AF868714

Structures for Nuclear Facilities

M.Y.H. Bangash

Structures for Nuclear Facilities

Analysis, Design, and Construction

Springer

Prof. Dr. M.Y.H. Bangash
Emeritus Professor of
Aerospace and Nuclear Structures
Consulting Engineer
39 Wontner Road
SW17 7QT London
United Kingdom

ISBN 978-3-642-12559-1 e-ISBN 978-3-642-12560-7
DOI 10.1007/978-3-642-12560-7
Springer Heidelberg Dordrecht London New York

Library of Congress Control Number: 2010934519

© 2011 M.Y.H. Bangash; London, UK
Published by: Springer-Verlag Berlin Heidelberg 2011
This work is subject to copyright. All rights are reserved, whether the whole or part of the material is concerned, specifically the rights of translation, reprinting, reuse of illustrations, recitation, broadcasting reproduction on microfilm or in any other way, and storage in data banks. Duplication of this publication or parts thereof is permitted only under the provisions of the German Copyright Law of September 9, 1965, in its current version, and permission for use must always be obtained from Springer-Verlag. Violations are liable to prosecution under the German Copyright Law.
The use of general descriptive names, registered names, trademarks, etc. in this publication does not imply, even in the absence of a specific statement, that such names are exempt from the relevant protective laws and regulations and therefore free for general use.

Cover design: WMXDesign GmbH, Heidelberg

Printed on acid-free paper

Springer is part of Springer Science+Business Media (www.springer.com)

Two well-known scientists, Dr. A.Q. KHAN and Professor Abdus Salam, have been chosen and this book is dedicated to them for their services to science and technology.

DR. A.Q. KHAN, a towering individual, has many achievements in the field of science and technology. The direction of his research reflects his deep passion for advanced materials. He deserves this dedication because of his deliverance of advanced nuclear technology to his country, Pakistan, and the establishment of advanced laboratories including the famous Kahuta Research Laboratories for existing and future practical implementations. His research is recommandable.

PROFESSOR ABDUS SALAM attained the pinnacle of his physics career when he shared a Nobel Prize with two other scientists on "Unification of Electromagnetism and the Weak Nuclear Force". The most revealing of it was the existence of a new particle at extreme energies – the conditions that would be similar to the few moments in the birth of the universe. These fleeting moments were confirmed under cosmic conditions at the CERN Accelerator. He had extended his hands to developing countries. He deserves this dedication.

Contents

Chapter 1
Nuclear Power Plant Facilities and Regulatory Guides

1.1 General Introduction

A number of nuclear power stations exist worldwide. This chapter concentrates on British, American, Canadian and European research and commercial reactors and nuclear power stations. The directory exists on the past, present and future planning for the NPS and are continuously being updated. In this chapter carefully selected nuclear stations with various design parameters have been clearly dealt with. British research and commercial nuclear power stations are first named. Various design and other parameters are given with reasonable artwork explaining the concepts behind the research and commercial reactor stations. Listings are provided for BWR, AGR, BHWR, PWR and FBR types and others in major countries of the world. Those countries that have designed and supplied the respective systems are named together with the suppliers. This chapter ends with relevant references for those who can carry out an in-depth study for initiating and planning of next generation of nuclear power stations. Owing to restrictions, some have been listed without explanatory notes and designs.

1.2 British Commercial Nuclear Power Station – Commissioned and Non-commissioned

1.2.1 General

The following well-known stations have been constructed and commissioned:

- Berkeley (275 MW) – decommissioned
- Bradwell (300 MW) – decommissioned
- Latina (200 MW)
- Dungeness A (550 MW)
- Dungeness B (500 MW)
- Oldbury (600 MW) – decommissioned
- Hinkley A (500 MW)

M.Y.H. Bangash, *Structures for Nuclear Facilities*,
DOI 10.1007/978-3-642-12560-7_1, © M.Y.H. Bangash 2011

- Hinkley B (550 MW)
- Hartlepool (550 MW)
- Wylfa (550 MW)
- Sizewell B PWR (1300 MW)

Ther are many others which are mentioned later on in the text.

1.2.1.1 Berkeley NPS

They have two units:

Unit 1 reactor on power: 12 June 1962
Unit 2 reactor on power: 2 November 1962

These are gas-cooled reactors, graphite moderated using natural uranium as fuel and are equipped with one-load charge/discharge machinery. Each reactor is associated with light heat exchangers which supply steam to 4No. 83 MW dual-purpose turbo alternators. Plate 1.1 shows the pictoral view of the Berkeley N.P.S. and Plate 1.2 gives an outline of the sectional elevation of this situation. The station was designed by TNPG – a consortium of engineering firms for the then Central Electricity Generating Board (CEGB).

Plate 1.1 A pictoral view of Berkeley nuclear power station

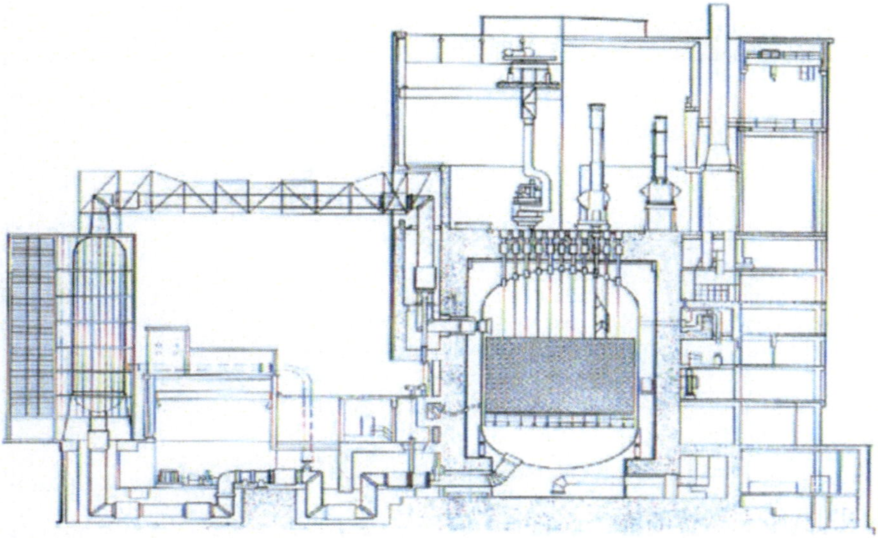

Plate 1.2 A sectional elevation of Berkeley nuclear power station

The station parameters:

Station net electric output		275 MW
Heat per reactor		560 MW
Moderator		Graphite
Diameter of active core		13.1 m
Height of active core		7.45 m
Fuel		Natural uranium solid rod in Magnox can
Rod diameter		27.9 mm
Rod length		0.48 m
No. of fuel channels		3265
No. of fuel elements per channel		13
Weight of fuel per reactor		230.8 t
Reactor vessel		Steel cylinder
Internal diameter		15.2 m
No. of standpipes	Control	133
	Charge	60
Blowers		8 per reactor
Coolant		Carbon dioxide
Reactor gas inlet temperature		160°C
Reactor gas outlet temperature		345°C
Power per blower		2.34 MW
Boilers		8 per reactor
Overall height		21.3 m
Diameter		5.3 m

(continued)

Fuel handling		On load
Turbo alternators		4
Speed		3000 rpm
Rating		80 MW
Steam conditions (at stop valve)		
HP	Flow	70.5 kg/s
	Pressure	21.8 kg/cm^2 abs
	Temperature	316°C
LP	Flow	32.6 kg/s
	Pressure	5.1 kg/cm^2 abs
	Temperature	316°C

Station status: Completed life service and decommissioned

1.2.1.2 Bradwell

Bradwell has two gas-cooled graphite-moderated reactors fuelled with natural uranium. There have been two units commercially operated:

Reactor 1 on power: 1 July 1962
Reactor 2 on power: 12 November 1962

Each reactor has six heat chambers supplying steam at two pressures to six 52.4 MW turbo alternators. For on-load refuelling universal charge and discharge machines are used. Plates 1.3 and 1.4 show the pictoral view and outline sectional elevation of the station, respectively.

Plate 1.3 A pictoral view of Bradwell nuclear power station (with compliments of CEGB)

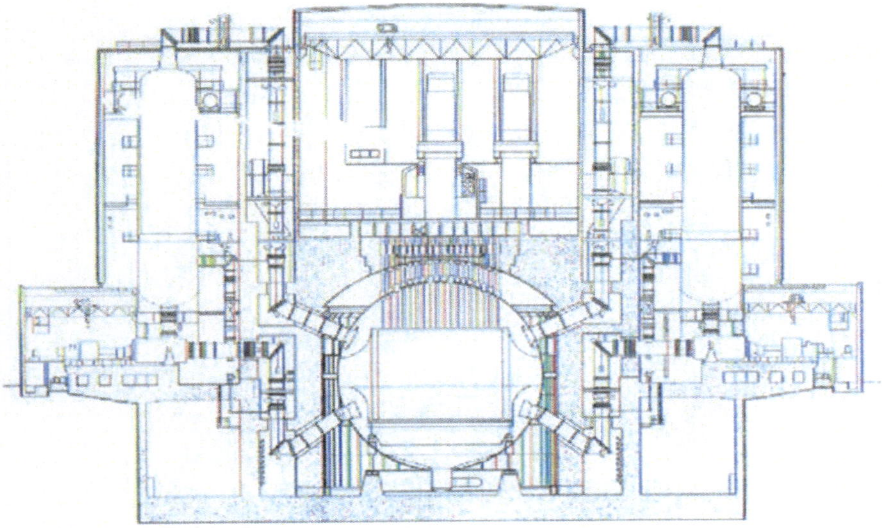

Plate 1.4 A sectional elevation of Bradwell nuclear power station (with compliments of CEGB)

The station parameters:

Station net electric output		300 MW
Heat per reactor		531 MW
Moderator		Graphite
Diameter of active core		12.2 m
Height of active core		7.90 m
Fuel		Natural uranium solid rod in magnox can
Rod diameter		29.3 mm
Rod length		0.91 m
No. of fuel channels		2564
No. of fuel elements per channel		8
Weight of fuel per reactor		239 t
Reactor vessel		Steel sphere
Internal diameter		20.3 m
No. of standpipes	Control	168
	Charge	88
Blowers		6 per reactor
Coolant		Carbon dioxide
Reactor gas inlet temperature		180°C
Reactor gas outlet temperature		390°C
Power per blower		2.47 MW
Boilers		6 per reactor
Overall height		28.1 m
Diameter	Top	5.8 m
	Bottom	6.1 m

(continued)

Fuel handling	On load
No. of charge machines	2 per reactor
Main turbo alternators	6
Speed	3000 rpm
Rating	52 MW
Auxiliary turbo alternators (blower drive)3	3
Speed	3000 rpm
Steam conditions (at stop valve)	
HP Flow	43.1 kg/s
Pressure	52.4 kg/cm^2 abs
Temperature	371°C
LP Flow	15.9 kg/s
Pressure	13.7 kg/cm^2 abs
Temperature	371°C

Station status: Completed life service

1.2.1.3 Latina

The power station is located some 40 miles south of Rome and was designed and constructed by TNPG in collaboration with refinements in the reactor design. It is a single refined gas-cooled graphite-moderated reactor which supplies steam to 3No.70 MW turbo alternator.

Reactor on power: 12 May 1963

Plates 1.5 and 1.6 show the pictoral view and outline sectional elevation of the structure, respectively.

The station parameters:

Station net electric output	200 MW
Reactor heat	705 MW
Moderator	Graphite
Diameter of active core	12.7m
Height of active core	7.88m
Fuel	Natural uranium solid rod in magnox can
Rod diameter	29.3mm
Rod length	0.91m
No. of fuel channels	2929
No. of fuel elements per channel	8
Weight of fuel	268t
Reactor vessel	Steel sphere
Internal diameter	20.3m
No. of standpipes	197
Blowers	6

Plate 1.5 Pictoral view of Latina (with compliments from TNPG a part of Mcalpine consultants)

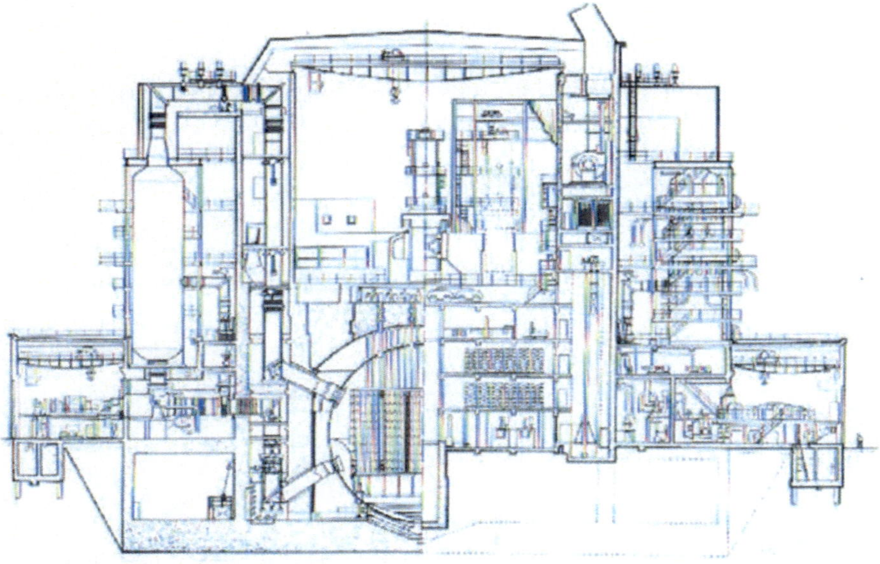

Plate 1.6 Outline of the sectional elevation of Latina

(continued)

Coolant	Carbon dioxide
Reactor gas inlet temperature	180°C
Reactor gas outlet temperature	390°C
Power per blower	3.06 MW
Boilers	6
Overall height	24.1m
Diameter	5.6m
Fuel handling	On load
No. of charge machines	2
Main turbo alternators	3
Speed	3000 rpm
Rating	70 MW
Auxiliary turbo alternators (blower drive)	2
Speed	2263 rpm
Steam conditions (at stop valve)	
HP Flow	57.6 kg/s
Pressure	52.4 kg/cm^2 abs
Temperature	371°C
LP Flow	21.1 kg/s
Pressure	13.7 kg/cm^2 abs
Temperature	371°C

Station status: Completed life service

1.2.1.4 Dungeness A

The TNPG, a consortium, installed two gas-cooled graphite-moderated type, using natural uranium as fuel with on-load refuelling. The number of gas circuits has been reduced to four back pressure steam turbines providing gas circulator drive. The turbine house contains 4No. 142.5 MW turbo alternators.

No. 1 reactor in power: 21 September 1965
No. 2 reactor in power: 18 November 1965

Plates 1.7 and 1.8 show the pictoral view and the outline elevation of the station, respectively.

The station parameters:

Station net electric output	550 MW
Heat per reactor	840 MW
Moderator	Graphite
Diameter of active core	13.9 m
Height of active core	7.47 m
Fuel	Natural uranium solid rod in magnox can
Rod diameter	27.9 mm

Plate 1.7 A pictoral view of the Dungeness A nuclear power station (with compliments of CEGB)

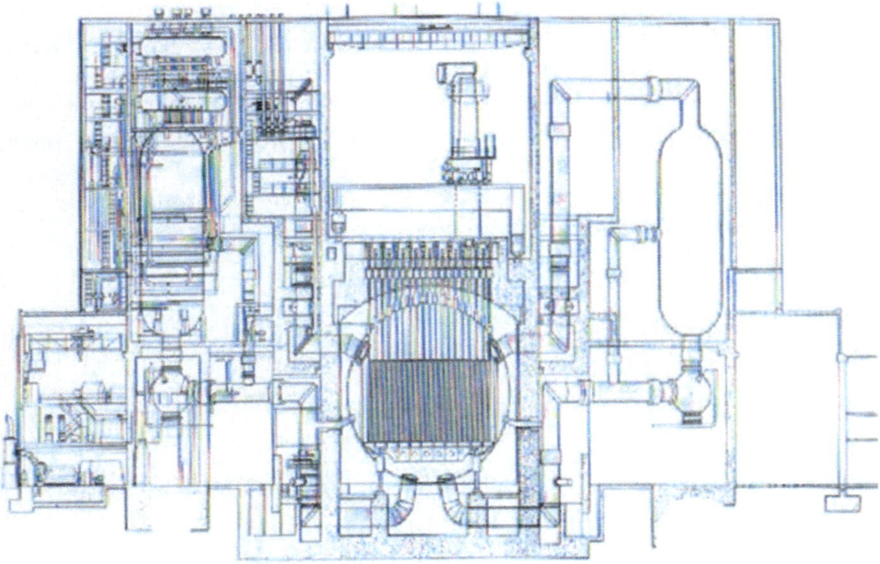

Plate 1.8 Outline of the sectional elevation of the Dungeness A nuclear power station (with compliments of CEGB)

(continued)

Rod length	0.972 m
No. of fuel channels	3876
No. of fuel elements per channel	7
Weight of fuel per reactor	300 t
Reactor vessel	Steel sphere
Internal diameter	19.1 m
Blowers	4 per reactor
Coolant	Carbon dioxide
Reactor gas inlet temperature	250°C
Reactor gas outlet temperature	410°C
Power per blower	7.16 MW
Boilers	4 per reactor
Overall height	22.9 m
Diameter	7.2 m
Fuel handling	On load
No. of charge machines	2 per reactor
Turbo alternators	4
Speed	1500 rpm
Rating	142.5 MW
Steam conditions	
Pressure	41.5 kg/cm^2 abs
Temperature	393°C

Station status: Under service operation

Commissioning date:
Reactor number 1: July 1965
Reactor number 2: 2 September 1965

The station is situated on the coast on the western side of Dungeness in Kent. There are two reactors, housed in spherical-type pressure vessels of 62.5 ft mean diameter, constructed of steel plates having a general thickness of 4 in. Each reactor has four boilers. Generating plant comprises four 1,500 rpm turbo alternators each with rated output of 142.5 MW at 0.8 power factor (lagging). The total guaranteed net output of the station in megawatts is planned as 550 MW which was expected to be achieved by the end of 1965.

Location:	Dungeness Point Kent
Reactor type:	Magnox
Reactor power:	840 MW (Th.), 275 MW (E) net from each of two reactors
Designer:	The Nuclear Power Group
Builder:	The Nuclear Power Group
Fuel element:	Natural uranium bars in Magnox cans
Data for each reactor	
Number of elements:	27,515 in 3932 channels. Total 300 t natural uranium
Moderator:	1500 t graphite

(continued)

Coolant:	Carbon dioxide, outlet temperature 410°C
Core size:	24 –sided prism: 13.83 m diameter, 7.39 m high
Specific power:	2.78 kW/kg uranium
Control:	91 bulk rods; 9 sector rods; 8 trim rods of boron steel in stainless steel sheath
Shielding:	Biological: 167.6–289.6 cm concrete

(With compliments from TNPG)

1.2.1.5 Dungeness B Advanced Gas-Cooled Reactor Station

The pressure vessel built in concrete, i.e. the PCPV, has been thoroughly investigated in this test.

1.2.1.6 Oldbury

An advancement is made on the gas-cooled reactor design. Each of the two reactor does have its core, boilers and gas circulators enclosed in a prestressed concrete pressure vessel. Natural uranium used as a fuel. In a turbine house unit generating sets are used, each one rated at 313 MW. It is four miles from Berkeley.

Both reactors on power in 1966/1967

Plates 1.9 and 1.10 show pictoral view and an outline elevation of the station, respectively.

Plate 1.9 A pictoral view of the Oldbury nuclear power station (with compliments of Mr. Ghalib of TNPG)

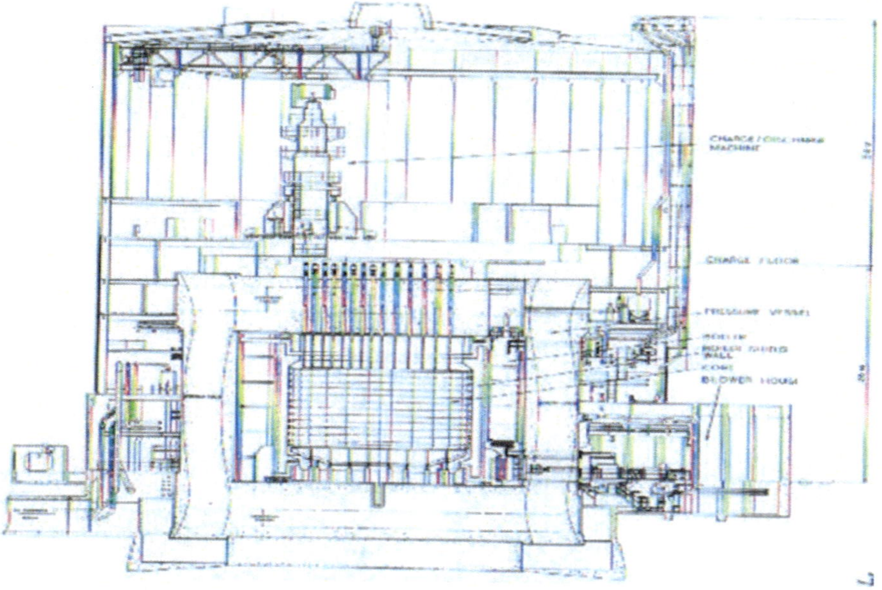

Plate 1.10 An outline sectional elevation of the Oldbury nuclear power station (with compliments of Mr. Ghalib of TNPG)

Station parameters:

Station net electric output	600 MW
Heat per reactor	892 MW
Moderator	Graphite
Diameter of active core	12.8 m
Height of active core	8.53 m
Fuel	Natural uranium solid rod in Magnox can
Rod diameter	27.9 mm
Rod length	0.972 m
No. of fuel channels	3308
No. of fuel elements per channel	8
Weight of fuel	292 t
Reactor vessel	Prestressed concrete cylinder
Internal diameter	23.5 m
Internal height	18.3 m
Blowers	4 per reactor
Coolant	Carbon dioxide
Blower drive	Back pressure steam turbine
Reactor gas inlet temperature	250°C
Reactor gas outlet temperature	411°C
Power per blower	4.9 MW

(continued)

Boilers	4 annular per reactor
Height over banks	11 m
Width of annulus	3.1 m
Fuel handling	On load
No. of charge machines	One per reactor
Turbo alternators	2
Speed	1500 rpm
Rating	313 MW
Steam conditions	
HP Flow	190 kg/s
Pressure	98.3 kg/cm^2 abs
Temperature	393°C
LP Flow	149 kg/s
Pressure	49.5 kg/cm^2 abs
Temperature	393°C

Station status: Originally completed life service

1.2.1.7 Hinkley

The station has a capacity of 1300 MW (c) twin reactor of AGR. It is similar to Oldbury for the internal arrangement of core and boilers. The significant features of the design are

(a) Gas pressure 42.5 kg/cm^2 (600 psia)
(b) Fully encapsulated gas coolers, removable under pressure
(c) Boiler shield walls of calcium hydroxide

The station is located in the north Somerset on the Bristol Channel.

Plates 1.11 and 1.12 show pictoral view and outline sectional elevation of the station, respectively.

Station parameters: Similar to Oldbury N.P.S.

Recent improvements in the design of the core have been given a mean fuel rating 30% higher than the figure published for Dungeness ‘B’. The reactors are refuelled on load. No axial or radial shuffling of the fuel is required during its life in the reactor.

The overall performance parameters of the station are as follows:

Electrical output (net)	2 × 630 MW (+ 40 MW from gas turbines)
Reactor heat output	1500 MW
Fuel rating	5.14 MW (e)/t
Thermal efficiency	42.0%

Station status: Completed life performance

Plate 1.11 A pictoral view of the Hinkley 'B' nuclear power station (with compliments of Mr. Ghalib of TNPG)

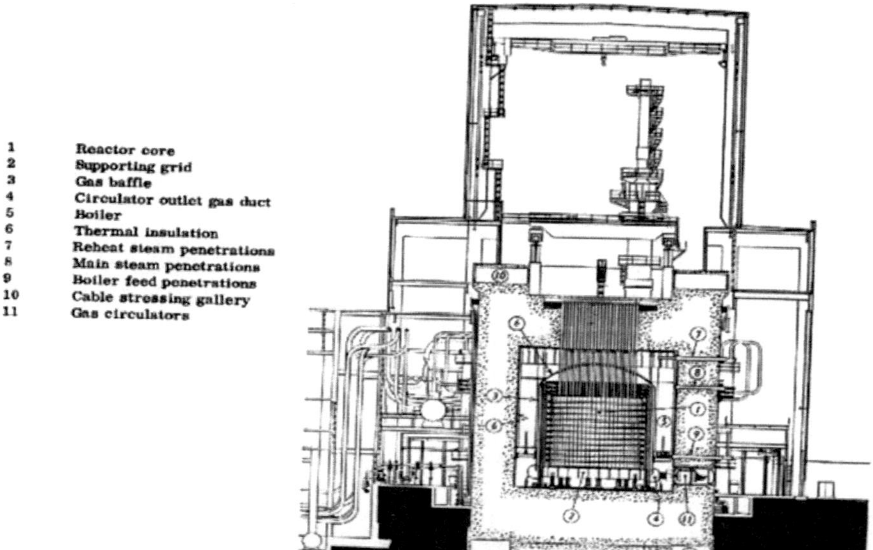

Plate 1.12 A sectional elevation of the Hinkley 'B' Nuclear power station (with compliments of Mr. Ghalib of TNPG)

1.2.1.8 Other British Reactors: Commercial and Research Types

General Information

The reactors mentioned in this section are collected from a number of sources. The information given herein is for research and advanced commercial reactors which are listed as under:

Each of the above station is supported by the respective station parameters. Some of them have completed their expected life with the possibility of demolition or replacement.

Gleep

Graphite Low-Energy Experiment Pile
Commissioning date: 15 August 1947
A thermal heterogeneous reactor, using natural uranium as fuel: graphite moderated and reflected, air cooled.

The reactor is formed from a graphite cube of 21 ft (6.4m) side width made of 682 channels. Natural uranium bars and aluminium clad, are placed in these channels to provide the fuel.5 ft (1.52m) thick barytes concrete forms biological shield. The listings of the other NP stations are first listed underneath.

Location:	A E R E Harwell Berkshire
Reactor type:	Thermal, heterogeneous
Reactor power:	Variable up to 4.5 kW (Th.)
Designer:	Ministry of Supply
Builder:	Ministry of Works
Fuel element:	Natural uranium bars in plain aluminium cans
Number of elements:	11,594 in 682 channels, total 32.5 t natural uranium
Moderator:	505 t graphite
Coolant:	Air at ambient temperature
Core size:	Cylindrical 552 cm diameter, 518 cm long
Specific power:	0.14 kW/t uranium
Control:	One safety rod and two sets of three shut-off rods; one fine, four coarse control rods of boron carbide in stainless steel annulus
Shielding:	Biological: concrete 152 cm thick. No thermal shield

(Data provided by the Late Lord Marshall of UKAEA 1989)

Bepo

British Experimental Pile Operation
Commissioning date: 5 July 1948
A thermal heterogeneous reactor with natural or low-enrichment uranium as fuel, graphite moderated and reflected, air cooled.

Research Reactors

- Gleep
- Bepo
- Lido
- Dido
- Pluto
- DMTR
- Horace
- Herald
- Zenith
- Vera
- Nestor
- Hero
- Daphne
- Dimple
- Zebra

- Hector
- Jund
- Tokaimura

Magnox Reactors

- Calder Hall
- Chapelcross
- Hunterston
- Hinkley Point A
- Trawsfynydd
- Dungeness A

AGR

- Sizewell
- Wylfa
- Hartlepool

$HTGFR_e$

- Dungeness B
- SGHWR, Winfirth
- DFR
- Dragon

The reacting core is a cylinder 20 ft diameter by 20 ft long and there are 20 fuel elements in each of 880 channels. Fuel is canned in aluminium, finned to improve heat transfer. Air is drawn at sub-atmospheric pressure through the core and is discharged from a high chimney. Shielding consists of a 6 in. cast iron thermal shield and $6\frac{1}{2}$ ft of barytes concrete.

Location:	A E R E Harwell Berkshire
Reactor type:	Thermal heterogeneous
Reactor power:	6 MW (Th.)
Designer:	Ministry of Supply
Builder:	Ministry of Works
Fuel element:	Natural or low-enriched uranium in finned aluminium cans
Number of elements:	16,558 in 830 channels total 39 t natural uranium
Moderator:	863 t graphite
Coolant:	Air outlet temperature 120°C
Core size:	Cylindrical 610 cm diameter, 610 cm long
Specific power:	0.15 kW/kg uranium
Control:	Two vertical shut-off rods: four horizontal control rods (falling under gravity) of hollow steel lined or filled with boron carbide
Shielding	Biological: concrete 198 cm thick. Thermal: cast iron 15.2 cm thick

(Data provided by the Late Lord Marshall of UKAEA 1989)

Lido

Commissioning date: 20 September 1956

Thermal heterogeneous swimming pool type, enriched uranium fuel, H_2O moderated and cooled.

The concrete tank has 7 ft thick walls and is 28 ft long, × 8 ft wide and 24 ft deep internally and contains light water. Three large aluminium 'windows' are fitted in the tank and these allow heavy shielding experiments to be set up outside. Three beam holes are also fitted. The reactor core is made up of uranium/aluminium plates clad in aluminium; it can be moved through the water into position on the inner side of any window. The water is circulated for cooling and cleaning.

Location:	A E R E Harwell Berkshire
Reactor type:	Thermal heterogeneous swimming pool
Reactor power:	100 kW (Th.)
Designer:	U K A E A
Builder:	Vickers Armstrong
Fuel element:	MTR-type curved plates 46% enriched uranium alloyed with aluminium; clad in aluminium
Number of elements:	22 standard elements, 4 control rod elements, a few partial elements, total 3.47 kg U-235
Moderator:	155,000 l demineralised water

(continued)

Coolant:	Water outlet temperature (core) 45°C
Core size:	Rectangular prism 46 cm square, 63.5 cm high
Specific power:	29 kW/kg U-235
Control:	Two safety rods, one shim rod; one regulating rod of cadmium strip clad in stainless steel, all rods operating in special fuel elements
Shielding:	The water is contained in a concrete tank 213 cm thick

(Data provided by the Late Lord Marshall of UKAEA 1989)

Dido Pluto DMTR

The authority has three *Material Testing Reactors* essentially of the same design: Dido and Pluto at Harwell and DMTR at Dounreay. Their main purpose is to test fuel, coolants and materials which may be used in future rectors, principally to study the effects of radiations under reactor conditions. Dido and Pluto are also used for the production of isotopes of high specific activity and for research in physics, chemistry and metallurgy.

Dido: Deuterium oxide-moderated reactor i.e. D_2O or DDO
Pluto: Loop Testing Reactor
Commissioning dates:
Dido: 7 November 1956
Pluto: 28 October 1957

Location:	Dido and Pluto: A E R E Harwell Berkshire
Reactor type:	Thermal heterogeneous, MTR
Reactor power:	15 MW (Th.)
Designer:	U K A E A
Builder:	Head Wrightson Processes Ltd.
Fuel element:	MTR-type plates 80% enriched uranium alloyed with aluminium; clad in aluminium
Number of elements:	Dido: 25 arranged in rows 4, 6, 5, 6, 4 Total 3.75 kg U-235
	Pluto: 26 arranged in rows 4, 6, 6, 6, 4 Total 3.9 kg U-235
Moderator:	10 nt D_2O
Coolant:	D_2O outlet temperature 53°C (bulk)
Core size:	Approximate vertical cylinder, 61 cm high Dido: equivalent diameter 84 cm Pluto: equivalent diameter 87.5 cm
Specific power:	Dido: 4,000 kW/kg U-235 Pluto: 3,850,000 kW/kg U-235
Control:	Coarse control/shut-off: six arms (Pluto seven arms) cadmium in stainless steel sheath, two vertical safety rods, one vertical fine control rod of cadmium in stainless steel tubes, D_2O reflector dumping
Shielding:	Dido: 0.65 cm boral, 10.2 cm lead (water cooled) surrounded by 152 cm barytes concrete Pluto: 0.65 cm boral, 10.2 cm lead (water cooled) about 45 cm, iron shot concrete (variable thickness) followed by not less than 120 cm of barytes concrete

(Data provided by the Late Lord Marshall of UKAEA 1989)

Horace

H_2O Reacot Aldermaston Critical Experiment
Commissioning date: May 1958
A zero energy reactor, H_2O moderated, with MTR-type fuel

The reactor core is a duplicate of the research reactor Herald core and is used for reactivity measurements required for the operation of Herald and the safety assessment of experiments.

Location:	A W R E Aldermaston Berkshire
Reactor type:	Thermal heterogeneous swimming pool
Reactor power:	5 MW (Th.)
Designer:	U K A E A
Builder:	A W R E Aldermaston
Fuel element:	80% enriched U-235 in uranium–aluminium alloy in aluminium plates
Number of elements:	Up to 70 in various configurations made up of elements, beryllium reflectors and mock-ups of incore rigs
Moderator:	Light water
Coolant:	Light water
Core size:	Maximum approximately 68.6 cm × 58.4 cm × 61 cm high
Specific power:	Zero energy
Control:	six flat absorbers, cadmium in stainless steel, with aluminium followers
Shielding:	Screening wall 366 cm high, 30.5 cm thick of aggregate concrete blocks

(Data provided by the Late Lord Marshall of UKAEA 1989)

Herald

Location:	A W R E Aldermaston Berkshire
Reactor type:	Thermal heterogeneous swimming pool
Reactor power:	5 MW (Th.)
Designer:	A E l-John Thompson Nuclear Energy Co. Ltd. and AWRE
Builder:	A E l-John Thompson Nuclear Energy Co. Ltd.
Fuel element:	Aluminium MTR type 14 fuel plates per element
Number of elements:	Up to 50 fuel elements, total 6 kg U-235
Moderator:	Light water
Coolant:	Light water up to 50°C
Core size:	68.6 cm × 58.4 cm × 61 cm high
Specific power:	800 kW/kg U-235
Control:	Six flat cadmium sheet absorbers in stainless steel with aluminium followers operating together
Shielding:	Biological: 61 cm iron shot concrete and 122 cm barytes concrete Thermal: 10.2 cm lead cast in aluminium segments

(Data provided by the Late Lord Marshall of UKAEA 1989)

Zenith

Zero Energy Nitrogen-Heated Thermal Reactor
Commissioning date: December 1959
A thermal, near-homogeneous, reactor with variable fuel-graphite moderated and reflected. Nitrogen is used as heating gas.

Zenith is designed for the study of the physics of high-temperature gas-cooled systems. Circulation of nitrogen gas over a 250 kW heater at the base of the core allows the core and side reflector regions to be heated to 800°C and 400°C respectively.

Location:	A E E Winfirth Dorset
Reactor type:	Thermal near-heterogeneous variable fuel
Reactor power:	Zero energy [max. 200 W (Th.)]
Designer:	UKAE and General Electric Company Ltd.
Builder:	General Electric Company Ltd.
Fuel element:	93% enriched uranium or plutonium alloy containing 97.5% Pu-239, 2.4% Pu-240 and 0.1% Pu-241
Number of elements:	235 fuel elements made of pallets in graphite tubes giving a variable core loading
Moderator:	Graphite (in the fuel element tubes)
Coolant:	Nitrogen is used as heating gas to give 800°C maximum core temperature
Core size:	Cylindrical 183 cm high, 122 cm diameter
Specific power:	Not applicable
Control:	1–10 motorised boron carbide safety rods; 3–6 motorised boron carbide shut-off rods; 1 fine, 1 coarse boron carbide control rod
Shielding:	1.91 cm thick mild steel pressure vessel in contained pit of 122 cm thick barytes concrete

(Data provided by the Late Lord Marshall of UKAEA 1989)

Vera

Versatile Experimental Reactor Assembly
Commissioning date: 22 February 1961

A zero-energy fast reactor built on a lattice plate which is separated into two halves for convenience and safety on loading. Fuelled with U-235 or Plutonium and designed for research in fast reactor physics, particularly for the purpose of checking nuclear data and calculation methods.

Location:	AWRE Aldermaston Berkshire
Reactor type:	Fast, zero energy
Reactor power:	10 W (Th.)
Designer:	UKAEA
Builder:	UKAEA
Fuel element:	U-235, square plates unclad, sprayed with lacquer. Pu-239 square plates clad in soft-soldered copper

(continued)

Number of elements:	Normally 40–60 in a maximum of 144 channels – the remainder, containing standard elements of reflector material
Moderator:	Variable (graphite, steel, polythene etc.) in square plates or blocks
Coolant:	None
Core size:	Maximum 55.8 cm cube. A typical core is a 35.6 cm isometric pseudo-cylinder
Specific power:	Not applicable
Control:	Four safety rods of standard fuel reflector elements, three control rods of similar elements
Shielding:	No integral shielding – the reactor is located in a shielded laboratory

(Data provided by the Late Lord Marshall of UKAEA 1989)

Nestor

Neutron Source Thermal Reactor
Commissioning date: September 1961
A thermal heterogeneous reactor, fuelled with highly enriched uranium, light water moderated and cooled, graphite reflected.

The Nestor reactor provides a source of neutrons and accordingly has a number of special features. It consists of an annular core structure containing highly enriched uranium–aluminium alloy as fuel plates immersed in water. Graphite fills the centre of the annulus and surrounds it externally. The graphite serves two important purposes. It reduces the amount of uranium required to obtain a self-sustaining chain reaction in the core. In addition, it screens the reactor from the experimental assemblies and ensures that the power level, control and safety of the reactor are not significantly affected by rearrangements in the experimental assemblies.

Location:	AEE Winfirth Dorset
Reactor type:	Thermal heterogeneous (modified Jason type)
Reactor power:	10 kW (Th.) rising to 30 kW in 1964
Designer:	Hawker Siddley/Nuclear Power Co. and UKAEA
Builder:	Hawker Siddley/Nuclear Power Co. and Turriff Construction Corporation Ltd.
Fuel element:	MTR-type plates containing 80% enriched U-235 in aluminium alloy clad in aluminium
Number of elements:	16–24 channel spaces not occupied by fuel elements are filled by aluminium-clad graphite blocks. Total 4.6 kg uranium
Moderator:	Demineralised water
Coolant:	Demineralised water at 35°C
Core size:	Fuel elements from an annulus 15.24 cm wide and 76.2 cm mean diameter in cylindrical aluminium tank
Specific power:	2.18 kW/kg U-235
Control:	Two safety plates of cadmium clad in stainless steel; two similar control plates. Water moderator can be dumped
Shielding:	76 cm of iron shot concrete

(Data provided by the Late Lord Marshall of UKAEA 1989)

Hero

Hot Experimental Reactor of 0 (zero) power
Commissioning date: 5 February 1962

This zero-energy reactor was designed to complement the Advance Gas-cooled Reactor at Windscale in providing physics information on the AGR system.

The graphite-moderated core is cylindrical, both height and diameter being 18 ft and the design is such that it can be dismantled and re-assembled so that it could be used to study a verity of core geometries but as planned the reactor is mainly used for testing lattices at the AGR lattice pitch. Carbon dioxide is circulated through the core and external heaters are provided which can heat the gas to give core temperature up to 500°C, thus enabling operating conditions of the temperature of the AGR to be simulated without the development of residual radioactivity, which would hamper experimental work.

Location:	Windscale Cumberland
Reactor type:	Head lattice test reactor
Reactor power:	3 kW (Th.)
Designer:	UKAEA
Builder:	UKAEA
Fuel element:	Enriched UO_2 (1.8% U-235) clad in stainless steel
Number of elements:	Variable 325 channels (maximum) are variable for use
Moderator:	Graphite, raised to 500°C by electrically heated CO_2
Coolant:	CO_2
Core size:	Vertical axis cylinder 549 cm high, 554 cm diameter
Specific power:	20×10^{-3} kW/kg
Control:	Variable; up to 6 boron steel safety rods, 2 fine control and 16 coarse control rods dependent on the experiment
Shielding:	122 cm high-density concrete

Daphne

Dido and Pluto Handmaiden for Nuclear Experiments
Commissioning date: 20 February 1962

A thermal heterogeneous reactor, fuelled with enriched uranium. D_2O moderated and cooled, D_2O and partial graphite reflectors.

Designed to provide basic physics support to Dido and Pluto and therefore is very similar to these reactors in the main design features. It can be loaded with fuel elements, rigs and beam tubes to stimulate Dido, Pluto or other reactors of this class. The reactor operates at powers up to 100 W and is shielded by 4 ft thick concrete blocks.

Location:	AERE Harewell Berkshire
Reactor type:	Thermal heterogeneous
Reactor power:	100 W (Th.)
Designer:	UKAEA
Builder:	UKAEA
Fuel element:	MTR-type plates 80% enriched uranium alloyed with aluminium; clad in aluminium l
Number of elements:	25 for Dido simulation (3.75 kg U-235), 26 for Pluto simulation (3.9 kg U-235)
Moderator:	D_2O at 7 t at ambient temperature
Coolant:	D_2O at ambient temperature
Core size:	Approximate vertical cylinder 61 cm high, equivalent diameter about 86 cm
Specific power:	27 kW/t U-235
Control:	Four safety rods cadmium in stainless steel sheath in hollow fuel elements, vertical fine control rod (same type). Control is by variation by D_2O level and/or reactor fine control rod. Signal-type coarse control arms (cadmium in stainless steel sheath) may be adjusted when sub-critical to stimulate coarse control/shut-off arms in Dido and Pluto
Shielding:	120 cm Portland concrete

(Data provided by the Late Lord Marshall of UKAEA 1989)

Dimple

Deuterium-Moderated Pile of Low Energy
Commissioning date at Winfirth: 18 June 1962
A thermal heterogeneous reactor, with fuel and moderator variable; removable graphite reflectors no coolant.

The Zero-Energy Reactor, Dimple, which was used at Harwell until late 1960, has been extensively modified and rebuilt at Winfirth. Dimple can now be used to study the physics of reactor systems moderated by light or heavy water or by an organic moderator. Containment and monitoring equipment have been provided to permit the use of plutonium-bearing fuels. A wide variety of coarse structures can be built into Dimple ranging from large pressure tube systems to small close packed systems moderated with H_2O or H_2O/D_2O mixtures, which are of interest in the field of marine propulsion. The reactor may be heated uniformly to about 90°C.

Location:	AEE Winfirth Dorset
Reactor type:	Thermal heterogeneous variable moderator and fuel
Reactor power:	100 W (Th.)
Designer:	UKAEA
Builder:	UKAEA
Fuel element:	Uranium or plutonium as required
Number of elements:	Variable

(continued)

Moderator:	Liquid; variable in amount and composition
Coolant:	None
Core size:	Variable: A maximum 259 cm diameter 249 cm high
Specific power:	Negligible
Control:	Safety rods: either two separate banks containing up to 10 rods per bank (variable in size and composition) or fast internal moderator dump into cavity within tank. Coarse and fine control rods of variable composition and number, the difference being achieved by different gear speeds
Shielding:	122 cm thick concrete block construction

(Data provided by the Late Lord Marshall of UKAEA 1989)

Zebra

Zero-Energy Breeder Reactor Assembly
Commissioning date: 19 December 1962

A fast reactor fuelled with plutonium and/or highly enriched uranium, enrichment variable; no moderator; reflector composition variable but commonly natural uranium; no coolant.
To date, experimental fast reactors have compromised very small highly enriched cores. For example, the core of the Dounreay experimental fast reactor is about 2 ft diameter and 2 ft high. For power producing fast reactor to be used in electricity generating stations much larger reactors with lower enrichment are envisaged.

The neutron physics of the reactor will be studied in Zebra.

Location:	AEE Winfirth Dorset
Reactor type:	Zero-energy fast breeder
Reactor power:	100 W (Th.) maximum
Designer:	UKAEA
Builder:	UKAEA
Fuel element:	Plates: initial experimental programme provided 350 kg U-235 as 93% enriched uranium, unclad; 150 kg U-235 as $36\frac{1}{2}$% enriched uranium, unclad; 100 kg plutonium clad in thin layers of copper
Number of elements:	Variable
Moderator:	None
Coolant:	None
Core size:	Variable
Specific power:	Variable
Control:	Nine control rods: five screw type operated from below; four cable type operated from above; two screw type used as safety rods; two groups of cable type used as shut-off rods: three screw-type rods used as regulating rods; one (of smaller worth) is used for fine control
Shielding:	106 cm Portland concrete; 2 heavy concrete doors roll back to give access to the top of the reactor

(Data provided by the Late Lord Marshall of UKAEA 1989)

Hector

Heated Experimental Carbon Thermal Oscillator Reactor
Commissioning date: 10 March 1963
A zero-energy thermal reactor; two-region core; graphite moderated and reflected with carbon dioxide heating and cooling.

Hector is designed for obtaining information on the physics of power reactors by means of experiments using small quantities of material. A sample of the material is introduced into the centre of the reactor and the observed rate of change of the reactor power gives a measure of the neutron absorbing properties of the sample. An alternative technique is to move the sample in and out of the reactor periodically and to observe the resultant fluctuation of the power level. The reactor lattice in the neighbourhood of the sample is adjusted so as to simulate the conditions existing in a power reactor.

The rector core consists of two regions. The outer region is an annulus of graphite containing fuel in the form of plates of an alloy of highly enriched uranium and aluminium. This is known as the 'driver' region. Within this annulus is a central region which may be heated to a temperature of up to 450°C by a stream of electrically heated carbon dioxide gas. A variety of assemblies of fuel and moderators may be lodged into this region.

Location:	AEE Winfirth Dorset
Reactor type:	Zero energy thermal
Reactor power:	100 W
Designer:	Fairey Engineering Ltd. to AEA specification
Builder:	Fairey Engineering Ltd.
Fuel element:	Central core region: variable. Outer core region: Dido-type flat plates
Number of elements:	Central core region: 128 channels Outer core region: 216 channels
Moderator:	Graphite
Coolant:	CO_2 used to heat the central region and cool the outer region
Core size:	Central: octagonal graphite stack 91.1 cm across flats by 285 cm high with outer diameter 359 cm and octagonal hole 123 cm across flats along axis
Specific power:	Very small
Control:	All control and safety rods are B.C packed in steel tubes; six safety rods and six shutdown rods; two coarse control rods and one fine control rod; up to eight temperature rods
Shielding:	122 cm concrete with shield doors which open to expose whole top surface of reactor vessel

(Data provided by the Late Lord Marshall of UKAEA 1989)

Juno

Commissioning date: April 1964

Juno is a dual-purpose assembly for the study either of compact liquid-moderated cores or of pressure tube-type liquid-moderated lattices in critical configurations. It may alternatively be used as an experimental system, driven external neutron sources, for the study of pressure tube-type cores.

It will be used for the study of the physics of the reactor systems moderated by light or heavy water supplementing the work of the reactor Dimple to which it bears many similarities.

When required, it may be modified to permit the use of plutonium-bearing fuel. A wide variety of core structures can be built into Juno ranging from critical cores of closed pack fuel assemblies moderated by light or heavy water (of interest of marine work) to large cores of Steam-Generating Heavy Water type. These may be studied exponentially if required.

When in use as a compact critical assembly, emergency shutdown is provided by a rapid internal dump system which is ideally suited for the study of close packed cores where the provision of control rods involves an unacceptable perturbation of the core.

Location:	AEE Winfirth Dorset
Reactor type:	Thermal, heterogeneous, variable moderator and fuel
Reactor power:	100 W (Th.)
Designer:	UKAEA
Builder:	UKAEA
Fuel element:	Uranium or plutonium as required
Number of elements:	Variable
Moderator:	Liquid; variable in amount and composition
Coolant:	None
Core size:	Variable – maximum 190 cm diameter, 350 cm high
Specific power:	Negligible
Control:	Fast shutdown: Internal moderator dump into cavity within tank for compact cores. Two banks of thermal absorbers may be fitted for pressure tube-type lattices. Fine control: Variation of moderator level or insertion of fine control rods
Shielding:	4 ft thick concrete block construction

(Data provided by the Late Lord Marshall of UKAEA 1989)

Tokai Mura Nuclear

Commissioning date: estimated July 1965

This Japan Atomic Power Company's Tokai nuclear power station is situated on the Pacific coast about 70 miles north-east of Tokyo. The station is a single reactor housed in a skirt-supported spherical pressure vessel having a mean diameter of

18.4 m (60.4 ft) and made from steel plates 92 mm (3.6 in.) and 80 mm (3.16 in.) thick. There are four steam-raising units feeding two 85 MW turbo generators. Also operating from the main HP steam circuit are four 8700 hp back-pressure turbines which drive the centrifugal coolant-circulators. Of particular interest are the special design features introduced because of the potential earthquake hazard. These include substantial reinforcement of the concrete foundation raft and superstructure; additional restraints at the top of the reactor vessel and steam-raising units; an earthquake-resistant support system for main gas ducts; a new type of core restraint structure and a specially designed brick and key system of core construction. In addition, there is a secondary shutdown system capable of operating even if damage to the core should prevent normal functioning of the control rods.

Location:	Tokai Mura, Ibaraki-ken, Japan	
Reactor type:	Magnox	
Reactor power:	585 MW (Th.), 157 MW (E) net from one reactor	
Designer:	GEC/Simon-Carves Atomic Energy Group	
Builder:	British General Electric Company of Japan Ltd. The First Atomic Power Industrial Group (Japan)	
Fuel element:	Natural uranium hollow rods in Magnox cans with graphite sleeves	
Number of elements:	16,357 in 2048 channels. Total 187 t natural uranium	
Moderator:	870 t graphite	
Coolant:	Carbon dioxide; outlet temperature 390°C	
Core size exclude reflector:	Cylinder mean diameter 11.72 m, height 6.63 m	
Specific power:	3.14 kW/kg uranium	
	9 Safety rods	boron steel Stainless clad mild steel
	63 Coarse (Shim) rods 10 Sector (Auto) control rods	Stainless clad
	16 flattering and trim rods	
Reactor control:	44 secondary shutdown Devices	boron steel balls.
Shielding:	Biological: 3.1 m concrete	

(With compliments of Japan Atomic Power, CO, Tokai, Japan 1970)
(Also to CEGB, England)

Calder Hall and Chapelcross

Commissioning date: Calder Hall: October 1956, Chapelcross: May 1959

Each station consists of four reactors housed in cylindrical mild steel pressure vessels which are 70/71 ft high and 37 ft in diameter. Six fuel elements are stacked vertically in each of the 1696 channels per reactor. Each reactor has four heat exchangers generating high- and low-pressure steam simultaneously. Generating plant comprises eight 3000 rpm turbo alternators.

Design study work began in 1951, and construction work on Calder site commenced in August 1953. The first reactor became critical in May 1956 and power was first fed to the National Grid in October 1956. Construction at

Chaplecross had begun in October 1955. Heat and electricity output had been increased in stages at both stations to well beyond the design figures. Present heat output per reactor is 235/240 MW compared with the design output of 180 MW, a 30% improvement. The turbines have been rebladed to utilise this additional power. The Calder reactors also supply the steam requirements of the adjacent Windscale factory for space heating and for the operation of the chemical plant.

In 1964 an average load factor in excess of 92% was achieved at both stations and full load was maintained for 98% of the time during the severe winter of 1962/1963. Up to 28 February 1965 the total electricity supplied to the national grid from each station has been as follows:

Calder 8280 million units
Chapelcross 7020 million units

Location:	Magnox
Reactor power:	Calder Hall: 235 MW (Th.) 45 MW (E) Net. Chapelcross: 235/240 MW (Th.) 45 MW (E) net from each of four reactors
Designer:	UKAEA
Builder:	Various contractors
Fuel element:	Natural uranium in Magnox cans
Data for each reactor	
Number of elements:	10,176 in 1696 channels
	Total 127 t natural uranium
Moderator:	650 t of graphite in 58,000 bricks
Coolant:	Carbon dioxide, outlet temperature 340°C
Core size:	Polygonal prism: 9.45 m diameter, 6.40 m high
Specific power:	1.35 kW/kg uranium
Control:	48 rods also used as safety rods (8 inner zone, 16 outer zone); any 4 can be selected as time-regulating rods of boron steel in stainless steel tubes
Shielding:	Biological: 213 cm concrete
	Thermal: 15.25 cm steel

(Data collected from Natural Grid 1956)

Hunterston

Commissioning date:
Reactor number 1: February 1964
Reactor number 2: July 1964

The South of Scotland Electricity Board's nuclear generating station at Hunterston, on the Ayrshire coast, is one of the first three civil stations – each with an electrical output of about 300 MW from two reactors – to be built under the government's nuclear power programme of 1957.

The design of the station embodies several features which are not available in other early stations and which should increase output, operational economy and safety.

These features include the charging and discharging of fuel from the bottom of the reactor vessels where the temperature is lower, the provision of separate fuel-handling machines at the top of the vessels if required and the provision of remotely adjustable rods for the maintenance of reactivity balance.

In addition each fuel element is separately contained and supported within a graphite sleeve – a refinement which eliminates bowing and the sticking of the elements and gives protection against the build-up of Wigner energy.

Each reactor at Hunterston is associated with a group of eight steam-raising units feeding three 60 MW turbo alternators. The station operates on the base load and feeds electricity into the 275 kV grid at Nielston, Renfrewshire.

Location:	Hunterston Ayrshire Scotland
Reactor type:	Magnox
Reactor power:	535 MW (Th.), 150 MW (E) net from each of two reactors
Designer:	GEC/Simon Carves Atomic Energy Group
Builder:	General Electric Company Ltd. (Main)
Fuel element:	Natural uranium bars in Magnox cans, with graphite sleeves
Data for each reactor	
Number of elements:	32,880 in 3288 channels. Total 251 t natural uranium
Moderator:	1200 t graphite
Coolant:	Carbon dioxide, outlet temperature 395°C
Core size:	Right cylinder: 13.5 m diameter, 7.01 m high
Specific power:	2.13 kW/kg uranium
Control:	8 safety rods; 130 coarse regulating rods; 18 fine regulating rods of stainless steel with 4% boron
Shielding:	Biological: 282 cm of concrete

(Data provided by GEC/Simon Curves Atomic Energy Group, U.K. (1972))

Hinkley Point 'A'

Commissioning Date:
Reactor number 1: February 1965
Reactor number 2: 2 April 1965

Hinkley point is located on the north coast of Somerset on the Bristol Channel, some 8 miles north-west of Bridgewater. The 'A' station consists of two reactors housed in steel spherical pressure vessels of 67 ft diameter, made of plates 3 in. thick. Each reactor has six boilers. Generating plant consists of six 3.5 MW turbo alternators and three 33 MW turbo alternators to provide auxiliary supplies. Total output of the station is 500 MW, which is fed into the 75 kV Supergrid system running between Melksham and Taunton.

Hinkley Point 'B'

Extensive land reclamation was carried out, and in view of the advantages of the site, the circulating water system was designed to cater for a further station,

Hinkley 'B' to be built at some later date. The Minister of Power has given his consent for 'B' station about 1000 MW output. It is identical to Hartleypool NPS.

Location:	Hinkley point Somerset
Reactor type:	Magnox
Reactor power:	971 MW (Th.), 250 MW (E) net from each of two reactors
Designer:	English Electric–Babcock & Wilcox–Taylor, Woodrow Atomic Power Group
Builder:	English Electric–Babcock & Wilcox–Taylor, Woodrow Atomic Power Group
Fuel element:	Natural uranium bars in Magnox cans
Data for each reactor	
Number of elements:	36,000 in 4500 channels. Total 376 t natural uranium
Moderator:	2475 t graphite
Coolant:	Carbon dioxide, outlet temperature 378°C
Core size:	24-Sided prism, 14.9 m diameter, 7.63 m high
Specific power:	2.55 kW/kg uranium
Control:	130 control rods in 12 zones; 15 safety rods of boron steel tubes
Shielding:	Biological: 22.8 cm + 297 cm concrete

(Data with compliments from Taylor Woodrow Atomic Power Group, Southall, Middlesex)

Trawsfynydd

Commissioning date:
Reactor number 1: January 1961
Reactor number 2: 2 March 1965

Trawsfynydd is situated on the artificial lake of that name created some 30 years ago as a storage reservoir for the Maentwrog hydroelectric power station and lies within the Snowdonia National Park, North Wales. The station consists of two reactors, housed in spherical-type pressure vessels made of steel, $3\frac{1}{2}$ *in.* thick, with an internal diameter each of 61 ft. Each reactor supplies heat to six boiler units. Generating plant comprises four 3000 rpm turbo alternators each with a rated output of 145 MW at 0.8 power factor (lagging). The total guaranteed net output of the station in megawatts sent is 500 MW.

Total cost of the power station, including the plant and the roadworking, etc., is estimated at £68.5 million, giving a capital cost of £137 per kilowatt sent out, and an estimated cost per unit of 0.96d.

Output from the station is fed into the 275 kV grid system, the main connection being to Connah' Quay and Ffestiniog. Some output will also reinforce the 33 kV main distribution system, North Wales.

Location:	Trawsfynydd Merionethshire Wales
Reactor type:	Magnox
Reactor power:	860 MW (Th.), 250 MW (E)net from each of two reactors
Designer:	Atomic Power Construction Ltd.

(continued)

Builder:	Atomic Power Construction Ltd. (now part of the United Power Company Ltd.)
Fuel element:	Natural uranium bars in Magnox cans
Data for each reactor	
Number of elements:	33,480 in 3720 channels Total 280 t natural uranium
Moderator:	2090 t graphite
Coolant:	Carbon dioxide, outlet temperature 394°C
Core size:	Cylinder: 13.56 m diameter, 7.31 m high
Specific power:	3.11 kW/kg uranium
Control:	185 control rods arranged in 9 sectors; 12 safety rods of boron steel in stainless steel tubes
Shielding:	Biological: 304 cm concrete

(Data provided by Atomic Power Construction Ltd, 1967)

Sizewell A

Commissioning date:
Reactor number 1: August 1965
Reactor number 2: February 1966

Sizewell is situated on the Suffolk coast between Aldeburgh and Southwold. The station has two reactors housed in spherical-type pressure vessels 63 ft 6 in. in diameter and formed from mild steel plates $4\frac{1}{8}$ *in.* thick. Each reactor has four boilers. Generating plant comprises two 3000 rpm turbo alternators, each with a rated output of 324.7 MW at 0.85 power facoter (lagging). The total guaranteed net output of the station on megawatts sent out is planned as 580 MW, estimated to be achieved by spring 1966.

Reactor type:	Magnox
Reactor power:	948 MW (Th.), 290 MW (E) net from each of two reactors
Designer:	English Electric–Babcock & Wilcox–Taylor Woodrow Atomic Power Group
Builder:	English Electric–Babcock & Wilcox–Taylor Woodrow Atomic Power Group
Fuel element:	Natural uranium bars in Magnox cans
Data for each reactor	
Number of elements:	26,600 in 3800 channels Total 321 t natural uranium
Moderator:	2237 t graphite
Coolant:	Carbon dioxide, outlet temperature 410°C
Core size:	24 – sided prism: 13.7 m diameter, 7.92 m high
Specific power:	2.96 kW/kg uranium
Control:	99 control rods in 9 zones; 8 safety rods of boron steel or mild steel
Shielding:	Biological: 304.8 cm concrete

(With compliment of Taylor Woodrow and English Electric, UK.)

Sizewell B PWR

It is the pressurised water reactor station. The containment vessel has been analysed in this text against all eventualities.

WYLFA

Commissioning date:
Reactor number 1: 1968
Reactor number 2: 1969

Wylfa is situated on the North-West coast of the Isle of Anglesey at WYLFA Head. The station is found on rocks and large blasting programme was entailed in its construction. Each of the two units comprises a prestressed concrete pressure vessel, reactor core and internal shielding, boiler and gas circulator. The pressure vessel has a spherical internal shape of 96 ft diameter and a minimum wall thickness of 11 ft. The boilers, one to each vessel, are single pressure one through type with four associated gas circulators. Generating plant consists of four 3000 rpm turbo alternators each with a rated output of 370.22 MW at 0.9 power factor (lagging). The total guaranteed net output of the station in megawatts sent out is planned as 1180 MW, the most powerful nuclear power station under construction anywhere in the world.

Location:	Wylfa Point Anglesey North Wales
Reactor type:	Magnox
Reactor power:	1,875 MW (Th.), 590 MW (E) net from each of two reactors
Designer:	English Electric–Babcock & Wilcox–Taylor Woodrow Atomic Power Group
Builder:	English Electric–Babcock & Wilcox–Taylor Woodrow Atomic Power Group
Fuel element:	Natural uranium bars in Magnox cans
Data for each reactor	
Number of elements:	49,200 in 6150 channels. Total 595 t natural uranium
Moderator:	3740 t graphite
Coolant:	Carbon dioxide, outlet temperature 414°C
Core size:	16–sided cylinder: 17.3.7 m diameter, 9.14 m high
Specific power:	3.16 kW/kg uranium
Control:	167 control rods in 16 zones; 18 safety rods of boron steel or mild steel
Shielding:	The prestressed concrete reactor pressure vessel is 353.3 cm thick

(With compliments of English Electric, WYLFA and Taylor Woodrow, Southall, Middlesex, 1966)

SGHWR

Steam-Generating Heavy Water Reactor (SGHWR)
Commissioning date:Late 1967
SGHWR is a steam-generating heavy water moderated reactor

The core of the reactor consists of a bank of pressure tubes which pass through vertical tubes in a calandria, or tank, containing the heavy water moderator. Fuel is contained within the pressure tubes and the heat generated is removed by light water passing up these tubes and being partially turned into steam. After separation into steam drum, most of the steam passes directly to a turbine; the rest is passed through eight super-heat channels which will raise the steam to an outlet temperature of about 1000°F. This steam is then mixed with main steam flow before being fed to the turbine.

About 13% of the light water which is circulated over the fuel elements in the most highly rated boiling channel is converted into steam to give a turbine a stop valve pressure of 900 psig. The heavy water moderator has its own cooling and purification circuit and the light water coolant has circuits for water treatment and the detection of fuel element can failure.

Construction began at Winfirth in 1963 and the reactor will have an electrical output of 100 MW.

Location:	Winfirth Heath Dorset
Reactor type:	Steam-generating heavy water moderator (SGHWM). Direct cycle pressure tube
Reactor power:	294 MW gross thermal. 100 MW gross electrical
Designer:	UKAEA
Builder:	Main contractors: Turriff, AEI, International Combustion, Fairey Aviation
Fuel element:	Boiling channels: UO_2 pellets clad in zircaloy-2. Superheat channels: UO_2 pellets clad in stainless steel
Number of elements:	Boiling: 3744 fuel pins, 36 pins per assembly. 1 assembly per channel, 104 boiling channels, one of which is connected to an independent cooling circuit for testing fuel. Total 21,100 kg uranium. Superheating: 184 fuel pins, 24 pins per assembly, 1 assembly per channel, 8 superheat channels. Total 950 kg uranium
Moderator:	Heavy water in calandria (some moderation also in boiling light water coolant)
Coolant:	Direct cycle light water, boiling channel outlet temperature 283°C, superheat channel outlet temperature 504°C
Core size:	Cylinder: 311 cm diameter (excluding D_2O reflectors), 366 cm long (excluding D_2O reflectors)
Control:	Shutdown: Boric acid solution in interlattice tubes, plus D_2O dump. Control: D_2O height for power control. Boric acid concentration in moderator for long-term reactivity changes. Flooding groups of interlattice tubes for spectral shift flux control
Specific power:	Boiling channels 19.7 kW/kg. Superheat channel 15.3 kW/kg
Shielding:	Water and steel plus concrete biological shield, 282 cm thick

DFR

Dounreay Fast Reactor
Commissioning Date: 14 November 1959

The reactor has been designed to establish the feasibility of the fast breeder system and to provide information and operating experience needed to design a prototype reactor for full-scale power production.

The reactor core is hexagonal, 21 in. high and $20\frac{1}{2}$ *in.* across the flats, containing 324 fuel channels, with a central hexagon 5 in. across the flats, which can accommodate experimental fuel pins or sub-assemblies. The core is surrounded by a breeder blanket of 1872 channels. A number of channels in both the core and the blanket are also used for experimental fuel elements. Control and shutdown are provided by 12 groups each of 10 fuel elements around the edge of the core, mounted in carriers which can be raised or lowered. Liquid sodium–potassium alloy is used as coolant and transfers heat from the fuel to secondary liquid metal circuit in 24 heat exchangers set around the reactor vessel within the biological shield. This field is in the form of a concrete vault 90 ft in diameter and 5 ft thick and occupies the lower half of the steel containment sphere, 135 ft in diameter.

The principal use of the rector is now as a testbed for plutonium-based fuels for the prototype power fast reactor, in relation to both performance and to the effects of high burn-up.

Location:	DERE Dounreay Scotland
Reactor type:	Fast breeder
Reactor power:	60 MW gross (Th.); 13 MW (E)
Designer:	UKAEA
Builder:	(Main) John Thompson Ltd., Motherwell Bridge and Engineering Co. Ltd., Whatlings Ltd. and others
Fuel element:	45.5% enriched U-235, seven rods per element with 15 cm natural uranium as a top blanket. Cladding niobium
Number of elements:	Variable according to the experimental programme. The driver charge coantains approximately 200 kg U-235
Moderator:	None
Coolant:	Liquid sodium potassium alloy (NaK) at 350°C (outlet)
Core size:	Hexagonal prism 52.5 cm across flats, 52.5 cm high
Specific power:	Approx. 300 kW/kg U-235
Control:	12 groups of fuel elements comprised of 6 regulating groups, 4 shut-down and 2 safety groups
Shielding:	Biological: concrete 152 cm thick. Vessel: 120 cm of borated graphite

Plate 1.13 DFR. Dounreay Fast Reactor. The photo provided by SSEB Scotland 1988

Dragon

Commissioning date: 22 October 1964

The chief purpose of the reactor experiment is to demonstrate the principles on which any high-temperature gas-cooled reactor must be based. The entire power station is now dismantled.

The main features of the reactor, arising from the high operating temperature in the core, are the use of helium as a coolant with the fuel elements made from impermeable graphite containing uranium and thorium dicarbide mixed with graphite in a suitable ceramic form. Each element consists of a seven-rod cluster, provision being made from fission products which escape from fuel inserts by passing a fraction of the total coolant through a system of fission traps, and coolant circuit is expected to be slightly radioactive.

The steel reactor vessel consists of main cylindrical portion 11 ft 6 in. diameters, containing the core and reflector, with a 'neck' 6 ft 3 in. diameter through which the fuel elements are inserted and removed. The whole reactor, with its shielding, primary coolant circuit and much of the ancillary plant, is enclosed in the containment vessel with an inner shell of steel 66 ft diameter, 81 ft high and an outer shell of concrete 110 ft diameter, 86 ft high with wall 2 ft thick.

Location:	AEE Winfirth Dorset
Reactor type:	High-temperature reactor project
Reactor power:	20 MW (Th.)
Designer:	OECD high-temperature reactor project
Builder:	Contractors in signatory countries
Fuel element:	Outer annular region (27 elements) pyrolytic carbon-coated U-235/ zirconium monocarbide spherical particles consolidated in graphite matrix, to form inserts, 30 per fuel rod. Inner central region (10 elements) pyrolytic carbon and silicon carbide-coated U-235 thorium carbide spherical particles. Zr:U-235 ratio 8:1 Th:U-235 ratio 10:1
Number of elements:	37 elements, 259 rods Total weight: 14 kg U-235 uniformly distributed
Moderator:	Graphite, integral with fuel elements
Coolant:	Helium, outlet temperature 750°C
Core size:	Equivalent diameter of assembly 107 cm. Length of core 160 cm
Specific power:	1.4 kW/kg
Control:	24 symmetrical granular boron carbide absorber rods are provided. Three or 4 are used as safety rods, 1 is an automatic regulator, 19 or 20 remainder can be used manually as regulators
Shielding:	*Primary:* 92 cm graphite reflector, 5 cm steel baffle, 5 cm steel pressure vessel, 38 cm thermal shield (alternate layers of steel and water) 175 cm concrete. *Secondary:* 61 cm (minimum) concrete

1.3 Reactor Power Stations Based on BWR, PHWR and PWR

1.3.1 Boiling Water Reactors (BWR)

All boiling water reactors in Scandinavia are of Asea-Atom design. Containment is based on the pressure suppression (PS) principle, i.e. at a major pipe rupture the steam is led from the upper part, dry well, through a number of pipes to the lower part of the containment, wet wall, with a water pool, where condensation takes place. This principle of the volume of the containment has been kept small, about one-fifth of that of a dry containment with the same design pressure.

The layout of the containments varies from plant to plant. The fuel element pools, together with the pool for the storage of reactor internals, rest on the containment proper, which is statically independent from the pool structure and from the surrounding outer reactor building which houses the reactor auxiliary systems. For Ringhals 1 the containment vessel has been built more integrally with the fuel element pools. In later designs the external circulation pumps have been replaced by internal circulation pumps, where the motor housing forms an integral part of the reactor vessel. Consequently, the layout of the containment can be simplified to cylinder with a solid partition slab between the dry-well and wet-well compartments.

he PS containments are all prestressed horizontally and vertically, the latter for internal gas pressure minus the overlying weight. The design pressures are between 460 and 550 kN/m^2 (absolute pressure). In most cases the fuel element pools, which are partially cantilevered, are provided with prestressing tendons in the side walls. The top of the containment over the reactor vessel has to be removable for the refuelling operation and is designed as a steel dome.

A further development of the layout of the containment and adjoining structural units is showing the Asea-Atom current design. Consideration has also been taken to earthquake and airplane crash loading. The structural connection between the fuel pools and the containment proper has been further simplified.

ALL BWR plants, except Barsebäck 1 and 2, are founded on rock. In Barsebäck 1 and 2 the bottom of the containment consists of a thick slab on hard moraine-clay.

Six BWR plants exist in Japan at present. Table 1.1 gives a summary of BWR stations and structures associated with them.

1.3.2 Pressurised Heavy Water Reactors (PHWR)

Generally this system is based on 'CANDU'. Table 1.1 shows six of them built in Canada and six of them built in India; one of them exists in Romania, one is constructed in South Korea and one is in existence in Pakistan.

1.3.3 Pressurised Water Reactors (PWR)

PWR nuclear plants, Ringhals 2, 3 and 4, are all standard Westlinghouse design, except for the arrangement of the liner. They are provided with a dry containment in the shape of a cylinder with a dome roof, both in prestressed concrete. The design pressure is 510 kN/m^2.

1.3.3.1 Design Details

A special feature in the design of the Scandinavian reactor containments is that the steel liner is usually embedded in the concrete. The internal concrete cover, approximately 250 mm, serves as an efficient protection for the liner against missiles during an accident while at the same time the liner is protected against corrosion. Furthermore, thermal expansion of the steel liner due to the elevated temperatures (155–180°C) during the initial phase of an accident is avoided, resulting in a saving of prestressing steel. The disadvantage of the embedment is that if leakage occurs during a test it may be difficult to locate and repair. For that reason the liner in some of the State Power Board plants has been provided with steel channels along all welds. The channels are connected in sections and with the outside, whereby leaking gas, if any, can be located and vented.

Table 1.1 BWR and PHWR Nuclear Power Stations in the World

Country	Name	Type	Nuclear system supplier	Power	No. of similar units onsite	First power	Design pressure	Start of erection
Japan	Kashiwazaki 4	BWR	Hitachi	1100	2	1993	0.416	1988
Japan	Kashiwazaki 6 ABWR	BWR	Toshiba Hitachi GE	1356	2	1995	0.416	1991
Japan	Kashiwazaki 1	BWR	Toshiba	1100	3	1984	0.385	1978
Japan	Shika 1 (NOTO)	BWR	Hitachi	540	1	1992	0.535	1988
Japan	Shimane 2	BWR	Hitachi	520	1	1988	0.535	1984
Japan	Tokai	BWR	GE ET Hitachi	1100	1	1978	0.385	1973
USA	Gra Gulf 1	BWR	General Electric	1305	1	1982	0.203	1974
USA	Hatch 2	BWR	General Electric	844	1	1978		1972
USA	Hope Creek	BWR	General Electric	1118	1	1986	0.528	1976
USA	La Salle 1	BWR	General Electric	1122	2	1984	0.410	1974
USA	Susquehanna 1	BWR	General Electric	1132	2	1982		1973
USA	WNP-2 Hanford	BWR	General Electric	1164	1	1984		1972
USA	Shoreham	BWR	General Electric	849	1		0.389	1972
USA	Clinton 1	BWR	General Electric	985	1	1987	0.204	1975
Switzerland	Leibstadt	BWR	BBC-GESTCO	1030	1	1984	0.203	1975
Taiwan	Kuosheng 1	BWR	General Electric	985	2	1982		1974
Taiwan	Lugmen	ABWR	General Electric	1350	2	2004		1999
Sweden	Forsmark 3	BWR	ABM Atom	1155	1	1984	0.60	1977
Sweden	Oskershamn 3	BWR	ABM Atom	1160	2	1985	0.60	1981
S. Korea	Wolsong 1	PHWR	AECL	679	4	1983	0.156	1979
Lituania	Ignalina 1	RBMK	Minatomenergo	1500	2	1986		1978
Mexico	Laguna Negra 1	BWR	General Electric	675	2	1988		1974

Table 1.1 (continued)

Country	Name	Type	Nuclear system supplier	Power	No. of similar units onsite	First power	Design pressure	Start of erection
Spain	Cofrentes	BWR	General Electric	990	1	1984		1975
Romania	Cernavoda	PHWR	Atomic Energy					
of Canada Ltd.	706	1	1996		1980			
India	Kaiga 1	PHWR	Nuclear Power					
Co. of India Ltd.	220	1	2000	0.273	1989			
India	Kaiga 2	PHWR	Nuclear PowerCo of India Ldt	220	1	2000	0.273	1989
India	Kakrapara 1	PHWR	Nuclear Power					
Co. of India Ltd.	220	2	1993	0.225	1985			
India	Narora 1	PHWR	Assoc. CanadaGe and AECL	220	2	1991	0.225	1976
India	Rajasthan 3	PHWR	Nuclear PowerCo of India Ldt	220	2	2000	0.273	1989
India	Tarapur 3	PHWR	Atomstrojex-port/ KWU	500	2		0.244	1998
Iran	Busher	PWR	Hitachi	1000	2			1995
Japan	Fukushima 1-4	BWR	Atomic Energyof Canada Ltd.	784	3	1978	0.49	1972
Japan	Fukushima 1-6	BWR	GE and Toshiba	1100	3	1979	0.385	1973
Japan	Fukushima 11-1	BWR	Toshiba	1100	4	1982	0.385	1975
Canada	Bruce A1	PHWR	Atomic Energy of Canada Ltd.	904	3	1976		1972

Table 1.1 (continued)

Country	Name	Type	Nuclear system supplier	Power	No. of similar units onsite	First power	Design pressure	Start of erection
Canada	Bruce B5	PHWR	Atomic Energyof Canada Ltd.	904	4	1984	0.291	1978
Canada	Darlington	PHWR	Atomic Energyof Canada Ltd.	935	4	1990	0.197	1982
Canada	Gentilly 2	PHWR	Atomic Energy					
of Canada Ltd.	675	1	1983	0.217	1979			
Canada	Pickering B5	PHWR	Atomic Energyof Canada Ltd.	540	4	1982	0.141	1974
Canada	Point Lepreau 1	PHWR	Atomic Energyof Canada Ltd.	680	1	1982	0.104	1975
Argentina	Emblase	PHWR	Atomic Energy of Canada Ltd.	648	1	1983		1974
China	Qinshan 3	PHWR	Atomic Energyof Canada Ltd.	700	2		0.224	1996

Note: Data collected from Nuclear Journals over a period of 10 years
Data collected from Atomic Energy of Canada, over a period of 10 years and are placed in the order

1.3.3.2 Design Criteria

The containment is designed for a sudden rupture of one (or in some cases two) of the main pipes. Diagram of transient pressures and temperatures in different parts of the containment for two conditions at Forsmark 1. It is conservatively assumed that 5% of the cladding metal in the fuel elements reacts with water causing generation of hydrogen gas. Maximum design temperature is 180°C.

Local reactions and missile forces and jet impingement during design basis accident are specified by the supplier of nuclear systems. Table 1.2 gives the PWR system and VVER adopted in several countries. This system has been widely used in the USA, France, Japan and Germany. In the Russian state, it is with small variation, named VVER. This system is adopted in the Eastern European countries. A typical pictoral view of the Tricastin Power Station based on PWR is shown in Plate 1.14.

Plate 1.14 Tricastin power station, PWR nuclear power station (Central France)

1.4 Additional Advanced Reactor Stations

1.4.1 General Introduction

Recent events in the energy field have emphasised the need for more thoughtful uses of energy and action on securing its transformation and its distribution. The fast breeder reactor has its natural, reasonable rate of development. The Dounreay prototype fast reactor has successfully ended its useful purpose in

Table 1.2 PWR and VVER Stations in the World

Country	Name	Type	Nuclear System Supplier	Power	No. of similar units onsite	First power	Design pressure	Start of erection
JAPAN	Genkai 4	PWR	Mitsubishi Atomic Power Indust.	1180	2	1997	0.500	1985
JAPAN	Ohi 1	PWR	Westinghouse Electric Corp.	1175	2	1978	0.540	1972
JAPAN	Ikata 1	PWR	Mitsubishi Atomic Power Indust.	566	2	1977	0.345	1973
JAPAN	Ikata 3	PWR	Mitsubishi Atomic Power Indust.	890	1	1981	0.389	1978
JAPAN	Mihama 3	PWR	Mitsubishi Atomic Power Indust.	826	1	1976	0.340	1972
ARMENIA	Medzamor 2	VVER	Minatomenergo	408	1	1995	0.200	1995
BELGIUM	Doel 3	PWR	Framaceco	970	2	1982	0.450	1975
BELGIUM	Tihange 2	PWR	Framaceco	960	1	1983	0.450	1976
BELGIUM	Tihange 3	PWR	Acecowen	1070	1	1985		1978
BRAZIL	Angra 1	PWR	Westinghouse	657	1	1982		1972
BULGARIA	Kozloduy 5	VVER	Atomenergoproekt	1000	2	1987		
CHINA	Qinshan II-1	PWR	CNEIC	600	2		0.450	1997
CHINA	Tiawan	VVER	Atomenergoexport	1000	4		0.500	1999
CZECH.REP	Dukovany	VVER	Skoda	440	4	1985		1974
CZECH.REP	Temelin	VVER		972	2		0.490	1982
FINLAND	Olkiluoto (TVO) 1	BWR	ASEA-Atom	820	2	1978	0.470	1973
FRANCE	Belleville 1	PWR	Framatome	1300	2	1987	0.520	1981
FRANCE	Blayais 1	PWR	Framatome	900	4	1981	0.500	1976
FRANCE	Bugey	PWR	Framatome	900	4	1978	0.500	1971
FRANCE	Cattenom 1	PWR	Framatome	1300	4	1986	0.520	1979
FRANCE	Chinon B1	PWR	Framatome	900	4	1982	0.500	1977
FRANCE	Chooz 1	PWR	Framatome	1455	2	1996	0.530	1984
FRANCE	Civaux 1	PWR	Framatome	1455	1	1997	0.530	1991
FRANCE	Civaux 2	PWR	Framatome	1455	1		0.530	1993
FRANCE	Cruas 1	PWR	Framatome	900	4	1980	0.500	1974
FRANCE	Dampierre 1	PWR	Framatome	900	4	1980	0.500	1974

Table 1.2 (continued)

Country	Name	Type	Nuclear System Supplier	Power	No. of similar units onsite	First power	Design pressure	Start of erection
FRANCE	Fessenheim	PWR	Framatome	900	2	1979	0.473	1972
FRANCE	Flamanville 1	PWR	Framatome	1300	2	1985	0.480	1979
FRANCE	Golfech 1	PWR	Framatome	1300	2	1990	0.520	1983
FRANCE	Gravelines 1	PWR	Framatome	900	6	1980	0.500	1974
FRANCE	Nogent 1	PWR	Framatome	1300	2	1987	0.520	1981
FRANCE	Paluel 1	PWR	Framatome	1300	4	1984	0.480	1977
FRANCE	Penly 1	PWR	Framatome	1300	2	1990	0.520	1983
FRANCE	St Alban 1	PWR	Framatome	1300	2	1985	0.480	1979
FRANCE	St Laurent B1	PWR	Framatome	900	2	1983	0.500	1978
FRANCE	Tricastin	PWR	Framatome	900	4	1980	0.500	1974
GERMANY	Brokdorf	PWR	KWU	1440	1	1986	0.750	1976
GERMANY	Emsland	PWR	KWU	1363	1	1982	0.630	1982
GERMANY	Grafenrheinfeld	PWR	KWU	1345	1	1981		1974
GERMANY	Grohnde	PWR	KWU	1430	1	1984		1976
GERMANY	Gundremmingen KRB II	BWR	SIEMENS	1344	2	1984	0.430	1976
GERMANY	Isar 1	BWR	AEG-KWU	907	1	1977		1972
GERMANY	Isar 2	PWR	KWU	1440	1	1988	0.630	1982
GERMANY	Krummel	BWR	AEG-KWU	1316	1	1983		1974
GERMANY	Mulheim Karlich	PWR	BBC-BBR	1302	1	1986		1975
GERMANY	Neckar 1	PWR	SIEM-KWU	840	1	1976	0.570	1972
GERMANY	Neckar 2	PWR	KWU	1365	1	1988	0.630	1982
GERMANY	Philppsburg 2	PWR	KWU	1424	1	1984		1977
GERMANY	Unterweiser	PWR	SIEMENS-KWU	1350	1	1978	0.580	1972
JAPAN	Ohi 3	PWR	Mitsubishi Atomic PoVVER Indust.	1180	2	1992	0.500	1987

Table 1.2 (continued)

Country	Name	Type	Nuclear System Supplier	Power	No. of similar units onsite	First power	Design pressure	Start of erection
JAPAN	Sendai 1	PWR	Mitsubishi Atomic PoVVER Indust.	890	2	1985	0.325	1981
JAPAN	Tomari 1	PWR	Mitsubishi Atomic PoVVER Indust.	579	2	1990	0.360	1984
JAPAN	Tsuruga 2	PWR	Mitsubishi Atomic PoVVER Indust.	1160	1	1987	0.500	1982
JAPAN	Takahama 3	PWR	Mitsubishi Atomic PoVVER Indust.	870	2	1984	0.360	1981
PAKISTAN	Chasma	PWR	China National Nuclear Corp.	300	1	2000		1993
RUSSIA	Balakovo 1	VVER	Minatomenergo (MAE)	1000	4	1985	0.490	1980
RUSSIA	Balakovo 5	VVER	Minatomenergo (MAE)	1000	1		0.490	1987
RUSSIA	Balakovo 6	VVER	Minatomenergo (MAE)	1000	1		0.490	1987
RUSSIA	Kalinin 1	VVER	Minatomenergo (MAE)	1000	2	1984	0.455	1977
CHINA	Daya Bay	PWR	Framatome	985	2	1993	0.520	1985
CHINA	Lingao	PWR	Framatome	985	2		0.520	1997
CHINA	Qinshan 1	PWR	Shanghai Boiler factory	300	1	1991	0.360	1984
RUSSIA	Kalinin 3	VVER	Minatomenergo (MAE)	1000	1		0.490	1988
RUSSIA	Kursk	RBMK	Minatomenergo	1000	4	1976		1972
RUSSIA	Novovoronej 5	VVER	Minatomenergo (MAE)	1000	1	1980	0.455	1974
RUSSIA	Novovoronej 6	VVER	Minatomenergo (MAE)	1000	1		0.490	1990
RUSSIA	Novovoronej 7	VVER	Minatomenergo (MAE)	1000	1		0.490	1990
RUSSIA	Rostov 1	VVER	Minatomenergo (MAE)	1000	2	2000	0.490	1984
RUSSIA	Smolensk 1	RBMK	Minatomenergo	1000	3	1982		1975
SLOVAKIA	Bohunice 1	VVER	Atomenergoproekt & Skoda	430	4	1978		1973
SLOVAKIA	Mochovce 1	VVER	Atomenergoproekt & Skoda	432	4			1983
SLOVENIA	Krsko	PWR	Westinghouse	664	1	1981		1974

Table 1.2 (continued)

Country	Name	Type	Nuclear System Supplier	Power	No. of similar units onsite	First power	Design pressure	Start of erection
SOUTH AFRICA	Koeberg	PWR	Framatome	965	2	1984	0.500	1976
SOUTH KOREA	Kori 3	PWR	Westinghouse Electric Corp	950	2	1985	0.520	1979
SOUTH KOREA	Uljin 1	PWR	Framatome	950	2	1988	0.500	1982
SOUTH KOREA	Yonggwant 1	PWR	KHIC-ABBCE	950	2	1986	0.520	1980
SPAIN	Almaraz 1	PWR	Westinghouse Electric Corp.	975	2	1981		1972
SPAIN	Asco 1	PWR	Westinghouse Electric Corp.	930	2	1983	0.480	1973
SPAIN	Asco 2	PWR	Westinghouse Electric Corp.	966	1	1985		1974
SPAIN	Trillo 1	PWR	KWU	1066	1	1988		1979
SPAIN	Vandellos 2	PWR	Westinghouse Electric Corp.	1009	1	1987		1976
SWEDEN	Ringhals 3	PWR	Westinghouse	915	2	1980	0.514	1972
SWITZERLAND	Gösgen	PWR	KWU	1020	1	1979	0.589	1973
TAIWAN	Maanshan 1	PWR	Westinghouse	915	2	1984		1977
UKRAINE	Khmelnitsky 1	VVER	Minatomenergo (MAE)	1000	2	1987	0.490	1981
UKRAINE	Khmelnitsky 2	VVER	Minatomenergo (MAE)	1000	1		0.490	1985
UKRAINE	Rovno 1	VVER	Minatomenergo (MAE)	402	2	1980		1976
UKRAINE	Rovno 3	VVER	Minatomenergo (MAE)	1000	2	1986	0.490	1981
UKRAINE	Rovno 4	VVER	Minatomenergo (MAE)	1000	1		0.490	1981
UKRAINE	Sud Ukraine 1	VVER	Minatomenergo (MAE)	1000	3	1982	0.490	1977
UKRAINE	Tchernobyl 3	RBMK	Minatomenergo	1000	1	1981		1977
UKRAINE	Zaporozhe 5	VVER	Minatomenergo (MAE)	1000	6	1989	0.490	1985
USA	Braidwood 1	PWR	Westinghouse	1175	2	1987	0.455	1975
USA	Byron 1	PWR	Westinghouse	1175	2	1985	0.445	1975
USA	Callaway-1	PWR	Westinghouse Electric Corp.	1234	1	1984		1976
USA	Catawba 1	PWR	Westinghouse	1205	2	1985	0.204	1974

Table 1.2 (continued)

Country	Name	Type	Nuclear System Supplier	Power	No. of similar units onsite	First power	Design pressure	Start of erection
USA	Comanche 1	PWR	Westinghouse Electric Corp.	1161	2	1993	0.445	1975
USA	Millstone 3	PWR	Westinghouse	1209	1	1986		1974
USA	Nine Mile Point 2	BWR	General Electric	1205	1	1987	0.411	1975
USA	Palo Verde 1	PWR	Combustion Engineering Co	1307	3	1985	0.514	1976
USA	Perry 1	BWR	General Electric	1250	1	1986	0.204	1974
USA	Rover Bend 1	BWR	General Electric	991	1	1985	0.204	1975
USA	San Onofre 2	PWR	Combustion Engineering Co	1127	2	1982	0.514	1974
USA	Seabrook 1	PWR	Westinghouse	1197	1	1989	0.459	1976
USA	Shearon-Harris 1	PWR	Westinghouse Electric Corp.	960	1	1987		1974
USA	South Texas	PWR	Westinghouse Electric Corp.	1315	2	1988		1975
USA	St Lucie 2	PWR	Combustion Engineering Co	882	1	1983	0.376	1976
USA	Summer 1	PWR	Westinghouse Electric Corp.	950	1	1982		1973
USA	Vogtle 1	PWR	Westinghouse Electric Corp.	1223	2	1987	0.459	1977
USA	Waterford 3	PWR	Combustion Engineering Co	1153	1	1985		1972
USA	Watts Bar 1	PWR	Westinghouse Electric Corp.	1218	1	1996	0.193	1973
USA	Wolf Creek	PWR	Westinghouse Electric Corp.	1214	1	1985		1977

Table 1.3 Fast Breeder Nuclear Power Plant

Country	Name	Type	Nuclear system supplies	Power (MW)	Unit	Status/ erection/ commission
India	FBTR Kalpakkam	FR	Dept. of Atomic Energy	15	1	1472/1985
Japan	Monju	FR	Power reactor and nuclear fuel development	280	1	1994/1985
Russia	BN350	FR		150	1	1973/1969
Russia	BOR360	FR		60	1	1973/1970
France	PHENIX	FR	EDF	253	1	1973/1971
USA	EBR II	FR	USAEC	62.5	1	1981/1975
USA	LMFBR	FR	TVA/AEC	975	1	1981/1976
Germany	KNK II	FR		60	1	1972/1969

Note: The Indian and Japanese reactors are installed using single-walled PCPR. The minatom of Russia reported installed power of 600 MW, the Japanese 280 MW stations.
Data collected from 15 years of Nuclear Engineering of Design, Elsevier, Holland.

Scotland. PHENIX in France went critical on 21 August 1973. The large prototype reactor SNR300 developed jointly by Germany, Belgium and the Netherlands did go critical on 1 January 1979. These projects gave around 800 MW_e. The Italian 130 MW_t PLC reactor does give considerable experience in this type of reactor system. Table 1.3 gives a brief of these reactors. In addition, British gas-cooled reactors are mentioned and other new diversified reactors are identified which are under planning.

1.4.2 Fast Breeder Reactor

Apart from the ones mentioned regarding fast reactors, it is therefore necessary to explain some technical aspects.

The design and construction and commissioning of PFR by the UKAEA on 14 December 1973 has in brief the following features:

- Reactor contained within a concrete vault situated between ground level.
- Possibility of raising and lowering the leak jacket during various stages of reactor construction.
- An existence of support lower around top of the vault and a system for jacking the roof structure offers a correct positioning for welding the diagrid support structures and primary tank to the roof, filling the top strake in solutions and welding the leak jacket to the roof. The diagrid was proof loaded to simulate the core and neutron shield rods.
- Commissioning sequence:
- Primary circuit
  - Heat and purge and prepare sodium fill
  - Dynamic running and sodium cleanup

- Vibration tests and functional testing of charge machine
- Fuel loading
- Zero energy test
- Hot dynamic running
- Run up to power

- Secondary circuits
  - All the above
  - Hot helium leak test
  - Commission water side
  - Hydrogen detection system tests

- Turbine and ancilliaries
  - Commissioning temporary boilers
  - Runup power

1.4.2.1 Fermi I Project

The Enrico Fermi project on fast breeder reactor was formally committed on 8 August 1955. In its 17-year history, the Fermi project has made significant contributions to fast rector technology. The following reports summed up Fermi I project.

(a) Alexanderson E.L. et al. Enrico-Fermi Atomic Power plant operating experience through 900 MW_t ANS-101, April 1967.
(b) Duffy I.G. et al. Operating experience with major components of the Enrico-Fermi Atomic power plant. IAEA-SM-127/40, Vienna, 1970.

1.4.2.2 British Gas-Cooled System

The advanced gas-cooled reactor systems are given in Table 1.4. The reactor pressure vessels are made either in prestressed concrete pressure vessels or in double PCCV.

The high-temperature gas-cooled reactors have been installed in Europe. Typical multi-cavity vessels have been thoroughly analysed in this Chapter.

Table 1.4 British system

Country	Name	Type	MW	No.	Erection
Great Britain	Heysham	AGR	660	4	1987
Great Britain	Sizewell B	PWR	1258	1	1995
Great Britain	Tomess PT 1	AGR	582	2	1987

1.4.3 New Diversified Systems

The following new systems have been planned.

1.4.4 EU-APWR

MHI has developed the EU-APWR based on 1538 MW planned for the PS using power station units 3 and 4 for the Japan Atomic Power Company.

1.4.4.1 PBMR

China is planning to develop the world's first commercially operated pebble bed nuclear power reactor of 195 MW_e station in Shandong Province.

1.5 Regulator Guides

The following are some of the useful guides for structures and appartances in nuclear power stations:

(a) US Nuclear Regulatory Commission

Title	Rev.	Issued year/ month
Net positive suction head for Emergency core cooling and containment Heat removal system pumps (safety guide 1)	...	70/11
Thermal shock to reactor pressure vessels (safety guide 2)	...	70/11
Assumption used for elevating the potential radiological consequence of a loss-of-coolant accident for boiling water reactors	...	70/11
	1	73/06
	2	74/06
Assumption used for elevating the potential radiological consequence of a loss-of-coolant accident for pressurised water reactors	...	70/11
	1	73/06
	2	71/03
Assumption used for elevating the potential radiological consequence of a steam line break accident for boiling water Reactors (safety guide 5)	...	71/03
Independent between redundant standby (onsite) power sources and between their distribution systems (safety guide 6)	...	71/03
Control of combustible gas concentrations in containment following a loss-of-coolant accident	...	71/03
	1	76/09
	2	78/11
Personnel selection and training	...	71/03
	1	75/09
	1-R	77/05
Selection, design and qualification of diesel generator units used as onsite electric power systems at nuclear power plants (for comment)	...	71/03
		78/11
Mechanical (cad weld) splices in reinforcing bars of category 1 concrete structures	...	71/03
	1	73/01
Instrument lines penetrating primary reactor containment (safety guide 11)	...	71/03

(continued)

Title	Rev.	Issued year/ month
Supplement to safety guide 11, backfitting considerations		72/02
Instruments for earthquakes	. . .	71/03
	1	74/04
Spent fuel storage facility design basis (for comment)	. . .	71/03
	1	75/12
Reactor coolant pump flywheel integrity (for comment)	. . .	71/1075/08
	1	
Testing of reinforcing bars for category 1 concrete structures	. . .	71/10
	1	72/12
Reporting of operating information–Appendix A technical specifications (for comment)	. . .	71/10
	1	73/10
	2	74/09
	3	75/01
	4	75/08
Protection of nuclear power plants against industrial sabotage	. . .	71/10
	1	73/06
Structural acceptance test for concrete primary reactor containments	. . .	71/10
	1	72/12
Non-destructive examination of primary containment liner welds safety Guide 19	. . .	71/12
	1	72/08
Comprehensive vibration assessment programme for reactor internals during preoperational and initial startup testing	. . .	71/12
	1	75/06
	2	76/05
Measuring, elevating and reporting radioactivity in solid wastes and releases of radioactive materials in liquid and gaseous effluents from light-water-cooled nuclear power plants	. . .	71/12
	1	74/06
Periodic testing of protection system actuation functions (safety guide 22)	. . .	72/02
Onsite meteorological programmes (safety guide 23)	. . .	72/02
Assumption used for elevating the potential radiological consequences of a pressurised water reactor gas storage tank failure (safety guide 24)	. . .	72/03
Assumption used for elevating the potential radiological consequences of a fuel handing accident in fuel handling and storage facility for boiling and pressurised water (safety guide 25)	. . .	72/03
Quality group classifications and standards for water-, steam- and radioactive-waste-containing components of nuclear power plants (for comment)	. . .	72/03
	1	74/09
	2	75/06
	3	76/02
Ultimate heat sink for nuclear power plants (for comment)	. . .	72/0374/0376/
	12	01
Quality Assurance programme requirements (design and construction) (for comment)	. . .	72/06
	1	78/03
Seismic design classification	. . .	72/06

(continued)

Title	Rev.	Issued year/ month
	1	73/08
	2	76/02
	3	78/09
Quality assurance requirements for the installation, inspection and testing of instrumentation and electric equipment (safety guide 30)	...	72/08
Control of ferrite content in stainless steel weld metal	...	72/08
	1	73/06
	2	77/05
	3	78/04
Criteria for safety-related electric power systems for nuclear power	...	72/08
plants	1	76/03
	2	77/02
Quality assurance programme requirements (operation)	...	72/11
	1	77/02
	3	78/02
Control of electric weld properties	...	72/12
In-service inspection of ungrouted tendons in prestressed concrete	...	73/02
containment structures	1	74/06
	2	76/01
Non-metallic thermal insulation for austenitic stainless steel	...	73/02
Quality assurance requirements for cleaning of fluid systems and associated components of water-cooled nuclear power plants	...	73/03
Quality assurance requirements for packaging, shipping, receiving,	...	73/03
storage and handling of items for water-cooled nuclear power plants	1	76/10
	2	77/05
Housekeeping requirements for water-cooled nuclear power plants	...	73/03
	1	76/10
	2	77/09
Qualification test for continuous-duty motors installed inside the containment of water-cooled nuclear power plants	...	73/03
Preoperational testing of redundant on-site electric power systems to verify proper load group assignments	...	73/03
(withdrawn – See 41 FR 11891, 3/22/.76)	...	...
Control of stainless steel weld cladding of low-alloy steel components	...	73/05
Control of the use of sensitised stainless steel	...	73/05
Reactor coolant pressure boundary leakage detection systems	...	73/05
Protection against pipe whip inside containment	...	73/05
Bypassed and inoperable status indication for nuclear power plant safety systems	...	73/05
Design limits and loading combinations of seismic category 1 fluid system components	...	73/05
Power levels of nuclear power plants	...	73/05
	1	73/12

(continued)

Title	Rev.	Issued year/ month
Control of preheat temperature for welding of low-llloy steel	...	73/05
(withdrawn – see 40 FR 30510, 7/21/75)	...	...
Design, testing and maintenance criteria for post-accident engineered-	...	73/06
safety-feature atmosphere cleanup system, air filtration and	1	76/07
adsorption units of light-water-cooled nuclear power plants	2	78/03
Application of the single-failure criterion to nuclear power plant protection systems	...	73/06
Quality assurance requirements for protective coating Applied to water-cooled nuclear power plants	...	73/06
Concrete placement in category 1 structures	...	73/06
Maintenance of water purity in boiling water reactor (for	...	73/06
comment)	1	78/07
Design limits and loading combinations for metal primary reactor containment system components	...	73/06
Qualification of nuclear power plant inspection, examination and testing personnel	...	73/08
Design basis floods for nuclear power plants	...	73/0876/0477/
	12	08
Design response spectra for seismic design of nuclear power plants	...	73/10
	1	73/12
Damping values for seismic design of nuclear power plants	...	73/10
Manual initiation of protective actions	...	73/10
Electric penetration assemblies in containment structures for	...	73/10
light-water-cooled nuclear power plants	1	77/05
	2	78/07
Quality assurance requirements for the design of nuclear power	...	73/10
Plants	1	75/02
	2	76/06
Material and inspections for reactor vessel closure studs	...	73/10
(withdrawn – see 42 FR 54478, 10/06/77	...	...
Installation of overpressure protection devices	...	73/10
Initial test programmes for water-cooled reactor power plants	...	73/11
	1	77/01
	2	78/08
Preoperational and initial startup testing of feedwater and condensate systems for boiling water reactor power plants	...	75/12
	1	77/01
Initial startup test programme to demonstrate remote shutdown	...	77/01
capability for water-cooled nuclear power plants	1	78/08
Standard format and content of safety analysis reports for nuclear	...	72/02
power plants	1	72/10
	2	75/09
	3	78/11
Welder qualification for areas of limited accessibility	...	73/12

(continued)

Title	Rev.	Issued year/ month
Spray pond piping made from fibreglass-reinforced thermosetting	...	73/12
resin	1	78/01
	2	78/11
Qualification test of electric valve operators installed inside the		
containment of nuclear power plant	...	74/01
Quality assurance terms and definitions	...	74/02
Physical independence of electrical system	...	74/02
	1	75/01
	2	78/09
Design basis tornado for nuclear power plants	...	74/04
Assumptions used for evaluating a control bed ejection accident		
for pressurised water reactors	...	74/05
Assumptions for evaluating habitability of a nuclear power plant		
control room during a postulated hazardous chemical release	...	74/06
Preoperational testing Of emergency core cooling system of	...	74/06
Pressurised water reactor	1	75/09
Preoperational testing of instrument air system	...	74/06
Shared emergency and shutdown electric systems for multi-unit	...	74/06
nuclear power plants	1	75/01
Sumps of emergency core cooling and containment spray systems	...	74/06
In-service inspection of pressurised water reactor steam generator	...	74/06
tubes	1	75/07
Code case acceptability – ASME section III design and fabrication	...	74/06
	...	76/03
Protection against low-trajectory turbine missiles	1	77/07
Quality assurance requirements for installation, inspection and	...	76/07
testing of mechanical equipments and systems	O-R	75/05
	...	76/07
Tornado design classification	1	78/04
Periodic testing of electrical power and protection systems	...	76/06
	1	77/11
	2	78/06
(withdrawn – see 42 FR 33387, 6/30/77)	...	...
Fire protection guidelines for nuclear power plants (for comment)	...	76/06
	1	77/11
Bases for plugging degraded PWR steam generator tubes (for comment)	...	76/08
Development of floor design response spectra for seismic design of	...	76/10
floor-supported equipment or components	1	77/07
Quality assurance requirements for control of procurement of items	...	76/10
and service for nuclear power plants	1	77/07
Overhead crane handling for nuclear power plants (for comment)	...	76/02
	...	75/11
Instrument set points	1	76/11
Thermal overload protection for electric motors on motor-operated	...	75/11

(continued)

Title	Rev.	Issued year/ month
valves	1	77/03
Qualifications for cement grouting for prestressing tendons in	...	75/11
containment structures	1	77/02
Periodic testing of diesel generator units used as onsite electric	...	76/08
power systems at nuclear power plants	1	77/08
Calculation of annual doses to man from routine releases of reactor	...	76/03
effluents for the purpose of evaluating compliance with 10 CFR Part 50, Appendix 1	1	77/10
Cost–benefit analysis for radwaste systems for light-water- cooled nuclear power reactors (for comment)	...	76/03
Methods of estimating atmospheric transport and dispersion of	...	76/03
gaseous effluents in routine releases rrom light-water-cooled reactors	1	77/07
Calculation of releases of radioactive materials in gaseous and	...	76/04
liquid effluents rrom light-water-cooled power reactors	1	77/05
Estimating aquatic dispersion of effluents from accidental and	...	76/05
routine reactor releases for the purpose of implementing Appendix 1	1	77/04
Guidance on being operator at the controls of a nuclear power plant	...	76/02
	1	76/01
Service limits and loading combinations for class 1 linear-type	...	76/11
component supports	1	78/01
Physical models for design and operation of hydraulic structures	...	77/03
and systems for nuclear power plants	1	78/10
An acceptable model and related statistical methods for the analysis	...	77/03
of fuel densification	1	78/03
Inspection of water-cooled structures associated with nuclear	...	77/04
power plants	1	78/03
Installation design and installation of large lead storage batteries for	...	77/04
nuclear power plants	1	78/10
Maintenance, testing and replacement of large lead storage	...	77/04
batteries for nuclear power plants	1	78/02
Service limits and loading combinations for class 1 plate-and-shell-	...	77/07
type component supports	1	78/10
Qualification test of electrical cables, field splices and connection	...	77/08
for light-water-cooled nuclear water plants (for comment)	...	77/09
Site investigations for foundations of nuclear power plant (for comment)	...	77/09
Loose-part detection programme for the primary system of light-water-cooled reactors (for comment)	...	77/09
Medical certification and monitoring of personnel requiring operator licences (for comment)	...	77/09
Normal water level and dischcarge at nuclear power plant (for comment)	...	77/09

(continued)

Title	Rev.	Issued year/ month
Material for concrete containments	...	77/11
	1	78/10
Fuel–oil systems are standby diesel generator (for comment)	...	78/01
Laboratory investigations of soils for engineering analysis and design of nuclear power plant (for comment)	...	78/04
Guidance for residual heat removal (for comment)	...	78/05
Design, testing and maintenance criteria for national ventilation Exhaust system air filtration and adsoprption units of light water-cooled nuclear power plants (for comment)	...	78/03
Containment isolation provisions for fluid systems (for comment)	...	78/04
Safety-related concrete structures for nuclear power plants (other than reactor vessels and containments) (for comment)	...	78/04
Design guidance for radioactive waste management systems, structures and components installed in light-water-cooled nuclear power plants (for comment)	...	78/07

(b) American Nuclear Society, Nuclear Standard Projects of Prime Interest to Structural Engineers

ANS 2	**Site elevation**
ANS 2.1	Guidelines for determining the vibratory found motions for the design earthquake for nuclear facilities
ANS 2.3	Standards for estimating tornadoes, hurricanes and other extreme wind parameters at power reactor sites
ANS 2.4	Guidelines for determining tsunami criteria for power reactor sites
ANS 2.7	Guidelines for assessing capability for surface faulting at nuclear power reactor sites
ANS 2.8	Standards for determining design basis flooding at power reactor sites
ANS 2.10	Guidelines for retrieval, review, processing and elevation of records obtained from seismic instrumentations
ANS 2.11	Guidelines for elevating site-related geotechnical parameters at power reactor sites
ANS 2.12	Guidelines for combining natural and external man-made hazards at power reactor sites
ANS 2.14	Determination of the shape of the response spectra for use in nuclearFacilities design
ANS 2.19	Guidelines for elevating site related parameters for independent spent fuel storage facilities

ANS 4	Criteria, control and dynamics
ANS 4.1	Design basis criteria for safety system in nuclear power generating stations
ANS 58.1	Plant design against missiles
ANS 58.2	Design basis for protection of nuclear power plants against effects of postulated pipe rupture
ANS 58.5	Probabilistic risk assessment

ANS 51	Pressurized water reactor
ANS 51.1	Nuclear safety criteria for the design of stationary pressurized water reactor plants
ANS 51.7	Single failure criteria for PWR fluid systems

ANS 52	Boiling water reactor
ANS 52.1	Nuclear safety criteria for the design of stationary boiling water reactor plants
ANS 58.9	LWR single failure criteria

ANS 53	High-cooled gas-cooled reactor management committee
ANS 53.1	Nuclear safety criteria for the design of stationary gas cooled reactor plants
ANS 53.6	Gas-cooled reactor plant containment system
ANS 53.21	Gas-cooled reactor plant secondary coolant systems

ANS 54	Liquid metal fast breeder reactor
ANS 54.1	LMFBR general design criteria
ANS 543	Principal design criteria for LMFBR containments
ANS 54.6	LMFBR safety classification and related requirements
ANS 54.10	Risk limit guidelines for LMFBR design
ANS 54.11	application of risk limit guidelines for LMFBR design
ANS 54.12	Event categorisation guidelines for LMFBR design

ANS 55	Fuel and radwaste
ANS 57.5	Away from reactor's spent dual-storage facilities

ANS 56	Containment
ANS 56.3	Overpressure protection of low-pressure system connected to the reactor coolant pressure boundary
ANS 54.6	Pressure/temperature transient analysis for LWR containments
ANS 56.8	Reactor containment leakage testing requirements
ANS 56.9	Environmental envelopes to be considered in safety-related equipment

Subsections NCA-NG, boiler and pressure vessel code section III, division I–nuclear power plant components

Boiler and pressure vessel code section III, division 2–subsection CB and CC concrete reactor vessels and containment

Section XI boiler and pressure vessel code–fuels for inservice inspection of nuclear power plant components

N626.3-1978 Qualification and duties of personnel engaged in ASME boiler and pressure vessel code section III divisions 1 and 2 certifying activities

N626.0-1974 Qualifications and duties for authorised nuclear Inspection

N626.1-1975 Qualifications and duties for authorised inservice Inspection

N626.2-1976 Qualifications and duties for authorised nuclear inspection (concrete)

N45.2/N45.2-1977 Quality assurance programme requirements for nuclear facilities QA-76-2

N45.2.5-1978 Supplementary Quality assurance requirements for installation, inspection and testing of structural concrete and structural steel, sails and foundations during the constructional phase of nuclear power plant QA-76-5

N45.2.6-1973 Qualifications of inspection, examination and testing personnel for the construction phase of nuclear power plants QA-76-6

N45.2.10-1973 Quality assurance Terms and definitions QA-76-10

N45.2.11-1974 Quality assurance requirements for the design of nuclear power plants QA-76-11

N45.2.20 Supplementary quality assurance requirements for subsurface investigation prior to construction phase of nuclear power plant

1.6 Boiling Water Reactors (BWR), Pressurised Heavy Water Reactor (PHWR) and Pressurised Water Reactor (PWR)

Since the publication of this book, many reactors of specific systems would have been planned and constructed throughout the world. Every offer has been made for the benefit of the readers under BWR, PHWR and PWR. They are constructed facilities.

Tables 1.1 and 1.2 give a list of well-known nuclear power stations. For others references can be made to various journals in the field of nuclear facilities.

References

1. Hojgaard, E., Kallero, K. E., and Nilsson, L. B. New Type of Reactor Containment of Prestressed Concrete. Report ANS Winter Meeting, 1968.
2. Gimstedt, Haga, I., Wassberg, R., and Nilsson, L. B. Articles on Oskarshamn 1 Nuclear Power Station. Nuclear Engineering International. 1970.
3. Lindbo, T. and Boye-moller, K. Design and construction of concrete containment for BWR. J 2/6, Second SMiRT Conference, Berlin, 1973.
4. Margen, P. H. and Menon, S. Preparation for exploitation of prestressed concrete reactor vessels for light water reactors. NUCLEX 72, Basel, 1972.
5. HAY, J. D. and EADIE, D. McD. The prestressed concrete pressure vessels for Hinkley Point 'B' and Hunterston 'B'. Paper No. SM.ll 1/31. Symposium on Advanced and High Temperature Gas-cooled Reactors. Jülich, Germany, October, 1968.

6. Burrow, R. E. D. and WILLIAMS, A. J. Hartlepool advanced gas-cooled reactor—reactor pressure vessel. Nuclear Engineering International. Vol. 14, No. 162, 1969.
7. Menon, S., Rasmussen, I., Tarandi, T., and Kraemer, W. Scandinavian PCRV development: present status and planned Work, 1973–75. H 3/9, Second SMiRT Conference, Berlin, 1973.
8. Burrow, R. E. D. and Brunton, J. D. Developments in large prestressing systems for nuclear pressure vessels. FIP Sixth International Congress. Prague, 1970.
9. Mckean, J. D., Tyrell, J., Lawson D. S., and Burrow, R. E. D. Heysham nuclear power station. Nuclear Engineering International Vol. 16, No. 186, 1971.
10. BS.4975. Specification for prestressed concrete pressure vessels for nuclear reactors. British Standard Institution, 1973.
11. TAN, C. A Study of the Design and Construction Practices of Prestressed and Reinforced Concrete Containment Vessels. The Franklin Institute Research Laboratories, Philadelphia, 1969, p. 506.
12. Kulka, F. and Wahl, H. W. American practices in the design of prestressed concrete containment structures. PCI Journal, 1968.
13. Schupack, M. Large post-tensioning tendons. PCI Journal, 1972.
14. Wahl, H. W. and Kosiba, R. J. Design and construction aspects of large prestressed concrete (PWR) containment structures. ACI Journal, 1969.
15. Lorenz, H. Design of the concrete containment vessel for the R. E. Ginna nuclear power plant. Nuclear Engineering and Design Vol. 6, 1967, pp. 360–366.
16. Preliminary Safety Analysis Report. Bellefonte Nuclear Plant, Docket No. 50438 and No. 50-439, Tennessee Valley Authority. Private Communications TVR 1990.
17. ACI-ASME Technical Committee on Concrete Pressure Components for Nuclear Service. Proposed standard code for concrete reactor vessels and containments, copyright 1973.
18. United Engineers and Constructors, Seabrook Station Containment Aircraft Impact Analysis, Prepared for Public Service Co. of New Hampshire, Seabrook, New Hampshire, Revision 1, January, 1975.
19. NRC Standard Review Plan Section 3.6.1, Plant Design for Protecting Against Postulated Piping Failures in Fluid Systems Outside Containment U.S. Nuclear Regulatory Commission, Office of Nuclear Reactor Regulation, 1975.
20. NRC Standard Review Plan Section 3.6.2. Determination of break locations and dynamic effects associated with postulated rupture of piping, U.S. Nuclear Regulatory Commission, Office of Nuclear Reactor Regulations, 1975.
21. ANS 58.2 Committee. Design basis for protection of nuclear power plants against effects of postulated rupture, ANS-N 176, 1977 (Draft).
22. CEGB Private Communications on British Nuclear Power Stations, 1979.
23. UKAEA Private Communications on British Nuclear Power Stations: Winfirth, 1984.
24. TNPG Group. Design of nuclear power stations belonging to TNPG consortium inclusive of Mccalpine Engineering 1979.
25. NRC list of regulatory guides related to nuclear power stations listed below and other guides:
26. The following NRC Regulatory guides apply in particular to containment structural design and construction:
27. Regulatory Guide 1.10—Mechanical (Cadweld) Splices in Reinforcing Bars of Category I Concrete Structures.
28. Regulatory Guide 1.15—Testing of Reinforcing Bars for Category I Concrete Structures.
29. Regulatory Guide 1.18—Structural Acceptance Test for Concrete Primary Reactor Containments.
30. Regulatory Guide 1.19—Non-destructive Examination of Primary Containment Liner Welds.
31. Regulatory Guide 1.29—Seismic Design Classification.

32. Regulatory Guide 1.35—In-service Inspection of Ungrouted Tendons in Prestressed Concrete Containment Structures.
33. Regulatory Guide 1 .46—Protection Against Pipe Whip Inside Containment.
34. Regulatory Guide 1.54—Quality Assurance Requirements for Protective Coatings Applied to Water-Cooled Nuclear Power Plants.
35. Regulatory Guide 1.55—Concrete Placement in Category I Structures.
36. Regulatory Guide 1.57—Design Limits and Loading Combinations for Metal
37. Primary Reactor Containment System Components.
38. Regulatory Guide 1.59—Design Basis Floods for Nuclear Power Plants.
39. Regulatory Guide 1.60—Design Response Spectra for Seismic Design of Nuclear Power Plants.
40. Regulatory Guide 1.61—Damping Values for Seismic Design of Nuclear Power Plants.
41. Regulatory Guide 1 .76—Design Basis Tornado for Nuclear Power Plants.
42. KWU. German nuclear power stations. Private Communications, 1982.
43. EDF. French nuclear power stations. Private Communications, 1983.
44. Asea. Atom. Private Communications (1981–1984) Stockholm, Sweden, 1981–1984.
45. AECL Pressurised Heavy Water Reactors: Candu stations, 1979–1984.
46. Japan (JNC) Private Communications Otawa, Tokyo. Private Communications on FBR stations, 1979–1985.
47. Swedish CNPW & CDL. Swedish Commercial Nuclear Power BWR Plants and Ancillaries. Private communications, Sweden. Stockholm, 1983–1987.
48. BNES. Fast Breeder Reactor Core and Fuel Structural Behaviour British Nuclear Energy Society, London. Conference held in Inverness 4.6 June 1990 Scotland.
49. BNES. Fast Reactor Power Stations. Proceeding of International conference, 1974, ICE Building, London, 1974.
50. Munce. J. F. The Architecture in the Nuclear Age. I Life Books Ltd, 1964.
51. I mechE Proceedings of the Symposium on Berkeley and Bradwell Nuclear Power Stations. The Institution of Mechanical Engineers, London, 1963.
52. APC SGHWR Atomic Power Constructions Ltd, Surrey, England, 1966.
53. BNES Steam Generating and Other Heavy Water Reactors. British Nuclear Society. Proceedings London, 1968.
54. CEGB Proof of Evidence on the Design of Sizewell-B PWR, 1982.
55. SMIRT Proceedings: (SMiRT 1–13 Vol. 8–10) Each Proceedings Structural Mechanics in Reactor Technology Proceedings No. 1–13. Nuclear Engineering & Design Journals, 1970–2001.
56. IJNE International Journals of Nuclear Engineering. 1969–2007, London, UK
57. JNE & D International Journal of Nuclear Engineering And Design. Elsevier, Holland, 1979–2008.

Chapter 2
Loads and Material Properties for Nuclear Facilities – A General Survey

2.1 Introduction

The probable failure assessment of structures for nuclear power facilities has bearings on the choice and postulation of the loads and load combinations, since the exact magnitude of the loads encountered in nuclear power plant design cannot be easily predicted. The loads are normally treated as random variables. These loads are generally defined in terms of probability of strength in different components/elements of structures for a nuclear facility. Together with the strength of characteristics of elements, it would be possible to determine the probability of the structure being able to perform the functions for which it has been designed. For obtaining reliable results a proper accounting of uncertainties practically at every stage of stress determination is necessary. The stress determination is the end product of

(a) the analysis and prediction of postulated even loads;
(b) the probability distribution of different variables involved causing the loads to occur. One form is the statistical sampling technique. The data and probability distribution will lead to the load to be considered.

In addition to the two load levels considered in conventional design, nuclear facilities are typically designed for third load level, termed the extreme load. Extreme loads include such natural phenomena as the maximum earthquake potential for a site considering regional and local geology, seismology, local foundation conditions, tornado wind and associated air-borne missile loads. It also includes postulated design basis accident loads consisting of high-energy system rupture that results in pipe break reaction and impingement loads, pipe whip and associated accident-generated missiles and pressurisation of building components, flooding and high thermal transients.

In the USA structures for nuclear facilities are designed for service load conditions. Three methods are recommended such as working stress design (WSD), factor load deign (FLD) and factor load reduction design (FLRD). In Europe the limit state design (LSD) is generally recommended using partial safety factors for materials and loads or actions. Where service load design parameters and material strength data for nuclear facilities are not given or

M.Y.H. Bangash, *Structures for Nuclear Facilities*,
DOI 10.1007/978-3-642-12560-7_2, © M.Y.H. Bangash 2011

Table 2.1 Service load parameters

Loads or stresses	Design load or range of actions	Notation
Dead load		D dead load
Reinforced/prestressed Concrete	$6.7-7.5\,kN/m^2$	
Structural steel	$24\,kN/m$	
Structural aluminium		
Structural wood		
I. S – snow load	Specified: ANSI A58.1 100-year interval or BS6399 Part 2 (see text table in this chapter)	L,LL Live or imposed or action
II. Construction	$1.73\,kN/m^2$	Co
III. Buoyancy	$0.91\,kN/m^2$	B
IV. Earth pressure (lateral)		E_p
(a) Active } Foundation below grade	$3.14-18.84\,kN/m^2/m$	
(b) Passive } Foundation below grade	$3.14-70.65\,kN/m^2/m$	
(c) At rest } Foundation below grade	$6.28-23.55\,kN/m^2/m$	
V. Piping equipment reaction	Depends on the piping analysis ⟶ variable	R_t
VI. Hatch for containment equipment	$67\,kN/m^2$	H_e
(a) Uniform load assumed	2000 kN	
(b) Concentrated load moving type		H_p
Note: Not concurrent with uniform lived load		
(c) Personnel hatch uniform load moving type	$67\,kN/m^2$ 44.5 kN	
VII. Linear for concrete containment		
(a) Concentrated load	6.67 kN	
(b) Line load	2.2 kN/m	
(c) Uniform load	$25\,kN/m^2$	
VIII. Uniform floor load		
(a) Reactor building operating deck	$1675\,kN/m^2$	
(b) Stairs, passage and escalator/moving walks	$67\,kN/m^2$(uniform) Plus concentrated from manufacturer B	
IX. Wind load	$F_w = q_{ref}.c_e(z_e).c_d.c_{fi}(A_{ref})$ where	W or F_w P
(a) Euro code and BS6399	z_e height above ground q_{ref} mean wind velocity pressure c_f derived force coefficient c_e exposure coefficient; c_d dynamic factor	

Table 2.1 (continued)

Loads or stresses	Design load or range of actions	Notation
	c_{fi} force coefficient	
	A_{ref} reference area of c_f	
	$v_{ref} = c_{DIR} \cdot c_{TEM} \cdot c_{ALT} \cdot v_{refo}$	
	v_{refo} basic value of reference wind velocity	
	c_{DIR} directional factor taken generally as 1.0	
	c_{TEM} seasonal factor taken generally as 1.0	
	c_{ALT} altitude factor taken to be 1.0 unless specified	
	q_{ref} = wind mean velocity pressure Xv_{ref} = Fw	
(b) Based on BS6399	V_s = site velocity = $V_o S_a S_d S_s S_p$	
	V_b = basic wind speed $\rightarrow v_{refo}$	
	S_a = altitude factor $\rightarrow c_{ALT}$	
	S_d = Direction factor $\rightarrow c_{DIR} = 1$	
	S_s = seasonal factor $\rightarrow c_{TEM} = 1$	
	S_p = probability factor = 1	
	q = wind dynamic pressure = kV_s^2	
	where k = 0.613	
	$V_e = V_s \times V_p$ – terrain factor	
	Example for $V_b = 45\,\text{m/s}; s_1 = 1; s_2 = 0.83$	
	Solution based on BS 6399	
	$V_b = V_b S_a S_d S_s S_p$;	
	V_b = basic wind speed = 23.5 m/s	
	$S_a = 1 + 0.001 \times 100 = 1.1$	
	$S_p = 1.0$	
	$S_s = 0.62$	
	$S_p = 1.0$	
	$V_s = 16.03\,\text{m/s}$	
	$V_e = SV_sS_b$	
	$S_b = S_cT_c[1 + (g_tS_tT_y) + S_h]$	
	$= 1.08(0.863)[1 + (2.52 \times 1.71 \times 1.38) + 0]$	
	$= 1.486$	
	$V_e = 16.03 \times 1.486 = 23.82\,\text{m/s}$	
	q = dynamic pressure	
	$= 0.613(23.83)^2 = 348\,\text{N/m}^2$	
P = net wind load on the surface	where	
	S_c = fetch factor	
$P = pA_{ref}$		
P = net pressure across the surface	S_h = topography factor	
	S_t = turbulence factor	
(c) American codes	g_f = gust factor	
	T_y = Turbulant adjust factor	
	T_c = fetch factor for adjustment	P_{cont}.
	ANSI A58.1 for exposure C Fig. 2 of the code	To
	Load based on 100-year recurrence wind, speed with gust factor and wind profile distribution	
X. Containments Internal Pressure	$p \not> 345\,\text{kN/m}^2$ for PWR	

Table 2.1 (continued)

Loads or stresses	Design load or range of actions	Notation
XI. Operating thermal load		
(a) Thermal gradients through the wall of containment	−11.2 (outside) + 6.672°C (inside) +5.5 (outside) + 6.672°C (inside)	
(b) Range of ambient temperatures at placement of concrete in containment	22.24 $< t <$ 50°C	
(c) Thermal gradient through reactor coolant compartment walls	±16.7°C gradient	V
(d) Range of ambient temperatures at placement of concrete	55.6°C $< t <$ 6.7°C	
(e) Thermal gradients through spent fuel pit walls	6.67°C max.mean temperature 4.5°C min.mean temperature −11.2 (outside) + 100°C (inside)	
XII. Resulting from the internal drop in containment (load)	0.13 kN/m^2	
XIII. Internal pressure for advanced cooled reactor vessels	$\ngtr$ 6900 kN/m^2	
XIV. Combination of Actions and Load Factors at the ultimate state (a) Permanent + variable (b) Permanent + wind + variable (c) Permanent + variable (d) Permanent + wind	G_k Q_k W or W_k 1.0 0 1.5 1.0 0 1.0 0 1.5 1.4 1.5 1.4 0 1.5	
XV. Prestressed concrete reactor pressure vessel Load combinations for elastic/work analysis	Loadings *Case 1* Prestress and ambient temperature *Case 2* Prestress 1.15 × design pressure + ambient temperature *Case 3** Prestress + design pressure + temperature *Case 4** Overload (prestress + increasing pressure + temperature) * Short- and long-term conditions apply	P_o $P^{/}$
$P' = P_0$(transfer force) − α ∞ = Losses in tendons P_o = (transfer force) $F_{rdu} = P$	$= \bar{\alpha} f_{ck} \left[\frac{A_{cl}}{A_{co}}^{0.5}\right] \leq 2 f_{ck}$ $\bar{\alpha}$ = depending upon adopted system range 0.67–0.85 A_{cl} = loaded area of the anchorage plate A_{co} = Anchorage area of the concrete block or block of material	

Table 2.2 Concrete stresses based on American and European codes

Stresses	Design stresses under range of actions	Notation
1. Concrete stresses sased on ACI 359		f'_c, f_{ck} Cylindrical Strength
Containment–primary		
(a) Membrane	$0.35f'_c$	
(b) Membrane plus bending	$0.45f'_c$	
containment–secondary		
(a) Membrane	$0.45f'_c$	
(b) Membrane plus bending	$0.60f'_c$	
concrete (after losses)		
Containments–primary		
(a) Membrane		
(b) Membrane plus bending		
Containment–secondary	$0.30f'_c$	
	$0.45f'_c$	
2. Euro code 2		
Concrete material properties		
f_{ck} = cylindrical strength	$25-60$ N/mm^{2}90 $\not>$ N/mm^2	
f_{ck}, σ_u = minimum cube Strength of concerete at 28 days	-41.34 kN/mm^2	
f_y, σ_y = yield strength	$-0.66\sigma_{cu}$	
σ_t = tensile strength	$+0.1\sigma_{cu}$	
E = elastic modulus	$+41.4$ kN/mm^2	
E_p = plastic modulus	$+0.476E$	
ε_{cu} = ultimate strains	0.0035	
ν = Poisson ratio	0.18	
α_T = coefficient of linear thermal expansion	8.0 μM/m°C	
K = thermal conductivity	1.75 W/m°C	
a' = coefficient of aggregates	0.65,0.68, 0.87,0.87	
ε_{ct} = shrinkage strain	200×10^{-6}	
Conventional steel		
$\bar{y} = \sigma_y$ = yield strength	4516 MN/mm^2(50ω) or T50	
E = elastic modulus	4400 (25ω) or T25	
E_p = plastic modulus	200 kN/mm^2	
α_T = coefficient of linear concrete thermal expansion	$0.1E = 20$ kN/mm^2	
f_{yk} = characteristic strength	500 N/mm^2	
Liner	12 mm + 10%	
t_s = thickness	19 mm + 5%	
	Up to 40 mm max to 400 mm up to	
$\bar{Y} = \sigma_y$ = yield strength	25 mm up to 25 m	
α_τ = coefficient of linear	3.4×10^5 kN/mm^2	
thermal expansion	10 μM/m°C	
K = thermal conductivity	41.6 W/m°C	
Prestress	Reference is made to the manufacturer's catalogues for various systems	
f_{mas}(at service)	$0.6-0.75f_{ck}$	
f_{mas}(at service)	$0.6f_{max} = P_o$transfer force : see catalogues for systems for accurate assessment	

Table 2.3 Structural steel

Service load parameters and stresses		
Loads or stresses	**Design load or range of actions**	**Notation**
1. Structural steel based on EC-3		
(a) Conventional steel design, load combinations		
Dead load	$1.46G_k$ or $Y_f F_k$	D
Dead load + restraining overturning	$1.0G_k$	
Dead + imposed load	$1.4G_k + 1.6Q_k$	
Dead + imposed + wind	$1.2(G_k + Q_k + W_k)$	
E_c' recent version		
F_d design action		
2. Structural steel Euro codes-3 (EC-3)	$\gamma_G = 1.35 : \gamma_Q = 1.5$	
(a) Grade shapes		
S275		Grade S
	F_y = design strength — f_y = stress, N/mm^2	
	$F_e 430 \leq 16$ — 275	
	≤ 40 — 275	
S355	$F_e 510 \leq 16$ — 355	
	≤ 40 — 355	
(b) Quenched tempered plates	$f_y = 690$ N/mm^2	
(c) Alloy bars – tension members	$f_y = 1030$ N/mm^2	
(d) High-carbon hard-drawn wires for cables	$f_y = 1700$ N/mm^2	
3. Load and resistance factor design (LRFD) (a) The design strength must equal or exceed the required strength R_u (b) Required strength and load combination for LRFD based on ASCE-7 Section 2.3	$R_u \leq \omega R_n$; R_n = safe working load × safety factor R_n = normal strength determined LRFD load combinations Ø = Resistance factor given by the Specification for a particular limit state 0.5	Ø
	$1.4D$ $1.2D + 1.6L + 0.5(L_r$ or S or $R)$ $1.2D + 1.6\ (L_r$ or S or $R) + (0.5L$ or $0.8\ W)$ $1.2D + 1.6\ W + 0.5L + 0.5\ (L_r$ or S or $R)$ $1.2D \pm 1.0E + 0.5L + 0.2S$ $0.9D \pm (1.6\ W$ *or* $1.0E)$ D = dead load L = live load due to occupancy L_r = roof live load S = snow load R = nominal load due to initial rainwater or ice W = wind load E = earthquake load	

Table 2.3 (continued)

Service load parameters and stresses		
Loads or stresses	Design load or range of actions	Notation
	LRFD $M_r = B_1 M_m + B_2 M_u = B_2 M_u$ $P_r = P_{nt} + B_2 P_{it} = B_2 P_u$ $B_1 \leq 1.05 \not> B_2$	Δ
(c) Simplified determination of required strength based on LRFD. Based on effective length method where P-δ factor is small (d) Stability design	$B_2 > 1.5$simplified method is not valid $\frac{\triangle 2nd}{\triangle 1st} \leq 1.5$ storey gravity load = minimum 2% $K = 1$ for braced frame K value to be determined for moment frames using sideway buckling analysis or $P_r \leq 0.5 P_y; \alpha = 1.0$ for LFRD P_r = required axial compressive strength P_y = member yield strength = AF_y	
(e) The required compressive strength contributing to stability (lateral) by flexural stiffness (f) Pin connected members i. Tensile strength ii. Shear rupture	$P_n = 2tb_{eff} \cdot F_u$ $\theta_r = 0.75$ for LFRD $P_n = 0.6 F_u A_{sf}$; $A_{sf} = 2t(a + d/z)$ a = shorter distance from the edge in mm d = pin diameter in mm t = place thickness in mm $b_{eff} = (2t + 16\,\text{mm})$ *Note*: For complete design specifications, a reference is made to steel construction manual (AISCE) 13th edition Dec 2005 ISBN 1-56424-055x or forward editions	

Table 2.4 Aluminium

Service load parameters and stresses		
Aluminium structures		Notation
Aluminium structures Characteristic values based on Euro code-9 f_u = ultimate strength f_o = 0.2% proof strength	L = longitudinal = $310 \frac{N}{mm^2}$ Temper $T6$ T = transfer = 260 N/mm^2 A = minimum elongation = 6%; Buckling class A	
(a) Alloy EN-AW 6082 100 mm thickness	$E = 74,000$ N/mm^2 $G = 27,000$ N/mm^2 v = Poisson's ratio = 0.3 α = coefficient = 23×10^{-6} per $^\circ C$ of linear expansion P = unit mass = 2700 kg/m^2	
(b) 6061	$T6/T651$ thickness < 12.5 $f_o = 110$ N/mm^2 $f_u = 205$ N/mm^2 $A_{so} = 12$	
(c) Bolts for 6082 T6 aluminium alloy	dia ≤ 6 $f_o = 250$ N/mm^2 $f_u = 320$ N/mm^2	

Table 2.5 Structural composites

Service load parameters and stresses		
Loads or stresses	Design load or range of actions	Notation
Structural composites based on EC-4		
(a) Design parameters b_e b_e = effective breadth L_z = distance between points of zero moment	$L_z/8 \ngtr$ half the distance of adjacent beam	b_e
Concrete slab stress Steel stress	$0.45f_{ck}$ $0.95f_y$	
p_v = shear capacity		P_v
Moment		M
$M_{api.RD} > M_{cd}$ Moment resistance of the steel beam	$W_{pl.}f_d$	
R_c = compressible resistance of slab	$0.85f_{ck}/\gamma_c \times b_{eff} \times b_c = 0.45f_{ckx}b_{eff}b_c$	R_c
f_{ck} = concrete cylindrical strength	$0.85f_{cu}$ or $0.8f_{k(cu)}$ γ_f = partial safety factor = 1.5	f_{ck}
R_s = compressive resistance of steel section	$f_d\ A_a$	$M_{pl,Rd}$
$M_{pl.Rd}$ = moment or resistance of composite beam	$R_s\left[\frac{h}{2} + h_c + h_p - R_s{}^{h_c}/2R_c\right]$	
$V_{pl.Rd}$ = shear resistance	$f_yA_v/(r_a\sqrt{3})$ $0.5V_{pl.RD} > V_sd$	
Shear connector Failure of concrete P_{Rd}	$0.29\alpha d^2\sqrt{Cf_{ck}E_c\gamma_r}$ or $R_d = \frac{0.8f_u\pi d^2}{4\gamma_v}$ shear failure of the steel at its weld	

available, the data given in this chapter shall be adopted by individual clients or their consultants. Tables 2.1, 2.2, 2.3, 2.4 and 2.5 give data on service load parameter and relevant acceptable stresses for the design of conventional structures for nuclear facilities, within the USA. In addition the Euro codes are mentioned for the design of conventional structures called ancilliary or auxiliary structures associated with nuclear facilities. Loads (actions) and stresses are tabulated from the US and the European codes. For detailed design a reference is made to the relevant codes where necessary.

2.2 Loads

2.2.1 Service Loads

Service load conditions are those loadings encountered during construction and in the normal operation of nuclear power facilities. A suggested summary of the list of loads is given below.

2.2.1.1 Dead Load (*D*)

Dead load is vertical load due to the weight of all permanent structural and non-structural components of a building, such as walls, floors, roofs and fixed service equipment as specified by the relevant codes and standards such as BS6399.

2.2.1.2 Operating Live Load (*L*)

Live load is the load superimposed by the used and occupancy of the building not including the wind load, earthquake load and impact load as specified by the relevant codes and standards.

2.2.1.3 Uniformly Distributed Loads (LL)

The live load is to be assumed in the design of building and other structures shall be the largest loads that can be expected to be produced by the intended use or occupancy, but in no case shall be less than the minimum uniformly distributed unit loads specified by the relevant codes and standards such as BS6399-1 to 3 or EC2, EC3.

2.2.1.4 Concentrated Loads (L_c)

Floors shall be designed to support safely a concentrated load simultaneously with the floor live loads. In European codes it is termed as knife edge loads.

2.2.1.5 Railroad Support (C_E)

For design purpose Cooper's E-72 loading should be used unless otherwise specified by intended use, such as support spent fuel cask handling car or other heavy equipment.

2.2.1.6 Truck Support (H_{20})

For design purposes, AASHO H-20-S16 loading should be used unless otherwise specified by intended use. The equivalent European track load can also be adopted instead.

2.2.1.7 Ordinary Impact Loads (*I*)

Machinery

The weight of machinery and moving loads should be increased to allow for impact. Some suggested values are 100% for elevator machinery; 20% for light machinery, shaft or motor driven; 50% for reciprocating machinery or power-driven units. All percentages should be increased or decreased as required by the design specification or manufacturer's recommendation.

Craneways

It is suggested that all craneways have their design loads increased for impact as follows: A vertical force equal to 25% of the maximum wheel load; a lateral force equal to 20% of the weight of trolley and lifter load only, applied one-half at the top of each rail; and longitudinal force of 10% at the maximum wheel loads of the crane applied at the top of rail. All percentages shall be increased or decreased if so recommended by the manufacturer or if otherwise specified by governing codes.

2.2.1.8 Construction Loads

Consideration shall be given to temporary large, heavy loads based on the 'Building Codes Requirements for Minimum Design Load in Building and Other Structures' (ANSI A58.1-1972)[1]. These provisions specifically exclude consideration of tornadoes. For extreme loads due to tornadoes, Section 3.3.2 should be referred to. Account should be taken of hurricane winds by comparison with the provisions of Section 3.3.3 for hurricane-susceptible sites. While using European codes a reference is made to BS6399, part 2 and Eurocode ENV1991-2-4 for wind loading. Table 2.1 part IX gives a brief relevant equation for the determination of loads caused by the wind.

2.2.1.9 Snow Loads (*S*)

Basic snow load requirements as a function of geographical area can also be found in ANSI Standard A58.1972. Table 2.1 gives a brief based on European code BS 6399-2-4.

Soil and Hydrostatic Pressure (E_p) and Buoyancy (*B*)

In designing nuclear facilities which are partly or wholly below grade, provision shall be made for the lateral pressure of adjacent soil, namely active pressure and at-rest pressure. The effect of dynamic pressure due to earthquake also should be given consideration. Due allowance shall be made for possible surcharge from fixed or moving loads. When a portion or the whole of the adjacent soil is below a free-water surface, computations shall be based on the weight of the soil diminished by buoyancy plus hydrostatic pressure. In the design of slabs below grade, the upward pressure of water, if any, shall be taken as the full hydrostatic pressure applied over the entire area. The hydrostatic head shall be measured from the underside of the slab. These are recommended also by the Nuclear Regulatory Commission (NRC) and ASCE of the USA.

Piping Equipment Reaction Load (R_o)

Piping system is attached directly to building structures through hangers, struts, restraints, anchors and snubbers. Hangers and struts are unidirectional, transmitting loads in one direction only. Hangers transmit only vertical loads. Restraints will transmit loads in any one or more of the coordinate directions. Anchors are capable of transmitting loads and moments in all three coordinate directions. Snubbers are unidirectional and transmit dynamic loads only.

Equipment loads include dead weight, restrained thermal expansion and dynamic effect such as pressure transients, changes in momentum, water and steam hammer in the equipment and earthquake. They also may include the effect of the restraint of attached piping. The effect of such phenomena must be considered in the design check.

Operating Pressure and Temperature (P_o, T_o)

In many cases compartments or sub-compartments within a structure which house highly radioactive pipes or equipment are maintained at lower pressure than the outside of the compartment in order to prevent out-leakage. Even though the differential pressure is not considerable, the magnitude should be determined and its effect evaluated particularly for steel structures which are more likely to experience external pressure buckling modes of failure.

2.2.2 *Operating Basis Earthquake (*E_o*)*

The respective nuclear organisation and regulatory commissions have criteria for the seismic design of nuclear power plants. The Operation Basis Earthquake (OBE) does consider the effect on a plant site during the operational life of the plant. Both local geology and seismology are related to specific characteristics of local subsurface materials.

Earthquakes can cause local soil failure, surface ruptures and structural damage of nuclear power plants. The most significant earthquake effects on plants or their structural components result from the seismic waves which propagate outwards in all directions from the earthquake focus. These different types of waves can cause significant ground movements up to several hundred miles from the source. The movements depend upon the intensity, sequence, duration and the frequency content of the earthquake-induced ground motions. For design purposes ground motion is described by the history of hypothesised ground acceleration and is commonly expressed in terms of response spectrum derived from that history. When records are unavailable or insufficient, smoothed response spectra are devised for design purposes to characterise the ground motion. In principal, the designers describe the ground motions in terms of two perpendicular horizontal components and a vertical component for the entire base of the nuclear power plants. A 3D analysis is essential using hybrid finite element non-linear method.

When the history of ground shaking at a particular site or the response spectrum derived from this history is known, plants' theoretical response can be calculated by various methods; these are described later.

The minimum acceptable acceleration for the OBE will be taken at least one-half of the Safe Shutdown Earthquake (SSE) acceleration. Sometimes OBE $<$ SSE have been permitted in some cases where SSE return period is such duration as not to be reasonably expected during the life of the nuclear power plant. If the vibratory ground acceleration of the site is equal to or greater than OBE acceleration, the US Federal Regulation makes it mandatory to shut the nuclear power plant for inspection.

2.2.2.1 Response Spectra

The main cause of the structural damage during earthquake is its response to ground motion which is in fact input to the base of the structure. To evaluate the behaviour of the plant under this type of loading condition knowledge of structural dynamics is required. The static analysis and design can now be changed to separate time-dependent analysis and design. The loading and all aspects of response vary with time which result in an infinite number of possible solutions at each instant during the time interval. For a design engineer the maximum values of the plant response are needed for the structural design.

The response may be deflection, shear, equivalent acceleration, etc: the response curves are generally similar with majority variations occurring in the vertical coordinates. The variation occurs with magnitude of the earthquake and the location of the recording instruments. Accelerations derived from actual earthquakes are surprisingly high as compared with the force used in designs and the main reason is the effect of different degrees of damping.

The recorded earthquake ground accelerations have no doubt similar properties to those of non-stationary random functions but owing to a lack of statistical properties related to such motions artificially generated accelerograms are used which are flexible for any duration.

The following three major aspects must be considered:

(a) Location of vibratory ground motion for OBE
(b) Direction of motion for OBE
(c) Vertical Acceleration associated with OBE

This book covers all aspects in detail when earthquake analysis of nuclear plant is considered. The reader is advised to examine the author's book on *Earthquake-Assistant Buildings* published by Springer-Verlag, Germany (2010) particularly using analyses and loadings with and without seismic devices.

2.2.3 *Extreme or Severe Loads*

These loads include extreme environmental conditions, such as tornadoes and the safe shutdown earthquake postulated to occur during the life of the facility. Also included are effects resulting from a postulated rupture of a high-energy system during normal operation, startup or shutdown of the plant or other postulated design basis accident.

2.2.3.1 Safe Shutdown Earthquake (E^1)

The Nuclear Regulatory Commission's Federal Regulation 10 CFR 100 Appendix A, entitled 'Siesmic and Geology Siting Criteria for Nuclear Power Plants' sets forth principal seismic and geological considerations which shall be used by the Commission in its evaluation of the suitability of proposed sites for nuclear power plants. Contained within this Appendix are definitions and procedures which are to be used as guidelines in establishing the various seismic input motion and potential faulting hazard for nuclear power plants, in the USA. Specifically, the safe shutdown earthquake (SSE) is defined as that earthquake which produces the vibratory ground motion for which structures, systems and components are important for safety of the structures or systems.

Required Investigations

Paragraph IV entitled 'Required Investigations.' in Appendix A of 10 CFR 100 sets forth the required geologic and seismic investigations that should be carried out to establish vibratory ground motion requirements and surface faulting. Sub-paragraph A entitled 'Required Investigation for vibratory Ground Motion' sets forth the specific investigations that should be carried out to establish the ground motion input associated with the SSE. Briefly, the items that should be considered to establish the ground motion input associated with SSE are as follows:

1. geologic conditions at the site;
2. tectonic structure determination;
3. identication of effects of prior earthquakes;
4. determination of static and dynamic characteristics of underlying materials;
5. historical listings of all earthquakes which may have affected the site;
6. correlation of epicentre;
7. determination of fault locations;
8. characteristics of faults in the vicinity.

Locations of Vibratory Ground Motion for SSE should be considered to be acting at the ground surface in the free field. The maximum acceleration of the vibratory ground motion for the SSE should be considered on the basis of evaluating the result of the investigation as stated above required investigation. The guideline shall be on this maximum acceleration as the largest possible acceleration at the site due to a postulated fault activity.

The direction of motion for SSE shall generally be assumed as resultant motion to correspond with one of the principal horizontal directions of the structure for the facility being analysed.

The vertical motion associated with SSE can be established on the basis of the information developed from the above-mentioned investigation. The value should not be less than 2/3 of the maximum horizontal ground acceleration of the SSE. The frequency strength is between 3.5 and 33 Hz. The vertical acceleration shall be equal in intensity to the horizontal component.

2.2.3.2 Tornado Loads (W_t)

Structures for the nuclear facility shall be designed to resist the maximum tornado load for a given plant site. The basis of the design shall be such that the safety class equipment remains functional; even a safe shutdown of the facility is accomplished in totality without endangering the plant. The AEC Regulatory Guide 1.76 recommends the design basis tornado.

The effects of a tornado that are manifested in structural damage are generated from three separate phenomena: wind, differential atmospheric pressure and missiles. These effects interact with structures and cause damage through three principal mechanisms:

1. pressure forces created by drag and lift as air flows around and over structure;
2. pressure forces created by relatively rapid changes in atmospheric pressure resulting in differential pressure between the interior and exterior of the building;
3. penetration, spalling and impact forces created by missiles.

Tornado missiles (Y_m): Tornado-generated missiles carry objects which are accelerated by the forces induced by the extreme wind speeds of the tornado. The parameters specified in the design basis tornado are translated into pressures and forces acting on the structures and its components. The important case is the real analysis that would be necessary to perform on the structure. The analysis is known as tornado structure interaction. In this analysis the load evaluated using a specific path width of the tornado field that experiences wind velocities $\geq$75 mph (120 km/h) is generally considered.

2.2.3.3 Hurricane Loads

A hurricane by definition is a cyclone storm having rotational wind velocities in excess of 70 mph (119 km/h). The dynamic strength of a hurricane builds up over water, but as it comes inland boundary layer drag forces cause a tremendous dissipation of the kinetic energy of the storm and the wind.

As regards wind distribution which is one aspect of hurricane loads, the maximum wind velocities generally occur to the right of the eye of the hurricane looking along the direction of its path. This is due to vectoral addition of the translational and rotational components of the wind. The following data can be adopted in the absence of specific data not available for the site under consideration:

1. α = Inclination of the direction of the wind = 20–30° (towards the centre of the hurricane)
2. Wind gust > the sustained wind by 30–50%
3. Hurricane diameter: 15 miles (24 km) to 100 miles (160 km)
4. Gale force wind: 40 mph (64 km/h occurring within 30 miles (560 km) to 400 miles (640 km)

Where sea swell surge and flooding occur, specific calculations would be required to algebraically evaluate the additional load occurring when considered along with other loads.

2.2.3.4 Tsunami Loads

Tsunami are long ocean water waves generated by mechanisms such as earthquakes or underwater explosions, which impinge on coastal areas. With regard to earthquakes tsunami appeared to be primarily associated with those tectonic movements having substantial vertical components of motion (dip-slip). The design of nuclear facilities to resist the effects of tsunami must be undertaken for all nuclear site adjacent to coastal areas, especially those bordering the Pacific Ocean. The basic criteria for tsunami are set forth in the NRC's Standard Format and the Contents of Safety Analysis Reports for Nuclear Power Plant. These silent feature must be known such as the

(i) Location relevant to the site
(ii) Magnitude
(iii) Tsunami wave height
(iv) Influence of harbour/break water and hydrography
(v) Records of the region with valuable statistics

The direct dynamic force of the moving tsunami wave impinging against structures of power facilities shall be algebraically added to the force produced due to the impact of the floating debris and water-borne missiles.

2.2.3.5 Missile Load (Y_m)

In nuclear facility design, safety class structures shall be protected against loss of functions due to postulated plant generated and extreme environmental missiles depending on aircraft crash should be considered.

The effect of missile impact on a target on the material and geometric properties of the impacting bodies. The phenomenon can be described in general as the formation of an impulse measured by the momentum exchange between the two bodies during the impact. Table 2.6 gives data on tornado and wind-generated missiles.

Table 2.6 Tornado and wind-generated missiles and their characteristics: wood, steel and concrete building components

Service load parameter and stresses

	Geometry				
Missile type	Diameter (mm)	Length (m)	Impact area (m^2)	Velocity (m/s)	Weight (kg)
Wooden plank	–	3.67	0.03	41.5	*56.7*
Wooden pole	200	3.67	0.03	*5.73*	94.8
Circular hollow sections in steel (average)	168.3	4	0.000026	70.2	60
Sign boards (average)	–	–	6.0	57	56
Steel I-beam light sections (average)	–	4	0.000032	40.5	100
Steel members channel sections (average)	–	3	0.000013	*50.5*	30
Steel members L-sections (average)	–	3	0.000015	45.5	36
Steel rafters T-sections (average)	–	3	0.000018	45.5	42
Steel rod	25	0.92	0.00049	75.6	3.63
Concrete lintels	–	3	0.025	60.5	1.8
Concrete sleepers	–	2.7	0.0031	75	0.2
Precast concrete beams or piles at delivery stage		9	0.09	60.5	19.44
Precast concrete wall panels		*5*	*11.5*	2.5	1380
Prestressed concrete pipes	400	–	–	–	1.1
	500	–	–	–	1.375
	600	–	–	–	1.65
	700	–	–	–	1.92
	800	–	–	–	2.2
	900	–	–	–	2.474
	1676	6	0.032	–	4.608
Prestressed concrete poles	–	17	0.0019	30.5	65.7
		12	0.0008	50.1	14.46
		9	0.000025	65.2	9.65

Missiles are usually classified by source as plant (accident)-generated missile and extreme environmental missiles. Typical plant-generated missiles include valve stems, valve bonnets, (caused by rupture of high-energy systems) and turbine discs and other rotating masses (caused by rupture of rotating parts). Extreme environmental missiles which are of major concern include tornado-generated missile and aircraft.

Table 2.7 gives a list of plant-generated missiles and their characteristics. They depend on their region, ranges of size, weight and impact velocity.

Tables 2.8, 2.9, 2.10, 2.11, 2.12, 2.13, 2.14 and 2.15 give various aircraft parameters and their characteristics and impact parameters.

Table 2.7 Plant-generated missile and their characteristics

Service load parameters and stresses			
Missile type	Weight (kg)	Impact area (cm^2)	Velocity (m/s)
Control rod mechanism or fuel	53	15.5	91.5
Disc 90° sector	1288	495	125
Disc 120° sector	1600	6573	156
Hexagon head bolts			
1.4 cm dia	0.20	1.54	250
2.0 cm dia	0.30	2.30	230
2.4 cm dia	0.37	2.84	189
3.3 cm dia	0.42	3.22	150
6.8 cm dia	0.97	7.44	100
Turbine rotor fragments			
High trajectory			
Heavy	3649	5805	198
Moderate	1825	3638	235
Light	89	420	300
Low trajectory			
Heavy	3649	5805	128
Moderate	1825	3638	162
Light	89	420	244
Valve bonnets			
Heavy	445	851	79
Moderate	178	181	43
Light	33	129	37
Valve stems			
Heavy	23	25	37.5
Moderate	14	9.7	20
Others			
30 cm pipe	33.7	260	68
12 cm hard steel disc	1.6	113	140
Steel washer	0.0005	3	250
Winfirth test missile	15.6	176	240

Table 2.8 Civilian sircraft

Service load parameters and stresses	
Data on civilian and military aircraft	
Civilian aircraft normally in service includes Concorde, Airbus, Boeing, Antonov, Ilyushin and Tupolov	
S = span; L = length; H = height; A_w = wing area; P_L = payload	
V = speed; W_a = weight at take-off or loading	
Basic parameters of Concorde	
Power Plant	
4 × 38,050 lb (169 kN)	
Rolls-Royce/Sneema Olympus	
593 Mk60 two-spool turbojet	
S (m)	25.61
L (m)	62.1
H (m)	12.19
A~ (m²)	358
$P_{l\sim}$(kg)	11,340
V (km/h)	2150
w_a (kg)	186,800

Tables 2.16, 2.17, 2.18, 2.19, 2.20, 2.21, 2.22 and 2.23 dictate again various military missiles with their characteristics and impact parameters.

For details and in-depth study references are made to the following authors:

(a) *Impact Explosion Analysis and Design*, Blackwell, 1993.
(b) *Manual of Numerical Methods in Concrete*, Thomas Tefford.
(c) *Shock, Impact & Explosion*, Springer, 2008.

2.2.3.6 Design Basis Accident Load

In addition to accident-generated missile loads there are several loading phenomena generated as the result of a design basis accident which normally includes all postulated high-energy system ruptures. Included in this category are all accident-induced pressure and temperatures, as well as high-energy fluid jet impingement and rupture reaction loads. The criteria for defining design basis high-energy system ruptures are found in the NRC Standard Review Plan.

Accident Pressure (P_a) and Temperature (T_a)

These pressures and temperatures are typically developed as a result of the blowdown of a high-energy system into a confined space. They typically include the containment design pressure and temperature as well as differential pressure and temperature across interior partitions or structures which house ruptured high-energy systems.

Table 2.9 (a) Data on the Airbus family and (b) data for Antonov aircraft

Type	Power plant	S (m)	L (m)	H (m)	A_w (m^2)	P_L (kg)	V (km/h)	W_a (kg)
(a) Data for Airbus family								
A300B2-100	2×51,000 lb (227 kN) GE CF6-50C turbofans	44.84	53.75	16.53	260	14,900	869	34,585
A300B2-200	2×51,000 lb (227 kN) GE CF6-50C turbofans	44.84	53.57	16.53	260	34,585	869	142,900
A300B2-100	2×51,000 lb (227 kN) GE CF6-50C turbofans	44.84	53.57	16.53	260	35,925	869	158,400
A300B4-.200	2×52,500 lb (233.5 kN) CF6-50C1 turbofans	44.84	53.57	16.53	260	35,600	869	165,900
A310-.202	2×48,000 lb (218 kN) GE CF6-80A turbofans	13.9	46.66	15.8	219	32,400	780	132,000

Table 2.9 (continued)

Type	Power plant	S (m)	L (m)	H (m)	A_w (m^2)	P_L (kg)	V (km/h)	W_a (kg)
(b) Antonov aircraft								
An-12	4 × 4000 ehp Ivchenko A1-20 K turboprops	38	37	9.83	119.5	10,000	550	54,000
An-22	4 × 15000 ehp Kuznetsov NK-12MA turboprops	64.4	57.8	12.53	345	80,000	679	250,000
An-24	2 × 2500 ehp Ivenchenlko A1-24 Seviiny 11 turboprops	29.2	23.53	8.32	74.98	13,300	450	21,000
An~26*	2 × 2800 ebp Ivenchenko A1-24T turboprops	29.4	23.8	8.575	74.98	5500	435	24,000
An-28	2 × 970 ehp Glushenkov TVD-1OB turboprops (similar to An-14)	21.99	12.98	4.6	39.72	1550	350	6100
An-72	2 × 14,330 1b (6500 kg) Lotarev D-36 turbofans	25.8	26.58	8.24	74.98	7500	720	30,500

An-30 and An-32 have similar status to An-26

Table 2.10 (a) Data for Boeing aircraft; (b) Data for the Ilyushin aircraft; and (c) Data for the Tupolev series of aircraft

Service load parameters and stresses								
Type	Power plant	S (m)	L (m)	H (m)	A_w (m^2)	P_L (kg)	V (km/h)	W_a (kg)
(a) Data for Boeing aircraft								
727-200	3 × 16000 lb (71.2 kN) Pratt and Whitney JT8D-17 turbofans	32.9	461	10.4	153.2	18,594	883	95,238
737-200	2 × 16,000 1b (71.2 kN) Pratt and Whitney JT8D-17 turbofans	28.3	30.5	11.4	91	15,422	775	53,297
767	2 × 44,300 1b (1.97 kN) Pratt and Whitney JT9D-7R 4A turbofans	47.24	48.46	15.38	200	40,224	800	128,030
757	2 × 37400 1b (166.43 kN) Rolls-Royce RB211-535C	37.95	47.32	13.56	181.25	71,530	899	29,8880
747-200B	4 × 50,000 lb (222 kN) Pratt and Whitney JT9D-7F (wet) turbofans	59.6	70.5	19.3	512	71,530	907	366,500
747-200B	4 × 53,000 lb (236 kN) Pratt and Whitney JT9D-7Q turbofans	59.6	70.5	19.3	512	69,900	907	373,300
747-200B	4 × 52,500 lb (234 kN) General Electric CF6-50E2 turbofans	59.6	70.5	19.3	512	69,080	907	373,300
767-200B	4 × 52,500 lb (234 kN) General Electric	76.65	70.5	19.3	512	69,080	907	373,300

Table 2.10 (continued)

Service load parameters and stresses

Type	Power plant	S (m)	L (m)	H (m)	A_w (m^2)	P_L (kg)	V (km/h)	W_a (kg)
	CF6-50E2 turbofans							
747-100B	4 × 52,500 lb (234 kN) General Electric CF6-50E2 turbofans	65.0	70.71	19.18	512	69,080	907	373,300
(b) Data for Ilyushin aircraft								
Ilyushin II-18	4 × Ivenchenko A1-20 M turboprops 4250ehp	37.4	35.9	10.17	140	14,000	625	64,000
Ilyushin II-62	4 × 25,000 lb (II3kN) Solovier 20-30-KU turbofans	43.2	53.1	12.4	280	23,000	860	165,000
Ilyushin II-76T	4× Solovier D.30KP turbofans. Each with 26,455 lb St (120,00 kg)	50.5	46.6	14.76	300	40,000	850	157,000
Ilyushin 11-86	4 × Kuznetsov turbofans, each with 28,635 lb St (13,000 kg)	48.06	59.5	15.81	320	42,000	900	206,000
(c) Data for Tupolev aircraft								
	2×21,385 lb (97 kN) Mikulin AM 3M500 turbojet	3454	25.85	11.9	174.4	900	800	76,000
TU-124	2 × 11,905 lb (54 kN) Soloviev D-20P turbofans	25.5	30.58	8.08	1.19	3500	800	26,300
TU-134	2 × 15,000 lb (66.5 kN) Solovjev D-30 turbofans	29	34.9	9	127	77,000	849	45,200
TU-144	4×44,000 lb St (20,000 kg) with Kuznetsoy NK-144 turbofans	28.8	6K7	12.85	438	14,000	2500	180,000
TU-154)	3 × 21,000 lb (93.5kN) Kuznetsov NK-8-2 turbofans	37.5	48	11.4	202	20,000	900	91,000

For 13767-200EH 46.55 48.46 16.155 For other details, reference is made to this Chapter
For B747-400 65.00 70.71 19.18 For other details, reference is made to this Chapter

Table 2.11 Aircraft information Boeing 767-200ER

Aircraft information

Boeing 767-200ER

General specifications

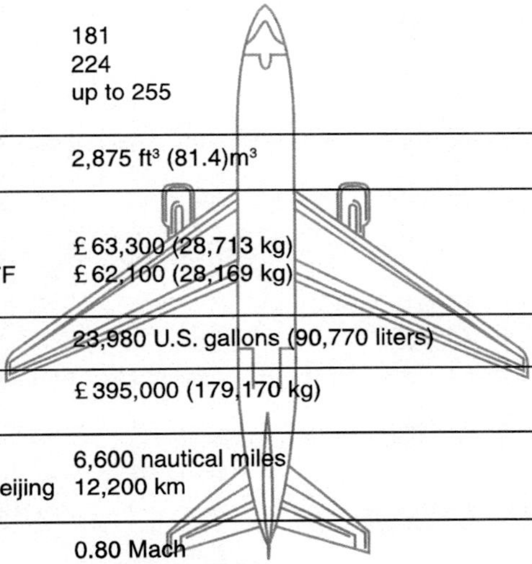

Passengers	
Typical 3-class configuration	181
Typical 2-class configuration	224
Typical 1-class configuration	up to 255
Cargo	2,875 ft³ (81.4)m³
Engines' maximum thrust	
Pratt & Whitney PW4062	£ 63,300 (28,713 kg)
General Electric CF6-80C2B7F	£ 62,100 (28,169 kg)
Maximum fuel capacity	23,980 U.S. gallons (90,770 liters)
Maximum takeoff weight	£ 395,000 (179,170 kg)
Maximum range	6,600 nautical miles
Typical city pairs: New York–Beijing	12,200 km
Typical cruise speed at 33,000 ft	0.80 Mach 530 mph (850 km/h)
Basic dimensions	
Wing span	156 ft 1 in. (47.6 m)
Overall length	159 ft 2 in. (48.5 m)
Tail height	52 ft (15.8 m)
Interior cabin width	15 ft 6 in. (4.7 m)

FEDERAL EMERGENCY MANAGEMENT AGENCY
Service load parameters and stresses

Note: This aircraft has been used in the Twin Tower collapse.

Table 2.12 Military aircraft

Service load parameters and stresses	S = span; L = length; H = height; A_w = wing area; P_L = payload; V = speed; W_a = weight at takeoff or loading	
	Data on the Tornado IDS and ADV aircraft	
	Power plant	
	Interdictor Strike (IDS) Turbo-Union RB 199-34R (101 or 103) after burning turbofan MK 8090 lb (3670 kg) to 15,950 lb (7253 kg) after burning thrust	Air Defense Variant (ADV) As for IDS, with MK 104
S (m)	8.60 max swept 13.90 max unswept	
L (m)	16.67	18.68
H (m)	5.95	5.95
$A\sim$ (m^2)	....	....
$P_{1\sim}$(kg)	9000	9000
V (km/h)	Mach 2 at high level Mach 1 at low level	Mach 2.2
w_a (kg)	28,000	28,000
Armament	4× MK 13/15 1000 lb (454.74 kg) bombs 2 AIM-9L missiles 8 MK 83 retarded bombs 2 CBLS-200 practice bomb containers 4 Kormoram ASM 8 × BL755 cluster bombs	
	Basic parameters for the F-5E and F-20 aircraft	
	Power plant	
	Engine 2GEJ 85-21 Engine 5000 lb (2268 kg) thrust each	GEF4O4-GE100 1800 lb (8164 kg) thrust each
S (m)	7.98 with missiles 8.53 without missiles	8.5 with missile

Table 2.12 (continued)

Service load parameters and stresses	S = span; L = length; H = height; A_w = wing area; P_L = payload; V = speed; W_a = weight at takeoff or loading			
L (m)	14.45		14.42	
H (in.)	4.07		4.10 (4.73 with wheels)	
A_w (m^2)	28.1		27.5	
P_1. (kg)	6350		7263	
V (miles/h)	850		1300	
W_a. (kg)	11,213.8		12,700	
Armament	Air-to-air 2 No. 20 mm guns and AIM 9 sidewinder missiles Air-to-ground 2 No. 20 mm guns and 9 bombs of 3020 kg			
	Data on the F-110 series F-16A and F-16B	Data on the F-16 series of aircraft F-16C and F-16D	F-16 N	TF-16 N
Power plant	Pratt and Whitney turbofan two shaft 24,000 lb (10,885 kg) thrust F100-PW-100	F100-PW-200F100-PW-200 F110-GE-100 25,000 lb (11,340 kg) thrust	F110-GE-100 25,000 lb (11,340 kg) thrust	F110-GE-100 25,000 lb (11,340 kg) thrust
S (m)	9.45 10.01 (with sidewinder)	9.45 10.01 (with sidewinder)	9.895 (without sidewinder)	
L (m)	14.52	15.03	15.1	
H (m)	5.01	5.09	5.1	
A_{s}., (m^2)	27.87	2787	27.87	
P_L (kg)	33,000 1b (14,969 kg)	37,500 lb (16,781 kg)	37,500 lb (16,781 kg)	
V (mile/h)	1300	1300	1300	
W_a (kg)	12,000 1b (5443 kg)	12,430 1b (5638 kg)	17,278 lb (7836 kg)	

Table 2.13 (a) Data on the F-15; (b) data on the F/A-18 Hornet and (c) data for the Grumman F-14 Tomcat

Service load parameters and stresses		
(a) Data on F-15	Power plant	
	2 No. Pratt and Whitney F-100-PW-220 each with 24,000 lb thrust	
S (m)	13.05	
L (m)	19.45	
H (m)	5.64	
A_w (m^2)	.....	
P_L (kg)	7000	
V (km/h)	2500	
w_a (kg)	20,000	
Armament	4 AIM-9L/M infrared-guided Sidewinder missiles; 4 AIM-7F/M radar-guided Sparrow missiles: 8 advanced medium-range air-to-air missiles (AMRAAMs); M-61 20 mm Gatling gun with 940 rounds of ammunition. Accommodates a full range of air-to-ground ordnance	
(b) Data on F/A -18 Hornet	Power plant	
	2 No. F404-UF'-400	
	Low bypass turbofan engines each in 1600 lb (70.53 kN) thrust and with a thrust/weight ratio of 8:1	
S (m)	11.43	
L (m)	17.06	
H (m)	4.7	
A_w (m^2)	37.2	
P_L (kg)	.....	
V (km/h)	2700	
w_a (kg)	24,402	
Armament	Up to 7711 kg maximum on nine stations: two wing-tips for sidewinder heat-seeking missiles; two outboard wings for air-to-ground ordnance; two inboard wings for Sparrow radar-guided missiles, air-to-ground or fuel tanks; two nacelle fuselages for Sparrow missiles or sensor pods: one centreline for weapons, sensor pods or tank. Internal 20 mm cannon mounted in nose	
(c) Data on F-14 Tomcat	Power plant	
	F-14A	F-14B, C
	2 × 20,900 lb (9480 kg) thrust Pratt and Whitney TF3O-1412A	2 × 28,090 lb (12,741 kg) thrust Pratt and Whitney F401-400
	Two shaft after-burning turbofans	
S (m)	11.630 (68° sweep)	Safely landing
	19.54 (20° sweep)	
L (m)	18.89	
H (m)	4.88	
A_w (m^2)	....	

Table 2.13 (continued)

Service load parameters and stresses	
P_L (kg)	17010
V (km/h)	Mach 2.3 or 1564 mph maximum speed 400–500 km/h cruise speed
w_a (kg)	27216
Armament	AIM-54 Phoenix missiles AIM-7 Sparrow missiles AIM-9 Sidewinder missiles

Table 2.14 (a) Comparison data of MIG aircraft and (b) Data on the British Aerospace Jaguar

Service load parameters and stresses					
Power plant	MIG-19 (Mikoyan)	MIG-21	MIG-23 (Flogger)	MIG-25 (Foxbat)	MIG-27
Engines	Single seater 2×600 lb (3040 kg) to 2×7165 lb (3250 kg) Kimov RD-39B turbojets	Single seater Range turbojet 11,240 lb (5100 kg) to 4150 lb (6600 kg) Tumanskey single shaft	Single seater 17,640 lb (8000 kg) to 25,350 lb (11,500 kg) thrust, 1 Tumansky turbofan	Single seater 27,000 lb (12,250 kg) thrust, 2 Tumansky R-266 after-burning turbojets	Single seater 17,640 lb (800 kg) to 25,350 lb (11,500) thrust, 1- Tumansky after-burning turbofan
S (m)	9	7.15	8.7 (72°sweep) 14.4 (16 sweep)	14 Foxbat A	8.7 (72°sweep) 14.4 (16 sweep)
L (m)	13.08 (S-5F)	14.35	16.15	22.3 (Foxbat A) 22.7 (Foxbat R) 23.16 (Foxbat U)	16.5
H (m)	4.02	4.5	3.96	5.6	4.6
A_w (m^2)	....	....	....	....	....
P_L (kg)	3760	4600	7050	14,970	9900
V (km/h)	Mach 1.3 or 1480 km/h (92 mph)	Mach 2.1 or 2070 km/h (1285mph)	Mach 1.1 or 1350 km/h (840 mph)	Mach 3.2 or 3380 km/h (2100 mph)	
w_a (kg)	9500	9800	15,000	34,930	17,750

Table 2.14 (continued)

Service load parameters and stresses		
	Power plant	
	2 No. Rolls-Royce Terbomeca Adour two shaft turbofans	
	7305 lb (3313 kg) to 8000 lb (3630 kg)thrust	
S (m)	8.69	
L (m)	15.4 –16.42	
H (m)	4.92	
A_w (m^2)	....	
P_L (kg)	6800	
V (km/h)	1450	
w_a (kg)	1550	
Armament and other data	2 No. 30 mm DFA 553 each with 150 rounds	
	5 No. pylons with total external loads of 4536 kg with guns	
	2 No. 30 mm Aden for its T-2 model	
	Matra 550 Magic air-to-air missiles	
Jaguar A and B and EMK 102 Audor engines		
Jaguar S	MK 104s, MK 108s } Audor engines	Using digital quadruplex fly-by-wire control system
Jaguar Act		
Jaguar FBW		

Jet Reaction (Y_r)

As a result of the postulated rupture of a high-energy system there develops an unbalanced differential pressure force plus a mass transfer momentum effect due to fluid being ejected from the rupture. In actuality an unbalanced external force develops on the system at each change in area and direction in the system. Typical reaction load characteristics due to a postulated rupture are given in the ANS N-176 guide.

Jet Impingement (Y_j)

As a result of a high-energy system rupture a high-energy fluid jet may form which would impinge on structures within its path. These structures in general would be designed 4o resist the momentum transfer resulting from the structure stopping the jet.

Reaction Load Due to Accident-Induced Differential Movement (R_a)

Many structures and components are supported by primary structures which would undergo deformation from an accident condition and thereby induce loads in the supported structure or component. Examples of this effect would be loads on piping systems attached to the containment, which would be induced when the containment expands due to accident pressure and temperature effects.

Table 2.15 Data on the Dassault aircraft

Service load parameters and stresses							
Type and power plant	S (m)	L (m)	H (m)	A_w (mm)	P_L (kg)	V (km/h)	w_a (kg)
Dassault Breguet Fl Single-seater multi-mission fighter, 7200 kg thrust, SNECMA Atar, 9 K-50 single shaft turbojet	8.4	15	4.5	–	7400	1472	14,900
Estendard IVM and WP Single-seater strike fighter, 4400 kg thrust, SNECMA Atar, 8B single shaft turbojet	9.6	14.4	4.26	–	5800	1083	10,000
Super Estendard single-seater strike fighter, 5110 kg thrust, SNECMA Atar 8 K-50 single shaft turbojet	9.6	14.31	4.26	–	6300	1200	11,500
Mirage 3 and 5 Single-seater or two-seater interceptor, trainer and reconnaissance aircraft, 6000 kg thrust, SNECMA Atar, 9B single shaft turbojet	8.22	15.5	4.25	–	6156	1390	12,000
Mirage 2000 Mirage 315 and F-l improved version of these aircraft with engines SNECMA turbofans	9	15	4.5	–	7800	2200	9000
Mirage 4000 SNECMA M53, single shaft bypass turbofan 8 stage axial compressor 2 × 14,500 lb (2 × 6579 kg) thrust	12	18.7	4.5	–	13000	2300	16,100
Armament Mirage 4000	Bombs						
Internal cannons 2 × 30 mm		Anti-runway Durandal up to 27					
4 long-range missiles		Clean or retarded (250 kg) up to 27					
4 air-to-ground missiles		Laser guided (250) up to 27					
2 air-to-surface missiles	Rockets 68 mm						

Table 2.16 Comparative data of some important combat aircraft

Type	Power plant	S (m)	L (m)	H (m)	P_L (kg)	V (km/h)	w_a (kg)
BAe Harrier GR-3 model T-Mark 4 model (AV-8A, TAV-8A) FRS-I model Single-seater, two-seater	1 × 21,500 lb (9752 kg) thrust Rolls-Royce Pegasus 103 two shaft turbofan (US designation F-400)	7.7	13.87 17.0 (GR-3)	3.43	6200	1.2 860	11,793
F-6 Sheriyang (or NATO's name FANTAN 'A') Single seater	2x Axial turbojets MD manufactured	10.2	15.25	3.35	4500	1 760	10,700
Rockwell B-I Four seater	4 × 30,000 lb (13,610 kg) General Electric F101-100 two shaft augmented turbofans	41.4	45.6	10.24	11,5670	0.85	179,170
Saab 37 AJ, JA Viggen SF, SH, SK versions Single seater	1 Svenska flying motor RMB; Pratt and Whitney two shaft 25,970 1b (11,790 kg) to 28,086 1b	10.6	AJ: 10.6 JA: 16.3	5.6	4500	2 1320	16,000
SU-9 Fishpot B SU-1 1 Fishpot C Single seater	(12,750 kg) thrust 1 Lyulka single shaft turbojet; 19,840 lb (9000 kg) thrust (SU-9) and 22,040 lb (10,000 kg) thrust (SU-11)	8.43	SU-9: 18.5 SU-11: 17.4	4.9	4540	0.95 1195	13,610
SU-15 (Flagen A to E models) Single seater	2x Tumonsky R-25 turhofans: 16,530 lb (7500 kg) thrust after burner (17)	9.5 D model	21.5	5	10,100 D 5100 A	2.3 1520	21,000 D 16,000 A
SU-17. SU-20 and SU-22 models Single seater	1-Lyulka AL-21 F-3 Thrust single shaft 17,200 lb (7800 kg) (20–22) AL-7F 22,046 lb (10,000 kg)	14 (28° sweep) 10.6 (62° sweep)	18.75	4.75	9000	1.05 798 To 2.17 1432	19,000

Table 2.17 Types of shells, bombs, rifles

Service load parameters and stresses								
Types of shells and bombs	Manufacturers	Dimensions	w_L(kg)					
Double base Shot gun	International	$d=1.25$ $t=0.150$	....	Porous disc	Short	....	....	Guns
Single or double base rim fire	International	$d=0.875$ $t=0.1$	....	Porous disc	Short	....	....	Guns
Double base revolver	International	$d=1$ $t=0.125$	....	Porous disc	Short	....	....	Guns
Single base rifle	International	$d=1.25$ $d=0.375$ $t=0.045$	....	Tubular	Short	....	....	Guns
GBU-15 HOBOS-guided bomb	Rockwell International USA	$L=3.75$ MK84 $S=112$ $d=46$ (M118) $S=132$ $d=61$	MK84 2240 1016 M118 3404	The forward guidance system is KMU-353 or electro-optical with TV camera or target-seeker optics	Medium	....	....	Free fall bomb from
Bofor	Sweden	.....	0.12	5500 kg guns	3.7 km	10,005	0.88	Trajectory fall
ZSU-23-4	USSR	$L=654$ $d=29.5$ 3400 rounds	20.5	SPAA vehicles 4×3 automatic	Short	7000	....	Guns
GBU-15(V)2 glide bomb (smart bomb)	USA	$L=388$		Imaging infrared	Short		1111	

d = diameter (mm); t = web thickness (mm); L = length (cm);
w_L = weight; S = span (cm); R = distance; V = velocity

Table 2.18 Missiles for armed forces (ICBM and ALCM) (USA, USSR)

Service load parameters

Types of missiles/rockets	Manufacturer/ country of manufacture	Dimensions	WL lb (kg)	Power plant and guidance	*R* miles (km)	V mach	Warhead manufacture	Mechanisms
AGM-86A air-	Boeing,	$L = 4.27$	1900	Williams Research F107-	760	0.6 –0.8	Thermo nuclear	B-52
launched cruise								
missile (ALCM)	Aerospace,	$S = 289$	862	WR-100 2. No shaft	1200	200 kt		
	Seattle, USA	$d = 64$		thrust; McDonnell Douglas				
				Tercom with inertial system				
LGM-30	Boeing	$L = 18.2$	Model II	I-Stage: Triokol TU-120	II stage:	1500 mph	Thermo nuclear	SG
Minuteman	Aerospace,	$d = 183$	70,116	(M55E), 2×10^5 lb	7000			
(ICBM) missile,	Ogden, USA		31,800	(9×10^5 kg) thrust	11,250		1—5 Mt	
located in silos			Model III	Il Stage: Hercules rocket	Ill stage:		II stage:	
			76,015	35×40^3lb (16×3 kg)	8000		AVCO MK	
			34,475	thrust	12,875		1IC	
				III-Stage: Aerojet 35 x			Ill stage: GE	
				10^3 lb (16 x 10^3 kg) thrust			MK-12 MIRV	
RS-121SS13	USSR	–	–	Solid fuel propellant; three stages	9400	0.8	Nuclear 75 kt	SG
Savage (ICBM)								
RS-16/SSI7	USSR			Liquid cold launch; two stages	11,000	0.8	3.6 Mt mode 2	SG
(ICBM)					1000		or MIRV	
							4×200kt	

Table 2.18 (continued)

Service load parameters								
Types of missiles/rockets	Manufacturer/ country of manufacture	Dimensions	WL lb (kg)	Power plant and guidance	*R* miles (km)	V mach	Warhead manufacture	Mechanisms
RS-20!SS18	USSR			Liquid propellant; two stages	10,000	1	5-lOMt	
(ICBM)							mode 2	
SS-23 Scalpel	USSR	–	–	Solid cold launch propellant:	10,000	1	Nuclear MIRV	Rail
(ICBM)				three stages			8–10x	mobile
SS-25 Sickle	USSR	–	–	Solid cold launch propellant;	10,500	1	Nuclear 550 kt	
(IRBM)				three stages				

Table 2.19 Missiles for armed forces (BGM, M, MK, RGM, etc.) (USA, France, Israel)

TOW 2 (BGM-71 D)	USA	$L = 140$ $d = 15$			21.5	Solid propulsion; wire manual or auto	3.75	Heat conventional	SSM anti-tank
Copperhead (M712)	USA	$L = 137$ $d = 15$			63.5	Cannon launched: laser homing	16	Heat 6.4 kg	SSM anti-tank
Gabriel	Israel	L	S	d		Two-stage solid		Conventional	SSM
MK I		335	135	34	430	propulsion; auto	18	100 kg	shipborne
MK II		341	135	34	430	pilot/command	36	100 kg	
MK III		381	135	34	560		36+	150 kg	
MK IV		381	135	34	560	Turbojet propulsion: inertial and active radar homing	200		
SS-N-3	USSR	$L = 1020$			4700	Solid boosters:	460–735	Nuclear and Conventional	SSM shipborne
Shaddock		$d = 86$				internal turbojet radio command			
(SRM)									
Harpoon	USA	$L = 384$			519	Solid booster turbojet cruise: inertial and radar	90	Conventional	SSM
(RGM-84A)		$S = 830$ $d = 34$							shipborne
Deadeye 5	USA	$L = 384$ $d = 12.7$			47.5	Solid propulsion; laser homing	24	–	Guided projectile; shipborne
Crotale	France	$L = 290$			80	Solid propulsion: radio command	18	15 kg	NSA
		$S = 54$ $d = 15$						conventiona	
Sadral	France	$L = 180$ $d = 16$			17	Solid propulsion; infrared homing	0.3–6	3 kg heat conventional	NSA

Table 2.20 Data and parameters on missile types

Service load parameters and stresses							
Barak missile	Israel Aircraft Industries (IAI)	$L = 217$ $d = 17$	86 –	Semi-active radar-horning missile with disposable launch canister, manually fitted to an 8-round launcher based on the MBT TCM-30 twin 30 mm anti-aircraft gun mounted	10	Conventional 2 nuclear 7 kg	NSA small patrol boats
SA-N-3 Goblet	USSR	$L = 620$ $S = 150$ $d = 33.5$	550	Solid propulsion: semi-active homing semi-active homing	55	80 kg conventional	NSA
SA-N-4	USSR	$L = 320$ $d = 21$	190	Solid propulsion: semi-active homing semi-active homing	14.8	50 kg	NSA
Roland 3	France and Italy	$L = 260$ $S = 50$ $d = 27$	85	Solid propulsion and command	8	9 kg conventional	ADM
SA-4 Gancf (SRM)	USSR	$L = 880$ $S = 290$	100,000	Ramjet and solid boosters; radio command	70	135 kg	ADM
SA-lO Grumble MMIO	USSR	$L = 700$ $S = 10$ $d = 45$	1500	Solid propulsion	100	Nuclear	ADM
SA-12 Gladiator	USSR	$L = 50$ $S = 350$ $d = 50$	2000	Solid propulsion: semi-active radar	80	150 kg conventional	ADM

Table 2.20 (continued)

Service load parameters and stresses								
SA-13 Gopher	USSR	$L = 220$	55	Solid propulsion:	10	4 kg	ADM	
				infra—red homing				
(HJ-8)		$S = 40$		infrared homing		conventional		
		$d = 12$						
BGM-109	General	$L = 5.56$	2200	Williams Research F107	1 727	0.72	Thermo-nuclear	SG
Tomahawk cruise missile	Dynamics	$S = 2.45$	1000	Turbofan 600 lb (272 kg)	2780	(550 mph/ 885 km/h)	nuclear	
	Convair.	$d = 0.53$	to	Thrust guidance			1000 lb	
(IRBM)	San Diego, USA		4000 1814	MD Tercom and inertial			(454 kg)	
CSS-2	China	–	–	Liquid propellant	120,00	0.8	Nuclear	SG
CSS-3				Stage 2			2 Mt	
(IRBM)				Stage 3			1–5 Mt	
CSS-1	China	–	–	Liquid propellant	1200	0.7	Nuclear	SG
CSS-2				Two stages	2700		20 kt	
(IRBM)							1 –3 Mt	
Sandal				Two stages	2000		1 Mt	
(IRBM)	–	–	–					
SSBS S-3	France	–		Solid propellant	3000	0.65	Nuclear	
				Two stages			1.2 Mt	
SS-N-20	USSR	–		Solid propellant	3000	0.65	Nuclear	
Sturgeon							1.2 Mt	
(1RBM)								

Table 2.20 (continued)

Service load parameters and stresses								
UGM-73	Lock heed	$L = 10.36$	65000	Solid motor Thiokol as first stage: Hercules as	3230	10	Lockheed	Submarine
Poseidon missile (IRBM)	Missiles and Space Company, California, USA	$d = 1.88$	–29500	second stage; guidance: inertial	5200		MIRV carrying 50 kt Rvs	launch
UGM-27	Lockheed,	$L = 9.45$	35000	Aerojet solid motor with jetavator control as first stage: Hercules motor missile with liquid injection as second stage	2875	10	MIRV	Submarine
Polaris (ICBM)	California, USA	$d = 1.37$	–15850		(4631)		200 kt	launch

Table 2.21 Missiles and rockets

Service load parameters and stresses

Type of missile/rockets	Manufacturer/country of manufacture	Dimensions	w_L (lb)	Power plant and guidance	R (km)	Warhead	Mechanisms
Blood hound	UK	$L=846$ $S=283$ $d=55$		Ramjet and solid boosters; semi-active radar homing	80	Conventional	ADM
Blow pipe	UK	$L=139$ $S=270$	20.7	Solid propulsion; radio command	Any short distance 3		
Javelin	UK	$L=40$ $d=8$		Solid propulsion; Saclos	4+	Conventional	ADM shoulder-fixed
Hawk MIM-23B	USA	$L=503$ $S=119$ $d=36$	627.3	Solid propulsion; semi-active Homing	40	Conventional	ADM
Stinger FIM-92A	USA	$L=152$ $S=14$ $d=7$	15.8	Solid propulsion; infrared homing	Short range 3	Conventional	ADM
MIM Hercules	USA	$L=1210$ $d=80$	4858	Solid propulsion; command	140	Nuclear and conventional	ADM
L-5B	China	$L=289$ $S=66$ $d=13$	85	Solid propulsion; infrared	5	Conventional	AAM
L-7	China	$L=25$ $S=66$ $d=16$	90	Solid propulsion; infrared	5	Conventional	AAM
Mistrel	France	$L=180$ $d=9$	18	Solid propulsion; infrared	3	3 kg conventional	AAM
Super 530	France	$L=354$	250	Solid propulsion	25	3 kg	AAM

Table 2.21 (continued)

Service load parameters and stresses

Type of missile/rockets	Manufacturer/country of manufacture	Dimensions	w_L (lb)	Power plant and guidance	R (km)	Warhead	Mechanisms
M Type (SRM)	China	$L=910$	6200	Solid propulsion; inertial	600	Nuclear	SSM
Frog-7 (SRM)	France	$L=910$ $d=85$	2300	Solid propulsion	70	Nuclear	SSM
SS-IC Scud B	USSR	$L=1125$ $d=85$	6370	Storable liquid; inertial	280	Nuclear	
SS-23 Spider (SRM)	USSR	$L=605$ $d=100$	3500	Solid propulsion; inertial	525 miles	Nuclear	SSM
Exocet MM10	France	$L=578$ $S=100$ $d=35$	850	Solid propulsion; two stages: inertial and active radar Homing	70	165 kg Conventional	SSM ship
RBS 15	Sweden	$L=435$ $d=50$	600	Turbojet boosters	150	Conventional	SSM ship
Arrow 8(Hj-8)	China	$L=99.8$ $S=47$ $d=10.2$	11.3	Solid propulsion; wire guided	0.1–3	Heat	SSM antitank
Spandrel	USSR	$L=100$ $d=16$	12 –18	Solid propulsion; semi-automatic	4	Heat	SSM antitank
Vigilant	UK	$L=107$ $S=28$ $d=11$	14	Solid propulsion; two stages: wire manual or auto	1.375	Heat 6 kg	
Dragon 1M47	USA	$L=74$ $S=33$ $d=13$	13.8	Multiple solid propulsion; wire manual	1	Heat	SSM antitank

Table 2.22 Missiles with warhead mechanisms

Service load parameters and stresses

Types of missiles/rockets	Manufacturer/ country ofmanufacture	Dimensions	$w_L$lb (kg)	Power plant and guidance	R	V_{mach} (mph)	Warhead	Mechanisms
		L S d						
C-601	China	738 280 92	2440	Liquid propulsion	100 km	–	400 kg	ASM
HY-4		736 280 7.6	1740	Turbofan propulsion	150 km		to	ASM
C-801		480 165 55	1025	Solid propulsion	150 km	–	500 kg	ASM
				All active radar			Conventional	
		L S d						
S-9	USSR	600 150 50	750	Solid propulsion:	150 km		Conventional	ASM
S-10		350 90 30	300	semi-active laser				
AS-II	USSR	$L = 350$ $S = 90$ $d = 30$	300	Solid propulsion: infrared	300 km min	–	Conventional	ASM
Sea Eagle	UK	$L = 414$ $S = 120$ $d = 40$	600	Turbofan inertial: active radar	110 km	–	Conventional	ASM
AGM-65	Hughes	$L = 246$	462 (210)	Thiokol TX-481 solid motor: AGM-65C laser guidance	8 miles	6–10	Mk 19 113 kg	ASM
Maverick	Aircraft, Tulson, USA	$S = 71$ $d = 44.5$			(13 km) to 14 miles (22.5 km)			
AGM-69A SRAM	Boeing	$L = 4127$	2230	LPC (Lockheed Company)	High 105 miles (170 km):	200	Conventional	ASM
	Aerospace,	$S = 89$	1010	Propulsion two-pulse solid motor:	low 35 miles (56 km)		Nuclear	Rocket
	Seattle,	$d = 44.5$		guidance: inertial			170 kT	

Table 2.22 (continued)

Service load parameters and stresses

Types of missiles/rockets	Manufacturer/ country ofmanufacture	Dimensions	$w_L$lb (kg)	Power plant and guidance	R	V_{mach} (mph)	Warhead	Mechanisms
	USA							
ASLAM	McDonnell Douglas, USA	$L = 4127$	2700	Internal rocket		6–10	170 kt nuclear and conventional	ASM
Missile MK III MK IV		$S = 89$ $d = 44.5$	1200	ramjet: guidance: inertial with Tercom				

Types of missiles/rockets	Manufacturer/country of manufacture	Dimensions	$w_L$lb (kg)	Power plant and guidance	R	Warhead	Mechanisms
550 Magic	France	$L = 275$ $S = 66$ $d = 16$		Solid propulsion; Infrared	5 km	Conventional	AAM
AIM-9	Naval	$L = 283$–	Up to	Rocket dyne, Thiokol,	2.5 + miles	XM 248	AAM
Sidewinder	Weapon Centre, Philco-Ford Ford Aero-space	2.91 $S = 56$– 64 $d = 12.7$	190 86	Bermite or Naval propellant; Single Grain solid MK 17, 36, 86; guidance: 9E to 9L, high	cruise 6 miles max	high explosive (150 lb)	(F-16, F-20, F-5E aircraft)
	General Electric			power servo-system, AM/			

Table 2.22 (continued)

Types of missiles/rockets	Manufacturer/country of manufacture	Dimensions	$w_L$lb (kg)	Power plant and guidance	R	Warhead	Mechanisms
	USA			FM IR conical scan head or Raytheon track via missile			
Alamo AA-10	USSR	$L=400$ $S=70$ $d=19$	200	Solid propulsion; semi-active or infrared	30 km	Conventional	AAM
AMBAAM	USA	$L=357$ $S=63$ $d=18$	150	Solid propulsion; command and inertial	12 km	Conventional	AAM
Falcon	USA	$L=213$ $S=639$ $d=29$	115	Solid propulsion; semi-active radar	8 km	Conventional	AAM
Phoenix	USA	$L=396$ $S=92$ $d=38$	454	Solid propulsion; semi-active radar	150 km	Conventional	AAM
Sparrow AIM-7 M	USA	$L=366$ $S=102$ $d=20$	227	Solid propulsion; infrared	140 km	Conventional	AAM

L = length (cm), S = wingspan (cm), d = diameter (cm), W_L = launch weight, R = range

Table 2.23 Missiles for armed forces (ADM and NSA types) (UK, USA)

Service load parameters and stresses

Types of missiles/ rockets	Manufacturer/ country of manufacture	Dimensions	$w_L$lb (kg)	Power plantand guidance	R miles (km)	V mach (mph)	Warhead	Mechanisms
MGM-31A	Martin,	$L = 10.51$	10,150	2 No. Thiokul	100	8	Nuclear	Mohile
Pershing missile	Orlando,	$S = 202$ $d = 101$	4600	solid motors: first stage – M105,second stage – M106; guidance:	160		400 kt	tactical system
	USA			Inertial	to 520 840			ADM/SSM
MGM-51C	Ford, New Post	$L = 1.14$	60	Amoco single-	3	800	Octal-	Battle tank,
Shillelagh missile	Beach, USA	$S = 29$ $d = 15.2$	-27	stage solid jetavators: guidance: optical tracking and infrared command link			shaped charge 15 lb (h.8 kg)	ADM/SSM
BGM-71 TOW	Hughes	$L = 1.17$	42	Quad boost	up to	620	Shaped	Vehicles,
Missile	Aircraft,	$S = 34$ $d = 14.7$	19	motor for recoil-less launch:	*3.75*		Charge	aircraft,
	Culver City,			guidance: with optical sighting and trailing wire			containing	ADM/SSM
	USA						5.3 lb (2.4 kg) high explosive	

Table 2.23 (continued)

Service load parameters and stresses

Types of missiles/ rockets	Manufacturer/ country of manufacture	Dimensions	w_L lb (kg)	Power plantand guidance	*R* miles (km)	*V* mach (mph)	Warhead	Mechanisms
Sea Dart Missile	British	$L = 4.4$ $d = 42$	550	Marconi radar	30	—	Conventional and nuclear	NSA patrol boats and ships
	Aerospace, UK			805 SW tracker with type 909 illuminator; guidance: semi				
Sea Wolf Missile	British Aerospace, UK	$L = 1.9$ $d = 18$	82	GWS 25 — Marconi radar (805547) search, tracking radars, Ferranti fire-control computer; guidance: radio command	30	—	Conventional and nuclear	NSA/SSM boats and Ships
Sea Cat Missile	Short Brothers, Belfast, UK	$L = 1.47$ $d = 19$	63	Radio command guidance with optical remote TV and radar- aided tracking	30	—	Conventional and nuclear	NSA/SSM boats and ships
Raytheon/ Martin	Martin Orlando	$L = 5.31$	3740	Thiokol TX-486	30	3–5	Nuclear and	Advanced
XMIM-104A	Division, USA	$S = 87$	1696	single-thrust solid	48		conventional	mobile
Patriot missile		$d = 41$		motor				system SAM

L = length (cm), S = wingspan (cm), d = diameter (cm), W_L = launch weight, R = range; V = speed

Helicopters

Helicopters are more vulnerable than aircraft in warfare. In peace time a helicopter may crash after losing a rotor or hitting objects such as offshore platforms, buildings, helipads or their surrounding structures. Table 73 gives useful data for some types of helicopters in the book “Shock, Impact and Explosion” by the authors published by Springer-Verlag (Germany) 2008.

2.2.3.7 Load Combinations

Based on American Standards

Load Combinations for Concrete Structures

Design load combinations for concrete structures are given in the following two industry standards, depending on the type of structure being designed.

1. Code for Concrete Reactor Vessels and Containments –ASME Section III, Division 2 and ACI 359-77. Prepared by Joint ACI-ASME Technical Committee on Concrete Pressure Components for Nuclear Service.
2. Standard for Design of Concrete Structures in Nuclear Service other than Pressure Retaining Components ACI-349.

   The loading equations found in these industry standards may not agree with NRC published guidelines. In such cases the designer should be assured the load combinations used are acceptable to the regulatory authorities.
3. Load Combinations for Steel Structures –For steel structures, a definitive industry standard (ANSI N690) is still used. In general, load combinations are acceptable if they are found in accordance with the following
4. For service load conditions, either the elastic analysis working stress design (WSD) methods of Part 1 of the AISC Specification or the plastic (limit) analysis load factor design (LRFD) methods of Part 2 of the AISC Specification may be used.

   If the elastic analysis WFD methods are used, the following load combinations should be considered:

   1. $D + L$
   2. $D + L + E_0$
   3. $D + L + W$

If thermal stresses due to T_0 and R are present, the following combinations should also be considered:

1a. $D + L + T_a + R_0$
2a. $D + L + T_0 + R_0 + E_0$
3a. $D + L + T_0 + R_0 + W$

Both cases of L having its full value or being completely absent should be checked.

If plastic (limit) analysis LRFD methods are used, the following load combinations should be considered:

1. 1.7D + 1.7L
2. 1.7D + 1.7L + 1.7E_0
3. 1.7D + 1.7L + 1.7 W

If thermal stresses due to T_0 and R_0 are present, the following combinations should also be considered:

lb. 1.3(D + L + T_0 + R_0)
2b. 1.3(D + L + E_0 + T_0 + R_0)
3b. 1.3(D + L + W + T_0 + R_0)

Both cases of L having its full value or being completely absent should be checked.

For factored load conditions which represent extreme loads the following load combinations should be considered:

Elastic analysis WSD methods are used:

4. $D + L + T_0 + R_0 + E^1_{ss}$
5. $D + L + T_0 + R_0 + W_t$
6. $D + L + T_a + R_a + P_a$
7. $D + L + T_a + R_a + P_a + 10(Y_j + Y_r + Y_m) + E_0$
8. $D + L + T_a + R_a + P_a + 10(Y_j + Y_r + Y_m) + E^1_{ss}$

If plastic (limit) analysis LRFD methods are used:

4. $D + L + T_0 + R_0 + E^1_{ss}$
5. $D + L + T_0 + R_0 + W_t$
6. $D + L + T_a + R_a + 1.5P_a$
7. $D + L + T_a + R_a + 1.25P_a + 1.0(Y_j + Y_r + Y_m) + 1.25E_a$
8. $D + +T_a + R_a + 1.0P_a + 1.0(Y_j + Y_r + Y_m) + 1.25E^1_{ss}$

European Codes

The most important codes indulging in nuclear facilities are EC2, EC3, EC8, EC9, etc. It is extremely difficult to specifically assign combination for nuclear facility. The best possible combination can be given after examining various design practices in Europe where European codes are used, all loads can be calculated using respective codes. The following combinations of various loads are given below:

$U = 1.4D + 1.7L$
$U = 0.75(1.4D + 1.7L \pm 1.87E)$
$U = 0.75(1.4D + 1.7L \pm 1.7\,W)$
$U = 0.9D \pm 1.43E$
$U == 0.9D \pm 1.3\,W$
$U = 1.4D + 1.7L + L^7H_{mep}$

$U = 0.9D + 1.7H$
$U = 0.75(1.4D + 1.4T_d + 1.7L)$
$U = 1.4(D + T_d)$

For reinforced concrete the following modifications are introduced where earthquakes are involved:

$U = 1.4(D + L \pm E)$
$U = 0.9D \pm 1.4E$

where U= required strength to resist factored loads or related internal moments and forces; D= dead loads or related internal moments and forces; L = live loads or related internal moments and forces; W = wind loads or related internal moments and forces; E = load effects of earthquake or related moments and forces; T_d= internal moments and forces due to differential settlement, creep, shrinkage or temperature effects; H_{mep} = moments or forces due to earth pressure.

Loads computed from the Euro codes can still be combined using American practices given above. Care should be taken that all industrial concerns have been consulted and approvals are obtained for the design of various structural elements of nuclear facilities.

2.3 Determination of Impulse/Impact caused by Aircraft and Missiles: Load (I)

2.3.1 General

An impactor in the form of an aircraft or a missile develops from initial velocity to a velocity caused by its movement under the action of its own weight or a booster's force. In any circumstances, if the kind of velocity is not vertical it will move in a curve and its flight can be evaluated in terms of horizontal and vertical components of displacement, velocity and acceleration. Alternatively, the directions are controlled in any specific direction from the control centre.

2.3.1.1 Direct Impulse/Impact and Momentum

An impulse is defined as a force multiplied by time, such that

$$F_1(t) = \int F dt \tag{2.1}$$

where F_1 (t) is the impulse, F is the force and t is the time. The momentum of a body is the product of its mass and its velocity:

$$\text{Momentum} = mv \tag{2.2}$$

where m is the mass and v is the velocity = dx/dt. Both velocity and momentum are vector quantities; their directions are the same. If a body is moving with a constant velocity, its momentum is constant. If velocity is to be changed, a force F must act on the body. It follows that a force F must act in order to change the momentum.

$$F = mdv/dt \tag{2.2a}$$

or

$$Fdt = mdv$$

Integrating both sides

$$\int_{t_z}^{t_2} = \int_{u}^{v} mdv$$
$$F_1(t) = m(v - u) \tag{2.3}$$

where u and v are the velocities at times t_1 and t_2, respectively. If the initial velocity $u = 0$, Eq. (2.3) becomes

$$I = mv \tag{2.3a}$$

Thus the impulse of a force is equal to the change in momentum which it produces.

2.3.1.2 Impacts/Collisions of Aircraft

When two solid aircrafts are in contact, they exert equal and opposite forces or impulses on each other and they are in contact for the same time. If no external force affects the motion, the total momentum in the specific direction remains constant. This is known as the *principle of conservation of linear momentum.* When two aircrafts m_1 and m_2, collide (Fig. 2.1), the mass ratios are then calculated from Eq. (2.1):

$$F_{11}(t) = m_1(v_1 - u_1) = \int F_1 dt$$
$$F_{12}(t) = m(v_2 - u_2) = \int F_2 dt \tag{2.4}$$

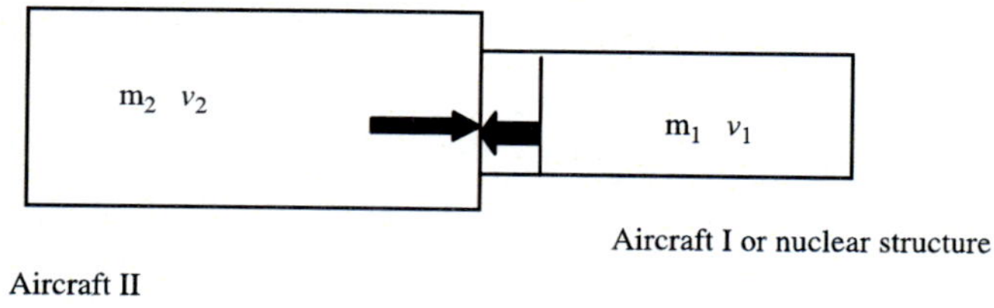

Fig. 2.1 Direct impact

Since $\int F_1 dt = \int F_2 dt = 0$, the relationship between the velocity change and mass becomes

$$m_2/m_1 = (\upsilon_1 u_1)/(\upsilon_2 - u_2) \tag{2.5}$$

During the collision process, although the momentum is conserved, there is a loss of energy on impact which is determined using the concept of the *coefficient of restitution, e,* which is defined as the relative velocity of the two masses after impact divided by the relative velocity of the two masses before impact. Before impact

$$e = (\upsilon_1 - \upsilon_2)/-(u_1 - u_2) = 0$$

When the relative velocity vanishes, and

$$e = (\upsilon_1 - \upsilon_2)/-(u_1 - u_2) = 1 \tag{2.5a}$$

there is no loss of relative velocity.
Where $e<1$, it is related to the loss in kinetic energy, and where $u_2 = 0$ (refer to Eq. (2.5a)

$$\begin{aligned} m_1(\upsilon_1 - u_1) + m_2(v_2) &= 0 \\ \upsilon_1 - \upsilon_2 &= eu_1 \end{aligned} \tag{2.6}$$

Hence

$$\upsilon_1 = u_1(m_1 - em_2)/(m_1 + m_2) \tag{2.6a}$$

$$\upsilon_2 = u_1[(1+e)m_1/(m_1 + m_2)] \tag{2.6b}$$

The original kinetic energy $(\mathrm{KE})' = \frac{1}{2}m_1u_1^2$
The final kinetic energy $(\mathrm{KE})'' = \frac{1}{2}(m_1v_1^2 + m_2v_2^2)$

$$(\mathrm{KE})' - (\mathrm{KE})'' = \frac{1}{2}m_1u_1^2 - \frac{1}{2}(m_1\upsilon_1^2 + m_2\upsilon_2^2) \tag{2.7}$$

Substituting the values of υ_1 and υ_2

$$(\mathrm{KE})' - (\mathrm{KE})'' = (\mathrm{KE})'[m_1(1-e^2)/(m_1 + m_2)] \tag{2.8}$$

The displacement resulting from a short-duration (τ) impact is given by

$$x = b(t - \tau) \tag{2.9}$$

where t is the time beyond τ. For dynamic analysis, the impact time is divided into ns small segments and, using Eq. (2.3a),

$$\begin{aligned} x &= \frac{1}{m}\sum_{0}^{n} \upsilon_n I_n(\mathbf{t} - \tau_n) \\ &= \frac{1}{m}\int_{0}^{t} F(t-\tau)d\tau \end{aligned} \tag{2.10}$$

The impact is divided into two phases such that in the first, from time t_1 to t_0, there will be compression and distortion until $(v_1 + v_2)$ are both reduced to zero (the both aircrafts moving together); in the second, the elastic strain energies in the aircraft are restored and are separated by a negative velocity, $-V_2 = (v_1 + v_2)$.

During the second phase the impulse relation between the aircraft ($F_T - F_{TO}$) will be proportional to F_{TO} and the coefficient or restitution e defined above is written as

$$e = (F_T - F_{TO})/F_{TO} \tag{2.11}$$

where F_T is the total impulse during the impact and F_{TO} is the impulse in phase one.

At time t_0

$$V_0 = \upsilon_{10} + \upsilon_{20} = \upsilon_1 + \left(\frac{F_{TO}}{m_1} + \upsilon_2 - \frac{F_{TO}}{m_2}\right) = 0 \tag{2.12}$$

hence

$$V = \upsilon_1 + \upsilon_2 = \left(\frac{1}{m_1} + \frac{1}{m_2}\right)F_{To} \tag{2.13}$$

Similarly, at time t_2 the relationship becomes

$$V_0 - V_2 = F_T\left(\frac{1}{m_1} + \frac{1}{m_2}\right) \tag{2.14}$$

Using Eq. (2.11), the expression given in Eq. (2.5a) may be written in the form

$$-\frac{V_2}{V}e \tag{2.15}$$

Equations (2.6), (2.6a) (2.6b) result from the above method. However, from Eq. (2.11) the total impulse is rewritten as

$$\begin{aligned} F_T &= \left(\frac{m_1 m_2}{m_1 + m_2}\right)(1+e)(\upsilon_1 + \upsilon_2) \\ & M(1+e)V \end{aligned} \tag{2.16}$$

where M is the equivalent combined mass of the aircrafts.

The changes in velocity after impact of the aircrafts are written as

$$\begin{aligned} \Delta V_1 &= \frac{M}{m_1}(1+e)(v_1+v_2) = \frac{M}{m_2}(1+e)V \\ \Delta V_2 &= \frac{M}{m_2}(1+e)V \end{aligned} \tag{2.17}$$

2.3.1.3 Oblique Impact

When two aircrafts collide and their axes do not coincide, the problem becomes more complex. With oblique impact, as shown in Fig. 2.2, two impulses are generated: the direct impulse, F_{T}, and the tangential impulse, F'_{T}. The latter is caused by friction between the impacting surfaces and by local interlocking of the two aircraft in the common surface. Let the angular velocity of the two aircrafts be θ_1 and θ_2, respectively. If $F'_{\mathrm{T}}/F_{\mathrm{T}} = \lambda'$ and the body's centre of gravity has a coordinate system X and Y,the components of the vector velocity, $v_1$and u_1, normal to the impact surface may be written as follows:

$x_1 - y_1$ system

$$v_1 = |\bar{v}_1| \cos\theta_1 \tag{2.18}$$

$$u = |\bar{v}_1 \sin\theta_1 \tag{2.18a}$$

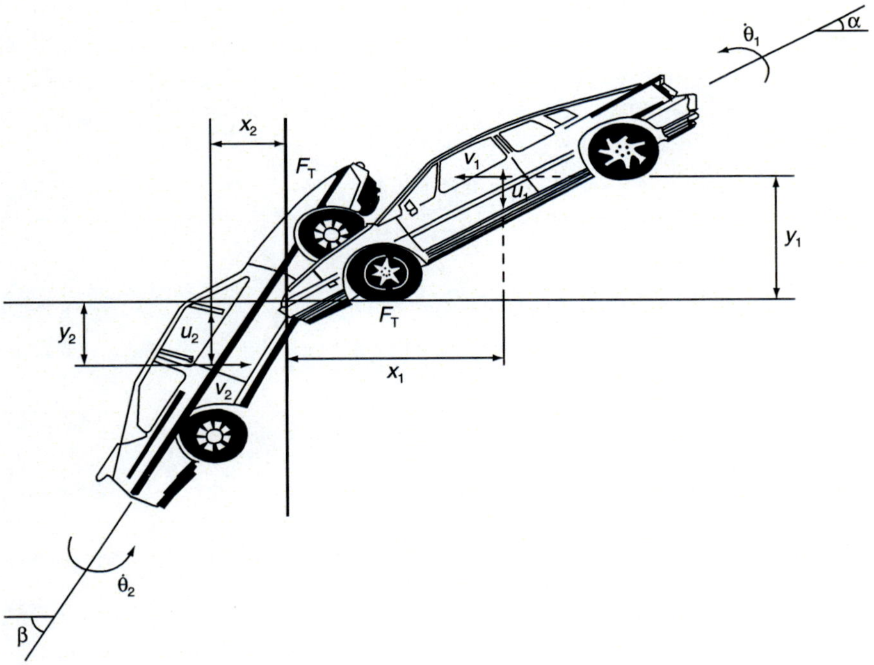

Fig. 2.2 Oblique impact of the bodies of two different objects

Similarly, υ_2 is written as

$$|\bar{\upsilon}_2| = V(\upsilon_2^2 + u^2{}_2) \tag{2.19}$$
$$\beta = \tan^{-1}(u_2/v_2) \tag{2.19a}$$

The momentum equations for the bodies are summarised below:

$$\left.\begin{aligned} m_1\upsilon'_1 - F_\mathrm{T} &= m_1\upsilon'_2 \\ m_1u'_1 - \lambda' F_\mathrm{T} &= m_1u'_2 \\ m_1R^2{}_1\theta'_1 + F_\mathrm{T}y_1 - \lambda' F_\mathrm{T}x_1 &= m_1R^2{}_1\theta'_2 \end{aligned}\right\} \quad \text{body 1} \quad (2.20)$$

where υ'_1, υ'_2, u'_1 and u'_2 are for t_1 and t_2.

$x_2 - y_2$ system

$$\left.\begin{aligned} m_2\upsilon''_1 - F_\mathrm{T} &= m_2\upsilon''_2 \\ m_2u''_1 - \lambda' F_\mathrm{T} &= m_1u'_2 \\ m_2R^2{}_2\theta'_2 + F_T y_2 - \lambda' F_\mathrm{T}x_2 &= m_2R^2{}_2\theta'_2 \end{aligned}\right\} \quad \text{body 2} \quad (2.21)$$

where $mR^2{}_1$ and $mR^2{}_2$ are the second moment of inertia about the vertical axis passing through the centre of gravity. The rate of approach and the sliding of the two surfaces at the point of contact can be written as

$$\Delta V_1 = \upsilon_1 + \upsilon_2 - \dot{\theta}_1 y_1 - \dot{\theta}_2 y_2 \tag{2.22}$$
$$\Delta V_2 = u_1 + u_2 - \dot{\theta}_1 x_1 - \dot{\theta}_2 x_2 \tag{2.23}$$

The addition to these equations is the restitution given by Eq. (2.15) in which when Eq. (2.22) is substituted and then, in the final equation, Eq. (2.20) is substituted, the value of F_T is evaluated as

$$F_T = \frac{V(1+e)}{C_1 - \lambda C_2} \tag{2.24}$$

where

$$C_1 = \frac{1}{m_1}\left(1 + \frac{y^2{}_1}{R^2{}_1}\right) + \frac{1}{m_2}\left(1 + \frac{y^2{}_2}{R^2{}_2}\right) \tag{2.24a}$$
$$C_2 = \left(\frac{x_1y_1}{m_1R_1^2} + \frac{x_2y_2}{m_2R_2{}^2}\right) \tag{2.24b}$$

Using Eqs. (2.20) and (2.21)

$$\begin{aligned} \upsilon'_\mathbf{2} &= \upsilon'_\mathbf{1} - (F_T/m_1) \\ u'_\mathbf{2} &= u'_\mathbf{1} - (\lambda' F_T/m_1) \\ \dot{\theta}_2 &= \dot{\theta}_1 + \frac{y_1 - \lambda' x_2}{m_1 R_1^2} F_\mathrm{T} \end{aligned} \tag{2.25}$$

$$\upsilon''_2 = \upsilon''_1 - \frac{F_T}{m_2}$$

$$u''_2 = u''_1 - \frac{\lambda' F_T}{m_2} \tag{2.26}$$

$$\dot{\theta}_2 = \dot{\theta}_1 + \frac{y_2 - \lambda' x_2}{m_2 R_2^2} F_T$$

Figure 2.3 shows plots for Eqs. (2.25) and (2.26). It is interesting to note that larger values of λ' show greater interlocking of the surfaces of the two aircrafts and with e reaching zero, a greater plastic deformation occurs.

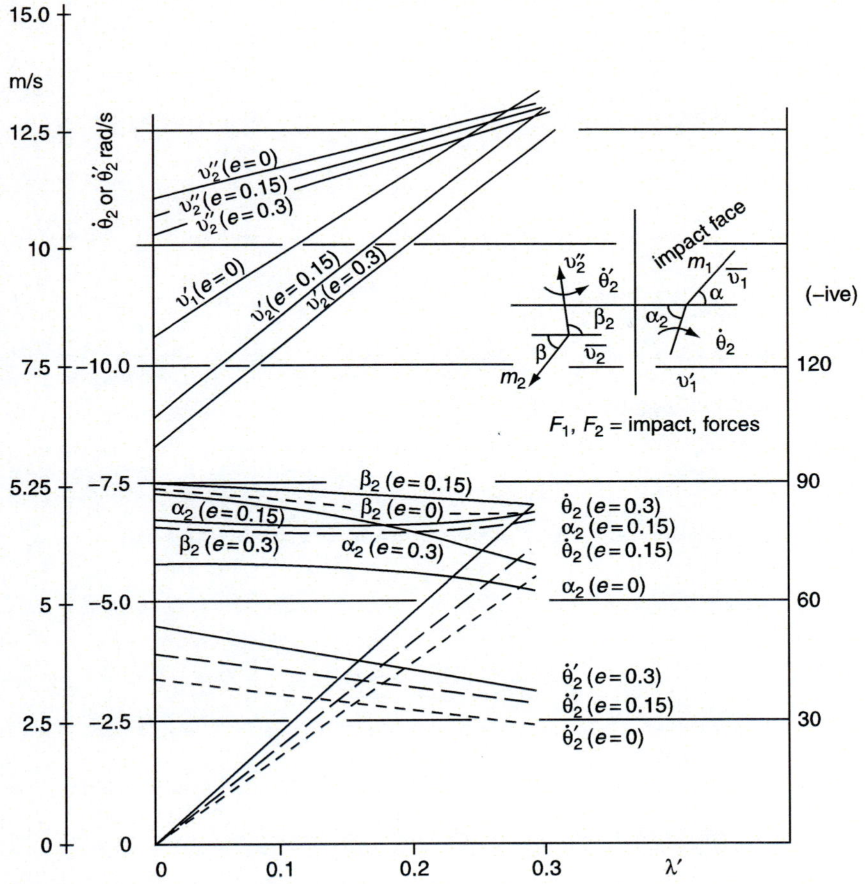

Fig. 2.3 Velocity versus λ' for oblique impact problems

2.3.1.4 Case Studies

(1) One aircraft impacting a rigid barrier, or containment vessel located with no angular velocity

$$\frac{1}{m_2} = 0, \quad v_1 = 0, \quad u_1 = 0, \quad \theta_1 = 0 \tag{2.27}$$

$$C_1 = \frac{1}{m_1}\left(1 + \frac{y_1^2}{R_1^2}\right); \quad C_2 = \left(\frac{x_1 y_1}{m_1 R_1^2}\right) \tag{2.27a}$$

$$v_2' = v_1'(y_1^2 - \lambda' x_1 y_1 - R^2)/\bar{\lambda} \tag{2.27b}$$

$$u'_2 = u_1' - v_1'\left(\frac{\lambda'(1+e)R^2}{\lambda}\right); \quad \dot{\theta}_1 = \frac{(1+e)(y_1 - \lambda' x_1)}{\lambda} \tag{2.27c}$$

$$\bar{\lambda} = y^2{}_1 - \lambda'^{x_1 y_1} + R^2 \tag{2.27d}$$

where

(2) Circular impactor with radius r_1.

$$x_1 = r_1 \quad \text{and} \quad y_1 = 0 \tag{2.28}$$

$$v_2' = e v_1' \tag{2.28a}$$

$$u_2' = u_1 - \lambda' v_1'(1+e) \tag{2.28b}$$

$$\dot{\theta}_1 = -v_1' \lambda' r_1 (1+e)/R^2 \tag{2.28c}$$

For a circular impactor, $R^2 = 2r_1^2/5$

$$\dot{\theta} = -v_1'(5\lambda'(1+e)/2r_1) \tag{2.28d}$$

(3) Inelastic collisions. The value of $e = 0$ in the above case studies:
Case study (1)

$$\begin{aligned} v'_2 &= v'_1(y^2{}_1 - \lambda' x_1 y_1)/\bar{\lambda} \\ u'_2 &= u_1' - v_1'(\lambda' R^2/\bar{\lambda}) \\ \theta_1 &= (y_1 - \lambda' x_1)/\bar{\lambda} \end{aligned} \tag{2.29}$$

Case study (2)

$$\begin{aligned} v_2' &= 0, u'_2 = u_1' - \lambda' v_1' \\ \theta_1 &= -v'_1 \lambda r_1/R^2 = -2.5 v_1' \lambda/r_1 \end{aligned} \tag{2.30}$$

(4) Where no interlocking exists, λ' in the above expressions.

This means the aircrafts do not interlock each other but their bodies have created damages.

2.3.2 *Aircraft Impact on Nuclear Structures – Peak Displacement and Frequency*

2.3.2.1 General

A great deal of work has been carried out on the subject of missile and aircraft impact. Tall structures are more vulnerable to civilian, wide-bodied jets or multi-role combat aircraft. A great deal of work on this subject will be reported later. In this section a preliminary analysis is given for the determination of peak displacement and frequency of a tall structure when subject to an aircraft impact. As shown in Fig. 2.4, the overall dimensions of the building are given. Let A be the base area and h be the maximum height of the building. According to the principle of the conservation of momentum, if m is mass and $\upsilon_1$is the velocity of the aircraft approaching the building, then using a linear deflection profile

$$I(t) = F_1(t) = m\upsilon_1 = \left(\frac{\rho A h}{2g}\right)\upsilon_{20} \tag{2.31}$$

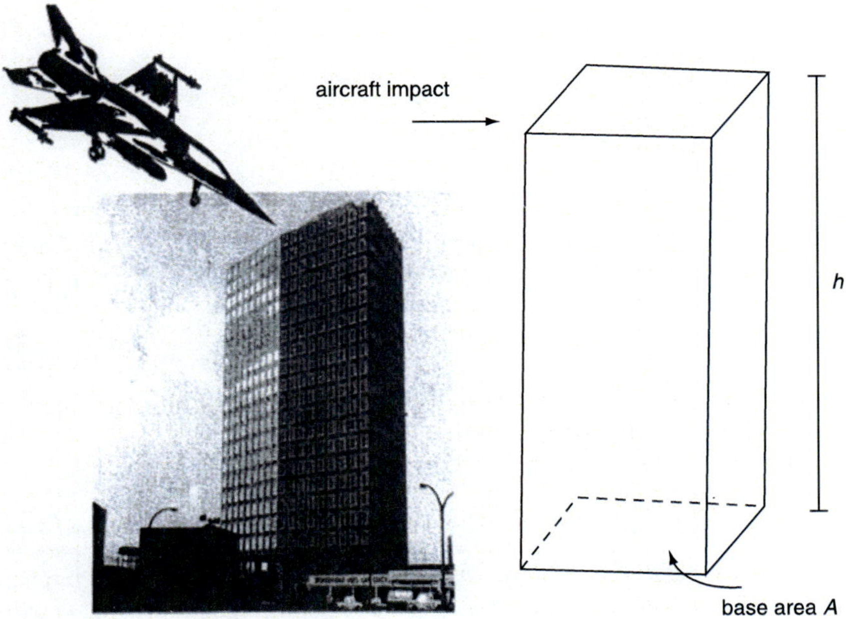

Fig. 2.4

where ρ is the density or average specific weight and υ_{20} is the velocity of the tip of the nuclear facility.

The initial velocity, υ_{20}, of the facility can thus be evaluated from Eq. (2.31). Free vibrations studied by the time-dependent displacement $\delta(t)$ is given by

$$\begin{aligned}\delta(t) &= \left(\frac{\upsilon_{20}}{\omega}\right)\sin \omega t \\ &= [\upsilon_{20}/(2\pi/T)]\sin \omega t \\ &= [\upsilon_{20}/\sqrt{(k_s/m_s)}]\sin \omega t\end{aligned} \tag{2.32}$$

where ω is the circular frequency and k_s and m_s are the equivalent nuclear facility stiffness and mass, respectively.

Using Eq. (2.31) for υ_{20} and $\sin \omega t = 1$ for $\delta_{max}(t)$, the *peak dynamic displacement*, $\delta_{max}(t)$, is given by

$$\delta_{max}(t) = m\upsilon_1 gT/\pi\rho Ah \tag{2.32a}$$

The equivalent point load generated for the peak dynamic displacement is given by Eq. (2.32a). If that load is $F_1(t)$, then work done is equal to the energy stored and

$$F_1(t) \times \delta_{max}(t) = \frac{1}{2}k_s\delta_{max}^2(t) \tag{2.33}$$

for which

$$F_1(t) = \frac{1}{2}k_s\delta_{max}(t) \tag{2.33a}$$

While momentum is conserved, a portion of energy of the aircraft is lost on impact. The loss of energy E_1 is then written as

$$E_1 = \frac{1}{3}(\rho Ah/mg)(\upsilon_{20}/\upsilon_1)^2 \tag{2.34}$$

Equation in case study (1) and Eq. (2.29) for inelastic collisions are applied with and without the interlocking parameter, λ'.

The velocity of the new target for the ideal plastic impact is given by

$$\dot{u}_t = [(m_b(t) + m_t)\dot{u}_i^- + m_t\dot{u}_i^-]/[(m_b(t) + 2m_t)] \tag{2.35}$$

Again the superscripts + and – indicate just after and just before impact. Wolf et al. (3.169) tested their work on rigid and deformable targets. Data used in their work are reported below:

Rigid target
Boeing 707-320
$m_a = 127.5$ Mg
$m_w = 38.6$ Mg included in m_a
$\varepsilon_y = 2 \times 10^{-3}$; $\varepsilon r = 5 \times 10^{-2}$
Deformable target
Impact area $= 37.2–45.1\ m^2$

Private Communication Feb 1993.
Also reported in the Author's book "Impact & Explosion 1993" published by Blackwell Science, Oxford 1993.

The tables given for aircrafts and other impacts give data to be used for load–time function or relationship for rigid and deformable targets. The elastic and inelastic systems have been included.

The method of Wolf et al. was idealised into 3D Finite Element method using programs BANG and ISOPAR. Both flexible and rigid targets of 15,000–350,000 isoparametric elements with 750,000 hybrid mixed elements for the aircraft were adopted. The force time–function relationships were combined and they are plotted for various aircrafts. The comparative study is given for these aircraft in (Fig. 2.5). This graph is readily available for the respective impact or crash analysis of any structure. These graphs can be improved by analysing other types of aircrafts.

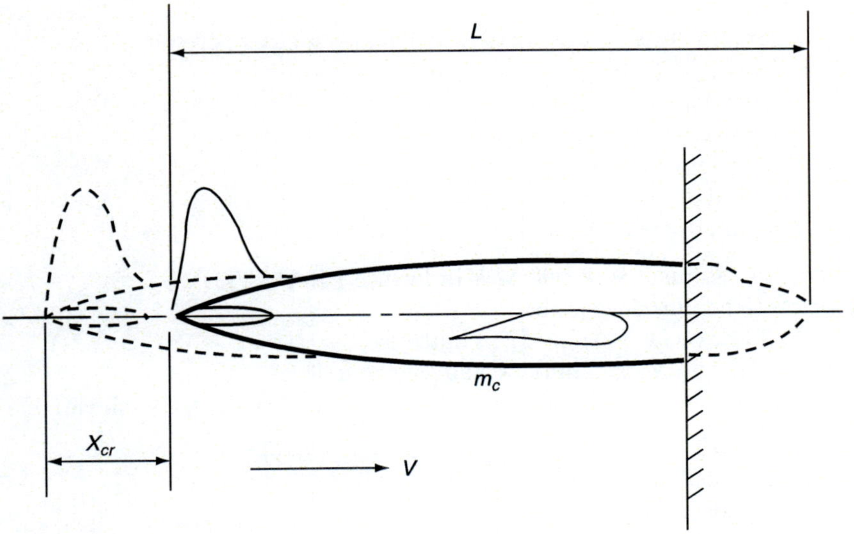

Fig. 2.5 Model aircraft impacting against a rigid surface

2.3.3 Finite Element Applications

This subject has been dealt with in much greater detail in Chapter 3 using dynamic finite element technique. The analysis given under Sections 2.3.1 and 2.3.2 are taken as basis for the finite element approach. Various load–time functions for different aircrafts have been evaluated so as to suite specific aircraft crash analysis, particularly on containment vessels. A list of containment vessels adopted for various nuclear power stations is given in a tabulated form in Tables 2.24 and 2.25. They can be treated as test examples on the lines given in a sample design example on the TVA containment vessel. The design pressures are given in Tables 2.24 and 2.25 with changed dimensions, loads and material properties; new calculations can be made for the existing vessels and for containments of future BWR and PWR nuclear stations. The dynamic analysis given in chapter 3 can also be used when these containments are accurately analysed under environmental and other anticipated loads such as aircraft and missile crashes, fire and explosion, earthquakes and other hazards.

2.3.4 Additional Data on Containment Parameters

2.3.4.1 General Introduction

Tables 2.24 and 2.25 show containments with internal pressures. Some of them are chosen in this section as test examples for readers who wish to test the work in this text.

(a) *Doel 4*: Status P.W.R 1041 MW in Belgium-FRAMATONE
Double walled-double dome resting on piles
$R = 21.90$ m inside cylinder
Space between walls = 3.34 m
Total height = 55 m
Spherical domical space = 3.34 m
Wall and dome thickness = 0.8–1.3 m

(b) *Tricastin*: Status P.W.R 900 MW in France-FRAMATONE
$R = 18.5$ m inside cylinder
Wall thickness = 0.9 m; 12 m buttresses
Total height = 59.5 m thick above ground level
Base slab = 5 m, base slab = 55.2 m with keys, 1.5 m keys depth variable

(c) *Civaux*: Status P.W.R 1400 MW in France-FAMATONE
Double wall type
Double dome type
Total height of walls = 545 m
Spaces = 3.34 m along the walls
Base slab = 50.90 m

(d) *Sizewell B*: Status P.W.R 1258 MW in UK. WESTING HOUSE
Dimensions given as an example in the text

Table 2.24 Nuclear Power Plants: Containment types and design pressures

Country	Name	Type	Containment type	Design pressure
Argentina	Embalse	PHWR	Steel and RC	
Armenia	Medzamor 2	WER	Reinforced concrete (RC)	0.200
Belgium	Doel 3	PWR	Double PC/RC	0.450
Belgium	Tihange 2	PWR	Double PC/RC	0.450
Belgium	Tihange 3	PWR	Double PC/RC	
Brazil	Angra 1	PWR	Steel and RC	
Bulgaria	Kozloduy 5	WER	Prestressed concrete (PC)	
Canada	Bruce A1	PHWR	Reinforced concrete (RC)	
Canada	Bruce B5	PHWR	Reinforced concrete (RC)	0.291
Canada	Darlington	PHWR	Prestressed concrete (PC)	0.197
Canada	Gentilly 2	PHWR	Prestressed concrete (PC)	0.217
Canada	Pickering B 5	PHWR	Reinforced concrete (RC)	0.141
Canada	Point Lepreau 1	PHWR	Prestressed concrete (PC)	0.104
China	Daya Bay	PWR	Prestressed concrete (PC)	0.520
China	Lingao	PWR	Prestressed concrete (PC)	0.520
China	Qinshan 1	PWR	Prestressed concrete (PC)	0.360
China	Qinshan 3	PHWR	Prestressed concrete (PC)	0.224
China	Qinshan II-1	PWR	Prestressed concrete (PC)	0.450
China	Tianwan	WER	Double PC/RC	0.500
Czech. Rep	Dukovany	WER		
Czech. Rep	Temelin	WER	Prestressed concrete (PC)	0.490
Finland	Olkiluoto (TVO) 1	BWR	Prestressed concrete (PC)	0.470
France	Belleville 1	PWR	Double PC/RC	0.520
France	Blayais 1	PWR	Prestressed concrete (PC)	0.500
France	Bugey	PWR	Prestressed concrete (PC)	0.500
France	Cattenom 1	PWR	Double PC/RC	0.520
France	Chinon B1	PWR	Prestressed concrete (PC)	0.500
France	Chooz 1	PWR	Double PC/RC	0.530
France	Civaux 1	PWR	Double PC/RC	0.530

Table 2.24 (continued)

Country	Name	Type	Containment type	Design pressure
France	Civaux 2	PWR	Double PC/RC	0.530
France	Cruas 1	PWR	Prestressed concrete (PC)	0.500
France	Dampierre 1	PWR	Prestressed concrete (PC)	0.500
France	Fessenheim	PWR	Prestressed concrete (PC)	0.473
France	Flamanville 1	PWR	Double PC/RC	0.480
France	Golfech 1	PWR	Double PC/RC	0.520
France	Granvelines 1	PWR	Prestressed concrete (PC)	0.500
France	Nogent 1	PWR	Double PC/RC	0.520
France	Paluel 1	PWR	Double PC/RC	0.480
France	Penly 1	PWR	Double PC/RC	0.520
France	St Alban 1	PWR	Double PC/RC	0.480
France	St Laurent B1	PWR	Prestressed concrete (PC)	0.500
France	Tricastin	PWR	Prestressed concrete (PC)	0.500
Germany	Brokdorf	PWR		0.750
Germany	Emsland	PWR	Steel and RC	0.630
Germany	Grafenrheinfeld	PWR		
Germany	Grohnde	PWR		
Germany	Gundremmingen KRB II	BWR	Reinforced concrete (RC)	0.430
Germany	Isar 1	BWR		
Germany	Isar 2	PWR	Steel and RC	0.630
Germany	Krummel	BWR		
Germany	Mulheim Karlich	PWR		
Germany	Neckar 1	PWR	Steel	0.570
Germany	Neckar 2	PWR	Steel and RC	0.630
Germany	Philppsburg 2	PWR		
Germany	Unterweiser	PWR	Steel	0.580
Great Britian	Heysham	AGR	Prestressed concrete (PC)	
Great Britian	Sizewell B	PWR	Double PC/RC	0.445
Great Britian	Tomess PT 1	AGR	Prestressed concrete (PC)	
Hungary	Paks 1	VVER		
India	FBTR Kalpakkam	FR		
India	Kaiga 1	PHWR	Double PC/RC	0.273
India	Kaiga 2	PHWR	Double PC/RC	0.273
India	Kakrapara 1	PHWR	Double PC/RC	0.225
India	Narora 1	PHWR	Double PC/RC	0.225
India	Rajasthan 3	PHWR		0.273

Table 2.24 (continued)

Country	Name	Type	Containment type	Design pressure
India	Tarapur 3	PHWR	Double PC/RC	0.244
Iran	Busher	PWR	Steel	
Japan	Fukushima 1-4	BWR	Steel	0.490
Japan	Fukushima 1-6	BWR	Steel	0.385
Japan	Fukushima II-1	BWR	Steel	0.385
Japan	Genkai 4	PWR	Prestressed concrete (PC)	0.500
Japan	Hamaoka 2	BWR	Steel	0.492
Japan	Hamaoka 3	BWR	Steel	0.535
Japan	Ikata 1	PWR	Steel	0.345
Japan	Ikata 3	PWR	Steel	0.389
Japan	Kashiwazaki 4	BWR	Steel	0.416
Japan	Kashiwazaki 6 ABWR	BWR	Reinforced concrete (RC)	0.416
Japan	Kashiwazaki 1	BWR	Steel	0.385
Japan	Mihama 3	PWR	Steel	0.340
Japan	Monju	FR		
Japan	Ohi 1	PWR	Steel	0.540

Table 2.25 Nuclear Power Plants–PWR, PHWR, RBMK, WER (Containments and design pressures)

Country	Name	Type	Containment type	Design Pressure
Japan	Ohi 3	PWR	Prestressed concrete (PC)	0.500
Japan	Sendai 1	PWR	Steel	0.325
Japan	Shika 1 (NOTO)	BWR	Steel	0.535
Japan	Shimane 2	BWR	Steel	0.535
Japan	Takahama 3	PWR	Steel	0.360
Japan	Tokai 2	BWR	Steel	0.385
Japan	Tomari 1	PWR	Steel	0.360
Japan	Tsuruga 2	PWR	Prestressed concrete (PC)	0.500
Lituania	Ignalina 1	RBMK		
Mexico	Laguna Negra 1	BWR		
Pakistan	Chasma	PWR	Prestressed concrete (PC)	
Romania	Cernavoda	PHWR		
Russia	Balakovo 1	WER	Prestressed concrete (PC)	0.490
Russia	Balakovo 5	WER	Prestressed concrete (PC)	0.490
Russia	Balakovo 6	WER	Prestressed concrete (PC)	0.490
Russia	Kalinin 1	WER	Prestressed concrete (PC)	0.455
Russia	Kalinin 3	WER	Prestressed concrete (PC)	0.490
Russia	Kursk	RBMK		
Russia	Novovoronej 5	WER	Prestressed concrete (PC)	0.455
Russia	Novovoronej 6	WER	Prestressed concrete (PC)	0.490
Russia	Novovoronej7	WER	Prestressed concrete (PC)	0.490

Table 2.25 (continued)

Country	Name	Type	Containment type	Design Pressure
Russia	Rostov 1	WER	Prestressed concrete (PC)	0.490
Russia	Smolensk 1	RBMK		
Slovakia	Bohunice 1	WER		
Slovakia	Mochovce 1	WER		
Slovenia	Krsko	PWR	Steel and RC	
South Africa	Koeberg	PWR	Prestressed concrete (PC)	0.500
South Korea	Kori 3	PWR	Prestressed concrete (PC)	0.520
South Korea	Uljin 1	PWR	Prestressed concrete (PC)	0.500
South Korea	Wolsong 1	PHWR	Prestressed concrete (PC)	0.156
South Korea	Yonggwang 1	PWR	Prestressed concrete (PC)	0.520
Spain	Almaraz 1	PWR		
Spain	Asco 1	PWR	Prestressed concrete (PC)	0.480
Spain	Asco 2	PWR		
Spain	Cofrentes	BWR		
Spain	Trillo 1	PWR		
Spain	Vandellos 2	PWR		
Sweden	Forsmark 3	BWR	Prestressed concrete (PC)	0.600
Sweden	Oskarshamn 3	BWR	Prestressed concrete (PC)	0.600
Sweden	Ringhals 3	PWR	Prestressed concrete (PC)	0.514
Switzerland	Gösgen	PWR	Steel	0.589
Switzerland	Leibstadt	BWR	Steel and RC	0.203
Taiwan	Kuosheng 1	BWR		
Taiwan	Lugmen	ABWR	Reincorced concrete (RC)	
Taiwan	Maanshan 1	PWR		
Ukraine	Khmelnitsky 1	WER	Prestressed concrete (PC)	0.490
Ukraine	Khmelnitsky 2	WER	Prestressed concrete (PC)	0.490
Ukraine	Rovno 1	WER		
Ukraine	Rovno 3	WER	Prestressed concrete (PC)	0.490
Ukraine	Rovno 4	WER	Prestressed concrete (PC)	0.490
Ukraine	Sud Ukraine 1	WER	Prestressed concrete (PC)	0.490
Ukraine	Tchernobyl 3	RBMK		
Ukraine	Zaporozhe 5	WER	Prestressed concrete (PC)	0.490
USA	Braidwood 1	PWR	Prestressed concrete (PC)	0.445
USA	Byron 1	PWR	Prestressed concrete (PC)	0.445
USA	Callaway -1	PWR	Reinforced concrete (RC)	
USA	Catawba 1	PWR	Steel and RC	0.204
USA	Clinton 1	BWR	Reinforced concrete (RC)	0.204
USA	Comanche 1	PWR	Reinforced concrete (RC)	0.445

Table 2.25 (continued)

Country	Name	Type	Containment type	Design Pressure
USA	Gran Gulf 1	BWR	Reinforced concrete (RC)	0.203
USA	Hatch 2	BWR		
USA	Hope Creek	BWR	Steel	0.528
USA	La Salle 1	BWR	Prestressed concrete (PC)	0.410
USA	Millstone 3	PWR	Reinforced concrete (RC)	
USA	Nine Mile Point 2	BWR	Reinforced concrete (RC)	0.411
USA	Palo Verde 1	PWR	Prestressed Concrete (PC)	0.514
USA	Perry 1	BWR	Steel	0.204
USA	River Bend 1	BWR	Steel	0.204
USA	San Onofre 2	PWR	Prestressed concrete (PC)	0.514
USA	Seabrook 1	PWR	Reinforced concrete (RC)	0.459
USA	Shearon-Harris 1	PWR	Reinforced concrete (RC)	
USA	Shoreham	BWR	Reinforced concrete (RC)	0.389
USA	South Texas	PWR	Prestressed concrete (PC)	
USA	St. Lucie 2	PWR	Steel and RC	0.376
USA	Summer 1	PWR	Prestressed concrete (PC)	
USA	Susquehanna 1	BWR	Reinforced concrete (RC)	
USA	Vogtle 1	PWR	Prestressed concrete (PC)	0.459
USA	Waterford 3	PWR	Steel and RC	
USA	Watts Bar 1	PWR	Steel and RC	0.193
USA	WNP-2 Hanford	BWR	Steel	
USA	Wolf Creek	PWR	Prestressed concrete (PC)	

(e) *GenKai* 4: P.W.R 1180 MW in Japan-MITSUBISHI
$R = 22.150$ m
Total height of walls = single-type 43 m wall thickness = 0.75–1.3 m
Single dome height = 22.6 m
Dome thickness = 1 m
Dome radius = 22.650
Base slab inclusive gallents = 44.30 m
Thickness varies from 10.2 m to 15.8 m

(f) *Kaiga-I*: P.H.W.R 220 MW in India (Double Wall Type)
$R=$ radius inside = 21.28 m
Total height of the wall = 48.235 m
Inner space of walls = 2.0 m

Inner space of dome = 2.0 m
Radius to centroid = 33.57 m inner
Radius of the dome spherical = 39.60 outer
Base slab (without keys) thickness = 3.5 m
Base slab (with keys) thickness = 5.5 m
Base slab total dimension = 49.0 m
Key base = 8.5 m

Note: LOCA ranges in all from 0.3 to 0.35 MPa. Exceptional causes LOCA = 0.47–0.60 MPa related to BWR types.

References

1. ANSI A58.l-1972. Building Code Requirements for Minimum Design Loads in Buildings and Other Structures. American National Standards Institute, 1972.
2. Shah, H. C. et al. A Study of Seiimic Risk for Nicaragua, Part 1, The J. A. Blume Earthquake Engineering Center, Report No. 11, Dept. of Civil Engineering, Stanford University, 1975.
3. Wiggins, J. H. Procedure of Determining Acceptable Risk Ground Motion Design Criteria. Technical. Report No. 75–1229, J. H. Wiggins Company, Redondo Beach, CA, 1975.
4. Der Kureghian, A. and Ang, A. H-S. A Line Source Model for Seismic Risk Analysis, Structural Research Series No. 419, University. of Illinois at Urbana-Champaign, Urbana, II, October 1975.
5. Specification for the Design, Fabrication and Erection of Structural Steel for Buildings. American Institute of Steel Construction, 1969.
6. Uniform Building Code, 1976 edn, Vol. 1, International Conference of Building Officials, 1976.
7. ACI Standard Building Code Requirements for Reinforced Concrete (ACI-3 18-71). American Concrete Institute, 1970.
8. ASME Boiler and Pressure Vessel Code Section III, Division I. Rules for Construction of Nuclear Power Plant Components. American Society of Mechanical Engineers, 1974.
9. ACI-ASME Joint Technical Committee. Code for Concrete Reactor Vessels and Containments. ASME Boiler and Pressure Vessel Code Section III, Division 2 and AC I Standard 359-74, 1975.
10. ACI Standard Code Requirements for Nuclear Safety Related Concrete Structures (ACl-34977). American Concrete Institute, 1975.
11. Specification for Steel Railway Bridges, Cooper E-72 Loading. American Railway Engineering Association, 1972.
12. Standard Specification for Highway Bridges, H-20-S16 Loading. American Association of State Highway Officials.
13. Federal Regulation 10 CFR Part 100, Reactor Site Criteria, Appendix A. Seismic and Geological Siting Criteria for Nuclear Power Plants. November 1973.
14. Tong, W. H. Seismic Risk Analysis for Two-Sites Case. Publication No. R75-22, Department of Civil Engineering, M.I.T., May 1975.
15. ANSI A58.l-1972, Building Code Requirements for Minimum Design Loads in Buildings and Other Structures, American National Standards Institute.
16. Newmark, N. M. Overview of Seismic Design Margins. Paper presented at A.I.F. Workshop on Reactor Licensing and Safety, December. 1974.

17. Hsieh, T., Okrent, D., and Apostolakis, O. E. On the Average Probability Distribution of Peak Ground Acceleration in the U.S. Continent Due to Strong Earthquakes. UCLA-ENG-75 16, University of California at Los Angeles, March 1976.
18. Anderson, D. L., Chariwood, R. G., and Chapman, C. B. On seismic risk analysis of nuclear plants safety systems, Canadian Journal of Civil Engineering, Vol. 2, No. 4, 1975, pp. 558–571.
19. St. Amand, P. Two proposed measures of seismicity, Bulletin of the Seismological Society of America., Vol. 46, 1961.
20. Howell, B. F., Jr., et al. Integrated and frequency band magnitude, two alternative measures of the size of an earthquake, Bulletin of the Seismological Society of America. , Vol. 60, No. 3, 1970.
21. Blume, J. A. An engineering intensity scale, Bulletin of the Seismological Society of America., Vol. 60, No. 1, 1970.
22. Blume, J. A. and Monroe, R. E. The Special Matrix Method of Predicting Damage from Ground Motion, John A. Blume & Associates, Research Division, San Francisco, CA, for Nevada Operations Office, U.S. Atomic Energy Commission, 1971.
23. Trifunac, M. D. and Brady, A. G. On the correlation of seismic intensity scales with the peaks of recorded strong ground motion, Bulletin of the Seismological Society of America., Vol. 65, No. 1, 1975, pp. 139–162.
24. Richter, C. F. An instrumental earthquake scale, Bulletin of the Seismological Society of America., Vol. 25, 1935.
25. Aki, K. Generation and Propagation of G Waves from Niigata Earthquake of June 16, 1964, part 2, Estimation of Earthquake Moment, Released Energy and Stress-Strain Drop from G-Wave Spectrum, Bulletin of the Earthquake Research Institute, Tokyo University, Vol. 44, 1966 pp. 73–88.
26. Cornell, C. A. and Vanmarcke, E. H. The major influences on seismic risk. Proceedings of the 4th World Conference on Earthquake Engineering, Santiago, Chile, 1969.
27. Algermissen, S. 1. Seismic risk studies in the United States. Proceedings of the 4th World Conference on Earthquake Engineering, Vol. 1, Santiago, Chile, 1969.
28. Algermissen, S. T. and Perkins, D. M. Earthquake risk studies in the branch of seismicity and risk analysis. Presented at USGS Earthquake Studies Advisory Panel, Colorado School of Mines, Golden, CO, June 4–5, 1975.
29. Liu, S. C. and Fagel, L. W. Seismic risk analysis– comparison of three different methods. for seismic regionalization, Bulletin of the Seismological Society of America., Vol. 65, No. 4, 1975, pp. 1023–1027.
30. Esteva, L. Seismic prediction, a bayesian approach. Proceedings of the 4th World Conference on Earthquake Engineering, Vol. 1, Santiago, Chile, 1969.
31. Mortgat, C. P. Bayesian Approach to Seismic Hazard Mapping. Ph.D. Thesis, Stanford University, 1976.
32. Esteva, L. Regionalization Seismica de Mexico para fines de Ingenieria. Report 246, Institute of Engineering, National University of Mexico, Mexico City, 1970.
33. Esteva, L. Seismic risk and seismic design input for nuclear power plants. Seismic Design for Nuclear Power Plants, ed. R.J. Hansen, MIT Press, Cambridge, MA, 1970, pp. 438–483.
34. Cloud, W. K. and Perez, V. Unusual accelerograms recorded at Lima, Peru, Bulletin of the Seismological Society of America., Vol. 61, 1970, pp. 663–640.
35. Schnabel, P. and Seed, H. B. Accelerations in rock for earthquakes in the western United States, Bulletin Seismological Society America, Vol. 63, 1973, pp. 501–516.
36. Gupta, I. N. and Nuttli, O. W. Spatial attenuation of intensities for central U.S. earthquakes, Bulletin Seismological Society America, Vol. 66, No. 3, 1976, pp. 743–751.
37. Seed, H. B., Murarka, R., Lysmer, J., and Idriss, I. M. Relationships of maximum acceleration, maximum velocity, distance from source and local site conditions for moderately strong earthquakes, Bulletin Seismological Society America, Vol. 66, No. 4, 1976, pp. 1323–1342.

38. Cough, R. W. Earthquake analysis by response spectrum superposition, Bulletin Seismological Society America, Vol. 52, No. 3, 1962.
39. Blume, J. A., Sharpe, R. L., and Dalal, J. S. Recommendations for Shape of Earthquake Response Spectra. John A. Blume and Associates, Engineers, San Francisco, CA, February 1973 (USAEC Report No. WASH-1254).
40. Mohraz, B., Hall, W. J., and Newmark, N. M. A Study of Vertical and Horizontal Earthquake Spectra. Nathan N. Newmark Consulting Engineering Services, Urbana, IL, 1972 (USAEC Report No. WASH-1255).
41. Seed, H. B., Ugas, C., and Lysmer, J. Site Dependent Spectra for Earthquake Resistant Design, Report No. EERC-74-12, U.C. Berkeley, 1974.
42. McGuire, R. K. Seismic Design Spectra and Mapping Procedures Using Risk Analysis Based Directly on Oscillator Response. Branch of Seismicity and Risk Analysis, USGS, Denver, CO, 1975.
43. Trifunac, M. D. Preliminary empirical model for scaling Fourier amplitude spectra of strong ground acceleration in terms of earthquake magnitude, source-to-station distance, and recording site conditions, Bulletin Seismological Society America, Vol. 66, No. 4, 1976, pp. 1343–1373.
44. Dobry, R. et al. Influence of magnitude, site conditions and distance on significant duration of earthquakes. The 6th World Conference on Earthquake Engineering, New Delhi, India, January 1977.
45. Bolt, B. A. Duration of strong motion. Proceedings of the 5th World Conference on Earthquake Engineering, Paper 292, Rome, 1973.
46. NRC Standard Review Plan Section 2.5.2.11.3, Vibratory Ground Motion Acceptance Criteria, U.S. Nuclear Regulatory Commission, Office of Nuclear Reactor Regulation, June 1975.
47. Shakal, A. F. and Toksoz, M. N. Earthquake hazard in New England, Science, Vol. 195, 1977, pp. 171–173.
48. Thom, H. C. S. New distributions of extreme winds in the United States, Journal of the Structural Division, ASCE, Vol. 94, No. ST7, Proc. Paper 6038, 1968, pp. 1787–1801.
49. Simiu, E. Probabilistic models of extreme wind speeds: uncertainties and limitation. Proceedings of the 4th International Conference on Wind Effect on Structures and Buildings, London, England, September 1975, pp. 53–62.
50. Russell, L. R. Probability distributions for hurricane effects, Journal of Waterways, Harbors and Coastal Engineering Division, ASCE, Vol. 97, No. WW 1, 1971, pp. 139–154.
51. Tryggvason, B. V., Surry, D., and Davenport, A. G. Predicting wind-induced response in hurricane zones, Journal of the Structural Division, ASCE, Vol. 102, No. STI2, 1976, pp. 2333–2351.
52. Sklarin, J. Probabilities of Hurricanes. Dept. of Civil Engineering, Research Report, MIT, Cambridge, MA, February 1977.
53. Fujita, T. T. Estimates of Areal Probability of Tornadoes from Inflationary Reporting of their Frequencies. Satellite and Mesometeorology Research Project (University of Chicago), SMRP Research Paper No. 89, October 1970.
54. Thom, H. C. S. Tornado probabilities, Monthly Weather Review, Oct–Dec 1963, pp. 730–736.
55. Pautz, M. E. (ed.). Severe Local Storm Occurrences, 1955–1967, U.S. Department of Commerce, Environmental Science Services Administration, ESSA Tech. Memo. VBTM FCST 12, September 1969.
56. USNRC, Regulatory Guide 1.76, Design Basis Tornado for Nuclear Power Plants, Directorate of Regulatory Standards, April 1974.
57. Garson, R. C., Catalan, J. M., and Cornell, C. A. Tornado design winds based on risk, Journal of the Structural Division, ASCE, Vol. 101, No. ST9, 1975, pp. 1883–1897.
58. Hoecker, W. H., Jr. Wind speed and air flow patterns in the Dallas tornado of April 2, 1957, Monthly Weather Review, Vol. 88, No. 5, 1960, pp. 167–180.

59. Hoecker, W. H., Jr. Three dimensional pressure patterns of the Dallas tornado and some resultant implications, Monthly Weather Review, Vol. 89, No. 12, 1961, pp. 533–542.
60. Wen. Y. K. and Ang, A. H-S. Tornado risk and wind effect on structures. Proceedings of the 4th International Conference on Wind Effect on Buildings and Structures, London, England, September 1975, pp. 63–73.
61. Wen, Y. K. Note on analytical modeling in assessment of tornado risks. Proceedings of Symposium on Tornadoes, Assessment of Knowledge and Implications for Man, June 1976, Texas Tech University, Lubbock, TX.
62. Wen, Y. K. Dynamic wind loads on tall buildings, Journal of the Structural Division, ASCE, Vol. 101, No. ST1, 1975, pp. 169–185.
63. Kuo, H. L. Axisymmetric flow in the boundary layer of a maintained vortex, Journal of Atmospheric Sciences, Vol. 28, No. 1. 1971, pp. 20–41.
64. McDonald, J. R., Minor, J. E., and Mehta, K. C. Tornado generated missiles. Specialty Conference on Structural Design of Nuclear Power Plant Facilities, Chicago, IL, December 17–18, 1973, Proceedings (Publ.: ASCE), Vol. II, pp. 543–556.
65. McDonald et al. Tornado generated missiles. Int. Conf. Nuclear Plant Facilities, Dec. 1973 ASCE, Vol. 11, pp. 543–556.
66. Abbey, R. F. and Fujita, T. T. Use of tornado path length and gradations of damage to assess tornado intensity probability. proceedings of the 9th Conference on Severe Local Storms, Norman, Oklahoma, 1975.
67. Abbey, R. F. Risk probabilities associated with tornado windspeeds. Proceedings of the Symposium on Tornadoes, Assessment of Knowledge and Implication for Man, June 1976. Texas Tech University, Lubbock, TX.
68. American Nuclear Society, ANS 24, Draft American National Standard, Tsunami Guidelines at Power Reactor Sites, ANSI N515, December 1975.
69. Wiegel, R. L. Earthquake Engineering. Prentice-Hall, Englewood Cliffs, NJ, 1970.
70. Gumbel, E. J. Statistical Theory of Extreme Values and Some Practical Applications, National Bureau of Standards, Applied Mathematics Series, No. 33, Superintendent of Documents, U.S. Government Printing Office, Washington, 1954.
71. Gumbel, E. J. Statistics of Extremes. Columbia University Press, New York, NY, 1958.
72. Wall, I. B. Probabilistic assessment of flooding hazard for nuclear power plants, Transactions of the American Nuclear Society, Vol.16, No.211, 1973; Nuclear Safety, Vol. 15, No. 4, 1974, pp. 399–408.
73. Linsley, R. K., Kohler, M. A., and Pauthus, J. L. H. Hydrology for Engineers, 2nd edn. McGraw-Hill, New York, NY, 1975.
74. Shane, R. M. and Lynn, W. R. Mathematical model for flood risk evaluation, Journal of the Hydraulics Division, ASCE, Vol. 90, No. HY6, 1964, pp. 1–20.
75. Linsley, R. K. and Franzini, J. B. Water Resources Engineering, 2nd edn. McGraw-Hill, New York, NY, 1972.
76. Linsley, R. K., Kohler, M., and Paulhus, J. L. H. Applied Hydrology. McGraw-Hill, New York, NY, 1949.
77. U.S. Weather Bureau. Rainfall Frequency Atlas of the United States for Durations from 30 Minutes to 24 Hours and Return Period from 1 to 100 Years, Weather Bureau Technical Paper No. 40, Washington, D.C., 1964.
78. U.S. Weather Bureau. Seasonal Variation of the Probable Maximum Precipitation East of the 105th Meridian for Areas from 10 to 1000 Square Miles and Durations of 6, 12, 24 and 48 Hours, Hydrometeorological Report No. 33, U.S. Weather Bureau, 1956.
79. Browzin, B. S. Statistical safety factor concept for determining the probable maximum flood. Specialty Conference on Structural Design of Nuclear Power Plant Facilities, Chicago, IL, December 1973; Proceedings (Publ. ASCE), Vol. II, pp. 557–570.
80. USNRC Standard Review Plan, Section 2.4.3, Probable Maximum Flood (PMF) on Streams and Rivers, U.S. Nuclear Regulatory Commission, Office of Nuclear Reactor Regulation, September 1975.

81. American Nuclear Society. Draft American National Standard, Guidelines for Determining Design Basis Flooding at Power Reactor Sites, ANSI N 170.
82. USNRC Standard Review Plan, Section 3.5.1, U.S. Nuclear Regulatory Commission, Office of Nuclear Reactor Regulation, September 1975.
83. Electric Power Research Institute. Tornado Missile Risk Analysis, EPRI NP- 154, Project 616-I, Topical Report 1, January 1975.
84. Johnson, B. Tornado Missile Risk Analysis, Study Being Conducted for Boston Edison Power Company by Science Applications, Inc., Private Communication, 1976.
85. Bush, S. H. Probability of damage to nuclear components due to Turbine Failure, U.S.A.E. C., ACRS, and Pacific Northwest Laboratories, Battelle Memorial Institute, November 6, 1972.
86. Bush, S. H. Probability of damage to nuclear components due to turbine failure, Nuclear Safety, Vol. 14, No. 3, 1973, pp. 187–201.
87. Emmert, H. D. Investigation of large steam-turbine spindle failure, Mechanical Engineering, Mechanical Engineering, 1956, p. 46.
88. Rankin, A. W. and Seguin, B. R. Report of the investigation of the turbine-wheel fracture at Tanners Creek, Mechanical Engineering, 1956, p. 47.
89. Schabtach, C., Fogleman, E. L., Rankin, A. W., and Winne, D. H. Report of the investigation of two generator-rotor fractures, Mechanical Engineering, p. 48, 1956.
90. Peterson, R. E., Mochel, N. L., Conrad, J. D., and Gunther, D. W. Large rotor forgings for turbines and generators, Mechanical Engineering, p. 49, 1956.
91. Lindley, A. L. G. and Brown, F. H. S. Failure of a 60-MW steam turbo-generator at Uskmouth Power Station, Proceedings of the Institution of Mechanical Engineers, Vol. 172, 1958, p. 627.
92. Kalderon, D. Steam turbine failure at Hinkley Point 'A', Proceedings of the Institution of Mechanical Engineers, Vol. 186, 1972, p. 341. Hinkley Point 'A' Power Station, Proceedings of the Institution of Mechanical Engineers, Vol. 186, 1958, p. 379.
93. Codier, E. O. Reliability growth in real life. Proceedings of the IEEE Annual Symposium on Reliability, Boston, January 1968;– Duane, J. T. General Electric Company Report DF62 MD300.
94. Downs, J. E. Hypothetical Turbine Missiles—Probability of Occurrence, General Electric Co., Turbine Dept., Memo Report, March 1973.
95. Analysis of the Probability of the Generation and Strike of Missiles from a Nuclear Turbine, Westinghouse Electric Corp., Steam Turbine Division Engineering, March 1974.
96. Hagg, A. C. and Sankey, G. O. The containment of disk burst fragments by cylindrical shells, Transactions of the ASME, Journal of Engineering for Power, Vol. 96, 1974, pp. 114–123.
97. Report Covering the Effects of a Turbine Accelerating to Destructive Overspeed, Westinghouse Electric Corp., Report 196/381 A, 1974.
98. Report Covering the Effects of a High Pressure Turbine Rotor Fracture and Low Pressure Turbine Disc Fractures at Design Overspeed, Westinghouse Electric Corp., Report 196/381 B, 1974.
99. Turbine Missile Analysis, Allis-Chalmers Power Systems Inc., Report Contained in Grand Gulf PSAR (Docket No. 50-416), Amendment 6, April 1973.
100. Downs, J. E. Hypothetical Turbine Missile Data, 43-Inch Last Stage Bucket Units, General Electric Co., Turbine Dept., Memo Report, March 1973.
101. Semanderes, S. N. Methods of Determining the Probability of a Turbine Missile Hitting a Particular Plant Region, WCAP-786I, Westinghouse Electric Corp., 1972.
102. Semanderes, S. N. Method of determining missile impact probability, Transactions of the American Nuclear Society, Vol. 15, No. 1, 1972, p. 401.
103. O'Connell, W. J., Bascheire, R. J. Design applications of turbine missile impact Analysis. Proceedings of the 2nd ASCE Specialty Conference on Structural Design of Nuclear Plant Facilities, New Orleans, LA, December 1975 (publ. ASCE), pp. 541–561.

104. O'Connell, W. J. and Fortney, R. A. Turbine missile impact analysis: a detailed treatment, Transactions of the American Nuclear Society, Vol. 19, 1974, p. 231; EDS Nuclear Inc., Report, October 27, 1974.
105. Swan, S. W. and Meleis, M. A method of calculating turbine missile strike and damage probabilities, Nuclear Safety, Vol. 16, 1975, pp. 443–451.
106. Bhattacharya, A. K. and Chaudhuri, S. K. The Probability of a turbine missile hitting a particular region of a nuclear power plant, Nuclear Technology, Vol. 28, 1976, pp. 194–198.
107. Johnson, B. et al. Analysis of the Turbine Missile Hazard to the Nuclear Thermal Power Plant at Pebble Springs, Oregon, Science Applications Inc. Report to the Portland General Electric Company, PGE-2012, January 1976.
108. Hornyik, K. Hazards to nuclear plants from surface traffic accidents, Nuclear Technology, Vol. 25, 1975, pp. 651–657.
109. USNRC Standard Review Plan, Section 3.5.1.6, Aircraft Hazards, U.S. Nuclear Regulatory Commission, Office of Nuclear Reactor Regulation, September 1975.
110. Census of U.S. Civil Aircraft, Office of Management Systems, Federal Aviation Administration, Department of Transportation.
111. Analysis of Aircraft Accident Data, U.S. Civil Aviation, National Transportation Safety Board, Department of Transportation.
112. Chelapati, C. V., Kennedy, R. P., and Wall, I. B. Probabilistic assessment of aircraft hazard for nuclear power plants, Nuclear Engineering and Design, Vol. 19, 1972, pp. 333–364.
113. Wall, I. B. and Augensteifl, R. C. Probabilistic assessment of aircraft hazard to a nuclear power plant – I, Transactions of the American Nuclear Society, Vol. 13, 1970, p. 217.
114. Chelapati, C. V. and Wall, 1. B. Probabilistic assessment of aircraft hazard for nuclear power plants – II," Transactions of the American Nuclear Society, Vol. 13, 1970, p. 218.
115. Wall, I. B. Probabilistic assessment of aircraft risk for nuclear power plants, Nuclear Safety, Vol. 15, No. 3, 1974.
116. Long Island Lighting Co., Shoreham Station (Docket 50-322), Amendment No. 3, February 1969.
117. Solomon, K. A. Hazards associated with aircrafts and missiles, Transactions of the American Nuclear Society, Vol. 23, 1976, p. 312.
118. Hornyik, K. Airplane crash probability near a flight target, Transactions of the American Nuclear Society, Vol. 16, 1973, p. 209.
119. Hornyik, K. and Grund, J. E. Evaluation of air traffic hazards at nuclear power plants, Nuclear Technology, Vol. 23, 1974, pp. 28–37.
120. NRC Regulatory Guide 1.70, Standard Format and Content of Safety Analysis Reports for Nuclear Power Plailts, LWR Edition, Revision 3, NUREG-75/094, U.S. Nuclear Regulatory Commission, Office of Standards Development, September 1975.
121. Aircraft crash probabilities, Nuclear Safely, Vol. 17, 1976, pp. 312–314.
122. Hahn, G. J. and Shapiro, S. S. Statistical Models in Engineering. Wiley, New York, NY, 1967.
123. Julian, O. G. Synopsis of first progress report of committee on safety factors, Journal of the Structural Division, Proceedings of the ASCE, Vol. 83, No. ST4, 1957.
124. Kececiouglu, D. Reliability analysis of mechanical components and systems, Nuclear Engineering and Design, Vol. 19, 1972, pp. 259–290.
125. Report of ASCE-ACI Joint Committee on Ultimate Strength Design, Proc. Sep. No. 809, ASCE, Vol. 81, October 1955.
126. Sexsmith, R. O. Reliability Analysis of Concrete Structures, Technical Report No. 83, Dept. of Civil Engineering, Stanford University, Stanford, CA, August 1967.
127. Andersen, P. The Resistance to Combined Flexure and Compression of Square Concrete Members, University of Minnesota Engineering Experiment Station Technical Paper No. 29, University of Minnesota, 1941.

128. Veist, E. and Hognestad, E. Sustained load strength of eccentrically loaded short reinforced concrete columns, ACI Proceedings, Vol. 52, 1956, pp. 727–755.
129. Chelapati, C. V. and Wall, I. B. Probabilistic assessment of seismic risk for nuclear power plants, Transactions of the American Nuclear Society, Vol. 12, 1969, p. 684; Holmes and Narver, Inc. Report, December 1969.
130. Hadjian, A. H. and Hamilton, C. W. Probabilistic frequency variations of concrete structures. 2nd International Conference on Structural Mechanics in Reactor Technology, Berlin, 1973; Paper K-3/5.
131. Stevenson, J. D. and Moses, F. Reliability analysis of framed structures, Journal of the Structural Division, ASCE, Vol. 96, No. STI 1, Proc. Paper 7692, 1970.
132. Riera, J. O. On the stress analysis of structures subjected to aircraft forces. Nuclear Engineering and Design. North Holland Publishing, Amsterdam, Holland, 1968, p. 8.
133. Long Island Lighting Co., Shoreham Station (Docket 50-322), Amendment No. 3, February 1969.
134. United Engineers and Constructors, Seabrook Station Containment Aircraft Impact Analysis, Prepared for Public Service Co. of New Hampshire, Seabrook, New Hampshire, Revision 1, January 1975.
135. NRC Standard Review Plan Section 3.6.1, Plant Design for Protecting Against Postulated Piping Failures in Fluid Systems Outside Containment, U.S. Nuclear Regulatory Commission, Office of Nuclear Reactor Regulation, March 1975.
136. NRC Standard Review Plan Section 3.6.2, Determination of Break Locations and Dynamic Effects Associated with Postulated Rupture of Piping, U.S. Nuclear Regulatory Commission, Office of Nuclear Reactor Regulations, March 1975.
137. ANS 58.2 Committee, Design Basis for Protection of Nuclear Power Plants Against Effects of Postulated Rupture," ANS-N 176, January 1977 (Draft).
138. Marcal, P. V. Finite element analysis with material Nonlinearities theory and practice. Pressure Vessel and Piping Design and Analysis, Vol. 1. American Society of Mechanical Engineers, 1972, p. 486.
139. AEC Standard Review Plan Section 3.8.1, Concrete Containment, U.S. Atomic Energy Commission, Directorate of Licensing, 1974.
140. AEC Standard Review Plan Section 3.8.3, Concrete and Steel Internal Structure of Steel and Concrete Containments, U.S. Atomic Energy Commission, Directorate of Licensing, 1974.
141. ASME code Section 3 Division 2. ACI code for PCCV and PCRV, Article CC 6000, 1989.
142. Axcell, Containment cooling by water film, The Nuclear Engineer, Vol. 42, Number 1, 2001.
143. Basu, K. Containment construction using HPC (i\EB) India. Proceedings of seminar on containments of nuclear reactors in conjunction with 15tn SMIRT, Seoul, 1999.
144. Chataigner, Granger, Touret et al. Containment leakage under accident conditions. Proceedings of Seminar on Containment of Nuclear Reactors in Conjunction with 15th SMIRT, Seoul, 1999.
145. Costaz. Le béton hautes performances (HPC), RGN No 4, 1992.
146. Costaz (EDF Septen), 15–22 November, Infrastructures for energy, Keynote, IFTA Seminar on Civil Engineering and Architecture, 1992.
147. Costaz, Picaut, Chataigner. Delayed Phenomena Analysis from French PWR 900 MW containment. Monitoring comparison with foreseen design values. SMIRT 10, 1989.
148. Danish, Hansen, Liersch, Peter. Testing of a concrete containment model coated with a composite liner. SMIRT 14, Vol. 5 Div. H concrete structures. HW/2, 1997.
149. Danisch, Hansen. Status report on German containment composite liner research. Proceedings of Seminar on Containment of Nuclear Reactor in Conjunction with 15th SMIRT, Seoul, 1999.
150. De Larrare, Ithuralde, Acker, Chauvel. HPC for NPP, 2nd International Conference on Utilisation of HSC, Berkeley, ACISP, 1990.

151. FIP – Fédération Internationale de la Précontrainte, state-of-art report: Design and construction of prestressed concrete reactor vessels, June 1978.
152. Granger, Labbé. Mechanical and leak tightness behaviour of a containment mock-up under severe accident – SMIRT 14, Vol. 5 Div. H concrete structures. HW/l, 1997.
153. Guinet, Decelle, Lancia, Barré. Design and erection of a large mock-up of containment under severe accident conditions – SMIRT 14, Vol. 5 Div. H concrete structures. HW/3, 1997.
154. Hessheimer, Pace, Klamerus, Matsumoto, Costello Instrumentation and testing of a PCCV model –. SMIRT 14, Vol. 5 Div. H concrete structures. H03/4, 1997.
155. IAEA – TECDOC-1025. Assessment and management of ageing of major nuclear power plant components important to safety: concrete containment buildings, June 1998.
156. Irving, Hinley, and Palfrey. The design, analysis and testing of the 1/10 scale model of the sizewell B containment building. International Conference on Structural Design For Hazardous Loads, Institution of Structural Engineers, Brighton, UK, 17–19 April 1991.
157. Jolivet, Richli, A seismic foundation system for nuclear power stations, SMIRT 4, Vol. K (b), 1977.
158. Kikuchi, Kobanashi, Ichizono. SIT, Kashiwazahi Kariwa, RCCV. Structural Integrity test of reinforced concrete containment at Kashiwazaki-Kariwa. SMIRT 14. Vol. 5 Div. H concrete structures. H03/5 - H03/6, 1997.
159. Klymov (State Research Institute of Building Construction, Kiev). Monitoring of stressed-strained state and forces in cables of prestressed containment shells of NPPs. Joint Wano/OECD-NEA. Workshop, Civaux, Poitiers, France, 1997.
160. Klymov (State Research Institute of Building Construction). Results and Problems of Monitoring of Prestressed Containment of Nuclear Power Plants. OECD-NEA. Workshop, Tractebel-Brussels, Belgium, 2000.
161. Kuroda et al. Recent advances in concrete containment vessels in Japan, Nuclear Engineering and Design, Vol. 140, 1993.
162. Kuroda, Hasegawa. Recent topics on concrete containment vessels in Japan: verification of SIT and assurance of safety margins, Japan. Proceedings of Seminar on Containment of Nuclear Reactors in conjunction with 5tii SMIRT, Seoul, 1999.
163. Libmann. Elements de Süreté Nucléaire – IPSN. Published by Lavoisier, France, 1997.
164. Maliavine Atom Energoproject Russian Federation. Prestress losses in containments of VVER 1000 Units. Joint Wano/OECD-NEA. Workshop, Civaux, Poitiers, France, 1997.
165. McFarlane, Smith, Davies, Millustry. In-service monitoring of AGR and PWR Nuclear Safety related Structures in the UK, Institution of Nuclear Engineers, September 1996, Cambridge, UK.
166. McFarlane, Smith. Plant Life Management. European Commission, Directorate General, FISA 99 Symposium, Luxembourg, 1999.
167. Martinet, Guinet, Roussel, Granger. Prestress losses in NPP containments: the EDF experience. Joint Wano/OECD-NEA. Workshop, Civaux, Poitiers, France, 1997.
168. OECD-NEA Nuclear Regulation. Regulatory aspects of life extension and upgrading of MPPs. CNRA Special Issues, Meeting 2000 Report, January 2001.
169. Picaut, Sidaner, Pastor, Lorteau. The dome of Daya Bay containment, design and construction, SMIRT 11, 1991.
170. Proceedings of seminar on containment of Nuclear Reactor in Conjunction with 15th SMIRT, Seoul. Thomas Jueger Seminar, 1999.
171. RCCG, Design and construction rules for civil works of PWR Nuclear Islands.
172. Roy, Guinet, Barré. Concrete monitoring analysis during the prestressing of a containment and comparison with foreseen design values – SMIRT 14. Vol. 5 Div. H concrete structures, HO 1/2, 1997.
173. Saito, H. et al. Experimental study on RCCV of ABWR plant. SMIRT 10, 1989.

174. Setogawa, Sakarai, Funokoshi, Yemaguchi, Koike, Kamei, and Hagro. Structural Integrity tests of prestressed concrete containment vessels at Ohi NPP, SMIRT 12, Vol H, 1993.
175. SFEN, Proceedings of International Conference on the EPR Project SMIRT 14, August 97, Vol. 5, Div. H concrete structures. H 04/3 PSA of typical 900 MW French nuclear containment. Grangier, Chataigner, November 1995.
176. Smith, L.M. and Taylor, M.F. The Long term In-Service Performance of Corrosion Protection to Prestressing Tendons in AGR PCPVS, Joint Wano/OECD-NEA Workshop, Civaux, Poitiers, France, 1997.
177. Smith, De Marneffe, Mathet, Contri. Civil Engineering Instrumentation on Nuclear Power Plants – An International Perspective, Institution of Nuclear Engineers, November 2000.
178. Tamura, Watanabe, Kato, Yamaguchi, Koike, Komatsu. A delayed phenomena evaluation of prestressed concrete containment vessel at Tsuruga Unit 2 power station. SMIRT 11, 1991.
179. UNSRC Reg. guide 1.35 ISI ungrouted tendons January 1976
180. USNRC Reg. guide 1.90 ISI grouted tendons August 1977
181. Xu Yao Zhang. Containment structure monitoring and prestress losses. Experience from Daya Bay NPP. Joint Wano/OECD-NEA. Workshop, Civaux, Poitiers, France, 1997.
182. Eurocode-3, BSI, 1999, Revision 2002, 2009.
183. Johnson, R.P. and Anderson, D. Designers' Guide to EN 1994-1-1 Eurocode-4. Design of composite steel and concrete structures for buildings. Thomas Telford, 2004.
184. Eurocode-3, Design of Steel Structures. BSI 2004.
185. Shahrooz, B.M. (ed.). Composite Construction in Steel and Concrete. ASCE, New York, 2002, No IV, pp. 584–595.
186. BSI. Design of Concrete Structures.
187. BSI. Design of Structures for Earthquake Resistance, BSI, London, BSEN, 1994.

Chapter 3
Dynamic Finite Element Analysis

3.1 Introduction

A great deal of work has been published on finite element techniques.

This chapter presents the linear and non-linear dynamic finite element analysis intended to be used for nuclear facilities. Plasticity and cracking models are included. Solid isoparametric elements, panel and line elements are included which represent various materials. Solution procedures are recommended. Programs ISOPAR, F-BANG and other computer packages are recommend for the dynamic non-linear analysis of structures for nuclear facilities with and without cracking.

3.1.1 Finite Element Equations

A 3D finite element analysis is developed in which a provision has been made for time-dependent plasticity and rupturing in steel and cracking in materials such as concrete. The influence of studs, tugs and connectors is included. Concrete steel liners and studs are represented by solid isoparametric elements, shell elements and line elements with or without bond linkages. To begin with, a displacement finite element is adopted.

The displacement field within each element is defined in Fig. 3.1 as

$$\{x\} = [N]\{x\}^{\mathrm{e}} = \sum_{i=1}^{n}(N_i[\boldsymbol{I}]\{x\}_i) \tag{3.1}$$

The strains and stresses can then be expressed as

$$\epsilon\{x\}^{\mathrm{e}} = \sum_{i=1}^{n}([\boldsymbol{B}_i]\{x\}_i) = [\boldsymbol{D}]\{\sigma\} \tag{3.2}$$

In order to maintain equilibrium with the element, a system of external nodal forces $\{\mathrm{F}\}^{\mathrm{C}}$ is applied which will reduce the virtual work (dW) to zero. In the

M.Y.H. Bangash, *Structures for Nuclear Facilities*,
DOI 10.1007/978-3-642-12560-7_3, © M.Y.H. Bangash 2011

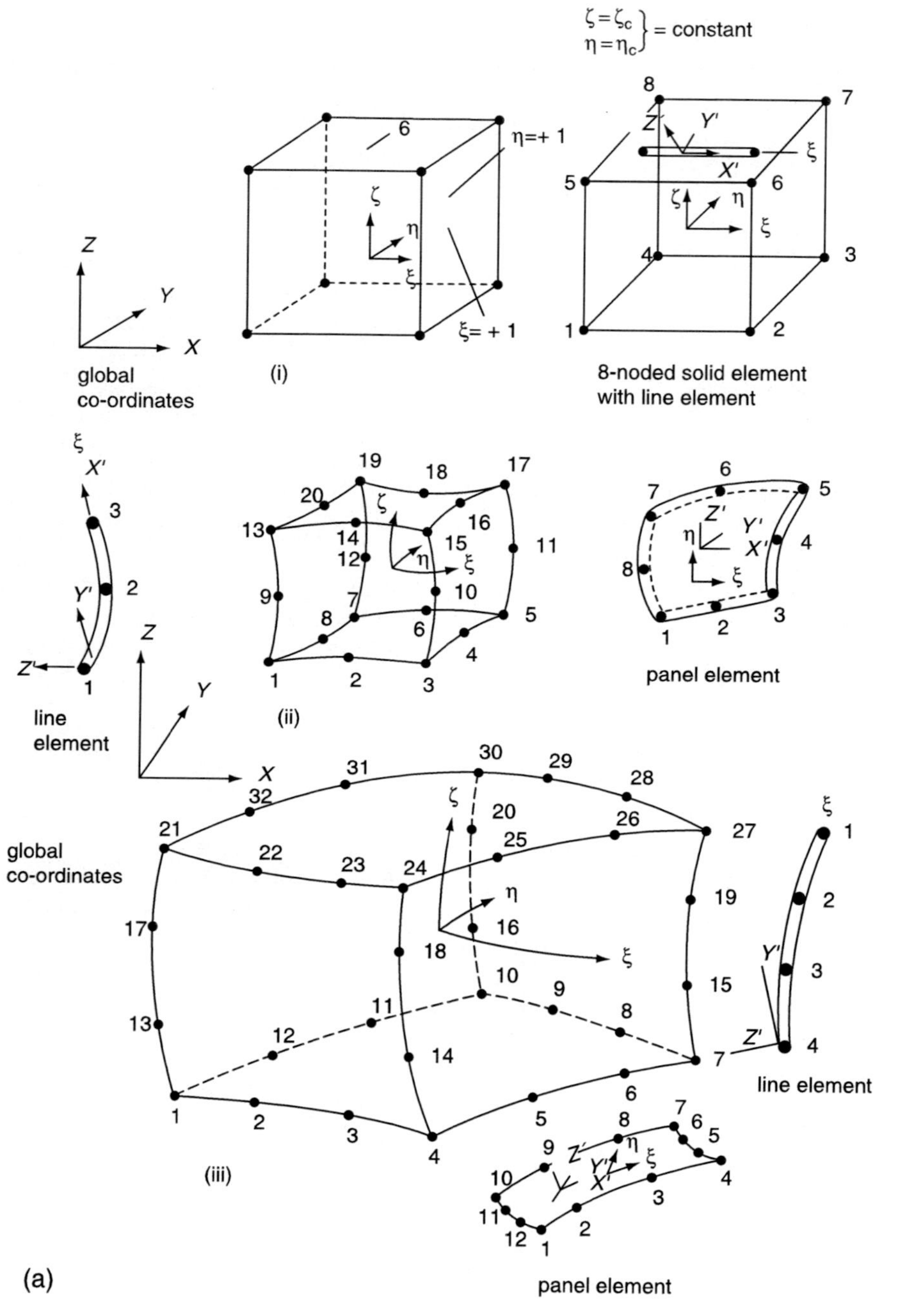

Fig. 3.1 (a) Isoparametric elements: (i) parent element, 3D isoparametric derived element; (ii) solid element (20-noded); (iii) 32-noded solid element. **(b)** Line elements within the body of the solid isoparametric elements (ISS – isoperametric solid element)

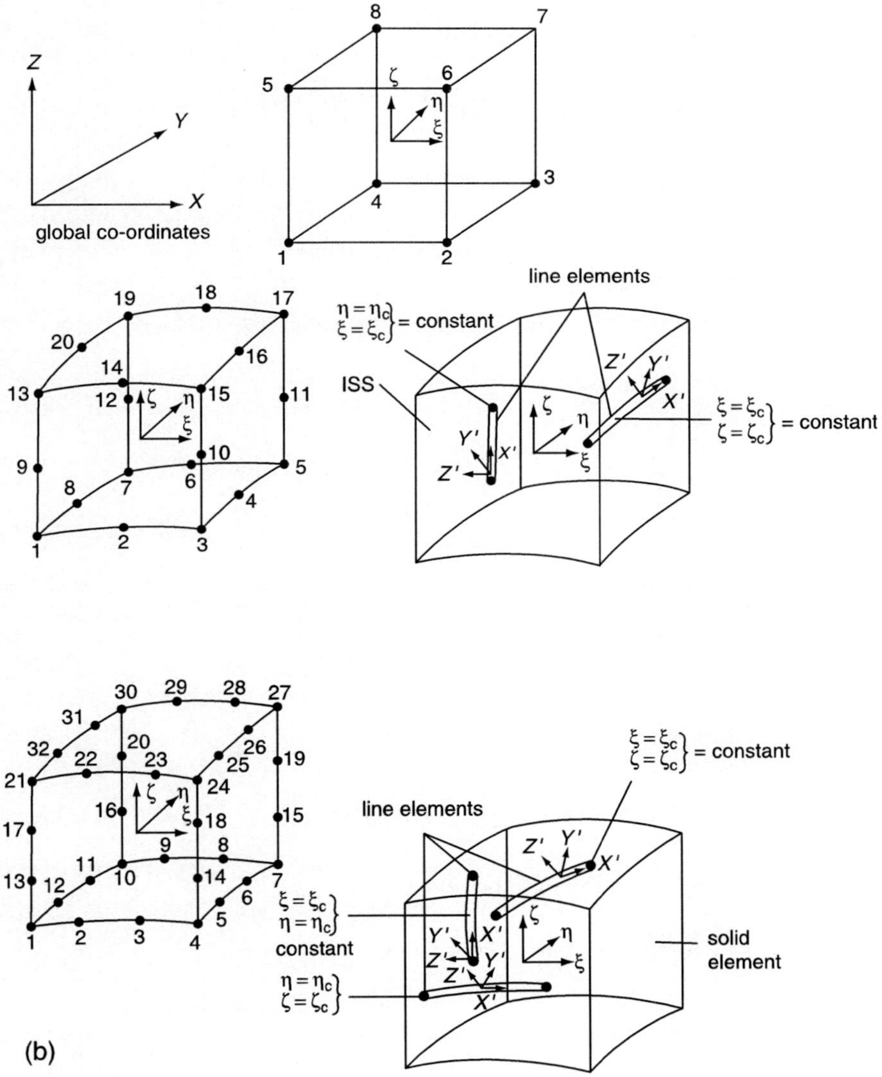

Fig. 3.1 (continued)

general equilibrium equation, both Eqs. (3.1) and (3.2) are included. The final equation becomes

$$(\{d\delta\}^{e})^{T}\{F\}^{e} = (\{d\delta\}^{e})^{T}\int_{\text{vol}}[B]^{T}\{\sigma\}dV \tag{3.3}$$

In terms of the local coordinate { $\{\xi,\eta,\zeta\}$ } system, Eq. (3.3) is written as

$$\{F\}^{e} = \int_{\text{vol}}[B]^{T}[D]\{\epsilon\}d\xi, d\eta, d\zeta \det[J]\{x\}^{e} \tag{3.4}$$

The force–displacement relationship for each element is given by

$$\{F\}^{e} = [K]^{e}\{u\}^{e} + \{F_{b}\}^{e} + \{F_{s}\}^{e} + \{F_{\sigma}\}_{i}^{e} + \{F_{\varepsilon}\}_{c}^{e} \tag{3.5}$$

where the element stiffness matrix is

$$[K_{c}] = \int_{\text{vol}}[B]^{T}[D][B]dV \tag{3.5a}$$

The nodal force due to the body force is

$$\{F_{b}\}^{e} = -\int_{\text{vol}}[N]^{T}\{G\}dV \tag{3.5b}$$

The nodal force due to the surface force is

$$\{F_{s}\}^{e} = -\int_{s}[N]^{T}\{p\}ds \tag{3.5c}$$

The nodal force due to the initial stress is

$$\{p_{\sigma}\}_{i}^{e} = \int_{\text{vol}}[B]^{T}\{\sigma_{o}\}dV \tag{3.5d}$$

The nodal force due to the initial strain is

$$\{p_{\epsilon}\}_{i}^{e} = -\int_{\text{vol}}[B]^{T}[D]\{\epsilon_{o}\}dV \tag{3.5e}$$

Equations (3.4) and (3.5) represent the relationships of the nodal loads to the stiffness and displacement of the structure. These equations now require modification to include the influence of the liner and its studs. The material compliance matrices $[D]$ are given in Tables 3.1 and 3.2. The numerical values are given for various materials or their combinations in Tables 3.3, 3.4, 3.5 and 3.6. These values of the constitutive matrices are recommended in the absence of specific information.

Table 3.1 Material compliance matrices $[D]$ with constant Poisson's ratio

(a) $[D]$ for steel components (isotropic)
Constant Young's modulus and Poisson's ratio

$$[D] = \frac{E}{(1+v)(1-2v)}\begin{bmatrix} 1-v & v & v & 0 & 0 & 0 \\ v & 1-v & v & 0 & 0 & 0 \\ v & v & 1-v & 0 & 0 & 0 \\ 0 & 0 & 0 & \frac{1-2v}{2} & 0 & 0 \\ 0 & 0 & 0 & 0 & \frac{1-2v}{2} & 0 \\ 0 & 0 & 0 & 0 & 0 & \frac{1-2v}{2} \end{bmatrix}$$

(b) $[D]$ for concrete: variable E and constant v

$$[D] = \begin{bmatrix} D_{11} = \frac{E_1(E')^3 - E_{cr}}{E''} & D_{12} = \frac{vE_1E_2(E')^2 + E_{cr}}{E''} & D_{13} = \frac{vE_1E_3(E')^2 + E_{cr}}{E''} & D_{14} = 0 & D_{15} = 0 & D_{16} = 0 \\ & D_{22} = \frac{E_2E_3(E')^2 + E_{cr}}{E''} & D_{23} = \frac{vE_2E_3(E')^2 + E_{cr}}{E''} & D_{24} = 0 & D_{25} = 0 & D_{26} = 0 \\ & & D_{33} = \frac{E_3(E')^3 - E_{cr}}{E''} & D_{34} = 0 & D_{35} = 0 & D_{36} = 0 \\ & & & D_{44} = G_{12} & D_{45} = 0 & D_{46} = 0 \\ & & & & D_{55} = G_{23} & D_{56} = 0 \\ & & & & & D_{66} = G_{31} \end{bmatrix}$$

$$E_{cr} = v^2 E_1 E_2 E_3 E'$$
$$E' = (E_1 + E_2 + E_3)/3$$
$$E'' = (E')3 - 2E_1E_2E_3v^2 - E'v(E_1E_2 + E_1E_3 + E_2E_3)$$
$$G_{12} = E_{12}/2(1+v)$$
$$E_{12} = (E_1 + E_2)/2$$
$$G_{23} = E_{23}/2(1+v)$$
$$E_{23} = (E_2 + E_3)/2$$
$$G_{31} = E_{31}/2(1+v)$$
$$E_{31} = (E_3 + E_1)/2$$

Table 3.2 [*D*] with variable Young's modulus and Poisson's ratio for concrete and other materials

$$\begin{bmatrix} D_{11} = \frac{(1-\nu_{23}\nu_{32})}{\overline{E}} E_1 & D_{12} = \frac{(\nu_{12}+\nu_{12}\nu_{32})}{\overline{E}} E_2 & D_{13} = \frac{(\nu_{13}+\nu_{12}\nu_{23})}{\overline{E}} E_3 & D_{14} = 0 & D_{15} = 0 & D_{16} = 0 \\ D_{21} = \frac{(\nu_{21}+\nu_{23}\nu_{31})}{\overline{E}} E_1 & D_{22} = \frac{(1-\nu_{13}\nu_{31})}{\overline{E}} E_2 & D_{23} = \frac{(\nu_{23}+\nu_{13}\nu_{21})}{\overline{E}} E_3 & D_{24} = 0 & D_{25} = 0 & D_{26} = 0 \\ D_{31} = \frac{(\nu_{31}+\nu_{21}\nu_{32})}{\overline{E}} E_1 & D_{32} = \frac{(\nu_{32}+\nu_{12}\nu_{31})}{\overline{E}} E_2 & D_{33} = \frac{(1-\nu_{12}\nu_{21})}{\overline{E}} E_3 & D_{34} = 0 & D_{35} = 0 & D_{36} = 0 \\ D_{41} = 0 & D_{42} = 0 & D_{43} = 0 & D_{44} & D_{45} = 0 & D_{46} = 0 \\ D_{51} = 0 & D_{52} = 0 & D_{53} = 0 & D_{54} = 0 & D_{55} & D_{56} = 0 \\ D_{61} = 0 & D_{62} = 0 & D_{63} = 0 & D_{64} = 0 & D_{65} = 0 & D_{66} \end{bmatrix}$$

$$\overline{E} - 1 - \nu_{12}\nu_{21} - \nu_{13}\nu_{31} - \nu_{23}\nu_{32} - \nu_{12}\nu_{23}\nu_{31} - \nu_{21}\nu_{13}\nu_{32}$$

$$E_1\nu_{21} = E_2\nu_{12} \quad D_{44} = G_{12}$$

$$E_2\nu_{32} = E_3\nu_{23} \quad D_{55} = G_{23}$$

$$E_3\nu_{13} = E_1\nu_{31} \quad D_{66} = G_{13}$$

The values of G_{12}, G_{23} and G_{13} are calculated in terms of the modulus of elasticity and Poisson's ratio as follows:

$$G_{12} = \frac{1}{2}\left[\frac{E_1}{2(1+\nu_{12})} + \frac{E_2}{2(1+\nu_{21})}\right] = \frac{1}{2}\left[\frac{E_1}{2(1+\nu_{12})} + \frac{E_1}{2\left(\frac{E_1}{E_2} + \nu_{12}\right)}\right]$$

$$G_{23} = \frac{1}{2}\left[\frac{E_2}{2(1+\nu_{23})} + \frac{E_3}{2(1+\nu_{32})}\right] = \frac{1}{2}\left[\frac{E_2}{2(1+\nu_{23})} + \frac{E_2}{2\left(\frac{E_2}{E_3} + \nu_{23}\right)}\right]$$

$$G_{13} = \frac{1}{2}\left[\frac{E_3}{2(1+\nu_{31})} + \frac{E_1}{2(1+\nu_{13})}\right] = \left[\frac{E_3}{2(1+\nu_{31})} + \frac{E_3}{2\left(\frac{E_3}{E_1} + \nu_{31}\right)}\right]$$

For isotropic cases: $E_1 = E_2 = E_3 = E \quad \nu_{12} = \nu_{13} = \nu_{23} = \nu_{21} = \nu_{31} = \nu_{32} = \nu$

Table 3.3 Material properties of concrete, bovine, steel and composites

(1) *Concrete*

$\sigma_1 = \sigma_2$	$\nu_1 = \nu_{23} = 0.2$ for any value of σ_3 up to 500 bar
$\sigma_1 < \sigma_2$	$\nu_{13} > \nu_{23}$
$\sigma_1 = 0$	$\nu_{13} = 0.2$ to 0.4 for any value of σ_2
	$= 0.4$ for up to $\sigma_3 = 500$ bar

for 80°C temperature, the above values are increased by 35–50%

E_c(KN/mm^2)	24	30	35	40
ν	0.15–0.18	0.17–0.20	0.20–0.25	0.25–0.30

Alternatively

$\nu = 0.2 + 0.6(\sigma_{2/}\,\sigma_{cu})^4 + 0.4(\sigma_{1/}\,\sigma_{cu})^4$

Where σ_1, σ_2 and σ_3 are the pressures/stresses along the three principal axes and σ_{cu} is the ultimate compressive strees of concrete

(2) *Bovine material*

$E_1 = 11$–18 GPa: $E_2 = 11$–19 GPa; $E_3 = 17$–20 GPa

$G_{12} = 3.6$–7.22 GPa; $G_{13} = 3.28$–8.65 GPa;$G_{23} = 8.285$–8.58 GPa

$\nu_{12} = 0.285$–0.58: $\nu_{13} = 0.119$–0.31; $\nu_{23} = 0.142$–0.31

$\nu_{21} = 0.305$–0.58: $\nu_{31} = 0.315$–0.46; $\nu_{32} = 0.283$–0.46

(3) *Steel*

$E = 200$ GN/m^2; $\nu = 0.3$–0.33

(4) *Composite*

Hot-pressed silicone nitride (HPSN) versus tungsten carbide

$E = 320$ GPa	$E = 320$ GPa
$\nu = 0.26$	$V = 0.24$

Carbon fibre (reinforced epoxy with 60% fibres by volume)

	Longitudinal	Transverse
Tensile strength (σ_{tu})	1750 MPa	60 MPa
Compressive strength	1300 MPa	–
Tensile modulus (E_t)	138 GPa	9.1 GPa
Compressive modulus (E'_c)	138 GPa	9.1 GPa
Failure strain in tension (ε_{tu})%	1.34	0.8
Failure strain in compression (ε_{cu})%	0.85	2.9

Table 3.4 Material properties of additional composites

Steel indenter	Versus
$E = 200$ GN/m^2	↓
$\nu = 0.3$–0.33	
	(1) *Plexiglass*
	$E = 3.435$ GN/m^2
	$\nu = 0.394$
	(2) Laminate: thornel 300/5208 with fibres oriented (0, + 60, −60)
	$E_1 = 50$ GN/m^2; $E_2 = 11.6$ GN/m^2
	$G_{11} = 19$ GN/m^2; $G_{12} = 4.0$ GN/m^2
	$\nu_{11} = 0.31$ $\nu_{12} = 0.06$

Table 3.4 (continued)

(3) *Aluminium and FRPs*

	Aluminium	BFRP[a]	GFRP[b]	CFRP[a]	CFRPL[c]
E(GN/m^2)					
ν	70	78.7	7.0	70	180
	0.3	0.32	0.30	0.30	0.28

[a]Quasi-isotropic
[b]Random mat
[c]Unidirectional

(4) *Graphite/epoxy*

(Web-stiffened foam sandwich panels with orthotropic facing and a number of four equally embedded stiffeners in a polyurethane (PU) core)

$E_1 = 120.7$ GN/m^2; $E_2 = 7.93$ GN/m

$G_{12} = G_{23} = G_{13} = 5.52$ GN/m^2

$\nu_{12} = 0.30$

Polyurethane foam

$E = 0.0431$ GN/m^2; $G_2 = 0.017$ GN/m^2; $\nu = 0.267$

(5) *Boron/epoxy composites*

$E_1 = 219.8$ GN/m^2; $E_2 = 21.4$ GN/m^2; $\nu = 0.208$

$E_{\mathrm{p1}} = 2.41$ GN/m^2; $E_{p2} = 0.04$ GN/m^2;

$G_{\mathrm{p}} = 0.008$ GN/m^2; Pat plastic level

$\sigma_{\mathrm{yt}} = 1.1$ GN/m^2

(6) *Layers of woven roving and chopped strand mat*

$E_1 = 14.5$ GN/m^2; $\sigma_{\mathrm{yt}} = 215$ N/mm^2; $\nu = 0.21$

(7) *Other materials*

Type	E_1 (GN/m^2)	E_2 (GN/m^2)	G (GN/m^2)	ν_{12}
CSM/polyester	8	8	3	0.32
WR/polyester	15	15	4	0.15
Glass fibre/polyester	25	25	4	0.17
UD glass/polyester	40	10	4	0.3
UD Kevlar/epoxide	76	8	3	0.34
UD carbon/epoxide	148	10	4	0.31
GY70/epoxy(celion with graphite fibre)	102	7.0	4.14	0.318
MODMORE II/epoxy HMS/E (with graphite fibres)	76.8	9.6	5.83	0.305
T300/E (thornel 300/ epoxy with graphite fibres)	54.86	12	5.83	0.30
GL/E (glass/epoxy)	30.3	149.9	5.84	0.32
Carbon fibre (60% volume)				
Reinforced epoxy compound:				

Table 3.4 (continued)

	Longitudinal	Transverse
E_l (GN/m^2)	140.0	9.00
E_c(GN/m^2)	140.0	9.00
σ_{tu} (GN/m^2)	1.8	0.06
σ_{cu} (GN/m^2)	1.3	0.27
V	0.3	0.02

Table 3.5 Material properties for brick and stone masonry and soil/rock

(1) *Brick masonry*

Brick Strength $f_b = 20-70$ N/mm^2 $E = 300f_b - 2000$

$f_b > 70$ N/mm^2 $E = 100f_b + 12{,}750$

Brick Strength (MN/m^2)	Mortar	Mortar mean cube Strength (MN/m^2)	Wall thickness (m^2)	Wall Strength (MN/m^2)
92	1:1:3	19.30	102.5	18.40
46	1:1:3	13.70	102.5	15.65
46	1:1:6	5.94	102.5	10.48

(3) Soil/rock

	$E \times 10^2$ MN/m^2	v	Density, ρ (kg/m^3)
Fine sand	57.456	0.35	
Silty clay	48.84	0.40	
Silty sand	47.88	0.35	
Plastic clay	3.56	0.40	
Silt stone	8.4	0.30	2622
Limestone	114.0	0.25	2671
Alluvial clay	5.0	0.20	
Clay (embankment fill)	20.0	0.20	1517
Saturated soil	200.0	0.30	
Jointed rock	150.0	0.25	
Sandstone	255.0	0.11	

For high plasticity, the frictional angle $\o_c' = 18°$
For low plasticity, the frictional angle $\o_c' = 25°$
For rocks, $\o_c'$ ranges between 20° and 30°
The adhesion coefficient c is around 1 kN/m^2

Table 3.6 Material properties of timber

Basic stresses and moduli					
Strength group	Bonding (N/mm^2) Parallel to grain	Tension (N/mm^2)	Compression to grain (N/mm^2) Parallel	Perpendicular	E_{min} (N/mm^2)
S_1	37.5	22.5	24.4	7.5	13,800
S_2	30.0	18.0	20.0	6.0	11,900
S_3	24.0	14.4	17.9	4.8	10,400
S_4	18.7	11.2	15.5	3.7	9200
S_5	15.0	9.0	13.3	3.0	7800
Dry-grade stresses and moduli					
Grade/species	Bonding (N/mm^2) Parallel to grain	Tension (N/mm^2)	Compression to grain (N/mm^2) Parallel	Perpendicular	E_{min} (N/mm^2)
SS/Douglas fir	6.2	3.7	6.6	2.4	7000
GS/Douglas fir	4.4	2.6	5.6	2.1	6000
SS/Redwood	7.5	4.5	7.9	2.1	7000
Whitewood					
GS/Corsican pine	5.3	3.2	6.8	1.8	5000
GS/European pine	4.1	2.5	5.2	1.4	4500
Plywood: all stresses and moduli are multiplied by the following factors:					
Grade/glued laminated	Bonding (N/mm^2) Parallel to grain	Tension (N/mm^2)	Compression to grain (N/mm^2) Parallel	Perpendicular	E_{min} (N/mm^2)
LA/4	1.85	1.85	1.15	1.33	1.0
LB/10	1.43	1.43	1.04	1.33	0.9
LB/20 or more	1.48	1.48			
LC/10	0.98	0.98	0.92	1.33	0.8
LC/20 or more	1.11	1.11			
	Permissible stresses				
	8 N/mm^2		–	–	8700
12 mm ply thickness	5 N/mm^2		–	–	7400

If the stiffness matrix $[K_c]$ for typical elements is known from Eqs. (3.4) and (3.5) as

$$[K_c] = \int_{vol} [\boldsymbol{B}]^T [\boldsymbol{D}][\boldsymbol{B}] \boldsymbol{d}\,\mathbf{Vol} \tag{3.6}$$

the composite stiffness matrix $[K_{TOT}]$, which includes the influence of liner and stud or any other material(s) in association, can be written as

$$[\boldsymbol{K}_{\mathrm{TOT}}] = [\boldsymbol{K}_{\mathrm{c}}] + [\boldsymbol{K}_{\mathrm{l}}] + [\boldsymbol{K}_{\mathrm{s}}] \tag{3.7}$$

where $[K_{\mathrm{l}}]$ and $[K_{\mathrm{s}}]$ are the liner and stud or connector matrices. If the initial and total load vectors on the liner/stud assembly and others are $[F_{\mathrm{T}}]$ and $[R_{\mathrm{T}}]$, respectively, then Eq. (3.4) is rewritten as

$$\{\boldsymbol{F}\}^{\mathrm{e}} + \{\boldsymbol{F}_{\mathrm{T}}\} - \{\boldsymbol{R}_{\mathrm{T}}\} = [\boldsymbol{K}_{\mathrm{TOT}}]\{\boldsymbol{x}\}^{*} \tag{3.8}$$

The displacement $\{x\}^*$ is different from $(x\}$ in Eq. (3.4), since it now includes values for both unknown displacements and restrained linear boundaries. Hence $\{x\}^*$ is defined in matrix form as

$$\{\boldsymbol{x}\}^{*}_{x,y,z} = \begin{Bmatrix} x_{\mathrm{un}x} \\ x_{\mathrm{un}y} \\ x_{\mathrm{un}z} \\ x_{\mathrm{b}x} \\ x_{\mathrm{b}y} \\ x_{\mathrm{b}z} \end{Bmatrix} = \begin{Bmatrix} x_{\mathrm{un}} \\ x_{\mathrm{b}} \end{Bmatrix} \tag{3.9}$$

where x_{un} and X_{b} are displacement values in unrestrained or unknown conditions and restrained conditions. Similarly, the values for $\{F_{\mathrm{T}}\}$ and $\{R_{\mathrm{T}}\}$ can also be written as

$$\begin{aligned} \{\boldsymbol{F}_{\mathrm{T}}\} &= \begin{Bmatrix} F_{\mathrm{un}} \\ F_{\mathrm{b}} \end{Bmatrix}_{x,y,z} \\ \{\boldsymbol{R}_{\mathrm{T}}\} &= \begin{Bmatrix} R_{\mathrm{un}} \\ R_{\mathrm{b}} \end{Bmatrix}_{x,y,z} \end{aligned} \tag{3.10}$$

The quantities for the liner corresponding to unknown displacements can be written as

$$[\boldsymbol{K}_{l}]\{\boldsymbol{x}_{\mathrm{un}}\}_{x,y,z} = \{\boldsymbol{F}_{\mathrm{un}}\}_{x,y,z} \tag{3.11}$$

The shear force τ acting on studs or any other type is evaluated as

$$\{\boldsymbol{\tau}\} = [\boldsymbol{K}_{\mathrm{s}}]\{\boldsymbol{x}_{\mathrm{un}}\}_{\boldsymbol{x},\boldsymbol{y},\boldsymbol{z}} \tag{3.12}$$

Table 3.7 gives the $[K_{\mathrm{s}}]$ matrix modified to include the stiffness of the liner.

Table 3.7 Stiffness matrix and load vector

$$[K_b]_{6\times 6} = \pi dL \begin{bmatrix} (l^2E_\mathrm{h} + pE_\mathrm{v} + \gamma^2E_\ell) & (lmE_\mathrm{h} + pqE_\mathrm{v} + \gamma sE_\ell) & (lnE_\mathrm{h} + \gamma t\,E_\mathrm{v}) & (-l^2E_\mathrm{h} - p^2E_\mathrm{v} - \gamma^2E_\ell) & (-lmE_\mathrm{h} - pqE_\mathrm{v} - \gamma sE_\ell) & (-nE_\mathrm{h} - \gamma tE_\mathrm{v}) \\ & (m^2E_\mathrm{h} + q^2E_\mathrm{v} + s^2E_\ell) & (mnE_\mathrm{h} + stE_\ell) & (-lmE_\mathrm{h} - pqE_\mathrm{v} - \gamma sE_\ell) & (-m^2E_\mathrm{h} - q^2E_\mathrm{v} - s^2E_\ell) & (-mnE_\mathrm{h} - stE_\ell) \\ & & (n^2E_\mathrm{h} + t^2E_\mathrm{v}) & (-lnE_\mathrm{h} - \gamma tE_\mathrm{v}) & (-mnE_\mathrm{h} - stE_\ell) & (-n^2E_\mathrm{h} - t^2E_\mathrm{v}) \\ & & & (l^2E_\mathrm{h} + p^2E_\mathrm{v} + \gamma^2E_\ell) & (lmE_\mathrm{h} + pqE_\mathrm{v} + \gamma sE_\ell) & (lnE_\mathrm{h} + \gamma tE_\mathrm{v}) \\ & \text{symmetrical} & & & (m^2E_\mathrm{h} + q^2E_\mathrm{v} + s^2E_\ell) & (mnE_\mathrm{h} + stE_\ell) \\ & & & & & (n^2E_\mathrm{h} + t^2E_\mathrm{v}) \end{bmatrix}$$

Component stiffnesses using springs:
E_h = horizontal stiffness
E_v = vertical stiffness
E_ℓ = longitudinal stiffness

$$\{\Delta P^e\}_{6\times 1} = \pi dL \begin{Bmatrix} -l\Delta\sigma_\mathrm{h} & -p\Delta\sigma_\mathrm{v} & -\gamma\Delta\sigma_\mathrm{l} \\ -m\Delta\sigma_\mathrm{h} & -q\Delta\sigma_\mathrm{v} & -s\Delta\sigma_\mathrm{l} \\ -n\Delta\sigma_\mathrm{h} & -t\Delta\sigma_\mathrm{l} & \\ l\Delta\sigma_\mathrm{h} & +p\Delta\sigma_\mathrm{v} & +\gamma\Delta\sigma_\mathrm{l} \\ m\Delta\sigma_\mathrm{h} & +q\Delta\sigma_\mathrm{v} & +s\Delta\sigma_\mathrm{l} \\ n\Delta\sigma_\mathrm{h} & +t\Delta\sigma_\mathrm{l} & \end{Bmatrix}$$

πdL = perimeter of the steel

$$\left.\begin{matrix} l, m, n \\ p, q, \gamma \\ s, t \end{matrix}\right\} = \text{direction cosines}$$

3.2 Steps for Dynamic Non-linear Analysis

The solutions of Eqs. (3.6), (3.7), (3.8), (3.9), (3.10), (3.11) and (3.12) require a special treatment such as under any increment of dynamic loading, stresses, strains and plasticity are obtained in steel, concrete and composites such as the liner and its anchorages and other similar materials. An additional effort is needed to evaluate the rupture of the steel or other material when cracks develop, especially in concrete beneath the liner or its anchorages.

The dynamic coupled equations are needed to solve the impact/explosion problems and to assess the response history of the structure, using the time increment δt. If $[M]$ is the mass and $[C]$ and $[K]$ are the damping and stiffness matrices, the equation of motion may be written in incremental form as

$$[M]\{\ddot{x}(t)\} + [C_{\text{in}}]\{\dot{x}(t)\} + [K_{\text{in}}]\{\delta(t)\} = \{R(t)\} + \{F_1(t)\} \tag{3.13}$$

where $F_1(t)$ is the time-dependent load including impact/explosion load. If the load increment of F_1(t) is $\delta P_n(t)$, where n is the nth load increment, then

$$P_n(t) = P_{n-1}(t) + P_n(t) \tag{3.13a}$$

and hence $\{R(t)\} = \{\delta P_n(t)\}$, which is the residual time-dependent load vector. The solution of Eq. (3.13) in terms of $t + \delta t$ for a δt increment becomes

$$[M]\{\ddot{x}(t+\delta t)\} + [C_{\text{in}}]\{\dot{x}(t+\delta t)\} + [K_{\text{in}}]\{\delta R(t+\delta t)\} + \{\delta P(t+\delta t)\} \tag{3.14}$$

where 'in' denotes initial effects by iteration using the stress approach; $\delta P(t + \delta t)$ represents the non-linearity during the time increment δt and is determined by

$$\{\sigma\} = [D]\{\varepsilon\} - \{\varepsilon_o\} + \{\sigma_o\} \tag{3.15}$$

The constitutive law is used with the initial stress and constant stiffness approaches throughout the non-linear and the dynamic iteration. For the iteration

$$\{x(t+\delta t)\}_i = [K_{\text{in}}]^{-1}\{R_{\text{TOT}}(t+\delta t)\}_i \tag{3.16}$$

The strains are determined using

$$\{\varepsilon(t+\delta t)\}_i = [B]\{x(t+\delta t)\}_i \tag{3.17}$$

where $[B]$ is the strain displacement. The stresses are computed as

$$\{\sigma(t+\delta t)\}_i = [D]\{\varepsilon(t+\delta t)\}_i + \{\sigma_o(t+\delta t)\}_{i-1} \tag{3.18}$$

where $\{\sigma(t+\delta t)\}$ is the total initial stress at the end of each iteration. All calculations for stresses and strains are performed at the Gauss points of all elements.

The initial stress vector is given by

$$\{\sigma_o(t+\delta t)\}_i = f\{\varepsilon(t+\delta t)\}_{i-1} - [D]\{\varepsilon(t+\delta t)\}_i \tag{3.19}$$

Using the principle of virtual work, the change of equilibrium and nodal loads $\{\delta P(t+\delta t)\}_i$ is calculated as

$$F_1(t+\delta t) = \{\delta P(t+\delta t)\}_{i\mathrm{TOT}} \tag{3.20}$$

$$= \int_{-1}^{+1}\int_{-1}^{+1}\int_{-1}^{+1} [B]^{T''}\{\delta\sigma_o(t+\delta t)\}_i d\xi d\eta d\zeta$$

$$\sigma_o(t) = \{\sigma_o(t+\delta t)\}_i = 0$$

where $d\zeta, d\eta$ and $d\zeta$ are the local coordinates and T″ is the transpose. The integration is performed numerically at the Gauss points. The effect load vector $F_1(t)$ is given by

$$\begin{aligned} F_1(t+\delta t) &= \{\delta P(t+\delta t)\}_{i\mathrm{TOT}} \\ &= -[\delta C(t)in](\{x(t+\delta t)\}_i - \{x(t)\}) \\ &\quad - [\delta C(t+\delta t)]_i(\{x(t+\delta t)\}_i - [\delta K(t)_{in}])(\{x(t+\delta t)\}_i - \{x(t)\}_i) \\ &\quad - [\delta K(t+\delta t)]_i\{x(t+\delta t)\}_i \end{aligned} \tag{3.21}$$

The Von Mises criterion is used with the transitional factor f^*_{TR} to form the basis of the plastic state, such as shown in Fig. 3.2:

$$f^*_{\mathrm{TR}} = \frac{\sigma_y(t) - \sigma_{y-1}(t)}{\sigma(t+\delta t)_i - \sigma(t+\delta t)_{i-2}} \tag{3.22}$$

The elasto-plastic stress increment will be

$$\{\delta\sigma_i\} = [D]_{\mathrm{ep}}\sigma(t+\delta t)_{i-1}(1 - f^*_{\mathrm{TR}})\{\delta\varepsilon\} \tag{3.23}$$

If $\{\sigma(t+\delta t)\} < \sigma_y(t)$, it is an elastic limit and the process is repeated. The equivalent stress is calculated from the current stress state where stresses are drifted; they are corrected from the equivalent stress–strain curve.

The values of $[D]_{\mathrm{ep}}$ and $[D]_{\mathrm{p}}$ are derived using plastic stress/strain increments.

In the elasto-plastic stage, the time-dependent yield function is $f(t)$. It is assumed that the strain or stress increment is normal to the plastic potential $Q(\sigma, K)$. The plastic increment, for example, is given by

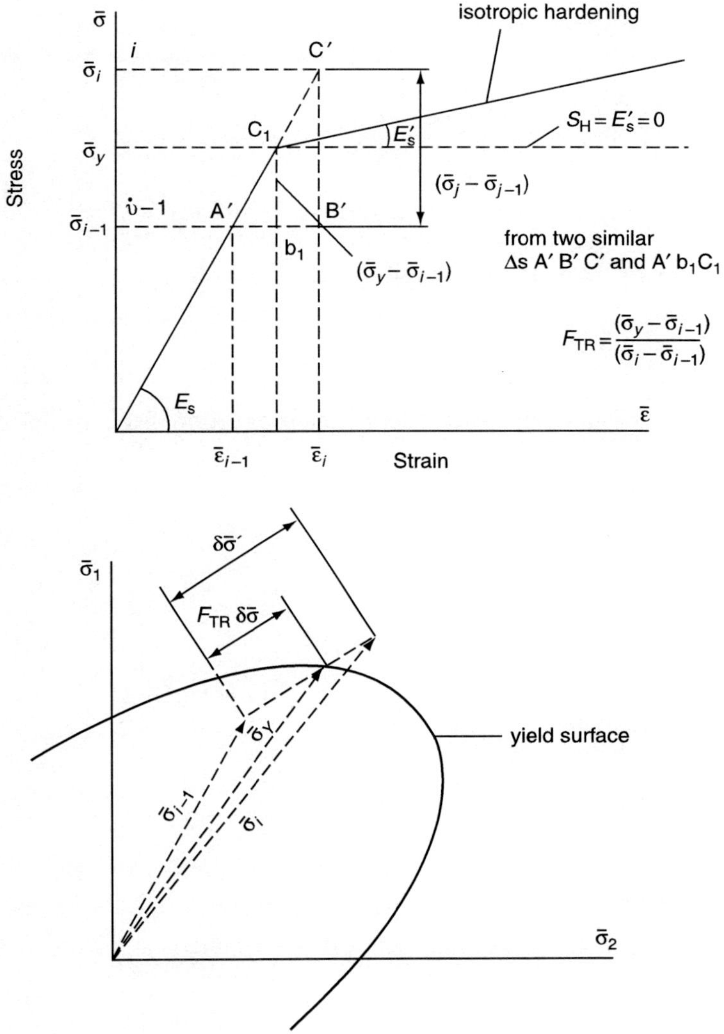

Fig. 3.2 Transitional factor and plastic point

$$\delta\varepsilon\sigma(t+\delta t)p = \frac{\partial Q}{\partial \sigma} = \lambda b \tag{3.24}$$

where λ = proportionality constant >0

$$b \approx \frac{\partial Q}{\partial \sigma(t+\delta t)}$$

When $f(t) = Q$

$$\delta\varepsilon(t+\delta t)_{\mathrm{p}} = \lambda a$$

$$a = \frac{\partial f}{\partial \sigma(t+\delta t)}$$

Therefore,

$$df = [\partial f/\partial \sigma(t+\delta t)] \quad d\sigma(t+\delta t) + (\partial f/dK)dK \tag{3.25}$$

If A is the hardening plastic parameter, then

$$A = \frac{1}{\lambda} + (\partial f/dK)dK$$

An expression can easily be derived for the proportionality constant λ

$$\lambda = \frac{a^{T''} D\delta\varepsilon(t+\delta t)}{[A + a^{T''} Db]} \tag{3.26}$$

hence

$$\delta\varepsilon(t+\delta t)p = b\lambda$$

The value of the elasto-plastic matrix $[D]_{\mathrm{ep}}$ is given by

$$[D]_{\mathrm{ep}} = D - \frac{Dba^{T''} Db}{[A + a^{T''} Db]} \tag{3.27}$$

The value of the plastic matrix $[D]_{\mathrm{p}}$ is given by

$$[D]p = \frac{Dba^{T''} D}{[A + a^{T''} Db]} \tag{3.28}$$

where $[D]$ is the compliance matrix for the elastic case. The elasto-plastic stress increment is given by

$$\{\delta\sigma_i\}_{\mathrm{t}} = [D]_{\mathrm{ep}}\{\sigma_i\}_{\mathrm{t}}^{Y*}(1 - F_{\mathrm{TR}}{}^*)\{\delta \in\}_{\mathrm{t}} \tag{3.29}$$

For the sake of brevity, $\{\delta\sigma_i\}_{\mathrm{t}} = \delta\sigma\ (t+\delta t)$ for the ith point or increment and other symbols are as given above. The total value becomes

$$\{\delta\sigma_i\}_{\mathrm{TOT}} = \{\sigma_i\}_{\mathrm{t}}^{Y*} + \{\delta \in\}_{\mathrm{t}} \tag{3.30}$$

If $\{\sigma_i\}_{\mathrm{t}} < \sigma_{\mathrm{yt}}$ it is an elastic point and $\{\sigma_i\}_{\mathrm{t}} = \{\sigma_i'\}_{\mathrm{t}}$. The process is repeated. Looking at the plastic point in the previous iteration, it is necessary to check for unloading when $\sigma \geq \sigma_y$; the unloading will bring about the total stress $\{\delta\sigma_{i-1}\}_t + \{\delta\sigma_i'\}_i$, and set $\{\sigma_y\}_t = \{\delta\sigma_{i-1}\}_t$. Then loading at this point gives

$$\{\delta\sigma_i\} = [D]_{ep}\{\sigma_{i-1}\}_t\{\delta \in\}_t \tag{3.31}$$

The total stress is then written as

$$\{\delta\sigma_i\}_{TOT} = \{\sigma_{i-1}\}_t\{\delta\sigma_i\} \tag{3.32}$$

Stresses are calculated using the elasto-plastic material matrix, which does not drift from the yield surfaces, as shown in Fig. 3.2. Stresses are corrected from the equivalent stress–strain curve by

$$\{\sigma_{corr} = \{\sigma_{i-1}\}_t + K\{\delta \in_P\}_t \tag{3.33}$$

where $\{\delta \in_P\}_t = \sqrt{\frac{2}{3}}\{\sqrt{(}\delta\varepsilon_{ij}^p\delta\varepsilon_{ij}^P\}_r$ = equivalent plastic strain increment. K is the strain-hardening parameter such that $\{\delta \in_p\}_t = \lambda$. The equivalent stress is calculated from the current stress state as shown below:

$$\{\sigma_i\}_{eq} = f\{(\sigma_i)\}_t \tag{3.34}$$

$$\text{the value of } \frac{\sigma_{corr}}{\sigma} \text{ is a factor} \tag{3.35}$$

Therefore the correct stress state on the yield surface is given by

$$\{\sigma_i = \text{factor} \times \{\sigma_i\} \tag{3.36}$$

A reference is made to Fig. 3.2 for evaluating this factor as f^*_{TR}.

3.2.1 Buckling State and Slip of Layers for Composite Sections

Within the above stages, there can be a possibility of plastic buckling of the liner or any embedded anchors or layers. The buckling matrix is developed so that at appropriate stages the layer/liner/anchor system is checked against buckling. The plastic buckling matrix is given below as

$$(K^i + \lambda_c K_G^i)F_T = 0 \tag{3.37}$$

where K^i is the elasto-plastic stiffness matrix as a function of the current state of plastic deformation using the above steps and K_G is the geometric stifness matrix:

$$\lambda_c = 1 + E_{ps} \tag{3.37a}$$

Table 3.8 T″ transformation matrices

$$[T''_{\varepsilon}] = \begin{bmatrix} l_1^2 & m_1^2 & n_1^2 & l_1m_1 & m_1n_1 & l_1n_1 \\ l_2^2 & m_2^2 & n_2^2 & l_2m_2 & m_2n_2 & l_2n_2 \\ l_3^2 & m_3^2 & n_3^2 & l_3m_3 & m_3n_3 & l_3n_3 \\ 2l_1l_2 & 2m_1m_2 & 2n_1n_2 & (l_1m_2+l_2m_1) & (m_1n_2+m_2n_1) & (l_1n_2+l_2n_1) \\ 2l_2l_3 & 2m_2m_3 & 2n_2n_3 & (l_2m_3+l_3m_2) & (m_2n_3+n_2m_3) & (l_2n_3+l_3n_2) \\ 2l_1l_3 & 2m_1m_3 & 2n_1n_3 & (l_1m_3+m_1l_3) & (m_1n_3+m_3n_1) & (l_1n_3+n_1l_3) \end{bmatrix}$$

$$[T''_{\sigma}] = \begin{bmatrix} l_1^2 & m_1^2 & n_1^2 & 2l_1m_1 & 2m_1n_1 & 2l_1n_1 \\ l_2^2 & m_2^2 & n_2^2 & 2l_2m_2 & 2m_2n_2 & 2l_2n_2 \\ l_3^2 & m_3^2 & n_3^2 & 2l_3m_3 & 2m_3n_3 & 2l_3n_3 \\ l_1l_2 & m_1m_2 & n_1n_2 & (l_1m_2+l_2m_1) & (m_1n_2+n_1m_2) & (l_1n_2+l_2n_1) \\ l_2l_3 & m_2m_3 & n_2n_3 & (l_2m_3+l_3m_2) & (m_2n_3+n_2m_3) & (l_2n_3+l_3n_2) \\ l_1l_3 & m_1m_3 & n_1n_3 & (l_1m_3+l_3m_1) & (m_1n_3+m_3n_1) & (l_1n_3+n_1l_3) \end{bmatrix}$$

where E_{ps} represents accuracy parameters.

Where composite layers, liner and studs are involved, the incremental slips from the nodal displacements are assessed in the following manner:

$$\{\Delta S_i\}_{x,y,z} = [T'']\{d\delta_i\} \tag{3.38}$$

where S_i is a slip at node i and T'' is the transformation matrix given in Table 3.8. The total slip at iteration is given without subscripts as

$$\{S_i\} = \{S_{i-1}\} + \{S_i\} \tag{3.39}$$

The strains are computed as

$$\{\varepsilon_i\}t = \{\varepsilon_{i-1}\}t + \{\delta\ \in_i\}t \tag{3.40}$$

Thes incremental stress $\{\sigma_B\}$ between the studs and concrete or between any composite materials for the ith node can then be computed as

$$\{\delta\sigma_{Bi}\}_t = [Ks]\{\sigma_{Bi-1}\}_t\{\delta S_i\}_t \tag{3.41}$$

The total stresses are

$$\{\delta\sigma_{Bi}\}_{TOT} =]\{\sigma_{Bi-1}\}_t +]\{\delta\sigma_{Bi}\}_t \tag{3.42}$$

If $|\delta S_i| > S_{max}$ the bond between the stud and the concrete or any composite material is broken and the pull-out occurs, i.e. $\{\sigma_B\} = 0$ and S_{max} has a value which is maximum. If $|\delta S_i| < S_{max}$ the value $\{S_i\}$ is calculated. The procedure is linked with the general finite element work discussed already in the non-linear dynamic cases for impact and explosion.

3.2.2 Strain Rate Effects Based on the Elastic–Viscoplastic Relationship for Earth Materials Under Impact and Explosion

It is assumed that for each dynamic loading increment, the strain rate ε_{ij} can be expressed as the sum of the elastic and viscoplastic components:

$$\{d\varepsilon_{ij}\}_{\mathrm{t}} = \{d\varepsilon_{ij}\}_{\mathrm{e}} + \{d\dot{\varepsilon}_{ij}\}_{\mathrm{vp}} \tag{3.43}$$

Where the subscripts e and vp denote the elastic and viscoplastic components, respectively. The elastic strains are related to the stress rate σ_{ij} by

$$\{d\dot{\varepsilon}_{ij}\}_{\mathrm{e}} = \frac{1}{9K} \times \left(\frac{dJ_1}{dt}\right)\delta_{ij} + \frac{1}{2G} \times \frac{dS_{ij}}{dt} \tag{3.44}$$

where $J_1 =$ deviatoric stress: first invariant; $S_{ij} = \{\sigma_{ij}\ \frac{1}{3}\ J_1\delta_{ij}\}_{\mathrm{t}}$; $\delta_{ij} =$ Knonecker delta; $K =$ elastic bulk modulus; G = elastic shear modulus.

The shear modulus is expressed in terms of the invariant J_2', where

$$J_2' = \frac{1}{2} S_{ij} S_{ij} \tag{3.44a}$$

Then

$$K = \frac{K_i}{1 - K_1}\left[1 - K_1 \mathrm{e}^{K_2 J_1}\right] \tag{3.44b}$$

$$G = \frac{G_i}{1 - G_1}\left[1 - G_1 \mathrm{e}^{-G_2 J_2'}\right] \tag{3.44c}$$

where K_{i}, K_1, G_{i}, G_1, K_2 and G_2 are material constants. The values of J_2' and Ss are given below:

$$J_2' = \frac{1}{2}(S_x^2 + S_y^z + S_z^2) + \tau_{xy}^2 + \tau_{yz}^2 + \tau_{zx}^2 \text{ is the second variant} \tag{3.44d}$$

$$S_x = \sigma_x - \sigma_m,\ S_y = \sigma_y - \sigma_m,\ S_z = \sigma_z - \sigma_m \tag{3.44e}$$

The linear values of K and G are

$$K = \frac{E}{3(1 - 2v)}; \quad G = \frac{E}{2(\mathbf{1} - 2v)} \tag{3.44f}$$

Tables 3.1 and 3.2 are used for variable properties of E and v.

The components of the viscoplastic strain rate are calculated using the above-mentioned plastic flow rule for rate-sensitive material:

$$\{\dot{\varepsilon}_{ij}\}_{\mathrm{vp}} = \gamma[f(\sigma_{\mathrm{D}}/B)\sigma_{\mathrm{D}}/\delta\sigma_{ij}]_{\mathrm{t}} \tag{3.45}$$

where γ = viscosity parameter:

$$f(\sigma_{\mathrm{D}}/B)_{\mathrm{t}} = f\left(\sigma_{\mathrm{s}} - \frac{\beta}{B}\right)$$

σ_{s} = static yield stress; B = material parameter:

$$\beta = f^{-1}\left\{\frac{1}{\gamma}\left[\frac{\frac{1}{3}\{d\dot{\varepsilon}_{KK}\}_{\mathrm{vp}} + (2\dot{e}^{-1})^2_{\mathrm{vp}}}{(\partial\sigma_{\mathrm{s}}/\partial J_1)^2 \frac{1}{2}((\partial\sigma_{\mathrm{s}}\partial J_2')^2)}\right]\right\} \tag{3.45a}$$

$$(\dot{e}^{-1})_{\mathrm{vp}} = \left[\frac{1}{2} d\dot{e}_{ij} d\dot{e}_{ij}\right]^{\frac{1}{2}}_{\mathrm{vp}} \tag{3.45b}$$

and is the square root of the second invariant of the viscoplasic strain rate.

Using this bulk modulus approach for soils, the time-dependent stress–strain relation is given in Table 3.9. With reference to *rocks,* the failure strength of the rock is defined in exactly the same way as described earlier; the values for E and ν will vary. Nevertheless, the various alternative failure models given in Table 3.10 for rocks are related in terms of strain rates by

$$\overline{M} = \frac{\sigma_{\mathrm{dyn}}}{\sigma_{\mathrm{s}}} = 1 + c\log\frac{\dot{\varepsilon}}{\dot{\varepsilon}_{\mathrm{s}}} \tag{3.46}$$

where σ_{dyn} = dynamic stress; σ_{s} static stress; $\dot{\varepsilon}$ strain rate (dynamic); $\dot{\varepsilon}_{\mathrm{s}}$ = strain rate (static); c = constant.

Table 3.9 Bulk modulus model for earth

$$\begin{Bmatrix} \delta\sigma_x \\ \delta\sigma_y \\ \delta\sigma_z \\ \delta\tau_{xy} \\ \delta\tau_{yz} \\ \delta\tau_{zx} \end{Bmatrix}_{\mathrm{t}} \begin{bmatrix} (K+\frac{4}{3}G) & (K-\frac{2}{3}G) & (K-\frac{2}{3}G) & 0 & 0 & 0 \\ & (K+\frac{4}{3}G) & (K-\frac{2}{3}G) & 0 & 0 & 0 \\ & & (K+\frac{4}{3}G) & 0 & 0 & 0 \\ & & & G_{12} & 0 & 0 \\ & \mathrm{sym} & & & G_{23} & 0 \\ & & & & & G_{13} \end{bmatrix} \begin{Bmatrix} \delta\epsilon_x \\ \delta\epsilon_y \\ \delta\epsilon_z \\ \delta_{xy} \\ \delta_{yz} \\ \delta_{zx} \end{Bmatrix}_{\mathrm{t}}$$

or in short $\{\delta\sigma\} = [D]\{\delta\varepsilon\}$

Where $[D]$ is the required material matrix

$$G_{12} = G_{23} = G_{13} = G = G_{\mathrm{e}} - \alpha \log\frac{J_2}{J_2^{\mathrm{e}}} \text{ for } J_2 > J_2^{\mathrm{e}}$$

$$G = G_e \text{ for } J_2 \leq J_2^{\mathrm{e}}$$

In the case where the soil/rock is orthotropic, the values of G_{12},G_{23} and G_{13} are given as indicated in Table 3.2

Table 3.10 Numerical models for rocks

(1) Sandstone
$\tau = 1538 + \sigma \tan \varnothing$ (1)
where τ and σ = shear and compressive stresses, respectively
$\varnothing = 29°15'$
(2) Rupture of sandstone: Mohr failure envelope
$\tau_{max}/\sigma_{cu} = 0.1 + 0.76(\sigma_m/\sigma_u)^{0.85}$ (2)
where σ_{cu} is the uniaxial compressive stress at rupture under pure shear $\sigma_1 = -\sigma_3$
(3) Realistic rock including friction
$\alpha = (\sigma_1 + \sigma_2 + \sigma_3) + (\sigma_1 - \sigma_2)^2 + (\sigma_2 - \sigma_3)^2 + (\sigma_3 - \sigma_1)^2 = K^*$ (3)
where $\alpha = \sqrt{6} \tan \varnothing / \sqrt{(9 + 12 \tan \varnothing)}$
$K^* = \sqrt{6c}/\sqrt{9 + 12 \tan \varnothing}$
$\varnothing$ = angle of friction
c = cohesion
A generalised Mohr coulomb criterion is written as
$\tau^2 = [\sqrt{n+1} - 1][\sigma_{tu}^2 - \sigma\sigma_{tu}]$ (4)
Where σ, t = normal stress and shear stress on the fractured plane
σ_{tu} = uniaxial tensile strength
$n = \sigma_{cu}/\sigma$ = brittleness
σ_{cu} = uniaxial compressive strength
Equation (4) can be expressed in terms of σ'_m mean stress and σ_s the maximum shear stress by
$\sigma_s = \sigma_{tu} - \sigma_m$ for $\sigma_{tu} > \sigma_{mo} > \sigma_m$
$\sigma_s = \tau_o \sqrt{[(1 - \sigma_m/\sigma_{tu}) - (\tau_o/2\sigma_{tu})^2]}$ for $\sigma_{mo} < \sigma_m$ (5)
where $\sigma_m(\sigma_1 + \sigma_2)/2$, $\sigma_s = (\sigma_1 - \sigma_2)/2$,
$\sigma_{mo} = \sigma_{tu} - \tau_0^2/\sigma_{tu}$ (6)
$\tau_o = [\sqrt{n+1} - 1]\sigma_{tu}$
The stress state is assessed for σ_m and σ_s from the failure surface as
$= \sigma_s/\sigma_{s(critical)} \geq 1$
representing the failure condition. If $a = \sigma_s/\sigma_{s(critical)}$, Eq. (5) is satisfied

The range of strain rate is $\dot{\varepsilon} = 10^{-5}\ \mathrm{S}^{-1}$. The dynamic failure criterion can then be written as

$$\sigma_t^1 = \tau_o X_3 + \sigma \left[\frac{\sigma_{cu}^2 X_1^2 - 4\tau_o^2 X_3^2}{4\sigma_{cu} X_1 \tau X_3} \right] \quad \text{for } \sigma \not< 0$$

$$\sigma_t^2 = \tau_0^2 X_3^2 + \sigma \frac{[\tau_o^2 X_3^2 - \sigma_{tu}^2 X_2^2]}{\sigma_{tu} X_2} - \sigma^2 \quad \text{for } \sigma \not> 0 \tag{3.47}$$

where

$$X_1 = \tfrac{3}{40}\overline{M} + \tfrac{1}{25}\overline{M}^2$$

$$X^2 = \tfrac{3}{100}\overline{M} + \tfrac{7}{1000}\overline{M}^2$$

$$X^3 = \tfrac{1}{40}\overline{M} + \tfrac{1}{100}\overline{M}^2$$

τ_o = octahedral shear stress under static loads

In the case of *brick material,* Khoo and Hendry relationships given below are used in the above failure models and strain rate simulations. The non-linear principal stress relationship (biaxial) is given by

$$\sigma_1/\sigma_{\mathrm{cu}} = 1 + 2.91(\sigma_2/\sigma_{\mathrm{c}})^{0.805} \tag{3.48}$$

where σ_1 = major principal stress; σ_2 = minor principal stress; σ_{cu} = uniaxial compressive strength.

The brick-failure envelope with the mortar triaxial strength curve is given by the polynomials

$$\sigma_{\mathrm{t}}/\sigma_{\mathrm{tu}} = 09968 - 2.0264(\sigma/\sigma_{\mathrm{cu}}) + 1.2781(\sigma/\sigma_{\mathrm{cu}})^2 - 0.2487(\sigma/\sigma_{\mathrm{cu}})^2 \tag{3.49}$$

$$\sigma_3/\sigma_{\mathrm{cu}} = -0.1620(\sigma_1/\sigma_{\mathrm{cu}}) + 0.1126(\sigma_1/\sigma_{\mathrm{cu}})^2 - 0.0018(\sigma/\sigma_{\mathrm{cu}})^3 \tag{3.50}$$

where $\sigma/\sigma_{\mathrm{cu}}$ = ratio of compressive strength; $\sigma_{\mathrm{t}}/\sigma_{\mathrm{tu}}$ = ratio of tensile strength; $\sigma_{\mathrm{t}} = \alpha\sigma_3$ where α = 0.15 and 0.40 for mortars of 1:1/4:3 and 1:1:6, respectively.

3.2.3 Finite Element of Concrete Modelling

A number of modelling methods are available for simulation into the finite element method. On impact and explosion work, methods such as the edochronic, Ottoson and Blunt crack have been widely used. They are covered in this section. The bulk modulus model of Table 3.9 is reviewed to include cracking with and without aggregate interlocking. On the basis of the endochronic concept, which is widely reported the following equations apply:

$$\{\delta\sigma_{x,y,z}\}_{\mathrm{t}} + \{\delta\sigma^{\mathrm{p}}_{x,y,z}\}_{\mathrm{t}} = [D^*_T]\{\varepsilon^p_{x,y,z}\}_{\mathrm{t}} \tag{3.51}$$

where the superscript p denotes stresses in the plastic case. Table 3.11 gives details of uncracked and cracked cases for Eq. (3.51). When cracks in three directions are open the concrete loses its stiffness, then

$$[D^*_{\mathrm{T}}] = 0 \tag{3.52}$$

Stresses $\{\{\sigma_i\}_{\mathrm{t}}\}\sim$ are checked against the cracking criteria. For example, if there is one crack normal to the X-direction, the concrete can no longer resist any tensile stress in that direction, then

$$\delta\sigma^*_x = 0$$

Table 3.11 Cracks using endochronic theory

Uncracked matrix

$$\begin{Bmatrix} \delta\sigma_x + \delta\sigma_x^p \\ \delta\sigma_y + \delta\sigma_y^p \\ \delta\sigma_z + \delta\sigma_z^p \\ \delta\tau_{xy} + \delta\tau_{xy}^p \\ \delta\tau_{yz} + \delta\tau_{yz}^p \\ \delta\tau_{zx} + \delta\tau_{zx}^p \end{Bmatrix}_t = \begin{bmatrix} D_{11} & D_{12} & D_{13} & 0 & 0 & 0 \\ & D_{21} & D_{23} & 0 & 0 & 0 \\ & & D_{33} & 0 & 0 & 0 \\ & & & \beta' D_{44} & 0 & 0 \\ & & & & \beta' D_{55} & 0 \\ & & & & & \beta' D_{66} \end{bmatrix} \begin{Bmatrix} \delta\sigma_x \\ \delta\sigma_y \\ \delta\sigma_z \\ \delta\gamma_{xy} \\ \delta\gamma_{yz} \\ \delta\gamma_{zx} \end{Bmatrix}_t$$

Where

$D_{11} = D_{22} = D_{33} = K + \frac{4}{3}G \quad \beta = \text{aggregate inter locking} \approx \frac{1}{2} \text{ to } \frac{3}{4}$

$$D_{12} = D_{13} = D_{23} = K - \frac{2}{3}G$$

$$D_{44} = G_{12} = \frac{1}{2}\left[\frac{E_1}{2(1+\nu_{12})} + \frac{E_2}{2(1+\nu_{21})}\right]$$

$$D_{55} = G_{23} = \frac{1}{2}\left[\frac{E_2}{2(1+\nu_{23})} + \frac{E_3}{2(1+\nu_{32})}\right]$$

$$D_{66} = G_{13} = \frac{1}{2}\left[\frac{E_3}{2(1+\nu_{31})} + \frac{E_1}{2(1+\nu_{13})}\right]$$

The values of E and ν are given in Tables 6.3, 6.4, 6.5 and 6.6

Cracked matrix

$$\sigma_1\text{-direction } [D]^* = \begin{bmatrix} 0 & 0 & 0 & 0 & 0 & 0 \\ 0 & \left(D_{22} - \frac{D_{12}^2}{D_{11}}\right) & \left(D_{23} - \frac{D_{12}D_{12}}{D_{11}}\right) & 0 & 0 & 0 \\ 0 & \left(D_{23} - \frac{D_{31}D_{21}}{D_{11}}\right) & \left(D_{33} - \frac{D_{13}D_{13}}{D_{11}}\right) & 0 & 0 & 0 \\ 0 & 0 & 0 & \beta' D_{44} & 0 & 0 \\ 0 & 0 & 0 & 0 & D_{55} & 0 \\ 0 & 0 & 0 & 0 & 0 & \beta' D_{66} \end{bmatrix}$$

$$\sigma_2\text{-direction } [D]^* = \begin{bmatrix} \left(D_{11} - \frac{D_{21}}{D_{22}}\right) & 0 & \left(D_{13} - \frac{D_{12}D_{23}}{D_{22}}\right) & 0 & 0 & 0 \\ 0 & 0 & 0 & 0 & 0 & 0 \\ \left(D_{31} - \frac{D_{21}D_{32}}{D_{22}}\right) & 0 & \left(D_{33} - \frac{D_{23}^2}{D_{22}}\right) & 0 & 0 & 0 \\ 0 & 0 & 0 & \beta' D_{44} & 0 & 0 \\ 0 & 0 & 0 & 0 & \beta' D_{55} & 0 \\ 0 & 0 & 0 & 0 & 0 & D_{66} \end{bmatrix}$$

$$\sigma_3\text{-direction } [D]^* = \begin{bmatrix} \left(D_{11} - \frac{D_{13}^2}{D_{33}}\right) & \left(D_{12} - \frac{D_{13}D_{23}}{D_{23}}\right) & 0 & 0 & 0 & 0 \\ \left(D_{21} - \frac{D_{31}D_{32}}{D_{33}}\right) & \left(D_{22} - \frac{D_{23}D_{23}}{D_{33}}\right) & 0 & 0 & 0 & 0 \\ 0 & 0 & 0 & 0 & 0 & 0 \\ 0 & 0 & 0 & D_{44} & 0 & 0 \\ 0 & 0 & 0 & 0 & \beta' D_{55} & 0 \end{bmatrix}$$

then

$$D_{11}\delta\varepsilon_x^* + D_{12}\delta\varepsilon_y^* + D_{13}\delta\varepsilon_z^* = \delta\sigma_x^{\mathrm{p}*}$$

$$\delta\varepsilon_x^* = \frac{\delta\sigma_x^{\mathrm{p}*}}{D_{11}} - \frac{D_{12}}{D_{11}}\delta\varepsilon_y^* - \frac{D_{13}}{D_{11}}\delta\sigma_y^* \tag{3.53}$$

In a similar manner, examples for shear terms can be written as

$$\begin{aligned} \delta\tau_{xy}^* + \delta\tau_{xy}^{\mathrm{p}*} &= \beta' D_{44}\gamma_{xy}^* \\ \delta\tau_{yz}^* + \delta\tau_{yz}^{\mathrm{p}*} &= D_{55}\delta\gamma_{yz}^* \\ \delta\tau_{zx}^* + \delta\tau_{zx}^{\mathrm{p}*} &= \beta' D_{66}\delta\gamma_{zx}^* \end{aligned} \tag{3.54}$$

3.2.3.1 Blunt Crack Band Propagation

The smeared crack concept, rather than the isolated sharp inter-element crack concept described above, is gaining ground. Here the element topology does change. The smeared crack band of a blunt front is that in which one can easily select cracks in any direction without paying a penalty, even if the crack direction is not truly known. Bazant et al. and Bangash introduced the equivalent strength and energy variation which are utilised for crack propagation once it is initiated within the element. The equivalent strength criterion is used for crack propagation by specifying an equivalent stress within the surrounding elements of an existing crack at which cracking should be propagated. The expression for the equivalent strength σ_{eq} is given (see Fig. 3.3) as

$$\sigma_{\mathrm{eq}} = C[EG_f/W(1 - 2\,V\sigma_2^0/\sigma_1^0)]^{1/2} \tag{3.55}$$

where C = a constant dependent on the choice of elements; E = elastic modulus; V = Poisson's ratio; $W = A/\delta\alpha = A/r\ \cos\alpha$; A = area of the element at the front.

The band length is specified as $a + \Delta a/2$

In the initial state prior to cracking, the strain energy U_0 is based on the principal stresses σ_2^0, where σ_1^0 is the largest tensile stress. After crackng, of becomes 0 and $\sigma_2^1 = 0$. Which is used for the current value of U_1. The change of strain energy $\delta U = U_0 - U_1$ is equated to the crack length $\Delta a \times G_f$, where U is the total strain energy in a cracked body and δU is the energy released by the structrure into the element which cracks. The crack direction within an arbitrary grid is given by

$$\theta_{\mathrm{A}} = \frac{1}{3}(\theta_{n-1} + \theta_n + \theta_{n+1}) \tag{3.56}$$

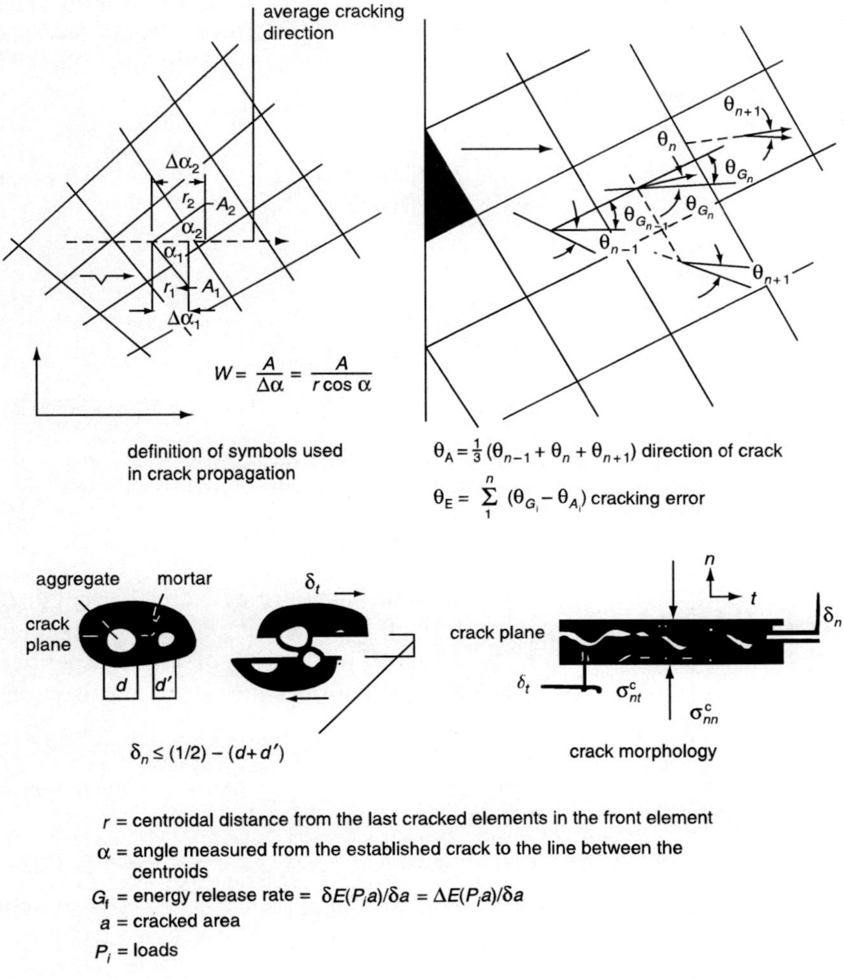

Fig. 3.3 Blunt crack propagation

where θ_{A} is the average crack direction, θ_{n-1} and θ_n are the cracking angles of the next to last cracked element and θ_{n-1} is the impending cracking angle of the element adjacent to the crack front. Since in the arbitrary grid the cracking direction is specified by the accumulated error of the cracked direction, the accumulated cracking error θ_{E} is given by

$$\theta_{\mathrm{E}} = \sum_{i}^{n} (\theta_{\mathrm{G}i} - \theta_{\mathrm{A}i}) \tag{3.57}$$

where n is the number of cracked elements and θ_G is the actual average crack propagation angle within each element. A better formulation by Gamborov and Karakoc, reported by Bangash, is given of the tangent shear modulus G^{CR} whenever cracking is initiated:

$$G^{CR} = \frac{\sigma^c_{nt}}{\varepsilon^{CR}_{nn}} K \frac{1}{r(a_3 + a_4 \mid r \mid^3)} \tag{3.58}$$

or

$$G^{CR} = \frac{\sigma_o}{\varepsilon^{CR}_{nn}} K\{1 - [2(P/D_a)\varepsilon^{CR}_{nn}]^{1/2}\}$$

where

$$K = \frac{a_3 + 4a_4 \mid \gamma \mid^3 - 3a_3 a_4 \gamma^4}{(1 + a_4\gamma^4)^2} \tag{3.59}$$

and a_3,a_4= coefficients as a function of the standard cylindrical strength f'_c; τ_0=crack shear strength (ranging from 0.25 to 0.7 f'_c); D_a = maximum aggregate size (up to 4 mm); P_c = large percentage of crack asperities; $\gamma = \delta_t/\delta_n$; δ_t, δ_n crack displacements along the normal and tangential directions (Figs. 3.3 and 3.4).

Curves have been plotted showing a decrease in the value of G^{CR} when a crack opening linearly increases at increasing shear. A constitutive law in which a confinement stress within the rough crack model is given by

$$\sigma^c_{nn} = -a_1 a_2 \frac{\delta_t \sigma^c_{nt}}{(\sigma^2_n + \sigma^2_t)^q} \tag{3.60}$$

where σ^c_{nn}= interface normal stress; σ^c_{nt}= interface shear stress; a_1, a_2= constant (a_1, a_2 = 0.62); q = a function of the crack opening; taken to be 0.25.

For different types of crack dilatancy δ_n/σ^c_{nt} (Fig. 3.3) the tangent shear modulus G^{CR} is plotted against the ratio r of the crack displacement. The value of σ^c_{nt} is given by

$$\sigma^c_{nt} = \tau_o\left[1 - \sqrt{\left(\frac{2P}{D_a}\in^{CR}_{nn}\right)}\right] r \frac{a_3 + a_4 \mid r \mid^3}{1 + a_4\gamma^4}, \quad r = \gamma^{CR}_{nt}/\varepsilon^{CR}_{nn} \tag{3.61}$$

where p is the crack spacing and CR is cracked concrete

$$\varepsilon^{CR}_{nn} = \delta_n/P = \text{strain against } \sigma^c_{nn}$$

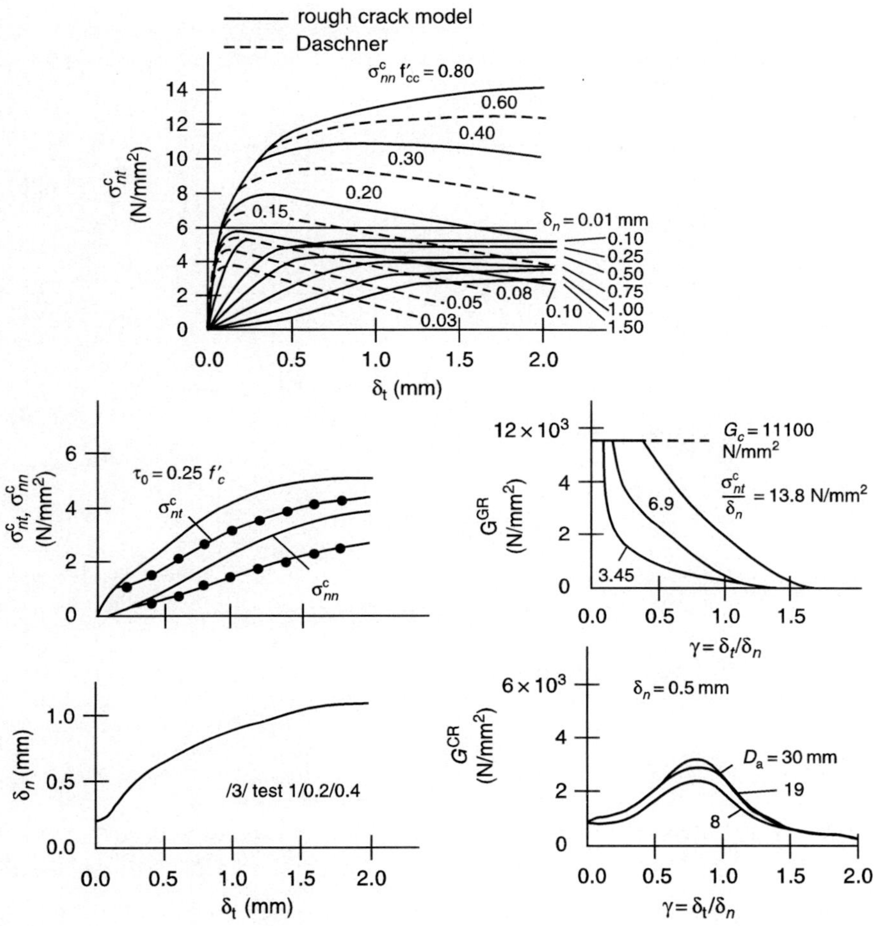

Fig. 3.4 Crack displacement versus tangent shear modulus[1.149]

The σ^{c}_{nn} values have been computed using points common to the curves of crack opening and constant confinement stress.

3.2.3.2 Ottoson Failure Model

The Ottoson four-parameter model has a smooth but convex surface with curved meridians determined by the constants a and b.

The analytical failure surface is defined by

$$f(I_1, J_2, J) = a\frac{J_2}{(f'_c)^2} + \lambda\frac{\sqrt{2}}{f'_c} + b\frac{l}{f'_c} - 1 = 0 \quad (3.62)$$

where

$$I_1 = \sigma_x + \sigma_y + \sigma_z = \text{the first invariant of the stress deviator tensor} \tag{3.62a}$$

$$\begin{aligned} J_2 &= \text{the second invariant of stress deviator tensor} \\ &= \frac{1}{2}\left(S_x^2 + S_y^2 + S_z^2\right) + \tau_{xy}^2 + \tau_{yz}^2 + \tau_{zx}^2 \end{aligned} \tag{3.62b}$$

$$J = \cos 3\theta = 1.5\sqrt{3}(J_3/J_2) \tag{3.62c}$$

$$\begin{aligned} J_3 &= \text{the third invariant of the stress deviator tensor} \\ &= S_x S_y S_z + 2\tau_{xy}\tau_{yz}\tau_{zx} - S_x\tau_{yz}^2 - S_y\tau_{zx}^2 - S_z\tau_{xy}^2 \end{aligned} \tag{3.62d}$$

$$\begin{aligned} S_x &= \sigma_x - I_1/3 \\ S_y &= \sigma_y - I_1/3 \\ S_z &= \sigma_z - I_1/3 \end{aligned} \tag{3.62e}$$

$\lambda = \lambda(\cos 3\theta) > 0$, a and b are constant

$$\lambda = K_1\left(\frac{1}{3}\cos^{-1}(K_2\cos 3\theta)\right) \quad \text{for } (\cos 3\theta)\ \leq 0$$

$$\lambda = K_1\cos\left(\frac{\pi}{3} - \frac{1}{3}\cos^{-1}(K_2\cos 3\theta)\right) \quad \text{for}(\cos 3\theta) \leq 0$$

k_1, k_2 a and b are material parameters $(0 \leq K_2 \leq 1)$
f_c' = uniaxial compressive cylinder strength for concrete $= 0.87\sigma_{cu}$
σ_t = uniaxial tensile strength for concrete

Table 3.12 lists some of the relevant parameters.

The knowledge of the mechanical properties of concrete and the reinforcement (conventional and prestressing steel) at high strain rates is essential for rational application of materials in those constructions where impact and explosion loadings can be expected. The usual magnitude of the strain rate $(d\varepsilon/dt) = \dot{\varepsilon}$ for all concrete structures is of the order of 5×10^5/s in the range of the ultimate load.

For reinforcement, the range is between 10^5 and 10 ^{2}/s. Table 3.13 gives relevant data. Figure 3.5 gives experimental stress–strain relationships for reinforcement for various strain rates. The theoretical expression in Eq. (3.46) was used.

Table 3.12 Ottoson's failure model for concrete

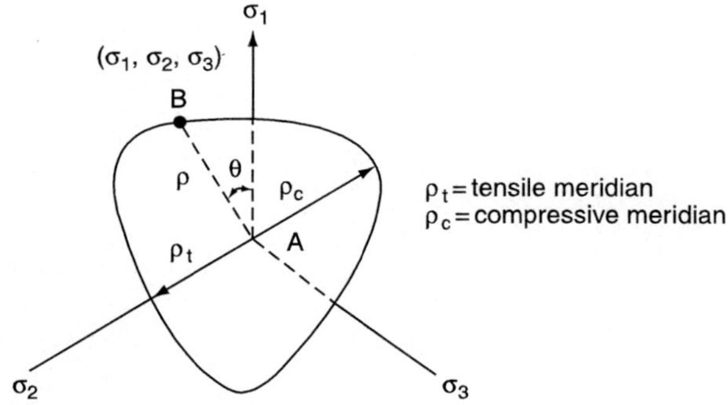

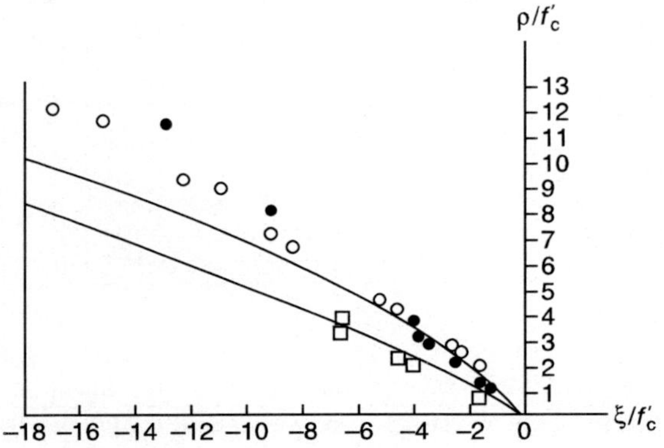

Comparison for the four-parameter model

Four-material Parameters($k = \sigma_t/\sigma_c$)

K	*a*	*b*	k_1	k_2
0.08	1.8076	4.0962	14.4863	0.9914
0.10	1.2759	3.1962	11.7365	0.9801
0.12	0.9218	2.5969	9.9110	0.9647

Values of the function($k = \sigma_t/\sigma_c$)

K	λ_t	λ_c	$\lambda_c\lambda_t$
0.08	14.4925	7.7834	07378
0.1	11.7109	6.5315	0.5577
0.12	9.8720	5.6979	0.5772

Table 3.12 (continued)

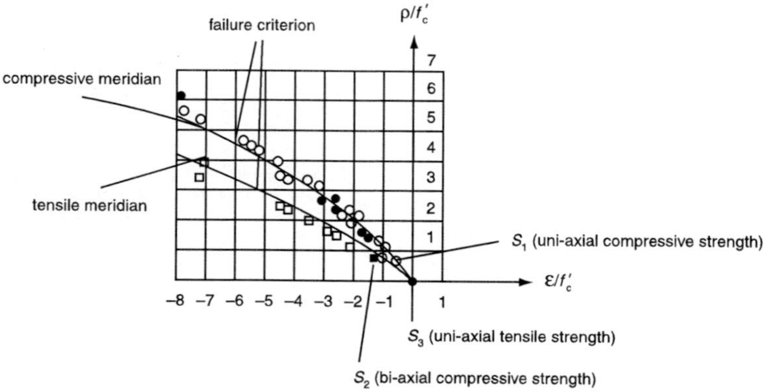

Determination of material parameters (S_1, S_2, S_3, S_4 = failure stresses)
The following three failure states were represented:
(1) Uniaxial compressive strength, $f_c'(\theta = 60°)$, Uniaxial tensile strength, $\sigma_t\ (\theta = 0°) = Kf_c'$
(2) Biaxial compressive strength, $\sigma_1 = \sigma_2 = -1.16\sigma_{ci}$; $\sigma_3 = 0\ (\theta = 0°)$
(3) The triaxial state $(\xi/f_c', \rho/f_c') = (-5, 4)$ on the compressive meridian $(\theta = 60°)$

Table 3.13 Strain rate for concrete and reinforcement

The relationship between the fracture strain ε_f and the strain rate ε is given by

$$\varepsilon_t = \dot{\alpha}^{1/3} \tag{1}$$

for plain concrete $\alpha = 206$; for reinforced concrete $\alpha = 220$.
Another expression given by DELFT for plain concrete is

$$\varepsilon_t = 100 + 109\dot{\varepsilon}^{1/2} \tag{2}$$

In Eq. (1), for say $\dot{\varepsilon}/s = 35$, the value of ε_f for reinforced concrete will be 740×10^{-6}.
For $\dot{\varepsilon}/s = 30$, the value of ε_f for plain concrete will be 650×10^{-6}.
For fibre-reinforced concrete, the influence of the strain rate upon the tensile strength for concrete is given by

$$\sigma_t = \alpha + \beta \log_e \dot{\varepsilon} \tag{3}$$

$\alpha = 0$ for no fibres, i.e. plain concrete
$\sigma = 1.7 + 0.0364 \log_e \dot{\varepsilon}$ for 3% fibres
$\sigma = 1.87 + 0.0424 \log_e \dot{\varepsilon}$
For low, intermediate and high strain rates, DELEFT gives an expression:

$$\sigma_t = \alpha + \beta N \tag{4}$$

where N is the number of fibres/reinforcements

	Low	Intermediate	High
α	3.32	4.87	5.49
β	1.85×10^{-3}	2.85×10^{-3}	6.3×10^{-3}

Table 3.13 (continued)

For the fracture energy, G_t as stated in the endochronic theory will be modified as follow:

$$G_t = \alpha + \beta N$$

	Low	Intermediate	High
α	12.72	22.90	29.200
β	0.12	0.18	0.211

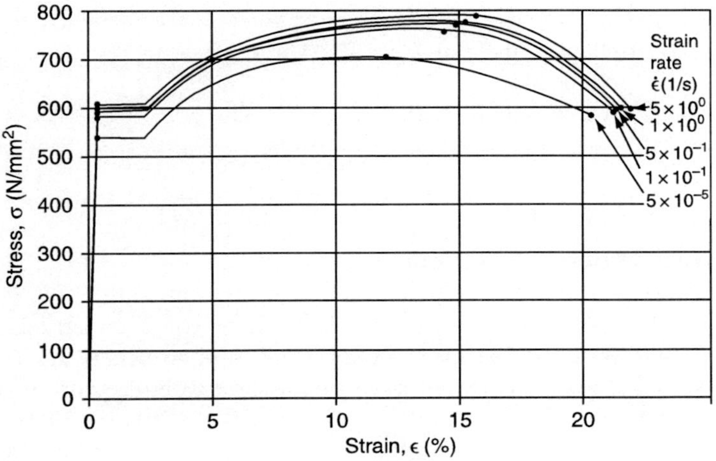

Fig. 3.5 Stress–strain relationships for various strain rates

3.3 Ice/Snow Impact

Chapter 2 gives a thorough survey on data regarding the effect of ice floes. When floating ice sheets move under the influence of strong winds and currents, a seagoing vehicle or a semi-submersible will be subject to an impact given by

$$F_1(t) = F_{1o}(t) + F_1'(t) \tag{3.63}$$

where $F_{1o}(t)$ and $F_1'(t)$ $F_1'(t)$ are constants and fluctuating values of the ice impact force, respectively. The value of $F_1'(t)$ is given by

$$F_{IO}(t) = S_1 S_2 S_3 f(\dot{\varepsilon}) TW \times h\sigma_c \text{ (or } f_c'\text{)} \tag{3.64}$$

where S_1 = contact factor around the member during crushing; S_2 = shape factor S_f of the impactor; S_3 = temperature factor which is $(1 - 0.012T)/(1 - 0.012T_s)$ (T: 0.5°C> − 20°C) ($T_s = -10$°C; σ_c compressive strength of ice measured at a strain rate of 5 × 10^4/s which is $\dot{\varepsilon}_o$; W = transverse width of the member; h = ice sheet thickness

$$f(\dot{\varepsilon}_o) = (\dot{\varepsilon}/\dot{\varepsilon}_o)\gamma; \qquad \phi = \alpha_1 + \alpha_2(h/w)^{1/2} \tag{3.64a}$$

$\dot{\varepsilon} = \dot{x}/4\,W$ where$\dot{x}$ = ice flow velocity; γ = empirical coefficient dependent on strain rate; a_1,a_2 = factors dependent on the ice thickness/diameter ratio of a member.

Depending on the type of impact (direct or angular) and the stiffnesses of members and ice floes, the global equation of motion, (3.13), will be influenced by roll, pitch and yaw motions (θ_s, Φ_s, Ψ_h) and surge, sway and heave motions (θ_r, Φ_p, Ψ_y), respectively. Generally, the values of θ_r, Φ_p, $\Psi_y \theta_s$, Φ_s and Ψ_h range as shown below:

$\theta_r = -0.012$ to 0.04×10^{-3} rad; $\theta_s = -0.75$ to 6 m
$\Phi_p = -0.012$ to 0.04×10^{-3} rad; $\Psi_s = -1.5$ to 3 m
$\Psi_y = -0.012$ to 0.04 rad; $\Psi_n \approx \theta_s$

3.4 Impact Due to Missiles, Impactors and Explosions: Contact Problem Solutions

Contact problems have been solved using Hallquis-contact method. Many similar methods are using the spring concept (given in Appendix A). In addition, at the time of impact constraints are imposed on global equations such as Eqs. (3.13), (3.14), (3.15), (3.16), (3.17), (3.18), (3.19), (3.20), and (3.21). Hallquist et al. developed a useful concept of *master*and *slave nodes* sliding on each other. As shown in Fig. 3.6, slave nodes are constrained to slide on master segments after impact occurs and must remain on a master segment until a tensile interface force develops. The zone in which a slave segment exists is called a *slave zone*. A separation between the slave and the master line is known as *void*. The following basic principles apply at the interface:

(1) Update the location of each slave node by finding its closest master node or the one on which it lies.
(2) For each master segment, find out the first slave zone that overlaps.
(3) Show the existence of the tensile interface force.

Constraints are imposed on global equations by a transformation of the nodal displacement components of the slave nodes along the contact interface. Such a transformation of the displacement components of the slave nodes will eliminate their normal degrees of freedom and distribute their normal force components to the nearby master nodes. This is done using explicit time integration, as described later under solution procedures. Thereafter impact and release conditions are imposed. The slave and master nodes are shown in Fig. 3.6. Haliquist et al.[3.40] gave a useful demonstration of the identification of the contact point, which is the point on the master segment to the slave node n_s and which finally becomes non-trivial. As shown in Fig. 3.6, when the master segment t is given the parametric representation and $\hat{t}$ is the position vector

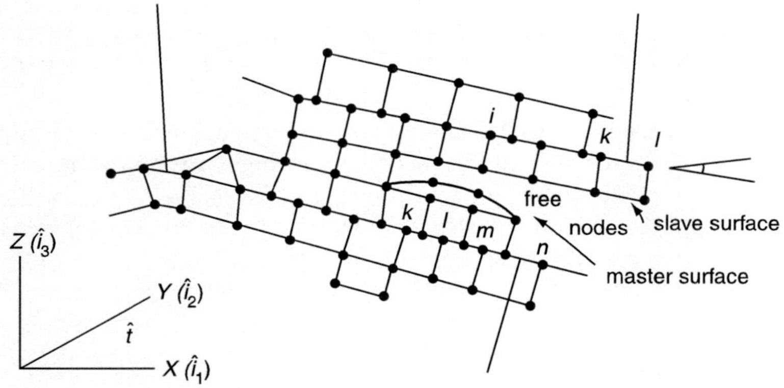

Fig. 3.6 Hallquist contact method (modified by Bangash)

drawn to the slave node n_s, the contact point coordinate must satisfy the following equations:

$$\frac{\partial \hat{r}}{\partial \xi}(\xi_c, \eta_c) \times [\hat{t} - \hat{r}(\xi_c, \eta_c)] = 0$$

$$\frac{\partial \hat{r}}{\partial \eta}(\xi_c, \eta_c) \times [\hat{t} - \hat{r}(\xi_c, \eta_c)] = 0 \tag{3.65}$$

where (ζ_c, η_c) are the coordinates on the master surface segment S_i. When penetration through the master segment S_i occurs the slave node n_s (containing its contact point) can be identified using the interforce vector f_s added, then

$$f_s = -lK_i n_i \quad \text{if } l<0 \tag{3.66}$$

to the degrees of freedom corresponding to n_s and

$$f_m^i = N_i(\xi_c, \eta_c) f_s \quad \text{if } l<0 \tag{3.67}$$

where

$$l = \hat{n}_i.[\hat{t} - \hat{r}(\xi_c, \eta_c)] < 0 \tag{3.67a}$$

A unit normal

$$\hat{n}_i = \hat{n}_i(\xi_c, \eta_c); \quad \hat{t}_i = \hat{n}_i \sum_{j=1}^{n} N_J (F_1)^j(t) \tag{3.67b}$$

$$K_i = f_{si} K_i A^2{}_i / V_i \tag{3.67c}$$

where $(F_1)^j(t)$ = impact at the jth node; k_i = stiffness factor for mass segment S_i; K_i, V_i, A_i = bulk modulus, volume and face area, respectively; F_{si} = scale factor normally defaulted to 0.10; $N_i = 1/4(1 + \xi\ \xi_i)(1 + \eta\ \eta_c)$ for a 4-node linear surface.

Bangash extended this useful analysis for others such as 8-noded and 12-noded elements. On the basis of this theory and owing to the non-availability of the original computer source, a new sub-program CONTACT was written in association with the program ISOPAR. CONTACT is in three dimensions; the values of N_i for 8- and 12-noded elements are given in Table 3.14.

3.5 High Explosions

The pressure P is generally defined as a function of relative volume and internal energy. Assuming $F_1(t)$ is the final surface load, the pressure P must replace $F_1(t)$ in relevant equations by taking into consideration the surface volume on which it acts. All equations defining relevant detonation pressures P must first be evaluated. They are then first applied as stated, in Eqs. (3.13), (3.14), (3.15), (3.16), (3.17), (3.18), (3.19), (3.20), and (3.21).

The cause of explosions can be nuclear (air burst or underground). The pressures, which are time dependent, can then act as surface loads on the body of the element concerned or at nodal points of the element as concentrated loads derived on the basis of shape functions. It is essential to choose a proper time-aspect ratio as it will affect the type of solution procedure adopted. The interaction between the loads and the structure can be considered and the method shown in Section 3.5 must be included.

3.6 Spectrum Analysis

Spectrum analysis is an extension of the mode frequency analysis, with both base and force excitation options. The response spectrum table is generally used and includes displacements, velocities and accelerations. The force excitation is, in general, used for explosions and missile aircraft impact. The masses are assumed to be close to the reaction points on the finite element mesh rather than the master degrees of freedom. The base and forced excitations are given below. For the base excitation for wave

$$\gamma_i = \{\psi_i\}_R^{T''}[M]\{b\} \tag{3.68}$$

For the impact excitation

$$\gamma_i = \{\psi_i\}_R^{T''}[M]\{F_1(t)\} \tag{3.69}$$

Table 3.14 N_i for 8- and 12-noded elements

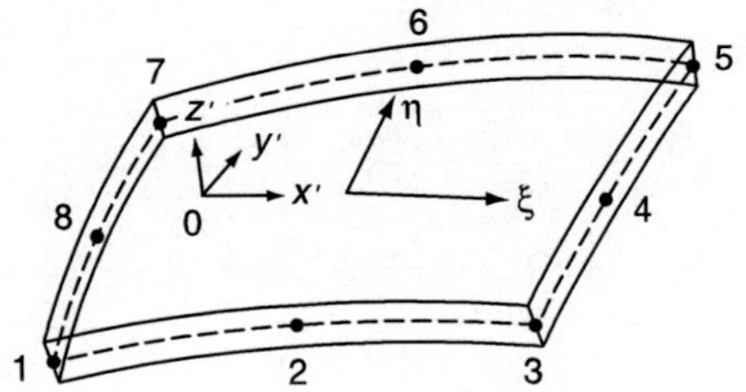

Eight-noded membrane element

Node i	Shape functions $N_i(\xi,\eta)$	Derivatives	
		$\frac{\partial N_i}{\partial \xi}$	$\frac{\partial N_i}{\partial \xi}$
1	$\frac{1}{4}(1-\xi)(1-\eta)(-\xi-\eta-1)$	$\frac{1}{4}(1-\eta)(2\xi+\eta)$	$\frac{1}{4}(1-\xi)(2\eta+\xi)$
2	$\frac{1}{2}(1-\xi^2)(1-\eta)$	$-\xi(1-\eta)$	$-\frac{1}{2}(1-\xi^2)$
3	$\frac{1}{4}(1+\xi)(1-\eta)(\xi-\eta-1)$	$\frac{1}{4}(1-\eta)(2\xi-\eta)$	$\frac{1}{4}(1+\xi)(2\eta-\xi)$
4	$\frac{1}{2}(1-\eta^2)(1+\xi)$	$\frac{1}{2}(1-\eta^2)$	$-\eta(1+\xi)$
5	$\frac{1}{4}(1+\xi)(1+\eta)(\xi+\eta-1)$	$\frac{1}{4}(1+\eta)(2\xi+\eta)$	$\frac{1}{4}(1+\xi)(2\eta+\xi)$
6	$\frac{1}{2}(1-\xi)(1+\eta^2)$	$-\xi(1+\eta)$	$\frac{1}{2}(1-\xi^2)$
7	$\frac{1}{4}(1-\xi)(1+\eta)(-\xi+\eta-1)$	$\frac{1}{4}(1+\eta)(2\xi-\eta)$	$\frac{1}{4}(1-\xi)(2\eta-\xi)$
8	$\frac{1}{2}(1-\eta^2)(1-\xi)$	$-\frac{1}{2}(1-\eta^2)$	$-\eta(1-\xi)$

Table 3.14 (continued)

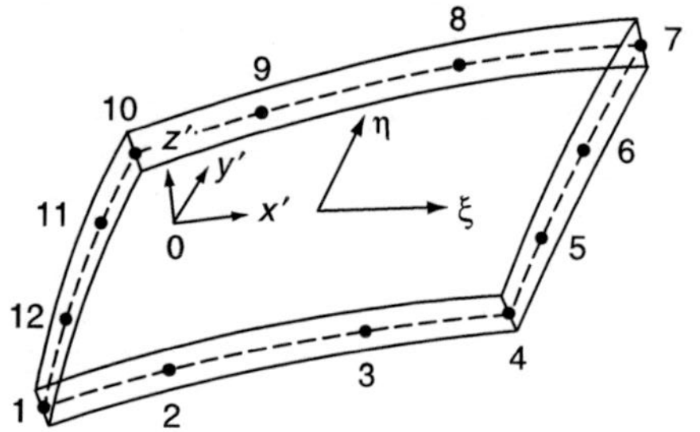

Twelve-noded membrane element

Node i	Shape function $N_i(\xi,\eta)$	Derivatives	
		$\frac{\partial N_i}{\partial \xi}$	$\frac{\partial N_i}{\partial \eta}$
1	$932(1-\xi)(1-\eta)[\xi^2+\eta^2-\frac{10}{9}]$	$\frac{9}{32}(1-\eta)[2\xi-3\xi^2-\eta^2+\frac{10}{9}]$	$932(1-\xi)[2\eta-3\eta^2-\xi^2+\frac{10}{9}]$
2	$932(1-\xi)(1-\xi^2)(1-\eta)$	$932(1-\eta)(3\xi^2-2\xi-1)$	$-932(1-\xi)(1-\xi^2)$
3	$\frac{9}{32}(1-\eta)(1-\xi^2)(1+\xi)$	$\frac{9}{32}(1-\eta)(1-2\xi-3\xi^2)$	$-\frac{9}{32}(1-\xi^2)(1+\xi)$
4	$\frac{9}{32}(1+\xi)(1-\eta)[\xi^2+\eta^2-\frac{10}{9}]$	$\frac{9}{32}(1-\eta)[2\xi+3\xi^2+\eta^2-\frac{10}{9}]$	$\frac{9}{32}(1+\xi)[2\eta-3\eta^2-\xi^2+\frac{10}{9}]$
5	$\frac{9}{32}(1+\xi)(1-\eta^2)(1-\eta)$	$\frac{9}{32}(1-\eta^2(1-\eta)$	$\frac{9}{32}(1+\xi)(3\xi^2-2\eta-1)$
6	$\frac{9}{32}(1+\xi)(1-\eta^2)(1+\eta)$	$\frac{9}{32}(1-\eta^2(1+\eta)$	$\frac{9}{32}(1+\xi)(1-2\eta-3\xi^2)$
7	$\frac{9}{32}(1+\xi)(1+\eta)[\xi^2+\eta^2-\frac{10}{9}]$	$\frac{9}{32}(1+\eta)[2\xi+3\xi^2+\eta^2-\frac{10}{9}]$	$\frac{9}{32}(1+\xi)[2\eta+3\eta^2+\xi^2-\frac{10}{9}]$
8	$\frac{9}{32}(1+\eta)(1-\xi^2)(1+\xi)$	$\frac{9}{32}(1+\eta)(1-2\xi-3\xi^2)$	$\frac{9}{32}(1+\xi^2)(1+\xi)$
9	$\frac{9}{32}(1+\eta)(1-\xi^2)(1-\xi)$	$\frac{9}{32}(1+\eta)(3\xi^2-3\xi-1)$	$\frac{9}{32}(1-\xi^2)(1-\xi)$
10	$\frac{9}{32}(1-\xi)(1+\eta)[\xi^2+\eta^2-\frac{10}{9}]$	$\frac{9}{32}(1+\eta)[2\xi-3\xi^2-\eta^2+\frac{10}{9}]$	$\frac{9}{32}(1-\xi)[2\eta+3\eta^2+\xi^2-\frac{10}{9}]$
11	$\frac{9}{32}(1-\xi)(1-\eta^2)(1+\eta)$	$\frac{9}{32}(1+\eta)(1-\eta^2)$	$\frac{9}{32}(1-\xi)(1-2\eta-3\eta^2)$
12	$\frac{9}{32}(1-\xi)(1-\eta^2)(1-\eta)$	$-\frac{9}{32}(1-\eta)(1-\eta^2)$	$\frac{9}{32}(1-\xi)(3\eta^2-2\eta-1)$

where $\{\psi_i\}_R$ = the slave degree-of-freedom vector mode; M = mass; {b} = unit vector of the excitation direction; $\{F_1(t)\}$ = an input force vector due to impact and explosion.

The values of $\{\psi_i\}_R$ = are normalised and the reduced displacement is calculated from the eigenvector by using a mode coefficient $\{M\}$

$$\{x_i\} = [M_i]\{\psi\}_i \tag{3.70}$$

where $\{x\}_i$ = reduced displacement vector and $[M_i]$ = mode coefficient and where

(a) for velocity spectra

$$[M_i] = [\dot{x}_{si}]\{\gamma_i\}/\omega_i \tag{3.71}$$

where $\dot{x}_{si}$ = spectral velocity for the ith mode;

(b) for force spectra

$$[M_i] = [Fsi]\{\gamma\}/\omega^2 t \tag{3.72}$$

where F_{si} = spectral force for the ith mode;

(c) caused by explosion P or impact $F_1(t)$

$$[M_i] = [\ddot{x}_{si}]\{\gamma_i\}/\omega_i^2 \tag{3.73}$$

where $\ddot{x}_{si}$ = spectral acceleration for the ith mode;

(d)

$$[M_i] = [x_{si}]\{\gamma\}/\omega_i^2 \tag{3.74}$$

$\{x_i\}$ may be expanded to compute all the displacements, as in

$$\{x_{\hat{\gamma}}\}_i = [K_{\gamma\prime\gamma\prime}]^{-1}[K_{\gamma\prime\gamma\prime}]\{x_i\}R \tag{3.75}$$

where $\{x'_\gamma\}_i$ = slave degree of freedom vector of mode i; $[K_{\gamma'\gamma'}][K_{\gamma'\gamma}]$= sub-matrix parts; γ, γ' = retained and removed degrees of freedom.

The impact/explosion load is then equal to

$$\left[[K_{\gamma\,\gamma}] - [K_{\gamma'\,\gamma}][K_{\gamma\,\gamma'}]^{-1}[K_{\gamma'\,\gamma}]\right]\{x_\gamma\} = \left[\{F_\gamma\} - [K_{\gamma\,\gamma\prime}][K_{\gamma\,\gamma\prime}]^{-1}\{F_{\gamma\prime}\}\right]$$

or

$$[\overline{K}][\overline{\ddot{x}}] = \{\overline{F}_1(t)\} \tag{3.76}$$

where

$$[\overline{K}] = [K_{\gamma\gamma}] - [K_{\gamma\gamma'}]\left[K_{\gamma'\gamma'}\right]^{-1}[K_{\gamma'\gamma}] \tag{3.77}$$

$$\{\overline{F}_1(t)\} = \{F_\gamma\} - [K_{\gamma\gamma'}][K_{\gamma'\gamma'}]^{-1}\{F_{\gamma'}\} \tag{3.78}$$

$$\{\overline{x}\} = \{x_\gamma\} \tag{3.79}$$

and $[K]$ and $\{F_1(t)\}$ are generally known as the substructure stiffness matrix and the impact load vector, respectively.

3.7 Solution Procedures

Three types of solution procedures are available for impact and explosion analysis, namely time-domain, frequency-domain and modal analysis.

3.7.1 Time-Domain Analysis

The following steps are adopted using a direct implicit integration procedure

3.7.1.1 Initialisation

(1) The effective stiffness matrix is

$$[K_o^*] = \left(\frac{6}{\tau^2}\right)[M] + \left(\frac{3}{\tau}\right)[C_o] + [K_o] \tag{3.80}$$

(2) Triangularise $[K_o^*]$

For each time step, calculate the displacement $\{x_{t+\tau}\}$

- Constant part of the effective load vector

$$\{R_\tau^* + \tau\} = \{R_\tau\} + \theta(\{R_{t+\delta t}\} - \{R_t\}) + \{F_t\} + [M] \times \left(\frac{6}{\tau^2}\right)\{x_t\} + \left(\frac{6}{\tau}\right)\{\dot{x}_t\}$$
$$+ 2\{\ddot{x}_t\} + [C_o]\left(\left(\frac{3}{\tau}\right)\{x_t\} + 2\{\dot{x}_t\} + \left(\frac{\tau}{2}\right)\{x_t\}\right) \tag{3.81}$$

- Initialisation, $i = 0$, $\{\delta F^i_{t\to t+t}\} = 0$
- Iteration

(a) $i \to i+1$
(b) Effective load vector $\{R^*_{t+\tau\mathrm{TOT}}\} = \{R^*_{t+\tau}\} + \{\delta F^i_{t\to t+\tau}\}$
(c) Displacement $\{x^i_{t+\tau}\}[K_0^*]\{x^i_{t+\tau}\} = R^{*i}_{t\to+\tau\mathrm{TOT}}$

(d) Velocity $\{\dot{x}^i_{t+\tau}\} + (\frac{3}{\tau})(\{x^i_{t+\tau}\}) - \{x_t\} - [\delta k^i_{t\to t+\delta t}] - 2\{\dot{x}_t\} + (\frac{\tau}{2}\{\ddot{x}_t\})$

(e) Change of initial load vector caused by non-linear behaviour of the material =

$$\begin{aligned}\{\delta F^i_{t\to t+\tau}\} = &- [\delta C_{o\to t}]\{\dot{x}^i_{t+\tau}\} - \{\dot{x}_t\}\omega - [\delta C^i_{t\to t+\tau}]\{\dot{x}^i_{t+\tau}\} \\ &\times [\delta K_{o\to t}](\{x^i_{t+\tau}\} - \{X_t\}) - [\delta K^i_{t\to t+\delta t}]\{X^i_{t+\tau}\}\end{aligned} \tag{3.82}$$

In fact, $\{\delta F^i_{t\to t+\tau}\}$is calculated using the initial-stress method.

(f) Iteration convergence

$$||\{\delta F^i_{t\to t+\tau}\} - \{\delta F^{i-1}_{t\to t+\tau}\}||/\{\delta F^i_{t\to t+\tau}\} < \text{tol} = 0.01 \tag{3.83}$$

or, analogously, on stresses.

Calculation of Velocity and Acceleration

Calculate the new acceleration $\{\ddot{x}_{t+\delta t}\}$, velocity $\{\dot{x}_{t+\delta t}\}$, displacement $\{x_{t+\delta t}\}$ and initial load $\{F_{t+\delta t}\}$

$$\{\ddot{x}_{t+\delta_t}\} = \left(\frac{6}{\theta\tau^2}\right)(\{x_{t+\tau}\} - \{x_t\}) - \left(\frac{6}{\theta\tau}\right)\{\dot{x}_t\} + \left(1 - \left(\frac{3}{\theta}\right)\right) \tag{3.84}$$

$$\{\ddot{x}_t\}\{\dot{x}_{t+\delta t}\} = \{\dot{x}_t\} + \left(\frac{\tau}{2\theta}\right)\{\ddot{x}_t\} + \{\ddot{x}_{t+\delta t}\} \tag{3.85}$$

$$\{x_{t+\delta t}\} = \{x_t\} + \left(\frac{\tau}{\theta}\right)\{\dot{x}_t\} + \left(\frac{\tau^2}{6\theta_2}\right)(2\{\ddot{x}_t\} + \{\ddot{x}_{t+\delta t}\}) \tag{3.86}$$

$$\{F_{t+\delta t}\} = \{F_t\} + \{\delta F^i_{t\to t+\tau}\} \tag{3.87}$$

Calculation by Quadratic Integration

When the velocity varies linearly and the acceleration is constant across the time interval, appropriate substitutions are made into Eq. (3.13), giving

$$\begin{aligned}[f_1[M] + f_2[C_t] + [K'_t]]\{x_t\} = &\{F_1(t)\} + [M](f_3x_{t-1}) - f_4(x_{t-2}) + f_5(x_{t-3}) \\ &+ [C](f_6\{x_{t-1}\} - f_7\{x_{t-2}\} + f_8\{x_{t-3}\})\end{aligned} \tag{3.88}$$

where f_1, f_2 are functions of time. This results in an implicit time integration procedure. The only unknown is $\{x_t\}$ at each time point and this is calculated in the same way as in static analysis. Equation (3.88) is then written as

$$\left(\frac{2}{\delta t_0 \delta t_0}[M] + \frac{\delta t_0 + \delta t_1}{\delta t_0 \delta t_0}[C] + [K_t']\right)\{x_t\}$$
$$= \{F_1(t)\} + [M]\left(\frac{2}{\delta t_0 \delta t_1}\{x_{t-1}\} - \frac{2}{\delta t_1 \delta t_0}\{x_{t-2}\}\right) \quad (3.89)$$
$$+ [C_t]\left(\frac{\delta t_0}{\delta t_0 \delta t 1}\{x_{t-1}\} - \frac{\delta t_0}{\delta t_0 \delta t_1}\{x_{t-2}\}\right)$$

where

$\delta t_0 = t_0 - t_1$ and t_0 = time of current iteration
$\delta t_1 = t_1 - t_2$ and t_1 time of previous iteration
$\delta t_2 = t_2 - t_3$ and t_2 time before previous iteration
$\delta t_2 = \delta t_0 - \delta t_1$ and $t_0 - t_2$ and t_3 = time before t_2

Calculation by Cubic Integration

Equation (3.88) becomes cubic and hence is written as

$$(f_1[M] + f_2[C_t] + [K_t']\{x_t\})$$
$$= \{F_1(t)\} + [M](f_3\{x_{t-1}\} - f_4\{x_{t-2}\} + f_5\{x_{t-3}\}) \quad (3.90)$$
$$+ [C](f_6\{x_{t-1}\} - f_7\{x_{t-2}\} + f_8\{x_{t-s}\})$$

where f_1 to f_8 are functions of the time increments; these functions are derived by inverting a 4 × 4 matrix.

For clear-cut solutions, the size of the time step between adjacent iterations should not be more than a factor of 10 in non-linear cases and should not be reduced by more than a factor of 2 where plasticity exists.

3.7.2 Frequency-Domain Analysis

The original equation of motion is reproduced as

$$[M]\{\ddot{x}\} + [C]\{\dot{x}\} + [K]\{x\} = \{J_F\}\{F\} \quad (3.91)$$

where $\{J_{\mathrm{F}}\}$ is a vector with all components zero except the last one, which is 1.

The terms $[K]$ and $[C]$ shall be frequency dependent. The value of $\{F\} = [K_{\mathrm{N}}]\{x_{\mathrm{s}}\}$ can be taken for solutions of rigid rock problems. If the excitation with frequency w assumes the form $\mathrm{e}^{\mathrm{i}\omega t}$ then

$$\dot{x} = i\omega x_{s'}; \quad \ddot{x}_s = -\omega^2 x_{s'}; \ \{x\} = i\omega\{x\}$$
$$\text{and } \{x\} = -\omega^2\{x\} \quad (3.92)$$

Equation (3.91) can thus be written as

$$([K] + i\omega[C] - \omega^2[M])\{x\} = \{J_F\}K_n x_s \tag{3.93}$$

For a given value of w, a set of algebraic equations is solved using any numerical scheme. The displacement of a mass can be written as

$$\{x\} = ([K] + i\omega[C] - \omega^2[M])^1\{JF\}K_n x_s \tag{3.94}$$

From displacements, accelerations, velocities, strains and stresses can be computed. The amplification function (AF) for each frequency x_1/x_s may be derived. Repeated solutions of Eq. (3.92) are necessary for a proper definition of this function. If the fast Fourier transform is used then AF must be tabulated at each frequency interval.

3.7.3 Runge–Kutta Method

It is an accurate method of time integration and is explicit in nature. Table 3.15 summarises this method. This method of higher order is a robust algorithm used to solve non-linear equations but may have problems which have discontinuous coefficients which take place spatially. Some coefficients change discontinuously as the load increases. Strain localisation will be difficult to produce in numerical simulations. Non-linear equations have bifurcation. The non-linear equation of stochastic elasto-plasticity is more suitable for finding the most unstable solution compared with the non-linear equation of deterministic elasto-plasty since the coefficients change continuously.

Table 3.15 The Runge–Kutta method

$x^{n+1} = x^n + \frac{\delta t}{6}\left(f^{(1)} + 2f^{(2)} + 2f^{(3)} + f^{(4)}\right)$	(3.95) (a)
where $f^{(1)} = f(x^n)$, $f^{(2)} = f(x^n) + \frac{\delta t}{2}f^{(1)}$	
$f^{(3)} = f(x^n) + \frac{\delta t}{2}f^{(2)}$	
$f^{(4)} = f(x^n) + \delta t f^{(3)}$	
Some computations are needed to calculate $f^{(1)}$ to $f^{(4)}$	
$f = kx$	(3.95) (b)

3.7.4 Keierleber Method

The author developed implicit with θ-operators covering linear, quadratic and cubic with θ-operators developed by Keierleber C.W. University of Nebraska (2003). The author found the Keierleber C.W. approach of these solution procedures extremely flexible and easily adaptable with both programs ISOPAR and F. BANG-2.

(a) Linear Implicit Method

$$\begin{Bmatrix} \ddot{x}_i + 1 \\ \dot{x}_i + 1 \\ x_i + 1 \end{Bmatrix} = [A] \begin{Bmatrix} \ddot{x}_i \\ \dot{x}_i \\ x_i \end{Bmatrix} + \{L\}(r_i\theta) \tag{3.96}$$

where

$$r_{i+\theta} = \frac{F_{i+\theta}}{m}, \{L\} = \begin{Bmatrix} \beta \\ \frac{\Delta t}{2}\beta \\ \frac{\Delta t^2}{6}\beta \end{Bmatrix} \tag{3.97}$$

$$[A] = \begin{bmatrix} \gamma\beta & \frac{1}{\Delta t}(\phi\beta) & \frac{1}{\Delta t^2}(-\omega^2\Delta t^2\beta) \\ \frac{\Delta t}{2}(1+\gamma\beta) & \frac{1}{2}(2+\phi\beta) & \frac{1}{2\Delta t}(-\omega^2\Delta t^2\beta) \\ \frac{\Delta t^2}{6}(2+\gamma\beta) & \frac{\Delta t}{6}(6+\phi\beta) & \frac{1}{6}(6-\omega^2\Delta t^2\beta) \end{bmatrix} \tag{3.98}$$

$$\beta = \frac{6}{6\theta + 6\xi\omega\theta^2\Delta t + \omega^2\theta^3\Delta t^2}$$

$$\gamma = \frac{6 + 6\theta - 12\xi\omega\theta\Delta t + 6\xi\omega\theta^2\Delta t - 3\omega^3\theta^2\Delta t^2 + \omega^2\theta^3\Delta t^2}{6}$$

$$\phi = \omega^2\Delta r^2\theta + 2\xi\omega\Delta t$$

$$\beta = \frac{6}{6\theta + 6\xi\omega\theta^2\Delta t + \omega^2\theta^3\Delta t^2}$$

(b) Quadratic Implicit with θ Method Operators

$$\begin{Bmatrix} \ddot{x}_{i+1} \\ \dot{x}_{i+1} \\ x_{i+1} \end{Bmatrix} = [A] \begin{Bmatrix} \ddot{x}_i \\ \dot{x}_i \\ x_i \end{Bmatrix} + \{L\}(r_i\theta) \tag{3.99}$$

where

$$r_{i+\theta} = \frac{F_{i+\theta}}{m}, \{L\} = \begin{Bmatrix} \beta \\ 0 \\ \frac{5\Delta t}{12}\beta \\ \frac{\Delta t^2}{8}\beta \end{Bmatrix} \tag{3.100}$$

$$[A] = \begin{bmatrix} \gamma\beta & \rho\beta & \frac{1}{\Delta t}(\phi\beta) & \frac{1}{\Delta t}^2(-\omega^2\Delta t^2\beta) \\ 1 & 0 & 0 & 0 \\ \frac{\Delta t}{12}(8+5\gamma\beta) & \frac{\Delta t}{12}(-1+5\rho\beta) & \frac{1}{12}(12+5\phi\beta) & \frac{5}{12\Delta t}(-\omega^2\Delta t^2\beta) \\ \frac{\Delta t^2}{24}(10+3\gamma\beta) & \frac{\Delta t^2}{24}(-1+3\rho\beta) & \frac{\Delta t}{8}(8+\phi\beta) & \frac{1}{8}(-\omega^2\Delta t^2\beta) \end{bmatrix} \tag{3.101}$$

$$\begin{aligned} \beta &= \frac{24}{12(\theta+\theta^2)+4\xi\omega\Delta t(3\theta^2+2\theta^3)+\omega^2\Delta t^2(2\theta^3+\theta^4)} \\ \gamma &= \frac{-12(1-\theta^2)-8\xi\omega\Delta t(3\theta-\theta^3)-\omega^2\Delta t^2(6\theta^2-\theta^4)}{12} \\ \rho &= \frac{12(\theta-\theta^2)-4\xi\omega\Delta t(3\theta^2-2\theta^3)-\omega^2\Delta t^2(6\theta^2-\theta^4)}{24} \end{aligned} \tag{3.102}$$

$$\phi = -2\xi\omega\Delta t - \omega^2\Delta t^2\theta$$

(c) Cubic Implicit with θ Method Operators

$$\begin{Bmatrix} \ddot{a}_{i+1} \\ \ddot{a}_i \\ \ddot{a}_{i+1} \\ \ddot{a}_{i+1} \\ a_{i+1} \end{Bmatrix} = [\mathrm{A}] \begin{Bmatrix} \ddot{a}_i \\ \ddot{a}_{i+1} \\ \ddot{a}_{i+2} \\ \ddot{a}_1 \\ a_{i+1} \end{Bmatrix} + \{L\}(r_i\theta) \tag{3.103}$$

where

$$r_{i+\theta} = \frac{F_{i+\theta}}{m}, \quad \{L\} = \begin{Bmatrix} \beta \\ 0 \\ \frac{3\Delta t}{8}\beta \\ \frac{19\Delta t^2}{180}\beta \end{Bmatrix} \tag{3.104}$$

Note: Keierleber Method

Keierleber C. W. developed a method of implicit method with θ opeators. The author developed linear implicit, quadratic implicit and cubic implicit all with θ operators. They are based on time integration approach. Plate 3.1 gives a summary of them.

3.7.5 Additional Solution Procedure

The following additional solution procedures are recommended. They are included in the programs ISOPAR and F. BANG-2.

3.7.6 Newton–Raphson Method

This concept benefits from the Taylor expression. Similar to the direct method given in Plate 3.1 and tabulated below, the Newton–Raphson as stated earlier can be viewed at for comparison: A reference is made to Plate 3.2. Additional solution procedures

Direct integration method (Plate 3.1)	Newton–Raphson method (Fig. 3.7)
$\Psi(u) = Ku + F = 0$ (3.105)	$\Psi(u) = Ku + F = 0$
where $K = K(u)$	$= \Psi[u^n] + [\frac{d\Psi}{du}]\{\Delta u^n\} = 0$ (3.108)
$u_1 = -[K_o]^{-1}.f$ where	with $\{u^{n+1}\} = \{u^n\} + \{\Delta u^n\}$ (3.109)
$K_o = K(u_o)$ (3.106)	
For nth iteration, one can write	In the above the derivative represents the
$u^n = [K^{n-1}]^{-1}.F$ (3.107)	*tangential matrix*
The process is terminated	The improved value u^{n+1} can be obtained as
when the error $e = u^n - u^{n-1}$ becomes	$\{\Delta u^n\} = -[k^n]_{T''}^{-1}\Psi^n$
sufficiently small usually expressed in	$= -[K^n]_{T''}^{-1}\{F^n + F\}$ where T'' represents
terms of NORM	transpose (3.110)

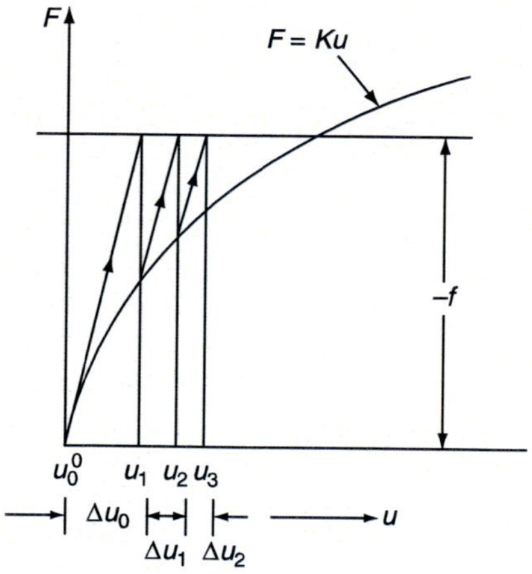

Plate 3.1 Direct method

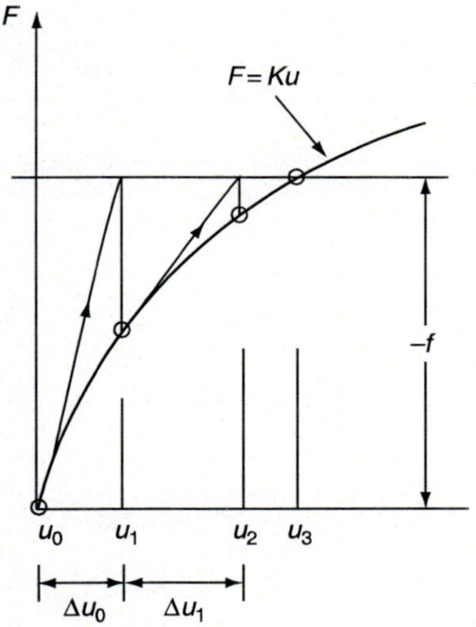

Fig. 3.7 Newton–Raphion method

3.7.7 Modified Newton–Raphson Method

Since there is a difficulty of having to solve a completely new sytem of equations at each iteration an approximation can be introduced [(Fig. 3.8; Plate 3.2)]. This can be done to

$$[K]^n{}_{T''} = [K^o]_{T''} \tag{3.111}$$

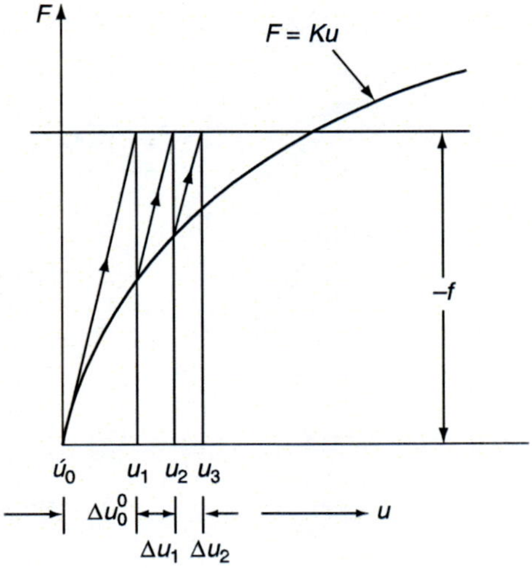

Fig. 3.8 Modified Newton–raphion method

Plate 3.2 Large deflection of a column

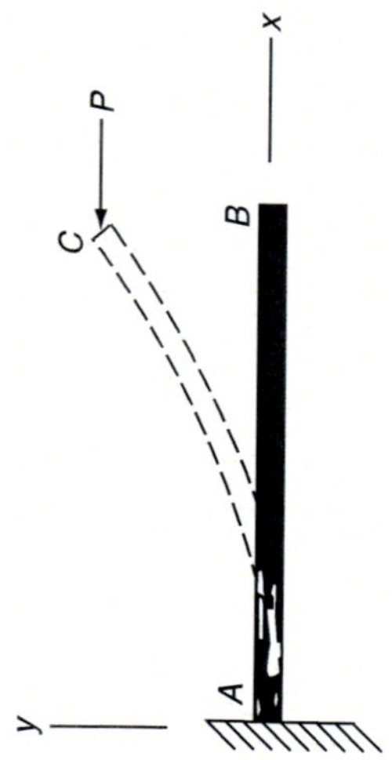

and a simple solution known as 'resolution' of the same system of equations is repeatedly used.

3.7.7.1 Incremental Method

This method is elaborately explained in the text [(Fig. 3.9)]. It is realised that the solution for $\{u\}$ is known when the load term is zero. Once the starting point is known, it will be useful to study the u as F is incremented. For small increments of F, convergence is highly likely. For the loading process, the intermediate computed results would be useful information. Thus the method begins with

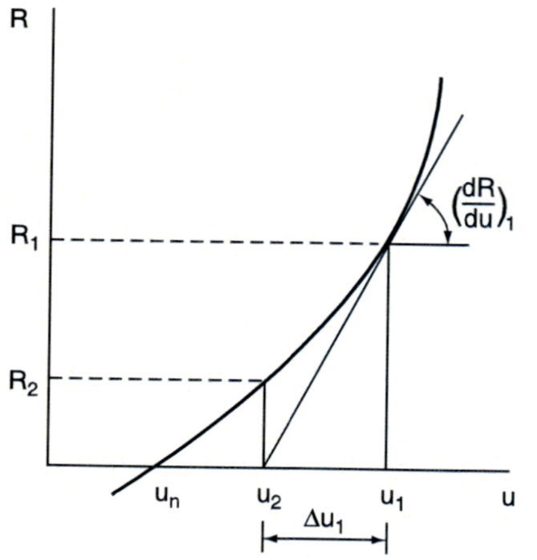

Fig. 3.9 Incremental method

$$\{F\}\{u\} + \{\lambda Fo\} = 0 \tag{3.112}$$

Differentiation is now needed. Assuming it is an exact differentiation, then

$$\frac{d\{F\}}{d\{u\}} \cdot \frac{d\{U\}}{d\{\lambda\}} + \{F_o\} = [K]_{T''} \frac{d\{u\}}{d\{\lambda\}} + \{F_o\} = 0 \tag{3.113}$$

In simplified form

$$\frac{dU}{d\lambda} = -[K_{T''}\{U\}]^{-1}\{F_o\} \tag{3.114}$$

Here one can identify the tangential matrix. To simplify the equation such as the above one, Euler's method is used which states

$$\{U_{m+1}\} - \{U_m\} = -[K_{T''}]\{U_m\}^{-1}\{F_o\}\{\Delta\lambda - m\} = -[K_{T''}]^{-1}\{\Delta F_m\} \tag{3.115}$$

where

$$\lambda_{m+1} = \lambda_m + \Delta\lambda_m \tag{3.116}$$

or

$$F_{m+1} = F_m + \Delta F_m \tag{3.117}$$

Improved integration schemes in the text can be used.

3.8 Geometrically Non-linear Problems in Finite Element

3.8.1 Introduction

Geometrically non-linear problems are assumed to be those associated with large displacements and large strains. In the presence of large displacements, the structure subjected to static or dynamic loads alters shapes and hence changes distributions. Large displacements normally affect stress–strain relationship. Equilibrium equations are written for deformed geometry. In dynamic condition or under dynamic loading, the deformed geometry can be included in the dynamic non-linear equilibrium equations. The linear analysis forms the basis and there is no need to revise these equations. During deformation the conditions are displaced which are known as moving coordinates and they cause solution problems. In Plate 3.2, the stiffness matrix of AB changes to AC position and $[K]$ is not constant. The matrix $[K]\{\Delta U\} = \{\Delta R\}$ where $\{\Delta R\}$ are the loads applied to the distorted nodes of the distorted elements and effectively are the arrayed *unbalanced* forces. The main idea is that for a typical solution, procedures described in Section 3.8, the interactive solution, will always be looking for or seeking the configuration which conforms to or with $\{\Delta R\} = 0$. One way is to account for rigid body displacement where a distortion of the load axis x' is introduced and for example the distortion member angle φ can be written for element AB as nodes 1 and 2:

$$\phi = tan^{-1}\frac{Y_L}{x_L} \tag{3.118}$$

If an element is distorted, it can also be expressed in the local coordinate system.

Various parameters are written as

$$\begin{aligned}\delta_2 = L - L_o &= \sqrt{x_2^2 + y_2^2} - L_o \\ &= \sqrt{(x_o + F_4 - F_1)^2 + (y_o + \theta_5 - \theta_2)^2 - L_o}\end{aligned} \tag{3.119}$$

And θ_1 and θ_2 are evaluated as

$$\theta_1 = M_3 - (\phi - \phi_o) = M_3 - [\tan^{-1}\frac{Y_L}{x_L} - \theta_o] \tag{3.120}$$

$$\theta_1 = M_6 - (\phi - \phi o) = M_6 - [\tan^{-1}\frac{Y_L}{x_L} - \theta_o] \tag{3.121}$$

Forces $F_{G(1,2)}$ applied at nodes 1 and 2 by the distorted element are written in local coordinates

$$\{F_{G(1,2)}\} = [K]\{\delta\} \tag{3.122}$$

where

$$\delta = \begin{bmatrix} 0 \\ 0 \\ \theta_1 \\ \delta_2 \\ 0 \\ \theta_2 \end{bmatrix} \tag{3.123}$$

Several computer packages exist to solve geometrically distorted structures, mostly based on the following criteria (Fig. 3.10):

1. Establish initially local coordinates using local displacements
2. Compute element distortions in local coordinates
3. Establish $[K]$ the stiffness matrix; $[K]\{\delta\}$ in local coordinates
4. Transfer $[K]$ and $\{F_G\}$ to global coordinates

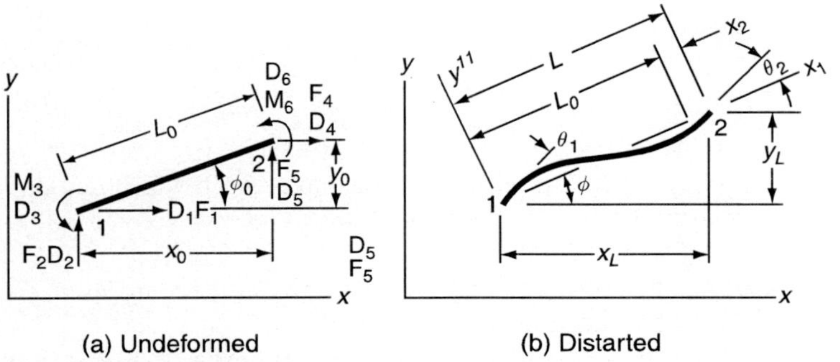

Fig. 3.10 Geometrically non-linear problems

5. The transfer technique is followed for all the distorted elements under new distributed loads. Repeat for all
6. Assemble global structural matrices $[K] = \sum[K]$ and $[R]_G = \sum\{F_G\}$
7. Compute the $\{\Delta R\}$, the unbalanced forces for the vector of the applied loads and $\{R\}$
8. Solve for the displacement increment $\{\Delta u\}$ from the structural equations $[K]\{\Delta u\} = \{\Delta R\}$
9. Add increments $\{\Delta u\}$ to global displacement $\{\delta\}$ accumulated in previous interactions.
10. Update estimate of the equilibrium configuration.
11. Apply convergence criteria given in the text.
12. When it does not converge go to step 1.

3.9 Finite Element Analysis of Explosion in Nuclear Facilities Using the Method of Explosive Factor

In order to stimulate detonation controlling of the release of chemical energy, a factor is needed to multiply the equations of high explosives given in books. The finite element method would require at the initial stage a lighting time t_{1l} for each element. Assuming the detonation velocity is v_D, the value of t_{1l} will be computed as follows:

Distance from the centre line of the detonation point to the centre of the element divided by v_D

The explosive factor f_{exp} between two points 1 and 2 will be

$$f_{exp} = \{f_{1,exp}, f_{2,exp}\}$$

such that

$$f_{1,exp} = \frac{2(t - t_{1l})v_D}{3(v_e/\mathbf{A}_{emax})} \quad \text{for } t > t_{1l} \tag{3.124}$$

$$= 0 \qquad \text{for } t \leq t_{1l} \tag{3.125}$$

$$f_{2,\exp} = \frac{1-v}{1-v_{\text{oj}}} \tag{3.126}$$

where v_s = detonation velocity; A_e max maximum area on which detonation or burn occurs; t = current time; v_{cJ} Chapman-Jouguet relative volume; v = current volume.

The value of $f_{\exp}$ has several steps towards unity and according to Wilkins M.L.[1] spread the burn front over several elements.

After reaching unity, $f_{\exp}$ is held constant and if it exceeds 1, it is reset to 1. According to the author the high explosive material be have as an elastic perfectly plastic solids prior to detonation. Hence, it will be necessary to update the stress tensor to an elastic stress $*S_{ij}{}^{n+1}$ such that

$$* S_{ij}{}^{n+1} = S_{ij}{}^{n} + S_{ip}\Omega_{pj} + S_{jp}\Omega_{pi} + 2G\dot{\varepsilon}'_{ij}\,dt \tag{3.127}$$

where G is the shear modulus and $\dot{\varepsilon}'_{ij}$ is a deviatoric strain rate. The von Mises yield condition is given by

$$\phi = J_2 = \frac{\sigma_y^2}{3} \tag{3.128}$$

where the second stress invariant, J_2, is defined in terms of the deviatoric stress components as

$$J_2 = \frac{1}{2} S_{ij} S_{ij} \tag{3.129}$$

and the yield stress is σ_y. If yielding has occurred, i.e. $\phi \geq 0$ the deviatoric trial stress is scaled to obtain the final deviatoric stress at time $n + 1$:

For detailed investigation in this field a reference is made to the following publication by the author: '*Manual of Numerical Methods in Concrete*. Thomas Telford, London, 2001.'

The value of S_{ij}^{n+1} can be written as

$$S_{ij} = \frac{\sigma y}{\sqrt{3J_2}} S_{ij}{}^{n+1} \tag{3.130}$$

If $\phi \leq 0$ then

$$S_{ij}{}^{n+1} = S_{ij}{}^{n+1} \tag{3.131}$$

---

[1] [Wilkins M.L.,"Calulations of Elastic Plastic Flow," Meth, Comp. Phys. 3 (Academic Press), 211–263 (1964).]

Before detonation pressure is given by the expression

$$P^{n+1} = K(\frac{1}{y^{n+2}} - 1)$$

where K is the bulk modulus. Once the explosive material detonates

$$S_{ij}^{n+1} = 0 \qquad (3.132)$$

and the material behaves like a gas.

For practical application a reference is made to the author's book on *Explosion-Resistant Buildings*. Springer Verlag 2006.

3.9.1 Good Achievement of the Explosive Burn

It will be necessary to look into the following points in order to achieve good explosive burn:

1. Where the impact occurs, the FE mesh must be kept constant.
2. The characteristic element dimension must be found by checking all explosive elements for the largest diagonal.
3. The detonation points, if possible, must be within or at the boundary of the explosive.

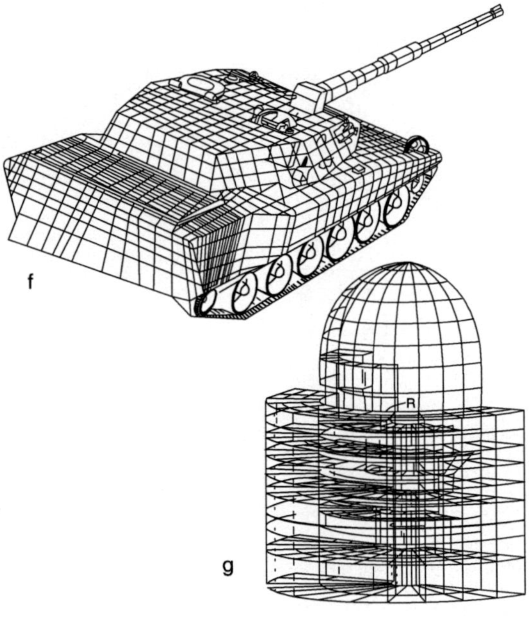

Fig. 3.11 3D global mesh schemes of a tank and PWR nuclear power station

4. Check always the computed lighting time for the explosive material. The lighting time I program LS-DYNA is kept at a negative number. This is true in program BANG F-FIRE

The line of detonation must have sufficient number of detonation points in order to visualise the line of fire. Some of the global mesh schemes such as Tank and PWR power station are shown in Fig. 3.11.

3.10 Finite Element Method Schemes

Several finite element analysis packages are available together with their mesh–generating schemes. Although the subject is beyond the scope of this book, however the following packages are mentioned for the reader's in-depth studies:

Various mesh-generation schemes are available as front ended to various computer packages such as mentioned above. Plate 3.3 just shows a few of such examples. The following are known packages are listed below.

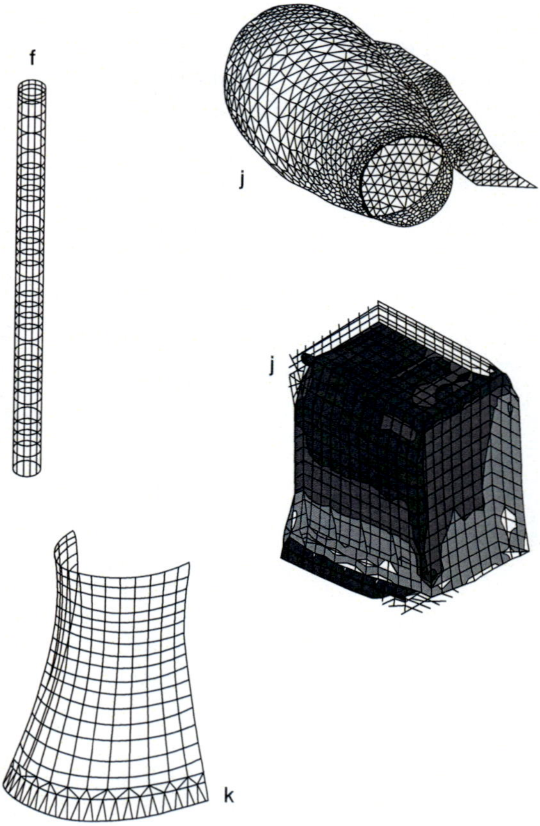

Fig. 3.12 3D mesh-generating schemes of nuclear structures

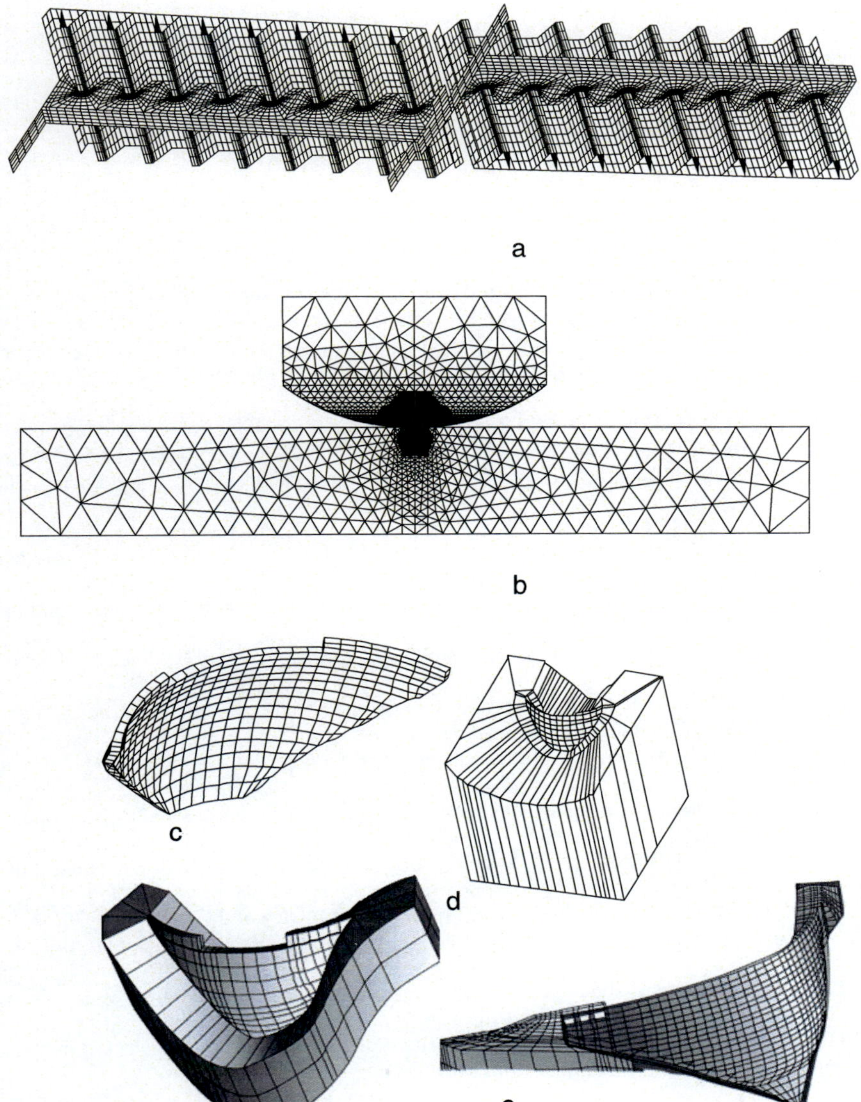

Plate 3.3 Finite Element Mesh Schemes for structures for nuclear facilities

ANSYS	IDEAS
ABACUS	OAYES
DYNA 3D	NASTRAN
ASAS	MARC
LUSAS	F. BANG
Ls DYANA	F. ISOPAR

References

1. Akyuz. F.A., and Merwin. J.A. Solution of non-linear problems of elastoplasticity by finite element method. J. AIAA, 6,1968.
2. Bangash, M.Y.H. The Automated three-dimensional cracking analysis of prestressed concrete vessels, Proc. 6th Int. Conf. Struct. Mech. Reactor Technology. Paper H3/2 Paris 1981
3. Bangash, M.Y.H. The structural Integrity of concrete containment vessels under external impact. Proc. 6th Inter. Conf. Struc. Mech. React. Technology Paper J7/6, Paris, 1981.
4. Bangash, M.Y.H. Reactor Pressure Vessel Design and Practice. Prog. Nucl. Energy, 10, 69–124, 1982
5. Bangash, M.Y.H. The simulation of endochronic model in the cracking analysis of PCPV. Proc. 9th Int. Conf. Struct. Mech. Reactor Tech. vol 4 pp 333–40, Lausanne, Switzerland 1987.
6. Bangash, M.Y.H. Manual of numerical methods in concrete. Thomas Telford, London. 2001 p. 918.
7. Bangash, M.Y.H. Explosion Resistant Buildings, Springer Verlag, Heidelberg, 2006, p. 783
8. Bangash, M.Y.H. Shock, Impact and Explosion. Structural Analysis and Design. Springer Verlag, Heidelberg 2009 pp.1365
9. Bangash, M.Y.H. Earthquake Resistant Buildings. Springer Verlag, Heidelberg. Inpress to be published in 2010
10. Banthia, N.P. Impact, resistance of concrete Ph.D. Thesis University of British Columbia, Canada 1987.
11. Bathe. K. J. and Wilson. E. L. Stability and accuracy analysis of direct integration methods. Earthquake Engineering Structural Dynamics 1, 1973, pp. 283–291.
12. Bathe, K. J. *Finite Element Procedures*. Prentice Hall, Englewood Cliffs, NJ, 1996.
13. Burden. R. L. and Faires. J. D. Numerical Analysis. Prindle, Weber and Schmidt, Boston, 1985.
14. Craig R.R. Structural dynamics: An introduction to computer methods, Wiley, New York, 1981
15. Hallquist. J.O. et al. Sliding interfaces with contact-impact in large scale Lagrangian computations. Comp. Methods. Appl. Mech. Eng., 51, 107–37 1985.
16. Hildebrand, F. B. Introduction to Numerical Analysis. McGraw Hill, New York, NY, 1956.
17. Hughes-Hallet, D., et al. Calculus. Wiley, New York, NY, 1994.
18. Keierleber, C. W. Higher-Order Explicit and Implicit Dynamic Time Integration Methods. University of Nebraska, Lincoln. NE, 2003.
19. Phillips D.V. Zienkiewicz O.C. Finite Element Non-linear Analysis of Concrete Structures. Proc. Inst. Civ. Eng. Res Theory. 61, 1976.
20. Tedesco, J. W., McDougal. W. G., and Ross, C. A. Structural Dynamics, Addison-Wesley Longman, Menlo Park, CA, 1999.

Chapter 4
Steel Pressure Vessels for Nuclear Facilities

4.1 General Introduction

Steel pressure vessels have been used to contain reactor designed for various nuclear power facilities in the world. The dimensions, loadings and materials used are dependent on the type of reactor systems adopted in the nuclear facilities. Some are core left standing covered by radiation shielding or embedded in thick concrete walls and slabs. A typical reactor pressure vessel for PWR is shown in Figure 4.1. Table 4.1 indicates design limit stresses for various pressure parts. This chapter begins with the general design criteria for pressure vessels. These vessels are first treated as thin shell surfaces. The equations of equilibrium of the element shell of revolution have been established. Equations are derived for cylindrical shell surfaces of the vessel. Membrane solution of domical surfaces are given and they are tabulated for a comparative study. Spherical, conical and elliptical dome parts of the vessels have been examined. Step-by-step equations have been established with design examples. Since mostly these vessels in the current environment have Toro spherical domes surfaces, previous equations have been readjusted to cater for these types of surfaces.

A comprehensive accident analysis of PWR loss-of-coolant (LOCA) type accident is generally required. A 3D finite element technique given in Chapter 3 has been used to investigate this effect of thermal shock load causing partial or complete rupture of the vessels. Various results have been produced and they are plotted. Jet impingement forces on PWR steel vessel components have been carefully examined. The R6 method of structure mechanics is examined with failure diagram. At the end, the state-of-the-art software for 3D fracture mechanics simulation in the reactor vessel has been introduced, known as ZENCRACK. Finite element interfaces, mesh procedure, adaptive remeshing and load handling in fracture mechanics have been thoroughly explained and sample applications are given to demonstrate the ZENCRACK by Zentech International Ltd., London.

M.Y.H. Bangash, *Structures for Nuclear Facilities*,
DOI 10.1007/978-3-642-12560-7_4, © M.Y.H. Bangash 2011

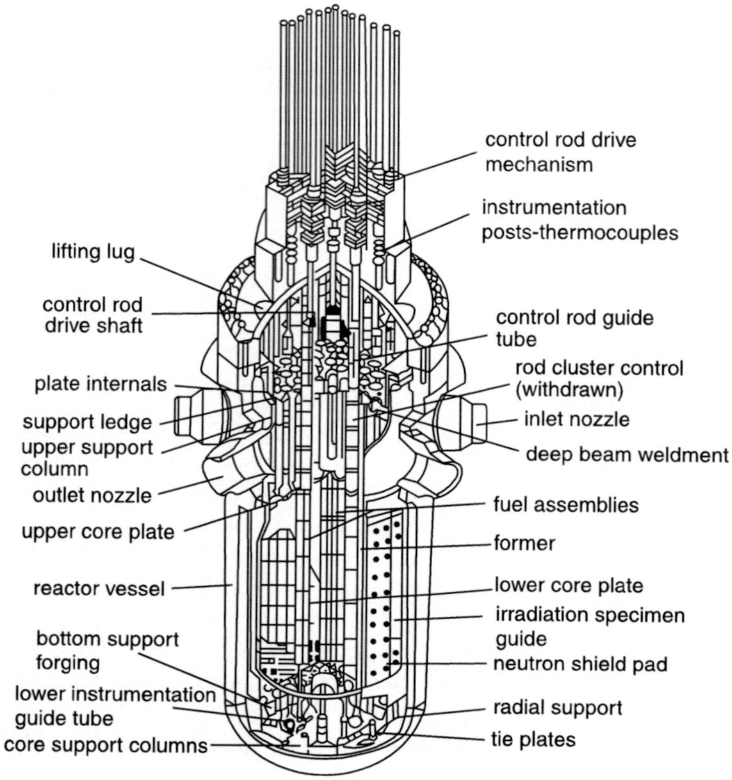

Fig. 4.1 Cut-away diagram showing the PWR steel vessel and internals (former CEGB Sizewell B inquiry). (From Bangash)[3.3]

4.2 General Design Criteria for Pressure Vessels

4.2.1 Introduction

ASME provides necessary formulae to compute vessel sizes and membrane stresses of the vessel due to internal and external pressures; it leaves to the designer to carry out analytical procedures for computing stresses due to other loads including combinations of loads.

However, it is a general practice to provide detailed stress analysis for the vessel components outside the Code approved details using either the maximum-stress or the maximum-shear theory of failure and to select allowable stresses for design conditions other than normal operations or for computed stresses other than direct membrane or direct membrane plus primary bending Code stresses. CEGB R6 method based on fracture mechanics is highly recommended. This method later on has been thoroughly discussed in this chapter.

Table 4.1 Design limit stresses for pressure parts

DESIGN CONDITION	COMBINATION OF DESIGN LOADS: DESIGN PRESSURE	VESSEL WEIGHT	WIND OR SEISMIC LOAD	TEMPORARY OVERLOAD	LOCAL MECHANICAL LOAD	THERMAL EXPANSION LOAD	MAXIMUM ALLOWABLE STRESS	NOTES:
1. Field erection		×	×				1.2 S_a	1
2. Operating	×	×	×				S_a	2, 5
3. Operating with thermal expansion	×	×	×			×	$1.25(S_a + S_{atm})$	2, 3, 5
4. Short-term mechanical overload	×	×		×			$1.33S_a < S_y$	2
5. Short-term mechanical overload with thermal expansion	×	×		×		×	$1.50(S_a + S_{atm})$	2
6. Operating with local mechanical load (bending)	×	×	×		×		$2S_a < S_y$	2, 5
7. Hydrotest	test pressure						S_{atm} × test factor	4

S_a = basic code allowable stress value at design metal temperature
S_{atm} = basic code allowable stress value at atmospheric metal temperature

Notes

1. Computed stresses are based on the full uncorroded thickness.
2. Computed stresses are based on the corroded thickness.
3. The criterion used here is the approach used in the ANSI B31.3 Petroleum Refinery Piping, that is, to limit the maximum calculated principal stress to the allowable stress range rather than using the shear theory of failure. The value $1.25(S_a + S_{atm})$ provides a safety margin against the possibility of fatigue due to localized stresses and other stress conditions (see Fig. 2.1). Obviously, stresses must be computed both with and without thermal expansion, since allowable stresses are much smaller for conditions without thermal expansion. As an additional safety precaution, the computed stresses are usually based on the modulus of elasticity E at room temperature [8].
4. Computed stresses are based on test thickness at test temperature. Since water pressure is a short-term condition, the allowable stresses for structural parts such as supports are frequently increased by a factor of 1.2. The upper limit of stress in the vessel shell during a hydrotest of pressure parts is not specified by the Code. However, it is a good engineering practice to limit the maximum membrane stress in any part of the vessel during a hydrotest to 80 percent of the yield strength. Thermal expansion stresses and local mechanical stresses will be absent and need not be considered.
5. This allowable stress limit is for statically stressed vessels. This means that the number of stress cycles applied during the life of vessel does not exceed several thousands. For cyclic conditions these allowable stresses would have to be substantially reduced and preferably a fatigue analysis made. Usually, in service the operating loads.are raised gradually up to their maximum values and maintained for some time and they are not reapplied often enough to make a fatigue analysis necessary.

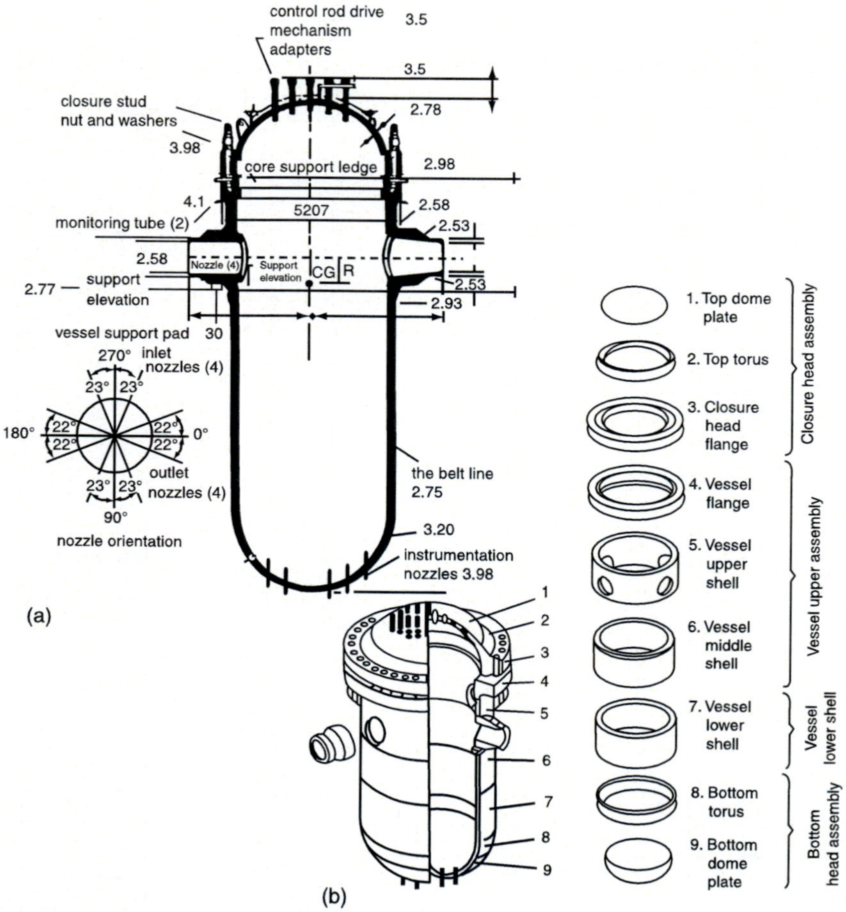

Fig. 4.2 (**a**) The reactor vessel of a 4-loop PWR system (safety factors computed using the finite-element method). (**b**) Diagram showing sections from which the PWR vessel may be fabricated when sections are either rolled plate or forged material (*shaded parts* are common to both). (Courtesy of Marshall[5.155]) (From Bangash[3.3])

Table 4.1 gives the design limit stresses for the pressure parts of the steel vessel and they are kept as BENCH MARK for checking linear. Non-linear and fracture/cracking analysis based on computerized numerical methods such as finite element and boundary elements techniques. Basically the stresses, as they occur in the vessel shells, are divided into three distinct categories: primary, secondary and peak.

Primary stress is produced by steady mechanical loads, excluding discontinuity stresses or stress concentrations. Its main characteristic is that it is not self-limiting. Primary stress is divided into two subcategories. It is imposed on the vessel by the equilibration of external and internal mechanical forces. Any yielding through the entire shell thickness will not distribute the stress, but will result in gross distortions, often carried to failure. General primary stress is divided into primary membrane stress and primary bending stress; the limit design method shows that a higher stress limit can be applied to the primary bending stress than to the primary membrane stress.

Local primary stress is produced by the design pressure alone or by other mechanical loads. It has some self-limiting characteristics. If the local primary *stress* exceeds the yield point of the material, the load is distributed and carried by other parts of the vessel.

The basic characteristic of *secondary stress* is that it is self-limiting. Minor yielding will reduce the forces causing excessive stresses. Secondary stress can be divided into membrane stress and bending stress, but both are controlled by the same limit stress intensities. Typical examples of secondary stress are thermal stresses and local bending stresses due to internal pressure at shell discontinuities.

Peak stress is the highest stress at some local point under consideration. In case of failure, peak stress does not generate any noticeable distortion, but it can be a source of fatigue cracks, stress-corrosion and delayed fractures. According to the shear theory of failure yielding in a member under loads begins if the maximum induced shear stress equals the yield shear stress developed in a test sample under simple tension. The maximum shear stress τ at the point under consideration equals one-half the largest algebraic difference between any two out of three principal stresses $(\sigma_1, \sigma_2, \sigma_3)$ at that point. It occurs on each of two planes inclined 45° from these two principal stresses. For $\sigma_3 > \sigma_2 > \sigma_1$, maximum $\tau = (\sigma_3 - \sigma_1)/2$. Twice the maximum shear stress τ is by definition the equivalent intensity combined stress or stress intensity.

There is a procedure for the computation of *stress intesities*.

1. *The maximum stress intensity* is computed on the basis of P_m, the sum of all primary membrane stress components which must not exceed a stress intensity equal to $\sigma_m \times k$ and k, a stress intensity factor ranges 1.0–1.2 for different load combinations such that when applied to $\sigma_m = \sigma_y$ or $\leq \frac{1}{3}\sigma_u$, while $\sigma_m =$ basic design stress value intension of the vessel materials.
2. *Local membrane stress intensity*. The maximum allowable stress intensity for S derived from σ_L stresses is $1.5\sigma_m$ times the factor k when applicable.
3. *Primary membrane* (σ_m or σ_L) *plus primary bending* (σ_b) *stress intensity*. The maximum stress intensity S based on σ_m or σ_L stresses plus the bending stress (σ_b) is limited to $1.5\sigma_m$ time the factor k when applicable.
4. *Primary plus secondary stress intensity*. The maximum stress intensity S *as* based on the primary or local membrane stresses plus the primary bending stress plus the secondary stress (σ_m *or* $\sigma_L + \sigma_b + \sum \tilde{Q}$) cannot exceed the

value of $3\sigma_m$. All stresses may be computed under operating conditions, usually less severe than the design conditions ($3\sigma_m \leq 2\sigma_y$)

5. *Peak stress intensity*. If fatigue analysis is required for cyclic operating conditions, the maximum stress intensity S must be computed from the combined primary, secondary and peak stresses (σ_m *or* $\sigma_L + \sigma_b + \sum \tilde{Q} + \sum \tilde{F}$) under operating conditions. The allowable value σ_a for this peak stress intensity S is obtained by the methods of analysis for cyclic operations with the use of the fatigue curves.

The symbols below define various stresses:

$\sigma_m =$ sum of all general primary membrane stress components
$\sigma_L =$ sum of all local primary membrane stress components
$\sigma_b =$ sum of all primary bending stress components
$\sum \tilde{Q} =$ sum of all secondary (membrane and bending) stress components
$\sum \tilde{F} =$ sum of all peak stress components

Note: In ASME codes σ is defined as "S" but this symbol is adopted for uniformity for suitable computerized analysis as done elsewhere in the text.

4.3 Stress Analysis of Vessel Shell Components

4.3.1 Shape and Curvature

At any point of a shell surface there are two principal curvatures: $1/R_1$, and $1/R_2$. The Gaussian or *mean* curvature $1/R_1$, $1/R_2$ is used for classification.

- *Synclastic* or *elliptical surfaces* have $1/R_1R_2 > 0$, i.e. both curvatures are at the same sign. Examples are domes, elliptic paraboloids ('elpers').
- *Anticlastic* or *hyperbolic surfaces* with $1/R_1R_2 < 0$ are of the saddle type; curvatures at a point are of different sign. They are sensitive to the way they are supported and to various secondary effects, such as temperature, shrinkage.
- Some anticlastic surfaces can be generated by straight tines (ruled or warped surface): the *conoid* (one system of straight generators), the *hyperboloid* and the *hyperbolic paraboloid* ('Hypar') both with two systems of straight generators. They are easier to form than synclastic surfaces, but are subject to large displacements if free at the edges.
- Surfaces of *zero Gaussian curvature* — i.e. *cylindrical shells* — are probably most affected by the bending effect, and the membrane theory can give good approximation only in exceptional cases.
- *Folded plate structures* — stripes of plate jointed at the edges — are not really shells at all, but in functioning and in the way they can be treated are quite close to cylindrical shells. The edges of the dome shell of the reactor vessel will, sometimes, develop the shape of the folded plates.

Kind of surface

- *Developable surfaces* can be bent into other shapes without stretching — e.g. cylindrical and conical shells.
- *Non-developable surfaces,* i.e. of double curvature.

4.3.2 Boundary and Edge Conditions

- Closed shells
- Open shells with various edge conditions:

  Free for clamped continuous support
  Single supports
  Edge beams
  Ribs
  Ties and traverses

Boundary and edge conditions modify the behavior of the shell, due to disturbance or perturbations originating from the edges.

4.3.3 Generalized Analysis of Thin Shell Surfaces of the Reactor Vessels

4.3.3.1 Assumptions Made in the Theory of Elastic Thin Shells

1. Direct stresses perpendicular to the middle surface can be neglected.
2. Points on a normal distribution to the middle surface before the deformation will be on a straight line after the deformation has taken place.
3. This line will remain perpendicular to the deformed middle surface.
4. Displacements are small compared to the thickness of the shell.

Stress resultants of a shell of the reactor vessel:

A system of coordinates, i.e. x, y, is defined on the middle surface, so that the lines x = const meets lines y = const at right angles (Gaussian coordinates). The element is out from the shell along pairs of adjacent coordinates lines, and the cuts are made so that four sides of the element are normal to the middle surface. The coordinates z is measured positive inwards, x, y, z, from a right-handed triad.
(*Note:* Some writers define z positive outwards, which results in an interchange of x and y axes to preserve the right-hand rule).

Stress resultants are defined as *forces* (or bending moments) *per unit of length* as in plate theory. Figure 4.3 shows the shell element with the stresses. Figure 4.4 shows the membrane stress resultants. The tensile forces are positive, and the

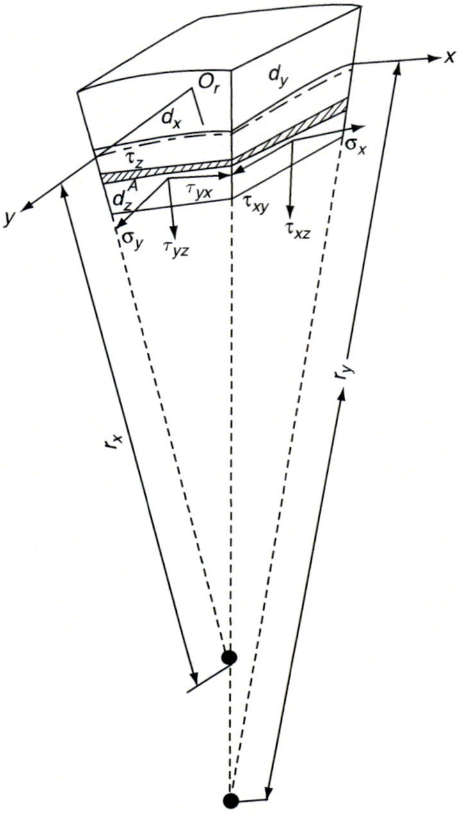

Fig. 4.3 Membrane stresses

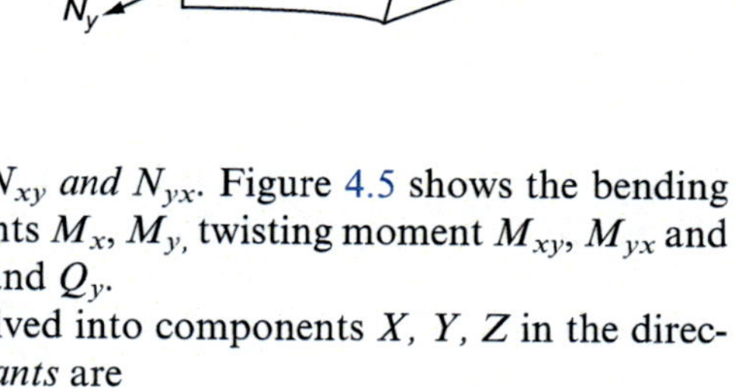

Fig. 4.4 Membrane stress resultants

figure shows positive shear forces N_{xy} *and* N_{yx}. Figure 4.5 shows the bending stress resultants, i.e. bending moments M_x, M_y, twisting moment M_{xy}, M_{yx} and transverse (radial) shear forces Q_x and Q_y.

The applied external forces revolved into components X, Y, Z in the directions for the *membrane stress resultants* are

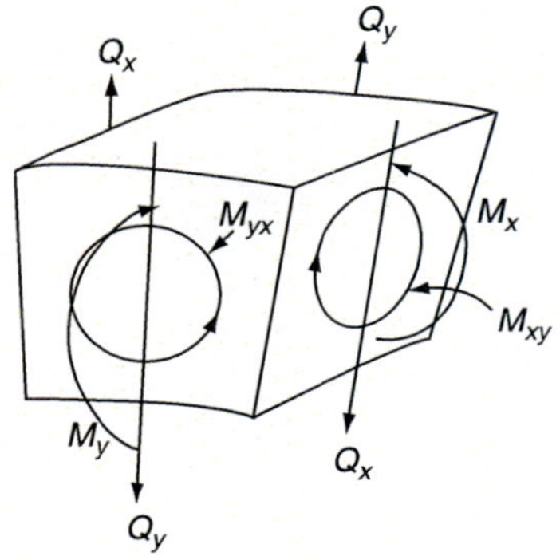

Fig. 4.5 Bending stress resultants

$$N_x = \int_{\frac{-t}{2}}^{\frac{t}{2}} \sigma_x\left(1 - \frac{z}{r_y}\right)dz; \quad N_y = \int_{\frac{-t}{2}}^{\frac{t}{2}} \sigma_y\left(1 - \frac{z}{r_x}\right)dz \tag{4.1}$$

$$N_{xy} = \int_{\frac{-t}{2}}^{\frac{t}{2}} \tau_{xy}\left(1 - \frac{z}{r_y}\right)dz; \quad N_{yx} = \int_{\frac{-t}{2}}^{\frac{t}{2}} \tau_{yx}\left(1 - \frac{z}{r_x}\right)dz \tag{4.2}$$

The second terms in these expressions $Z/r_y, Z/X$ are due to the trapezoidal form of the sides of the element: as the thickness t is small compared with the radii of curvature, these terms are omitted.

$\tau_{xy} = \sigma_{yx}$ are complementary shear stresses, but shear forces N_{xy} and N_{yx} are only equal when $r_x = r_y$.

4.3.3.2 Assumptions of the Membrane Theory for Shells

The stresses are uniformly distributed over the thickness of the shell, from which it follows that $N_{xy} = N_{yx}$ (by dropping the second term τ_{yx} and τ_{xy} can be taken in front of the integration sign). The membrane stress resultants are therefore

- all the other six stress resultants are assumed to be zero;
- the three equations of equilibrium of forces are sufficient to determine the three unknowns, N_x, N_y and N_{xy}.

The *requirements* for the use of the membrane theory are

1. The middle surface must be continuous (no sudden changes of curvature).
2. No sudden changes of the thickness are allowed.
3. Reactions at the edges must be directed tangentially to the middle surface, and support arrangements must be made to allow this.

The geometry of a dome, as a shell of revolution, is shown in Fig. 4.6 and with notation in Figs. 4.7 and 4.8.

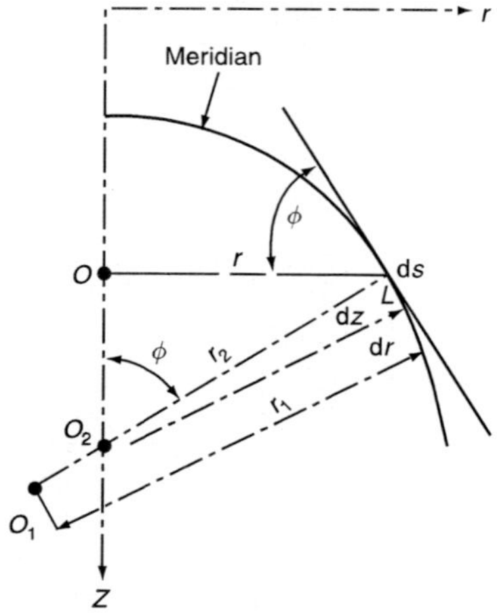

Fig. 4.6 Geometry of dome

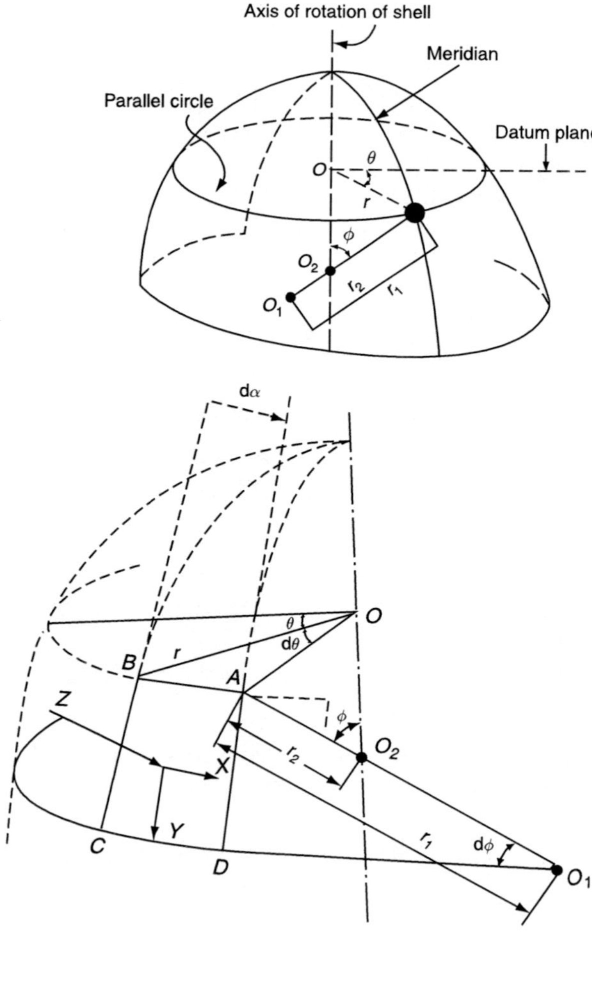

Fig. 4.7 Element in the middle surface of a shell of revolution

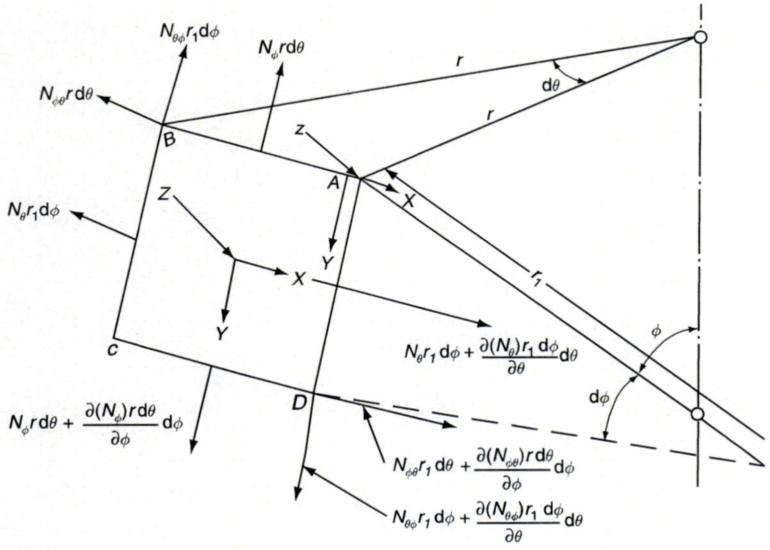

Fig. 4.8 The membrane stress resultants and the load components

4.3.4 *The Equations of Equilibrium of the Element of a Shell of Revolution*

The equation for forces in the x direction (along tangent to a parallel circle)
Geometry:

Tangents to meridians *at* A and B intersect at angle a (Fig. 4.9a)

$$r_2 \tan \phi \mathrm{d}\alpha = AB = r\mathrm{d}\theta$$

$$\mathrm{d}\alpha = \frac{r\mathrm{d}\theta}{r_2 \tan \phi}$$

But,

$$r = r_2 \sin \phi$$

$$\mathrm{d}\alpha = \mathrm{d}\theta \cos \phi$$

Required: component of $N_{\theta\phi}$, $R_{\mathrm{d}\theta}$ in the x direction

$$2 \times N_{\theta\phi} r_1 \mathrm{d}\phi \sin\left(\frac{\mathrm{d}\alpha}{2}\right) = N_{\theta\phi} r_1 \mathrm{d}\phi \mathrm{d}\alpha$$

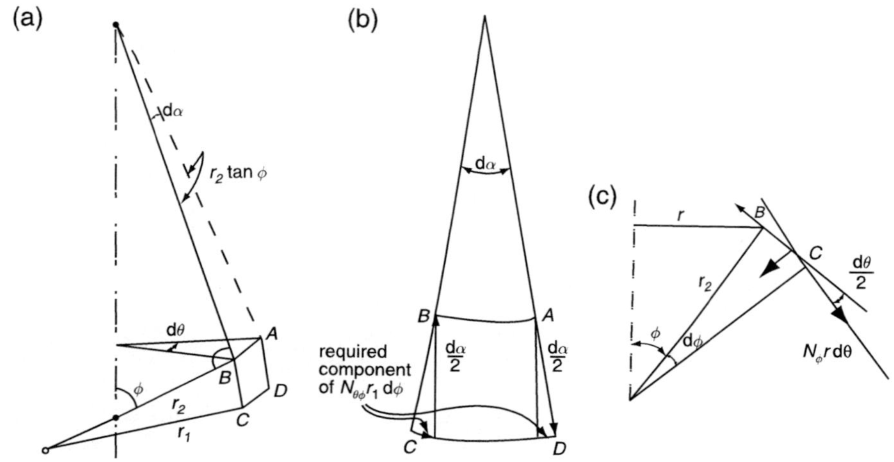

Fig. 4.9 (**a**) Tangents to meridian. (**b**) Equilibrium diagram. (**c**) Elements of meridian

But, $\mathrm{d}\alpha = \mathrm{d}\theta \cos \phi$

$$N_{\theta\phi} r \mathrm{d}\theta + \frac{\partial N_{\theta\phi}}{\partial \phi} r \mathrm{d}\theta \mathrm{d}\phi - N_{\theta\phi} r \mathrm{d}\theta + N_\theta r_1 \mathrm{d}\phi + \frac{\partial N_\theta}{\partial \theta} r_1 \mathrm{d}\phi \mathrm{d}\theta - N_\theta r_1 \mathrm{d}\phi \\ + N_{\phi\theta} r_1 \mathrm{d}\phi \mathrm{d}\theta \cos \phi + X r_1 \mathrm{d}\phi r \mathrm{d}\theta = 0 \quad (4.3)$$

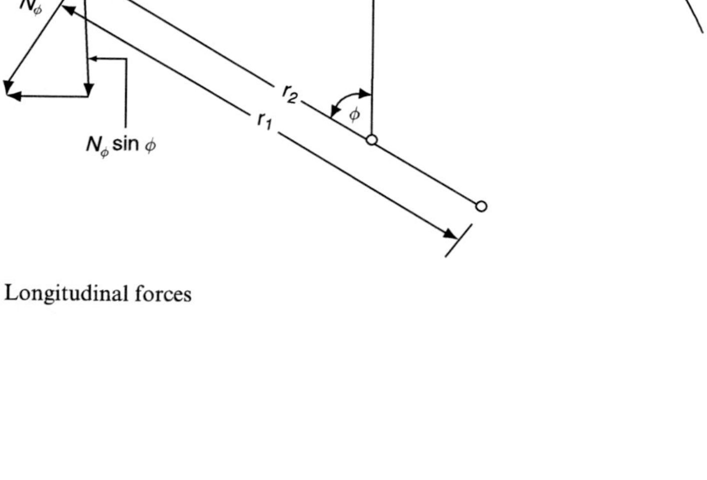

Fig. 4.10 Longitudinal forces

r = radius of curvature of parallel circle
r_1 = radius of curvature of meridian
r_2 = length of normal to surface, to axis of rotation
r_1, r_2 = principal radii of curvature at A (Meusnier theorem)

$$AB = \mathrm{r}d\theta$$

$$AD = r_1\, d\phi$$

Total forces acting on the sides of the element are shown in the figure as $AB \neq CD$. Coordinate axes are

x = along tangent to parallel circle
y = along tangent to meridian
z = normal to the shell surface

And after reduction, as

$$\frac{\partial}{\partial\phi}(rN_{\phi\theta}) + r_1\frac{\partial N_\phi}{\partial\phi} + r_1 N_{\phi\theta}\cos\phi + Xrr_1 = 0 \tag{4.4}$$

The equation for forces in the y direction (along tangent to meridian).

Required: components of total force on AD and BC $N_\theta \mathrm{r}d\phi$ has component along the radius of the parallel circle:

$$2N_\theta rd\phi\sin\left(\frac{\mathrm{d}\theta}{2}\right) = 2N_\theta r_1\mathrm{d}\phi\mathrm{d}\theta$$

Resolving this in the y and z directions:

Component in y direction $N_\theta \mathrm{r}\mathrm{d}\phi\mathrm{d}\theta\cos\phi$
Component in z direction $N_\theta r_1\mathrm{d}\phi\mathrm{d}\theta\sin\phi$

$$\frac{\partial}{\partial\phi}(rN_\phi) + r_1\frac{\partial N_{\theta\phi}}{\partial\theta} + r_1 N_\theta\cos\phi + Yrr_1 = 0 \tag{4.5}$$

The equation for forces in Z direction (i.e. normal to the middle surface of the shell):

Component due to N_ϕ in the Z direction:

$$N_\phi r\mathrm{d}\theta\sin\left(\frac{\mathrm{d}\phi}{2}\right) = N_\phi r_1\mathrm{d}\theta\mathrm{d}\phi$$

Component due toN_θ
The complete equation of this equilibrium is

$$N_\theta r_1\mathrm{d}\phi\mathrm{d}\theta\sin\phi + N_\phi r\mathrm{d}\theta\mathrm{d}\phi + Zr_1 r\mathrm{d}\theta\mathrm{d}\phi = 0$$

where $r = r_2 \sin \phi$ and $r_1 = r_2 \sin \phi \mathrm{d}\theta \mathrm{d}\phi$

$$\frac{N_\phi}{r_2} + \frac{N_\theta}{r_2} + Z = 0 \tag{4.6}$$

Identity of cross shears

$$N_{\phi\theta} = N_{\theta\phi} \tag{4.7}$$

Equations (4.4), (4.5), (4.6) and (4.7) allow the calculation of the membrane forces of a shell of revolution, also for asymmetrical loading.

4.3.5 Shells of Revolution: Axisymmetric Loading

With loads symmetric about the vertical axis of rotation, Eqs. (4.4), (4.5), (4.6) and (4.7) simplify considerably

1. Stresses are independent of θ, and all partial derivatives with respect to θ disappear.
2. $N_{\phi\theta} = N_{\theta\phi} = 0$, otherwise they would produce asymmetrical deformation with respect to the vertical axis.
3. There is no variation of other stress resultants with θ.
4. The component X of the load must vanish; it would produce twist about the axis and shear deformation.

Therefore, Eq. (4.1) is identically satisfied and disappears. Hence, Eq. (4.5) simplifies to

$$\frac{\mathrm{d}}{\mathrm{d}\phi}(rN_\phi) + rN_\theta \cos \phi + Yrr_1 = 0 \tag{4.8}$$

Eq. (4.6) is unchanged

$$\frac{N_\phi}{r_1} + \frac{N_\theta}{r_2} + Z = 0$$

The differential Eq. (4.8) and the common Eq. (4.6) are sufficient to determine meridian force N_ϕ and the hoop force N_θ for an axisymmetrically loaded shell of revolution.

Procedure in solving Eqs. (4.8) and (4.6)

From Eq. (4.6)

$$N_\theta = -Y_2\left(Z + \frac{N_\phi}{r_1}\right) \tag{4.9a}$$

Substituting into Eq. (4.8) using $r = r_2 \sin\phi$, and multiplying the equation with $\sin\phi$ gives

$$\frac{d(N_\phi r_2 \sin\phi)}{d\phi}\sin\phi + N_\phi r_2 \sin\phi\cos\phi = -r_1 r_2 (Y\sin\phi + Z\cos\phi)\sin\phi$$

The left-curved side can be written as $d(N_\phi r_2 \sin^2\phi)/d\phi$ hence

$$\frac{d(N_\phi r \sin\phi)}{d\phi} = -rr_1(Y\sin\phi + Z\cos\phi) \tag{4.9b}$$

The integration with ϕ is variable

$$N_\phi = -\frac{1}{r\sin\phi}\left[\int -rr_1(Y\sin\phi + Z\cos\phi)d\phi + C\right]$$

where the constant C is found from boundary conditions. The hoop force is obtained later from Eq. (4.6).

The membrane forces of an axisymmetrically loaded dome can be obtained much faster and more directly. Instead of dealing with the equilibrium of the element of the shell, consider the equilibrium of the dome above a certain parallel circle.

If $\bar{W}$ is the vertical resultant of the applied load and $2\pi r N_\phi \sin\phi$ is the total vertical component around the parallel circle of the meridian force N_ϕ, then

$$2\pi r N_\phi \sin\phi - \bar{W} = 0 \tag{4.10}$$

where $N_\phi = \bar{W}/(2\pi r \sin\phi)$.

The hoop force N_θ can now be calculated (see Fig. 4.9).

4.3.6 Cylindrical Shell Surfaces of the Vessel

4.3.6.1 Types of Cylindrical Shell Surfaces

Cylindrical shells — shells in which either the directrix or the generatrix is a straight line.

Membrane forces on the cylindrical shell part of the vessel under different loads are summarized in Table 4.2.

Table 4.2 Membrane forces in cylindrical shell surfaces under different loads

Loading	N'_ϕ	N'_x	$N'_{\phi t}$
per m² of surface 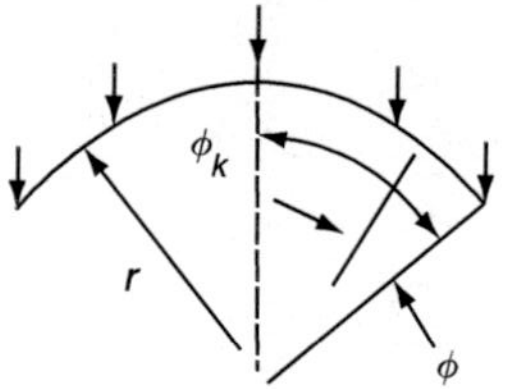	$-r\cos(\phi_k-\phi)$	$-\frac{x}{r}(L-x)\cos(\phi_k-\phi)$	$-(L-2x)\sin(\phi_k-\phi)$
per m² projection 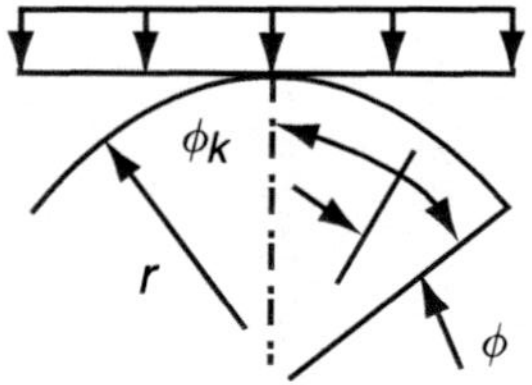	$-r\cos^2(\phi_k-\phi)$	$\frac{3}{2}\frac{x}{r}(L-x)$ $[1-2\sin^2(\phi_k-\phi)]$	$-\frac{3}{2}(L-2x)\sin(\phi_k-\phi)$ $\cos(\phi_k-\phi)$
p_w per m² projection 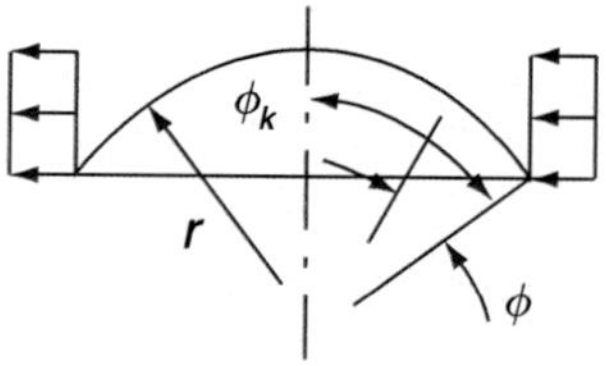	$-p_w r\sin(\phi_k-\phi)$	$-p_w\frac{x}{2r}(L-x)\sin(\phi_k-\phi)$	$p_w\left(\frac{L}{2}-x\right)\cos(\phi_k-\phi)$

4.4 Membrane Solution of Domical Surfaces

As the solution of a 3D elasticity problem is tough, shell theory may be viewed as a 3D subset of elasticity valid for certain classes of structures. For this subset we shall develop appropriate kinematics, constitutive and equilibrium relations.

Now consider an element at the dome in the middle surface of the shell of revolution as shown in Fig. 4.11 where R is the radius of curvature of the parallel circle and a is the radius of curvature of the meridian.

The membrane stress, stress resultants and the load components can be designed as P_x, P_y, P_z The coordinate axes, as in Section 4.2.2, are

x = along tangent to parallel circle
y = along tangent to meridian
z = normal to the shell surface

Equations of equilibrium of the element of a shell of revolution. As shown in Figs. 4.11, 4.12 and 4.13, the equations are given below. Equations of forces in the x direction.

Refer to Fig. 4.12: the tangents to meridians at A and B interact at an angle $d\theta$

$$AE = BE = a \tan \phi$$

$$AB = a \tan \phi d\theta \text{ and } AB = r \times \theta P \tag{4.11}$$

This required component of $N_{\theta\phi} \times a d\theta$ in the x direction is

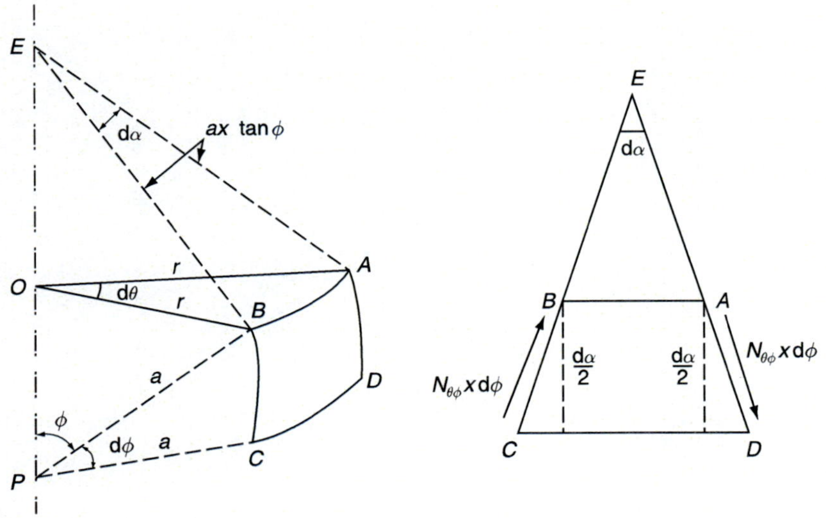

Fig. 4.11 Side element equilibrium

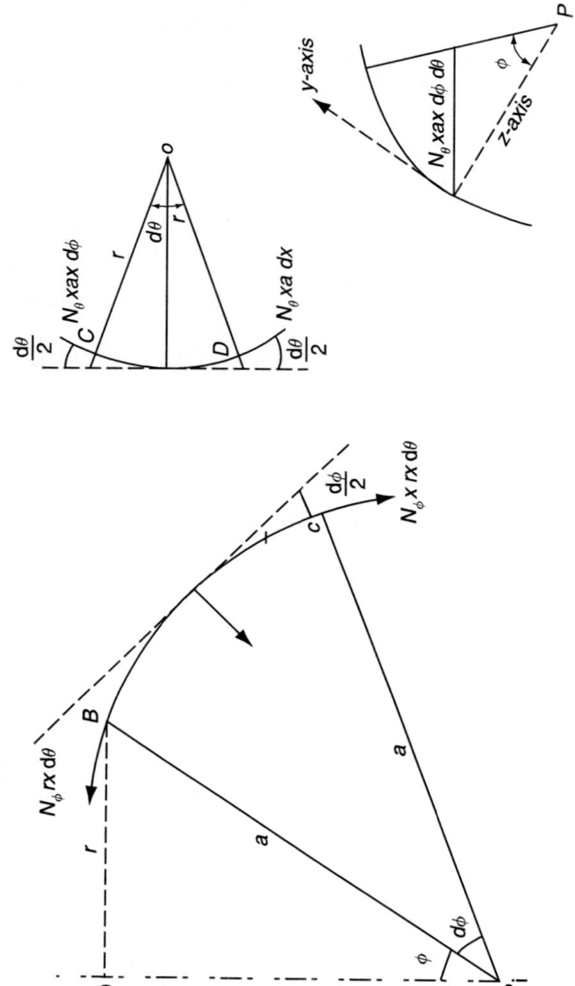

Fig. 4.12 Curve on longitude

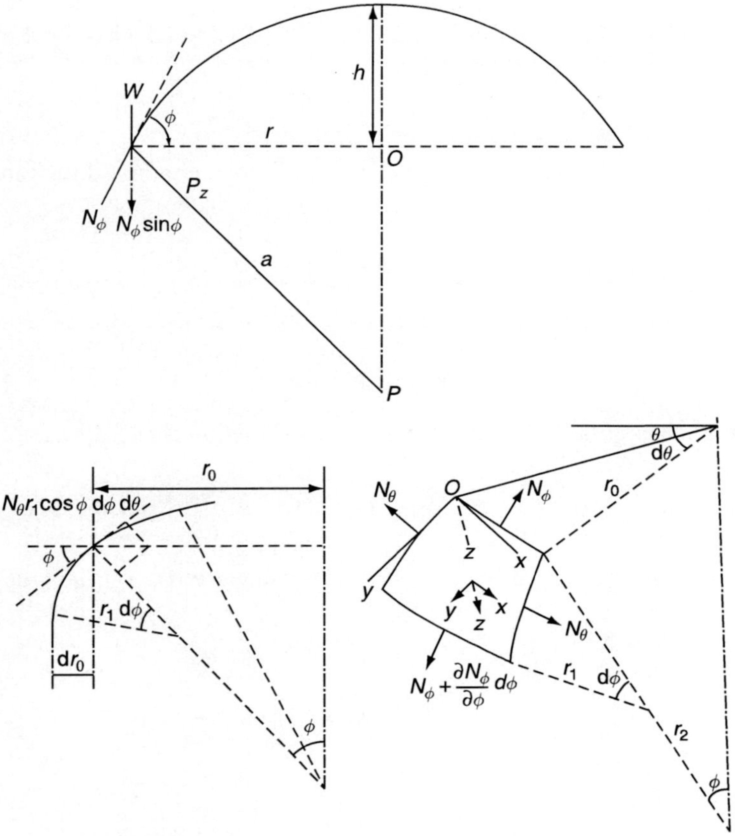

Fig. 4.13 Dome geometry

$$2 \times N_{\theta\phi} \times a\mathrm{d}\theta \sin\left(\frac{\mathrm{d}\alpha}{2}\right) = N_{\theta\phi} \times a\mathrm{d}\phi\mathrm{d}\theta = N_{\theta\phi} a \cos\phi \times \mathrm{d}\theta\mathrm{d}\phi = 0 \quad (4.12)$$

Therefore, the complete equation of equilibrium in the x direction is

$$\left(N_{\theta\phi} r\mathrm{d}\theta + \frac{\mathrm{d}N_{\theta\phi}}{\mathrm{d}\theta} \times r\theta\mathrm{d}\phi\right) - N_{\theta\phi}.r\mathrm{d}\theta + \left(N_{\theta}.a\mathrm{d}\phi + \frac{\mathrm{d}N_{\phi}}{\mathrm{d}\theta}.a\mathrm{d}\theta\mathrm{d}\phi\right) - N_{\theta}.a\mathrm{d}\phi$$
$$+\left(N_{\phi} a \cos\phi.\mathrm{d}\theta\mathrm{d}\phi\right) + P_x.a\mathrm{d}\phi.r\mathrm{d}\theta = 0$$

Divide all by $\mathrm{d}\theta\mathrm{d}\phi$ and the equation becomes

$$\frac{\mathrm{d}(N\phi\theta.r)}{\mathrm{d}\phi} + a.\frac{\mathrm{d}N_{\theta}}{\mathrm{d}\theta} + aN_{\theta\phi}.\cos\phi + P_x.ra = 0 \quad (4.13)$$

Equation of forces in the y direction
Refer to Fig. 4.13 the total force on AD and BC, N_θ and $a\mathrm{d}\phi$ has a component along the radius of the parallel circle:

$$2N_\theta a\mathrm{d}\phi\sin(\mathrm{d}\theta/2) \approx N_\theta \mathrm{d}\phi\mathrm{d}\theta$$

Resolving this in the y and z directions, one has Element in the y direction and is as follows:

$$N_\theta.a\mathrm{d}\phi\mathrm{d}\theta\cos\phi \tag{4.14a}$$

Element in the z direction is as follows:

$$N_\theta.a\mathrm{d}\phi\mathrm{d}\theta\sin\phi \tag{4.14b}$$

The complete equation of equilibrium in the y direction is as follows:

$$\left[N_\phi r\mathrm{d}\theta + \mathrm{d}(N_\phi)r\mathrm{d}\theta\frac{\mathrm{d}\phi}{\mathrm{d}\phi}\right] - N_\phi r\mathrm{d}\theta + \left[N_{\theta\phi}a\mathrm{d}\phi + \frac{\mathrm{d}(N_{\theta\phi})}{\mathrm{d}\theta}a\mathrm{d}\phi\mathrm{d}\theta\right] - N_{\theta\phi}a\mathrm{d}\phi$$
$$-N_\theta.a\mathrm{d}\phi\mathrm{d}\theta\cos\phi + P_\mathrm{y}a\mathrm{d}\phi r\mathrm{d}\theta = 0$$

Divide all by $\mathrm{d}\theta\mathrm{d}\phi$ we have

$$\frac{\mathrm{d}}{\mathrm{d}\theta}rN_\phi + a\frac{\mathrm{d}(N_{\theta\phi})}{\mathrm{d}\phi} - N_\theta a\cos\phi + P_\mathrm{y}ra = 0 \tag{4.15}$$

Equation of forces in the z direction
Required component of forces *on AB* and *CD* $N_\phi r d\theta$: component of $N_\phi r d\theta$ in the z direction is as follows:

$$2N_\phi r\mathrm{d}\theta\sin\left(\frac{\mathrm{d}\phi}{2}\right) \approx N_\phi r\mathrm{d}\theta\mathrm{d}\phi$$

Combining this with the component of z of $N_\theta a d\phi$ from (4.20a), the complete equation of equilibrium is as follows:

$$N_\theta a\mathrm{d}\phi\sin\phi + N_\phi r\mathrm{d}\theta\mathrm{d}\phi + P_z ra\mathrm{d}\theta\mathrm{d}\phi = 0$$

But $r = a\sin\phi$ therefore

$$\frac{1}{a(N_\theta + N_\phi)} + P_z = 0 \tag{4.16}$$

Assuming that the stresses are uniformly distributed over the thickness of the shell:

$$N_\theta = N_{\phi\theta} \tag{4.17}$$

Equations (4.19) and (4.21), (4.22) and (4.23) allow the calculation of the membrane forces of a shell of revolution. Hence with loads symmetric about the vertical axis of rotation, these four equations simplify considerably:

1. Stresses are independent of $\o$ and all partial derivatives with respect to ϕ disappear.
2. $N_{\theta\phi} = N_{\phi}\theta = 0$, otherwise they would produce asymmetrical deformation with respect to the vertical axis.
3. There is no variation of other stress resultants with θ.
4. The component P of the load must vanish; it would produce twist about the axis and shear deformation.

Therefore, (4.19) is identically satisfied and disappears. Hence, Eq. (4.21) becomes

$$\frac{\mathrm{d}}{\mathrm{d}\phi}(rN_{\phi}) - aN_{\theta}\cos\phi + P_{y}ra = 0 \tag{4.18}$$

Equation (4.22) remains the same:

$$\frac{1}{a(N_{\theta} + N_{\phi})} + P_{z} = 0$$

Therefore

$$N_{\theta} = -(P_{z}a + N_{\phi}) \tag{4.19}$$

For an axisymmetrically loaded dome, the meridional forces N_{ϕ} can be obtained much faster and directly by considering the equilibrium of the part of the dome above a certain parallel circle.

Then, if V is the vertical resultant of the applied load and $2\pi rN_{\phi}\sin\phi$ is the total vertical component around the parallel circle of the meridional force N_{ϕ}

$$2\pi rN_{\phi}\sin\phi + V = 0 \quad N_{\phi} = -\frac{V}{2\pi r\sin\phi} \tag{4.20}$$

Let W = weight per unit area. Surface area above parallel circle.

$$S_{\phi} = 2\pi aH = 2\pi a^{2}(1 - \cos\phi) \tag{4.21}$$

Therefore, the meridional force is as follows:

$$N_{\phi} = -W_{a}/(1 + \cos\phi) \tag{4.22}$$

Therefore, the hoop force is as follows:

$$N_\theta = W_a(1 - \cos\phi - \cos^2\phi)/(1 + \cos\phi) \tag{4.23}$$

The load on $r\mathrm{d}\phi$ about the axis

$$= r\mathrm{d}\phi.2\pi r \sin\phi w'(\phi - \phi_o) \tag{4.24}$$

Total load is 0–1 about axis

$$W_u = 2\pi r^2 w'(\sin\phi_1 - \sin\phi_o) - \cos\phi_1(\phi_1 - \phi_o)$$

Determination of meridional thrust T and hoop force H
The reader is referred to Figs. 4.14, 4.15 and 4.16 for computing the value of T and H.

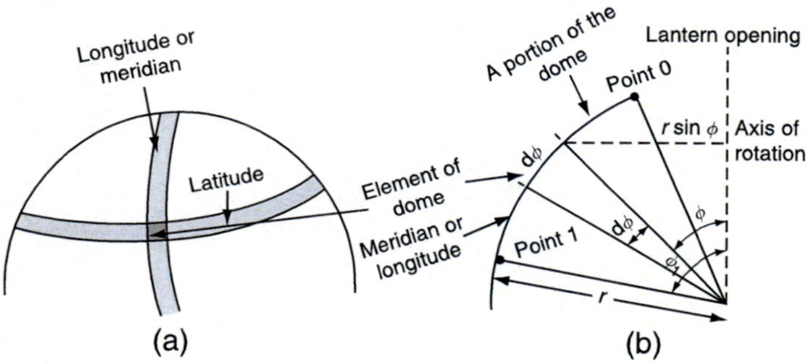

Fig. 4.14 Longitude and latitude of a spherical dome

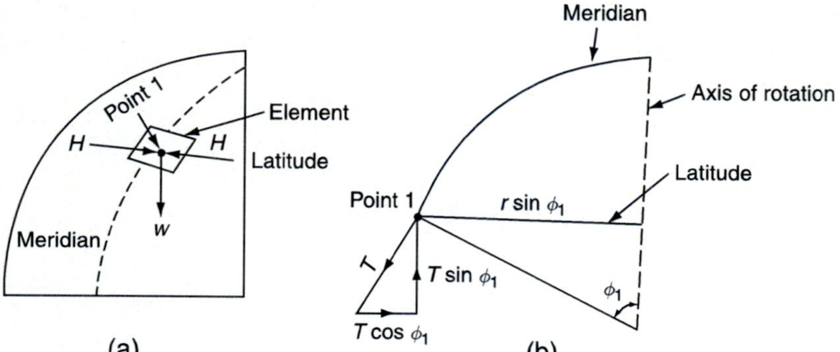

Fig. 4.15 Meridional thrust and hoop or circumferential forces

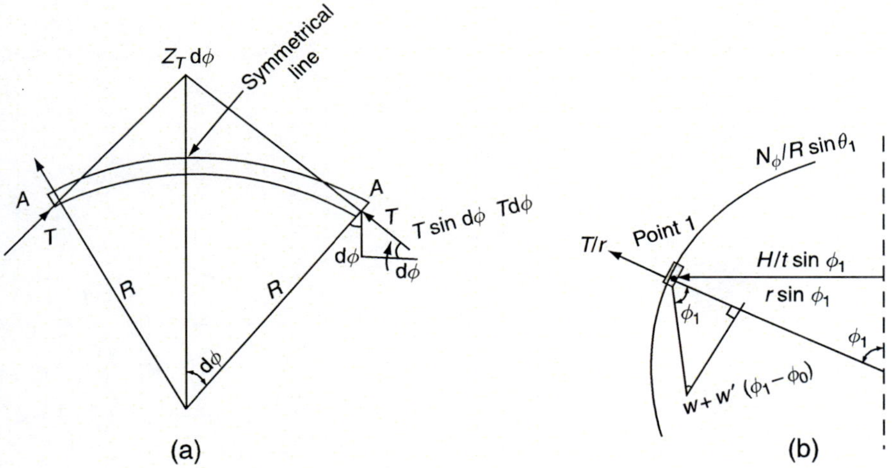

Fig. 4.16 Forces in equilibrium

$$W_u = \text{sum of vertical componets } T$$

$$W_u = 2\pi r\sin\phi_1 \times T\sin\phi_1 \tag{4.24a}$$

$$T = \frac{W_u}{2\pi r\sin^2\phi_1}$$

If the dome is continuous along the circle of latitude through 1, a circular ring through that point is subjected to a unit radial force of $(T \times \cos\phi_1)$. Therefore

$$S = \text{ring tension} = (W_u\cos\phi_1)/(2\pi \times \sin\phi_1) \tag{4.25}$$

Total radial component
Length of arc $A - A' = (2T \times \mathrm{d}\phi)/(2r \times \mathrm{d}\phi) = T/r$ therefore,

$$H = -T + [w + w'(\phi_1 - \phi_o)] \times r\cos\phi_1 \tag{4.26}$$

4.4.1 Elliptical Dome-Shaped Surfaces of the Vessel

The following analysis is carried out for the elliptical dome (Fig. 4.17). Equation of ellipse

$$\frac{X^2}{a^2} + \frac{Y^2}{b^2} = 1 \tag{4.27}$$

where $X = \text{a}\sin\phi$ and $Y = b\cos\phi$. Consider a ring produced by rotating an element ds about the Y axis of the dome.

Fig. 4.17 Elliptical dome

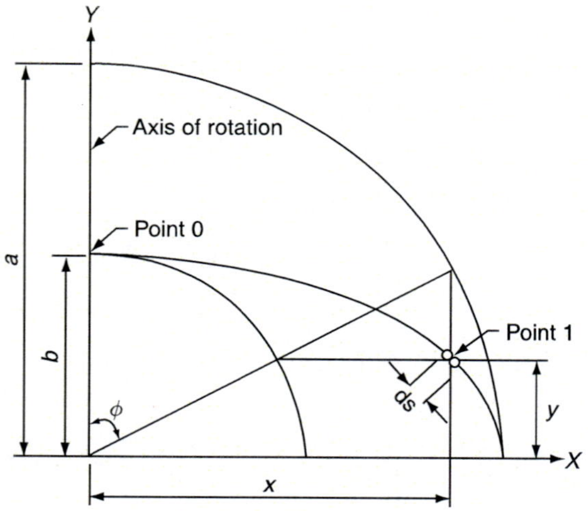

$$ds = \sqrt{[(dx)^2 + (dy)^2]} = \sqrt{[(a^2\cos^2\phi + b^2\sin^2\phi)]}d\phi \tag{4.28}$$

Weight of element ring is as follows

$$\begin{aligned} dw &= 2\pi \times X \times W \times ds \\ &= 2\pi \times w \times a^2 \sin\phi\sqrt{(1 - [((a^2 - b^2)/a^2)\sin^2\phi] \times d\phi)} \end{aligned} \tag{4.29}$$

Let $k^2 = (a^2 - b^2)/a^2$

W_u = total load between points 0–1

$$W_u = dw = 2\pi \times a^2 w\left[\cos\phi(\sqrt{1 - k^2\sin^2\phi}) - \frac{1}{2}k(1 - k^2) \times \log k\cos\sqrt{}(1 - k^2\sin^2\phi)\right]$$

Let $\cos\phi = y/b = g$ and $\sin^2\phi = 1 - \cos^2\phi = 1\, g^2$

Substitute into Eq. (4.35)

$$\begin{aligned} W &= 2\pi \times a^2 w\left[\frac{1}{2} + \frac{1}{2}k(1 - k^2) \times \log(1 + k) \right. \\ &\left. - \frac{1}{2}g\left(\sqrt{1 - k^2(1 - g^2)} - \frac{1}{2}k(1 - k^2) \times \log k + \sqrt{1 - k^2(1 - g^2)}\right)\right] \\ &= (2\pi a^2 w)C \end{aligned} \tag{4.30}$$

where C is the quantity in the bracket.

To find the meridional thrust

$$\begin{aligned} T &= \frac{W}{2\pi \sin \phi x} \\ &= \frac{2\pi \times a^2 \mathrm{w} C}{2\pi \times a(\sqrt{(1 - g^2 \sin \phi)}} \\ &= wCa^2 \sqrt{\frac{1 - k^2(1 - g^2)}{b(1 - g^2)}} \\ &= \frac{wa^2}{b} \times \frac{CQ}{1 - g^2} \end{aligned} \tag{4.31}$$

where $Q = \sqrt{1 - k^2(1 - g^2)}$

$$\begin{aligned} H &= \frac{-Tx}{R \sin \phi} + \frac{wx \cos^2 \phi}{\sin \phi} \\ &= \frac{-wx^2}{2R \sin^2 \phi} + \frac{wx \cos^2 \phi}{\sin \phi} \\ &= \frac{wa^2}{2b} \times \frac{2g^2 - 1}{Q} \end{aligned} \tag{4.32}$$

If the dome is continued along the circle of latitude through point 1, an edge member must be provided along that circle and that member is subjected to ring tension, S

$$S = \frac{Wa}{2\pi \times b} \times \frac{g}{\sqrt{1 - g_1}} \tag{4.33}$$

4.4.2 Torispherical Shell Surfaces of the Reactor Vessel Top Dome

The torispherical shell surface is formed when a toroidal segment is inserted between a spherical and a cylindrical cap. The dome surface is always used as well for the reactor vessel.

Main analysis

The geometry of this shell is shown in more detail in Fig. 4.18.

In terms of the radii of curvature of the sphere a_1, the toroid a_3 and the cylinder a_2, the meridional angle at the sphere—toroid junction is given by

$$\sin \phi_1 = \frac{a_2 - a_3}{a_1 - a_3} \tag{4.34}$$

In practice, the radius of the toroid is selected so as to provide smooth transitions at both ends.

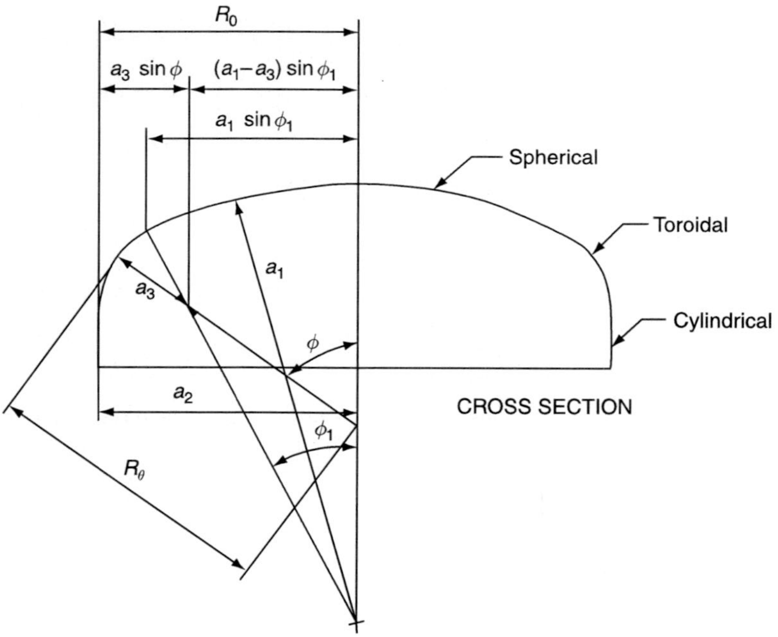

Fig. 4.18 Torospherical shell geometry

To derive the membrane theory stress resultants, one again uses the overall equilibrium method. At a general section within the toroid, $\phi_1 < \phi < \pi/2$, the resultant axial load consists of two parts: $P_I =$ the load on the spherical cap, and $P_2(\phi) =$ the load between ϕ_1 and ϕ on the toroid. For a uniform positive pressurep, the magnitude of the first force is the pressure multiplied by the projected base area of the cap.
Toroidal-shaped surfaces

$$P_1 = p\pi(a_1 \sin \phi_1)^2 \tag{4.35a}$$

For the second force, first compute the horizontal radius of the toroid

$$R_0 = (a_1 - a_3) \sin \phi_1 + a_3 \sin \phi \tag{4.35b}$$

Whereupon the magnitude equals the pressure multiplied by the area of the projected annulus,

$$\begin{aligned} P_2(\phi) &= p\pi[R_0^2 - (a_1 \sin \phi)^2] \\ &= p\pi[(a_1 - a_3)^2 \sin^2 \phi_1 + a_3^2 \sin^2 \phi + 2(a_1 - a_3)a_3 \sin \phi_1 \sin \phi - a_1^2 \sin^2 \phi_1] \end{aligned}$$

which reduces to

$$P_2(\phi) = p\pi a_3(\sin\phi - \sin\phi_1)[a_3(\sin\phi - \sin\phi_1) + 2a_1 \sin\phi_1] \tag{4.35c}$$

Summing Eqs. (4.35a) and (4.35c)

$$P(\phi) = p\pi\bar{p}(\phi)$$

where,

$$\begin{aligned}\bar{p}(\phi) = {} & (a_1 \sin\phi_1)^2 + a_3(\sin\phi - \sin\phi_1) \\ & \times [a_3(\sin\phi - \sin\phi_1) + 2a_1 \sin\phi_1]\end{aligned} \tag{4.35d}$$

$$\begin{aligned}N_\phi &= \frac{1}{2\pi R_0 \sin\phi}\bar{p}(\phi) \\ &= \frac{p\bar{p}(\phi)}{2[(a_1 - a_3)\sin\phi_1 + a_3 \sin\phi]\sin\phi}\end{aligned} \tag{4.36a}$$

To compute the circumferential stress resultant, the equation for N_θ is

$$N_\theta = R_\theta\left(p - \frac{N_\phi}{R_\phi}\right) \tag{4.37}$$

From Eq. (4.37) since $R_0 = R_0 \sin\phi$

$$R_\theta = \frac{a_2 - a_3}{\sin\phi} + a_3 = (a_1 - a_3)\frac{\sin\phi_1}{\sin\phi} + a_3 \tag{4.38}$$

From Fig. 4.18

$$R_\phi = a_3$$

Substituting (6.55a) and (6.56) into (6.55b) gives

$$\begin{aligned}N_\theta = {} & p\left[\frac{(a_1 - a_3)\sin\phi_1 + a_3 \sin\phi}{\sin\phi}\right] \\ & \times \left\{1 - \frac{\bar{Q}(\phi)}{2a_3 \sin\phi[(a_1 - a_3)\sin\phi_1 + a_3 \sin\phi]}\right\}\end{aligned} \tag{4.39}$$

Equation (4.36a) and (4.39) constitute the membrane theory solution for the so-called toroidal knuckle portion of the torospherical head, called Toroidal Knuckle:

4.5 Nuclear Reactor: Accident Analysis

4.5.1 PWR Loss-of-Coolant Accident (LOCA)

4.5.1.1 Introduction to LOCA

The important steel reactor pressure vessel analysis is to check them against loss-of-coolant accident (LOCA), especially in PWR. A loss-of-coolant accident (LOCA) occurs as a result of a penetration to the main coolant boundary such that the primary circuit water is released through the break to the containment area, causing a rapid decrease in the pressure and temperature of the primary coolant Fig. 4.19. This will give an impact thermal shock load. The streamline break accident (SLBA) occurs as a result of a complete and partial rupture of a steam line inside the containment vessel. A rapid cool-down and depressurization of the primary circuit normally take place. In order to restore

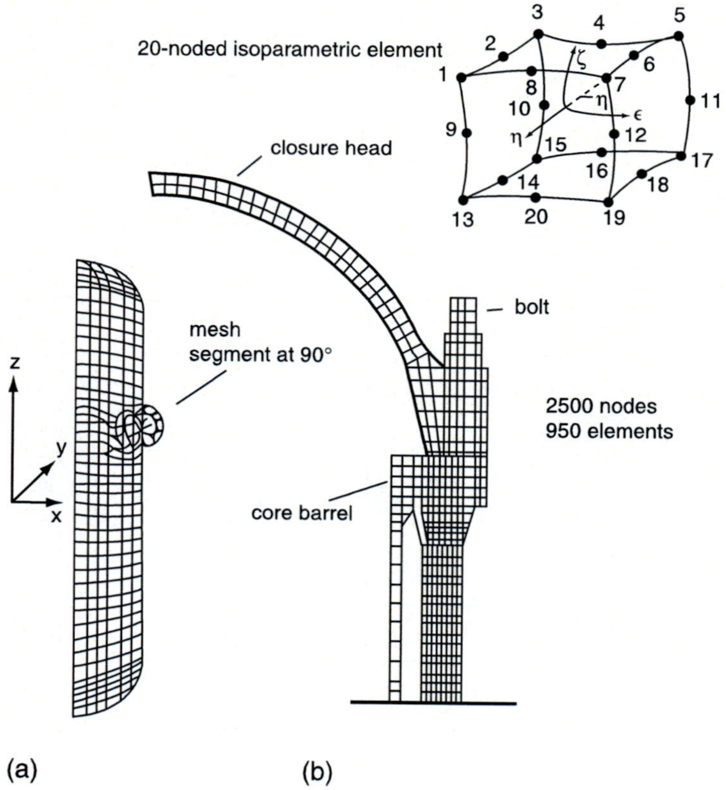

Fig. 4.19 Finite-element mesh generation of the vessel (**a**) generalized vessel mesh. (**b**) Closure head—wall flange region. (From Bangash)

the reactor coolant pressure, a pressure loading unconnected with LOCA is required.

4.5.1.2 Description of the PWR Vessel and Its Materials

Figure 4.1 shows a typical PWR vessel for sizewell B. The vessel steel must possess high toughness and strength coupled with adequate weldability in thick sections, with generally low-alloy steels containing manganese, nickel and molybdenum. The material grade for plates is SA-533 B alloys, and SA-508 alloys for forging. Both must be in quenched and tempered conditions. The suitability of these steels rests on the mechanical properties such as yield stress, ultimate tensile strength, elongation to fracture and charpy impact energy affected by thermal aging, strain aging and neutron irradiation. The vessels are made out of thick-section plates of up to 360 mm or from ingots of over 200,000 kg. The ingots generally develop cavities of up to 3 mm in the ν-segregation regions. These are healed by hot working processes. Both plates and forgings are welded. Figure 4.2 shows vessel fabrications.

Table 4.1 shows the data used in the 3D finite-element analysis. Figures 4.3 and 4.20 show the finite-element mesh generation scheme of the vessel, wall nozzles and closure heads. These are the important sections and locations in the reactor vessels from the point of view of fracture assessment. Figure 4.21a, b shows the pressure—time and temperature—time relationships. Figures 4.22–4.23 indicate various stresses in different zones due to a LOCA Table 4.3. The defect size in each region is checked using the R6 method of the former CEGB (UK) which is given in Tables 4.4–4.6.

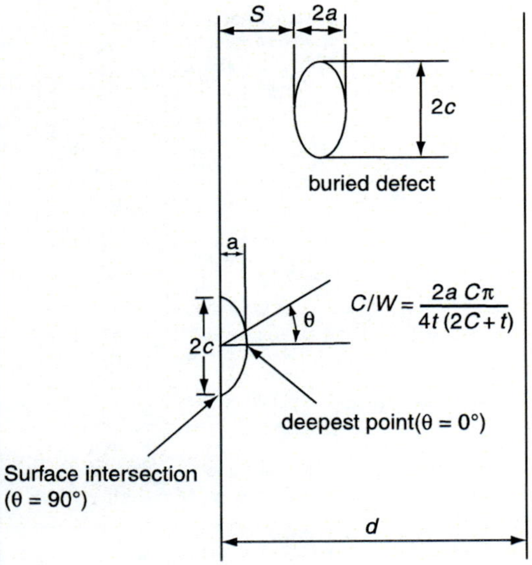

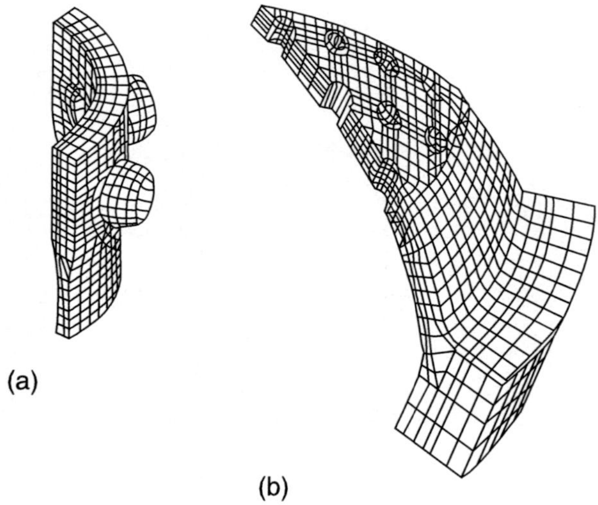

Fig. 4.20 Finite-element mesh generation of (**a**) wail nozzles and (**b**) closure head. (From Bangash)

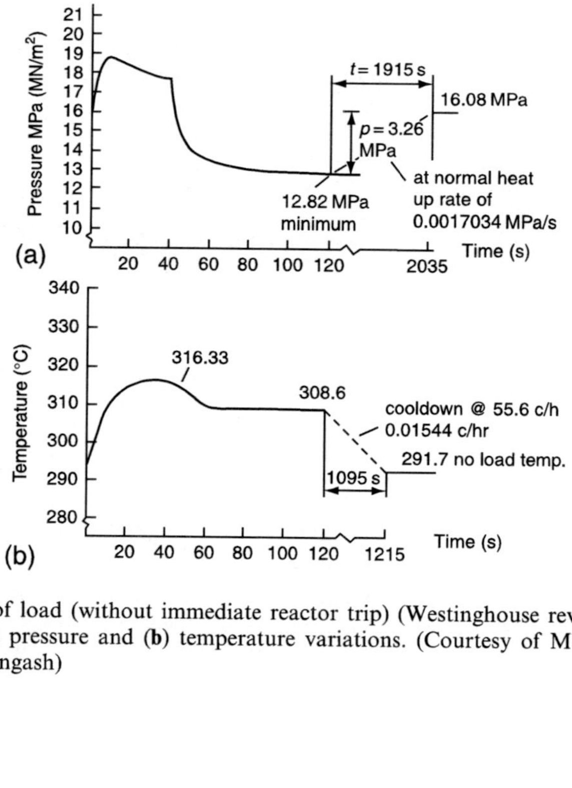

Fig. 4.21 Loss of load (without immediate reactor trip) (Westinghouse revision no 5). (**a**) Reactor coolant pressure and (**b**) temperature variations. (Courtesy of Marshall of U.K. AEA) (From Bangash)

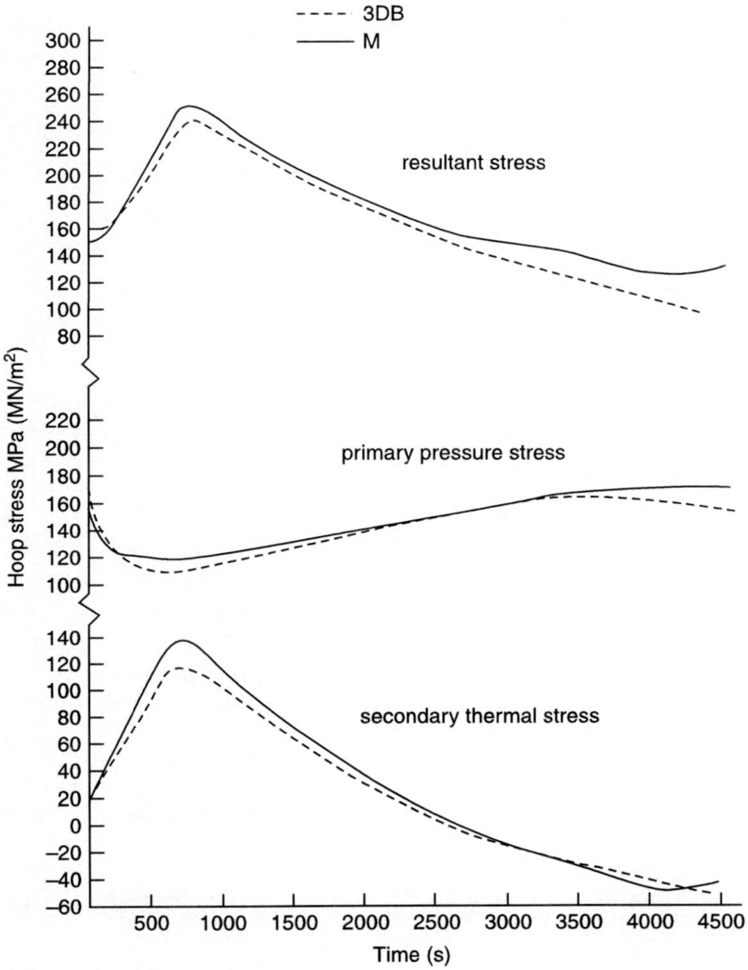

Fig. 4.22 Belt-line reactor trip from full power. Time variation of hoop stress in vessel material from clad/vessel interface. 3DD: three-dimensional finite-element analysis (Bangash); M. Marshal[5.155] ; (From Bangash[3.3])

4.5.1.3 Nuclear Containment Under Hydrogen Detonation

Hydrogen detonation has become an important issue after the Three-Mile-Island accident. The hydrogen burning occurred approximately 10 hours into the accident. The steam reacting with the Zircaloy cladding and the oxidation of the overheated steel vessel interiors created large quantities of hydrogen. This can also occur due to interaction of the molten core. In order to predict the wall pressures due to such detonations, non-linear gas dynamics equations for the entire volume of the containment vessel have to be solved. In the current analysis of the Sizewell B containment vessel, it is assumed that the wall pressure

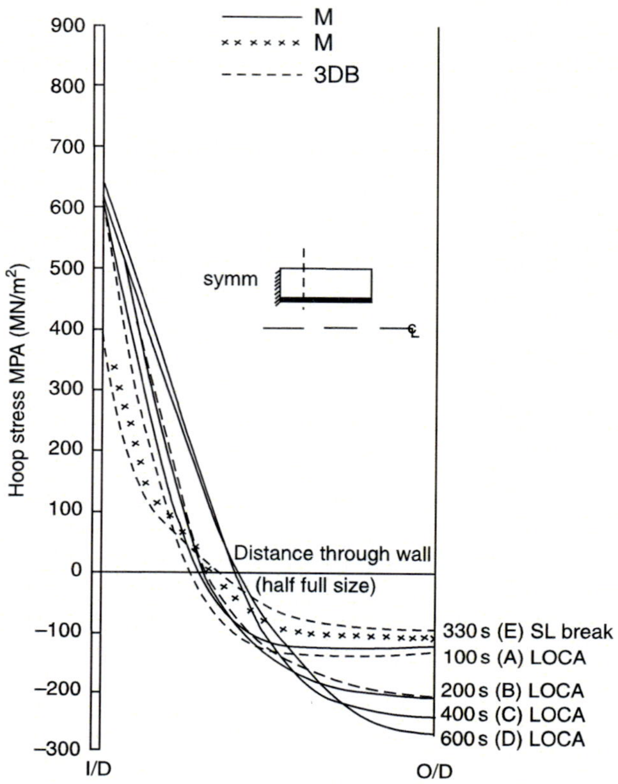

Fig. 4.23 Secondary thermal stress at various durations: belt-line region. LOCA condition. 3DB: Bangash: M: Marshall.[5.155] (From Bangash[3.3])

P_0 is proportional to the (containment pressure P. The vessel parameters are given in Table 4.1 and Figure 4.1. It is assumed that the detonation starts approximately at the mid-height of the containment vessel. The spherical shock front generated obliquely converges at the dome, causing a strong reflection around the apex. The containment finite-element mesh scheme is unchanged. Bond slip and shear slip for the reinforced elements are considered. The following additional input data have been included:

Figure 4.34 a, b shows the non dimensional relationship for the pressure P/P_o and $(t/R)v_{so}$ for the Sizewell B vessel at the apex of the dome and at the mid-height. The parameters R is the radius of the containment vessel. Figures 4.35 to 4.37 illustrate the stress-time histories fopr the containment wall, dome springing and dome apex due to detonation.

E.4 Jet impingement forces on PWR steel vessel components

The safety of nuclear installations such as the pressure vessel and its piping systems requires strict measures. In the event of steel failure, the safety rules

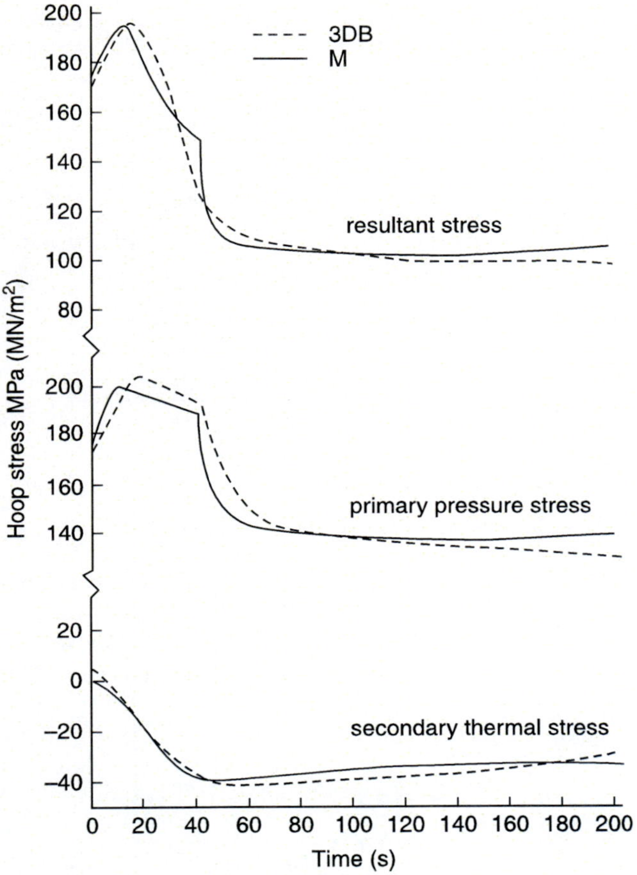

Fig. 4.24 Time variation of hoop stress in vessel material from clad/vessel interface (belt-line loss of load). 3DB: Bangash; M: MarshalE[5.155](From Bangash[3.3])

require the assessment of the jet impinging forces on the vessel nozzle areas. In the current analysis it was assumed that the transition from the cross-section of the discharging pipe to that of the outlet is abrupt. It is imperative to evaluate the structural behaviour under the jet impinging forces. The vessel chosen for this analysis was the Sizewell B PWR steel vessel. All parameters and finite-element mesh schemes are kept the same as in Figure 4.1. The following data is used:

At any height/diameter ration ($z/D = z/737$), the pressure ration $R_{p'} = p'T/P_w$ is computed, where p' is the saturation pressure at temperature T and P_w is the vessel pressure: $z/D = 0.85, 0.95$, 1, 2, 3, 4, 5 and 10: $p'T/P_w$ ranges between 0 and 1 and there is a total of 20 times steps (t), with time interval;s pf 0.24 s.

The jet impinging force against $P'T/P_w$ for failure conditions is given in Figs. 4.28 and 4.29

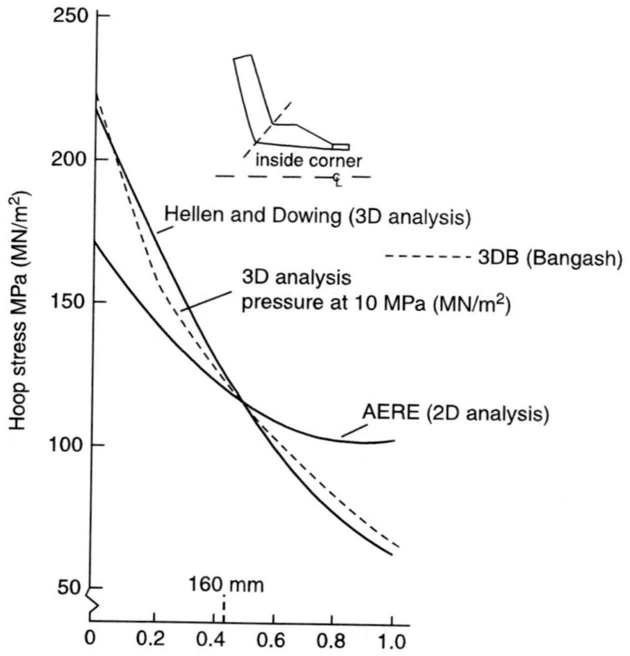

Fig. 4.24 (a) Comparative study of results for the inside corner of an inlet nozzle. From Bangash[3.3]

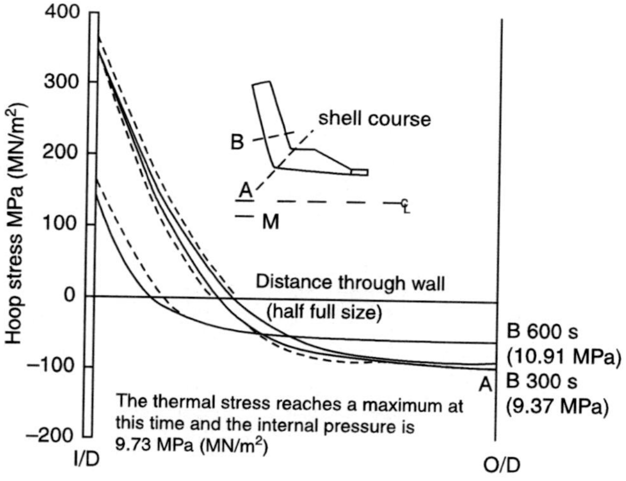

Fig. 4.25 Secondary thermal stresses at various durations for an inlet nozzle shell course large streamline break, M: Marshall,[5.155] (From Bangash[3.3])

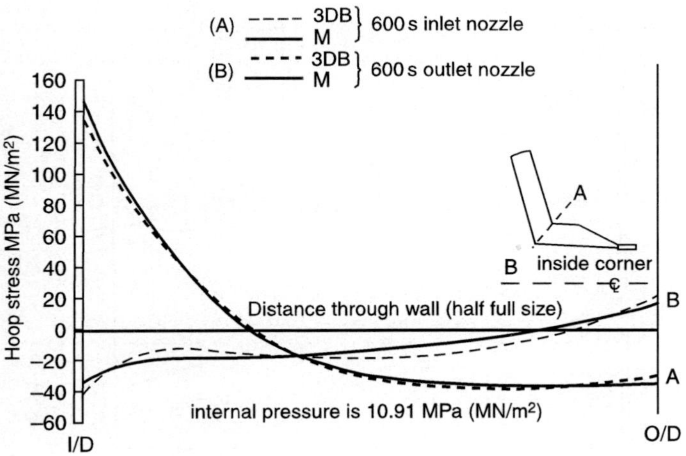

Fig. 4.26 Inlet nozzle inside corner: reactor trip from full power. 3DB: Bangash: M: Marshall[5.55] (From Bangash[3.3])

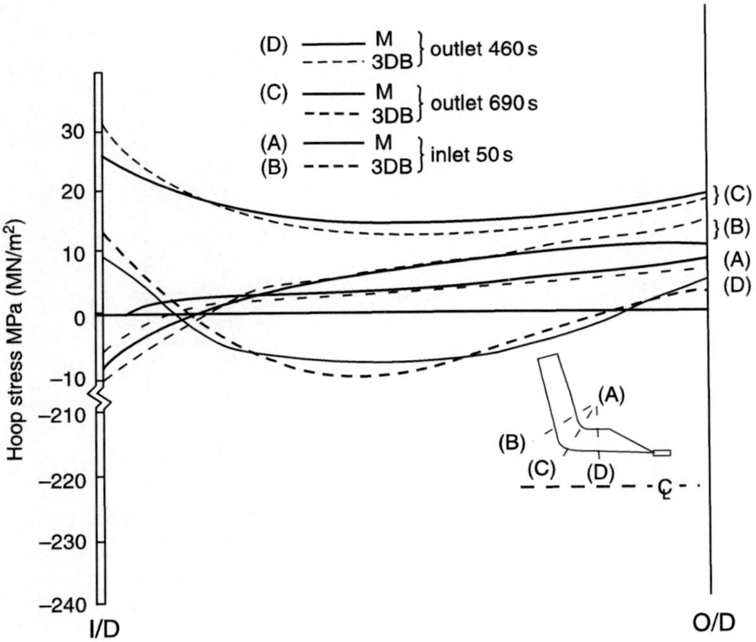

Fig. 4.27 Inlet nozzle inside corner; loss of power, secondary thermal stresses. 3DB: Bangash; M: Marshall[5.155] (From Bangash[3.3])

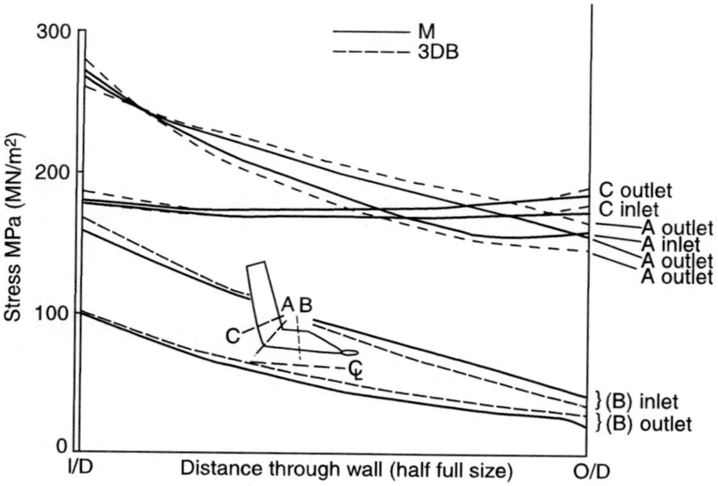

Fig. 4.28 Inlet nozzle hoop stress plotted through the wall at the sections shown. Basic pressure 17.24 MPa (MN/m^2). 3DB: Bangash: M: Marshall[5.155] (From Bangash[3.3])

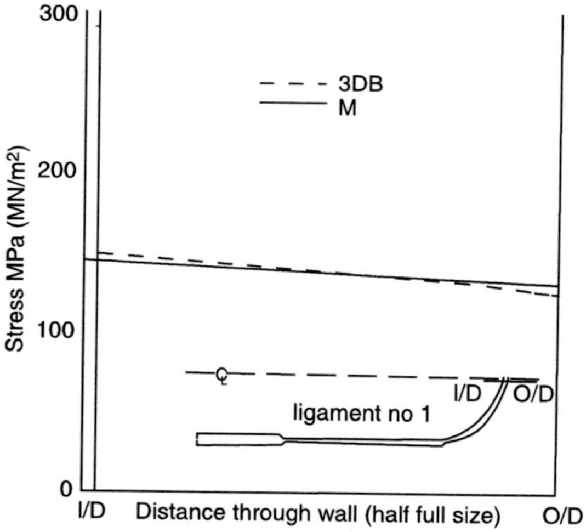

Fig. 4.29 Main hoop stress plotted through the wall at the sections shown. Basic pressure load of 17.24 MPa (MN/m^2). 3DB: Bangash: M: Marshall[5.155] (From Bangash[3.3])

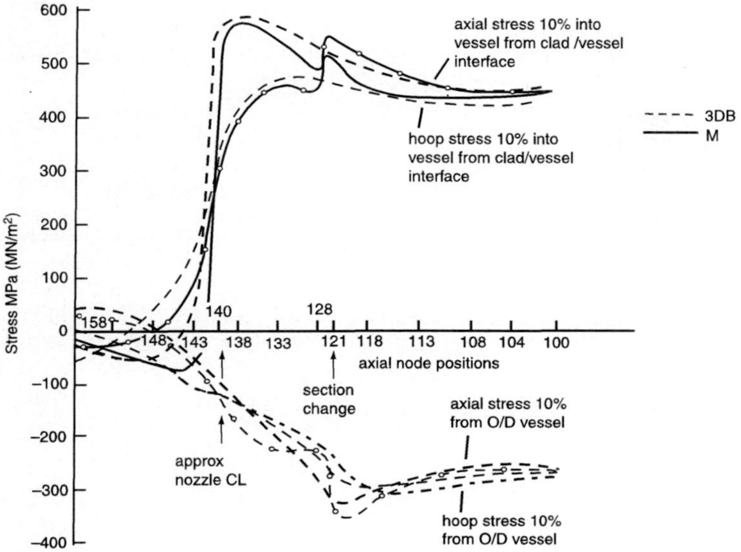

Fig. 4.30 Axial and hoop stress plotted along the nozzle course and belt-line region for a large LOCA at 600 s. 3DB: Bangash: M: Marshall[5.155] (From Bangash[3.3])

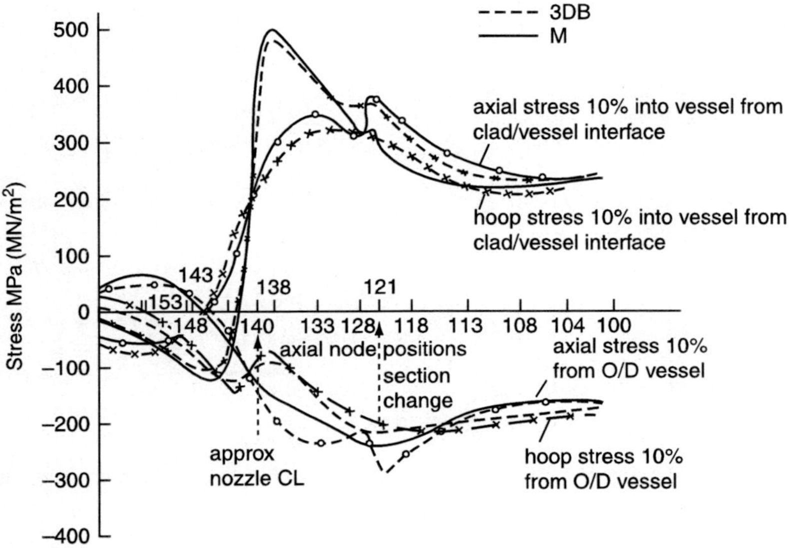

Fig. 4.31 Axial and hoop stress plotted along the nozzle course and belt-line region for a large LOCA at 2000 s, 3DB: Bangash: M: Marshall[5.155] (From Bangash[3.3])

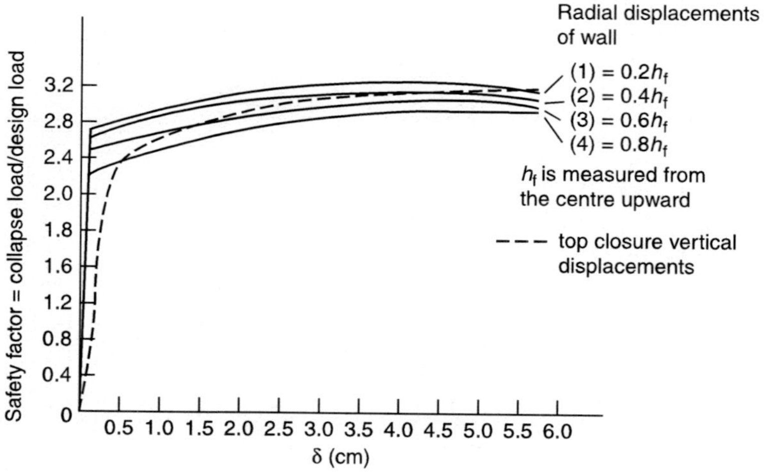

Fig. 4.32 Safety factors versus displacement (δ). (From Bangash[3.3])

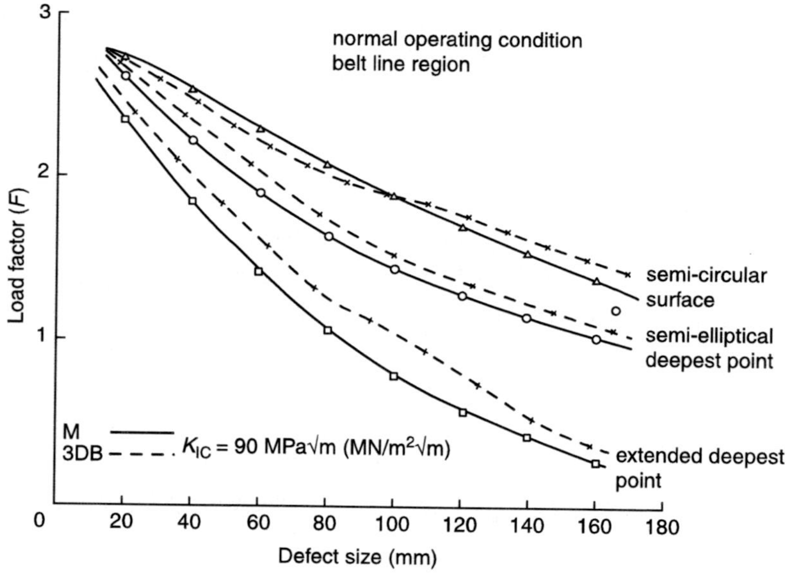

Fig. 4.33 Load factor versus defect size. 3DB: Bangash; M: Marshall[5.155] (From Bangash[3.3])

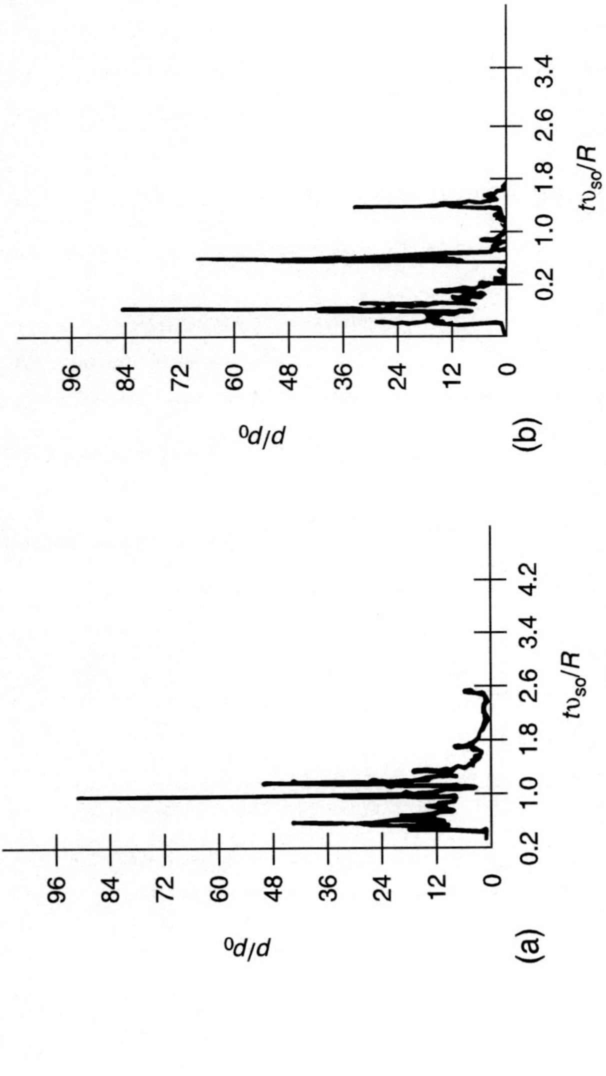

Fig. 4.34 (**a**) pressure at the mid-height; (**b**) pressure at the apex of the dome

Material

Liner: $\sigma_y = 390$ MPa

Strain rate = 0.1/s

Pressures and shock front

Detonation strength $q/\bar{R}T_0 = 17$ (base) and 23 (mid-height)

where q = energy release/unit volume

$\bar{R}$ = gas constant = 8.31 kJ/kg-mole K

T_0 = absolute temperature

Shock front

ν_{so} = shock wave velocity = 2000 m/s

t = 0,50, 100, 200, 250, 300 μs

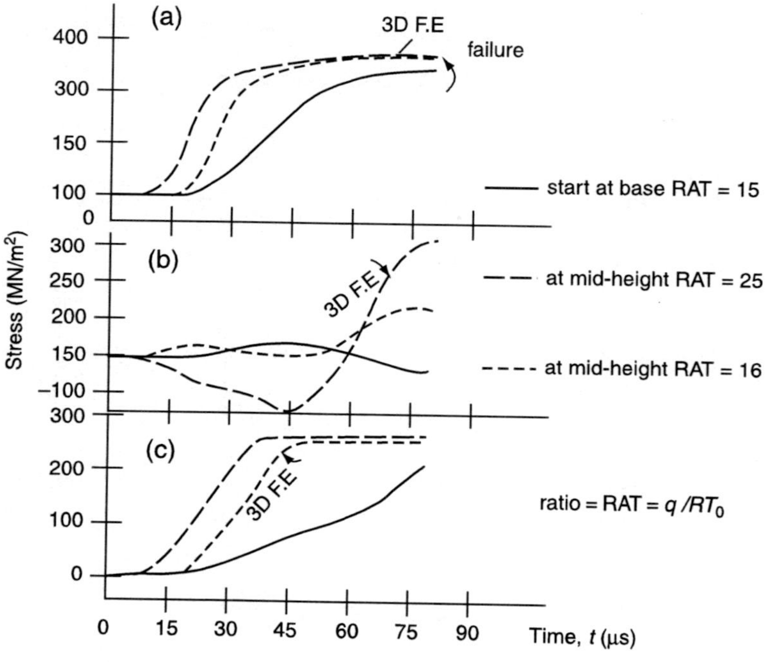

Fig. 4.35 Stress histories at 40% of the cylinder, with variation of steel properities included: (**a**) inside hoop bars; (**b**) inside vertical bars; (**c**) seismic diagonal bars.

Table 4.3 Vessel material properties and parameters

Proposed parameters for Sizewell B:

Vessel overall height	13660 mm
Inside diameter	4394 mm
Wall thickness opposite core	215 mm
Wall thickness at the flange	500 mm
Normal clad thickness	6 mm
Thickness of the dome top	178 mm
Thickness of the dome bottom	127 mm
Inside diameter of inlet nozzle	700 mm
Inside diameter of outlet nozzle	737 mm
Number of closure studs	54 (each 1466 mm high) (nut 268×203) (washer 268×38)
Diameter of closure studs	173 mm
Dry weight of the pressure vessel	434.8×10^3 Kg
Normal operating pressure	15.98 MPa
Design pressure	17.13 MPa
Initial hydraulic pressure	21.43 MPa
Normal operating inlet temperature	288°C
Normal operating outlet temperature	327°C
Design temperature	343°C
No load temperature	292°C
Design life	40 years at 80% load factor
$E_s = C\times E\times10^5$ MPa	C varies with temperature
(1) Material SA533B	C = 0.218, Mn = 1.367, Ni = 0.547, Mo = 0.547, Si = 0.236, Cr = 0.074, P = 0.009, S = 0.014
(weight per cent)	C = 0.117
(2) Submerged arc-welding	C = 0.16, Mn = 2.20, Ni = 0, Mo = 0.6, Si = 0.05, Cr = 0, I = 0.025, S = 0.035
(electrode wire content)	C = 0.15

(3) Mechanical properties	Yield stress (N/mm^2)	Ultimate tensile stress (N/mm^2)	Charpy test (minimum value)
(a) SA533B plates	at 20°C 345	at 20°C 555	
	at 400°C 280	at 400°C 520	34 J at 4.4°C flanges, shell
(b) SA508 forging	at 400°C 280	at 20°C 555	rings and nozzles average or
		at 400°C 430	values 150, 138, 100,
(c) Weld metal	483	593	respectively

Table 4.3 (continued)

(d) Stresses after 288°C irradiation unirradiated	These are given before from 3(a) to 3(c) Add 2% of the above values	
(e) Fracture toughness of plates	$K^{Ic} = 106$ N/mm^2,$\surd m = 153$	
(f) Bonding material	Yield = 900–1050 N/mm^2; ultimate tensils stess = 1050 N/mm^2; impact energy = 60.9–81.2 J	
(4) Mechanical properties	Yield stress (N/mm^2)	Ultimate tensile stress (N/mm^2) Charpy test (minimum value)
(a) SA533B plates	at 20°C 345 at 400°C 280	at 20°C 555 at 400°C 520 34 J at 4.4°C
(b) SA508 forging	at 400°C 280	at 20°C 555 at 400°C 430 flanges, shell or rings and nozzles
(c) Weld metal	483	593 average values 150, 138, 100, respectively
(d) Stresses after 288°C irradiation unirradiated	These are given before from 3(a) to (3c) Add 2% of the above values	
(e) Fracture toughness of plates	$K_{Ic} = 106$ N/mm^2, $Vm = 153$	
(f) Bonding material	Yield = 900-1050 N/mm^2; ultimate tensile stress = 1050 N/mm; impact energy = 60.9-81.2 J	

Table 4.4 R6 method of fracture assessment

The failure assessment diagram for the R6 method (courtesy of CEGB, UK).

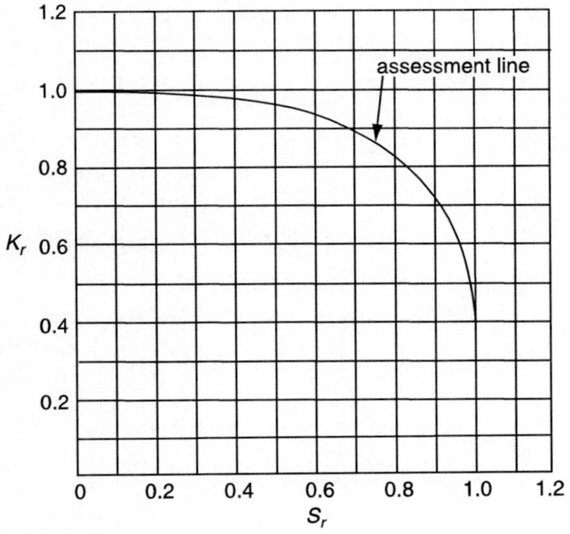

Figure represents K_r–S_r diagram

$$K_r = S_r\left\{\tfrac{8}{\pi^2}\log_e \sec\left(\tfrac{\pi}{2}S_r\right)\right\}^{-1/2} \quad (4.40)$$

$$K_r = K_r^P / K_{IC} \quad (4.41)$$

where K_r is a measure of how close the vessel is to linear clastic failure = 0.59 and $K_I{}^p$ is the stress intensity factor due to σ stresses and is given by

$$K_I^P = Y\sigma\sqrt{\pi a} = 39.42\,\text{Mpa}\sqrt{\text{m}}$$

where Y = magnification factor when applying unflaw stresses obtained to postulated flawed vessel; a Y value of 1.25 has been taken

σ = applied stress

a = crack height

K_{IC} = fracture toughness, with a lower limit of 170 MPa $\sqrt{\text{m}}$ at 288°C

Defect length $2C$ parallel to component surface

Defect depth S distance between nearest edge of the defect and component surface (normal distance)

Defect height $2a$ distance between nearest and the furthest extremities of a defect normal of the surface (buried $2a$, surface breaking a)

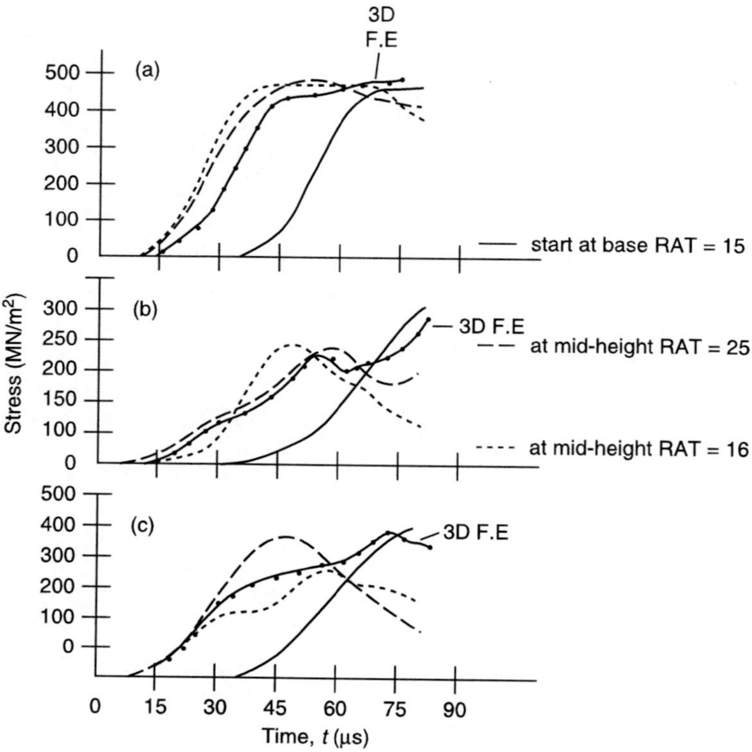

Fig. 4.36 Stress histories at hte dome springline, with variation of steel properties included: (**a**) inside hoop bars; (**b**) inside vertical bars; (**c**) seismic diagonal bars

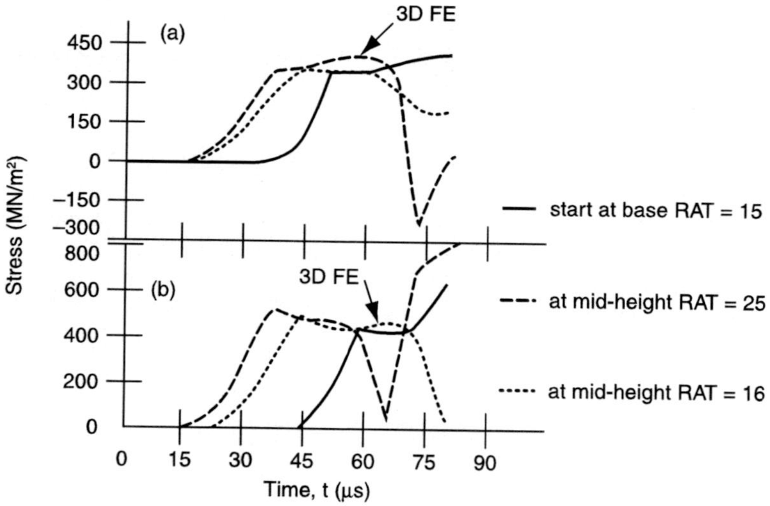

Fig. 4.37 Strees histories 10° from the dome apex, with variation of steel properties included: (**a**) inside hoop bars; (**b**) inside meridional bars

Nozzles and pipe
Pw = 17.13 MPa; initial hydraulic pressure = 21.42 MPa design
Temperature: 327° C
Discharging pipe diameter: 133 mm (120 Elements)
Outside nozzle diameter: 737 mm (390 Elements 20 noded)
Inside nozzle diameter: 730 mm (390 Elements 20 noded)
Hydraulic coefficient of resistance: 0.37 to 0.82
Nozzle-structure distance: 330 mm

Finite Element Data
400 elements in the cylindrical part
20 noded isoparametric Elements
351 dome part
mixiture of 8-noded and 82 noded elements

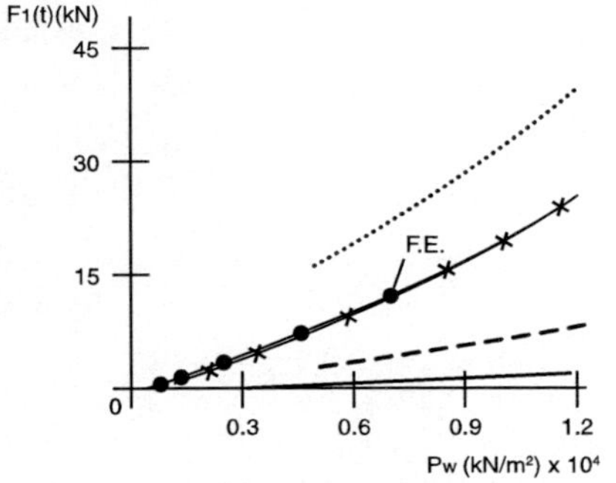

symbol	ND	z/D
········	65	0.5
—●—●— F.E.	50	0.5
—✕—✕—	25	1.0
— — —	10	0.5

Fig. 4.38 Measurements for jet impingement forces on a flat plate due to the discharge of saturated pressurized water from circular nozzles of different diameters

4.6 State-of-the-art Software for 3D Fracture Mechanics Simulation (ZENCRACK)

(Provided by R. Chandwani, Managing Director ZENTEC Corp. London). Zencrack is an advanced engineering analysis tool for 3D fracture mechanics assessment and crack growth simulation. The program uses finite element analysis to allow calculation of fracture mechanics parameters such as energy release rate and stress intensity factors. This is achieved by automatic generation of focused cracked meshes from uncracked finite element modules. A general mixed-mode crack growth capability allows non-planar crack growth prediction for fatigue and time-dependent load conditions via automated adaptive meshing techniques.

Overview

Zencrack provides flexibility with two levels of simulation capability – Standard and Professional.

For industries where static loading is important the Standard version can be used to evaluate stress intensity factors using energy release rate and nodal

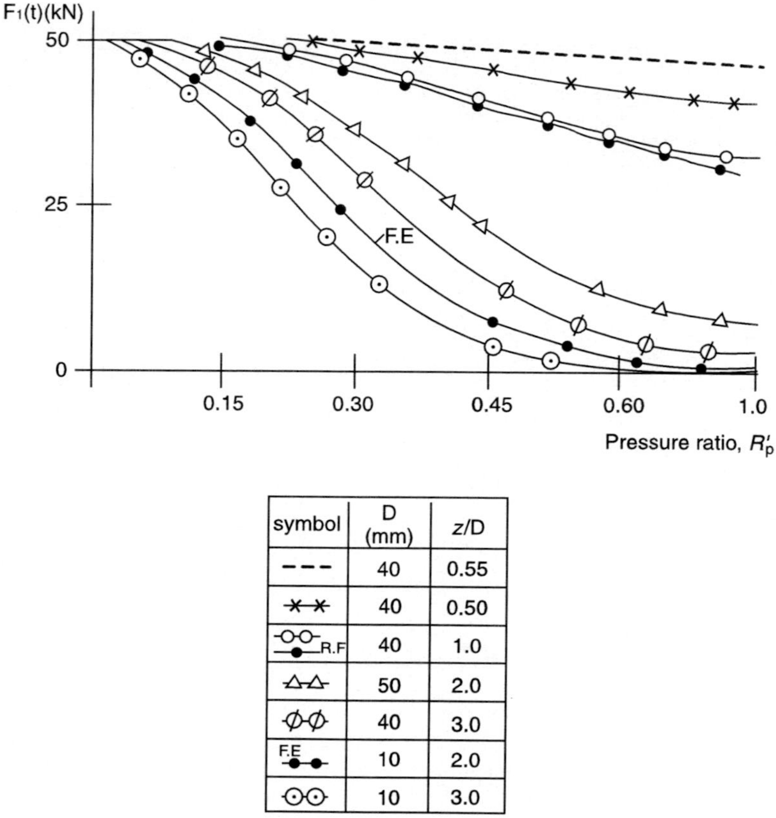

Fig. 4.39 R'_p versus $F_1(t)$

Table 4.5 Semi-elliptic surface breaking floor configuration: sample calculations

For $0.1 < a/2c < 0.5$ and $a/t \leq 0.8$. When $a/2c < 0.1$ then $c/w = a/t$ (extended crack).
Therefore the S_r value is given by

$$S_r = \{1/\bar{\sigma}(1-c/w)^2\}\{[\sigma_{bc}/4+\sigma_{mc}(c/w)]\}+[(\sigma_{bc}/4+\sigma_{mc}(c/w))^2+\sigma^2_{mc}(1-c/w)^2]^{1/2} \quad (4.42)$$

$a/c13$ if $a = 70$ mm: $c = 210$ mm

$0.1 < 7/42 < 0.5$ is satisfied

$$c/w = 2ac\pi/4t(2c+t) = 2(70)(210)\pi/4(216)(420+216) = 0.168$$

$$\sigma bc = 6m/t^2 = 7.56/, \text{MPa} \quad (4.43)$$

With $a/c\frac{1}{3}$. $\sigma_{mc} = 171.5$ MPa, if $a = 70$ mm and $c/w = 0.168$ = area of flaw/area of rectangle.

$$S_r = [7.56E/4 + 175.5E6(0.168)] + [(7.56E6/4 + 171.5E6(0.168)^2$$
$$+ (171.5)^2(1-0.168)^2|^{1/2}/440E6(1-0.168)^2$$

or $S_r = 0.58$

and $K_I^P = Y\sigma_{mc}\sqrt{\pi a} = 1.25(171.5E6)\sqrt{\pi(0.07)} = 100.53\,\text{MPa}\,\sqrt{\text{m}}$ (4.44)

where $K_r = K_I{}^P/K_{IC} = 100.53E6/170E6 = 0.59$

σ_{bc} = clastically calculated bending stress evaluatued over the gross section containing the flaw
σ_{mc} = clastically calculated tensile stress evaluated over the gross section containing the flaw

Table 4.6 Failure assessment of crack heights due to LOCA

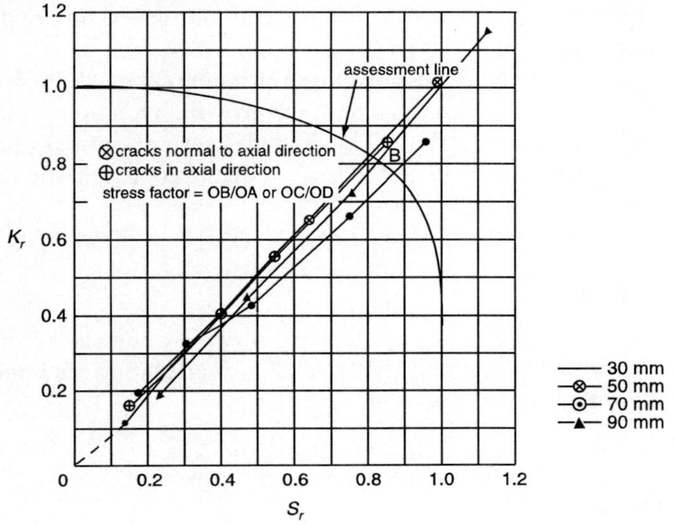

Computed data: normal operating conditions, belt-line region S_r and K_r values for flaws normal to the hoop stress.

Crack height	Effective ration	Calculated values	
a (mm)	c/w	S_r	K_r
70	0.168	0.58	0.59
90	0.234	0.70	0.67
110	0.301	0.86	0.74
130	0.371	1.10	0.81

Normal operating conditions, belt-line region S_r and K_r values for flaws normal to hoop stress.

Crack height	Effective ratio	Calculated values	
a (mm)	c/w	S_r	K_r
70	0.168	0.25	0.23
90	0.234	0.30	0.26
110	0.301	0.38	0.29
130	0.371	0.48	0.32
150	0.440	0.64	0.34
170	0.510	0.88	0.36

displacement methods. For thermal transients, the instantaneous stress intensities may be evaluated through the transient to steady state conditions. Collapse analyses may be undertaken to allow, for example, generation of data for failure assessment diagrams.

The second level of capability is introduced through Zencrack Professional and provides a facility for 3D non-planar crack growth prediction for cases of fatigue and time-dependent loading. This includes several options for crack growth data definition and a flexible "load system" approach for defining complex load spectra.

4.6.1 Application Areas

ZENCRACK can be applied in any industry in which knowledge of crack behaviour, crack growth prediction and residual life calculation are important. Zencrack is relevant in many situations, e.g.

- Post-failure, forensic and accident investigations
- Parametric studies of different crack sizes in a component
- Leak before break studies
- Design of experiments
- Repair assessments
- Determination of inspection periods
- Determination of residual life
- Assessment of brittle-ductile failure using failure assessment diagrams

ZENCRACK takes an uncracked finite element mesh and inserts one or more user-specified cracks before submitting the model for analysis. Results are processed automatically and the mesh updated ready for the next finite element analysis of the advanced crack position.

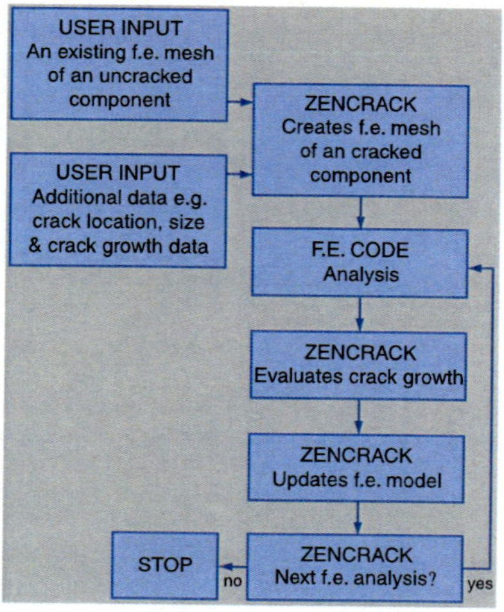

4.6.2 Sample Applications

- Roller support including significant mode II and mode III effects and the use of crack face contact
- Crack at a bolt hole in an engine disk including temperate effects.

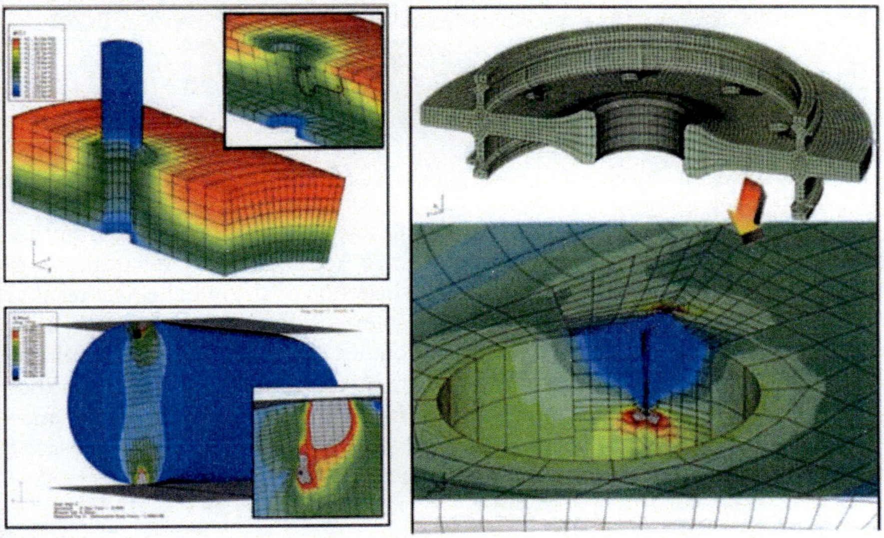

4.6.3 F.E. Interfaces

By interfacing to commercial finite element codes rather than using a proprietary finite element solution, Zencrack is able to take advantage of the many man-years of development within these codes and their associated pre and post-processors.

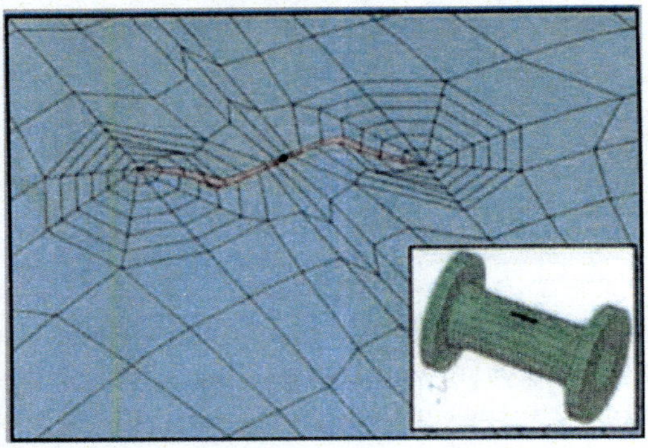

4.6.4 A Typical Crack Growth Analysis Procedure

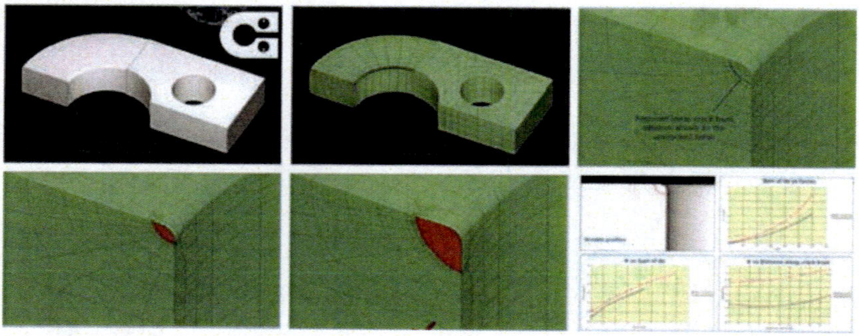

A typical crack growth analysis begins with a geometry model created in the user's pre-processor. This is meshed as an "uncracked model" in preparation for the insertion of the defect(s) by Zencrak. The insertion of the initial defect results in a cracked mesh ready for analysis in the interfaced finite element code. The analysis results are extracted for a fully automatic crack growth simulation in which re-meshing technologies allow the crack to be advanced through the structure. Typical results from an analysis include crack growth profiles and plots such as crack growth vs load cycles and stress intensity factor vs crack size. This is a useful method also when the vessel cracks under LOCA.

4.6.5 Meshing Procedure

The generation of 3D finite element models suitable for analysis of cracks requires special attention in and around the crack region where focused rings of hex elements are used. In addition, the initial crack front may be straight, a simple elliptic section, or a general curve in space.

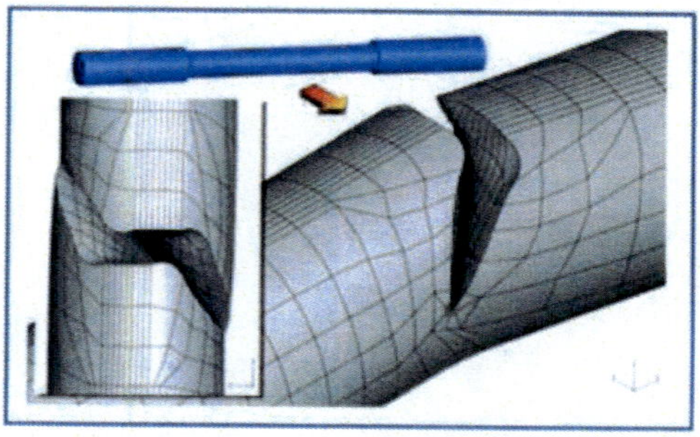

To address these issues, Zencrack removes the onus of modelling the crack region from the analyst and requires instead that an uncracked mesh is supplied. A "crack-block" approach is then used to introduce one or more crack fronts into the uncracked mesh. The term crack-block refers to a collection of brick elements stored as a unit cube. The arrangement of these noded brick elements in a user-supplied uncracked mesh by crack-blocks. During the mapping process to introduce the crack-blocks the user can control the size and shape of the generated crack-block. Crack-blocks can be connected together to form distinct crack fronts of the required size in the cracked mesh.

4.6.6 Adaptive Re-meshing

In a crack growth analysis the shape of the crack that develops is a function of the geometry, loading and material properties. The re-meshing of the crack-blocks and if necessary the modification of the surrounding mesh that takes places automatically to allow the correct crack shape development in the structure. No assumptions are forced onto the developing crack shape – it is a consequence of the structure, loading and material. Any local mixed mode effects arising at the crack front are embodied in the stress and displacement solution and are handled during the growth and re-meshing procedure.

4.6.7 Load Handling

The loading requirements in a fracture mechanics analysis vary considerably depending upon the goal of the analysis. Some possible examples include:

- Parametric study of crack sizes to determine stress intensity factors
- Evaluation of j-integrals at multiple load increments in a non-linear analysis

Chapter 5
Concrete Reactor Pressure Vessels

This chapter attempts to review the state of the art of methods of analysis and design of the concrete reactor pressure vessels and their components. Existing vessels have been examined for elastic, inelastic and cracking conditions. The results obtained from the analysis given in this chapter and in the appendices are well collaborated with those available for the experimental tests on models and from site monitoring of similar structures of vessels. The text is provided with an up-to-date comprehensive bibliography.

5.1 Introduction

The state-of-the-art review of the prestressed concrete reactor vessels covers, in this chapter, historical development, the available methods of analysis and design procedures of vessels and their elements such as concrete, prestressing systems, conventional steel, vessel steel liner and anchorages including penetration liners. Each section is presented with a clear-cut discussion on its contents.

Several methods of analysis and design are described in detail. The author has, for the first time, presented detailed analyses based on Finite Element and Limit State concepts. A suite of computer programs has been developed by the author to evaluate the entire operational history, the inelastic and cracking conditions and the safety margins of the concrete vessels. The methods have been applied to existing vessels chosen built for gas-cooled reactors at Oldbury, Dungeness B, Hartlepool/Heysham and Fort St. Vrain. The results obtained from these analyses are in good agreement with those available from published data.

The analytical and design tools thus available in this text have made it possible to investigate fairly accurately the safety margins of other vessels adopted for pressurised water reactors (PWR), boiling water reactors (BWR) and fast breeder reactors (FBR).

Nomenclature

$2h_I$	internal height
R_I, R_E	internal and external radii

M.Y.H. Bangash, *Structures for Nuclear Facilities*,
DOI 10.1007/978-3-642-12560-7_5, © M.Y.H. Bangash 2011

P_G	internal gas pressure
P_{GB}	boiler gas pressure
R_P	standpipe radius
R_B	radius of boiler/circulator
H_B	Height of boiler/circulator
$\left.\begin{matrix}\varepsilon'_{ps}\\ \varepsilon'_{pc}\\ \varepsilon^{e}_{pb}\end{matrix}\right\}$	circumferential strains at hinges
$\bar{P}_c$	total cap pressure
$\bar{x}$	distance at which $\bar{P}_c$ acts
F_V	total vertical prestressing load
F'_V	total modified vertical prestressing load
a	distance from the outside face of the vessel to F_V or F'_V
R'	distance from the centre line of the vessel
x_p, x_b, x_w	depths of compression hinge at top, bottom and at equator
ε_{pa}	$\varepsilon_{ce} + \varepsilon_u$
ε_{ce}	concrete stress at the tendon level divided by E_C
ε_u	average strain at the level of the tendon at ultimate conditions
x	neutral axis depth
d	depth of the section
ε_{pc}	effective prestressing strain after losses
ε_0	$\sqrt{\sigma}_{\mathrm{cu}/5000}$
σ_{cu}	concrete stress at ultimate
F_{bst}, F'_{bst}	longitudinal forces in tension and compression reinforcement
T_F	total force
F_{BL}	liner force
C_C	concrete force
b	average breadth $(R_E+R_t)\sin\theta$
A'_{sb}	area of bonded steel in compression
σ_{yb}	yield stress of bonded steel
2θ	angle of a segment
A_{Sb}	area of the bonded steel in tension
A_{SL}	area of the liner
σ_{yl}	yield steel of the liner
D_w	unit weight of concrete
σ^P_{ij}, σ^P_{ij}	plastic stress and strain
f	stress function
$\mathrm{d}\lambda$	constant of proportionality
C	factor in plasticity rule
E_t	tangent modulus
Z	coefficient for strain hardening
P_{HR}	total radial compression
P_{Hr}	radial compression $= P_{HR/2RE^{\sin\Phi}}$
ε_z	strain profile at a particular position

K'	strength reduction factor from wire/strand bands or from the circumferential tendons
l_c	crack depth
ω_{th}, ω_{tb}	crack width top and bottom
$\hat{x}$	typical value of x_t or x_b
ω_{el}	longitudinal crack
ω_{eh}	horizontal crack
α_c	cap rotation factor
δ_z'	deflection of cap at the yield zone
n_1, n_2	concrete block parameters
β	measure of a distance
σ_t	tangential stress
Q	shear force
N	axial load
τ_{RZ}	uniform or parabolic shear stress
Φ'	angle of frictional resistance for concrete
C'	apparent cohesion for concrete
X_1, X_2	dimension on a cap plug
$\bar{S}$	terms for the uniform or parabolic shear stress
ε_{pp}	strain due to rigid plastic rotation of the barrel wall
ε_{cpv}	strain due to rigid plastic rotation of the cap
E_o, E_s	Young's moduli for concrete and steel
ε_{pu}	strain of tendons or band in the circumferential direction
A_c	area of concrete
A_{sp}	area of prestressing steel
δ	displacement in a respective direction
Z	position of a hinge from equator
K_1, K_2, K_3	coefficients
P_{w1}, P_{w2}	total pressure on various zones of the barrel wall
M_{HT}, M_{HR}, M_E	moments at top, bottom and equatorial hinges
A_{hp}	area of circumferential prestressing tendons or bends
n_b	number of boilers/circulators
m_ϕ	a plastic moment per unit length of the radial yield line
$P_{H\varphi}$	pressure due to circumferential tendons
H'	slope of the effective plastic stress
n	series of values or nth term
D_p	plasticity matrix
σ_{in}	initial stress
σ_{e-p}	clasto-plastic stage
σ_{TOT}	total stress
$\bar{y}$	yield
$\{\psi\}$	residual force vector
α	alpha-constant used for acceleration
$\bar{A}$	acceleration factor
$[T'']$	transformation matrix

ε_{cu}	ultimate or critical strain
σ_{cv}, ε_{cv}	stress and strain normal to the closed crack
τ	shear stress
τ_{cv}	shear stress on crack interface
$\hat{tf}$	transfer stress
$\hat{F}$	transfer force
$\Delta\varepsilon_{su}$	plastic strain in steel
$\Delta\varepsilon_s^p$	plastic strain increment
X, Y, Z	global co-ordinates
p_i	shape function
ξ, η, ζ	local system co-ordinates
x_n, y_n, z_n	nodal co-ordinates
J	Jacobian
u, v, w	displacements
a_n	polynomial coefficients
$[c]$	matrix relating a_n and u
det $[J]$	determinant of a Jacobian
ε	strain at any point
$[B_i]$	matrix relating strains and displacements
σ	stress
δ	generalized displacement
δ^*	virtual displacements
$[S_{ijkl}]$	material compliance tensor
W_e	external work done
U	internal energy
$[K]$	stiffness matrix
$\{F\}$	nodal forces
$\{F_p^e\}$	nodal forces at elastoplastic stages
$\bar{f}$	body forces and any other loads
$\{\hat{g}_{x,n}\}$	vector containing the values of the body load at nodal pints in a given direction
R_i	residual load
L_F	loading function
S_H	the yield of work hardening function

5.2 Historical Development, Existing Analyses and Scope of Research

5.2.1 *General*

The use of nuclear energy in the production of electrical power involves substantial structural systems comprised of pressure vessels that house the reactor. Most of the light water (LWR) and pressurised water reactors (PWR) in operation at present are sheltered by two major structural units: the primary container which is a steel pressure vessel for holding the coolant systems pressure and a secondary container providing a second protective shield. An alternative to the use of liquid coolants is a pressurised gas system. The coolant in this case is either carbon dioxide or helium. The current British and French reactors, using these coolants, are called advanced gas-cooled reactors (AGR) and high-temperature gas-cooled reactors (HTR or HTGCR). Efficient operation of these nuclear power plants requires a large electrical output. This requirement calls for a large nuclear core space and much larger primary pressure vessels and supporting structures of sizeable dimensions. The inherent difficulties that would be involved in the fabrication and transportation of the pressure vessel steel units of the size necessary for such a reactor have led to the need for a more versatile type of structure. The vessel must, undoubtedly, be built on site and meet the serviceability and safety requirements. The solution adopted has been the use of prestressed concrete reactor vessels (PCRV).

One incentive for adopting PCRV is the economic advantages to be expected. These advantages arise, for example, from

(a) the ability to contain large reactors with high pressures and temperatures;
(b) simplification of plant layout;
(c) the fact that a highly developed steel fabrication industry is not necessary.

Attractive features from a safety point of view include

(a) physical isolation of the steel prestressing tendons and reinforcement from sources of heat and radiation and from the primary coolant;
(b) the high degree of redundancy in the prestressing systems.

The shapes used, so far, for PCRVs have varied from a cylinder bounded by two inverted, non-prestressed, hemispherical domes, torospherical domes to spherical shells. The current trend in PCRV configuration is the use of thick-walled cylinders, the ends of which are closed by flat slabs known as caps. The boilers and circulators are either housed within the main cavity or within the thickness of the walls; the latter is known as a *multi-cavity-type vessel*. This study is concerned with such vessels also apart from other shapes.

The first application of prestressed concrete as a reactor vessel was in the Marcoule G-2 and G-3 installations in France. These are horizontally placed concrete cylinders with concave domes as caps. The vessel main cylinder is wrapped with prestressing cables describing an arc of 2700 which are finally anchored to prestressed concrete foundations under the vessel. The domes are not prestressed, resulting in a substantial loss of vessel volume. The standpipe and the control rod areas are badly situated for the charge machine. The French have recognised these limitations of the Marcoule vessels. However, in the meantime the requirements for high-quality steel and more rigorous inspections for both British and French reactor vessels have reawakened interest in concrete vessels. Consequently, in planning the construction of the EDF-3 reactor, the French designers decided to re-examine the application of prestressed concrete to reactor vessels. The design selected provides a vessel with a cylindrical shape and with prestressing cables placed in layers parallel and perpendicular to the long axis of the vessel. Supplemental prestressing steel is also provided at the corners where the stress concentrations are high.

The British interest in concrete pressure vessels began with some experimental work done by the then General Electric Company and the Simon-Carves group on cylindrical vessels prestressed circumferentially by wire winding. Based on these tests and other tests carried out by the then United Kingdom Atomic Energy Authority (UKAEA), Waters and Barrett [1] published design studies for cylindrical and spherical vessels containing advanced gas-cooled reactors. They carried out a comparative study of steel and concrete pressure vessels and concluded that there were significant economic and technical attractions to adopting such vessels. They then presented a detailed design philosophy in which particular attention was paid to the short- and long-term effects of tendon loads on concrete, the premature failure of the liner and the optimum choice of the conventional reinforcement as an anti-crack steel. Apart from highlighting design and constructional problems, they suggested a list of research programmes. At a time when no suitable code of practice and no completely reliable mathematical tools existed for analysing and designing such structures, their recommendations prompted many UK firms with interests in nuclear technology to carry out research and development in this field. The first order placed by the then Central Electricity Generating Board (CEGB) was for the Oldbury Power Station. The then Nuclear Power Group (TNPG) proposed a vessel design that was a hybrid of the French Marcoule and EDF-3 concepts. The vessel is of cylindrical shape with flat top and bottom caps, and it utilises alternate layers of helical cables which wrap the vessel in reverse directions. These cables are anchored in groups in specially provided top and bottom galleries. Both longitudinal and lateral prestressing loads are produced from this single prestressing system.

By the time of completion of this vessel a great deal of knowledge had been gained from both the British and the French experience in this field. Another concept was under consideration at the CEGB, namely a spherical vessel containing the entire reactor internals. Owing to immediate requirements for more electrical output, the then CEGB gave the go-ahead to a group member of the British Nuclear Design Company to design and construct such a vessel for the Wylfa power plant. This vessel has a spherical inner surface and the prestressing is provided by a six-sided polyhedron made up of cables anchored to buttresses along the vessel circumference. Each tendon is laid perpendicular to the preceding one. Since the prestressing tendons in the Wylfa Nuclear Power Station are nearly straight, the possibility of large frictional effects is remote. The prestressing loads are uniform and this is partly due to its spherical shape. Compared with cylindrical vessels, the Wylfa spherical vessel effectively utilises the entire prestressing system, and the peak stressing is considerably less.

In the meantime, a number of variations on the French EDF-4 (now Bugey) reactor were under consideration. It was decided to have a vertical cylindrical pressure envelope divided into two chambers: the upper one supports the reactor core while the lower one contains the boilers or heat exchangers. These affect the size of the internal cavity and the prestressing system. Complete data for these vessels were published by Marsh and Melese based on detailed literature surveys and the preliminary designs carried out by the Frankline Institute.

Both the Frankline Institute [2] and the Oak Ridge Laboratory [2] in the USA made a start on the research and development required for the Fort St. Vram high-temperature gas-cooled reactor. The choice was made of a cylinder with flat top and bottom caps. Both in the wall and in the cap the prestressing was by means of cables of variable curvature.

In the meantime, in the UK a great deal of experience was gained from the Oldbury and Wylfa vessels. Problems associated with the plant layout and the construction of the Wylfa vessel were such that the British designers had to rethink the use of cylindrical vessels. The then CEGB finally approved the cylindrical vessel of the Dungeness B Power Station using large tendons in the vertical and in the circumferential directions. The circumferential tendons are anchored on specially formed buttresses. This time, using a minimum number of large tendons, a greater concentration of forces has been achieved for the limited available space in the vessel.

However, the Oldbury contractors carried out optimisation studies on the UK Hinkley B and Hunterston B power stations. Slight changes in the vessel parameters were inevitable. This time true helices were adopted for the prestressing tendons rather than the 'barrelised' type adopted for the Oldbury vessel.

At this stage, the concept of multi-cavity vessels was under active consideration. Boilers and circulators occupied a large space within the main cavity. The dimensions of the main cavity were reduced by sheltering the boilers and circulators within the thickness of the wall. Burrow of Taylor Woodrow construction Ltd. reported on providing the wire-winding system for the circumferential_prestressing loads on the vessel, thus leaving the main vessel thicknesses for the large capacity longitudinal tendons only. Such a vessel was accepted for the AGR system adopted for both the Hartlepool and Heysham power stations. In the last few years substantial progress has been made in the development of such vessels and much experimental data have been produced from both model tests and on-the-spot measurement of the prototype vessel.

5.2.2 Problems Associated with Vessels

Although the multi-cavity vessel has much to offer, it is by no means an easy structure to model. Many designers and structural analysts have tried to tackle the various problems associated with these structures. In many cases they have claimed some achievements but for some reason they have not made them public. However, most are united on the design criteria for such vessels and these are

(a) the vessel should be designed for elastic response to all possible combinations of loads during operation;
(b) it must show a progressive mode of failure under increasing gas pressures with large deformation to warn against impending failure; and
(c) the vessel must have an acceptable safety margin against failure.

In order to meet the above criteria satisfactorily, one has to first validate the performance of the individual components forming such structures. The major components are

(a) concrete,
(b) prestressing tendons,
(c) the liner and other penetrations, and
(d) bonded primary and secondary reinforcements or conventional steel.

The problems associated with each one of these are enormous. For example, a certain amount of evidence must be available to demonstrate that for short- and long-term loadings (under multiaxial compressive stresses) concrete can safely withstand higher compressive stresses than are generally acceptable under uniaxial loading. For all service load conditions, including startup and shutdown, both initially and at the end of the vessel's life, the stress–strain characteristics for the concrete should take account of the age, temperature and time under load. It is also considered that limited cracking may be accepted provided due regard is paid to any significant redistribution of the stresses which may arise, due to lack of integrity and leak tightness of the liner. Where local concentrations of stress occur, due to the presence of embedments or other discontinuities in the vessel geometry, these should be assessed individually. In such cases, due regard should be paid to the effects of increased creep rates or tensile cracking on the distribution of stresses in the vessel concrete and the influence which they may have on the strains in the vessel liner.

Stresses, strains, deflections and cracking in the vessel should be analysed for all relevant combinations of mechanical and thermal loads which can arise under normal service and ultimate conditions. In such cases the prestressing forces play a great part. The tendon forces adopted in each analysis should include allowances for the most severe effects of friction and loss of prestress. A proper method of analysis together with short- and long-term experimental tests is required to design tendon systems for cyclic loads produced by gas pressures. These requirements are translated in terms of range of stress or strain cycles in the steel liner, the concrete or the other components in the nuclear islands. Beyond the elastic range, the method of assessing the ultimate behaviour of prestressing tendons and their anchorages is very important for predicting a safety margin for the vessel. This, of course, depends on the time at which the incident occurs to the vessel and the rate of increased gas pressure and whether or not the tendons are grouted.

The liner undoubtedly represents a vital safety element of the vessel. Indeed, it serves the fundamental purpose of forming an impervious barrier to the cooling gas. The reactor's operability depends on its integrity. The vessel will collapse, not at the ultimate collapse pressure coincident with the failure of the prestressing tendons but at the liner failure pressure which might be lower. This would require knowledge of the topology and magnitude of the cracking on the internal face of the concrete in contact with the liner. It would be necessary, therefore, to have a clear idea of the cracking mode of the vessel for increasing pressures up to the collapse pressure. Moreover, the liner integrity is also dependent upon the performance of the liner anchorages and cooling pipes, buckling stability and fatigue in general areas, the insulation and temperature distribution.

The choice and the distribution of the conventional bonded steel reinforcement in the vessel main areas and around the penetrations are extremely important should initial cracking occur due to moisture migration and unpredictable shrinkage and creep in concrete. The bonded steel reinforcement is needed in areas where extreme stresses under serviceability conditions cannot be avoided. Above all, the progressive failure of vessels depends on the amount and the distribution of such reinforcements.

Now turning to the vessel behaviour under extreme loads, it is necessary to provide some means of limiting the effect on the vessel of an excessive rise in internal operating pressure. The magnitude and rate of a postulated pressure rise can be determined only by reference to the characteristics of the particular reactor contained. It is common practice to provide automatic venting devices, such as safety valves, for this purpose. An alternative is to design the vessel so that it is self-venting by partial structural failure. It is generally recognised that this alternative cannot be relied upon at the present method of design/construction.

It is customary and advantageous to make provision to verify the state of the vessel. This may be done by installed instrumentation and/or by periodic in-service inspection. Such measures, which are taken in a manner appropriate to the particular situation, serve to verify the vessel integrity and to confirm the design criteria. However, this does not rule out the main problem of numerically assessing such design criteria. Under any circumstances the design philosophy is based on the recognition of two or more modes in the vessel response to increasing pressure, and this cannot be met by simple experimental models.

The design objective is to ensure that a particular response to the imposed load can be achieved in each mode and that this behaviour is consistent with the appropriate operational and predetermined fault conditions.

Over a range of pressures and temperatures, including normal operating conditions, the vessel will respond to short-term variations in pressure in an elastic manner. This facilitates machine analysis of stress and strain in the vessel. Long-term stresses and strains are affected by shrinkage and creep of the concrete, relaxation of tendons and possibly fatigue. In this range of response the effects of short- and long-term behaviour can be combined to demonstrate that stresses and strains are limited to acceptable values.

Beyond the elastic range the response becomes increasingly inelastic and nonlinear. The vessel would not be expected to enter this phase except under the most severe overpressure fault conditions. In this phase the vessel is stable but may experience permanent damage. It is in this phase of vessel response that some limit states occur.

The ultimate load condition, in which the vessel is incapable of sustaining any further increase of internal pressure, is a further limit state. Evaluation of the ultimate load provides a measure of the factor of safety above design conditions.

5.2.3 *Vessel Layouts and Finite Element Mesh Schemes*

Typical layouts of the Dungeness B, Oldbury and HTGCR vessels are given in Figs. 5.1, 5.2 and 5.3, respectively. Figures 5.4, 5.5, 5.6 and 5.7 show the finite element mesh layout of the Dungeness B, Oldbury, Hartlepool and HTGCR vessels, respectively.

The mesh layouts take into consideration the vessel penetrations, prestressing configurations and other features such as the existence of buttresses, galleries and the liner plugs and cooling pipes.

5.2.4 *Design Analysis*

The objective of this analysis is to demonstrate that reactor vessels will meet both the serviceability and ultimate limit state conditions, taking into consideration the gas increasing pressure.

5.2.4.1 Service Conditions

Under service conditions it is important to know the technique and precision with which the vessel geometry can be reproduced by analytical means. A number of key loading cases are given below.

(a) Initial prestressing loads: Analysis includes losses due to friction and elastic shortening and creep; this is to see that no abnormal stresses develop in the vessel concrete.
(b) Proof test: In this analysis the vessel is tested at the commissioning stage for changes in strains and deflection due to proof pressure loading.
(c) Normal operating conditions: This is intended to observe the vessel operating normally under extreme loads caused by the combination of prestress, design pressure and temperature gradient together with the effects of applicable environmental loadings. This can be at an early-life or late-life operating condition.

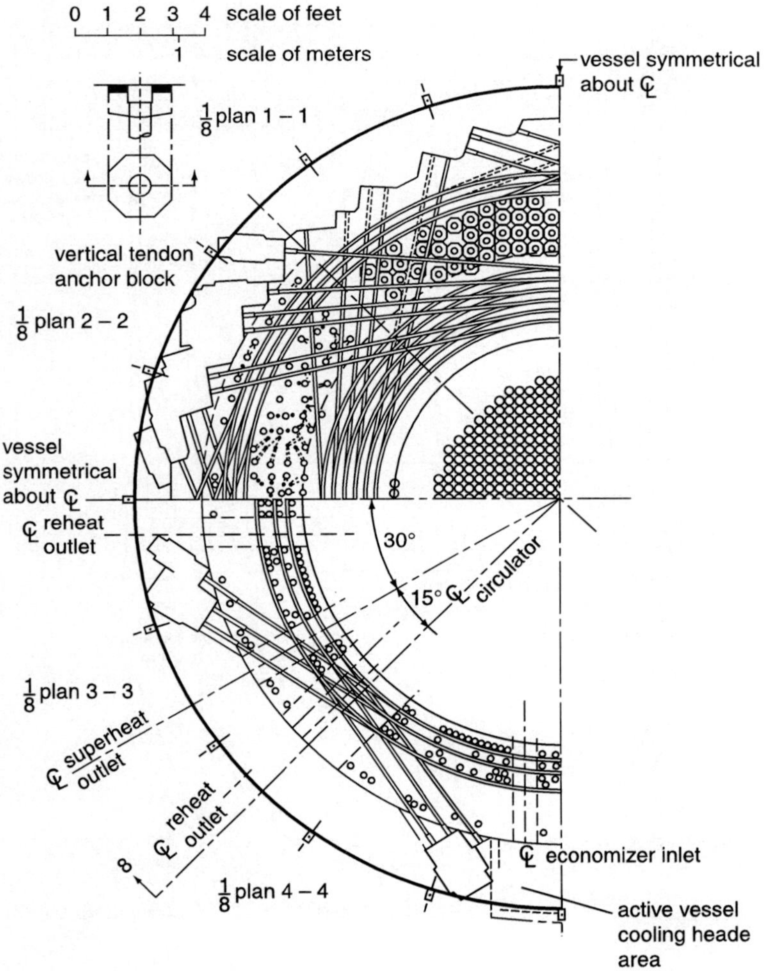

Fig. 5.1 Dungeness B vessel prestressing layout (plan)

(d) Start-up and shut-down conditions: These are transient and quasi-transient conditions normally imposed on the steady conditions of normal operation. Together they reproduce the extremes of the early-life shut-down/start-up cycle. The late-life conditions are also examined by changing the material properties and temperature transient conditions.
(e) Deviations from operating conditions: A number of inaccuracies arise from the simple treatment of loading cases; these mainly concern the effect of concrete creep on stresses, unloading and cooling effects on the vessel and the allowance for low-probability loadings arising from accident or fault conditions. Analyses are required of these uncertainties.

a

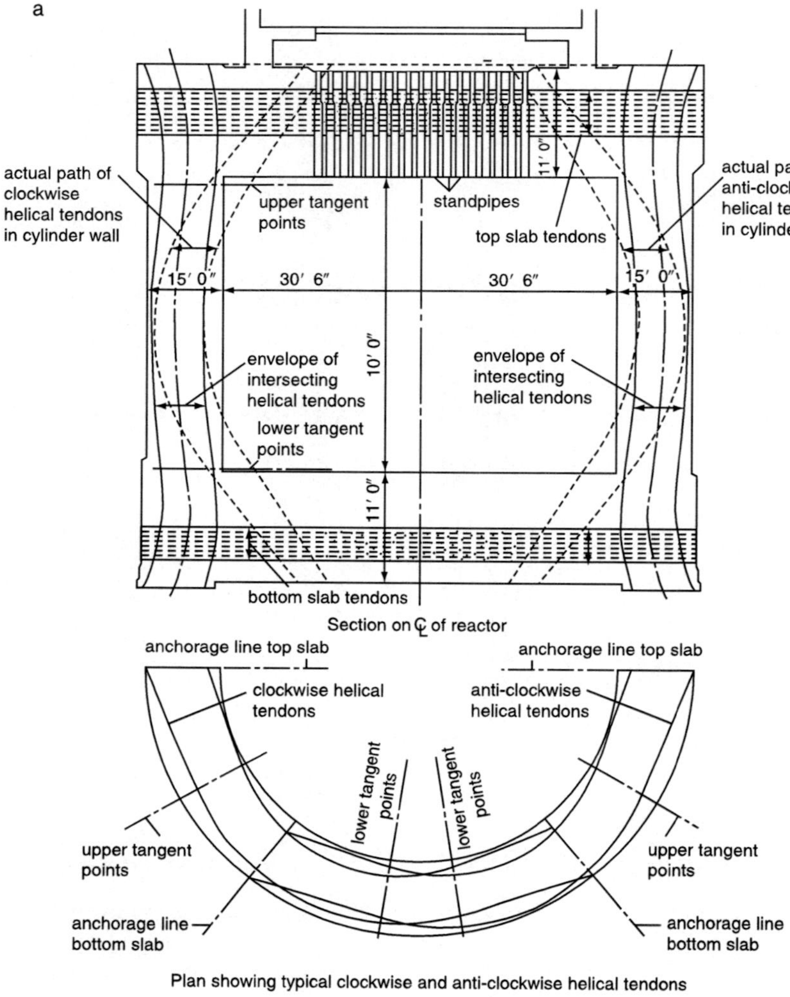

Fig. 5.2 (**a**) Principal prestressing systems (patented) (*Note*: All dimensions in imperial units; 1 ft = 0.3048 m,. 1 in. = 25.4 mm). (**b**) Model for Oldbury nuclear power station – layout of prestressing cables and cable profiles (*Note*: All dimensions in imperial units; 1 ft = 0.3048 m, 1 in. = 25.4 mm)

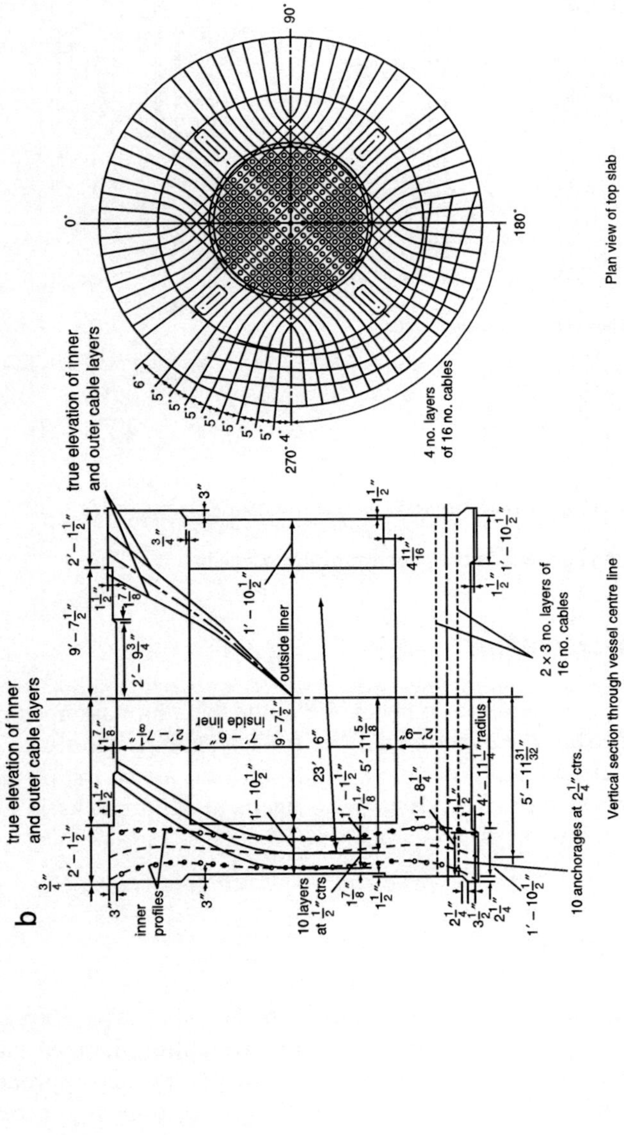

Fig. 5.2 (continued)

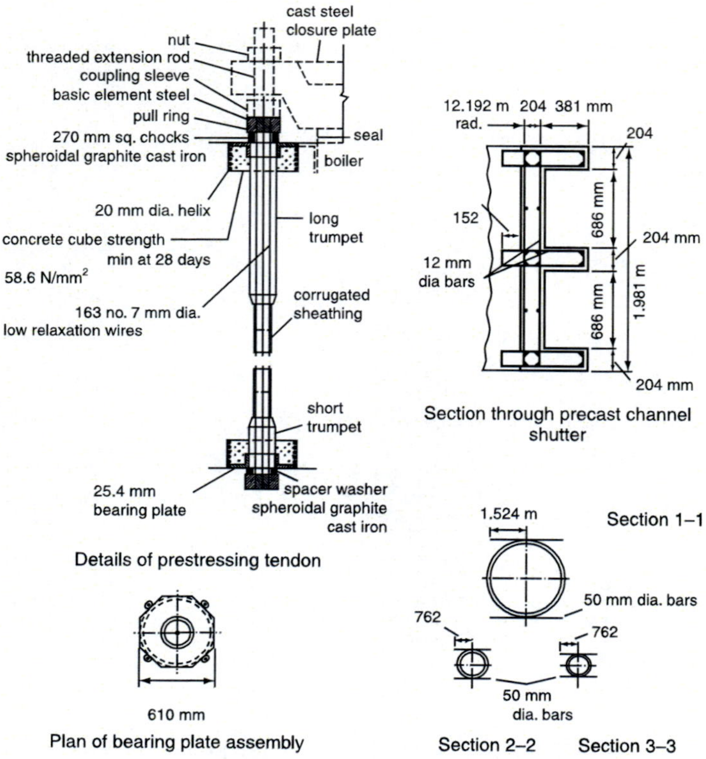

Fig. 5.3 Vertical and circumferential prestressing precast channels

5.3 Ultimate Conditions

A validity analysis is required to check the mode of failure assumed or calculated or derived from the experimental tests. This analysis should cover cases such as the intact and ruptured liner, failure of local areas, the possibility of shear failure in top caps or at cavity/wall ligaments and cracks in hot spots (areas around cooling pipes).

5.4 Methods of Analysis

Many methods are available to designers. The most common [3–7] are based on the finite element, finite difference or dynamic relaxation, lumped parameter and limit state methods. This book gives earlier the step-by-step approach of the finite element method. In some service and fault conditions it is necessary to consider the influence of external hazards and environmental conditions. Major

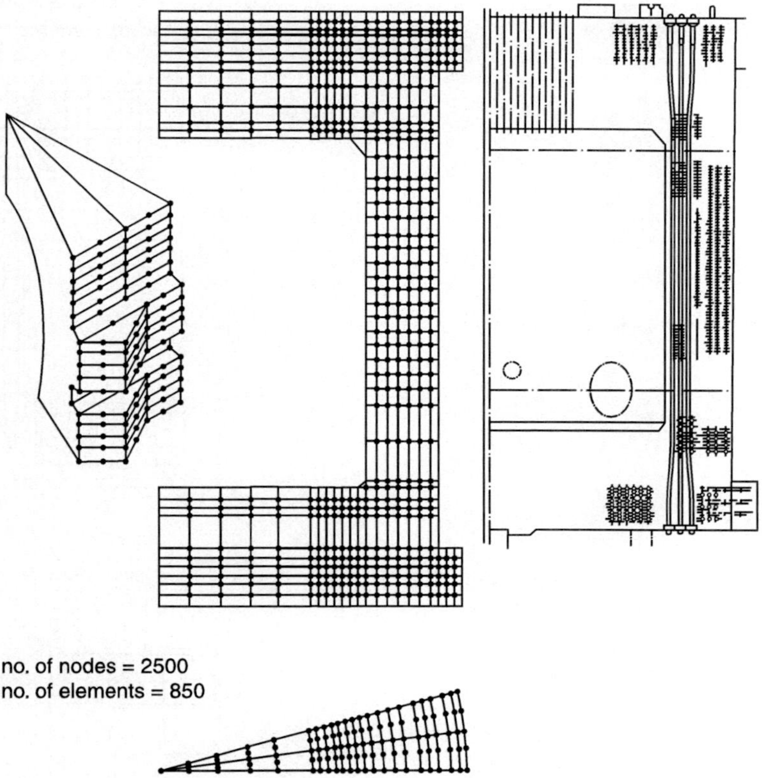

Fig. 5.4 Finite element idealisation of Dungeness B vessel (*Note*: For clarity certain nodes are not shown)

external hazards are seismic disturbances, wind, missiles and aircraft crashes. These hazards are considered for the case of containment vessels.

5.5 Model Testing

The purpose of a model test is to verify the methods of analysis and to provide a visible physical demonstration of the adequacy of the design requirements. As described earlier, model tests have been carried out to verify the service and ultimate state behaviour of vessels and to assess prestressing, reinforcement and liner requirements. The scale chosen for a specific model depends on the objective of the test and on the reduction in size of important vessel components. For a complete vessel a suitable scale varies from 1:25 to 1: 10, and scales of 1:25 and 1:50 have been used for the investigation of the top slab behaviour. A large number of models for shear modes of failure have been tested with

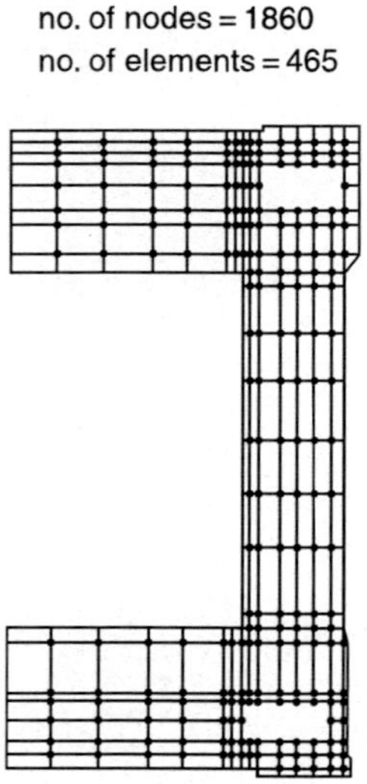

Fig. 5.5 Finite element idealization of Oldbury vessel

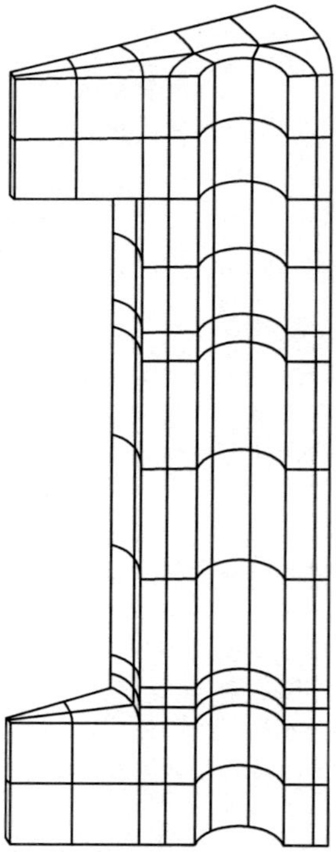

Fig. 5.6 Prestressed concrete cylindrical multi-cavity reactor vessel (HTR) – three-dimensional finite element (first rough meshing)

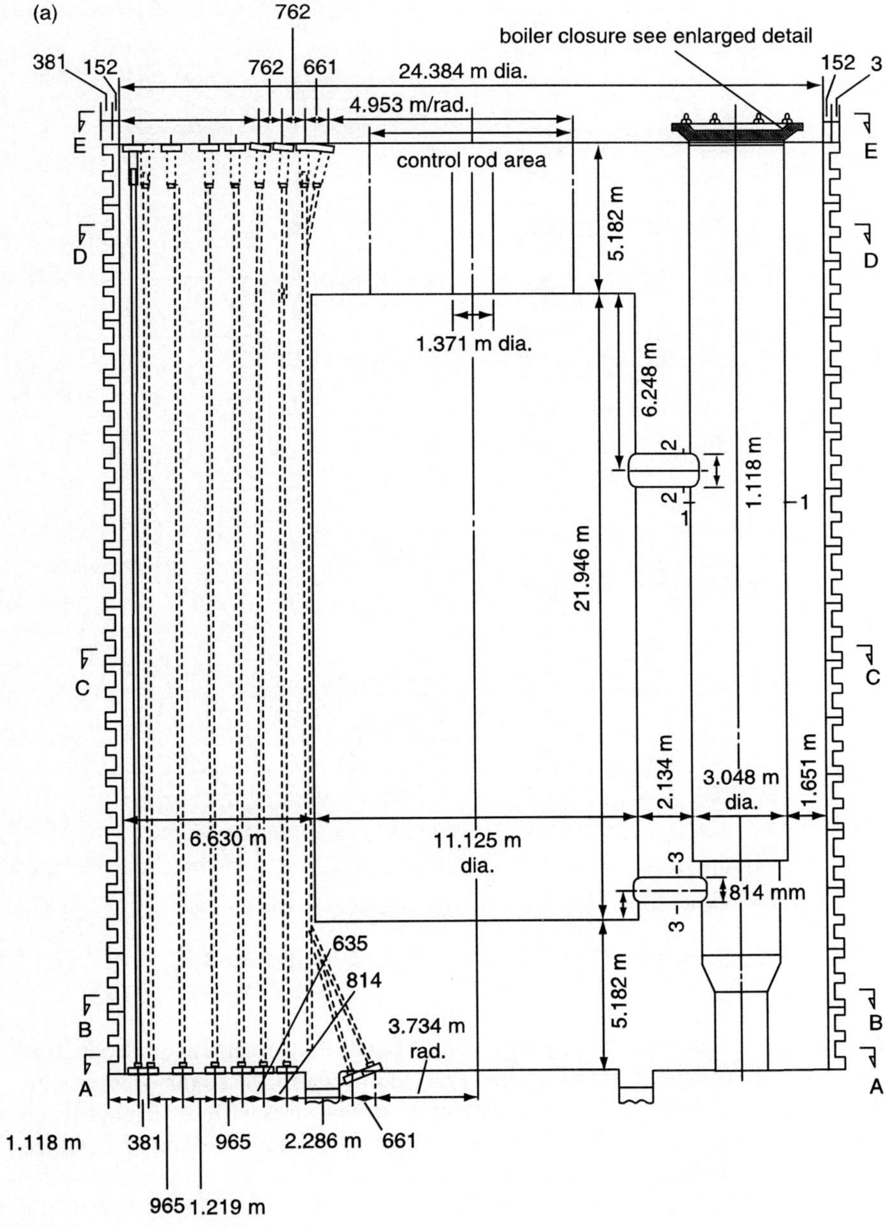

Fig. 5.6 (a) Prestressed concrete cylindrical multi-cavity reactor vessel, HTR, Section with tendous and precast channels. Prepared by Bangash for 3D Analysis Based on Finite Element

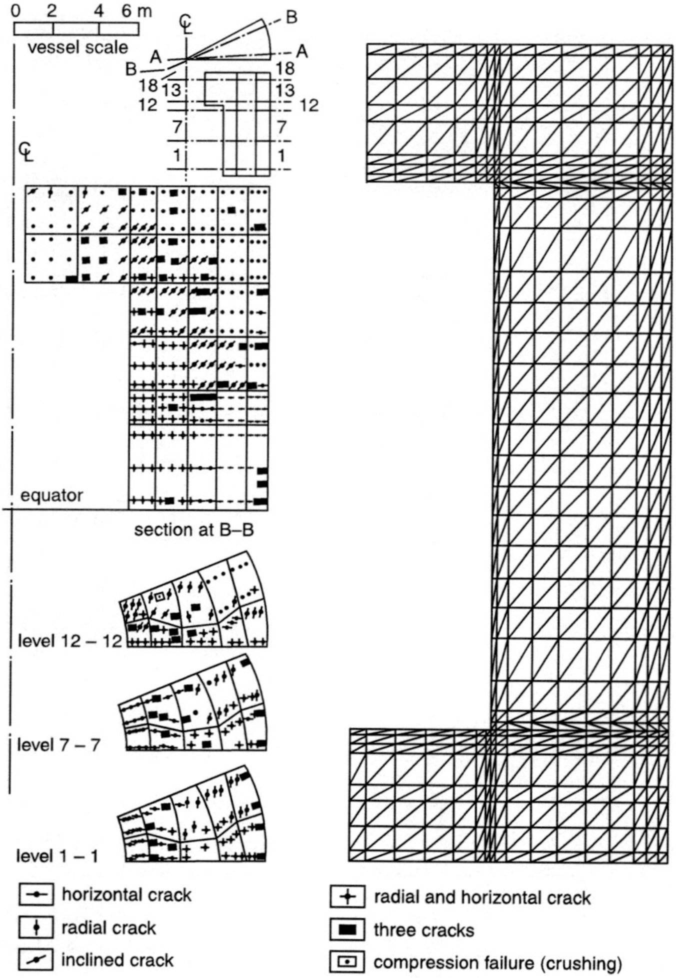

Fig. 5.7 Prestressed concrete cylindrical 2D finite element pressure vessel

varying geometries, materials, prestressing loads, reinforcement and penetrations. Some pneumatic tests have been carried out using gas as the pressurising medium. Most of these tests produced progressive failures with sufficient means available for detecting the elastic limit, the onset of cracking and the failure mode. Many individual tests have been carried out to assess the integrity of the liners, penetrations, closures and thermal protection systems. The finite element analysis and constitutive model discussed earlier are applied to validate the proposed analysis.

5.6 Analysis of Results

(a) *Dungeness* B *vessel*: Figures 5.8 and 5.9 show gas pressure versus vessel deflection for the top cap. Here only 12 increments of pressure are plotted. For various pressures the trend in the incremental deflection is fairly regular and no abrupt change in the deflections between any two increments has been found. Finite element analysis using three concrete numerical models, namely the four-parameter, hypoelastic and endochronic models, has been carried out. Simultaneously the model has been analysed. Figure 5.10 shows the pressure–deflection relation for two areas of the barrel wall.

The experimental tests have been carried out on two pressure vessel models, namely that of Dungeness B and Oldbury, both cylindrical with top and bottom flat caps, with different prestressing systems, indicate different modes of failure. These will be the reference models of the limit state analysis which has been developed in the text and which will be finally provided by using 3D hybrid finite element analysis under increasing gas load pressure. The Oldbury will be given a detailed assessment later on in the text.

Most of the other ultimate load analyses of cylindrical PCRV follow general procedure as commented earlier.

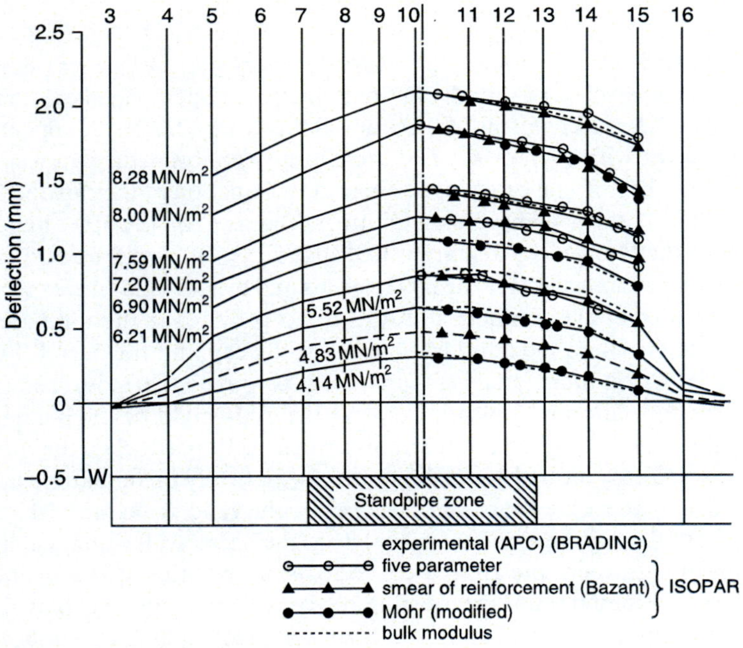

Fig. 5.8 Deflection versus pressure: 4.14–8.28 MN/m^2

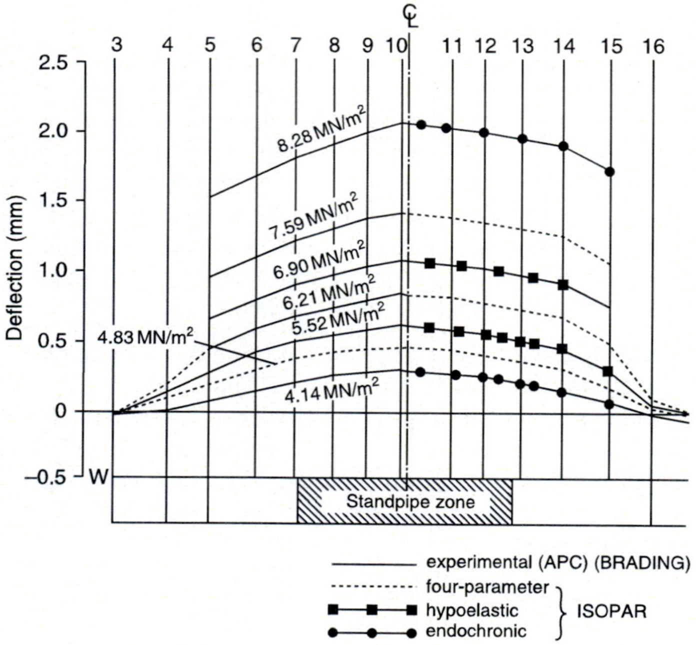

Fig. 5.9 Deflection versus pressure: 4.14–8.28 MN/m^2

The reported ultimate load analysis for Oldbury and Dungeness B vessels considers each vessel to be initially broken up into an open-ended cylinder and two end caps. The mechanisms of these initial break-ups, however, appear to be in disagreement with each other. For the Oldbury analysis, the cap was separated from the barrel wall by a plastic hinge at the junction of the cap and barrel (see Fig. 5.11). A shear-compression failure mechanism, on the other hand, was adopted for the Dungeness B analysis (see Fig. 5.12). Of even greater disparity between the two analyses were the mechanisms and modes of failure experiment adopted to predict the ultimate pressure or to prove that the ultimate load factor was higher than the adopted factors for the cap and the barrel. For the barrel wall, the Oldbury analysis apparently favoured the modes of failure through the formation of a plastic hinge at the mid-height of the barrel. This mode of failure has aroused considerable criticism. In particular, the mode of failure was pointed out by Morice [8] to be kinematically incompatible in a 3D vessel without extensive cracking. It is probably valid to assume that with extensive radial cracking along the barrel wall, the barrel wall would be divided into strips of equivalent prestressed concrete beams positioned in a cylindrical array by the circumferential prestressing or equivalent hooping cables. Under such a condition, assuming that the liner remained intact (a hypothetical assumption commonly used for convenience of analysis), the pressure force

Fig. 5.10 Deflected shape – sides of model, ultimate condition

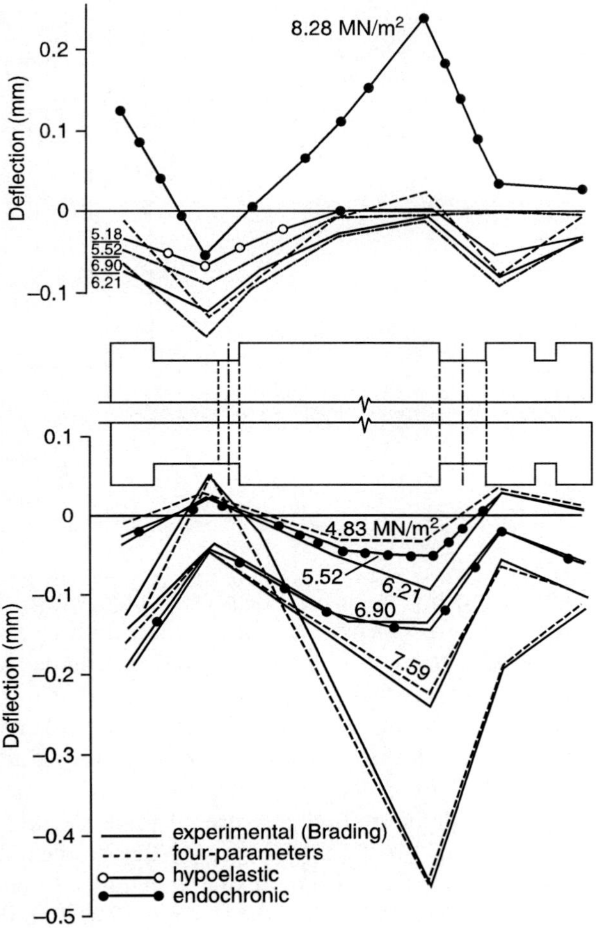

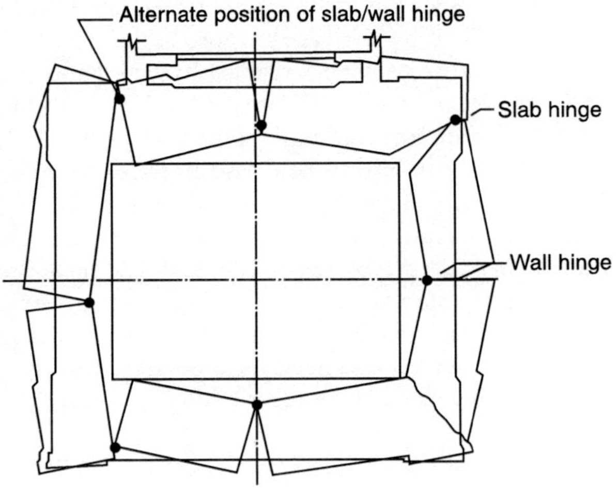

Fig. 5.11 Oldbury POPV mode of failure (with complements of McAlpine and Sons and the authors listed)

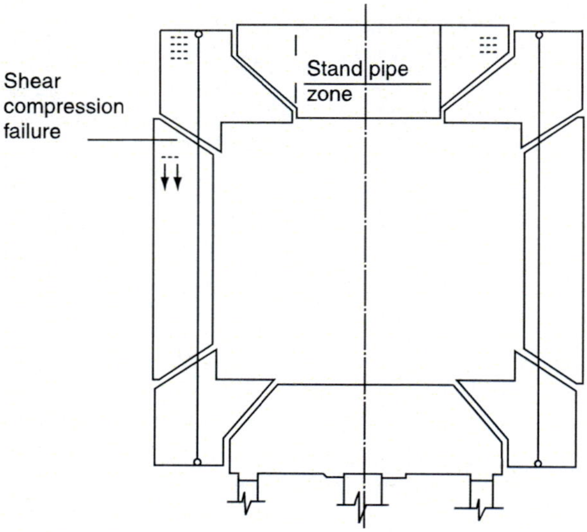

Fig. 5.12 Dungeness POPV mode of failure (with complements of the Atomic Power Construction and the authors listed)

would be entirely resisted by the prestressing system. The approach adopted in such a case is to consider the ultimate load analysis of the barrel on the basis of the rupturing strength of the tendons. The ultimate load analysis for the barrel of Dungeness B in fact adopted a similar approach with the exception that the membrane action of the wall of the barrel was taken into account. The tensile strength of the membrane and tendons, with due allowance for tendon efficiency, was checked in the analysis to be greater than the circumferential force associated with an internal pressure 2.5 times the design pressure.

Zudans and Tan [9] similarly used the simpler ultimate load analysis by assessing the ultimate strength of the cylinder solely on the strength of the prestressing tendons to resist the circumferential and axial components of the pressure force. A series of tests [10–14] have been carried out successfully on individual vessel components and scaled models. Some of them are worth mentioning. To crown all, the purpose of the model test is to verify the methods of analysis and to provide a visible physical demonstration of the adequacy of the design requirements. As described earlier, model tests have been carried out to verify service and ultimate state behaviour of the vessel [126], to assess prestressing and reinforcement and liner requirements [15]. The scale chosen for a specific model did depend on the objective of the test and on the reduction in size of important vessel components. For a complete vessel a suitable scale varied from 1:25 to 1:10, scales 1:25 and 1:50 have been used for the investigation of the top slab behaviour. A large number of models for shear modes of failure have been tested [13–15] with varying geometries, materials, prestressing loads, reinforcement and penetrations. Some tests have been carried out pneumatically using gas as the pressurising medium [13–15]. Most of these tests produced progressive failures with sufficient means available for detecting the elastic limit, the onset of cracking and the failure mode. Many individual tests have been carried out to assess the integrity of the liners, penetrations, closures and thermal protection systems [142–168].

5.7 Fundamental Elements of the Concrete Pressure Vessels

As stated earlier the major fundamental elements of the vessel are

(a) Prestressing systems
(b) The steel liner
(c) The bonded reinforcement
(d) The embedded elements
(e) The concrete

5.7.1 Prestressing Systems

The layout of the prestressing system is dependent on the shape of the vessel. The safety characteristics of the vessel can only be claimed if the integrity of the prestressing tendon and achorage system is correctly established. Several tests [1–4, 6, 8–52] have been carried out to firmly establish the strength and deformation characteristics of these systems. These include reliable tests on cables, wires, anchorages and ducts. Both short- and long-term data have been obtained in stress–strain behaviour, losses and ultimate load-carrying capacity of longitudinal, circumferential, helical and radial tendons. The latest vessels built in the UK are provided with a wire-winding system in which the wires or strands are wound under tension into a steel-lined R.C. channel either formed in the vessel walls or made of precast concrete. The band comprises a series of layers of wire/strand anchored at both ends. In this way maximum circumferential pressures are exerted within a very limited area. The advantages of this system are fully discussed [200] in the literature.

In many vessels only unbonded systems have been adopted. This choice is based on a condition that all tendons must be inspected for defects including those due to corrosion. The advantages or disadvantages of bonded and unbonded tendons are fully discussed [200] by the author.

Figures 5.1, 5.2, 5.3, 5.4 and 5.5 show the prestressing systems adopted or recommended for the several existing and future vessels. Table 5.1 shows a summary of equations in case of wire-winding for obtaining axial, circumferential, radial and boundary pressures. Details and geometry of each vessel are given in several references [13–15].

Figures 5.2, 5.3, 5.4, 5.5 and 5.6 show typical layouts of prestressing systems for existing prestressed concrete vessels.

Table 5.1 Circumferential prestre due to wire/strand winding

$P_n = np_H \tan\theta$ — odd number } of wires

$P_n = \frac{(n-1)}{2} P_H \tan\theta$ — even number } of wires

$P_n = np_H$; $F_s = F_q + \frac{3}{2}\sec\theta$

$P_q = \frac{1}{2} P_H \sec\theta$

$P_H = \Sigma \frac{k_1 T_W}{l} \quad \frac{l}{\Sigma R_1}$

$\Sigma \bar{R}_1 = \frac{1}{R_1} + n_L . \sqrt{3\phi/2}$

WIRE WINDING

k - factor for friction

ϕ - wire/strand diameter

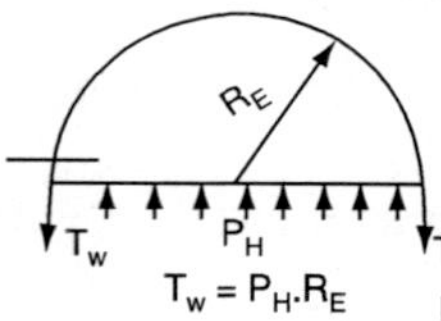

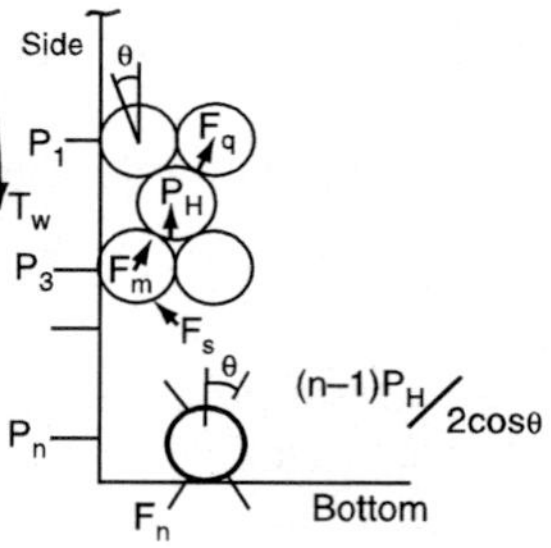

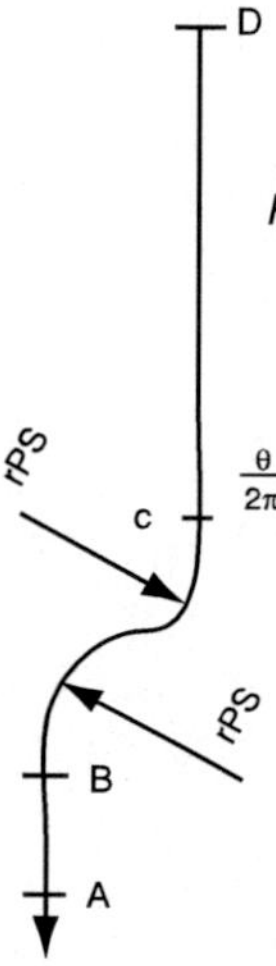

Kinked longitudinal prestressing

P_{HV} = horizontal force due to curved vertical tendon

$= \Sigma(F_{\theta 1} \sin\theta_1 + F_{\theta 2}\sin\theta_2 + \ldots) a_s \frac{\theta}{2\pi \bar{f}}$

$\bar{f}$–rotation factor a_s – losses; θ's angles subtended to a vertical

$\frac{\theta}{2\pi} F_V$ = Total net vertical force excluding losses

Axial force / radian / area $= \frac{P_K \cdot n_b \sin\theta}{\pi \tau_{PS} W_b}$; P_k = Tendon's characteristic load

Boundary Pressures $= \frac{\phi}{\frac{\pi}{4} \cdot Rg \cdot l}$

n_b = Total tendons in a bandwidth

w_b = a bandwidth

ϕ = Total load in the grid line

l = Grid line

R_g = Radius of grid line

$\bar{\theta}$ = angle subtended by circumferential tendons

H = Zonal height

P

Circumferential tendons

$P_H = P_K \cdot n_b \cdot \bar{\theta} / 2\pi w_b \cdot H$

Helical tendons

$\hat{y}$ – position vector $= xi + yj + zk$

$= a\cos\theta_i + a\sin\theta_j + zk$

Straight helix :

$$\frac{d^2 \dot{v}}{ds^2} = (-a\cos\theta_i - a\sin\theta_j) \quad \frac{1}{\sqrt{a^2 + b^2}}$$

Barrelled helix :

$$\frac{d^2 \dot{v}}{ds^2} = \cos\alpha \frac{da}{dz} (-\sin\alpha) [(a' \cos\theta - a\sin\theta d\theta)i + a' \sin\theta + a\cos\theta \cdot d\theta)) + K] + [a'' \cos - 2a' \sin\theta \cdot d\theta - a\cos\theta (d\theta)^2 - a\sin\theta d^2\theta)i - ((a'' \sin\theta + 2a' \cos\theta d\theta - (\sin\theta (d\theta)^2) + \cos\theta d^2\theta)]\cos^2 d = \bar{K}\dot{n}$$

Tendon normal load $P_N = P_K / rPS \cdot n$

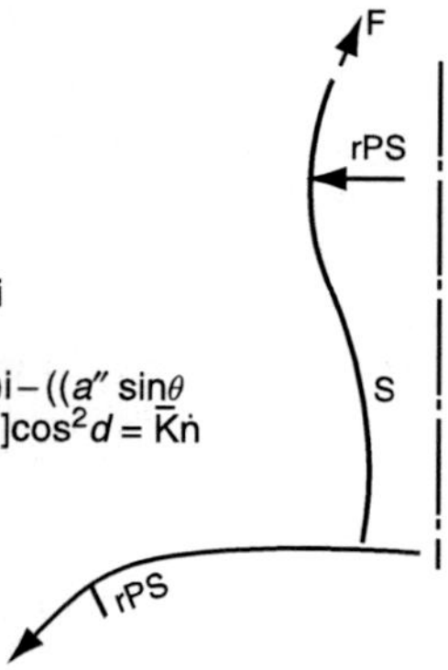

5.7.2 *Liner*

The liner in general is to protect concrete from excessive reactor temperatures and radiation developed from inside. It carries cooling pipes circulating water at suitable intervals. The liner is anchored to the concrete by means of plugs, etc. The details are given later on under a separate caption.

5.7.3 *Bonded Reinforcement*

Prestressing systems would be able to avoid completely a premature failure either in the primary zones or in local areas. Under such conditions greater emphasis can be laid on the control of cracking by using good quality bonded steel.

5.7.3.1 Primary and Secondary Reinforcement

Bonded reinforcements are divided into two main categories:

i. Primary type: they reinforce radially, circumferentially and longitudinally the walls and caps, say, of a cylindrically shaped vessel.
ii. Secondary type: these are mainly for local areas such as zones below anchorages, buttresses or wire/strand winding channels.

5.7.3.2 Factors Affecting the Quality and Layout of Bonded Reinforcements

Elastic, ultimate load and construction analyses, if properly carried out, do give sufficient information about the quantity and disposition of bonded reinforcement. Nevertheless they are not the only means of achieving the right objective. Several factors need to be considered prior to these analyses.

The shape of the vessel, whether spherical or cylindrical with flat, haunched or curved closures, is of importance.

The type and layout of a reactor system including boilers, circulators, standpipes and control rods and penetrations required for instrumentations and other equipment can also affect the reinforcements. Substantial changes in the layout and design would be necessary if boilers, circulators and their closures were located within the walls rather that in the annular space. The same is true of the manner in which standpipes and control rods are arranged in the caps.

Prestressing system layout is another factor which is to be considered in the conventional reinforcement design. Generally prestressing systems come first. The tendons will have to go into the vessel first and their positions are determined by analyses and practical requirements. Thus the bonded steel design layout will be dependent on how a typical vessel has been stressed. The layout and hence the design would be different if the vessel were stressed with a helical wire-winding system, tendons describing horizontal arcs and which are anchored on buttresses and finally a sequential cable system.

Other factors are the anchoring of liners, positioning of cooling pipes behind the liners, charge machine tract and tendon anchorage and their protection methods.

Vessel construction methods, which include the selection of lifts and bays and existing joints in between and methods of liner erection in relation to concrete pours, must be considered.

The final factors are material inherent properties which include the choice of reinforcement, its short- and long-term behaviour in the vessel under various combinations of loads, temperature, radiation and corrosion.

5.7.3.3 Choice of Bonded Reinforcement

Generally for thick concrete vessels the choice of bonded steel in the UK is based on its high tensile strength and its interaction with the concrete. In thick sections it most certainly gives a higher bond length and an improved degree of crack control. Several tests have been carried out on GK 60 deformed bars [218] or equivalent in Eurocodes. Bond tests carried out on the 50 mm dia. bars indicate that using these reinforcements no hooks are essentially required in the PCRV. Bars used in the UK vessels are 50 and 32 mm as primary reinforcements and 18 and 12 mm as secondary reinforcements.

5.7.3.4 Procedure for the Design of Bonded Reinforcement

The step-by-step procedure is recommended for the final disposition of primary and secondary reinforcements in the PCRV. The reactor layout and the influence runs on the effects on vessel and the allowance of low probability loadings arising from accident or fault conditions. Analyses are required to cover these uncertainties.

In certain zones such as embedded items areas would create where stresses in concrete are unacceptable; reinforcement may be arranged to reduce these stresses. Due to temperature hot spots may occur in concrete zones which could be unacceptable. The stresses could be high. By providing some of these local reinforcement will reduce stresses to acceptable values in concrete. The most appropriate technique is to use advance analytical means such as finite element. Stress trajectories with and without reinforcement could be developed. These stress trajectories would determine the sizes and the zones to which these reinforcement could form shapes. The extent of such reinforcement will have bond lengths between 24 and 48 diameters of the bar. Obviously these reinforcement would be evaluated under prestressing anchorages too to avoid cracks under stressing loads.

When the final layout is obtained for these bonded reinforcement, it will be necessary to include them in the hybrid finite element analysis, either placed on solid element nodes or within the concrete body linking them to the adjacent nodes of the solid elements. Alternatively, these reinforcement can be spread within the body of the solid elements. These methods are discussed in Chapter 3 under Finite Element Analysis.

5.7.4 Embedded Elements

Based on the vessel layout, certain components may be embedded to save vessel area for the reactor. These embedded elements such as shown in

Table 5.2 Embedded elements

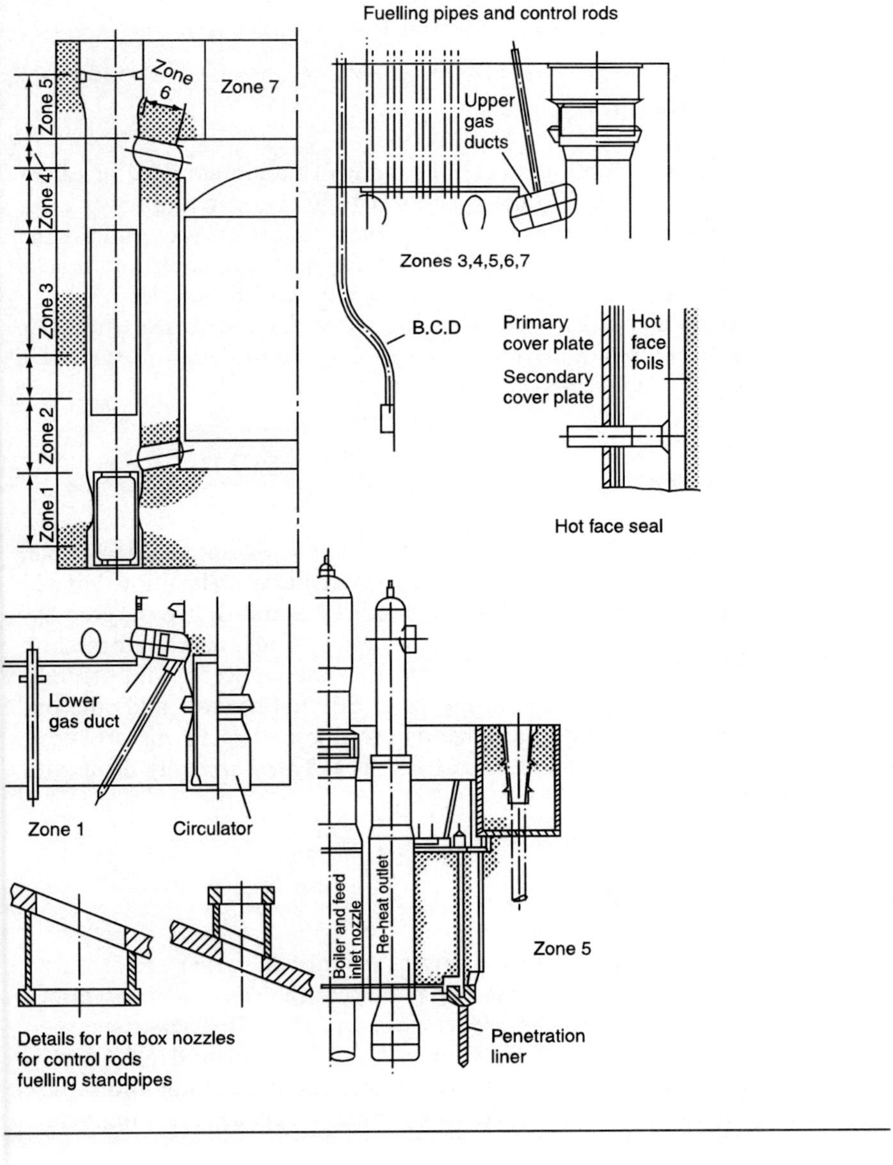

Table 5.2 can produce stress concentration in areas quiet unacceptable. They should be thoroughly analysed using numerical techniques such as finite element.

5.7.5 *The Concrete*

A separate analysis is given on the behaviour of the reactor concrete in this chapter.

5.8 Ultimate Conditions

A validity analysis is required to check the mode of failure assumed or calculated or derived from the experimental tests. This analysis should cover cases such as the intactness and rupture of the liner, failure of local areas, the possibility of shear failure in top caps or at cavity/wall ligaments and cracks in hot spots (areas around cooling pipes) and computer subrouteries.

Several chapters give a list of analyses required for the vessels and their components. The ultimate limit state analysis is given in detail on the bases tests and carried out and discussed in Section 5.7.

5.9 Methods of Analysis

Many methods are available to the designers. The most common are [14, 15] based on the finite element, finite difference or dynamic relaxation lumped parameter and the limit state methods. Appendices A and B give the step-by-step approach of the finite element analysis, Chapter 3 and limited state method used for these vessels is given in this chapter. In some service and fault conditions it is required to consider the influence of external hazards and environmental conditions. Major external hazards are due to seismic disturbances, wind/local generated missiles and aircraft crashes. These are fully dealt with by a number of researchers or designers [231].

5.10 Analysis of Results

The linear, non-linear and cracking analyses mentioned earlier are useful to predict the operational and overload behaviours of the vessels. Four vessels, namely the Dungeness B, Oldbury, Hartepool and Fort St. Vrain have been analysed. Figures 5.13, 5.14, 5.15, 5.16, 5.17, 5.18, 5.19 and 5.20 give a summary of the load displacement relations, modes of cracking and failure and the safety margins. Cracks predicted by the finite element and limit state

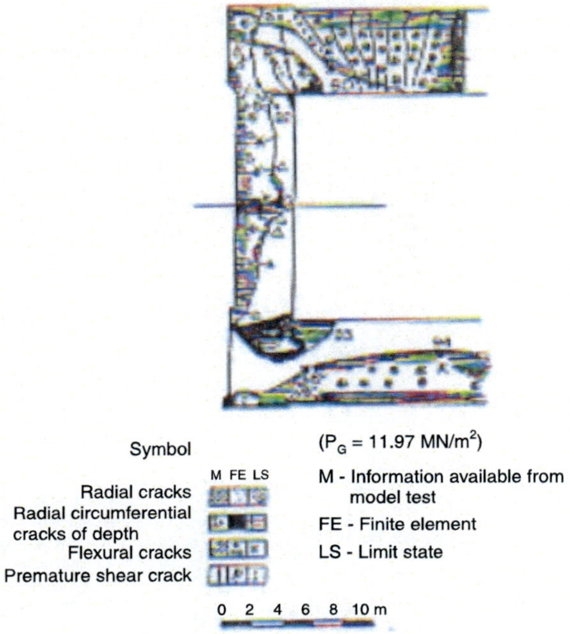

Fig. 5.13 Oldbury prestressed concrete reactor pressure vessel

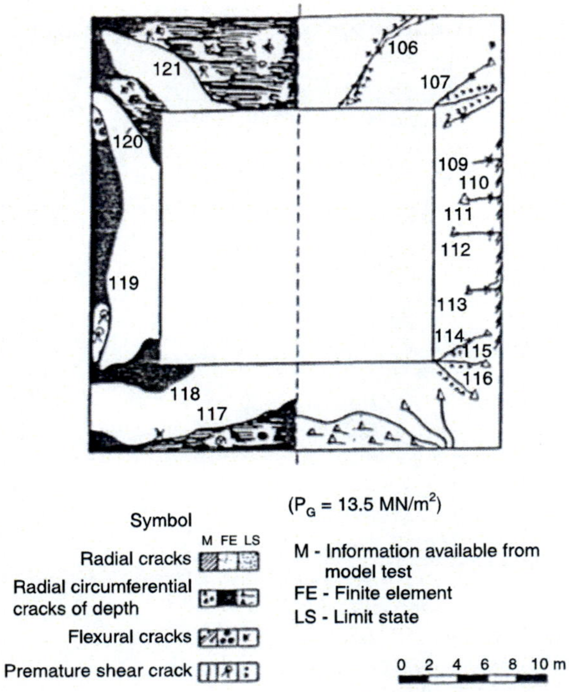

Fig. 5.14 Dungeness B prestressed concrete reactor vessel

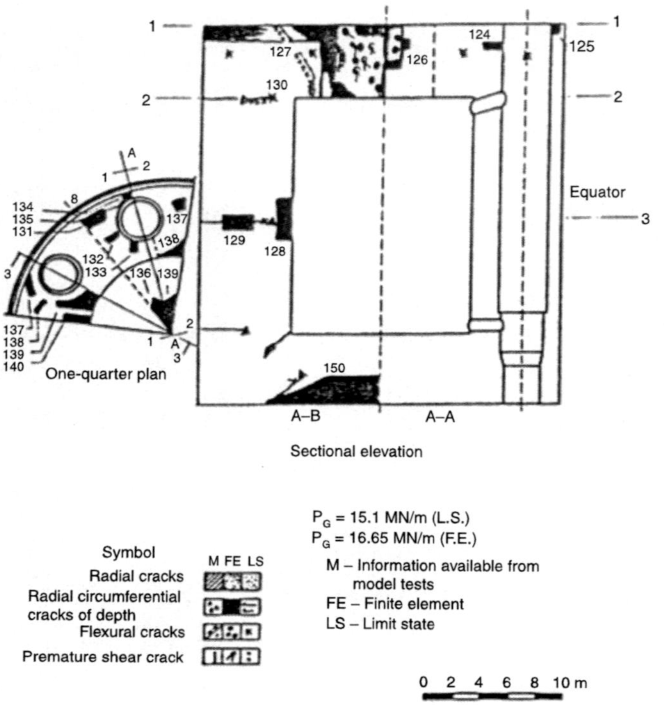

Fig. 5.15 Hartlepool prestressed concrete reactor pressure vessel

Fig. 5.16 Fort St. Vrain prestressed concrete reactor vessel (P_G = 18 HN/m^2)

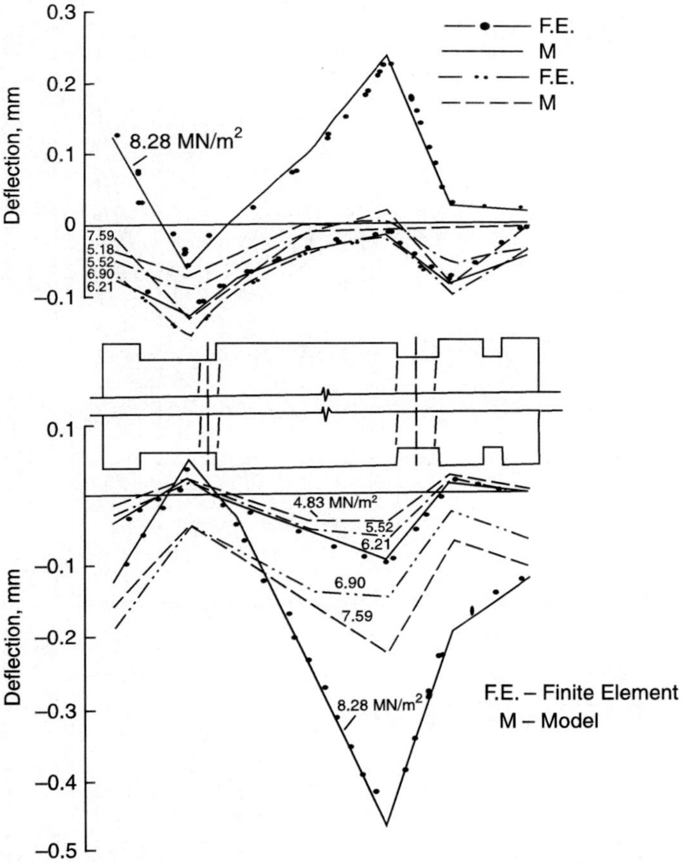

Fig. 5.17 Deflected shape. Sides of model – ulimate run

analyses for these vessels are compared with each other and wherever possible with those predicted by experimental tests. The comparative study shows that theoretical and experimental results are in good agreement.

In two such vessels conditions under normal operations, 2.5 P_{GD} and failure conditions are particularly examined – principal stresses are examined at suitable pressure increments right up to 2.5 P_{GD} and beyond. The complete plug failure occurred at 4:1 P_{GD} (13.5 MN/m^2) and the top cap (of the Dungeness B) and the wall. It was the shear compression failure. At 2% ultimate strain in the tendons, the possible failure of longitudinal tendons, a group of circumferential tendons did fail. From this model testing, central core plug failure at the top cap (shear compression type) occurred at 13.8 MN/m^2. Two vertical tendons failed at 10.2 MN/m^2.

It is concluded that the failure of this vessel has occurred owing to the complete removal of the central core or plug in the top cap at a pressure

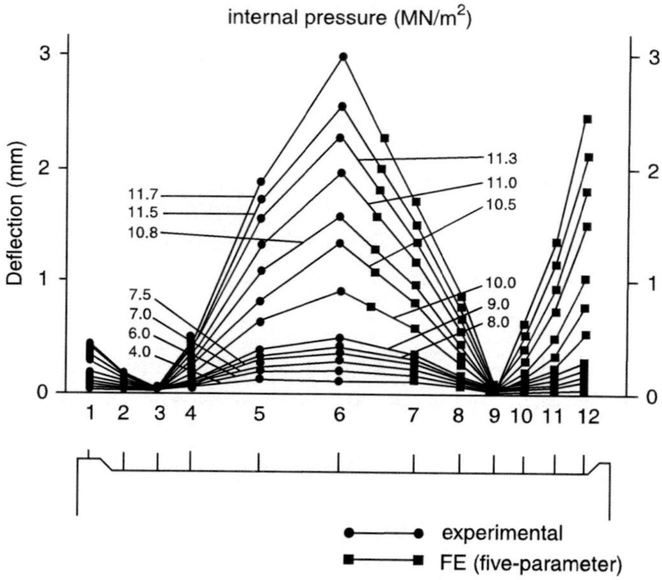

Fig. 5.18 Horizontal deflection of wall (ultimate pressure) – Oldbury vessel

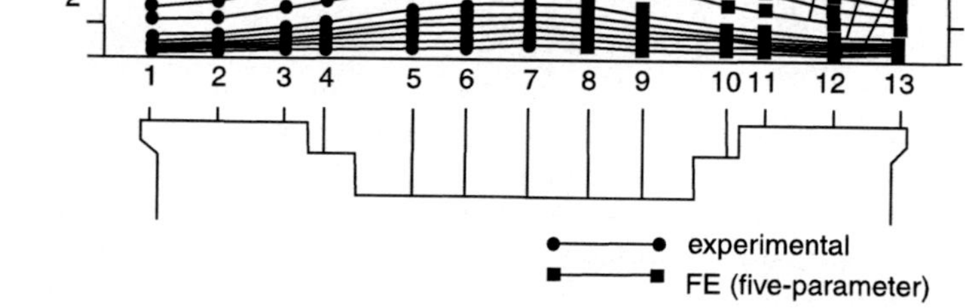

Fig. 5.19 Vertical deflection of top cap (ultimate pressure) – Oldbury vessel

around four times the design pressure. The FE and the M analyses are in good agreement with each other and with most of the published data from the experimental tests. Throughout the vessel there is no premature failure of the liner.

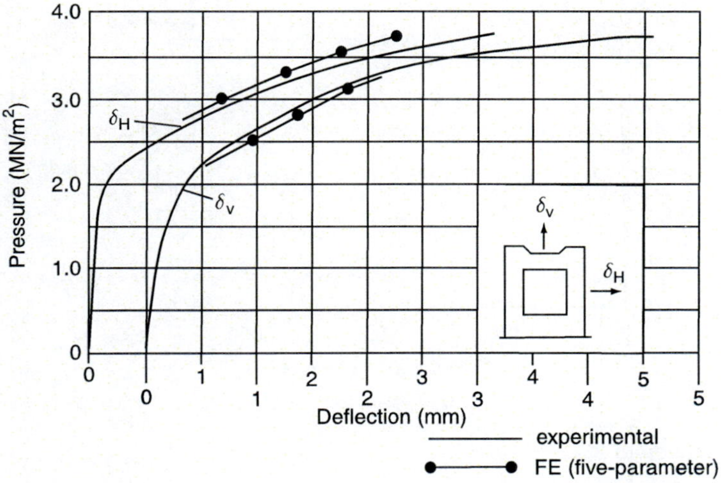

Fig. 5.20 Presure–deflection curye of Oldbury vessel [418: experimental]

5.10.1 Oldbury Vessel

For the model test of the Oldbury vessel, gas pressure versus deflections is plotted in Figs. 5.18 and 5.19. The results from finite element analysis have been plotted for comparison in the same figures. Because of great variations in tendon layout, the pressure versus deflection curves are given but at a pressure of 4.825 MN/m^2 (1.815P_{GD}). This must have been produced even earlier in the model, since such a crack cannot suddenly appear at such a marginal pressure difference of 0.065P_{GD}. The results from FE clearly indicate that there is a plastic zone appearing at the equator.

Raising the pressure to 5.32 MN/m^2 (1.995P_{GD}) FE indicates that the vessel top cap plastic zone has been abruptly increased and additional flexural cracks have emerged. The plastic zone in this case has reached the areas surrounding the central hole in the standpipe area and the prestressing galleries. Model tests show similar effects in the top cap. The results from M give two surface flexural cracks and various positions or locations. Therefore each curve takes into account the contribution it has received from the exact position of tendons, conventional steel and the liner in the original layout. In order to test these results, some of them are plotted in Fig. 5.20 along with those from the model tests (four tests) carried out by Eddie et al. and others. Both the horizontal and vertical deflections are in good agreement. However, in Fig. 5.21 although the computed axial strains are in good agreement with those measured on the prototype, the hoop or circumferential strains are wide apart. The only explanation that can be offered is the difference between the exact chosen tendon layout of the vessel parameters and that of the prototype.

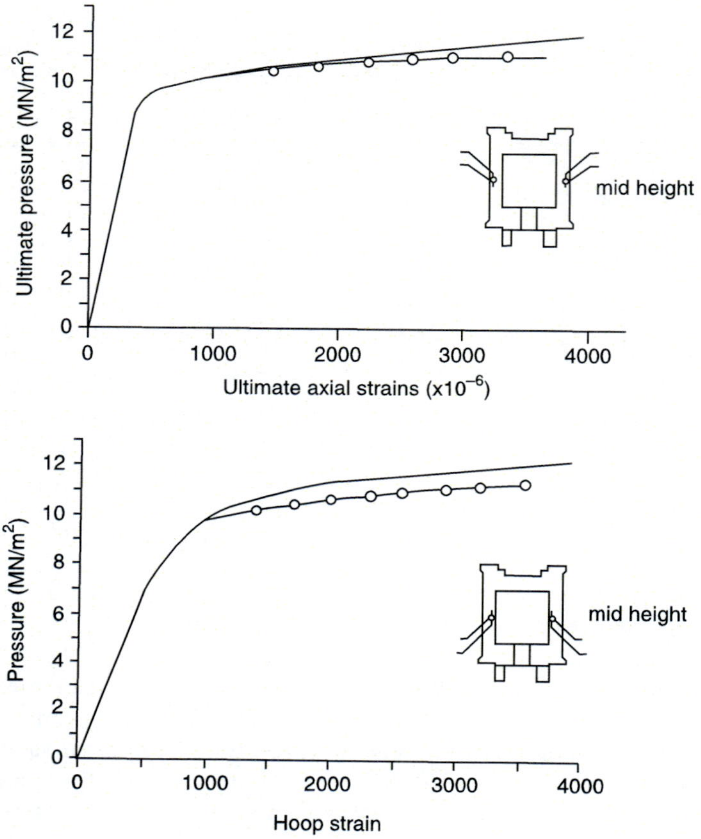

Fig. 5.21 Hoop and axial strains at wall mid-height – Oldbury reactor vessel (conditions at ultimate pressure) (——, finite (five-parameter): —O—, experimental)

The percentage difference is still within 5–8%. A further comparison is given in Fig. 5.21 which is based on the monitoring of the vessel's performance for a period of 10 years.

Figures 5.22 and 5.23 show plots for various principal stresses. It is interesting to note that in Fig. 5.24 the effect of removal of cracked concrete at 1.81 times the design pressure is shown. Figures 5.25 and 5.26 show the principal stresses at 2.5 times design pressure. No cracks have been observed for a pressure range of 2.66–3.75 MN/m^2. At a pressure of 3.99 MN/m^2 ($1.455P_{GD}$) a plastic zone in the top cap centre has been predicted by FE. Both FE and M show a plastic zone and a minor crack at the top haunch. Radial hair cracks together with a plastic zone have also been predicted by FE. However, M assumes a crack width at the equator and finally proves that it is the only crack that exists in the wall at this pressure.

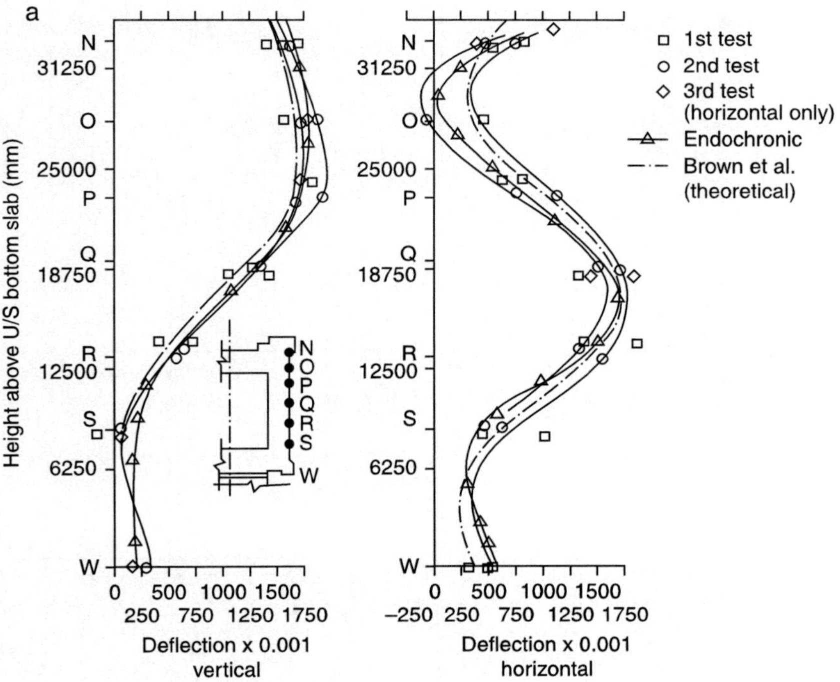

Fig. 5.22 (**a**) Wall horizontal and vertical deflections. (**b**) Top slab horizontal and vertical deflections. (**c**) Circumferential stresses – operating conditions only (finite element analysis)

At a pressure of 4.665 MN/m^2 (1.75P_{GD}), a wider and deeper plastic zone appeared (mark 6) in the top cap. It is interesting to note that the top haunch plastic zone has also been increased. In addition, a minor plastic zone has been predicted by FE at the bottom cap haunch. Moreover, at the upper part of the side wall a wider radial crack has occurred and this crack has also been confirmed by the model tests.

Both top and bottom haunch plastic zones of this vessel predicted by FE are fairly large. Here there is some disagreement between FE and M on the type of crack. The top haunch M results show both flexural and premature shear cracks. On the other hand, the results from FE clearly indicate the possibility of cracks caused by the principal tensile stresses caused by shear. However, there is an agreement between the two analyses on the initiation of flexural cracks in the bottom cap haunch. The magnitude of these cracks is not greater than 0.007 mm. Practically there is no change in the flexural crack. The bottom cap has an enlarged plastic zone.

Reports suggest that a horizontal-cum-radial crack has appeared above the circulator penetration, but the published data have not confirmed it. In the

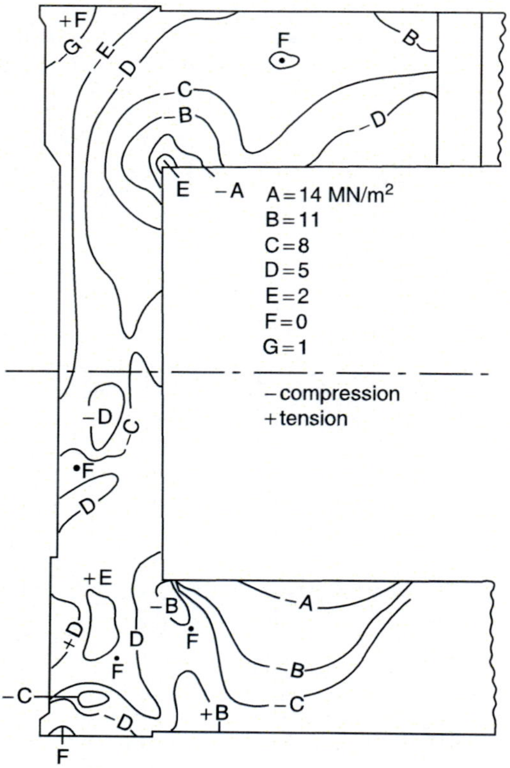

Fig. 5.23 Principal stress 1 – normal operation

barrel wall M shows flexural cracks in both radial and circumferential directions. Examination shows that cracks in radial directions do exist at pressures between 4.664 and 5.32 MN/m^2. They have been superimposed for comparison. The results obtained from FE show plastic zones and surface cracks. Obviously there is a slight discrepancy. It seems that FE does not agree with M up to a pressure of 5.32 MN/m^2 for the creation of plastic zones or surface cracks. The difference is in the modelling of the layouts. No test data are available for this vessel or any other cylindrical vessels for such cracks appearing at 1.995 times the design pressure. However, there seems to be some agreement in this area at pressures above 5.32 MN/m^2.

Carrying on in the same way, many failure zones have been developed at additional incremental pressures. As shown in Fig. 5.13 at P_G, ultimate = 1197 MN/m^2 (4.5P_{GD}), both model tests and FE results are in agreement. The vessel top cap and top corner have been separated.

This failure mechanism is different from that predicted for the Dungeness B vessel, although they both have an identical shape and reactor system layout but with different prestressing system and gas design pressure.

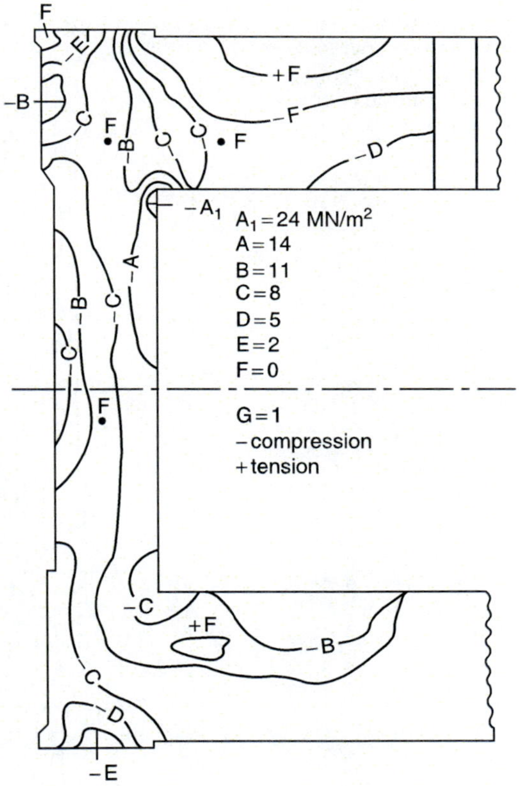

Fig. 5.24 Principal stress 2 – normal operation

5.10.2 High-Temperature Gas-Cooled Reactor Vessel (HTGCR)

Figure 5.22 and 5.23 show gas pressure versus deflection, and Figs. 5.24 and 5.25 show principal stresses at normal operation. Figures 5.26 and 5.27 show principal stresses at 2.5 times the design pressure. Although this vessel layout is different in many respects to that of the Hartlepool reactor, the stress flow in many areas is identical. The stress concentrations around the boilers and at the junction between the caps and the walls are similar in both cases. However, the quantities do differ. It is because of these and some dissimilar stress contours that the final failure modes in both vessels differ.

Figures 5.6 and 5.7 show 3D and 2D finite element mesh generation schemes of a prestressed concrete vessel designed for a high-temperature gas-cooled reactor. Both internal and external loads have been computed in accordance with the method given in Chapter 3. The program CREEP, which

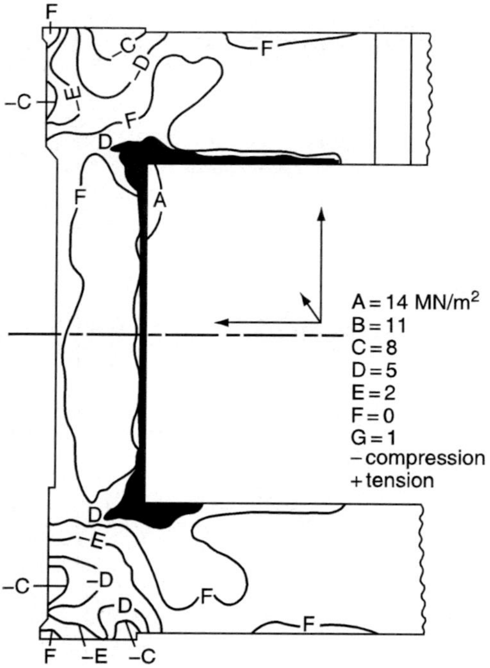

Fig. 5.25 Minimum principal stress – pressurisation after long-term shut-down showing effect of removal of cracked concrete (A)

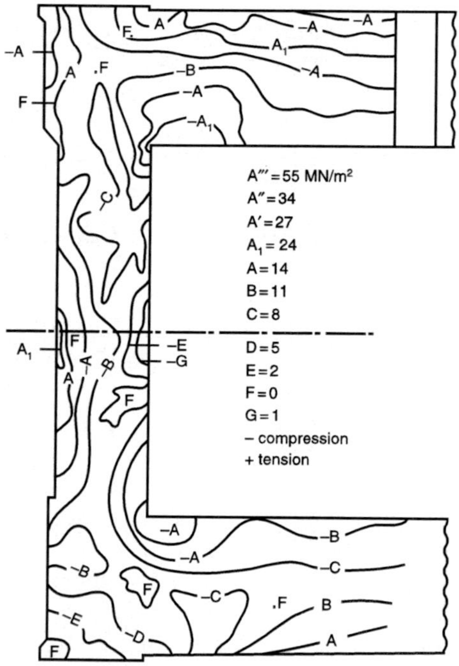

Fig. 5.26 Maximam principal stresses at 2.5 P_{GD}

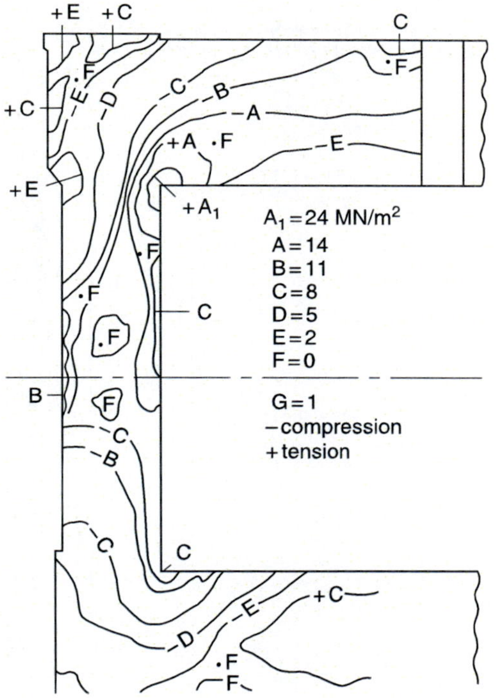

Fig. 5.27 Minimam principal stresses at 2.5 P_{GD}

is a part of the program ISOPAR, gives results for when the vessel undergoes the effect of creep. First a normal operation condition (prestress + pressure + temperature) is considered. This vessel is analysed with and without creep. The axial, radial and circumferential stresses have been computed. The influence of creep when the Young's modulus is constant or variable is considerable. In some local areas, because of creep the stresses change significantly with passage of time. Figure 5.28 shows stresses at the top cap with and without creep for up to 20 years. Figure 5.28 shows the behaviour of the vessel under operating and thermal shut-down conditions with and without the influence of creep.

The vessel is then taken to elasto-plastic and cracking conditions. As discussed later on, the relationships between gas load, prestressing strains, and deflections and cracks with and without the influence of creep have been established. These results (Fig. 5.29) show that where the influence of creep is not considered, the load-carrying capacity is overestimated by as much as 25%. With creep, considerable changes were discovered in the cracking pattern of the vessel and the overstress conditions in local areas (standpipes and boiler/circulator penetrations).

Fig. 5.28 Deflection curve of wall HTGCR reactor vessel (— —, experimental; —○—, FE (four-parameter); — • —, FE (five-parameter))

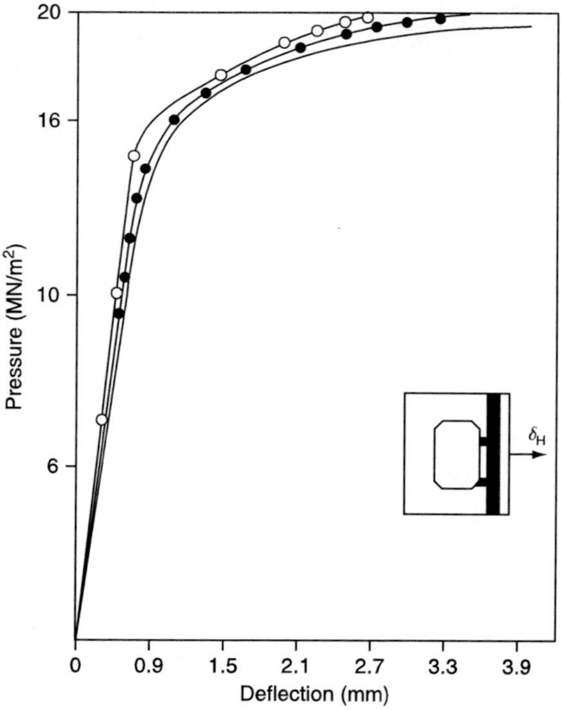

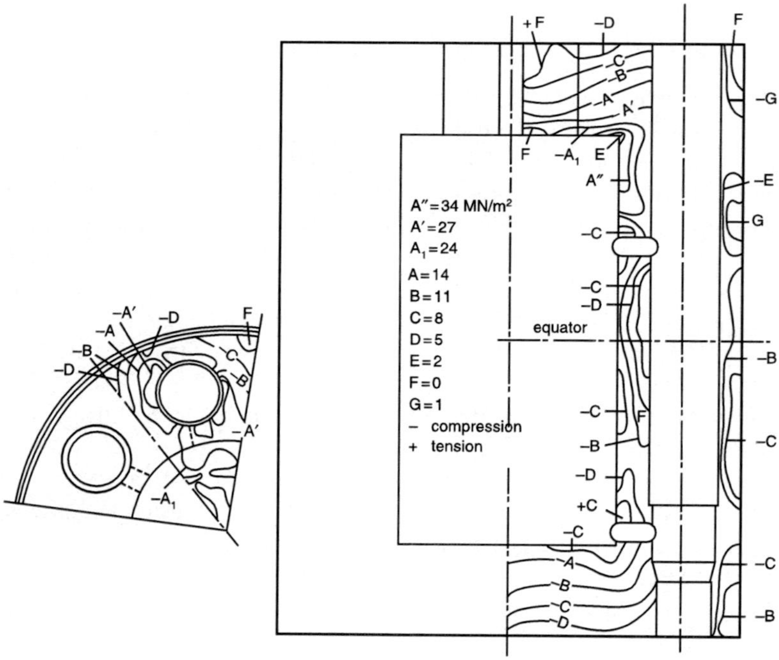

Fig. 5.29 Maximum principal stresses at normal operation

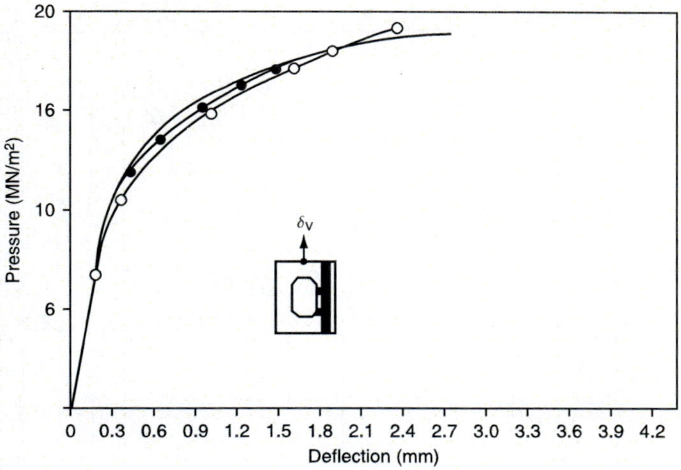

Fig. 5.30 Pressure–deflection curve for top cap of HTGCR reactor vessel (—— Experimental: —○—, FE (four-parameter); —•—, FE (five-parameter))

Ahmad has idealised the same vessel, as shown in Fig. 5.30. He has used the vessel bonded and unbonded and produced deformations (Figs. 5.30, 5.31, 5.32 and 5.33) for normal operation and for 40 years of creep. He has also produced a graph (Fig. 5.34) showing deformation for this vessel at various pressures. He has included creep in his analysis. This vessel has also been considered for elasto-plastic and cracking effects by Bangash. The deformation of the vessel at the cracking stage under the influence of creep and with the steel liner anchored to concrete is shown in Fig. 5.35.

The stress trajectories at normal operation and at 2.5 times design pressures can be obtained in the same manner as for vessels discussed previously. The vessel design pressure P_{GD} here is 5.68 MN/m^2. At a pressure of 7.48 MN/m^2 (1.315P_{GD}), this vessel's behaviour (Fig. 5.36) is identical to that of the Hartlepool vessel. The only change is the plastic zone developed around the gas inlet duct under mark 6.

At a pressure of 9.28 MN/m^2 (1.635P_{GD}) the behaviour of this vessel (Fig. 5.37) in most zones is identical to that of the Hartlepool vessel [423, papers H3/5, H5/3] and others. When the pressure was raised to 6 MN/m^2 the only difference is in the position of cracks (marks 11, 12 and 13) at the top haunch and the extent of the plastic zones (marks 7, 14, 20, 21 and 22). The crack sizes have not been predicted in the top cap. The bottom cap has been affected only at the bottom haunch. Both FE and M are in agreement in most areas, as shown in Fig. 5.37. Raising the pressure to 11.08 MN/m^2 (1.95P_{GD}) there is a significant difference in the behaviour of this vessel (Fig. 5.38).

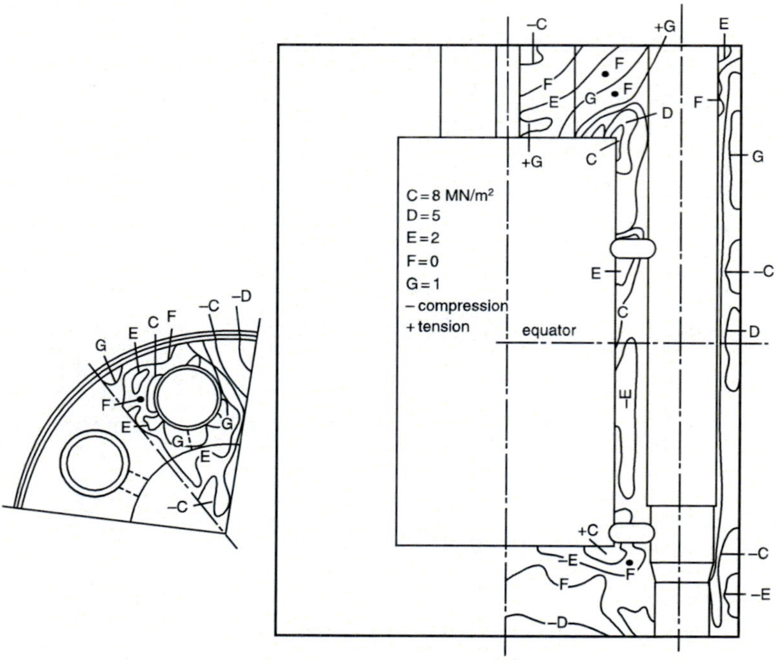

Fig. 5.31 Minimum principal stresses at normal operation

The HTGCR vessel failure mode is moving more towards flexural criteria than shear criteria. The initiation and propagation of cracks and their positions are very different. The reason is purely the dominant role that the barrel wall/cap slenderness ratio has played in producing flexural cracks. In plan around the boiler and circulator penetrations and in the standpipe area, most of the plastic zones are identical.

At a pressure of 12.88 MN/m^2 (2.24P_{GD}) the top cap mode (marks 46–53) is given in Fig. 5.39.

There is a similarity in the development of plastic zones between the two vessels in the bottom caps. In plan, the plastic patch zones are very similar between these two vessels. The only difference is the extent of cracks developed at the equator (mark 49) by FE. At this pressure in the top cap the plug failure (mark 60) is imminent. It is interesting to note that the crack sizes are identical in these zones. Another difference is in the sudden appearance of a plastic zone just above the gas inlet (mark 55) and a flexural crack in the circumferential direction in the top corner (mark 54) at the outside of the boiler penetration. There is also a slight shift in the position of the crack (mark 57) around the circulator. The crack widths have reached about 0.2 mm

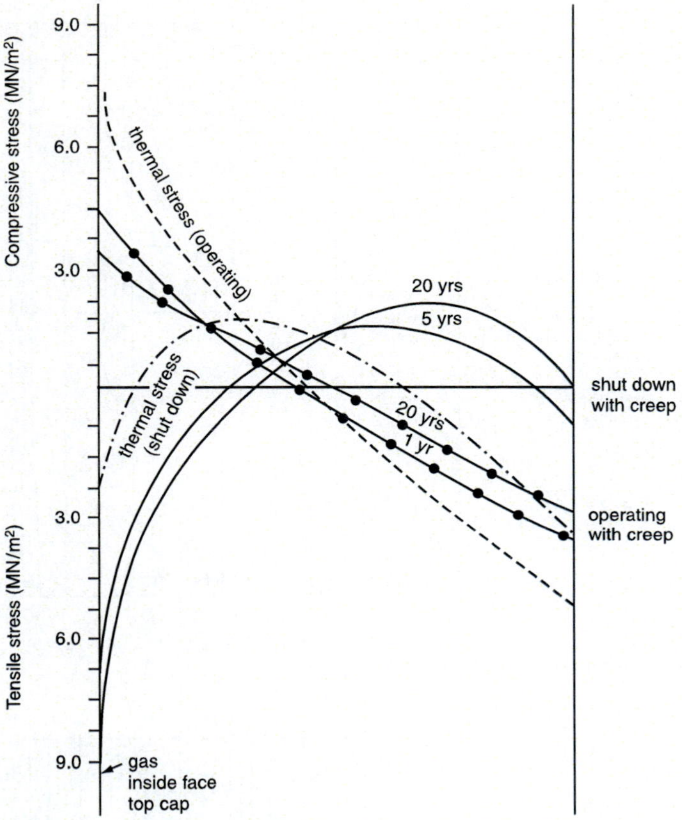

Fig. 5.32 Stresses through top cap – at operating temperature and shut-down

in most places. In many places strains in the tendons and in the bands do not exceed 0.009.

Raising the pressure to 14.68 MN/m^2 (2.258P_{GD}), sufficient flexural cracking in the barrel wall (Fig. 5.40) has occurred.

The nearest case that can be compared is the behaviour of the Hartlepool vessel under 2.5P_{GD}. In the top cap, plastic zones have identical locations.

However, in the HTGCR vessel no internal cracking has been predicted by LS around the boiler and the circulator penetrations.

In plan, the plastic zones are quite similar. There is a change in the plastic zone around the boiler penetration between sections (3)–(3) and (4)–(4). In the top cap a plug failure under mark 69 has been initiated. The strain and crack widths in most places do not exceed 0.01% and 30 mm.

At a pressure of 16.48 MN/m^2 (2.81 P_{GD}) the behaviour of this vessel (Fig. 5.41) is identical in many respects to the behaviour of the Hartlepool vessel [418–431] for 2.82 P_{GD}. In plan, the cracks and plastic zones are very

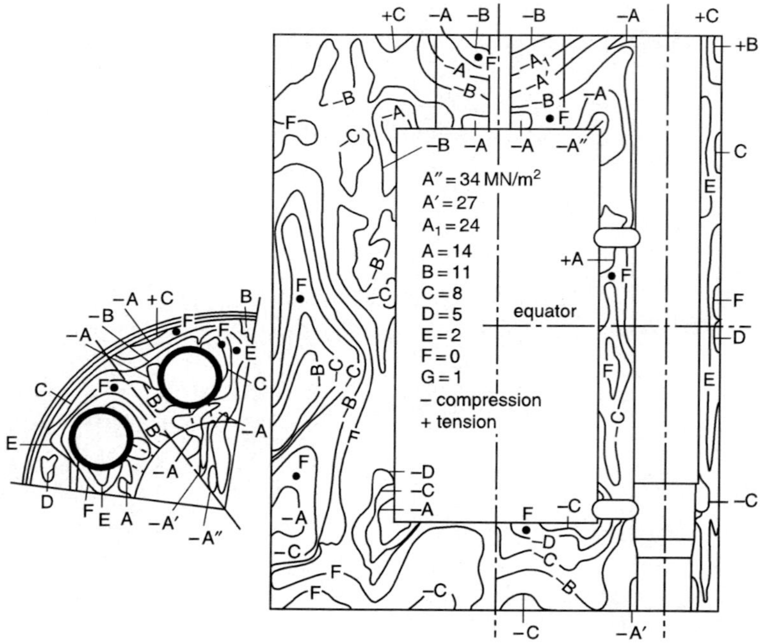

Fig. 5.33 Maximum principal stresses at $2P_{GD}$

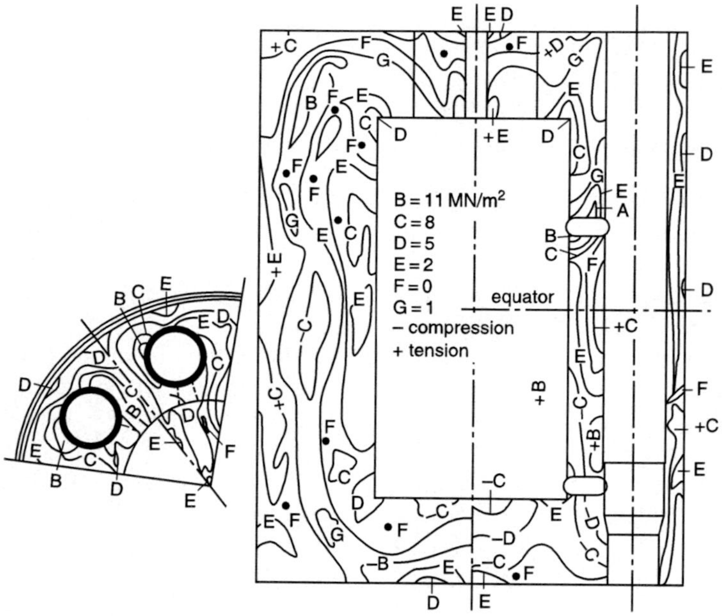

Fig. 5.34 Minimum principal stresses at $2P_{GD}$

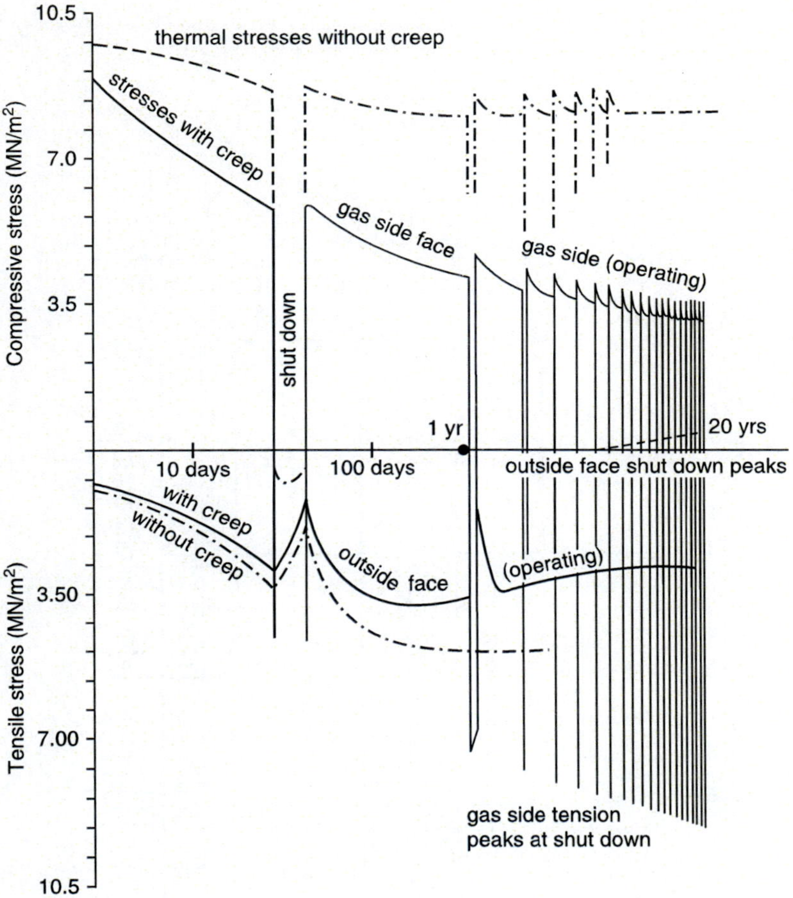

Fig. 5.35 Stresses in top cap showing inflection of shut-down

similar in both cases. In both the top and bottom caps flexural cracks have been extended in depth. Not enough change is visible of the plug failure line in the top cap. In addition, the barrel (Figs. 5.41 and 5.42) wall cracks and the haunch cracks have advanced considerably. These flexural cracks are more pronounced than in the case of the Hartlepool vessel.

The circumferential prestressing bands have reached the ultimate strain of 1.5% and the cracks in *most places* are around 50 mm. When the pressure is raised to 18.28 MN/m^2 (3.2 1P_{GD}), Fig. 5.43a shows the positions of plastic zones between sections (1)–(1) and (2)–(2).

Additional flexural cracks have appeared in the top cap. It seems that the influence of cracks due to shear compression and plug failure has been minimised. At the equator (marks 115 and 116) a larger plastic *zone* has been

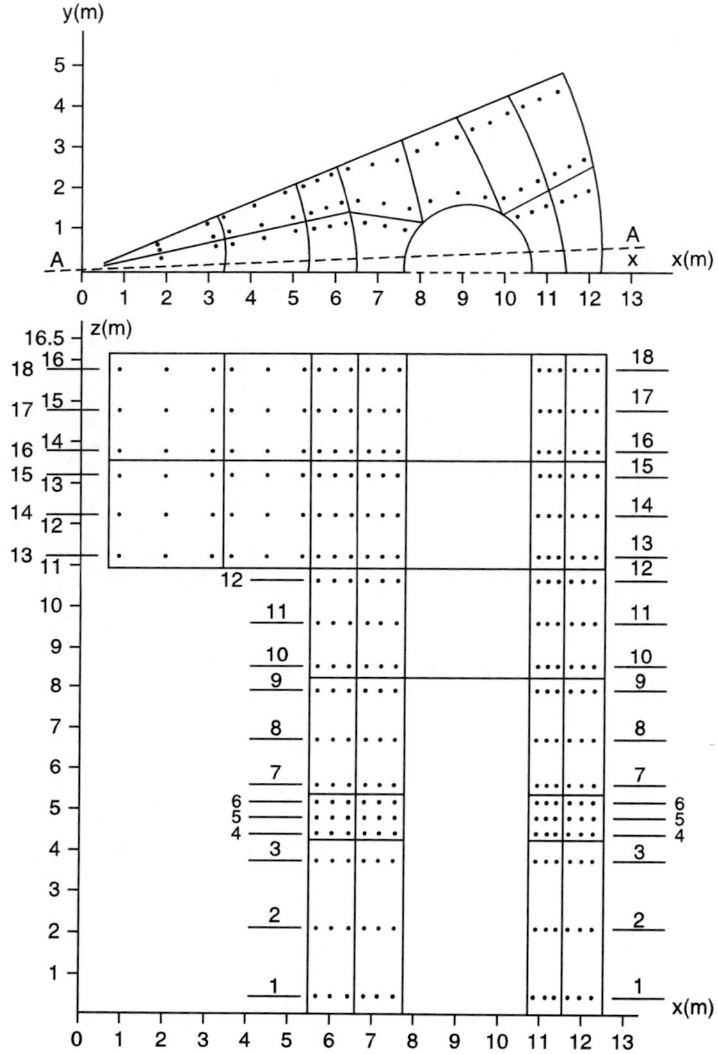

Fig. 5.36 Games point locations in the reactor vessel [262] (all dimensions in num)

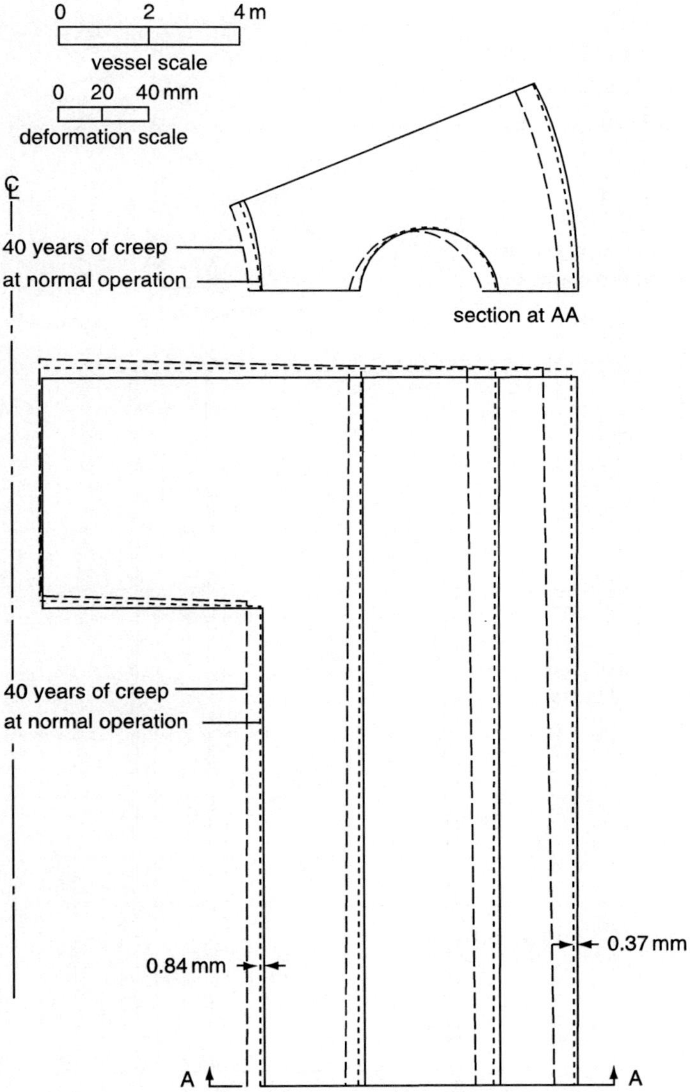

Fig. 5.37 Deformations under normal operation and after 40 years of creep [262] – vessel type, perfectly bonded

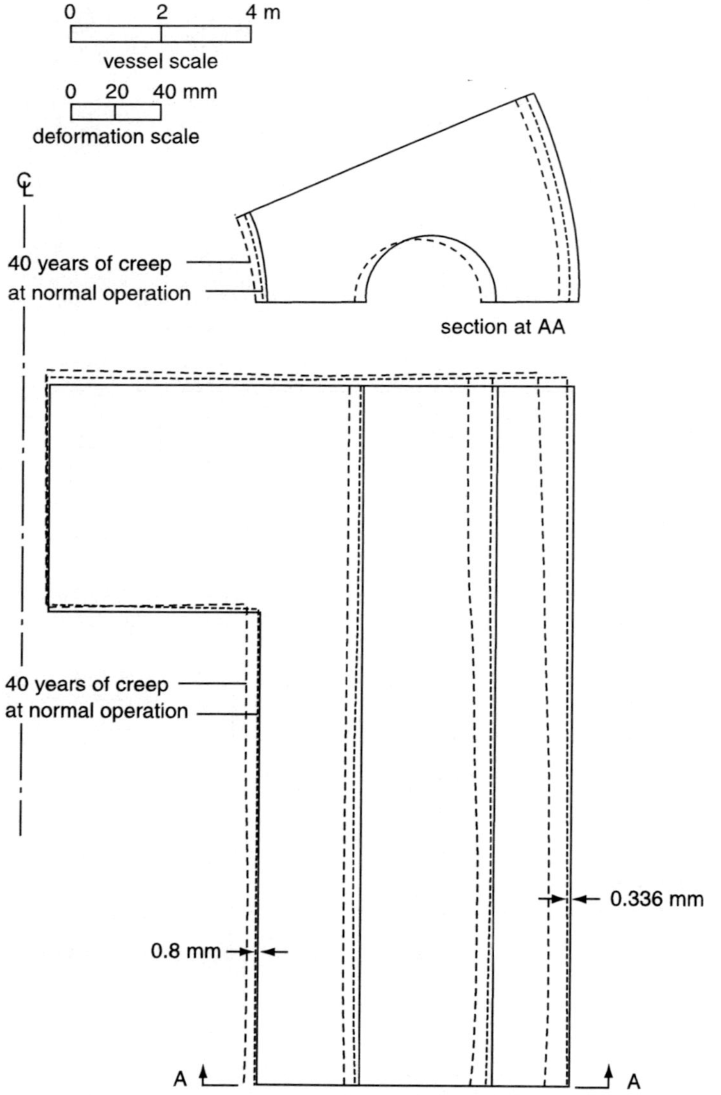

Fig. 5.38 Deformations under normal operation and after 40 years of creep [262] – vessel type, bonded (with Ahmlink element)

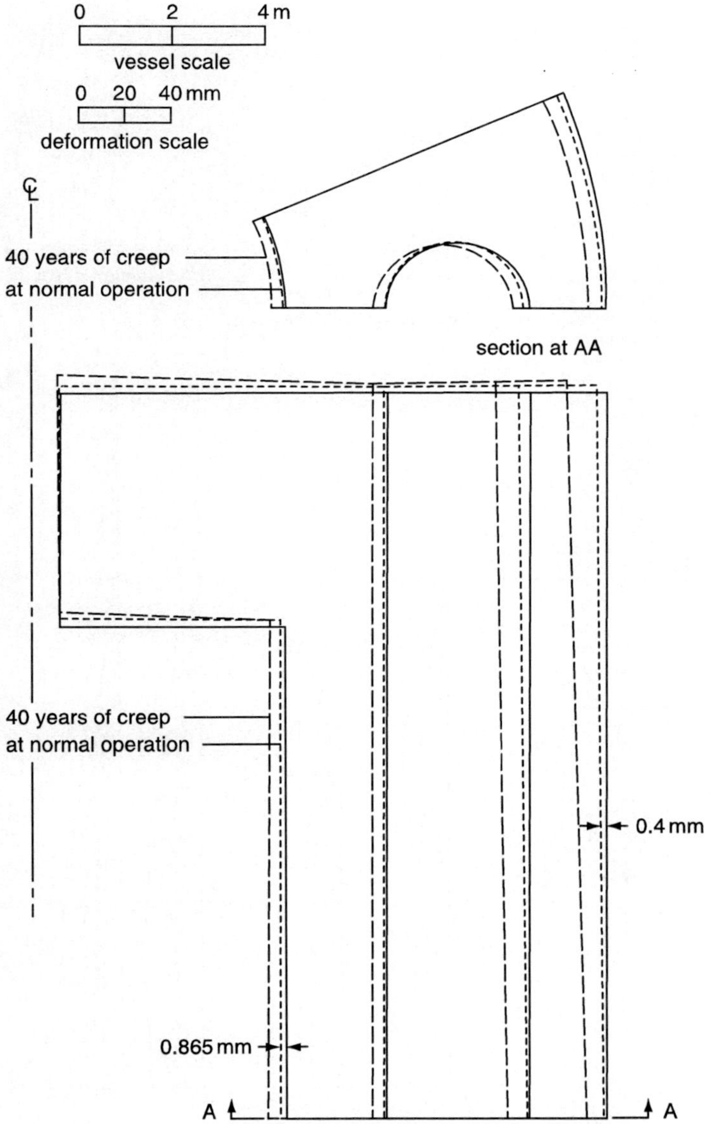

Fig. 5.39 Deformations under normal operation and after 40 years of creep [262] – vessel type, unbounded

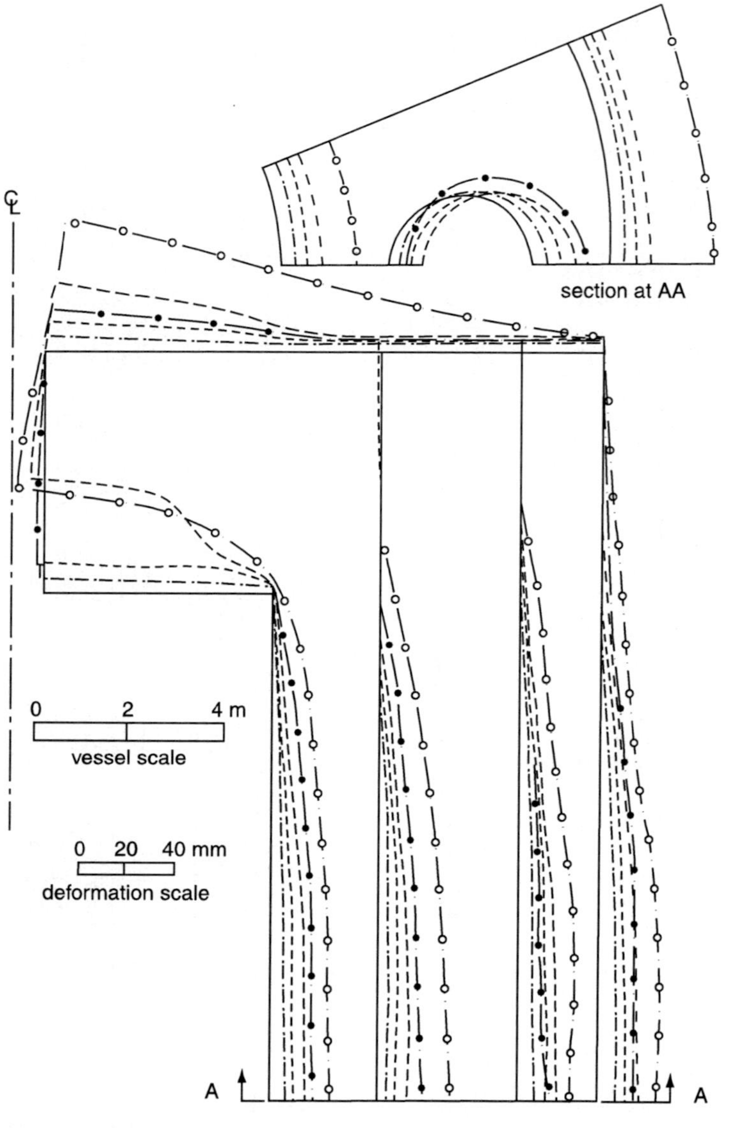

Fig. 5.40 Deformed shapes at various pressures [262, 749] – vessel type, bonded vessel with Ahmlink element (—, —, deformed shape at 3.0; ——, deformed shape at 3.3; ———, deformed shape at 3.5; —○ —, FE (four-parameter, Bangash); —• —, FE (five-parameter, Bangash))

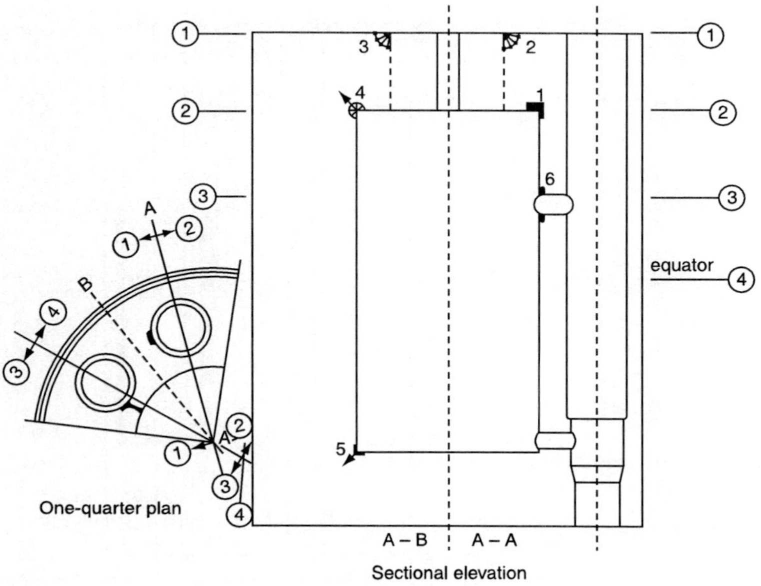

Fig. 5.41 High-temperature gas-cooled reactor pressure vessel (P_G = 7.48 MN/m^2) (see Fig. IX.118a for key)

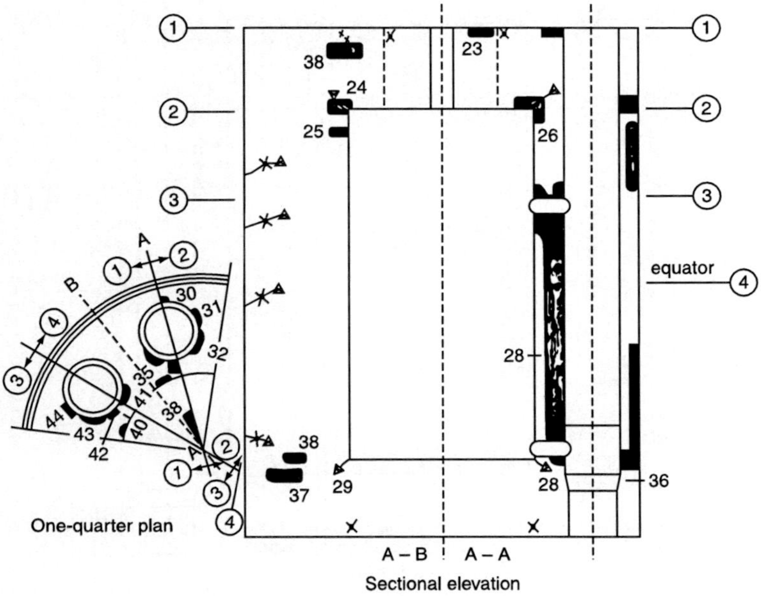

Fig. 5.42 High-temperature gas-cooled reactor pressure vessel (P_G = 11.08 MN/m$_2$) (see Fig. IX.118a for key)

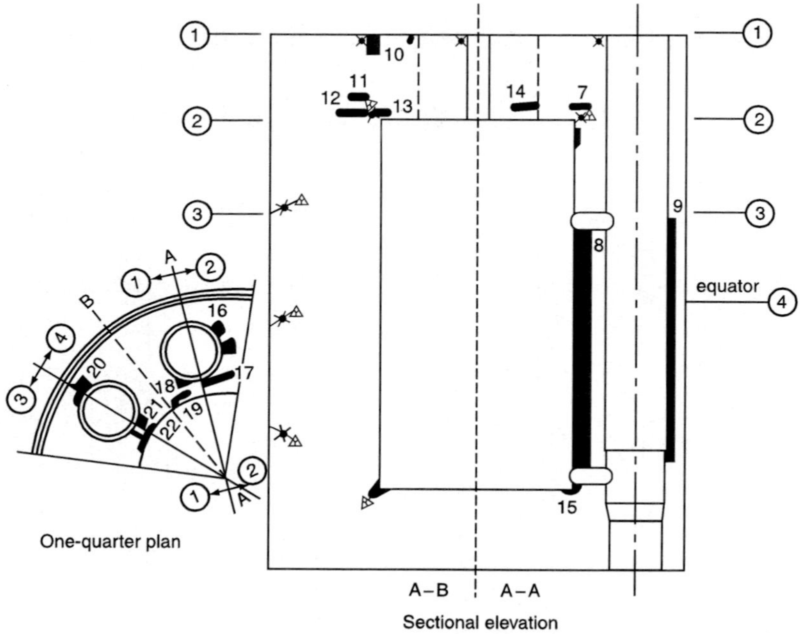

Fig. 5.43 High-temperature gas-cooled reactor pressure vessel (P_G = 9.28 MN/m^2) (see Fig. IX.118a for key)

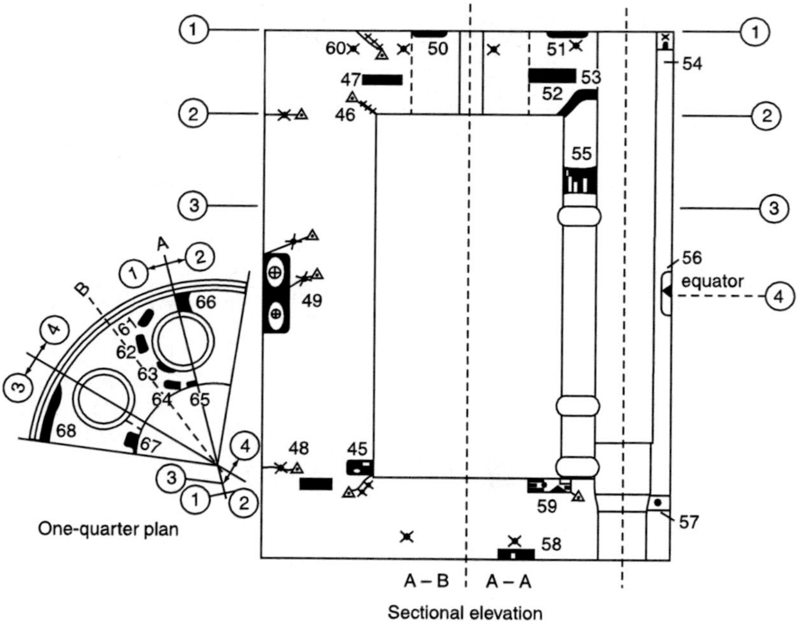

Fig. 5.44 High-temperature gas-cooled reactor pressure vessel (P_G = 12.88 MN/m^2) (see Fig. IX.118a for key)

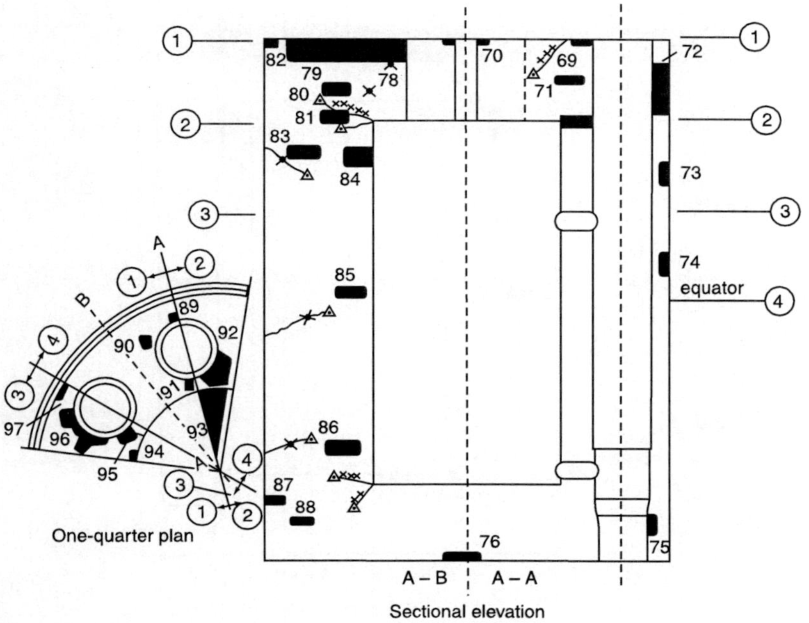

Fig. 5.45 High-temperature gas-cooled reactor pressure vessel (P_G = 14.68 MN/m^2) (see Fig. IX.118a for key)

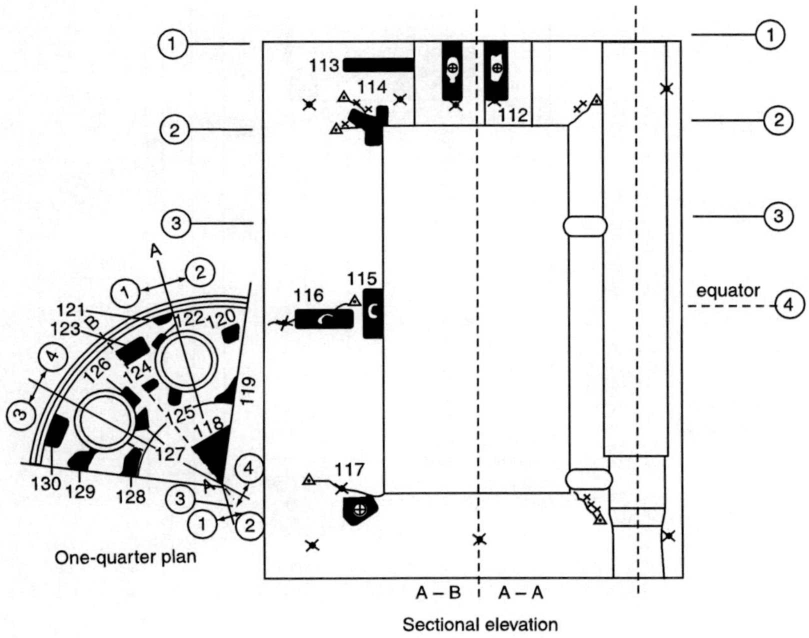

Fig. 5.46 High-temperature gas-cooled reactor pressure vessel (P_G = 18.28 MN/m^2) (see Fig. IX.118a for key)

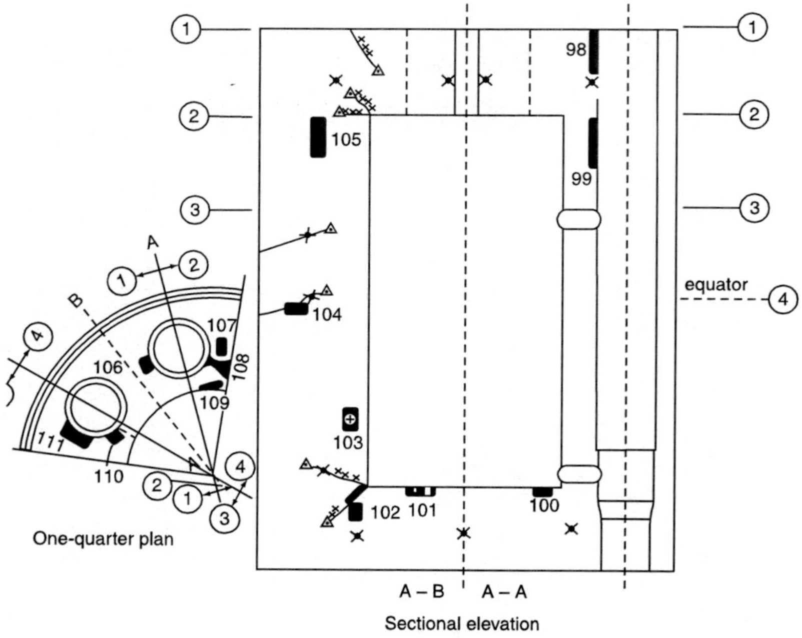

Fig. 5.47 High-temperature gas-cooled reactor pressure vessel (P_G = 16.48 MN/m^2) (see Fig. IX.118a for key)

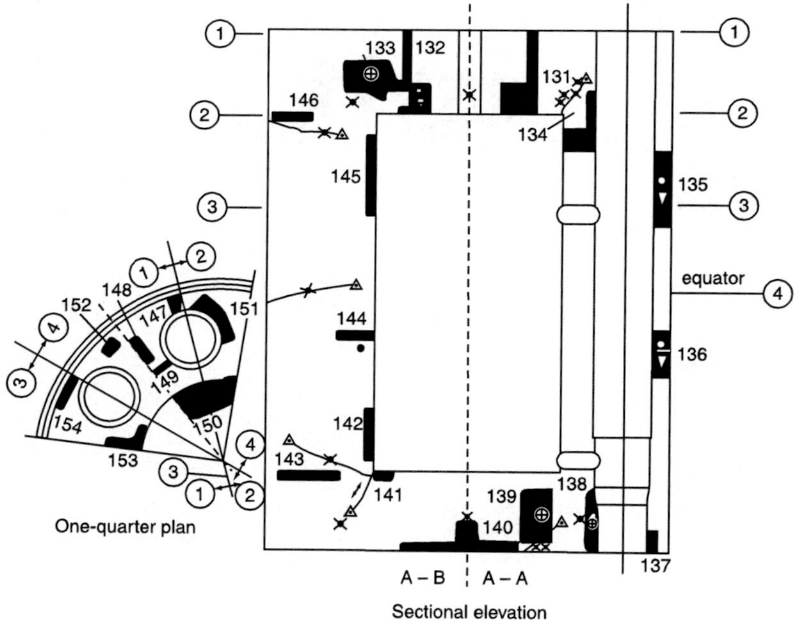

Fig. 5.48 High-temperature gas-cooled reactor pressure vessel (P_G = 20.08 MN/m^2) (see Fig. IX.118a for key)

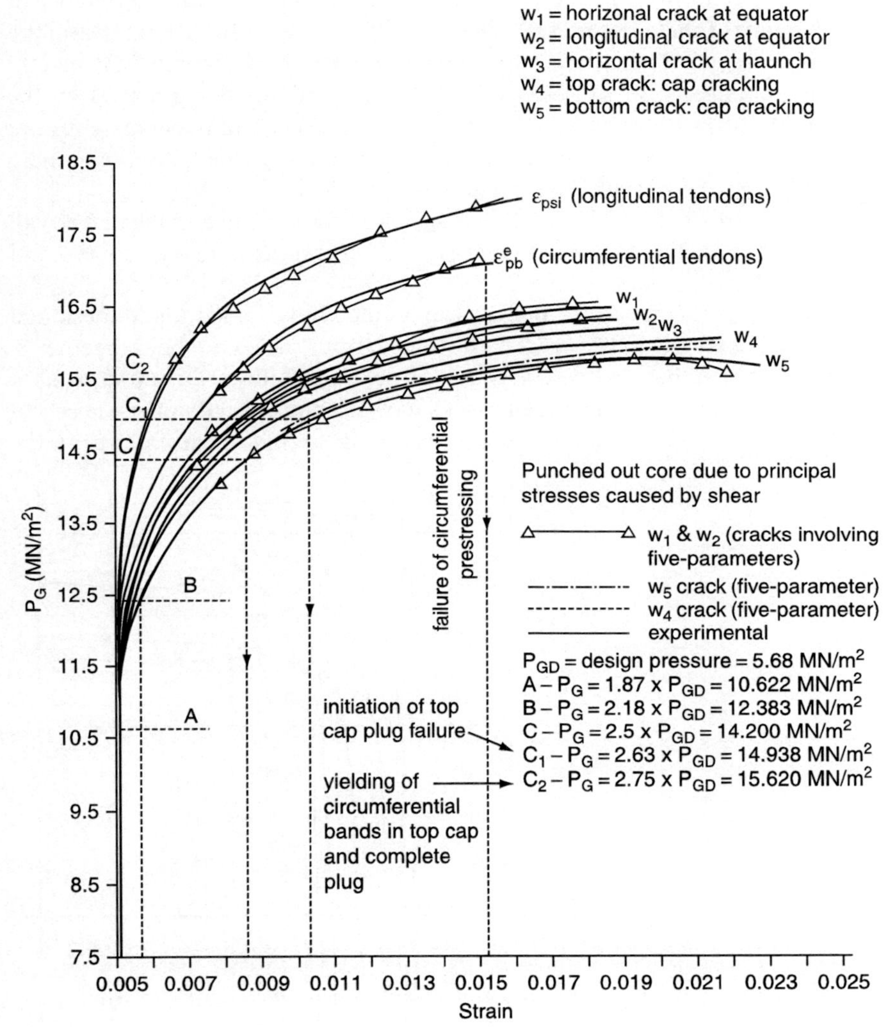

crack width top cap & wall (mm)	5	10	15	20	25	30	35	40	45	50	55	60	65	70	75	80	85	90	95	100
crack width bottom cap (mm)		8		16		24		32		40		48		56		64		70		78
top cap central deflection (mm) x 10		0.9		1.2		1.5		1.8		2.1		2.4		2.7		3.0		3.3		3.6
bottom cap central deflection (mm) x 10		0.6		0.9		1.2		1.5		1.8		2.1		2.4		2.7		3.0		3.3
wall horizontal deflection (mm) x 10		0.9		1.2		1.5		1.8		2.1		2.4		2.7		3.0		3.3		3.6

Fig. 5.49 High-temperature gas-cooled reactor pressure vessel and deflections

created. The bottom haunch crack (mark 117) position is another difference in the behaviour of these two vessels. The plastic zones also differ from those in the Hartlepool vessel. Their shapes and positions and crack sizes do not match. This again is due to the difference in the dominant role being played by the span/depth ratios. Longitudinal prestressing has failed and the crack sizes are approximately over 90 mm. Bonded reinforcement has yielded. Although there is sufficient yielding of the liner, it still has not failed.

At a pressure of 20.08 MN/m^2 ($3.55P_{GD}$) both the top cap and the barrel wall have failed in flexure one after another. All cracks around the boilers and circulators have widened.

Circumferential cracks around the top haunch have joined the longitudinal cracks. The bottom cap has cracked but the various pieces are held together.

It is concluded that the vessel has failed due to the failure of a strand band at a pressure of $3.21P_{GD}$ and enough cracks have been produced in the top cap to define a premature fiexural failure intervening in the shear failure in the

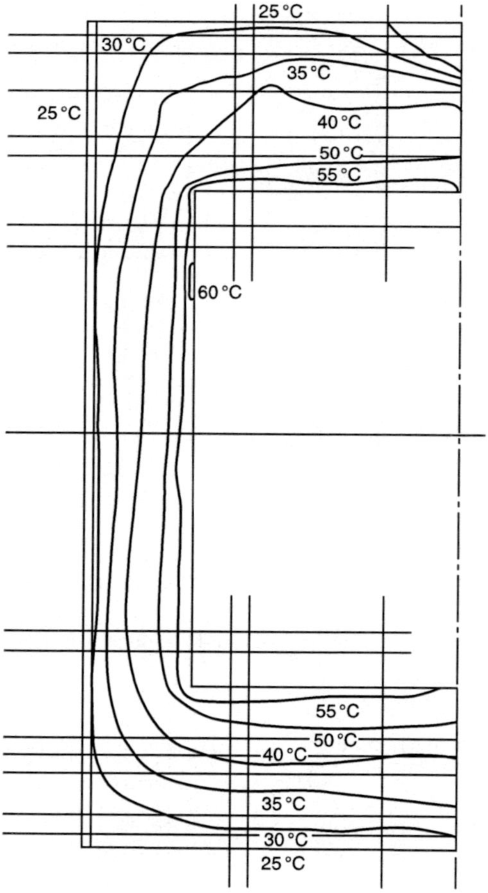

Fig. 5.50 Dungeness B type vessel – temperature distribution

top cap. Fllexural failure has also occurred at several places in the barrel wall. Figure 5.43a,b gives additional cases for failure modes but these are not governing cases. Figure 5.44 gives the deflection and cracking history of this vessel.

5.11 Thermal Analysis of Vessels

Thermal analysis has been carried out for all vessels noted in this chapter. Figures 5.50, 5.51, 5.52 and 5.53 give the result of temperature distributions for the vessels. Where thermal loads are to be included in the overall analysis they are computed from these figures and are put at the respective nodes of the elements as concentrated loads or patch loads. Alternatively strains are evaluated from these temperatures and they are included in the finite element formulations as initial strains. Stress trajectories can be drawn for the temperature-only case on the lines given for vessels in normal operation – cases which in fact include temperature effects.

5.12 Concrete Failure Theories

Several concrete theories exist which can easily be simulated into the overall various failure stages of the vessels. A summary of the important ones are given in Table 5.3 for detailed derivations and applications, a reference is made to the authors' following books:

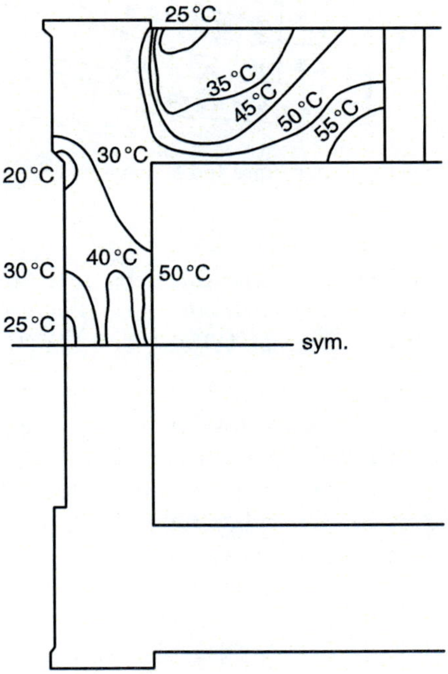

Fig. 5.51 Oldbury vessel – temperature distribution

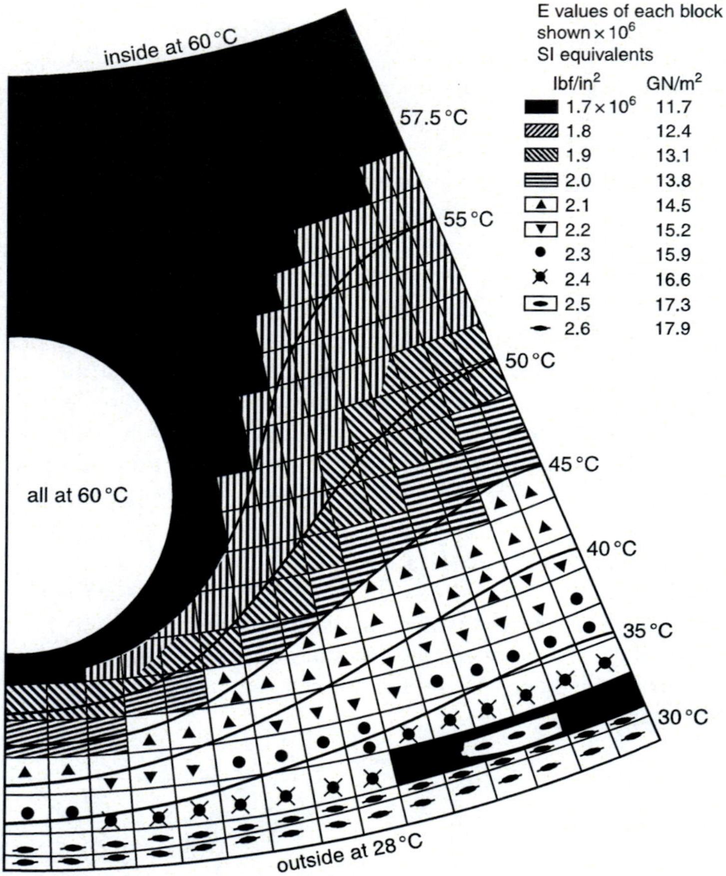

Fig. 5.52 Temperature and Young's moduli at equator – HTGCR vessel

Manual of Numerical Methods in Concrete
"*Modeling and Applications Validity by Experimental and Site Monitoring Data*", Thomas Telford, London (1999).

Tables 5.4, 5.5 and 5.6 give derivations of equations for, respectively, four-parameter (OTTOSON) model, five-parameter model, Bulk shear moduli model and Endochronic cracking model, which have been used in the existing vessels to present the concrete failure. Reference is made to above reference by the author on other failure criteria including bulk and shear moduli and Endochronic cracking models.

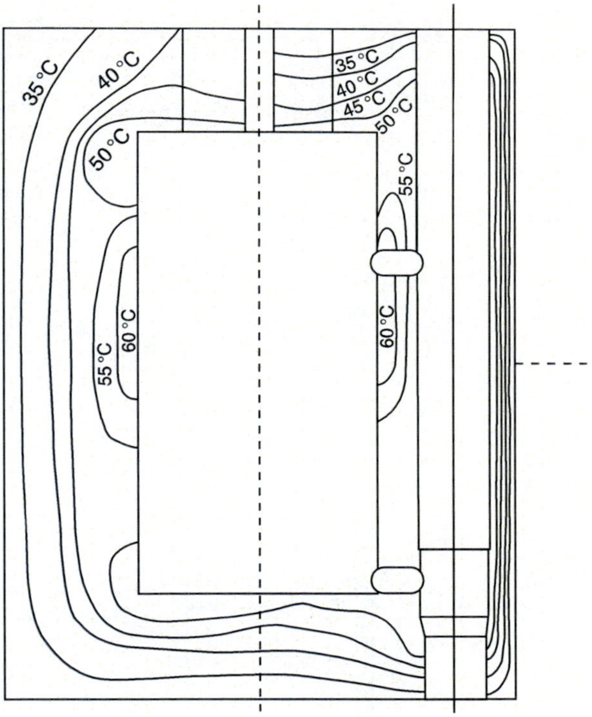

Fig. 5.53 HTGCR vessel – temperature distribution

5.13 The Contribution Made by the Steel Liner

Table 5.5 gives a brief on the analytical formulation of the steel liner which can easily be associated with the overall numerical modelling of the PCPV including the ultimate load and collapse conditions.

5.14 Prestressing Systems

Appendix C gives data on various prestressing systems.

5.15 Limit State Formulation

A reference is made to the future models (Fig. 5.11 for Oldbury and Fig. 5.12 for Dungeness); they definitely form the background for the development of the Limit State Analysis.

Several failure criteria [10–15] have been established from the experimental tests. Flexural failure is reported in the wall of the vessel. In the top cap both

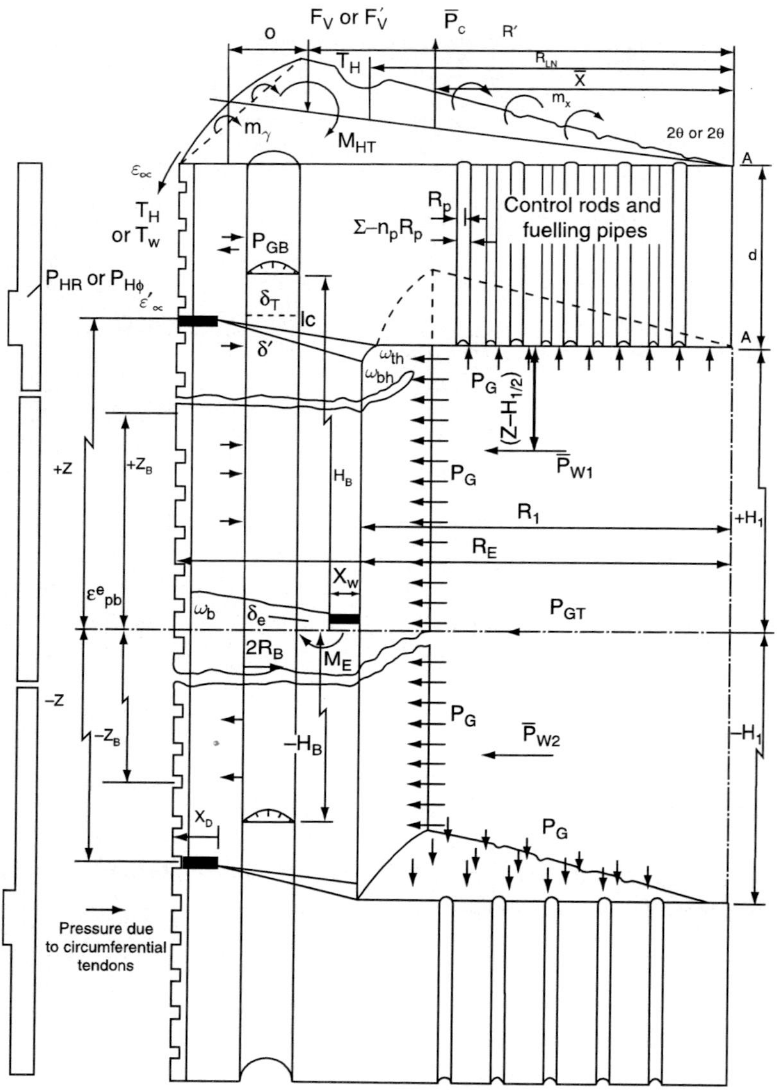

Fig. 5.54 Vessel and its parameters

flexural and plug failures have been reported. The central plug normally consists of a perforated zone due to the location control rods or standpipes. When the resistance of the compressive area in this region is generally reduced, the outside perimeter of the standpipes develops a circumferential plane of weakness. Under gas-increasing pressure a complete punched-out failure of the central core of the cap has occurred. This is due to the force caused by internal gas pressure

Table 5.3 Summary of concrete failure criteria

1. Krupfer and Rusch

$f(\sigma_{ij}) - A = 0 = \tau_{oct} + Bp - A$

where

$\tau_{oct} = \left(13\sigma_{ij}\sigma_{ij} - p^2\right)^{\frac{1}{3}} \quad i,j = 1,2,3$

$p = 13\sigma_{KK} \quad K = 1,2,3$

$$\frac{\partial f}{\partial \sigma_{ij}} = \sigma_{ij}(B\tau_{oct} - P)\delta_{ij}$$

2. Ottosen **Four-parameter model**

$f[(J_1, I_2)\cos 3\theta] = \frac{AI_2}{\sigma_y^2} + \lambda\frac{\sqrt{I_2}}{\sigma_y} + B\frac{J_1}{\sigma_y} - 1 = 0$

3. Argyris et al.

$f\left(A\frac{J_1}{\sigma_c} + (B - C\cos 3\theta_c)\frac{\sqrt{I_2}}{\sigma_c} - 1\right) = 0$

4. William et al.

(a) $(I_1, I_2, I_3) = 0 \qquad I_1 = \sigma_1 + \sigma_2 + \sigma_3;\ I_2 = \sigma_1\sigma_2 + \sigma_2\sigma_3 + \sigma_3\sigma_1$

$I_3 = \sigma_1\sigma_2\sigma_3$

(b) $\left(I\frac{1}{2_p}, I\frac{1}{2_c}\right) = 0 \qquad I_{2,c} = A_{01} + A_{11}J_1 + A_{22}J_1^2/\sigma_c$ tensile $\theta_c = 0$

$I_{2,c} = B_{01} + B_{11}J_1 + B_{22}J_1^2/\sigma_i$ comp $\theta_i = 60°$

5. Mohr's equations

$$f(z, r, \delta) = 0 \qquad z = \frac{1}{\sqrt{3}}(\sigma_1 + \sigma_2 + \sigma_3)$$

$$r = \frac{1}{\sqrt{3}}\sqrt{(\sigma_1 - \sigma_2)^2 + (\sigma_2 - \sigma_3)^2 + (\sigma_1 - \sigma_3)^2}$$

$$\cos\delta = \sigma_1 + \sigma_2 - 2\sigma_3/\sqrt{2}\sqrt{(\sigma_1 - \sigma_2)^2 + (\sigma_2 - \sigma_3)^2 + (\sigma_1 + \sigma_3)^2}$$

$$\tau_{ect} = \frac{1}{\sqrt{3}}\gamma; \quad \sigma_0 = \frac{1}{\sqrt{3}}.z$$

6. Launay et al.

$f(\sigma^{**} - 1) = 0$

where

$$\sigma^{**} = P^2\left[\frac{\cos^2\frac{3}{2}\theta_c}{P_c/J_1^2/3} + \frac{\sin^2\frac{3}{2}\theta_c}{P_r J_1^2/3}\right]; P = \sqrt{2}P_cP_t/\sqrt{(P_t^2 + P_c^2) + (P_t^2 - P_c^2)}\cos 3\theta_c$$

$$\cos 3\theta_c = \frac{3\sqrt{3}}{2}\left[I_3/(I_2)^{\frac{3}{2}}\right]; I_3 = \frac{(\sigma_1^3 + \sigma_2^3 + \sigma_3^3)}{3}$$

7. Bangash Y

Hypo-elastic theory

$f(\{\sigma\}, \{\varepsilon\}, \{\dot{\sigma}\}, \{\dot{\varepsilon}\}) = 0$

$d\sigma_{ij} = D_r(\sigma_{ij})d\varepsilon_{Kl} \qquad \delta =$ Kronecker delta

$D_r =$ Material property tensor

$$= (A_0 + A_{11}\sigma_{rr})\delta_{ij}\delta Kt + 12(A_{02} - 2A_{11}\sigma_{rr})(\delta_{iK}\delta_{ji} - \delta_{jK} - \delta_{il})$$
$$+ A_{13}(\sigma_{ij}\delta_{Kl} + \sigma_{Kl}\delta_{ij}) + 12A_{14}(\sigma_{jK}\delta_{il} + \sigma_{jl}\delta_{Kl} + \sigma_{iK}\delta_{ij} + \sigma_{il}\delta_{Kl})$$

Table 5.3 (continued)

$$d\sigma_{ij} = A_{01}\dot{\varepsilon}_{kk}\delta_{ij} + A_{02}\dot{\varepsilon}_{ij} + A_{11}\dot{\varepsilon}_{kk}\delta_{ij}$$
$$- 2A_{11}\sigma_{kk}\dot{\varepsilon}_{ij} + A_{13}(\sigma_{ij}\dot{\varepsilon}_{kk} + \sigma_{k1}\dot{\varepsilon}_{k1}\delta_{ij})$$
$$+ A_{14}(\sigma_{jk}\dot{\varepsilon}_{ik} + \sigma_{ik}\dot{\varepsilon}_{jk})$$

where δ's are Kroneckler deltas and A's are various hydrostatic constants to be determined by experiments

$K_{11} = A_{01} + A_{02} - A_{11}J_1 + 2(A_{13} + A_{14})\sigma_1$
K_{22}, K_{33} have all the terms of K_{11} except σ_1 is changed to σ_2 for K_{22} and σ_3 for K_{33}
$K_{44} = (A_{02} - 2A_{11}J_1) + A_{14}(J_1 - \sigma_1)$
$K_{55} = (A_{02} - 2A_{11}J_1) + A_{14}(J_1 - \sigma_2)$
$K_{66} = (A_{02} - 2A_{11}J_1) + A_{14}(J_1 - \sigma_3)$
$K_{12} = (A_{01} + A_{11}J_1) + A_{13}(\sigma_1 + \sigma_2) = R + A_{13}(\sigma_1 + \sigma_2)$
$K_{13} = R + A_{13}(\sigma_1 + \sigma_3)$
$K_{23} = R + A_{13}(\sigma_2 + \sigma_3)$
$K_{21} = R + A_{13}(\sigma_2 + \sigma_1)$ where $R = \dot{A}_{01} + A_{11}J_1$
$K_{31} = R + A_{13}(\sigma_3 + \sigma_1)$
$K_{32} = R + A_{13}(\sigma_3 + \sigma_2)$

The determinant in ( ) after expansion leads to

$$\Big[(A_{02} - 2A_{11}j_1)\Big\{-3A_{13}^2\left(2J_2 - \frac{J^3}{3}\right) + [3A_{01} + A_{02} + (A_{11} + 2A_{13})J_1]\Big\}$$
$$+ 2A_{14}\Big\{(A_{02} - 2A_{11}J_1)^2 J_1 + 2(A_{02} - 2A_{11}J_1)J_1(A_{01} + A_{11}J_1)$$
$$+ (2J_{1/3}^2 - (J_2 - J_{1/3}^3))[A_{14}(A_{01} + A_{11}J_1) + (A_{02} + 2A_{11}J_1)(A_{13} + 2A_{14})]$$
$$+ \left(\left(J_{3/3} - J_1\left(J_2 - \frac{J_1^3}{3}\right)\right)\Big/2\right)[2A_{14} + (6A_{13} + 2A_{14}) + 6A_{13}^2]$$
$$-A_{13}^2[J_{1/3}^3 + J_1(J_2 - J_{1/3}^3) - J_3]\Big\}\Big] \times [(A_{02} - 2A_{11}J_1)A_{14}(J_1 - \sigma_1)]$$
$$\times [(A_{02} - 2A_{14}J_1) + A_{14}(J_1 - \sigma_2)][(A_{02} - 2A_{11}J_1) + A_{14}(J_1 - \sigma_3)] = 0$$

for $\sigma_3 > \sigma_1$ or σ_2

$$\frac{d\sigma_3}{d\varepsilon_3} = \frac{\{(A_{01}+A_{02})+(-A_{11}+2A_{13}+2A_{14})\sigma_3-2A_{11}\sigma_1[A_{01}+\frac{1}{2}A_{02}+(2A_{13}+A_{14})\sigma_1]-[A_{01}+(A_{11}+A_{13})\sigma_3+(2A_{11}+A_{13})\sigma_1]^2\}}{A_{01}+\frac{1}{2}A_{02}+(2A_{13}+A_{14})}$$

The inclination of the strain path for $d\varepsilon$ is given as

$$dx' = \tan^{-1}\frac{1}{\sqrt{3}}\frac{(d\varepsilon p_1 - d\varepsilon p_2) + d\varepsilon p_1 - d\varepsilon p_3)}{(d\varepsilon p_2 - d\varepsilon p_3)}$$

7. Sulkand shear moduli noddle
8. Endochronic cracking model
9. Fifth para model

Table 5.4 Four-parameter model

An analytical failure model in which four parameters are included has been developed by Ottoson. This failure surface is described already. The same chain rule of partial differentiation is applied. The gradient $\{\partial f/\partial \sigma\}$ is given by Bangash:

$$c = \frac{\partial f}{\partial \sigma} = \frac{\partial f}{\partial I_1}\frac{\partial I_1}{\partial \sigma} + \frac{\partial f}{\partial J_2}\frac{\partial J_2}{\partial \sigma} + \frac{\partial f}{\partial \cos 3\theta}\frac{\partial \cos 3\theta}{\partial \sigma}$$

$$a_1c_1 + a_2c_2 + a_3c_3 \qquad (5.1)$$

$$a_1 = \frac{\partial f}{\partial I_2} = \frac{B}{f_c}$$

$$a_2 = \frac{\partial f}{\partial J_2} = \frac{\partial}{\partial J_2}\frac{AJ_2}{f_c^2} + \frac{\partial}{\partial J_2}\frac{\lambda J_2^{1/2}}{f_c^2} = \frac{A}{f_c^2} + \frac{\partial}{\partial J_2}\frac{\lambda J_2^{1/2}}{f_c^2}$$

$$\frac{\partial}{\partial J_2}\frac{\lambda J_2^{1/2}}{f_c^2} = \frac{\lambda}{f_c}\frac{1}{2}J^{-1/2} + \frac{J^{1/2}}{f_c}\frac{\partial(\lambda)}{\partial J_2} \qquad (5.2)$$

$$\frac{\partial(\lambda)}{\partial J_2} = \frac{\partial}{\partial J_2}\left\{K_1 \cos\left[\frac{1}{3}\cos^{-1}\left(K\frac{3\sqrt{3}}{2}\frac{\sqrt{3}}{3/J_2^2}\right)\right]\right\} \qquad (5.3)$$

where

$$\cos 3\theta \le 0 = -K_1 \sin(P)\frac{1}{3}\frac{-1}{\sqrt{1-t^2}}K_2\left(\frac{3}{2}\right)\sqrt{3}J_3\left(-\frac{3}{2}\right)J_2^{-6/2}$$

$$-K_1 \sin(P)\frac{1}{3J}\left(K_2\frac{3\sqrt{3}}{2}\frac{\sqrt{3}}{3/J_2^2}\right)\frac{1}{\sqrt{1-t^2}}\left(+\frac{3}{2}\right)$$

$$-K_1\frac{t}{\sqrt{1-t^2}}\sin(P)\frac{1}{2J} = -\frac{K_1}{2J}\frac{t}{\sqrt{1-t^2}}\sin(P)$$

$$t = K_2\frac{3\sqrt{3}}{2}\frac{J_3}{3/J_2^2}$$

$$\frac{\partial(\lambda)}{\partial J_2} = \frac{\partial}{\partial J_2}\left\{K_1 \cos\left[\frac{\pi}{3} - \frac{1}{3}\cos^{-1}\left(K_2\frac{3\sqrt{3}}{2}\frac{J_3}{3/J_2^2}\right)\right]\right\} \quad \cos\theta \le 0$$

$$= K_1(-\sin\ P)\left[-\frac{1}{3}\frac{(-1)}{\sqrt{1-t^2}}\left(-K_2\frac{3\sqrt{3}}{2}J_3\right)\left(-\frac{3}{2}\right)J_2^{-6/2}\right]$$

$$= \frac{-K_1}{T}(-\sin\ P)\left[\frac{1}{3}\frac{t}{\sqrt{1-t^2}}\left(-\frac{3}{2}\right)\right]\frac{-K}{2J}(\sin\ P)\frac{t}{\sqrt{1-t^2}} \qquad (5.4)$$

From this equation

$$a_2 = \frac{A}{f_c^2} + \frac{1}{J_2^{1/2}f_c}\left\{\lambda - K_1 \sin\left[\frac{1}{3}\cos^{-1}\left(K_2\frac{3\sqrt{3}}{2}\frac{J_3}{J_2^{3/2}}\right)\right]\frac{t}{\sqrt{1-t^2}}\right\}$$

$$\frac{A}{f_c^2} - \frac{1}{2J_2^{1/2}f_c}\left\{\lambda - K_1 \sin\left[\frac{\pi}{3}\frac{1}{3}\cos^{-1}\left(K_2\frac{3\sqrt{3}}{2}\frac{J_3}{J_2^{3/2}}\right)\right]\frac{t}{\sqrt{1-t^2}}\right\} \qquad (5.5)$$

$$a_3 = \frac{\partial}{\partial \cos 3\phi}\left(\lambda\frac{J_2^{1/2}}{f_c}\right) \quad \cos 3\theta \ge 0 \qquad (5.6)$$

$$\frac{J_2^{1/2}}{f_c}\frac{1}{3}K_1 \sin\left[\frac{1}{3}\cos^{-1}(K_2\cos 3\theta)\right]\frac{K_2}{\sqrt{1-t^2}}$$

$$= \frac{1}{3}\frac{K_1K_2J_2^{1/2}}{3f_c\sqrt{1-t^2}}\sin\left[\frac{1}{3}\cos^{-1}(K_2\cos 3\theta)\right] \quad \cos\theta \ge 0 \qquad (5.7)$$

$$a_3 = \frac{K_1K_2J_2^{1/2}}{3\sqrt{1-t^2}f_c}\sin\left[\frac{\pi}{3} - \frac{1}{3}\cos^{-1}(-K_2\cos 3\theta)\right] cos\, 3\theta \ge 0$$

where $t = K_2 \cos 3\theta$

$$J_2 = \left[\frac{1}{2}(S_x^2 + S_y^2 + S_z^2) + \tau_{xy}^2 + \tau_{yz}^2 + \tau_{zx}^2\right]$$

The values of c_1, c_2, c_3 are as follows:

Table 5.4 (continued)

$$C_1 = \frac{\partial I_1}{\partial\{\sigma\}} = \left\{ \begin{array}{c} 1 \\ 1 \\ 1 \\ 0 \\ 0 \\ 0 \end{array} \right\}$$

$$C_2 = \frac{\partial J_2}{\partial\{\sigma\}} = \frac{\partial J_2}{\partial S_x}\frac{\partial S_x}{\partial\{\sigma\}} + \frac{\partial J_2}{\partial S_y}\frac{\partial S_y}{\partial\{\sigma\}} + \frac{\partial J_2}{\partial S_z}\frac{\partial S_z}{\partial\{\sigma\}} + \frac{\partial J_2}{\partial}\frac{\partial \tau_{xy}}{\partial\{\sigma\}} + \frac{\partial J_2}{\partial \tau_{yz}}\frac{\partial \tau_{yz}}{\partial\{\sigma\}} + \frac{\partial J_2}{\partial \tau_{zx}}\frac{\partial \tau_{zx}}{\partial\{\sigma\}}$$

$$= S_x \left\{ \begin{array}{c} \frac{2}{3} \\ -\frac{1}{3} \\ \frac{1}{3} \\ 0 \\ 0 \\ 0 \end{array} \right\} + S_y \left\{ \begin{array}{c} -\frac{1}{3} \\ \frac{2}{3} \\ -\frac{1}{3} \\ 0 \\ 0 \\ 0 \end{array} \right\} + S_z \left\{ \begin{array}{c} -\frac{1}{3} \\ -\frac{1}{3} \\ \frac{2}{3} \\ 0 \\ 0 \\ 0 \end{array} \right\} + 2 \left\{ \begin{array}{c} 0 \\ 0 \\ 0 \\ \tau_{xy} \\ \tau_{yz} \\ \tau_{zx} \end{array} \right\} \qquad (5.8)$$

or

$$C_2 = \left\{ \begin{array}{c} \frac{1}{3}(2S_x - S_y - S_z) \\ \frac{1}{3}(2S_y - S_x - S_z) \\ \frac{1}{3}(2S_z - S_x - S_y) \\ 2\tau_{xy} \\ 2\tau_{yz} \\ 2\tau_{zx} \end{array} \right\} = \left\{ \begin{array}{c} S_x \\ S_y \\ S_z \\ 2\tau_{xy} \\ 2\tau_{yz} \\ 2\tau_{zx} \end{array} \right\} \qquad (5.9)$$

$$\cos 3\theta = \frac{3\sqrt{3}}{2} \times \frac{J_3}{J_2^{\frac{3}{2}}} \qquad (5.10)$$

$$C_3 = \frac{\cos 3\theta}{\partial J_3} = \frac{\cos 3\theta}{\partial J_3}\frac{\partial J_3}{\partial\{\sigma\}} + \frac{\cos 3\theta}{\partial J_2}\frac{\partial J_2}{\partial\{\sigma\}} \qquad (5.11)$$

From Eq. (5.11)

$$\frac{\cos 3\theta}{\partial J_3} = \frac{3\sqrt{3}}{2J_2^{3/2}} \; \frac{\cos 3\theta}{\partial J_2} = \left(\frac{3\sqrt{3}}{2} J_3\right)\left(\frac{-3/2}{J_2^{5/2}}\right) = -\frac{9}{4}\sqrt{3\frac{J_3}{J_2^{5/2}}} \qquad (5.12)$$

Now

$$J_3 = [S_x S_y S_z + 2\tau_{xy}\tau_{yz}\tau_{zx} - S_y\tau_{xy}^2 - S_y\tau_{yz}^2 - S_z\tau_{zx}^2]$$

$$\frac{\partial J_3}{\partial\{\sigma\}} = \frac{\partial J_3}{\partial S_x}\frac{\partial S_x}{\partial\{\sigma\}} + \frac{\partial J_3}{\partial S_y}\frac{\partial S_y}{\partial\{\sigma\}} + \frac{\partial J_3}{\partial S_z}\frac{\partial S_z}{\partial\{\sigma\}} + \frac{\partial J_3}{\partial \tau_{xy}}\frac{\partial \tau_{xy}}{\partial\{\sigma\}} + \frac{\partial J_3}{\partial \tau_{yz}}\frac{\partial \tau_{yz}}{\partial\{\sigma\}} + \frac{\partial J_3}{\partial \tau_{zx}}\frac{\partial \tau_{zx}}{\partial\{\sigma\}}$$

$$\frac{\partial J_3}{\partial S_x} = S_y S_z - \tau_{yz}^2, \quad \frac{\partial J_3}{\partial S_y} = S_x S_z - \tau_{xz}^2, \quad \frac{\partial J_3}{\partial S_z} = S_x Sy - \tau_{zx}^2 \qquad (5.13)$$

$$S_x = \frac{1}{3}(2\sigma_x - \sigma_y - \sigma_z) \qquad (5.14)$$

$$\frac{\partial S_x}{\partial \sigma} = \frac{1}{3}\left\{ \begin{array}{c} 2 \\ -1 \\ -1 \\ 0 \\ 0 \\ 0 \end{array} \right\}, \quad \frac{\partial S_y}{\partial \sigma} = \frac{1}{3}\left\{ \begin{array}{c} -1 \\ 2 \\ -1 \\ 0 \\ 0 \\ 0 \end{array} \right\}, \quad \frac{\partial S_z}{\partial \sigma} = \frac{1}{3}\left\{ \begin{array}{c} -1 \\ -1 \\ 2 \\ 0 \\ 0 \\ 0 \end{array} \right\} \qquad (5.15)$$

$$\frac{\partial J_3}{\partial \tau_{xy}} = 2\tau_{xy}\tau_{zx} - 2S_z\tau_{xy}, \quad \frac{\partial J_3}{\partial \tau_{yz}} = 2\tau_{xy}\tau_{zx} - 2S_x\tau_{yz}$$

$$\frac{\partial J_3}{\partial \tau_{xz}} = 2\tau_{xy}\tau_{yz} - 2S_y\tau_{xz} \qquad (5.16)$$

Table 5.4 (continued)

$$\frac{\partial\tau_{xy}}{\partial\{\sigma\}}=\begin{Bmatrix}0\\0\\0\\1\\0\\0\end{Bmatrix},\quad \frac{\partial\tau_{yz}}{\partial\{\sigma\}}=\begin{Bmatrix}0\\0\\0\\0\\1\\0\end{Bmatrix},\quad \frac{\partial\tau_{zx}}{\partial\{\sigma\}}=\begin{Bmatrix}0\\0\\0\\0\\0\\1\end{Bmatrix} \qquad (5.17)$$

$$\frac{\partial J_3}{\partial\{\sigma\}}=\begin{Bmatrix}\frac{1}{3}\left[2(S_yS_z-\tau_{yz}^2)-(S_xS_z-\tau_{xz}^2)-(S_xS_y-\tau_{xy}^2)\right]\\ \frac{1}{3}\left[-(S_yS_z-\tau_{yz}^2)+2(S_xS_z-\tau_{xz}^2)-(S_xS_y-\tau_{xy}^2)\right]\\ \frac{1}{3}\left[-(S_yS_z-\tau_{yz}^2)-2(S_xS_z-\tau_{xz}^2)+(S_xS_y-\tau_{xy}^2)\right]\\ 2(\tau_{yz}\tau_{zx}-S_z\tau_{xy})\\ 2(\tau_{xy}\tau_{zx}-S_x\tau_{yz})\\ 2(\tau_{xy}\tau_{yz}-S_y\tau_{xz})\end{Bmatrix} \qquad (5.18)$$

Equation (5.18) is further simplified as

$$\frac{\partial J_3}{\partial\{\sigma\}}=\begin{Bmatrix}\frac{1}{3}\left[2S_yS_z-S_xS_z-S_xS_y-2\tau_{yz}^2+\tau_{xz}^2+\tau_{xy}^2\right]\\ \frac{1}{3}\left[2S_xS_z-S_yS_z-S_xS_y-2\tau_{xz}^2+\tau_{yz}^2+\tau_{xy}^2\right]\\ \frac{1}{3}\left[2S_xS_y-S_yS_z-S_xS_z-2\tau_{xy}^2+\tau_{yz}^2+\tau_{zx}^2\right]\\ 2(\tau_{yz}\tau_{zx}-S_z\tau_{xy})\\ 2(\tau_{xy}\tau_{zx}-S_x\tau_{yz})\\ 2(\tau_{xy}\tau_{yz}-S_y\tau_{xz})\end{Bmatrix} \qquad (5.19)$$

From the flow rule of normality principle, the following relationship exists between the plastic strain increment and the plastic stress increment:

$$d\{\varepsilon\}=\lambda\frac{\partial f}{\partial\{\sigma\}} \qquad (5.20)$$

This equation can be interpreted as requiring the normality of the plastic strain increment vector to yield the surface in the hyper-space of n stress dimensions. As before $d\lambda$ is the proportionality constant.

For stress increments of infinitesimal size, the change of strain can be divided into elastic and plastic parts, thus (as before)

$$d\{\varepsilon\}=d\{\varepsilon\}_\epsilon+d\{\varepsilon\}_\text{p} \qquad (5.21)$$

The elastic increment of stress and strain is related to an isotropic material property matrix $[D]$ by

$$d\{\varepsilon\}_\epsilon=[D]^{-1}d\{\sigma\} \qquad (5.22)$$

From Equation (5.20) – (5.22) the following equation is established:

$$d\{\varepsilon\}=[D]^{-1}d\{\sigma\}+d\lambda\left\{\frac{\partial f}{\partial\{\sigma\}}\right\} \qquad (5.23)$$

The function stresses, on differentiation, can be written as

$$df=\frac{\partial f}{\partial\sigma_1}d\sigma_1+\frac{\partial f}{\partial\sigma_2}d\sigma_2+\ldots+0=\left\{\frac{\partial f}{\partial\{\sigma\}}\right\}^{T'}d\{\sigma\} \qquad (5.24)$$

Equation (5.24) together with (5.25) can be written in matrix form as

$$d\{\varepsilon\}=[D]_{\epsilon\text{p}}^{-1}d\{\sigma\} \qquad (5.25)$$

or

Table 5.4 (continued)

$$\begin{Bmatrix} d\varepsilon_x \\ d\varepsilon_y \\ d\varepsilon_z \\ d\gamma_{xy} \\ d\gamma_{yz} \\ d\gamma_{zx} \\ 0 \end{Bmatrix} = \begin{bmatrix} [2pt] & & & & & & \frac{\partial f}{\partial \sigma_x} \\ & & & & & & \frac{\partial f}{\partial \sigma_y} \\ & & & [D]^{-1} & & & \frac{\partial f}{\partial \sigma_z} \\ & & & & & & \frac{\partial f}{\partial \tau_{xy}} \\ & & & & & & \frac{\partial f}{\partial \tau_{yz}} \\ & & & & & & \frac{\partial f}{\partial \tau_{zx}} \\ \frac{\partial f}{\partial \sigma_x} & \frac{\partial f}{\partial \sigma_y} & \frac{\partial f}{\partial \sigma_z} & \frac{\partial f}{\partial \tau_{xy}} & \frac{\partial f}{\partial \tau_{yz}} & \frac{\partial f}{\partial \tau_{zx}} & 0 \end{bmatrix} \begin{Bmatrix} d\sigma_x \\ d\sigma_y \\ d\sigma_z \\ d\tau_{xy} \\ d\tau_{yz} \\ d\tau_{zx} \\ d\lambda \end{Bmatrix} \qquad (5.26)$$

Inversion of the above matrix $[D]^{[-1]}$ will give stresses

$$d\{\sigma\} = [D]_{\epsilon p} d\{\varepsilon\} \qquad (5.27)$$

The explicit form of the elasto-plastic material matrix $[D]_{sp}$ is given by

$$[D]_{\epsilon p} = [D] - [D]\left\{\frac{\partial f}{\partial \sigma_x}\right\}\left\{\frac{\partial f}{\partial \sigma_x}\right\}^T [D]\left\{\frac{\partial f}{\partial \sigma_x}\right\}^T [D]\left\{\frac{\partial f}{\partial \sigma_x}\right\}^{-1} \qquad (5.28)$$

The value of $(\partial f/\partial_{\sigma_x})$ has been evaluated above.
The rest of the procedure using finite element is described in Chapter 3.

Table 5.5 Five-parameter model

A full description of this model is given which predicts failure. The same failure surface in terms of tensile and compressive meridians is well defined. The average normal stress σ_a is given by

$$\rho(\sigma_a, \theta) = \frac{(c+t)}{v} \qquad (5.29)$$

$$c = 2(\rho_c^3 - \rho_c \rho_t^2)\cos\theta$$

$$t = (2\rho_c \rho_c - \rho_c^2)[\alpha_1'' + 5\rho_c^2 - 4\rho_c \rho_c]^{1/2}$$

$$v = \alpha_1'' + (\rho_c - 2\rho_c)^2$$

The chain rule of differentiation is used in order to evaluate the gradient $\partial f/\partial \sigma_{ij}$

$$\frac{\partial f}{\partial \sigma_{ij}} = \frac{\partial f}{\partial p}\frac{\partial p}{\partial \sigma_a}\frac{\partial \sigma_a}{\partial \sigma_{ij}} + \frac{\partial f}{\partial p}\frac{\partial p}{\partial \theta}\frac{\partial \theta}{\partial \sigma_{ij}} + \frac{\partial f}{\partial \tau_a}\frac{\partial \tau_a}{\partial \sigma_{ij}} \qquad (5.30)$$

Values for the terms in (5.30) are given below:

$$\frac{\partial f}{\partial p} = \frac{-\tau_a}{f_c}\frac{1}{\rho^2} \qquad (5.31a)$$

where

$$\frac{\partial \rho}{\partial \sigma_a} = \frac{1}{v}\left[\left(\frac{\partial c}{\partial \sigma_a}\right)\left(\frac{\partial t}{\partial \sigma_a}\right)\right] - (c+t)\frac{\partial v}{\partial \sigma_a} \qquad (5.31b)$$

$$\frac{\partial c}{\partial \sigma_a} = 2\cos\theta\left[(3\rho_c^2 - \rho_t^2)\frac{d\rho_c}{d\sigma_a} - 2\rho_c\rho_t\frac{dt}{d\sigma_a}\right]$$

$$\frac{\partial t}{\partial \rho_a} = \left[2p_c\frac{d\rho_t}{d\sigma_a} + 2(\rho_t - \rho_c)\frac{d\rho_c}{d\sigma_a}\right][\alpha_1'' + 5\rho_t^2 - 4\rho_c\rho_t]^{1/2} \qquad (5.31c)$$

$$+\frac{(2\rho_c\rho_t - \rho_c^2)\left[(+10\rho_t - 8\rho_t\cos^2\theta - 4\rho_c)\frac{d\rho_t}{d\sigma_a} + (8\rho_c\cos^2\theta - 4\rho_c)\frac{d\rho_c}{d\sigma_a}\right]}{2[\alpha_1'' + 5\rho_t^2 - 4\rho_c\rho_t]^{1/2}} \qquad (5.31d)$$

$$\frac{\partial v}{\partial \sigma_a} = (8\rho_t\sin^2\theta - 4\rho_c)\frac{d\rho_t}{d\sigma_a} + (8\rho_c\cos^2\theta + 2\rho_c - 4\rho_t)\frac{d\rho_c}{d\sigma_a} \qquad (5.31e)$$

Table 5.5 (continued)

From Eq. (5.30) the vectors ρ_c and ρ_t are evaluated.

In Eq. (3.31e) $d\rho_c/d_{\sigma_a}$ is taken as

$$\frac{d\rho_c}{d\sigma_a} = \sqrt{5b_1} + \frac{2\sqrt{5b_2\sigma_a}}{f'} \qquad (5.31f)$$

Similarly

$$\frac{d\rho_t}{d\sigma_a} = \sqrt{5a_1} + \frac{2\sqrt{5a_2\sigma_a}}{f'}$$

The value of the term $\partial_{\sigma_a}/\partial_{\sigma_{ij}}$ is already known and equal to $1/3S_{ij}$.

Once the gradient is known the elasto-plstic formulation described in Chapter 3 can be adopted.

Table 5.6 Non-linear bond linkage element

The Ahmlink element procedure described in ~i4—I can easily be included in the finite element analysis by adopting the following steps:

(a) Calculate the incremental slip from the nodal displacements:

$$\{\Delta S_i\} = [T''] \{\Delta U_i\} \qquad (5.32)$$

where *[T"j* is the transformation matrix and {~L,} are the element nodal displacements. The total slip at iteration *i* is calculated as

$$\{S_i\} = \{S_{i-1}\} + \{\Delta S_i\} \qquad (5.33)$$

(b) Calculate the incremental stress based on the bond stress at iteration *i* — I:

$$\{\Delta\sigma_{bi}\} = [E_b^t]\{\sigma_{bi-1}\}\{\Delta S_i\} \qquad (5.34)$$

Total stress is equal to

$$\{\sigma_b i\}_{TOT} = \{\sigma_{bi-1}\} + \{\Delta\sigma_{bi}\} \qquad (5.35)$$

(c) Check the state of the bond, i.e. whether the bond is broken or not, and calculate the stress accordingly.

If *IS~~* > S_{max}, set flag 'sat = I, i.e. bond is broken. At this point, the bond stress is instantaneously dropped to zero, i.e. $\{u_5\} = 0.0$, where $\{S_{max}\}$ is the maximum slip allowed.

If JS~~ < *S~*, calculate the bond stress which is compatible with the slip, S~. This is obtained by linear interpolation of a nonline r ~ond—slip curve. The curve is simulated by multi-linear lines. Figure V ~ gives the scheme for the linear interpolation. Let {*o*~,} be the bond stress compatible with the slip *S*. The difference between {ab.} and {*o*~} is treated as initial stress and this may be converted into nodal loads on the structure under consideration.

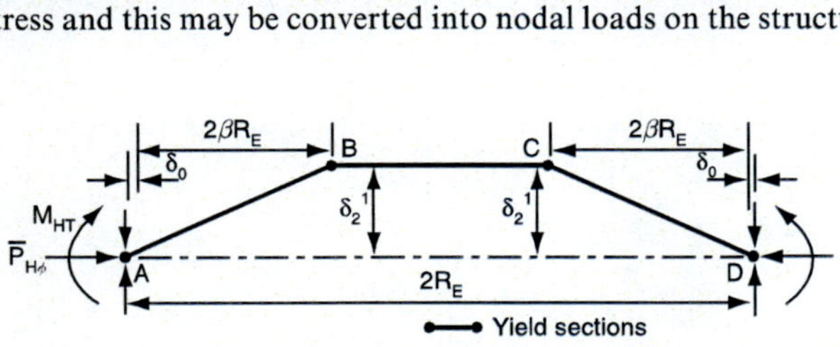

Table 5.6 (continued)

$\{\triangle\sigma_D\} = \{\sigma_{bi}\} - \{\sigma_b^r\}$ (5.36)

The correct stress is written as

$\{\sigma_{bi}\}_{cor} = \{\sigma_{bi}\} - \{\triangle\sigma_D\}$ (5.37)

(d) Total internal equivalent loads and residuals are calculated as

$\{F\}_{internal} = \pi dL[T]^{T''}\{\sigma_{bi}\} \quad \{R\} = \{F\}_{external} - \{F\}_{internal}$ (5.38)

The rest of the procedure is the same as for the other above cases.

Table 5.7 Analytical formulation of the steel liner

$[K]_{TOT}\{\delta\}^* + \{F_T\} - \{R_T\} = 0$

where

$[K]_{TOT} = [K_l] + [K_a]$

$\{\delta\}^* = \begin{Bmatrix} \delta_{un} \\ \delta_b \end{Bmatrix}; \ \{F_T\} = \begin{Bmatrix} F_{un} \\ F_b \end{Bmatrix}; \ \{R_T\} = \begin{Bmatrix} R_{un} \\ R_b \end{Bmatrix}$

$[K_{TOT}]$ = total stiffness matrix, $[K_l]$ = liner stiffness matrix, $\{K_j\}$ = a stud stiffness matrix, $\{F_T\}$ total initial load vector, $R\sim\}$ = total external load vector

= quantities corresponding to unknown displacement = quantities corresponding to restrained boundaries

[K],,] {~5~,} + ~ =0

{c} =[*B*]{*b4*

~*cr*} =[*D*] (~}—{e~})

= anchor shear forces

$F_{un} = \int_r [B]^T[D]\{\varepsilon_0\}dv = [B]^T[D]\{\varepsilon_o\}\det[J]d\xi d\eta d\zeta$

The plastic buckling matrix is given by

$(K + \lambda K_G)F_T = 0$

where

K = elasto-plastic stiffness matrix as a function of the current state of plastic deformation

K_G = initial stress geometric stiffness matrix.

The determinant $|K + \lambda K_G| = 0$

The essential equation is characteristically triangularised for the ith loading step as

$(K^i + \lambda_c K_G^i)F_T^i = 0$

$\lambda_c = 1 + E_{ps}E_{ps}$ – accuracy parameter

Table 5.8 Material matrices typical cracks in one, two, three directions

$$\sigma_1 - \text{Direction}\ [D]^* = \begin{bmatrix} 0 & 0 & 0 & 0 & 0 & 0 \\ 0 & \left(D_{22} - \frac{D_{12}^2}{D_{11}}\right) & \left(D_{23} - \frac{D_{18}D_{12}}{D_{11}}\right) & 0 & 0 & 0 \\ 0 & \left(D_{23} - \frac{D_{31}D_{21}}{D_{11}}\right) & \left(D_{33} - \frac{D_{13}D_{13}}{D_{11}}\right) & 0 & 0 & 0 \\ 0 & 0 & 0 & \beta' D_{44} & 0 & 0 \\ 0 & 0 & 0 & 0 & D_{55} & 0 \\ 0 & 0 & 0 & 0 & 0 & \beta' D_{66} \end{bmatrix}$$

$$\sigma_2 - \text{Direction}\ [D]^* = \begin{bmatrix} \left(D_{11} - \frac{D_{21}}{D_{22}}\right) & 0 & \left(D_{13} - \frac{D_{12}D_{23}}{D_{22}}\right) & 0 & 0 & 0 \\ 0 & 0 & 0 & 0 & 0 & 0 \\ \left(D_{31} - \frac{D_{21}D_{32}}{D_{22}}\right) & 0 & \left(D_{33} - \frac{D_{23}^2}{D_{22}}\right) & 0 & 0 & 0 \\ 0 & 0 & 0 & \beta' D_{44} & 0 & 0 \\ 0 & 0 & 0 & 0 & \beta' D_{55} & 0 \\ 0 & 0 & 0 & 0 & 0 & D_{66} \end{bmatrix}$$

$$\sigma_3 - \text{Direction}\ [D]^* = \begin{bmatrix} \left(D_{11} - \frac{D_{13}^2}{D_{33}}\right) & \left(D_{12} - \frac{D_{13}D_{23}}{D_{23}}\right) & 0 & 0 & 0 & 0 \\ \left(D_{21} - \frac{D_{31}D_{32}}{D_{33}}\right) & \left(D_{22} - \frac{D_{23}D_{23}}{D_{33}}\right) & 0 & 0 & 0 & 0 \\ 0 & 0 & 0 & 0 & 0 & 0 \\ 0 & 0 & 0 & D_{44} & 0 & 0 \\ 0 & 0 & 0 & 0 & \beta' D_{55} & 0 \end{bmatrix}$$

exceeding the vertical components of the shear resistance and normal stress on the inclined failure surface of the core. The vessel is assumed to deform. Several flexural and shear modes are considered. Cracks are assumed to develop at different positions in the wall and the cap. Strains, stresses and crack widths are computed under each mode. Several concrete strength criteria are included for comparison. Each failure mode is judged against the following criteria:

(a) exceeding the minimum rupturing strength of the tendons,
(b) exceeding concrete strength under complex loads,
(c) limiting crack sizes affecting the integrity of the vessel,
(d) rupturing strength of the liner crossing the crack. Any one of these cases, if reached, will define the failure of the vessel. The gas pressure that causes these is treated to have a minimum value.

5.15.1 Analysis of Flexural Failure

Generally the symmetry dictates and it is also evident from the experimental test [10–15] that one hinge should form at the equator and the position of any haunch hinge is determined by considering the equilibrium of the mechanism with various assumed hinge positions at the junction of the cap and the wall. Figure 5.28 shows a slice of the vessel between the two radial yield lines. The internal height of the main cavity is $2H_i$. The internal and external radii are R_I and R_E, respectively. For distinction the main cavity is subject to the internal gas pressure P_G and where boilers or circulators exist in the vessel walls the internal gas pressure is denoted by P_{GB}. As before the top cap standpipes/ control rods are shown as each having a diameter of $2R_p$. Where boilers/ circulators exist, the diameter of each boiler is denoted by $2R_B$ with an assumed pressurised height H_B above or below the equator. Let ε'_{ps}, ε'_{pc}, ε^e_{pc} be the circumferential strain at the hinge at any distance Z, the circumferential strain at the top of the cap and the circumferential strain at the equator, respectively. The total cap pressure $\bar{P}_c$ acts at a distance X from the central line of the vessel. The total vertical prestressing load F_v or F'_v defined earlier acts at a centroid at a distance 'a' from the outside face of the vessel or R' from the central line of the vessel. Under incremental pressure or during deformation, all the above parameters do vary, particularly the magnitude of the pressure and prestressing loads and their centroids for specific values of strains and deformations. As the vessel deforms and cracks are propagating, the depth of the compression block at the hinge such as x_p, x_b and x_w changes. It is essential to assume the shape of the compressive block before developing the equilibrium equations defining various incremental stages and ultimate conditions.

5.15.1.1 Concrete Stress Block

For the ultimate limit state conditions, the stress block given in Fig. 5.28 is adopted. The general flexural theory may be modified to include tendon strain ε_{pb} at ultimate conditions which is made up of the two parts, namely ε_{ps}, the effective tendon prestress after losses as dictated earlier, and the additional strains $\varepsilon_{pa} = \varepsilon_{ce} + \varepsilon_u$, as shown later on. The value of $\varepsilon_{ce} = (1/E_c)x$ the concrete prestress at the tendon level and the value of $\varepsilon_{\mathbf{u}}$ is the average concrete strain at the level of the tendon at ultimate conditions. This of course is true for a bonded tendon. For an unbonded tendon ε_{pa} is always less than the algebraic sum of ε_{ce} and ε_u. In general ε_{pa} is written as

$$\varepsilon_{pa} = m_1\varepsilon_e + m_2\varepsilon_u \tag{5.39}$$

For bonded tendon in the vessel $m_1 = m_2 = 1$. In the examples chosen where unbonded tendons are used, the values of m_1 and m_2 are assumed to be 0.5 and 0.25, respectively, as are normally assumed for heavy beams and slabs.

5.15.1.2 Barrel Wall

For the limit state flexural analysis of the barrel wall, Fig. 5.28 shows one-half of the barrel wall. The equations of equilibrium of forces on the wall have been computed and are given below. The axial force F'_v on segment 2θ is then related as

$$f_{\mathrm{v}}' = \frac{\theta}{\pi}\left[A_{\mathrm{sp}}E_{\varepsilon}\left[\varepsilon_{\mathrm{p}\varepsilon} + \frac{\pi p_{\mathrm{G}}(R_{\mathrm{I}}^2 + n_{\mathrm{o}}R_{\mathrm{B}}^2)}{E_{\mathrm{c}}A_{\mathrm{c}} + E_{\mathrm{s}}A_{\mathrm{sp}}}\right]\right] \tag{5.40}$$

$$+\frac{R_{\mathrm{E}}(\varepsilon_{\mathrm{po}}^{\varepsilon} - \varepsilon_{\mathrm{ps}}')R_{\mathrm{E}} - R_1 - \dfrac{2(F_{\mathrm{o}} + T_{\mathrm{F}})}{0.45\sigma_{\mathrm{cu}}(R_{\mathrm{E}} + R_1)\sin\theta}}{Z(H_1 + d)}$$

$$+\frac{R_{\mathrm{E}}(\varepsilon_{\mathrm{pc}}^{\varepsilon} - \varepsilon_{\mathrm{ps}}')\left(a - \dfrac{(F_{\mathrm{o}} + T_{\mathrm{F}})}{0.45\sigma_{\mathrm{cu}}(R_{\mathrm{E}} + R_1)\sin\theta}\right)}{(H_1 + d)(H_1 - d - z)}$$

$$+ D_{\mathrm{W}}\pi(R_E^2 - R_1^2) - \pi P_{\mathrm{G}}(R_{\mathrm{I}}^2 + n_{\mathrm{b}}R_{\mathrm{B}}^2)$$

Since $\boldsymbol{F'}_1$ is a function of $\mathrm{P_G}$ for iterative solutions on the computer, it will be convenient to separate constant terms from the F'_1 and P_0 dependent terms.

F_{V}' Dependent Terms

$$1 + \frac{\theta}{\pi}A_{\mathrm{sp}}E_{\mathrm{s}}\left[\frac{R_{\mathrm{E}}(\varepsilon_{\mathrm{pc}}^{*} - \varepsilon_{\mathrm{ps}}')}{Z(H_1 + d)}\left\{\frac{1}{0.225\sigma_{\mathrm{cu}}(R_{\mathrm{E}} + R_1)\sin\theta}\right\} + \frac{R_{\mathrm{E}}(\varepsilon_{\mathrm{pc}} - \varepsilon_{\mathrm{ps}}')}{(H_1 + d)(H_1 - d - Z)}\right.$$
$$\left. + \left\{\frac{1}{0.45\sigma_{\mathrm{cu}}(R_E + R_1)\sin\theta}\right\}\right]$$

or

$$K_1 = \left[1 + \frac{R_{\mathrm{E}}\frac{\theta}{\pi}A_{\mathrm{sp}}E_{\mathrm{s}}}{(\sigma_{\mathrm{cu}}(R_{\mathrm{E}} + R_1)\sin\theta)(H_1 + d)}\left\{\frac{\varepsilon_{\mathrm{pb}}^{\epsilon} - \varepsilon_{\mathrm{ps}}'}{0.225Z} + \frac{\varepsilon_{\mathrm{pc}}^{*} - \varepsilon_{\mathrm{ps}}'}{0.45(H_1 + dZ)}\right\}\right] \tag{5.41}$$

Equation (5.41) is common for all the vessels.

P_{G} Dependent Terms

$$\frac{\theta}{\pi}\left[A_{\mathrm{sp}}E_s\frac{\pi(R_{\mathrm{I}}^2 + n_{\mathrm{b}}R_{\mathrm{B}}^2)}{E_{\mathrm{c}}A_{\mathrm{c}} + E_{\mathrm{s}}A_{\mathrm{sp}}} - \pi(R_{\mathrm{I}}^2 + n_{\mathrm{b}}R_{\mathrm{B}}^2)\right] \tag{5.42}$$

or

$$K_2 = \theta(R_{\mathrm{I}}^2 + n_{\mathrm{b}}R_{\mathrm{B}}^2)\left[\frac{A_{\mathrm{sp}}.E_{\mathrm{s}}}{E_{\mathrm{c}}.A_{\mathrm{c}} - E_{\mathrm{s}}A_{\mathrm{sp}}} - 1\right] \tag{5.42a}$$

Equation (5.42a) can be used directly for the HTGCR and Hartlepool vessels. For the Dungeness B and Oldbury vessels, equation (5.42a) is reduced to

$$K_2 = -\theta.R_{\mathrm{I}}^2\left(\frac{E_{\mathrm{c}}A_{\mathrm{c}}.E_{sp}}{E_{\mathrm{c}}A_{\mathrm{c}} + E_{\mathrm{s}}A_{\mathrm{sp}}}\right) \tag{5.43}$$

Constant Terms

$$K_3 = \frac{\theta A_{sp} E_s}{\pi}\left[\varepsilon_{ps} + \frac{R_E}{(H_1+d)}\left\{\frac{(\varepsilon^{\varepsilon}_{pc}-\varepsilon'_{ps})\left(R_E + R_1\dfrac{T_F}{0.225\sigma_{cu}(R_E+R_1)\sin\theta}\right)}{Z} + \frac{R_E(\varepsilon^{\epsilon}_{pc}-\varepsilon'_{ps})\left(a - \dfrac{T_F}{0.45\sigma_{cu}(R_E+R_1)\sin\theta}\right)}{(H_1-d-Z)}\right\}\right] + D_w(R_E^2+R_1^2) \tag{5.44}$$

where

$$\sigma_{pb} = (\varepsilon_{ps} + \varepsilon_1 + \varepsilon_{pp} + \varepsilon_{cpv}) \times E_s \tag{5.45}$$

It is initially assumed that, $\sigma_{pb} \leq$ pu…where pu …. the circumferential strain when the circumferential bands or tendons yield.

Evaluation of Strain and Stresses

(a) ε_{ps} stress in tendons/E_s

(b)
$$\varepsilon_1 = \frac{[(P_G \pi R_1^2) + n_b \pi (P_{GB}) \pi R_B^2]}{E_c}.A_c + E_s.A_{sp} \tag{5.46}$$

where $A_c = \pi\left(R_E^2 - R_1^2 - n_b - \pi R_B^2\right)$
A_{sp} = area prestressing steel

(c) ε_{pp} and ε_{cpt}. Figure J.4 shows the top half of the barrel wall in a rotated position. Line $A-A$ is centroid of the vertical tendon. Various displacements are related which are given below:

$$\frac{\delta_\epsilon}{(R_E - F_1 - t - x_t)} = \frac{\delta_T}{(t - x_w)} = \frac{R_E(\varepsilon^{\epsilon}_{pb} - \varepsilon'_{ps})}{Z} \tag{5.47}$$

Therefore

$$\delta_\varepsilon + \delta_T = \frac{R_E(\varepsilon^{e}_{pb} - \varepsilon'_{ps})}{Z}\{R_E - R_1 - t - x_t + t - x_w\} \tag{5.48}$$

Hence

$$\varepsilon_{\mathrm{pp}} = \frac{\delta_e + \delta_{\mathrm{T}}}{H_1 + d} \tag{5.49}$$

where δ_e and δ_T are defined in Fig. 1.4 and thickness of the cap.

Then

$$\varepsilon_{\mathrm{cpo}} = \frac{\delta'}{H_1 + d} = \frac{R_{\mathrm{E}}(\varepsilon_{\mathrm{pc}} - \varepsilon'_{\mathrm{ps}})(a - x_{\mathrm{t}})}{(H_1 + d)(H_1 - d - Z)} \tag{5.50}$$

where a is the distance from the outside face to the centroid of the vertical tendons.

$\varepsilon_{\mathrm{ps}}$ = strain at transferless losses including those due to creep effect.

ε_1 = elastic strain due to gas pressure, $\varepsilon_{\mathrm{pp}}$ strain due to rigid plastic rotation of the barrel, ε_{cpt} strain due to rigid plastic rotation of the cap. The general equation for F' in the abbreviated form is now given by

$$F'_{\mathrm{v}} = \frac{K_2}{K_1} P_{\mathrm{G}} \frac{K_3}{K_1} \tag{5.50}$$

Now the pressure P_0 to hold forces due to prestressing tendons, bonded steel reinforcement, liner steel and concrete for specified values of Z, $\varepsilon^e_{\mathrm{pc}}$ and $\varepsilon'_{\mathrm{ps}}$ can be found by taking first moments about the haunch hinge zone at Z. Referring to the previous equations, the equating moments are

$$\overline{P}_{\mathrm{w1}} \text{ or } \overline{P}_{\mathrm{w2}}\left(Z - \frac{H_1}{2}\right) + P_{\mathrm{GB}}.2R_{\mathrm{B}}Z(2\sin\,\theta)\frac{Z}{2} =$$

$$M_{\mathrm{HT}} \text{ or } M_{\mathrm{HB}} + M_{\mathrm{E}} + \sum_{z_{\mathrm{B}} < z}(T(Z - Z_{\mathrm{B}})) + \sum\left((A_{\mathrm{sb}} + \sigma_{yb} + A_{\mathrm{SL}}\sigma_{yl})\frac{Z^2}{2}\right) \tag{5.51}$$

In this case $P_{08} = P_0$, the second term on the left-hand side of equation (~is zero for the Oldbury and the Dungeness B vessels. The value of T, the radial component of force from the circumferential tendons or bands, is given by

$$T = \sigma_{\mathrm{ps}} \times A_{\mathrm{hp}} \times 2\sin\,\theta$$
$$\overline{P}_{w1} \text{ or } \overline{P}_{\mathrm{w2}} = P_{\mathrm{G}}(2R_{\mathrm{I}}\sin\,\theta.H_1) \tag{5.52}$$

The expression for the moment M derived earlier is now the hinge moment M_{HT} or M_{HB} and M_{E} depending on the respective positions. Substituting equation (~~T into the above equations the following relationship between gas pressure and hinge moment is obtained:

$$M_{\mathrm{HT}} \text{ or } M_{\mathrm{HB}} + M_{\mathrm{E}} = C_{\mathrm{c}}(d_1 - \overline{x}) + F'_{\mathrm{bst}}(d_1 - d') - \left[\frac{K_2}{K_1}P_{\mathrm{G}} + \frac{K_3}{K_1}\right](d_e - e)$$

$$= \left(\frac{K_2}{K_1}P_{\mathrm{G}} + \frac{K_3}{K_1}\right)(d_1 - d_\varepsilon + e - \overline{x}) + T_{\mathrm{F}}(d_1 - \overline{x}) + F'_{\mathrm{bst}}(\overline{x} - d') \tag{5.53}$$

Now take the same segment 2θ representing the top and bottom cap. For computer simulation procedure the segmental angle is given a new rotation 2ϕ which will differentiate between the barrel wall rotation and the cap rotation and will also assess distinctly the compatibility between them. Hence for both the top and the bottom caps the segmental angle $2\phi = 2\theta$ is retained throughout in the calculations.

The cap is subject to disturbing forces and moments due to gas pressures. The restoring forces and moments are due to

(a) vertical compression of longitudinal prestressing tendons, bonded reinforcement and the liner causing a yield line moment m, of the transverse yield lines per unit length;
(b) radial compression due to circumferential prestress, bonded reinforcement and the internal pressure acting on the barrel wall.

The equilibrium equations can be established by taking moments about a line passing through the centroid of the longitudinal tendons. This is the point about which M_{HT} and $M_{\mathrm{R}_{,,}}$ have previously been calculated. Similar to the barrel wall the equilibrium equation for the caps can be derived.

Total pressure P_c (Fig. 1.4)

$$\begin{aligned} &P_{\mathrm{G}}\pi R_{\mathrm{I}}^2 . \frac{2\phi}{2\pi} + P_{\mathrm{GB}} n_{\mathrm{b}} \pi R_{\mathrm{B}}^2 \frac{2\phi}{2\pi} \\ &P_{\mathrm{G}}\phi R_{\mathrm{I}}^2 + P_{\mathrm{GB}} n_{\mathrm{b}} \phi R_{\mathrm{B}}^2 \end{aligned} \tag{5.54}$$

The centroid $\bar{X}$ has already been evaluated earlier. The relationship between circumferential tendon forces T_{H} and P_0 given for various cases can now be easily derived. This pressure $\bar{P}_{\mathrm{G}}$ is required to maintain the cap in equilibrium for specified values of Z, $\boldsymbol{t}$ and

As the cap is divided into radial and transverse yield lines, they have their individual plastic moments which help in restoring the cap to equilibrium. Let m_ϕ be the plastic moment per unit length of the radial yield lines. The total length of any two radial lines is $(2RE—n_p 2R_p)$ where n, is the number of penetrations each having a radius R_{p} crossing the yield lines. Similarly m_{r} is the plastic moment of the transverse yield lines per unit length with a total length of this type of yield line $2(R_{\mathrm{E}}—a)$. In addition there is a pressure $P_{\mathrm{A}\phi}$ due to circumferential tendons. The equilibrium equations can therefore be written as

$$\begin{aligned} &\phi P_{\mathrm{G}}(R_I^2 + n_{\mathrm{b}} R_{\mathrm{B}}^2)\left(R_\phi - a - \frac{2}{3} R_{\mathrm{I}} \frac{\sin\phi}{\phi}\right) \\ &= D_{\mathrm{w}} R_{\mathrm{B}}^3 \sin\phi \left(1 - a - \frac{\sin\phi}{\phi}\right) \\ &+ \left(\frac{K_2}{K_1} P_{\mathrm{G}} + \frac{K_3}{K_1}\right)(a) + T_{\mathrm{F}}(d_1 - \overline{x}) + F'_{\mathrm{bst}}(\overline{x} - d') \end{aligned} \tag{5.55}$$

$$
\begin{aligned}
&+2(R_{\mathrm{E}}-n_{\mathrm{p}}R_{\mathrm{p}}\sin\phi)\left[\sum_{Z_{\mathrm{B}=1}}^{n}\left\{(A'_{\mathrm{hp}}E_{\mathrm{s}}\varepsilon_{z}K')-P_{\mathrm{G}}(2n_{\mathrm{b}}R_{\mathrm{B}})H_{1}+d-(H_{\mathrm{I}}R_{\mathrm{I}})\right.\right.\\
&\qquad\left.+2\bar{A}_{\mathrm{sb}}\sigma_{yb}(d_{1}-\overline{x})+A''_{\mathrm{sb}}\sigma_{\mathrm{scy}}(d_{1}-d')\right\}\\
&\qquad-2\sin\phi\left\{\sum_{Z_{\mathrm{B}<1}}(T_{\mathrm{H}})+\overline{A}_{\mathrm{sb}}\sigma_{yb}Z\right\}\left(H_{1}+\frac{d}{2}-Z\right)\\
&\qquad\left.+\sum_{Z_{\mathrm{B}<1}}\left((T_{\mathrm{H}})(H_{1}+\frac{d}{2}-Z_{\mathrm{B}})\right)\right]\\
&\qquad+2\sin\phi P_{\mathrm{G}}(R_{\mathrm{I}}H_{\mathrm{I}}+2R_{\mathrm{B}}Z)\left(H_{1}+\frac{d}{2}-Z\right)\\
&\qquad+2R_{\mathrm{B}}(H_{\mathrm{B}}-Z)\left(H_{1}+\frac{d}{2}-\frac{H_{\mathrm{B}}-Z}{2}-Z\right)
\end{aligned}
\tag{5.56}
$$

The above term $P_{\mathrm{H}\phi}(\mathrm{e})$ is evaluated which is a moment of the radial forces about the centre line of the cap. These are due to shear forces at the hinge at Z and circumferential tendons and any other pressure forces acting above the position Z of the hinge. Using Figs. 5.29 and 5.30 the breakdown of these forces and moments is given below:

(a) Circumferential prestresses below the hinge $Z_{\mathrm{b}} \leq Z$

$$
M_{1}=-2\sin\phi\sum_{Z_{B<Z}}^{n}(T_{\mathrm{H}})\left(H_{1}+\frac{d}{2}-Z\right) \tag{5.57}
$$

(b) Circumferential prestresses below the hinge $Z_{\mathrm{b}} > Z$

$$
M_{2}=-2\sin\phi\sum_{Z_{\mathrm{B}}<Z}^{n}(T_{\mathrm{H}})\left(H_{1}+\frac{d}{2}-Z\right) \tag{5.58}
$$

(c) Boiler pressure P_{GB} $Z_{\mathrm{b}} \leq Z$

$$
M_{3}=+\sin\phi P_{\mathrm{GB}}2R_{\mathrm{B}}\left(H_{1}+\frac{d}{2}-Z\right) \tag{5.59}
$$

(d) Boiler pressure P_{GB} $Z_{\mathrm{b}} \geq Z$

$$
M_{4}=+2\sin\phi P_{\mathrm{GB}}2R_{\mathrm{B}}(H_{\mathrm{B}}-Z)\left(H_{1}+\frac{d}{2}-\frac{H_{\mathrm{B}}-Z}{2}-Z\right) \tag{5.60}
$$

(e) Gas internal pressure P_G

$$M_5 = 2\sin\phi P_G R_I H_I\left(H_1 + \frac{d}{2} - Z\right) \tag{5.61}$$

(f) Bonded reinforcement T_ϕ

$$M_6 = 2\sin\phi \bar{A}_{sb}\sigma_{yb} Z\left(H_1 + \frac{d}{2} - Z\right) \tag{5.62}$$

5.15.1.3 Crack Sizes

The major cracks are as follws:

- Horizontal cracks at haunch w_{th} and w_{bh}
- Longitudinal and horizontal cracks w_{el} and w_{eh}
- Substituting *t, l* and *e* represent a top, bottom and equator of the vessel. Before computing the crack width, the barrel wall and the cap rotation α_B and α_c, respectively, are evaluated. The relevant strain used for these rotations are ε^e_{po}, ε'_{ps} and ε_{pc}

$$\alpha_B = \left(\varepsilon^e_{pb} - \varepsilon'_{ps}\right) R_E Z \tag{5.63}$$

$$\alpha_c = \left(\varepsilon^e_{pb} - \varepsilon'_{ps}\right) R_E/(H_1 + d - Z) \tag{5.64}$$

Next the depth of the crack is evaluated. From the geometry (Fig. 5.28) it is easy to compare the crack depth l_c *given by*

$$l_c = [(R_E - R_I - \hat{x})^2 + (Z - H_I)^2]^{1/2} \tag{5.65}$$

where $\hat{x}$ is a typical value for x_t or x_b. Then crack width

$$w_{th} \text{ or } W_{tb} = (\alpha_B + \alpha_c).l_c = \frac{R_E}{(H_1 + d - Z)}\left\{(\varepsilon^e_{pb} - \varepsilon'_{ps})\frac{H_1 + d - Z}{Z} + (\varepsilon_{pc} - \varepsilon'_{ps})\right\} \tag{5.66}$$
$$x[(R_E - R_I - \hat{x})^2 + (Z - H_I)^2]^{1/2}$$

The longitudinal crack width at the equator w_{el} is computed as

$$W_e = 2\pi R_E(\varepsilon^e_{pb} - \varepsilon'_{ps})\frac{2\sin\theta}{2\pi} \tag{5.67}$$

The horizontal crack width at the equator w_{eh} is computed as

$$W_{\text{eh}} = R_{\text{E}}\left(\varepsilon^{e}_{\text{pb}} - \varepsilon'_{\text{ps}}\right)\left[\frac{H_1 + d}{H_1 + d - Z}\right] \tag{5.68}$$

5.15.1.4 Additional Mechanisms in Caps

Once the vessel starts developing tensile cracks, only aggregate interlock friction and bonded steel reinforcement prevents the formation of one large crack. In zones where the reinforcement is not sufficient additional cracks form in the cap area vertically [10–15] near the haunch and at the standpipe perimeter (central plug) due to cap fixture-cum-shear. Under pure flexural conditions, the additional hinges form in the cap, say, at any distance $2R_E\beta$ as shown in Fig. 5.55. It is assumed that lateral restraint is provided to the cap due to barrel and cap circumferential prestressing and no vertical translation occurs between the hinge and the external boundary or rim of the cap. This assumption demands that analysis should include the effect on the compressive membrane action of this partial restraint against lateral displacement at the rim and the radial strain in the plane of the cap. The radial strain e_7. is assumed to be the algebraic sum of Σ_c, the creep strain Σ_s, the shrinkage strain and Σ_e the elastic strain.

The total radial strain $\varepsilon_T = \varepsilon_c + \varepsilon_s + \varepsilon_e$ is assumed to be constant since elastic or membrane forces due to these are constant. Because of the value of C_T, the middle portion *BC* between the yield and cracked zone will be reduced by $2R_E\varepsilon_T(1-2\beta)$. This will cause an outward lateral displacement $\ddot{o}_0$ at the haunch boundaries *A* and *B* or *C* and *D* as shown in Fig. 5.55 owing to the same c_7 the dimensions *AB* or *CD* will decrease *AB* by $2R_E\beta(1-2\beta)$. Hence due to the total movement caused by ε_T and δ_o, the final reduced dimension R' will become

$$R' = [2R_E(\beta + \varepsilon_T(1 - 2\beta) + \delta_o)] \sec \alpha_c \tag{5.69}$$

where α_c is the cap rotation. -

Let the distances to the neutral axes from the compressed faces of the concrete at the sagging and hogging conditions be the two values of x, i.e. x_1 and x_2. These values of x_1 and x_2 can be evaluated in the same way as before. Due to the inclusion of membrane forces, the value of x may not be equal to x_1 or x_2, during incremental deformation. Assume $x_1 = n_1 d$ and $x_2 = n_2 d$, (Fig. 5.56). The equations for n_1 and n_2, F_e, M_e are written as

$$(1 - n_2)d\tan\alpha_c + 2R_E(1 - \varepsilon).\beta n_1 d\tan\alpha_c \tag{5.70}$$

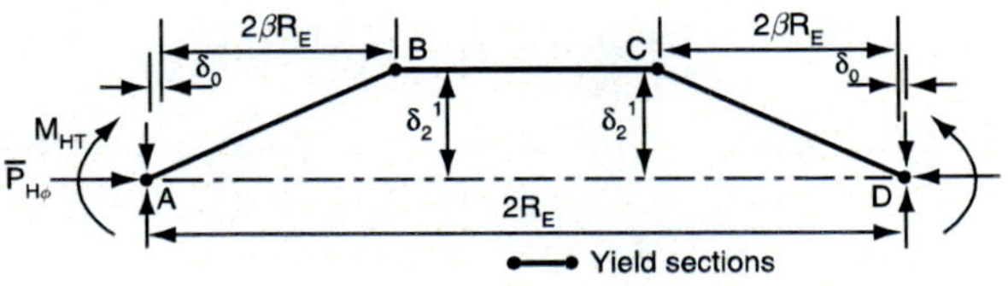

Fig. 5.55 Collapse mechanism

Fig. 5.56 Portion of a YIELD SECTION

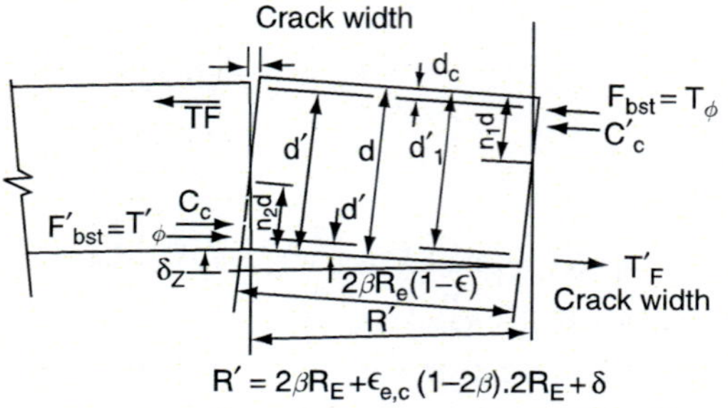

From Eqs. (5.69) and (5.70)

$$1 - (n_1 + n_2) = \frac{[2R_E\beta + 2R_E\varepsilon_T(1-2\beta) + \delta_o] - 2R_E(\beta - \varepsilon\beta)\cos\alpha_c}{d\sin\alpha_c}$$

or

$$\begin{aligned} n_1 + n_2 &= 1 - \frac{2R_E\{\beta - \varepsilon_T(1-2\beta) - (\beta - \varepsilon_T)\cos\alpha_c\} + \delta_o}{d\sin\alpha_c} \\ &= 1 - \frac{R_E\{\sin^2\frac{\alpha_c}{2} + \beta\epsilon\cos\alpha_c + \varepsilon_T(1-2\beta)\} + \delta_o}{d\sin\alpha_c} \end{aligned} \quad (5.71)$$

For membrane forces α_c and ε_T are very small values therefore it is assumed that

$$\sin\alpha_c = 2\sin\frac{\alpha_c}{2} = \delta'_z/\beta(2R_E) \quad (5.72)$$

where δ'_z is the deflection of the cap at yield zone. Hence Eq. (5.71) is written as

$$n_1 + n_2 = 1 - \left\{\frac{\delta'_z}{2d} + \frac{4R_E^2\beta}{(d)(\delta'_z)}\varepsilon_T\beta + \varepsilon_T - 2\beta\varepsilon_T + \frac{\delta_o}{2R_E}\right\} \quad (5.73)$$

If the crack occurs around the standpipe perimeter or the central plug the hinges at *B* and *C* will be at equal distance from *A* and *D*, respectively. In that case equation (5.73) is modified by replacing $(\varepsilon_T - 2\beta_{\varepsilon T} + \frac{\delta_0}{2R_E})$ with $(\varepsilon_T + \frac{\delta_0}{2R_E})$. Proceeding as above Eq. (B. 8 1) then assumes the following form:

$$n_1 + n_2 = 1 - \left\{\frac{\delta'_z}{2d} + \frac{4R_E\beta}{(d)(\delta'_z)}\left(\varepsilon_T + \frac{\delta_o}{2R_E}\right)\right\} \quad (5.74)$$

The equilibrium of horizontal forces acting on zone *AB*

$$F_\varepsilon = F'_2$$

or

$$C_c + T'_\phi - T_F = C'_c + T_\phi - T'_F \tag{5.75}$$

Therefore

$$n_2 - n_1 = \frac{T'_F - T_F - T_\phi + T'_\phi}{0.45\sigma_{co}.d} \tag{5.76}$$

$$n_1 = \frac{1}{4d}\left[2d - \delta'_z - \frac{\left(\varepsilon_T + \frac{\delta_o}{2R_E}\right).4R_E^2\beta}{\delta'_z} - \frac{T'_F - T_F - T_\phi + T'_\phi}{0.225\sigma_{cu}}\right] \tag{5.77}$$

$$n_2 = \frac{1}{4d}\left[2d - \delta'_z - \frac{\left(\varepsilon_T + \frac{\delta_o}{2R_E}\right).4R_E^2\beta}{\delta'_z} + \frac{T'_F + T_F - T_\phi + T'_\phi}{0.225\sigma_{cu}}\right] \tag{5.78}$$

Equations (5.77) and (5.78) indicate that the values of n_1 and n_2 do differ as stated earlier and the term $(\varepsilon_T + (\delta_0/2R_E))$ in the same equations merely signifies the reduction in the depth to the neutral axis at the cracked sections. For the sagging part of the cracked section, the forces C_c, T'_ϕ and T_F are statically equivalent to the membrane force F_e at mid-depth. Hence

$$F_\varepsilon = C_c + T'_\phi - T_F = 0.45\sigma_{cu}.n_1 d + T'_\phi - T_F. \tag{5.78}$$

Various values in Eq. (5.78) have already been defined. The moment M caused by these forces is computed as

$$\begin{aligned} M_\varepsilon &= 0.45\sigma_{cu}n_1 d\left(\frac{1}{2}d - K_3.x\right) + T'_\phi\left(\frac{1}{2}d - d'\right) + T_F\left(d_1 - \frac{1}{2}d\right) \\ &= 0.45\sigma_{cu}n_1 d\left(\frac{1}{2}d - K_3 n_1 d\right) + T'_\phi\left(\frac{1}{2}d - d'\right) + T_F\left(d_1 - \frac{1}{2}d\right) \end{aligned} \tag{5.79}$$

where K_3 is already defined.

The moment M_e then replaces all the terms in the previous equations, except the terms $F_v(d_e - e)$ or $F_v a$ respectively. Wherever necessary, equations are similarly adjusted to include the above terms. For Hogging moment M_e' for yield or cracked sections is similar to equation 5.79 and can be computed by replacing n_1 by n_2. The summation of moments for, say, AB or CD portions due to the above stress resultants at one end for a segment of 2ϕ will be given by

$$\begin{aligned}\sum M &= M_c + M'_e - F_e.\delta'_z \\ &= 0.45\sigma_{cu}\left\{\frac{1}{2}d^2(1-K_3) + \frac{1}{4}\delta'_z.d(2K_3-3)\right. \\ &+ R_E^2\beta\varepsilon_T(1-K_3+2K_3d-d) + \frac{1}{(8d)(\delta'_z)}[(\delta'_z)^4(2-K_3)] \\ &\left. - 16K_3\varepsilon_I^2\beta^2R_E^4\right\}(2\sin\phi) - \frac{K_3}{0.9\sigma_{cu}}(T'T_F - T_\phi T'_\phi)^2(2\sin\phi) \\ &+ T_\phi\left(\frac{1}{2}d - d_c - \frac{\delta'_z}{2}\right)(2\sin\phi) + T'_\phi\left(\frac{1}{2}d - d' - \frac{\delta'_z}{2}\right)(2\sin\phi) \\ &+ T'_F\left(d_1 - \frac{1}{2}d + \frac{\delta'_z}{2}\right)(2\sin\phi) \\ &+ T_F\left(d_1 - \frac{1}{2}d + \frac{\delta'_z}{2}\right)(2\sin\phi)\end{aligned} \tag{5.80}$$

From the barrel wall rotation caused by the eccentric prestressing about the haunch hinge the moment value is

$$M_{pB} = 2F_v\left(e + \bar{x} - \frac{d}{2}\right) = 2\sin\phi\left(\frac{K_2}{K_1}.P_G + \frac{K_3}{K_1}\right)\left(e + \bar{x} - \frac{d}{2}\right) \tag{5.81}$$

This term is algebraically added to Eq. (5.82). Hence M_{HT}, M_{HB} of previous equations are written in an abbreviated form as

$$M_{HT} \text{ or } M_{HB} = \sum M + M_{PB} \tag{5.82}$$

The new values of M_{HT}, M_H from Eq.(5.83) are then used and the intermediate equations are similarly adjusted accordingly for the redevelopement of the interaction equation.

5.15.1.5 Load Variation Due Radial Prestressing in Caps

In the Oldbury vessel a slight variation to the above equations is necessary for computing forces and moments. As stated in both the top and bottom caps of the Oldbury vessel the tendons are arranged horizontally in layers which form a square mesh in plan view. As shown in Figs. 5.55 and 5.56 both top and bottom caps are divided into slices. Equations are modified to include the term

$$\sum_{i=1}^{n}(P_{hi}.Y_i + F_{vi}.R_i).$$

The final equation is then written as

$$\phi P_G R_I^2 (R_E - a - \frac{2}{3} R_I \frac{\sin\phi}{\phi} = D_w R_E^3 \sin\phi\left(1 - a - \frac{\sin\phi}{\phi}\right) + \left(\frac{K_2}{K_1} P_G + \frac{K_3}{K_1}\right)(a)$$
$$+T_F(d_1 - \overline{x}) + F'_{\mathrm{bst}}(\overline{x} + d')$$

$$\pm\sum_{i=1}^{n}(P_{Hi}.Y + F_{vi}.R_i) + 2(R_E - n_p R_p \sin\phi)$$
$$\times\left[\sum_{Z_B=1}^{n}\left\{(A'_{\mathrm{hp}}E_s\varepsilon_z K') + P_G H_I R_I + 2A'_{\mathrm{sb}}\sigma_{yb}(d_1 - \overline{x}) + A''_{\mathrm{sb}}\sigma_{\mathrm{scy}}(d_1 - d')\right\}\right]$$

(The term $(-n_p R_p \sin\phi)$ is zero from bottom cap)

$$-2\sin\phi\left\{\sum_{Z_B<Z}(T_H) + A'_{\mathrm{sb}}\sigma_{yb}Z)\right\}\left(H_1 + \frac{d}{2} - Z\right)$$
$$+\sum_{Z_B>Z} T_H\left(H_1 + \frac{d}{2} - Z_B\right) + 2\sin\phi P_G - (R_I H_I)\left(H_1 + \frac{d}{2} - Z\right). \quad (5.83)$$

Equation (5.84) can then be recalled as an independent sub-routine in a computer program in solving this specific case.

Ultimate Limit State of Shear Analysis

Basic concept: The failure of concrete under complex loading has been adequately discussed in the references. In this section a further explanation is required regarding the so-called ‘shear failure’ under gas increasing pressure.

The concept of shear failure in thick sections of brittle material such as concrete is obscure and in many instances it could be misleading. One clear concept is that concrete failure can easily be put to the test if it is assumed that it is governed by the principal tensile and compressive stress caused by the so-called ‘shear’. The problem with this simple concept is the limitation on these stresses. In the prestressed concrete reactor vessels due to variations in loading conditions these principal stresses at any time at any point may vary from biaxial and triaxial compression to compression—tension—tension in any combination. These instant changes can bring about any kind of failure. For example, it may be pure flexural-cumnominal shear failure or principal tensile or compressive failure or by the so-called shear compression failure or in any combination of these. It must be borne in mind that the type of failure is directly related to the vessel overall layout. Before discussing the individual sample examples, it is necessary to know what the above-mentioned terms are and what effect they have on the prestressed concrete vessels.

(a) *Flexural-cum-Nominal Shear*

As evident from previous discussions, flexural failure occurs by the formation of principal tensile or bending cracks followed by the crushing of the compressive zone. This may have been preceded by the yielding or breaking of steel in the vessel. In zones where nominal shear governs, the failure will take place along planes parallel to the direction of the applied loads. In some zones a certain magnitude of shear resistance is provided side by side with the flexural failure case.

(b) *Principal Tensile Failure*

Failure can take place by inclined principal tensile cracks produced by inclined principal tensile stresses in zones of larger shear. These cracks can spread rapidly through the section across the compression zone. Unless conventional steel reinforcements are present in such zones, an internal redistribution of forces is not possible.

(c) *Principal Compressive Failure*

Depending upon the low span/depth ratio, there is a possibility for the vessel components, like a short deep beam, to fail directly in compression as a tied arch or truss.

(d) *Shear Compressive Failure*

This term is normally associated with shorter deeper beams. Failure takes place when the compression zone, after its reduction in size due to cracking, fails under combined action of triaxial compression and shear stresses. This failure type can happen at principal tensile cracks. The limitation to this shear compression is that the average compressive stress in the concrete has only been found at failure in pure flexure. As it is based on extreme concrete fibre reaching its limiting strain it is possible that 'shear' failure can occur on an inclined plane near the centre of a compression zone. In the vessel the haunch hinge is under a multiaxial state of stress due to prestressing bonded steel, liner and penetrations. This zone will in that case resist shear compression failure.

However, principal stresses and consequently the cracks they cause, may appear in a number of different positions depending upon the relative magnitudes of loads, moments and 'shear' carried by the section. Significant differences occur among isolated cap tests and full model tests. Reported ultimate load tests for Oldbury and Bungeness B vessels consider each vessel to be initially broken up into an open-ended cylinder and two end slabs. The mechanisms of this initial break-up, however, appear to be in disagreement with each other. For the Oldbury analysis, the slab was separated from the barrel by a plastic hinge at the junction of the slab and barrel.

A shear-compression failure mechanism on the other hand was adopted for the Dungeness B vessel analysis. Of even greater disparity between the two analyses were the mechanisms and modes of failure adopted to predict the ultimate pressure or to prove that the ultimate load factor was higher than the adopted factors for the slab and the barrel. For the barrel, the Oldbury analysis apparently favoured the modes of failure through the formation of a plastic hinge at the mid-height of the barrel. For Hartlepool (20 g, 24Aa, 24Ae) the one-tenth scale model tests carried out gave flexural cracking in top cap but with isolated cap tests plug failure similar to the Dungeness B has been observed which is claimed to be due to shear compression and inclined tensile cracking phenomena. Other tests have reported either flexural or shear plug failure.

Shear in Barrel Wall and Caps

The present analysis is based on gas increasing pressure and progressive incremental deformation. Any of the above-mentioned 'shear' failure conditions caused by the net difference of the instantaneous inward and outward vessel forces can be automatically handled provided sufficient mathematical tools and iterative techniques are available. A separate analysis is required for both the barrel wall and the caps. Concrete failure criteria have been discussed previously. Some of the concrete strength theories discussed in this chapter are reproduced in the following order:

(a) Mohr's theory based on octahedral shear and normal stresses.
(b) Biaxial compression state.
(c) Combined compression-tension state.
(d) Empirical for shear derived from actual vessel data using hypo-elastic concept.

Any one of the above is associated with a straight line and parabolic stress variation. The above equations can be arranged so that they can be linked up with those of the ultimate limit state of flexure for both wall and caps. For each incremental deformation of the vessel, these equations can be tested for any premature 'shear failure' intervening flexural failure.

Barrel 'Shear'

The shear forces at any position of the hinge $Z \gtrless Z_B$ for each vessel incremental deformation can be found by resolving inward forces due to prestressing liner and bonded reinforcement and outward radial forces due to gas pressures from the main cavity with and without gas pressures in the boiler penetrations. The net force divided by the area of the concrete at that position is the shear stress.

Cap 'Shear'

The elastic analyses of many vessels indicate that everywhere except at the centre of the cap the circumferential tensile stresses are greater than the radial stresses and they cause flexural mode of failure by breaking up the cap into a series of wedges with their apices at the centre of the cap. In addition to this mode of failure the resistance of the compressive area in the standpipe region is generally reduced and it finally develops a circumferential plane of weakness at the outside perimeter of the standpipes. The Dungeness B model tests show a complete punched-out failure of the central core of the cap. Failure has occurred when the force due to the internal gas pressure P_G exceeded the vertical components of the shear resistance and normal stress on the inclined failure surface of the core. Figure 5.30 shows the equilibrium forces on the central core having sloping surfaces. The normal force and the shear force components are represented by N and Q. It is assumed that they act on single straight lines of intersection between the failure surface and the radial plane of the cap. The angle of failure y is considered as the rupture angle corresponding to the minimum pressure required to punch out the core. At the point of sliding the shear resistance on the failure surface is dependent on the normal stress. Once the relationship between the shear resistance and the normal stress is established then it is not difficult to determine the failure pressure for the core. As indicated in this case the distances x_1 and x_2 which are the inner and outer radii of the central core are dependent on the angle y. Let Z_c be the vertical distance from the top of the cap to any point on the sloping surface at a horizontal distance R_x from the circumferential plane of weakness as shown in Fig. 5.31. The radial forces $\bar{P}_{H\phi}$ are known from previous equations. Let the shear stress in the R_x–Z_c plane be τ_{RZ} and the radial thrust on the inner core be $\bar{P}_{Hi}$, the minimum value of which is $\bar{P}_{H\phi}$. The upper value of $\bar{P}_{Hi}$ is given from Fig. 5.31 as

$$\overline{P}_{Hi} = \frac{R_E}{R_x}.\overline{P}_{H\phi} + \left(\frac{R_E}{R_x} - 1\right)\sigma_t \tag{5.84}$$

where σ_t is the tangential stress along the crack taken right through the thickness of the ring and is not less than $T_H/(R_E - R_I)\,(R_E + R_I)$ sin $\phi.\sigma_t$ can also be the concrete tensile stress. The forces on the central core are shown in Fig. 5.31 and when they are resolved vertically

$$P_{Gx} = \text{the total pressure on the core diameter}$$

$$x_1 = Q\sin\gamma - N\cos\gamma. \tag{5.85}$$

For simplification it is assumed that the shear and normal stresses acting at the inclined surface at distance (R_x , Z_c) from the top of the central core are of the octahedral type and are denoted, as before, by τ_o and σ_o, respectively. The above forces can then be written as

$$P_{Gx} = \pi(x_2 - d\cot\gamma)^2.P_G$$

$$Q = \int_{x_2 - d\cot\gamma}^{x_2} \tau_o.R_x \sec\gamma dR_x$$

$$N = 2\pi \int_{x_2 - d\cot\gamma}^{x_2} \sigma_o R_x \sec\gamma.dR_x \tag{5.86}$$

Substituting Eq. (5.86) into Eq. (5.85) those stresses are related to the internal gas pressure

$$P_G = \frac{2}{(x_{2-d\ \cot\ \gamma})^2} \int_{x_{2-d\cot\gamma}}^{x_2} (\tau_o \tan\gamma - \sigma_o)R_x dR_x \tag{5.87}$$

The equilibrium equation in the radial direction (Fig. 5.31) is given by

$$\sigma_o \sin\gamma + \tau_o \cos\gamma = \overline{P}_{Hi} \sin\gamma - \tau_{RZ} \cos\gamma. \tag{5.88}$$

Campbell–Allen et al. suggest straight line and parabolic stress variations for the shear stress τ_o influenced by concrete strength and maximum aggregate size. As stated previously at the point of sliding the relationship between τ_o and σ_o can be written using different concepts. Campbell–Allen et al. suggests Mohr envelops. For straight line stress variation

$$\tau_o = C' + \sigma_o \tan\phi' \tag{5.89}$$

where C' is apparent cohesion for concrete, ϕ' is the angle of frictional resistance for concrete. In terms of parameters largely dependent on the concrete strength and the aggregate size, Campbell–Allen et al.[20d] suggest an expression

$$\frac{\tau_n}{f_c'} = A(\sigma_n/f_c' + F_t/f_c')^k \tag{5.90}$$

where τ_n is the limiting shear resistance

σ_n normal stress on failure surface
f_c' unconfined concrete compressive strength
f_t the uniaxial tensile strength of concrete
A, K parameters.

In this section $\tau_n = \tau_0, \sigma_n = \sigma_0, K = K', f_t = \sigma_t$ have been adopted and therefore Eq. (5.90) can be rewritten as

$$\tau_o = B'(\sigma_o + \sigma_t)^k \tag{5.91}$$

where

$$B' = \frac{A.f_c'}{(f_c')^k}. \tag{5.92}$$

Now in addition to the above relationship τ_o, equations for shear reproduced below are considered for τ_o

Biaxial State of Stress

$$\tau_o = -\sqrt{2}\left(\frac{R_C - 1}{2R_C - 1}\right)\left(\sigma_o + \frac{1}{3}\sigma_o\right) \tag{5.93}$$

where R_c = biaxial strength/uniaxial strength of concrete.

Combined Compression Tension

$$\tau_o = -\sqrt{2}\left(\frac{L}{3} + \frac{2}{3}L_4\right) \tag{5.94}$$

where

$$L_3 = \frac{1 - R_t}{1 + R_t}; \quad L_4 = \frac{R_t}{1 + R_t}$$

Hypoelastic Concept

$$\frac{\sigma_3}{\sigma_c} = -0.053 - 0.92\left(\frac{\tau_o}{\sigma_c}\right)^{1.35} \tag{5.95}$$

From Equations (5.89) and (5.95) any one of the value of τ_o can be considered during the vessel incremental deformation or gas increasing pressure. Case (A) $\tau_o = C' + \sigma_o \tan \phi'$

Substituting τ_o from Eq. (5.98) into Eq. (5.88), Eq. (5.88) becomes

$$\sigma_o = \frac{1}{\sin\gamma + \cos\gamma \tan\phi'}\{\overline{P}_{Hi}\sin\gamma - \tau_{RZ}\cos\gamma - A'\cos\gamma\} \tag{5.96}$$

Substituting τ_o and σ_o into Eq. (5.87) then Eq. (5.87) is rewritten as

$$(x_2 - d\cot\gamma)^2 P_G + 2\int_{x_2-d\cot\gamma}^{x_2} \frac{\tau_{Rz}\cos\gamma(\tan\phi'\tan\gamma-1)}{(\sin\gamma+\tan\phi'\cos\gamma)} R_x d.R_x$$

$$= 2\int_{x_2-d\cot\gamma}^{x_2} \frac{\overline{P}_{Hi}\sin\gamma(\tan\phi'\tan\gamma-1)}{(\sin\gamma+\tan\phi'\cos\gamma)} R_x d.R_x \tag{5.97}$$

$$+ 2c'\int_{x_2-d\cot\gamma}^{x_2} \left[\tan\gamma - \frac{\cos\gamma(\tan\phi'\tan\gamma-1)}{\sin\gamma+\tan\phi'\cos\gamma} R_x d.R_x\right]$$

The value of $\bar{P}_{Hi}$ is linear to those of P_{HQ} and τ_{RZ} and can be assumed to vary linearly with P_G, the internal gas pressure. The left-hand side of Eq. (5.97) involves τ_{Rz} values and the integral on the left-hand side can be solved for two different shear stress distributions.

Case (A): Uniform Shear Stress T_{Rz}

The value $\tau_{Rz} = \left(\frac{1}{2d}\right)P_G R_x$ is substituted in the integral part of the left-hand side of Eq. (5.64) and is finally itegrated

$$\text{l.h.s} = (x_2 - d\ \cot\ \gamma)^2.P_G + \frac{\cos\gamma(\tan\phi'\tan\gamma-1)}{(\sin\gamma+\tan\phi'\cos\gamma)}\left[\frac{R_x^3}{3}\right]_{x_2-d\cot\gamma}^{x_2}$$

$$= (x_2 - d\cot\gamma)^2 + \frac{P_G\cos^2\gamma(\tan\phi'\tan\gamma-1)}{\sin\gamma(\sin\gamma+\tan\phi'\cos\gamma)} \tag{5.98}$$

$$(x_2^2 - x_2 d\cot\gamma - 0.334\cot^2\gamma) = P_G(\hat{\alpha}\hat{k})$$

where $\hat{\alpha}\hat{k}$ are the terms of Eq. (5.98) in the bracket after P_G is taken outside the bracket.

Now the right-hand side of Eq. (5.97) is evaluated. This is written as:

$$\text{r.h.s} = \overline{P}_{Hi}\frac{\sin\gamma(\tan\phi'\tan\gamma-1)}{(\sin\gamma+\tan\phi'\cos\gamma)}(2x_2 d - d^2\cot\gamma) \tag{5.99}$$

$$+ C'\left[\tan\gamma - \frac{\cos\gamma(\tan\phi'\tan\gamma-1)}{(\sin\gamma+\tan\phi'\cos\gamma)}\right](2x_2 d - d^2\cot\gamma)$$

In Eq. (5.99) $\bar{P}_{Hi}$ may have two values. One is

$$\overline{P}_{Hi} = \overline{P}_{H\phi} \tag{5.100}$$

The other is given by Eq. (5.84) when σ_o is compressive $\bar{P}_{Hi}$ must always be related as

$$\overline{P}_{Hi} \geq \cot\gamma(\tau_{Rz} + C') \tag{5.101}$$

where τ_{Rz} has a uniform or a parabolic shape.

Subject to modification given by Eq. (5.100) and equation (5.84) the gas minimum vertical pressure component P_G of the cap causing the parameter 'shear' failure is given by

$$\overline{P}_G = \text{Eq.}(5.100)/(\hat{\alpha}\hat{K}). \tag{5.102}$$

The value of γ in the above equations will then be the rupture angle.

Case (B) $\tau_o = B'(\sigma_o + \sigma_t)^k$

Using the parabolic stress variations, the value of

$$\tau_{Rz} = \frac{3P_G R_x Z_c (d - Z_c)}{d^3}. \tag{5.103}$$

The value of τ_{Rz} when substituted in the integral part of the left-hand side of Eq. (5.97) gives another expression very similar to Eq. (5.98):

$$\text{l.h.s} = (x_2 - d\cot\gamma)^2 P_G + \frac{P_G \cos^2\gamma(\tan\phi' \tan\gamma - 1)}{\sin\gamma(\sin\gamma + \tan\phi' \cos\gamma)}$$

$$\left[x_2^2 - x_2 d\cot\gamma - 0.3d^2\cot^2\gamma\right] \tag{5.104}$$

The difference between Eqs. (5.98) and (5.104) is very small and therefore hardly affects the ultimate pressure of the cap. Equation (5.104) is written in an abbreviated form as

$$\text{l.h.s} = P_G(\hat{a}_1\hat{k}_1). \tag{5.105}$$

Now substituting $\tau_o = B'(\sigma_o + \sigma_t)^k$ into Eq. (5.88) the equilibrium equation is written as

$$\sigma_o \sin\gamma + B'(\sigma_o + \sigma_t)^k \cos\gamma = \overline{P}_{Hi}\sin\gamma - \tau_{Tz}\cos\gamma \tag{5.106}$$

For a parabolic variation Eq. (5.105) is further written as

$$\sigma_o \sin\gamma + B'(\sigma_o + \sigma_t)^k \cos\gamma = \overline{P}_{Hi}\sin\gamma - \frac{3P_G R_x.Z_c(d - Z_c)}{d^3}\cos\gamma. \tag{5.107}$$

From Eq. (5.106)

$$\sigma_o = \overline{P}_{Hi} - B'(\sigma_o + \sigma_t)^k + \frac{3P_G R_x.Z_c(d - Z_c)}{d^3}\cot\gamma \tag{5.108}$$

Substituing σ_o and σ_t into Eq. (5.87)

$$P_G = \frac{2}{= (x_2 - d\cot\gamma)^2}\int_{x_2 - H\cot\gamma}^{x_2} [B'(\sigma_o + \sigma_t)^k \tan - \sigma_o] R_x d.R_x$$

$$= \frac{1}{(x_2 - d\cot\gamma)^2}[B'(\sigma_o + \sigma_t)^k \tan\gamma - \sigma_o(2x_2 d \cot \gamma - d^2 \cot^2\gamma] \qquad (5.109)$$

Since binomial terms exist in Eq. (5.109), the solution depends upon the variables such as $\sigma_0^2 \gtrless \sigma_t^2$.

Case (C) Biaxial Compression State

Equation (5.86) is utilised for the value of τ_o, hence the same equation is written as

$$\sigma_o \sin\gamma + \left[C_1\left(\sigma_o + \frac{1}{3}\sigma_o\right)\right]\cos\gamma = \overline{P}_{Hi}\sin\gamma - \tau_{RZ}\cos\gamma \qquad (5.110)$$

where

$$C_1 = \sqrt{2L_1} - \frac{\sqrt{2}(R_c - 1)}{2R_c - 1}.$$

Therefore

$$\sigma_o = \frac{1}{= (C_1\cos\gamma + \sin\gamma)}\left[\overline{P}_{Hi}\sin\gamma - \tau_{Rz}\cos\gamma - \frac{1}{3}C_1\sigma_c\cos\gamma\right]. \qquad (5.111)$$

The value of P_G is computed as

$$P_G = \frac{2}{= (x_2 - d\cot\gamma)^2}\int_{x_2 - d\cot\gamma}^{x_2}\left[C_1\left(\sigma_o + \frac{1}{3}\sigma_o\right) - \sigma_o\right]R_x.dR_x \qquad (5.112)$$

or

$$\begin{aligned} &P_G(x_2 - d\cot\gamma)^2 \\ &\quad = \left\{\left[+\frac{1}{3}C_1\sigma_c d\cot\gamma(2x_2 - d\cot\gamma)\right]\right. \qquad (5.113) \\ &\quad \left. + \frac{(c_1 - 1)}{c_1\cos\gamma + \sin\gamma}\left[\overline{P}_{Hi}\sin\gamma - \tau_{RZ}\cos\gamma - \frac{1}{3}c_1\sigma_c\cos\gamma\right]\right\}. \end{aligned}$$

(i) For a case of uniform shear stress distribution

$$\tau_{RZ} = \frac{1}{2}\frac{P_G R_x}{d}.$$

$$P_G = \frac{\overline{S}_1}{\left\{S_2 + \frac{(C_1 - 1)}{2d}R_x.\overline{S}_3\right\}}. \qquad (5.114)$$

(ii) For a case of parabolic stress distribution

$$\tau_{RZ} = \frac{3P_G R_x Z_c(d - Z_c)}{d^3}$$

$$P_G = \frac{\overline{S}_1}{\left\{\overline{S}_2 + \frac{3(C_1-1)(Z_c(d-Z_c))}{d} \cdot R_x \cdot \overline{S}_3\right\}} \tag{5.115}$$

where

$$\overline{S}_1 = \left\{\frac{1}{3}C_1\sigma_c d\cot\gamma(2X_2 - d\cot\gamma) + \frac{(C_1-1)}{C_1\cos\gamma + \sin\gamma} \times \left[\overline{P}_{Hi}.\sin\gamma - \frac{1}{3}C_1\sigma_c\cos\gamma\right]\right\}$$
$$\overline{S}_2 = (x_2 - d\cot\gamma)^2 \tag{5.116}$$
$$\overline{S}_3 = \frac{\cos\gamma}{(C_1\cos\gamma + \sin\gamma)}$$

Case (D): Combined Compression–tension State

$$+\tau_o = +\sqrt{2}\left(\frac{L}{3} + \frac{2}{3}L_4\right) \tag{5.117}$$

where

$$L_3 = \frac{1 - R_t}{1 + R_t} \text{ and } L_4 = \frac{R_t}{1 + R_t}$$

Substituting Eq. (5.117) into Eq. (5.88) the same equation is written as

$$\sigma_o \sin\gamma + \sqrt{2}\left(L_3 + \frac{2}{3}L_4\right)\cos\gamma = \overline{P}_{Hi}\sin\gamma - \tau_{RZ}\cos\gamma$$

or

$$\sigma_o = \overline{P}_{Hi} - \tau_{RZ}\cot\gamma - \sqrt{2}\left(L_3 + \frac{2}{3}L_4\right)\cot\gamma. \tag{5.118}$$

Similarly Eq. (5.118) is substituted in Eq. (5.87) and the value of P_G is arrived at in the following manner:

$$P_G(x_2 - d\cot\gamma)^2 = \left\{\sqrt{2}\left(L_3 + \frac{2}{3}L_4\right)\left(\frac{1}{\sin\gamma\cos\gamma}\right) - \overline{P}_{Hi} + \tau_{RZ}\cot\gamma\right\} \tag{5.119}$$
$$\left[2X_2 d\cot\gamma - d^2\cot^2\gamma\right].$$

Putting τ_{RZ} terms on the left-hand side

$$P_G(X_2 - d\cot\gamma)^2 - \tau_{RZ}(2x_2 d\cot^2\gamma - d^2\cot^3\gamma)$$
$$= \left\{\sqrt{2}\left(L_3 + \frac{2}{3}L_4\right)\left(\frac{1}{\sin\gamma\cos\gamma}\right) - \overline{P}_{Hi}\right\} \tag{5.120}$$
$$(2X_2 d\cot\gamma - d^2\cot^3\gamma).$$

(i) For a uniform shear stress distribution

$$\tau_{RZ} = \frac{1}{2}\frac{P_G.R_x}{d}$$
$$P_G = \frac{\bar{S}_4}{\bar{S}_2 + \frac{1}{2}\bar{S}_5 \cdot R_x}. \tag{5.121}$$

(ii) For a parabolic shear stress distribution

$$\tau_{RZ} = \frac{3P_G R_x Z_c(d - Z_c)}{d^3}$$
$$P_G = \frac{\bar{S}_4}{\bar{S}_2 + \dfrac{3Z_c(d - Z_c)}{d^2}\bar{S}_5 \cdot R_x} \tag{5.122}$$

where

$$\bar{S}_4 = (2X_2 d\cot\gamma - d^2\cot^2\gamma)\left\{\sqrt{2}\left(L_3 + \frac{2}{3}L_4\right)\left(\frac{1}{\sin\gamma\cos\gamma}\right) - \bar{P}_{Hi}\right\} \tag{5.123}$$
$$\bar{S}_5 = (2X_2 - \cot^2\gamma - d\cot^3\gamma).$$

Case (E) τ_o Value from the Hypo-elastic Analysis

From the hypo-elastic approach discussed earlier the expression for the deviatoric stresses arrived at is

$$\frac{\sigma_3}{\sigma_c} = -0.053 - 0.92\left(\frac{\tau_o}{\sigma_c}\right)^{1.35}. \tag{5.124}$$

Adopting the same convention

$$\left(\frac{\tau_o}{\sigma_c}\right)^{1.35} = 1.09\frac{\sigma_3}{\sigma_c} + 0.058$$
$$\tau_o = \sigma_c{}^{1.35}\sqrt{1.09\frac{\sigma_3}{\sigma_c} + 0.058}. \tag{5.125}$$

In the analysis σ_3 could also be σ_1 and σ_3 whichever is the higher compressive stress. Substituting Eq. (5.125)

$$\sigma_o = \bar{P_{Hi}} - \tau_{RZ}\cot\gamma - \left\{{\sigma_c}^{1.35}\sqrt{1.09\sigma_3/\sigma_c + 0.058}\right\}\cot\gamma \tag{5.126}$$

Now the value of P_G is calculated on the lines suggested above

$$P_G = \frac{2}{(X_2 - d\cot\gamma)^2}\int_{x_2 - d\cot\gamma}^{X_2}\left\{\left({\sigma_c}^{1.35}\sqrt{\frac{1.09\sigma_3}{\sigma_c} + 0.058}\right)\tan\gamma - \sigma_o\right\}R_x dR_x \tag{5.127}$$

or

$$P_G(\bar{S}_2) = \left\{(\bar{S}_6)\left(\frac{1}{\sin\gamma\cos\gamma}\right) + \tau_{RZ}\cot\gamma - \bar{P_{Hi}}\right\}(2X_2 d\cot\gamma - d^2\cot^2\gamma) \tag{5.128}$$

(i) For a uniform shear distribution

$$\begin{aligned} \tau_{RZ} &= \frac{1}{2}P_G\frac{R_x}{d} \\ P_G &= \frac{\bar{S}_7}{(\bar{S}_2 - \frac{1}{2}\bar{S}_5.R_x)} \end{aligned} \tag{5.129}$$

(ii) For a parabolic shear distribution

$$\begin{aligned} \tau_{RZ} &= \frac{3P_G R_x Z_c(d - Z_c)}{d^3} \\ P_G &= \frac{\bar{S}_7}{\bar{S}_2 = \frac{3Z_c(d-Z_c)}{d^2}.\bar{S}_5.R_x} \end{aligned} \tag{5.130}$$

where

$$\begin{aligned} \bar{S}_7 &= \left\{(\bar{S}_6)\left(\frac{1}{\sin\gamma\cos\gamma}\right) - \bar{P_{Hi}}\right\}(2X_2 d\cot\gamma - d^2\cot^2\gamma) \\ \bar{S}_6 &= {}^{1.39}\sqrt{(1.09\sigma_3/\sigma_c + 0.058)}.(\sigma_c) \end{aligned} \tag{5.131}$$

$$\begin{aligned} &\pm\sum_{i=1}^{n}(P_{Hi}.y_i + F_{vi}.R_i) + 2(R_E - n_p R_p\sin\phi) \\ &\times\left[\sum_{Z_B=1}^{n}\left\{(A'_{hp}E_s\varepsilon_Z K') + P_G H_I R_I\right.\right. \\ &\left.\left.+2A'_{sb}\sigma_{yb}(d_1 - \dot{x})A''_{sb}\sigma_{scy}(d_1 - d')\right\}\right] \end{aligned}$$

The barrel walls are constructed on top of bottom caps in lift and bays generally not exceeding 2 m. During this process it is impossible to avoid a congestion of embedded components. In these regions it is difficult to achieve a full compaction of concrete and hence the possibility of using high workability concrete mixes, secondary grouting procedures and super-plasticising admixtures cannot be ruled out. Individual bays in the walls and top caps are kept free of 4harp discontinuities. Around all embedded components including cooling pipes, aggregates with low coefficients of expansion and low long-term shrinkage are adopted. The sequence of concrete casting is chosen which is conductive to acceptable distributions of tensile strain as and when the vessel concrete cools.

5.16 Quality Assurance and Control

Quality assurance in any applied science and engineering project is an essential task. It is a disciplined approach in the provision of systems and components of nuclear power plants. The main objectives for the vessel are to ensure that each part of the vessel manufacture is assembled, inspected, tested and operated strictly in accordance with the performance requirements, reliability and safety codes and the designers' intent. The quality control is those quality assurance actions which provide a means to control measure and monitor the characteristics of the vessel throughout its life. The quality assurance in practice embodies formal documentation and procedures:

(a) Design control.
(b) Instructions, procedures and drawings.
(c) Identification and control of special monitoring processes including material equipment, inspection and test control.
(d) Operating and maintenance manual.
(e) Quality performance records of liners, penetrations, closure members, prestressing system, reinforcement and concrete under mechanical, thermal and creep loading effects.
(f) Records to monitor and control the instrumentation of the vessels. Types of instruments embedded are the strain, stress, deformation and temperature and crack gauges which determine the general behaviour of the vessel. Other special instruments load cells for monitoring prestress level; moisture gauges and gas detectors for concrete and the liner, respectively, need quality performance and control.
(g) Periodic checks for mechanical damage, variation of pressures around the cooling circuits and the safety valve accumulation pressure insulation and the difference between test and operating temperature are all part of surveillance and quality control.

5.17 Conclusions

In this chapter a step-by-step review of the literature pertaining to the analysis, design and model testing of the prestressed concrete reactor pressure vessel is given. The author has looked in some detail at methods of analysing the operational and overload behaviour of these vessels. Vessel component designs are given together with a relevant design data of some of the existing vessels. The methods of analysis discussed provide tools which could prove useful in the future development and monitoring of the existing vessels. The comparative study of analytical and model analysis validates the design and overload performance of these vessels fairly accurately. An emphasis is given on the monitoring and quality assurance of these vessels. Since any vessel has not completed its life, many more predictions in future will be made using site monitoring data. Using these analyses and in the light of new available data, the exact operational and overload performances can be safely predicted on the lines suggested earlier.

Acknowledgements *The* author acknowledges with thanks the help given by the CEGB, UKAEA and the Nuclear Consortia of UK, USA, West Germany and France.

References

1. Waters T. C. and Barrett N. T. Prestressed concrete pressure vessels for nuclear reactors. *Journal of British Nuclear Energy Society*, 1963.
2. Frankline Institute Laboratories, State of art of prestressed concrete pressure vessels for nuclear power reactors—A critical review of literature, USEAC Report ORNL-TM-812, Oak Ridge National Laboratory, 1964.
3. Rashid Y. R. Analysis of axisymmetric composite structures by the finite element. Nuclear Engineering Design, Vol. 3, 1966, pp. 163–182.
4. Rashid Y. R. Ultimate strength analysis of prestressed concrete pressure vessels. Nuclear Engineering Design Vol. 7, 1968, pp. 334–344.
5. Good Paster D. W. et al. Design and analysis of multi-cavity prestressed concrete reactor vessels, Nuclear Enq. Des. Vol. 46, 1978, pp. 101–107.
6. Orr R. S. and Holland D. A. Theoretical analysis of model spherical vessels, Conference on PCPV, London. Paper 33, 1967.
7. Bangash Y. et al. The influence of thermal creep on the operational behaviour of complex structures. International Conference Fundamental Creep & Shrinkage, Laussanne, Switzerland.
8. Morice P. B. Discussion on Group F papers on design and analysis of vessel structures, Conference on PCPV, London, 1967, pp. 397–434.
9. Zudans Z and Tans C. P. Feasibility study of prestressed concrete pressure vessel design, USAEC Report ORNLTM-813, Oak Ridge National Laboratory, 1967.
10. Papers on ultimate load tests, Conference on PCPV Model Techniques, British Nuclear Energy Society, Papers 3–14, 1968.
11. Sozen et al. Strength and behaviour of prestressed concrete vessels for nuclear reactors. Vols. 1, 2. University of Illinois USA, 1969.

12. ACI-BAM Concrete for nuclear reactors. Proceedings of mt. Seminar at Bundesantalt für Materialspriifungin, Berlin. ACI Special Publication SP-34. Vols. 1,11,111 Papers SP34-12/67168/69, 1970.
13. ACI-BAM concrete for nuclear reactors. Proceedings of mt. Seminar at Bundesanstalt für Marerialsprüfungin Berlin. ACI Special Publication SP-34, Vols. 1,11,111 Papers SP34-7/9/I0/70/39, 1970.
14. CEC and BAM Proceedings of the First International Conference on Structural Mechanics in Reactor Technology, 20–24, Berlin, Vol. 4, Reactor Pressure Vessels Part H Prestressed Concrete Pressure Vessels'. Papers H3/4–H5/8, 1971.
15. CEC and BAM Proceedings of the Second International Conference on Structural Mechanics in Reactor Technology, 844 M, 1973.
16. Marsh R. O. and Melese G. B. Prestressed concrete pressure vessels. Nucleonics Vol.23 No. 9, 1975, pp. 63–74.
17. Bender M. A status report on prestressed concrete reactor vessels, pressure vessel technology. Nuclear structural engineering. Vols. 1, 2, 1965, pp. 83–90, pp. 206–223.
18. Jaeger T. A. Note on stress analysis of prestressed concrete reactor vessels. Nuclear structural engineering I, 1965 pp. 133–136.
19. Tottenham H. and Kanchi M. B. (1966) Structural characteristics of cylindrical pressure vessels of medium thickness. Nuclear Engineering Design Vol. 4, 1966, pp. 177–192.
20. Kanchi M. B. Elastic theories of thick plates and shells, CE Dept. Report No. CEII-65, University of Southampton.
21. Reissner E. Stress-strain relations in the theory of thin elastic shells, J. Math. Physics, Vol. 31, 1952, p. 109.
22. Private communication from UKAEA, 1967.
23. Common D. K. and Hannah I. W. Specifications of concrete vessels for gas-cooled reactors. Paper 7, Conference on PCPV, London, 1967.
24. Finigen A. Ultimate analysis of Dungeness B vessels. Conference on PCPV. Paper 31 Group F, 1967.
25. Zbirohowskj-Kuscja K. and Carlton D. Analysis of vessel structures with particular reference to Wylfa, Conference on PCPV, Inst. of Civil Engrs, London. Paper 30, 1967.
26. Brown A. H. The Oldbury vessels, Conference PCPV. Paper 1 Group A, March (1967). Paper I Group A, (1967). (b) Harris A. J. and Hay J. D. Rupture design of the Oldbury vessels, Conference PCPV. Paper No. 29, Group F, 1967.
27. Gomez A. E. et al. Lumped parameter analysis of cylindrical prestressed concrete reactor vessels. University of Illinois, Chicago Vols 1, 2, USA, 1968.
28. Koerner R. J. Over pressure analysis of prestressed concrete pressure vessels, Nuclear Engineering and Design, 1970.
29. CEC and Transactions of the Third International Conference on Structural Mechanics in Reactor Technology, 1–5, London, Vol. 3, Reactor VesselsPart H. Structural Analysis of Prestressed Concrete Reactor Pressure Vessels, 1975.
30. Richart F. E. et al. A study of the failure of concrete under combined compressive Stresses. Engineering Experiment Station Bulletin No. 185, University of Illinois, Chicago, 1928.
31. Glucklich J. On the compression failure of plain concrete T and AM Report No. 215, University of Illinois, Chicago 1962.
32. Balmer G. G. Shearing strength of concrete under high triaxial stress—computation of Mohr's envelope as a curve. Struct. Research Lab. Report No. SP-23 USBR, 1949.
33. Westergaard H. M. On the resistance of ductile materials to combined stresses in two or three directions perpendicular to one another. Journal Franklin Institute 189, 1920, pp. 627–640.
34. Nadai A. Theory of flow and fracture of solids. Vol. 1, McGraw-Hill Newyork, NY, 1950.
35. Merkle J. G. An engineering approach to multiaxial plasticity. ORNL-4138, 1967.
36. Sandbye P. A plastic theory for plain concrete. Bygnings-statiske meddelelser 36, 41–62, Teknisk Forlag. Copenhagen, 1965.

37. Newman K. and Newman J. B. Failure theories and design criteria for plain concrete. Structure, solids, solid mechanics, Wiley Interscience, London, 1971 pp. 963–995.
38. Wästlund G. Nyc ron angaende betongens grundläggende hallfasthetsegenskabar. Betong Haft 3, Stockholm, 1937.
39. Bresler B. and Pister K. S. Failure of plain concrete under combined stresses. Transactions ASCE, Vol. 122, 1957.
40. McHenry D. and Karni J. Strength of concrete under combined tensile and compressive stresses. J.ACI. 1958.
41. Schleicher F. per spannungszustand an der fliessgrenze (plasticitatsbedingung). Zietschrift für Angewandte mathematik und mechanik. Band 6 Heft 3. Berlin, 1926.
42. Vile G. W. D. The strength of concrete under short term static biaxial stress. The structure of concrete and its behaviour under load—Proceedings Int. Conference, London, 1968 pp. 275–288.
43. Smith G. M. Failure of concrete under combined tensile and compressive stresses. Journal of Allergy and Clinical Immunology, Vol. 50, No. 2, 1953, pp. 137–147.
44. Bresler B. and Pister K. S. Strength of concrete under combined stresses. Journal of Allergy and Clinical Immunology Vol. 55, No. 3, 1958, pp. 321–345.
45. Bellamy C. J. Strength of concrete under combined stresses. Journal of Allergy and Clinical Immunology 367–380.
46. Akroyd T. N. W. Concrete under triaxial stress. Magazine of concrete research Vol. 13, No. 39, 1961, pp. 11–118.
47. Nishizawa N. Strength of concrete under combined tensile and compressive loads. Japan Cement Engineering Association Review, 1961 pp. 126–131.
48. Cambell Allen D. Strength of concrete under combined tensile and compressive loads. Constructional Review, Vol. 35, 1962, pp. 29–37.
49. Sigvaldason O. T. Failure characteristics of concrete, Ph.D Thesis, University of London, 1965.
50. Baker A. L. L. (a) An analysis of deformation and failure characteristics of concrete. Magazine of Concrete Research, Vol. 11 No. 33, 1959, pp. 119–128. (b) A criterion of concrete failure. Proc. Inst. Civ. Vol. 45, 1970, pp. 269–278.
51. Anson M. An investigation into a hypothetical deformation and failure mechanism for concrete. Ph.D Thesis, University of London, 1962.
52. Hopkins and Pruger. The load carrying capacities of circular plates, Journal of Mechanics and Physics Society 2 (1) 1954.
53. Bangash Y. Prestressed concrete reactor vessel Timesaving ultimate load analysis. J. Inst Nucl. Engg: Vol. 13, No. 4, 1972.
54. Bangash Y. A basis for the design of bonded reinforcement in the prestressed concrete reactor vessels Paper 7478S. Supplement (viii). Inst. of Cu. Enqrs 1972.
55. Smee D. J. The Effect of aggregate size and concrete strength on the failure of concrete under multi-axial compression, Civil Engineering Transaction, Inst of Enqineer's Australia. Vol. CE9, No. 2, 1967 339–344.
56. Private communication from The Nuclear Power Group Company Risely, Lancashire, 1968, U.K.
57. Corum J. M. and Smith J. E. Use of small models in design and analysis of prestressed concrete reactor vessels ORNL-4346, IJ.S.A, 197
58. Smith J. R. Problems in assessing the correlation between the obseived and predicted behaviour of models Pipcr 10. Model Techniques for Prestrcsse(l Concrete Pressure Vessels, liNES, London, 1969.
59. Treharne and I)avics. Tests on samples and calculation of standard deviations for bonded reinforcement Private Reprt, 1969.
60. Paul et al. Mortar models of prestressed concrete reactor vessels, J. St. Dir. ASC'E, February 1969.

61. Brown G. Advances in the gas-cooled reactor. l.ecture joint session of IEE. 1. MECH. E and liNES, University of Strathclyde, Glasgow I 6 Available from British Nuclear Energy Society, 1968.
62. Buchert K. P. et al. (1978) Analysis of reinforced concrete containment vessels considering concrete cracking Nuclear Eng. Des. 46. 101–107.
63. Argyris J. H. et al. Finite element analysis of inelastic structural behaviour. Nuclear Eng. Des Vols. 46, 23, 1978, pp. 5–262.
64. 6th Conference on SMIRT, Paris, Vol J (a), J (b), 1981.
65. 6th Conference on SMIRT, Paris, Vol. M, 1981.
66. 5th Conference on SMIRT, Berlin, Papers HI to H30, 1979.
67. Bangash Y. The automated three-dimensional cracking analysis of prestressed concrete vessels, 6th Conference on SMIRT, Paris, Paper H3/2, 1981.
68. Kotulla B. et al. Three-dimensional analysis of stresses and crack development in prestressed thick-walled cylinders, 6th Conference on SMIRT, Paris, Paper H3/1, 1981.
69. Cheung K. C. An assessment of long-term structural behaviour of an asymmetric multicavity PCRV, 6th Conference on SMIRT, Paris, August 1981, Paper H3/9, 1981.

Chapter 6
A Complete Manual Design Analysis of Concrete Containment Vessel (Building) Using American Practices and Codes

6.1 Introduction

A comprehensive analysis is related to these structures has been given in this text: Design loads, material properties and other design specifications and parameters have been thoroughly explained. In this chapter, preliminary design calculation for the containment based on limits state concept are given using U.S Regulations and codes. One typical case of BELLEFONTENUCLEAR PLANT of the TVA is considered. Design calculations for other kinds of PWR shall be carried out on the same line using new design parameters and guidelines. For detailed analysis a reference is made to the following comprehensive paper by author.

'Containment Vessel Design And Practice'
-Annals of Nuclear energy. Pergamon Press Ltd.,Vol1.No. 2, pp. 107–181(1982)

The geometry and other parameters are based on the information given by the TVA(USA). They have been taken into consideration. The drawings given by TVA are gracefully recorded. The author took a great pain to interpret the data given in Imperial units and then convert them using SI units. The calculations carried out seems to be a final check o the TVAdesign and in conformity to the safely analysis demanded by the NUCLEAR Regulatory Commission (NCR) of the USA and the additional work based on local requirements of the TVA. Based on the requisite guidelines (N4-50-D702) the design of the annulus and auxiliary structures is included. Both the barrel walls and the dome have been designed as prestressed concrete structures integrated fully. Hence a complete manual design makes this entire chapter extremely useful prior to.

6.2 Geometry And Design Parameters Adopted by the TVA

Figures 6.1 and 6.2 show the vertical presentation and plan of the containment which are self explanatory as far as the geometry is concerned.

Secondary Containment. This category I reinforced concrete structures consists of a cylinder wall having a 162 ft, (49.37 m), inside diameter which is

M.Y.H. Bangash, *Structures for Nuclear Facilities*,
DOI 10.1007/978-3-642-12560-7_6, © M.Y.H. Bangash 2011

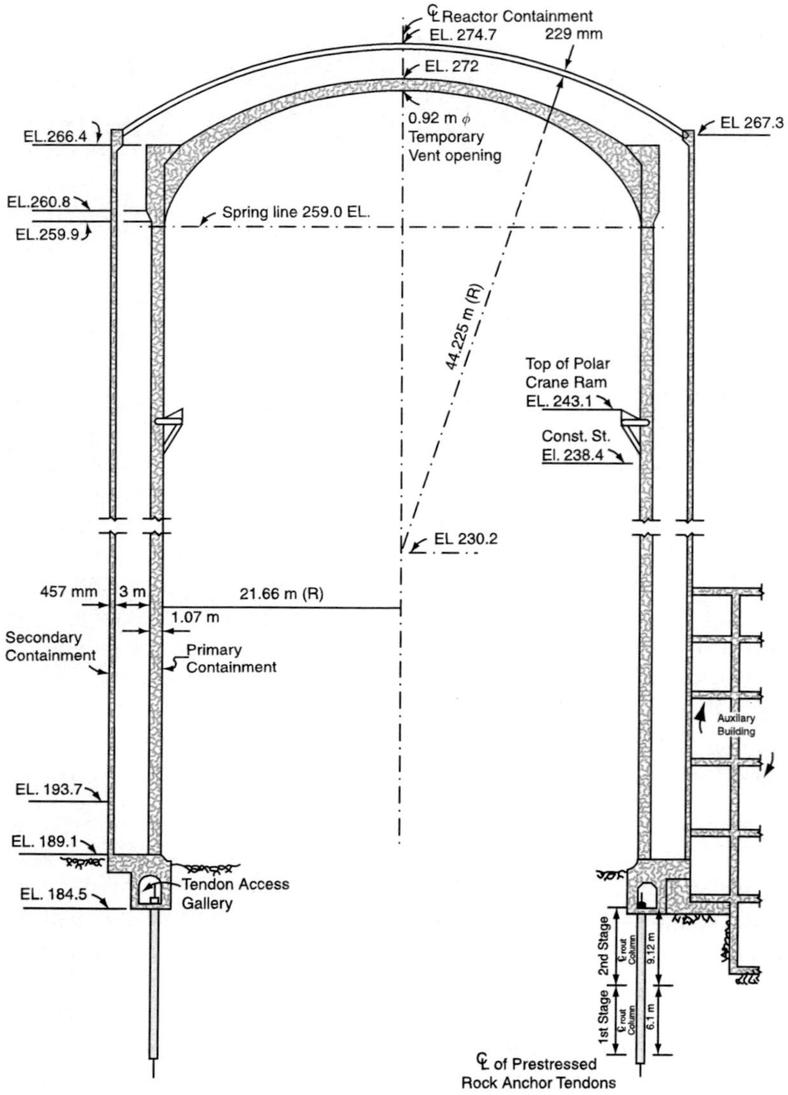

Fig. 6.1 Vertical presentation (with compliments from TVA)

capped with a ring girder and a 18-in., (457 mm), thick circular dome. The base of the cylinder wall is 2′6″, (0.725 m), thick at EL. 620.0 where it is anchored into an extension of the tendon gallery top slab. The wall decreases in thickness linearly from EL. 620.0 to EL. 635.0 where the wall is 18 in., (457 mm), thick. The 18 in., (457 mm), thick cylinder extends above EL. 635.0 approximately 237 ft, 72.214 m, to the ring girder. The top inside face of the dome is approximately 280 ft, (85.34 m), above the EL. 62.0 base slab.

Note 'C'. The typical dome tendon spacing is 978 mm measure horiz: the final prestressing force for each tendon group is 5453.25 KN/m.

Purpose of this structure is to minimize off-site radiation dosage which might occur during certain atmospheric conditions. The structure also serves to protect the mechanical and electrical systems, which pass through the annulus enroute to the Auxiliary Building, from catastrophic effects of earthquake, tornado, and flood.

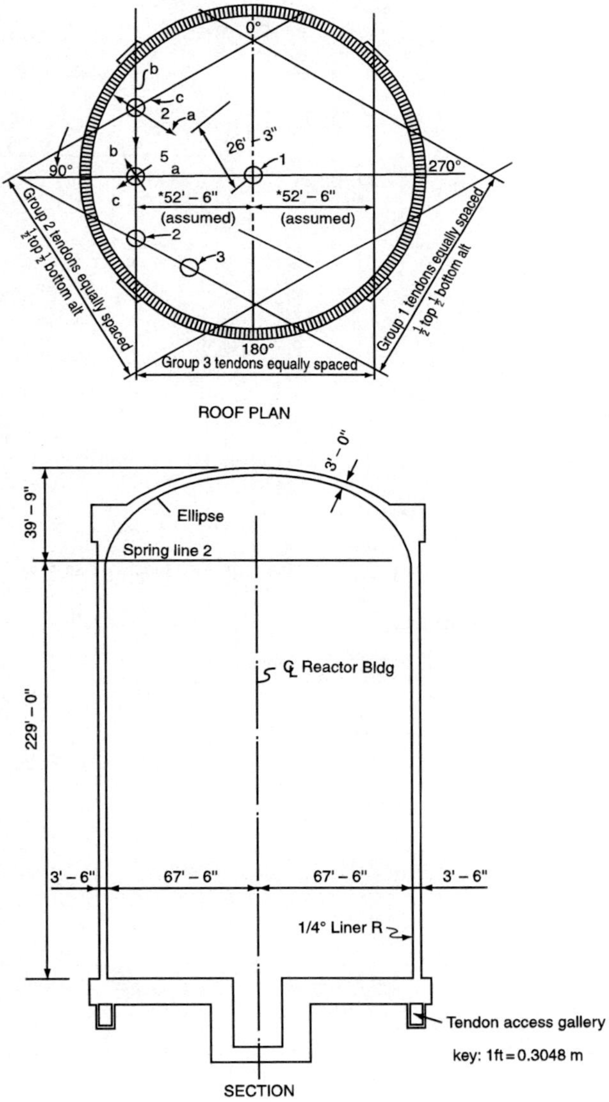

Fig. 6.2 1 Plan of the dome prestressing layout and elevations

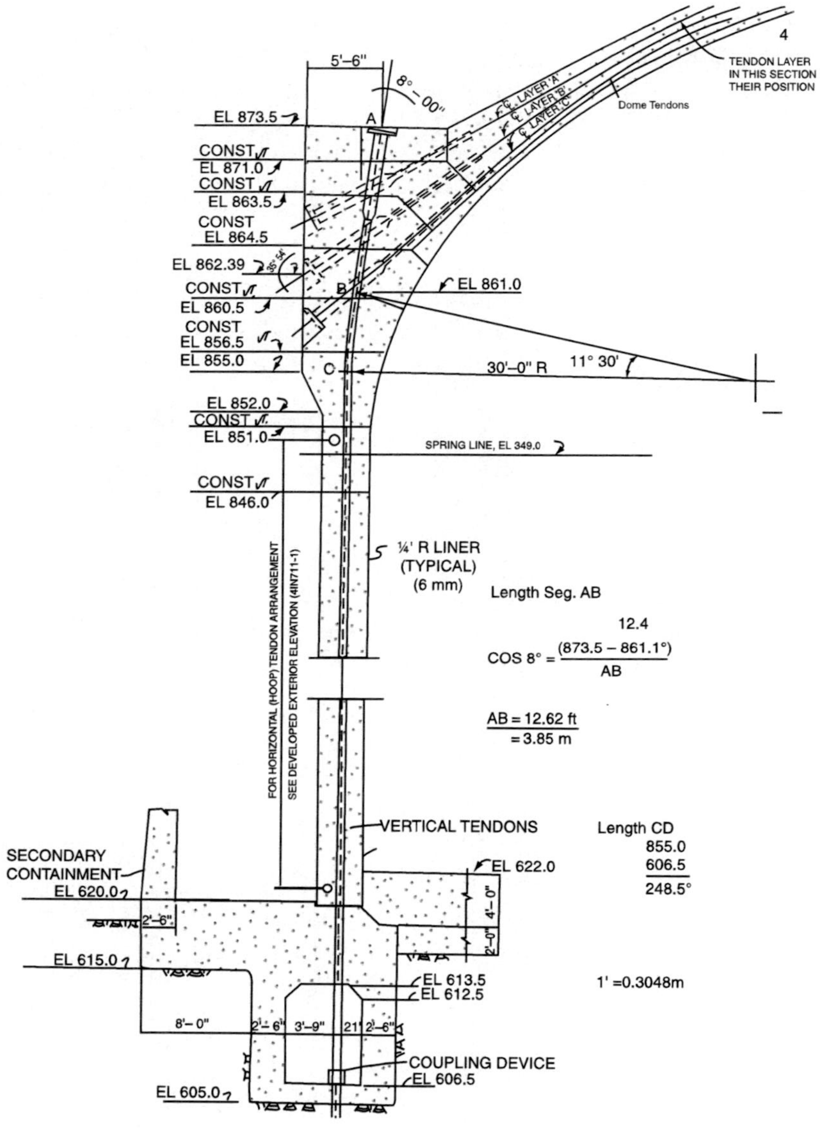

Fig. 6.2 (continued)

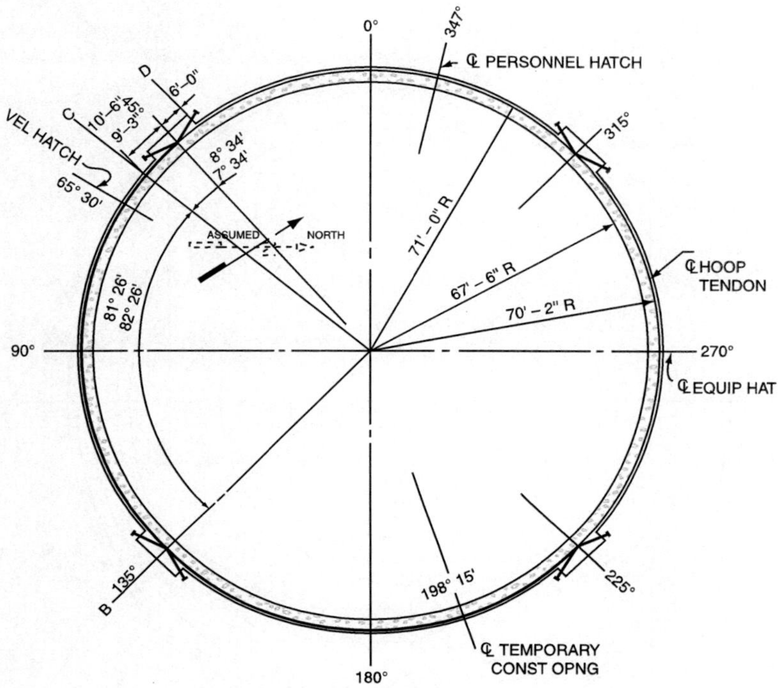

PLAN - REACTOR BLDG
SHOWING HOOP TENDON ARRANGEMENT
UNIT 1 AS SHOWN
UNIT 2 OPP HAND

Fig. 6.2 (continued)

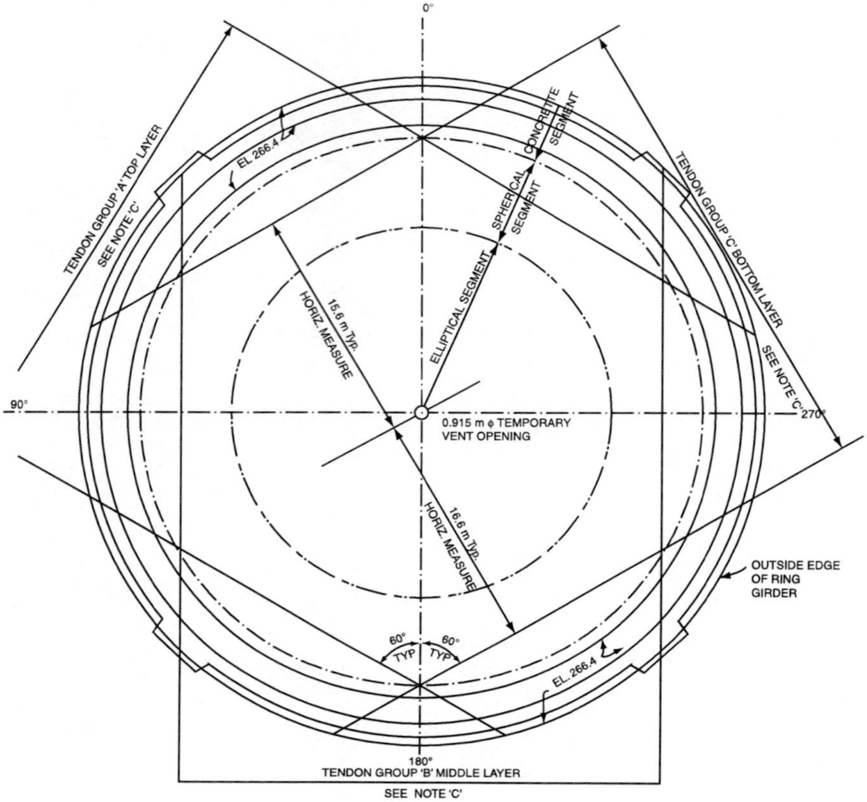

Fig. 6.2 (continued)

The TVA structure has been sealed with PVC seals to prevent ingress of water at construction joints below El 625.0. PVC seals shall also be provided across the expansion joints at the bottom slabs and the exterior walls of the Auxiliary Building. The seals into exterior walls shall extend to EI 646.5.

All doors and openings connecting the secondary containment and the Auxiliary Building have been sealed to maintain a vacuum pressure in the annulus between the primary and secondary containment.

Where this structure supports slabs in the annulus, care has be taken to insure minimum resistance from these slabs by the use of expansion joints between the cylinder wall and the slabs. In addition, low friction bearing pads have been used under the slab support steel at the steel brackets on the primary containment wall to insure that there is essentially no interaction between the primary and secondary containment.

Where the walls and slabs from the Auxiliary Building intersect the secondary containment, fiberous glass joint filler material has been used to separate the structures to allow for seismic movement. The material is intended to be held in place by the concrete pouring pressure. This material will provide adequate sealing in areas that are subjected to low pressure purging.

6.2.1 Annular Structures

6.2.1.1 El 666.0 Shield Slab

This 2-ft (0.61 m) thick slab extends from the refuelling transfer tube compartment slab at approximately the 90^0 azimuth around the annulus past 0° stopping at the 300° azimuth. Its basic function is to provide radiation shielding for personnel working in the annulus above El 646.0. The slab support steel has been connected to steel brackets on the secondary and primary containment walls. The slab support steel will be free to slide on low friction bearing pads at the steel support bracket on the primary containment wall. Openings have been made in this floor to allow purging underneath. The openings have been located so that someone walking on the 646.0 floor level will not be exposed, to a radiation stream.

6.2.1.2 Barrier Wall

This wail is located at the 0° azimuth and extends from El 620.0 to El 646.0 levels.

The barrier is designed to protect redundant electrical systems that are required for safe shutdown from the effects of fire, pipe jets, and missiles. The barrier shall have a minimum thickness of 2 ft. (0.61 m).

The barrier has been separated from the primary and secondary containment structures by expansion joints to allow free movement of each structure during earthquake. The barrier has been connected to the El 64.6.0 slab.

6.2.1.3 Primary Shield Door

This 3-ft thick concrete door is approximately 25′-5″(7.8 m) wide and 24′-6″ (7.47 m) high. It is located at the equipment hatch in the primary containment at the 270° azimuth. The basic function of the door is to provide radiation and tornado borne missile shielding for the equipment hatch opening.

6.2.1.4 Primary Shield Door Support

This structure is located under the equipment access hatch at the 270° azimuth. It consists of a wall, two columns, two beams, and a thin slab. The wall provides door support during normal plant operation. The 300-kip (1334 KN) door is mounted on a rail and opens toward a decreasing azimuth providing access to the primary containment during maintenance operations. A long beam supported by one column provides door support in the open position. The remaining beam and column allow the door to be moved slightly toward an increasing azimuth to allow for wheel maintenance. The thin slab provides a working surface for door operation and maintenance.

6.2.1.5 Compartment Around the Fuel Transfer Tube

This structure serves primarily as shielding against radioactivity around the refuelling transfer tube. It consists of two side walls and a top slab. The walls and slab are separated from the secondary and primary containments by an expansion joint to allow independent movement of these structures during earthquake. The top slab is concrete plug which can be removed to inspect the transfer tube.

6.2.2 Tendon Gallery and Annulus Maintenance Access

This structure is located at the 214° azimuth, it consists of a vertical shaft extending from grade at El 646.0 down to the horizontal passageways which enter the annulus at El 620.0 and the tendon gallery at El 606.5, respectively.

Steel ladders with safety devices shall extend between horizontal levels and the shafts can serve as a work access to the tendon gallery during tendon installation. After tendon installation a removable structural steel floor at El 620.0 has been installed to provide access to the annulus. The annulus access can be used to bring in the tendon jacks during surveillance testing. The surveillance tests will be performed at one year intervals for the first three years of plant life and at five year intervals thereafter. The annulus access can also serve as the maintenance access for servicing various mechanical systems located in the annulus.

Due to the infrequent use of the access structure, the top cover slab at El 646.25 has been bolted into place and will have lifting eyes such that it can be removed with a small mobile hoist.

The removable floor at El 620.0 shall have a man-hatch to provide access to the tendon gallery for inspection purposes. If it becomes necessary to take large tools or parts into the tendon gallery, the steel floor at El 620.0 may be lifted out to grade.

This structure has been connected to the tendon gallery by construction joints below El 620.0. Above El 620.0, an expansion joint shall be used, to allow independent, movement of this structure and the secondary containment during earthquake.

6.2.3 Applicable Design References

1. Bellefonte PSAR
2. Concrete Design Standards, Parts II and III ACI 1972. (Revised 2006)

3. General Criteria for Design of Civil Structures (N4-50-D702)
4. Civil Design Branch report, "Dynamic Earthquake Analysis of Secondary Containment Structure and Response Spectra for the Attached Equipment - Bellefonte Nuclear Plant (1972) (Revised 1985)
5. OEDC Quality Assurance Manual (1972) (Revised 2007)
6. American Concrete Institute (ACI) (Revised 2008)
   ACl 315-65- Manual of Standard Practice of Detailing Reinforced Concrete Structures
   ACI 318-71- Building Code Requirements for Reinforced Concrete
   ACI 347-68- Recommended Practice for Concrete Formwork
   ACI 305-72- Recommended Practice for Hot Weather Concreting
   ACI 211.1-70 - Recommended Practice for Selecting Proportions for Concrete
   ACI 6114-73- Recommended Practice for Measuring, Mixing, and Placing Concrete
7. American Society for Testing and Materials-1972 ASTM Standards (Revised 2008)
8. American Society of Civil Engineers (ASCE), Paper No. 3269, "Wind Forces on Structures" (1969)
9. Code of Federal Regulations, Title 29, Chapter XVII - "Occupational Safety and Health Standards"
10. NRC Regulatory Guides
    No. 1.10 - "Mechanical (Cadweld) Splices in Reinforcing Bars of Category I Concrete Structures"
    No. 1.55 - Concrete Placement in Category I Structures
11. TVA Construction Specifications
    G-2 - TVA General Construction Specification for Plain and Reinforced Concrete
    G-30 - TVA General Construction Specification - Fly Ash for Use as an Admixture in Concrete (1972) (Revised 2007)
    G-32 - TVA General Construction Specification - Bolt Anchors Set in Hardened Concrete (1972) (Revised 2007)
12. "TVA Report on Wind Analysis of Sequoyah Shield Building," by R.G. Domer and B. C. Giordano(1972) (Revised 2007
13. "THERMCYL - Thermal Stresses in Reinforced Concrete Cylinders," by R. G. Domer and R. C. Giordano available at TVA (1973)
14. "Design of Structures for Missile Impact," Bechtel Power Corporation Report BC-Top-9 (1972)
15. Bangash M.Y.H. Shock, Impact & Explosion
    -Springer Verlag, Heidelberg 2008.
16. Bangash M.Y.H. Manual of Numerical Methods in Concrete
    -Thomas Telford, London (2003)
17. Bangash M.Y.H. Earthquake Resistant Buildings-Analysis and Design Springer Verlag, Heidelberg 2010

6.2.4 *General Design Requirements and Loads Used by TVA. For Calculations*

6.2.4.1 Load Symbols and Definitions

- D = dead load of structure and equipment plus any other permanent loads excluding soil and hydrostatic pressure. An allowance is also made for future permanent loads.
- E^* = 1/2 safe shutdown earthquake resulting from horizontal rock acceleration of 0.09 g (0.09 g vertical).
- $*E$ = safe shutdown earthquake resulting from horizontal rock acceleration of 0.18 g (0.18 g vertical).
- L = live loads (loads which vary in intensity and occurrence) including movable equipment, soil and hydrostatic pressure loads.
- L_c = construction live loads (greater than normal operating live loads of a temporary or unusual nature) which occur during construction.
- R_a = pipe reactions under thermal conditions generated by the postulated pipe break and. including R_0.
- R_0 = pipe reactions during normal operating or shutdown conditions, based on the most critical transient or steady state conditions.
- T_a = thermal loads due to temperature differences under accident conditions,
- T_0 = thermal loads due to temperature differences under normal operating or shutdown conditions, based on the most critical transient or steady state condition.
- U = required ultimate load capacity of structures.
- W = wind load as defined
- W_t = tornado loadings including missiles as defined in the General Design Criteria.
- Y_j = jet impingement equivalent static load on a structure generated by the postulated break including an appropriate dynamic factor,
- Y_m = missile impact equivalent static load on a structure generated by or during the postulated break, pipe whipping, including an appropriate dynamic factor.
- Y_r = equivalent static load on the structure generated by the reaction of the broken high-energy pipe during the postulated break including an appropriate dynamic factor.

In determining an appropriate equivalent static load for Y_j, Y_m and Y_r, elasto-plastic behaviour may be assumed with appropriate ductility ratios and as long as excessive deflections will not result in loss of function.

6.2.5 *Earthquake Loads*

Earthquake effects on these concrete structures shall be determined by using the appropriate structural response prepared by the Civil Engineering Branch of the TVA or any other organisations.

6.2.6 Temperature Loads (Concrete Members)

The surface temperature and the gradient through the concrete members for normal operation and accident conditions shall be calculated by the Nuclear Systems Engineering Group of the Mechanical Engineering Branch. These gradients will be used with the THERMCYL computer program or a similar method to compute the thermal stresses. Currently many advanced computer programs are being used. Program ISOPAR or any other 3D-hybrid F.E. Non-linear Computer Program.

6.2.7 Jet Loads and Pipe Restraint Loads

Restraints for steam and other systems piping will be located by the Bellefonte Project Mechanical Design Sections and the Civil Engineering Branch will evaluate the design criteria for determining loads.

Jet forces on the structure from accidental pipe breaks shall be calculated by the following formulas:

- F = 1.26 K PA for steam or flashing liquid
- F = 2.0 K PA for subcooled nonflashing liquid
- F = total force for flexural design or shear design
- P = maximum operating pressure
- A = internal area of pipe
- K = dynamic load factor for flexure, peripheral shear, and diagonal tension.

Pressures for piping systems shall be obtained from the Project Mechanical Design Sections.

Three types of pipe breaks shall be postulated and are defined below:

- Circumferential rupture. This type of pipe break exists when a pipe breaks circumferentially and the ends at the break displace laterally such that a full diameter jet from each broken end is possible.
- Longitudinal split. The area of this type of break is equal to the internal cross-sectional area of the pipe and is two inside diameters long. The shape may be assumed to be a narrow rectangle, circular, or elliptical for analyses purposes.
- Through-wall leakage cracks are not postulated in high energy piping systems, except where these lines are routed in the vicinity of vital electrical equipment. Fluid flow from a through-wall leakage crack may be based on a circular opening with an area equal to an equivalent rectangular opening of one-half the pipe inside diameter in length and one-half the wall thickness in width and oriented parallel to the pipe flow axis.
- A jet shall be assumed to expand 10 degrees each direction from the sides of the break for purposes of determining an impact area. The load F shall be assumed to act uniformly over the impact area.

When designing for jet forces, local yielding is permitted which will reduce the dynamic effects of the force on the structures. The dynamic load factor, K, shall be determined by using a ductility ratio (Y_m/Y_{el}) as described in Structural Dynamics by J. M. Biggs, 1961, pp. 69–81 and 222–223 or elsewhere.

- Y_m = maximum permissible deflection of the structure
- Y_{el} = deflection of the structure at yield strain of the reinforcing steel

Y_{rn} and Y_{el} shall be determined by moment rotation concepts as outlined in "Limit Design for Structural Concrete," Journal of the PCA Research and Development Laboratories, May 1959, p. 14–23.

6.2.8 Load Combinations

The load combinations used by the TVA are identical to ones recommended in Chapter 2. Table 6.1 gives a brief on load combinations and load factors for secondary containment, the following special live loads have been adopted with appropriate load factors as given in table 6.1.

6.2.9 Secondary Containment

6.2.9.1 Special Live Loads

The following loads shall be treated as live loads using the appropriate load factors as given in Table 6.1. Snow. The structure shall be designed for a 20 psf (0.958KN/m^2) snow load on the dome.

Table 6.1 Load Combinations and Load Factors

1. Normal operating	$U = 1.14D + 1.7L + 1.3T_0 + 1.3R_0$
2. Seismic operating	$U = 1.14D + 1.7L + 1.9E + 1.3T_0 + 1.3R_0$
2a. Dead and seismic	$U = 1.2D + 1.9E$
3. Normal wind	$U = 1.4D + 1.7L + 1.7W + 1.3T_0 + 1.3R_0$
3a. Dead and wind	$U = 1.2D + 1.7W$
11. Extreme seismic	$U = D + L + T_0 + R_0 + E'$
5. Extreme wind	$U = D + L + T_0 + R_0 + W_t$
6. Extreme accident	$U = D + L + T_a + R_a$
7. Seismic/accident	$U = D + L + T_a + R_a + 1.25E + Y_r + Y_j + Y_m$
8. Extreme seismic/accident	$U = D + L + T_a + R_a + E' + Y_r + Y_j + Y_m$
9. Temporal7 construction	$U = 1.4D + 1.4L_c$
10. Probable maximum flood	$U = D + L$

Notes:

1. All components that constitute these structures shall be designed to resist all applicable loading combinations.

   With the exception of the load factors given in Table 6.1, this structure shall be designed in accordance with the ultimate strength design provision of ACI 318-71.
2. In combinations 6, 7, and 8, the maximum values of T_a, R_a, Y_j, Y_r, and Y_m' including an appropriate dynamic factor shall be used unless a time history analysis is performed to justify otherwise.
3. For combinations 7 and 8, local stresses due to the concentrated loads Y_r, Y_j, and Y_m, may exceed the allowable provided there will be no loss of function of any safety-related system.
4. Where L reduces the effects of other loads it shall be taken as the minimum possible value.
5. In addition to the above, combination 1 will also be designed for temperature and shrinkage, in place of 1.3 T_0, using TVA "Concrete Standards for Temperature and Shrinkage Reinforcement" as the criteria.

Water. Water pressure shall be assumed outside the Auxiliary Building to El 625.0 for design purposes. The dynamic effect on the water due to earthquake shall be considered in accordance with N4-50-D702.

Earth. Design for soil pressure shall be based on soil backfiled to El 646.0 in all areas that are not adjacent to the Auxiliary Building. Surcharge on the soil surface shall be 200 psf (9.58 KN/m^2) except where larger surface loads are known to exist, The dynamic effect on the earth due to earthquake shall be considered.

Vacuum. The structure shall he designed to resist a theoretical vacuum of 5.2 psf (0.2491 KN/m^2) during normal operation with a design value of 10.4 psf (0.477 KN/m^2) to allow for marginal error.

6.2.10 Annular Structures

6.2.10.1 El 646.0 Shield Slab

Since no equipment is supported by this floor it shall be designed for a 100 psf (4.79 KN/m^2) live load.

Analysis may be done by considering the floor as a simple span in the circumferential direction, Temperature and shrinkage steel shall be placed in the radial direction.

6.2.10.2 Barrier Wall

Analysis of this wall may be done by conventional methods.

6.2.10.3 Primary Shield Door

This door shall be designed to withstand the stresses produced by the credible tornado-borne missiles listed in the General Criteria for Design of Civil Structures, N14-50-D702. For further design considerations by others see Shield Door at Equipment Access Hatch Criteria, N4-MAD740C.

6.2.10.4 Primary Shield Door Support

This structure shall be designed to carry the moving door load, plus a 100 psf (4.79 KN/m^2) live load on the slab outside the door area. The wall shall be designed to take the vertical load that could possibly occur during the removal of a steam generator. A horizontal load of 5% of the weight shall be assumed to occur simultaneously with the vertical load on the wall.

This structure shall be analyzed by a conventional frame analysis taking account of moment distribution, flexibility and stiffness methods.

6.2.10.5 Compartment Around the Fuel Transfer Tube

The top slab shall be designed for a 100 psf (4.79 KN/m^2) live load.

6.2.11 Geometrical Analysis

6.2.11.1 Elliptical Dome of The Reactor Containment

General

Developmental Analysis and computer Program exist in this text in Chapter 3. In addition the formulation under this section has been provided by the TVA.

Geometric Analysis of Elliptical Dome

By concentric circles method, points on the elliptical curve of minor axis "*b*" and major axis "*a*" can be located graphically as followed:

Standard elliptic EQ

$$\frac{x^2}{a^2}+\frac{y^2}{b^2} \tag{6.1}$$

From figure

$$x = a\sin\phi \tag{6.1a}$$

$$y = b\sin\phi \tag{6.1b}$$

For $x - 14$ and $y - 5$

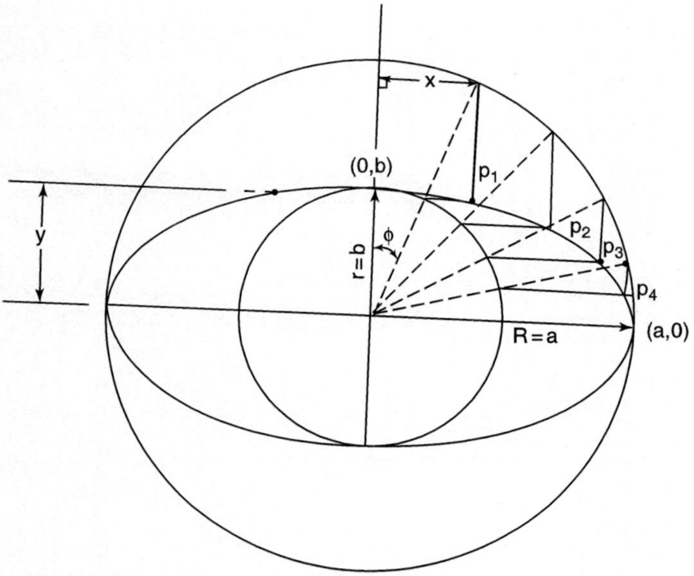

Fig. 6.3 Concentric circles for elliptical curve development

Elliptical dome

$$a = 67.5'(20.574\ m)$$
$$b = 39.75'(12.116\ m)$$

Method of Obtaining Arc Length of A Elliptical Curve

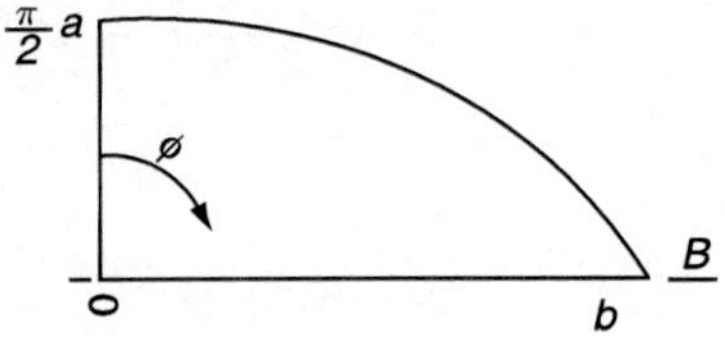

Fig. 6.4 Method of obtaining arc length of a elliptical curve

For length of a plane curve:

$$L_A^B = \int_A^B \sqrt{1 + \left(\frac{dy}{dx}\right)^2}\, dx \tag{6.2}$$

Equation of an ellipse can be written as

$$y^2 = b^2 - \frac{b^2 x^2}{a^2} \tag{6.2a}$$

Taking derivative

$$2ydy = -\tfrac{2b^2}{a^2} \times \mathrm{d}x$$
$$\frac{\mathrm{d}y}{\mathrm{d}x} = -\frac{b^2x}{ya^2}$$
$$\left(\frac{\mathrm{d}y}{\mathrm{d}x}\right)^2 = \left(\frac{b^2x}{ya^2}\right)^2 = \frac{b^4x^2}{y^2a^4} = \frac{b^4x^2}{\left(b^2 - b^2\frac{x^2}{a^2}\right)a^4} = \frac{b^2x^2}{a^4 - a^2x^2} \tag{6.2b}$$

Substituting

$$\begin{aligned} L_A^B &= \int_A^B \sqrt{1 + \frac{b^2x^2}{a^4 - a^2x^2}}dx \\ &= \int_A^B \sqrt{1 + \frac{a^4 - a^2x^2 + b^2x^2}{a^4 - a^2x^2}}dx \end{aligned} \tag{6.2c}$$

Letting $x = a\sin\phi$

$$\mathrm{d}x = a\cos\phi d\phi$$

$$L_A^B = \int_A^B \sqrt{\frac{a^4 - a^2(a\sin\phi)^2 + b^2(a\sin\phi)^2}{a^4 - a^2(a\sin\phi)^2}}(a\cos\phi \mathrm{d}\phi) \tag{6.2d}$$

$$L_A^B = \int_A^B \sqrt{\frac{a^4 + (b^2a^2 - a^4)\sin^2\emptyset}{a^4(1 - \sin^2\emptyset)}}a\cos\phi \mathrm{d}\phi \tag{6.2e}$$

Since

$$a = 67.5'(20.574\ m)$$
$$b = 39.75'(12.118\ \mathrm{m})$$
$$b = .58889\mathrm{a}$$
$$b^2 = .34679a^2$$

$$\begin{aligned} L_A^B &= \int_A^B \sqrt{\frac{a^4 + (.34679a^4 - a^4)\sin^2\phi}{a^4(\cos^2\phi)}}a\cos\phi \mathrm{d}\phi \\ &= \int_A^B \sqrt{\frac{a^4(1 - .65321\sin^2\phi}{a^4\cos^2\phi}}a\cos\phi \mathrm{d}\phi \\ & a\int_A^B \sqrt{1 - .65321\sin^2\phi \mathrm{d}\phi} \end{aligned} \tag{6.2f}$$

$$L_A^B = aE_1$$

where

$$E_1 = \int_0^{\pi/2} \sqrt{1 - 0.65321 \sin^2 \phi;\ d\phi} \tag{6.2g}$$

and

$$\frac{\sqrt{a^2 - b^2}}{a} = k$$
$$k = 0.65321$$
$$k = \sqrt{0.65321} = 0.80821$$
$$\sin\theta = k$$
$$\theta = \sin^{-1} k = \sin^{-1}(0.80821) = 53°55'$$
$$\text{use } \theta = 54°$$

Using CRC table (table269 page) $\theta = 53°55'\theta = 90°$
Handbook $E_1 = 1.2681$
One quarter length of an ellipse

$$L_0^{\pi/2} = aE_1 = 67.5 \times 1.2681$$
$$L = 85.59675\text{ft} \approx 85.6\,\text{ft}(26.06\text{ m})$$

By approximation

$$L = \frac{2\pi}{4}\sqrt{\frac{a^2 + b^2}{2}} = 87\text{ ft }(26.5176\text{ m})$$

Table 6.2 is prepared for the elliptical dome (CRC Math tables @ $E_1\ \theta = 54°$ and using interpolation.

Dome with Portions of Elliptical, Spherical and Others

General. Here analysis is presented where various segments of ellipse.

Geometrical Analysis.

Let

$$\begin{aligned} l_e &= \text{length of elliptical portion} \\ l_s &= \text{lenght of spherical portion} \\ l_r &= \text{length of remaining portion} \\ L_{\text{Total}} &= l_e + l_s + l_r \end{aligned} \tag{6.3}$$

Table 6.2 ϕ versus θ values of E_I

ϕ	$\theta = 54°$ E_I	ϕ	$\theta = 54°$ E_I
1	.0175	46	.7502
2	.0349	47	.7643
3	.0523	48	.7782
4	.0698	49	.7922
5	.0872	50	.8060
6	.1046	51	.8197
7	.1220	52	.8332
8	.1393	53	.8467
9	.1566	54	.8600
10	.1739	55	.8730
11	.1912	56	.8860
12	.2084	57	.8988
13	.2256	58	.9116
14	.2427	59	.9243
15	.2598	60	.9367
16	.2769	61	.9492
17	.2939	62	.9614
18	.3108	63	.9736
19	.3277	64	.9856
20	.3445	65	.9973
21	.3612	66	1.0093
22	.3779	67	1.0210
23	.3945	68	1.0326
24	.4110	69	1.0441
25	.4274	70	1.0555
26	.4439	71	1.0668
27	.4601	72	1.0780
28	.4763	73	1.0891
29	.4924	74	1.1001
30	.5084	75	1.1110
31	.5243	76	1.1221
32	.5401	77	1.1326
33	.5558	78	1.1433
34	.5714	79	1.1540
35	.5870	80	1.1646
36	.6024	81	1.1750
37	.6177	82	1.1856
38	.6329	83	1.1960
39	.6479	84	1.2064
40	.6629	85	1.2167
41	.6777	86	1.2254
42	.6925	87	1.2373
43	.7071	88	1.2475
44	.7216	89	1.2578
45	.7360	90	1.2681

With reference to calculation "length of Elliptical Dome", by DQL on 3-10-72.

$$l_{eA}^{B} = aE_1 = aE_{1B} - aE_{1A} = a(E_{1B} - E_{1A}) \tag{6.3a}$$

$$E = \int_0^0 \sqrt{1 - k^2 \sin^2 \phi} d\phi$$

$$k = \frac{\sqrt{a^2 - b^2}}{a} = \frac{\sqrt{69.125^2 - 41.375^2}}{69.125} = \frac{\sqrt{3066.375}}{69.125} \tag{6.3b}$$

$$k = \frac{55.37485}{69.125} = 0.8011 \tag{6.3c}$$

$$\theta = \sin^{-1} k = 53°14$$

$$\text{Limit } A = 90° - \tan^{-1}\left(\frac{40'}{41.375' - 7.64'}\right)$$

$$\phi = \mathrm{A} = 90° - 49.857 = 40.143°$$

$$\text{Limit } B = \phi = 90°$$

$$\begin{aligned} &\text{At } \phi = 40.143°, E_{1A} \approx .666 \\ &\text{At } \phi = 90°, E_{1B} = 1.275 \end{aligned} \tag{6.3d}$$

$$l_{eA}^{B} = (69.125)(1.275 - 0.666) = (69.125)(.609) = 42.097'(12.831\ m) \tag{6.3e}$$

$$\begin{aligned} l_s &= \frac{\pi R}{180°}(\theta_3) = \frac{(3.1416)(102.33)}{180°}(12.8815°) = 23.02'(7.165\ m) \\ l_s &= \frac{R' - 60'}{\cos \theta_4} \end{aligned} \tag{6.3f}$$

Where

$$\mathrm{R}' = \text{distance from £ of dome to external edge of ring girder} \tag{6.3g}$$

$$\begin{aligned} &= \frac{72.5' - 60'}{\cos(35.8975)} = \frac{12.50'}{0.8101} = 15.43'(4.7031\ m) \\ L_T &= 42.100' + 23.02' + 15.43' = 80.55'(24.552\ m) \end{aligned} \tag{6.3h}$$

Compute Elev. of Anchorage Point on Ring Girder:

$$\begin{aligned} h &= \sqrt{(15.43)^2 - (12.50)^2} = 9.05'(2.7858\ m) \\ E_1 &= 871.44 - 9.05 = 862.39 = Anch.elev. \end{aligned} \tag{6.3i}$$

6.2.11.2 Equations of Curvature for Elliptical Dome at specific Location

Figure 6.5 gives the curvature phenomenon of the elliptical dome at specific location.

Ellipsoid at right has the equation

$$\frac{x^2}{a^2} + \frac{y^2}{b^2} + \frac{z^2}{c^2} = 1$$

where

$$|x| \leq a, |y| \leq b, |z| \leq c$$

If a plane x-z cut through the ellipsoid @ $y = 0$, then the equation reduced to:

$$\frac{x^2}{a^2} + \frac{z^2}{c^2} = 1 \tag{6.4}$$

which is the equation of ellipse.

Further, if the plane $x - z$ cuts through the ellipsoid @ $y = y$, where $y_1 < b$, then the equation of an ellipsoid becomes:

$$\frac{x^2}{a^2} + \frac{y_1^2}{b^2} + \frac{z^2}{c^2} = 1$$
$$\therefore \frac{y_1^2}{b^2} \tag{6.4.1}$$
$$= constant$$

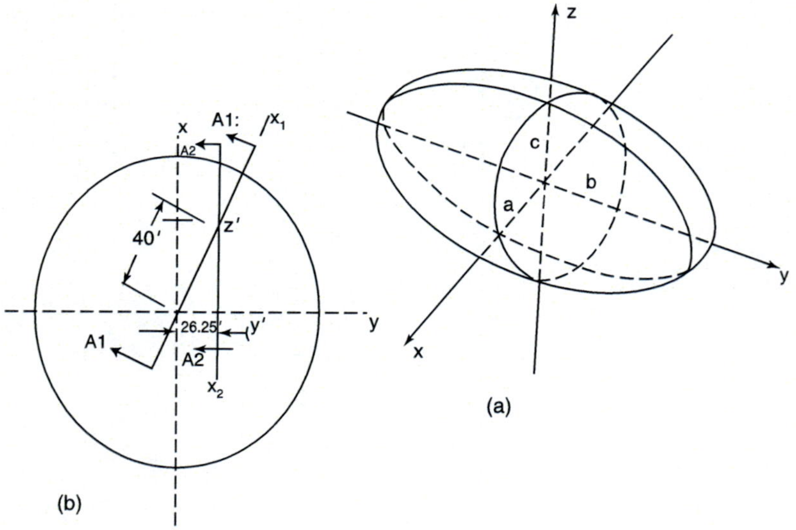

Fig. 6.5 A curvature at specific location of the elliptical domes. **(a)** Curvature is xyz. **(b)** Plan

$$\frac{x^2}{a^2} + \frac{z^2}{c^2} = 1 - \frac{y_1^2}{b^2} \tag{6.4.2}$$

or
$$\frac{x^2}{a^2(1 - \frac{y_1^2}{b^2})} + \frac{z^2}{c^2(1 - \frac{y_1^2}{b^2})} = 1 \tag{6.4.3}$$

Reference to Fig. 6.5b section $A1 - A1$ cut the ellipsoid at $Y' = 0$ then the section has the equation of the form:

$$\frac{x_1^2}{a^2} + \frac{z^2}{c^2} = 1.0$$

Let

$$a = 69.125'(21.069\ m)$$
$$c = 41.375'(12.61\ m)$$
$$x_1 = 40.0'(12.192\ m)$$

Then

$$z^2 = \frac{a^2c^2 - c^2x^{-2}}{a^2}$$
$$z^2 = \frac{69.125)^2(41.375)^2 - (41.375)^2(40)^2}{(69.125)^2}$$
$$z = 33.7441\text{ft}(10.285\ m)$$

Let section $A2 - A2$ cut the ellipsoid @ $y_1 = 26.25'$, the section has an equation:

$$\frac{x_2^2}{a^2(1 - \frac{y_1^2}{b^2})} + \frac{z^2}{c^2(1 - \frac{y_1^2}{b^2})} = 1 \tag{6.4.4}$$

where

$$x_2 = \sqrt{(40)^2 - (26.25)^2} = \sqrt{910.9375} = 30.1817(9.2\ m)$$
$$b = 69.125'(21.069\ m)$$

Solve for

$$\frac{x_2^2}{a^2\left(1 - \frac{y_1^2}{b^2}\right)} = \frac{910.9375}{(69.125)^2(1 - \frac{26.25^2}{69.125^2})} = \frac{910.9375}{(69.125)^2(1 - 14421)} = .2228$$

$$c^2\left(1-\frac{y_1^2}{b^2}\right) = (41.375)^2(1-.14421) = (41375)^2(.8558) = 1465.0$$

$$0.2228 + \frac{z^2}{1465.04} = 1.0$$

$$z^2 = (1-.2228)(1465.04)$$

$$z^2 = (9.7772)(1465.04) = 1138.6290$$

$$z = 33.7435(10.285\ m)$$

Radius of curvature of section $A2 - A2$:
Major axis $a = \sqrt{(69.125)^2 - (26.25)^2} = \sqrt{4089.2031} = 63.95(19.492\ m)$
Major axis:
Previously where $x^2 = 30.1871$:

$$0.2228 + \frac{z^2}{1465.04} = 1.0$$

where $x^2 = 0$

$$z^2 = 1465.04$$

$$c = z = \sqrt{1465.04} = 38.28'(11.67\ \mathrm{m})$$

Find radius of curvature @ pt. $z = 33.7441$

$$a = 63.95'(19.442\ m)$$

$$c = 38.28'(11.69\ m)$$

$$x_2 = 30.1817'(9.2\ m)$$

$$z = 33.7435'(10.28\ m)$$

$$R_2 = \text{ radius of curvature } = \frac{(\mathrm{a}^4z^2 + c^4x_2^2)^{1/2}}{c^2}$$

$$= \frac{[(63.95)^4(33.7441)^2 + (38.28)^4(30.1817)^2]^{1/2}}{(38.28)^2} = 98.8930'(30.1426\ m)$$

Radius of Curvature @ pt. $x^2 = 0, z = 38.28'(11.668\ m)$

$$R_2 = \frac{[(63.95)^4(38.28)^2 + 0]^{1/2}}{(38.28)^2} = 106.834'(32.563\ m)$$

Angle formed by C.G. Tendon with respect to the Horizontal Axis of section $-A2$:

$$x_2 = \sqrt{60^2 - 26.25^2} = \sqrt{2910.9375} = 53.953(16.4449\ m)$$

$$\cos\theta_2 = \frac{53.9531'}{98.893'} = .54539$$

$$\theta_2 = 56^\circ - 57'$$

$$\theta_4 = 90^\circ - (56^\circ - 56' - 00)$$

$$\theta_4 = 33^\circ - 04'$$

$$\sin\theta_1 = \frac{30.1817'}{98.893'} = .3052$$

$$\theta_1 = 17.77^\circ = 17^\circ - 46' - 13$$

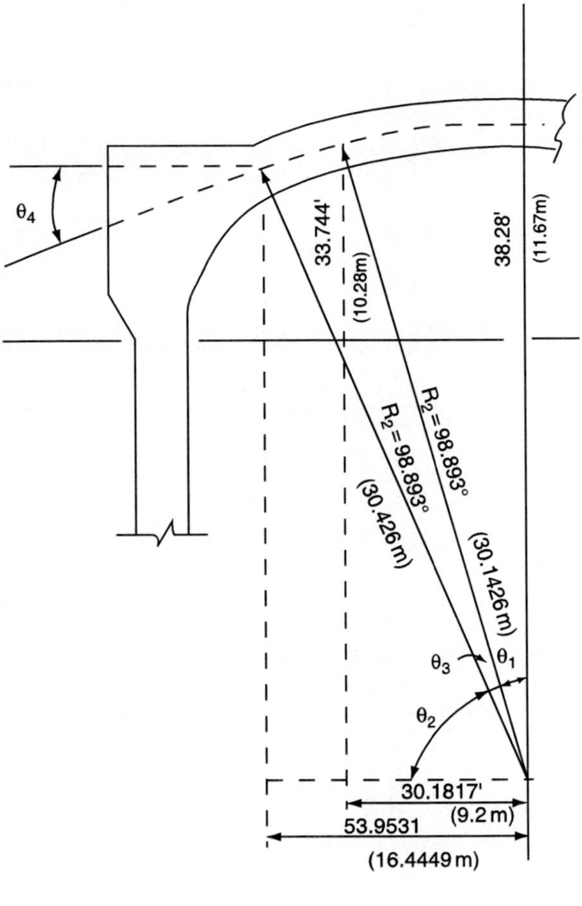

$$\theta_3 = 90^\circ - (\theta_2 + \theta_1)$$
$$\theta_3 = 90^\circ - (56^\circ - 57' - 00 + 17^\circ - 46' - 13)$$
$$\theta_3 = 90^\circ - (74^\circ - 43' - 13) = 15^\circ - 14'$$

Length of Tendon of Section A2–A2:

$$l_{eA}^{B} = a(E_B - E_A)$$

Knowing that

$$a = 63.95'(19.492\ m)$$
$$b = c = 38.28'(11.07\ m)$$
$$k = \frac{\sqrt{a^2 - b^2}}{a} = \frac{\sqrt{(63.95)^2 - (38.28)^2}}{63.95}$$

$$k = .8011.\ \theta = \sin^{-1}(.8011) = 53.75^\circ$$

$$\text{Limit } A = \phi = 90^\circ - \tan^{-1}\left(\tfrac{30.1817}{33.7441}\right) = 90^\circ - \tan^{-1}(0.8944)$$
$$= 90^\circ - 41.8^\circ = 48^\circ 11' 24$$
$$E_A \approx 0.7823$$

$$\text{Limit } B = \phi = 90^\circ$$

$$E_B = 1.2750$$
$$l_{eA}^{B} = 63.95(12.750 - .7820) = 63.95(.4930) = 31.53'$$
$$l_s = \frac{\pi R_2 \theta_3}{180^\circ} = \frac{(3.1416)(98.893)(15.2777)}{180^\circ} = 26.37'(8.038\ m)$$
$$l_r = \frac{12.65'}{\cos\theta_4} = \frac{12.63'}{\cos(33^\circ 030)} = \frac{12.5}{.838195} = 16.86'(4.9577\ m)$$
$$L_T = 31.53' + 26.37' + 16.26' = 74.16'(22.604\ m)$$

6.2.11.3 A Spheroid Equation

Similar to ellipsoid a spheroid has an equation of the form. A reference is made to Fig. 6.5d.

$$\frac{x^2}{a^2} + \frac{y^2}{b^2} + \frac{z^2}{c^2} = 1$$

where $a = b = c = r, x^2 + y^2 + z^2 = r^2$

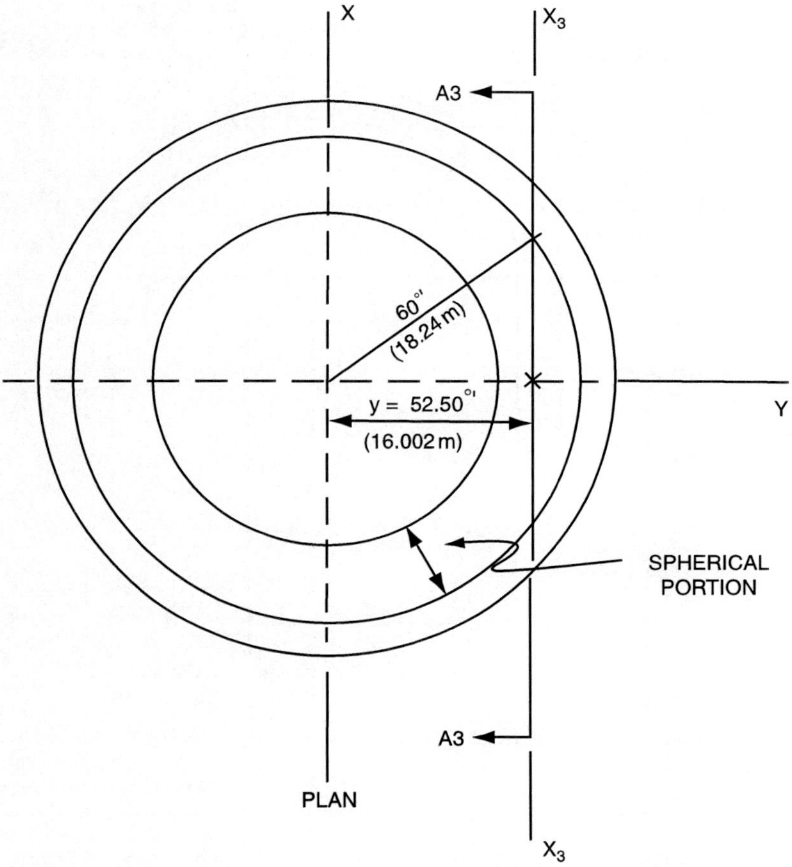

Fig. 6.5d Plan for spheroid

At any section where $y = y_1 = constant$
Now

$$x^2 = y_1^2 + z^2 = r^2$$
$$x^2 + z^2 = (r^2 - y_1^2)$$

Which is an equation of a circle

$$\begin{aligned} x_3 &= \sqrt{(60)^2 - (52.5)^2} \\ &= \sqrt{843.75} \\ &= 29.0473'(8.85362m) \end{aligned}$$

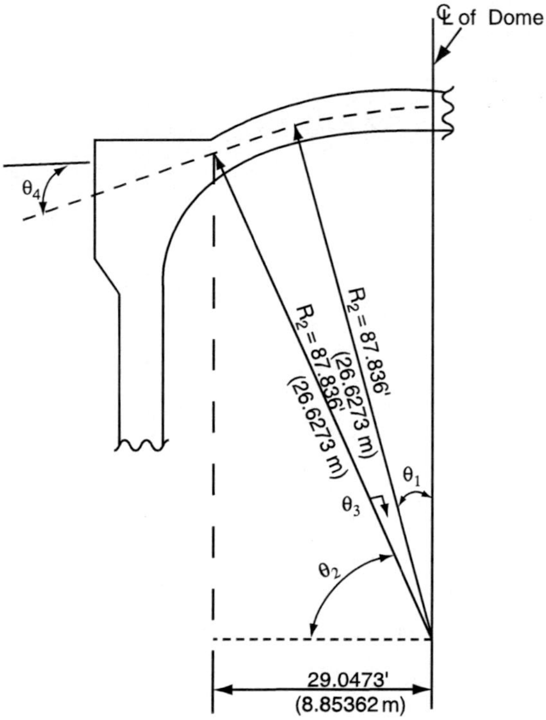

Fig. 6.5 (e) Section A2 – A2

We have

$$x_3^2 + z^2 = (r^2 - y_1^2)$$

$$r = R_2 = 102.33'(31.190\text{ m})$$

$$y_1 = 52.50'(16.002\text{ m})$$

$$z^2 = (102.33)^2 - (52.5)^2 - (29.0473)^2$$

$$z = \sqrt{6871.433} = 82.8941'(25.266\,m)$$

where $x_3 = 0$

$$z^2 = 102.33^2 = 52.5^2 = 7715.178'$$

$$z = \sqrt{7715.178} = 87.836'(26.6273\,m)$$

Radius of curvature @ $x_3 = 0, z = 87.836'(26.6273\text{ m})$

$$R_2 = \frac{[(29.0473)^4(87.836)^2 + (29.0473)^4(0)]^{1/2}}{(29.0473)^2} = 87.836'(26.6273\,m)$$

Radius of curvature @ $x_3 = 29.0473, z = 82.8941$

$$R_2 = \frac{[(29.0473)^4(82.8941)^2 + (29.0473)^4(29.0473)^2]^{1/2}}{(29.0473)^2} = 87.836'(26.6273\ \text{m})$$

Angle formed by C.G.Tendon with respect to Horizontal Axis of section A3–A3:

$$\cos\theta_2 = \frac{29.0473'}{87.836'} = .3307$$

$$\theta_2 = 10^\circ - 41' - 39''$$

$$\theta_4 = 90^\circ - (70^\circ - 41' - 18'')$$

$$\theta_4 = 19^\circ - 18' - 21''$$

6.2.11.4 Length of the Tendon of Section A3–A3 (DQL)

$$l_e = 0$$

$$l_s = \frac{\pi R_2 \theta_4}{180^\circ} = \frac{3.1416 \times 87.836 \times 19.306}{180^\circ} = 29.60\,ft$$

$$l_r = \frac{20.95}{\cos\theta_4} = \frac{20.95}{\cos(19.312)} = 22.20'$$

$$L_T = 29.60 + 22.20 = 51.80\,ft\,(15.79\,\text{m})$$

Similarly this distance can be computed through out.

6.2.11.5 Coordinates for the Tendon Length (DQL) section A3–A3 used possibly for the Infinite Element mesh scheme

Method used

Starting from £ of dome, determine coordinates x,y with respect to the origin of the elliptical curve at $L = 18''(457\,\text{mm})$

$$L = aE_1$$

$$E_1 = \frac{L}{a} = \frac{18''}{67.5 \times 12}$$

$$E_1 = .0222$$

Interpolating

$$\phi = 1.270^\circ$$

$$\sin\phi = \sin(1.2701)^\circ = 0.2217$$

$$x = a\sin\theta = (67.5)(.02217)$$

$$x = 1.4965'(0.456133m)$$

$$x = 17.9577''$$

$$y = \sqrt{b^2 - \frac{b^2x^2}{a^2}} = \sqrt{(39.75)^2 - \frac{(39.75)^2(1.4965)^2}{(67.5)^2}}$$

$$39.74022ft = 476.88275''*$$

Again $\phi R\,type\, y = b\cos\phi = 39.75\cos(1.2701) = 39.74165ft(12.113255\,m)$

476.89982″∗ checked.

6.2.11.6 TVA Response Curve for Containments Plates No. 6.1–6.3 Were Decided by he TVA Preliminary Earthquake Response Curves

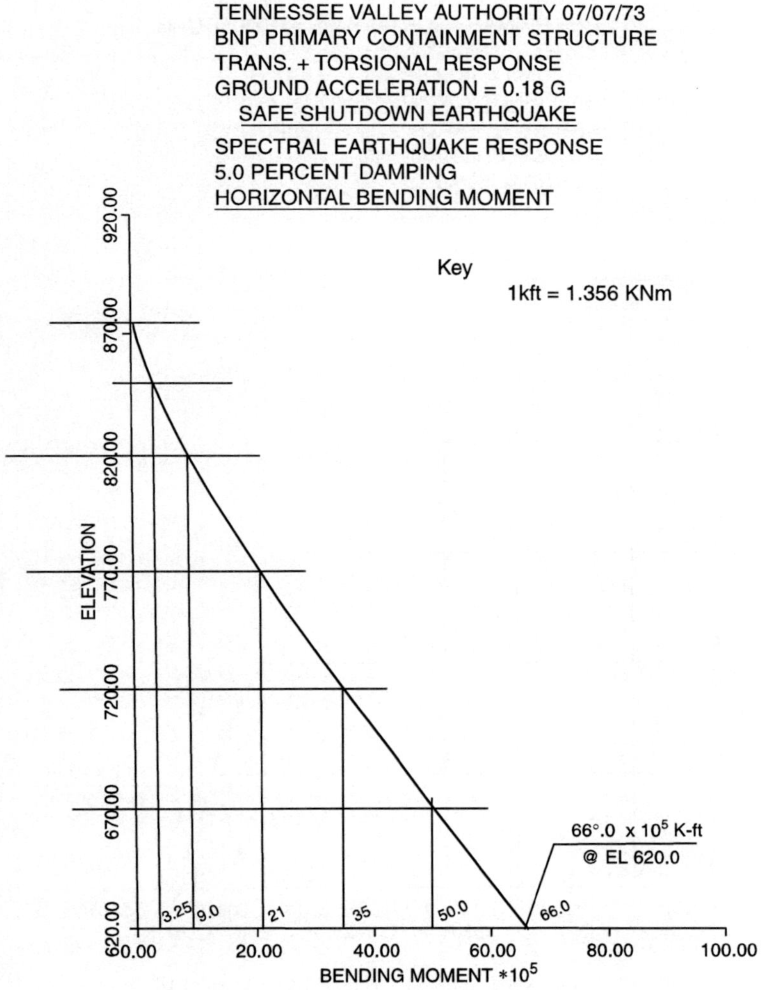

Plate No. 6.1 TVA vessel elevation versus bending moment

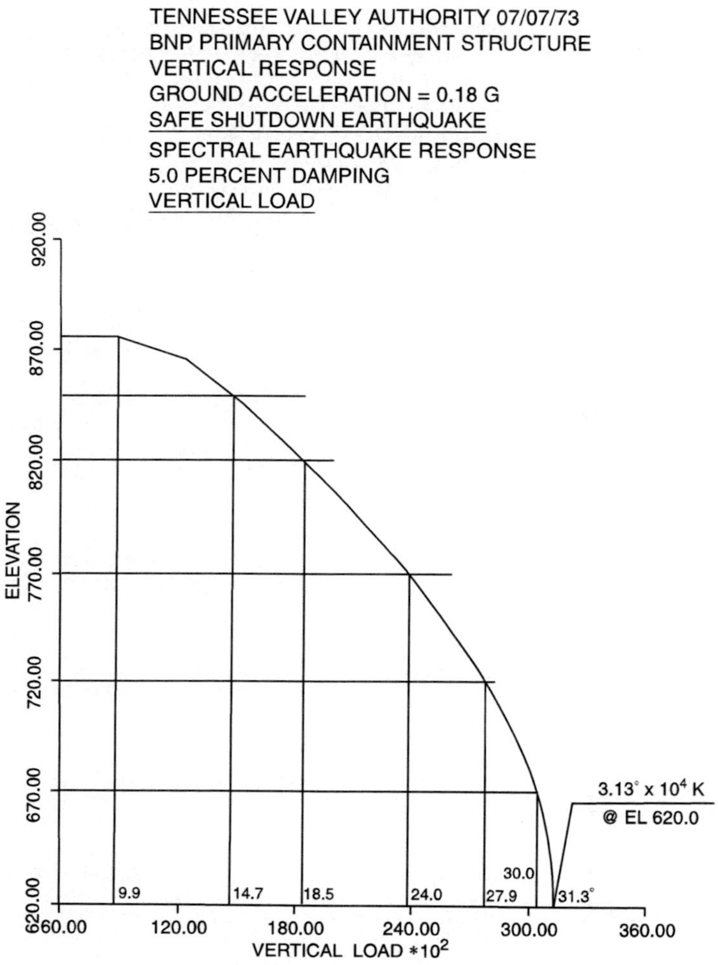

Plate No. 6.2 Elevation versus vertical loads. *Ground Acceleration* 0.18 G = r_g

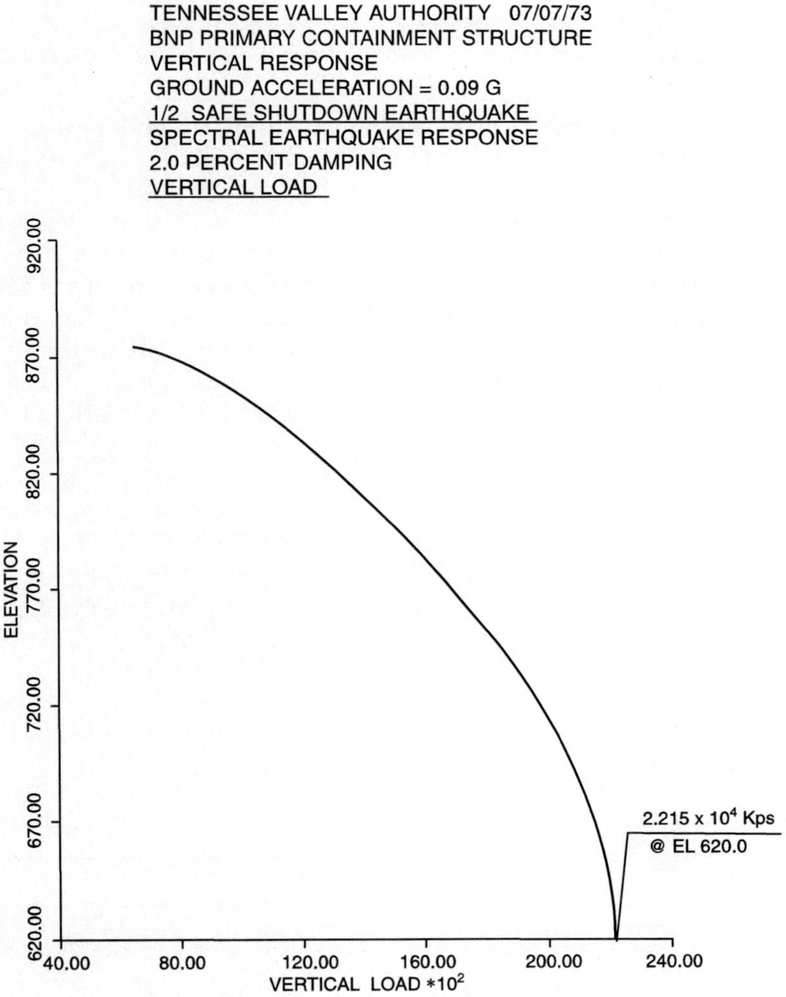

Plate No. 6.3 Load versus elevation. Ground Acceleration = 0.09 G1/2 safe shutdown

6.3 Design Calculations for Containment Structures

6.3.1 General Criteria

A great deal of substance available in this text to design both primary and secondary containments. The geometry material and loading are always dominant. The decision based on a number of variables has to be made regarding the usage of prestressed or reinforced concrete or perhaps both many primary and secondary containments have been designed in concrete using conventional steel reinforcement and prestressing system to resist loads. Here typical design parameters and prestressing systems adopted for the TVA Containment have been given. Any other geometry and design criteria for a typical vessel of a country must be adopted keeping in mind the methodology given in theses preliminary calculations adopted by the Tennessee Valley Authority (TVA). These calculations give a fair assessment of the containment prior to subjecting it to a more sophisticated analysis such as the dynamic finite element analysis involving the three dimensional behaviour under seismic impact, explosion and hurricane loadings in any combinations. These calculations cover both the dome and the walls of the TVA containments. Whatever criteria have been used, the calculation follow the TVA guidelines. Any changes needed for other containments must follow that country's codified methods. The author acknowledges the TVA support on the design calculations.

6.3.2 Types of Prestress Losses

Type of Loss	Value
1. Seating of anchorage	None (see comments 1 below)
2. Elastic shortening of concrete, in./in.	(see comment 2 below)
3. Creep strain of concrete, in./in. per psi	0.20×10^{-6}
4. Shrinkage strain of concrete, in./in.	100×10^{-6}
5. Relaxation of prestressing steel, psi	$8\% \, of 0.65 for \, 12,500 \, psi (8625 MN/\text{m}^2)$
6. Friction losses	$k = 0.0003, \mu = 0.13$

6.3.2.1 Comments on Types of Losses

1. There will be no allowance needed for seating losses in the end anchorage, since no slippage occurs in the anchorage during the transfer of the tendon force to the structure. In addition, the tendon force will be measured by making a life off reading after the anchorage has been seated.
2. The elastic shortening of the concrete due to the stressing of subsequent tendons results in a strain changes is considered in determining the effective force of a typical tendon. The average value of strain loss in the tendons is

determined by taking one –half the stress in the concrete under initial prestress and dividing it by the modulus of elasticity of the concrete(50 × 10^6 psi).

3. The above assumed values for the friction factors are somewhat larger than the factors measured on full-scale friction tests on the similar projects; however, the listed values are based on those used in similar containments.
4. Prior to the actual tensioning operation of the containment structures, friction tests will be performed to determined the actual friction factors and stressing data will then be revised as necessary.

6.3.3 Primary Containment Structures

6.3.3.1 Vertical Prestressing Losses (A)

a) Shrinkage losses

$$\Delta f_g = E. \in_s = (29 \times 10^3 \text{ksi})(100 \times 10^{-6}) = 2.90\,\text{ksi} = (20\ MN/\text{m}^2)$$

b) Steel relaxation losses

$$\Delta_1 f_g = 12,500\,psi(86.2MN/\text{m}^2)$$

c) sum *of* creep + shrinkage + relaxation

$$6.54 + 2.9 + 12.5 = 21.94\,\text{ksi}(151.28\,\text{MN/m}^2)$$

$$\text{Final } F = 0.7fc_{\text{pu}} - 21.94 = 146.06\,\text{ksi}(1007\,\text{MN/m}^2)$$

NOTE: These are losses may not be recovered by prestressing 0.7 f_s

$$\text{Initial prestress } f_{\text{in}} = f_f = 146.06(1007MN/\text{m}^2)$$

$$\text{creep} = 6.54(45\text{MN/m}^2)$$

$$\text{Shrinkage} = 2.90(19.99\ \text{MN/m}^2)$$

$$1/2(\text{Re}\mathit{laxation}) = 6.23\,\text{ksi}$$

$$\text{Total } 161.73\,\text{ksi}(1115\,\text{MN/m}^2)$$

d) Dome prestressing

Tendon at $26' - 2''$ (7.9748 m) from the £ dome

Segment	L	KL	α	$\mu.\alpha$	$KL + \mu\alpha$	$e^{-(KL+\mu\alpha)}$	Tx
A	0	0	0	0	0	1	1.0
AB	16.26'(4.96 m)	.00487	0.	0	.00487	.995±	0.995
BC	626.37'(8.376 m)	.00791'	.268	.0348	.0427	.958	0.954
CD	31.53'(9.610 m)	.00945	.306	.0394	.0489	.952	0.907

$$\alpha_{BC} = \frac{L_{SC}}{R} = \frac{26.37'}{98.89'} = 0.268$$

$$\alpha_{CD} = \frac{L_{CD}}{R_{AVG}} = \frac{31.53'}{\frac{106.83' + 98.89'}{2}} = .306$$

$$T_o = \frac{T_x}{.907} = 1.103 \therefore \text{Frictionloss} = 10.3\%$$

Segment	L	KL	α	$\mu.\alpha$	$KL + \mu\alpha$	$e^{-(KL+\mu\alpha)}$	Tx
A	0	0	0	0	0	1	1.0'
AB	22.2	.00666	0'	0	.00666	.9935	0.994
	29.60	.00888			.00888	.992	0.992
BC	29.61	.00888	0.337	0.0438	.0527	.949	0.994
	22.20	.00666			.0505	.951	0.943
CD	NO						

$$\alpha_{BC} = 29.60' = 0.337'$$

$$T_0 = \frac{T_x}{943} = 1.060' \therefore \text{Friction Loss} = 6.0\%$$

6.3.3.2 Layout of the Containment Dome Assumptions

The flat portion of the DOME-RING-GIRDER will be assumed to be 10'-3" (3.1242 m) wide in order to satisfy anchorage requirement. Therefore, the X-Y coordinates of points "A" are

$$X_A = 72.5 - 10.25 = 62.25(18.974\,\text{m})$$

$$Y_A = 903.5 - 879.0 = 24.5(7.4676\,\text{m})$$

6.3.3.3 Computations

Equation for roof

$$\frac{X^2}{(70.5)^2} + \frac{Y^2}{(42.75)^2} = 1$$

$$Y^2 = (42.75)^2 \frac{(42.75)^2}{(70.5)^2} X^2$$

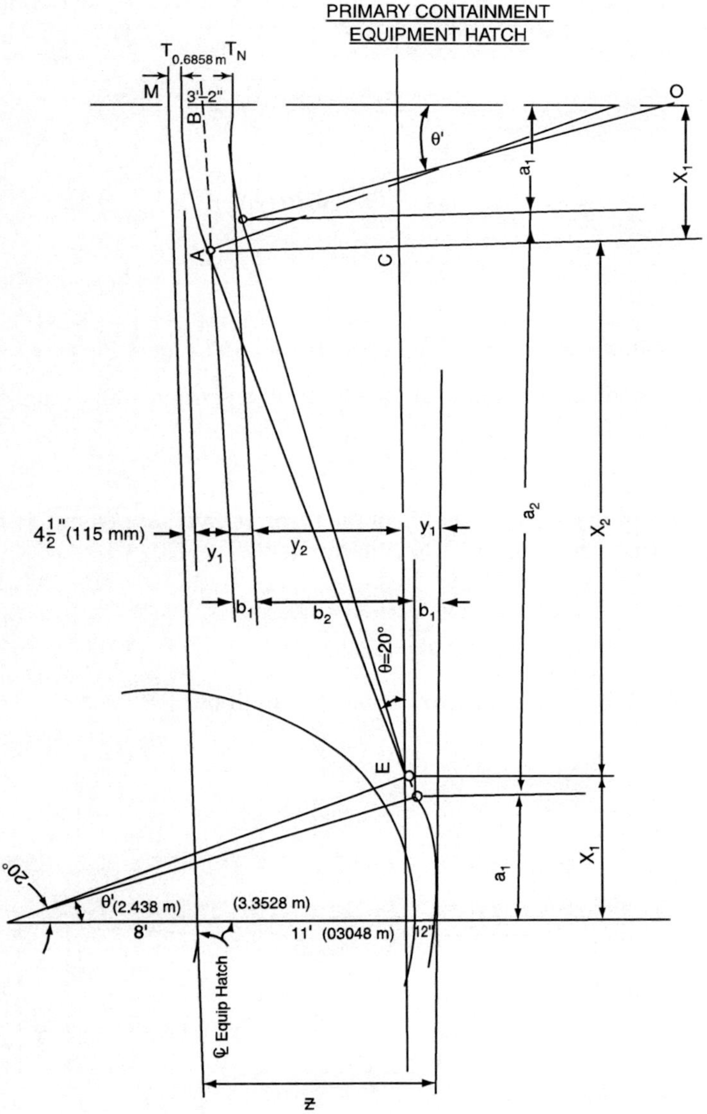

Fig. 6.6 Equipment hatch geometry

Slope to the curve above @ any point can be forced by determining the derivative.

$$y = \sqrt{(42.75)^2 \frac{(42.75)^2}{(70.5)^2} X^2}$$

$$Y = (1827.56 - 0.3677X^2)\frac{1}{2}$$

Defining $U = 1827.56 - 0.3677X^2$

6.3.3.4 Layout of Containment Dome Verifiction of Tangent Purpose

To verify the point of Tangency of the Dome-Ring-girder Intersection for final design notes.

Data

The roof of primary containment is in the form of an ellipse from £ R.B to the point of tangency in question. The equation is :

$$\frac{x^2}{(70.5)^2} = \frac{y^2}{(42.74)^2} = 1$$

Info. On Sketch Fig 6.6(a) primary containment dome

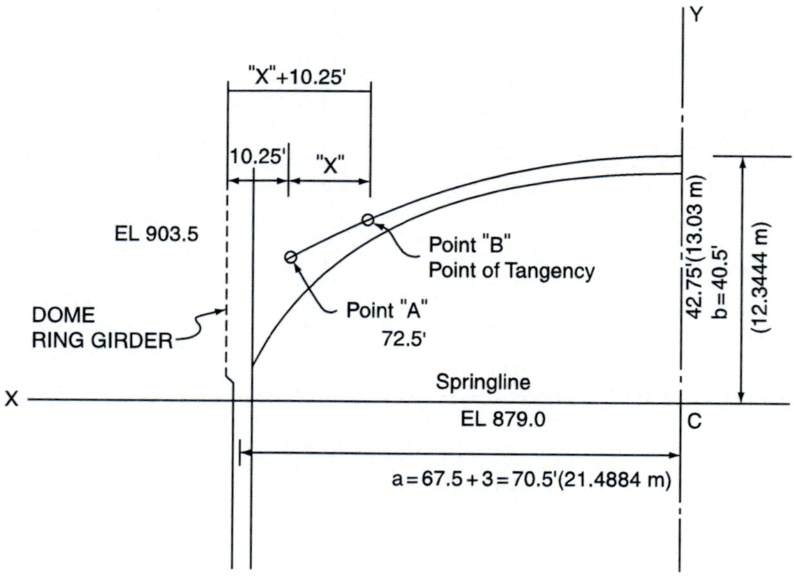

Fig. 6.6 (a) Dome verification for tangency (dome ring order)

Then

$$y = (U)^{1/2}$$

$$\frac{dy}{dx} = 1/2(U)^{1/2}\frac{d}{dx}$$

$$\frac{dy}{dx} = 1/2(1827.56 - 0.3677x^2)^{1/2} \cdot \frac{d}{dx}(1827.56 - 0.3677x^2)$$

$$\frac{dy}{dx} = 1/2(1827.56 - 0.3677x^2)^{1/2}(-0.7354x)$$

With above equation the Slope @ several points will be calculated
Trail # 1

$$"x" + 10.25' = 27.5'$$

$$x = 72.5 - 27.5 = 45.0'(13.176\,m)$$

$$\frac{dy}{dx} = 1/2(1827.56 - 0.3677(45)^2)^{1/2} \cdot (-0.7354 \times 45)$$

$$= (0.0151936561)(-0.7354)(45)$$

$$\text{Slope} = -0.5028036615$$

$$\text{when } x + 10.25 = 27.5'$$

The rectangular coordinates for the assumed point of tangency can be found by solving the equation of ellipse.

$$\frac{(45)^2}{(70.5)^2} + \frac{y^2}{(42.75)^2} = 1$$

$$y = 32.91'(10.031\,\text{m})$$

$$x = 45.00'(13.716\,\text{m})$$

From equation of straight line

$$Slope = m = \frac{y_2 - y_1}{x_2 - x_1}$$

Slope of trail # 1

$$Slope = \frac{32.91 - 24.5}{45.0 - 62.25} = \frac{8.41}{17.25} = 0.487536$$

$$0.487536\phi = 0.50280$$

Trail # 2

∴ Try point of tengancy @ 27.0′

$$x = 72.5 - 27.0 = 45.5'$$

$$\frac{dy}{dx} = 1/2(1827.56 \times 0.3677(45.5)^2)^{1/2} \bullet (-0.7354 \times 45.5)$$

$$= (0.0153117341)(-0.7354)(45.5)$$

$$Slope = -0.512341$$

Rectangular coordinates for assumed point of tangency.

$$x = 45.5(13.8684\,m)$$

$$y = 32.66(9.9538\,m)$$

The equation for straight line

$$Slope = \frac{32.66 - 24.5}{45.3 - 62.2} = \frac{8.16}{-16.75} = 0.48716$$

$$0.48716\phi = 0.512341$$

Trail # 3

$$''x'' + 10.25 = 28.5\,feet$$

$$x = 72.5 - 28.5 = 44.0$$

$$\frac{dy}{dx} = 1/2\,(1827.56 - 0.3677(44^2)^{-1/2} \times (-0.7354)(44)$$

$$= (0.014969198)(-0.7354)(44)$$

$$\text{Slope} = -0.4843663801$$

$$\text{When } x + 10.25 = 28.5\,\text{ft}$$

Solving equation of ellipse when

$$x = 44'.0(13.4112\,\text{m})$$
$$y = 33'.402(10.181\text{m})$$

From straight line equation

$$\text{Slope} = \frac{33.402 - 24.5}{44.0 - 62.25} = \frac{8.902}{16} = 0.48778081$$

Slope of ellipse $\approx$ Slope of line

$0.48436638 \approx 0.4877808$

Summary

$\therefore$ Tangent point to the ellipse is 28.5 ft from edge of any girder x = 18.25(5.5626 m)

6.3.3.5 Dome Layout & Design C.G. of Tendon

Elliptical curvature of C.G. ends at 40′ – 0″ (12.19m) from £ of Dome.

Equation of Ellipse, $\frac{x^2}{a^2} + \frac{y^2}{b^2} = 1$

For C.G. of Tendon

$$a = 67.5' + 1.625' = 69.125'\ (21.069\,\text{m})$$
$$b = 39.750' + 1.625' = 41.375'\ (1.986\,\text{m})$$

From Ron Howerton's oliveHi program, With $x = 40'$ $y = 33.74' + 849.00' = 882.74'(26.9059\,\text{m})$

(Elev: of transition from elliptical to circular curve) $El = 882.747'$ [26.9059 m]

Radius of curvature R_2 of elliptical dome @ $y = 40' = 102' - 4''[31.1432\,\text{m}]$

$$R_2 = ^{(a^4y^2+b^4x^2)^{1/2}} / b^2$$

$$a = 69.125(21.0693\,\text{m})$$
$$b = 41.375(12.611\,\text{m})$$
$$x = 40'$$
$$y = 33.74'$$

Table 6.3 Table of Offsets For Eliptical Roof of Reactor Containment Table .1471I Imperial units X 0.3048 = (m)

x	Y	y (approx)
5	39.6408	39′ – 8″
6	39.5926	39′ – 7″
7	39.5357	39′ – 6$^{1/2}$″
8	39.4698	39′ – 5$^{1/2}$″
9	39.3951	39′ – 4″
10	39.3114	39′ – 3″
12	39.1169	39′ – 1″
14	38.8858	38′ – 10$^{1/2}$″
16	38.6174	38′ – 8″
18	38.3108	38′ – 4″
20	37.9631	38′ – 11$^{1/2}$″
22	37.5800	37′ – 6″
24	37.1532	37′ – 2″
26	36.6836	36′ – 8″
28	36.1697	36′ – 2″
30	35.6083	35′ – 7″
32	35.0005	35′ – 0″
34	34.3405	34′ – 4″
36	33.6264	33′ – 8″
40	32.0187	32′ – 0″
42	31.1204	31′ – 1$^{1/2}$″
44	30.1471	30′ – 2″
46	29.0937	29′ – 1″
48	27.9511	28′ – 0″
50	26.7037	26′ – 8$^{1/2}$″
52	25.3492	25′ – 4″
54	23.8554	23′ – 10″
56	22.1993	22′ – 2″
58	20.3411	20′ – 4″
60	18.2104	18′ – 2$^{1/2}$″
62	16.7274	15′ – 9″
63	14.2831	14′ – 3$^{1/2}$″
64	12.6482	12′ – 8″
65	10.72	10′ – 9″

Alternative Tendon Arrangement for
Solving Anchorage Problems
key: 1″ = 25.4 mm

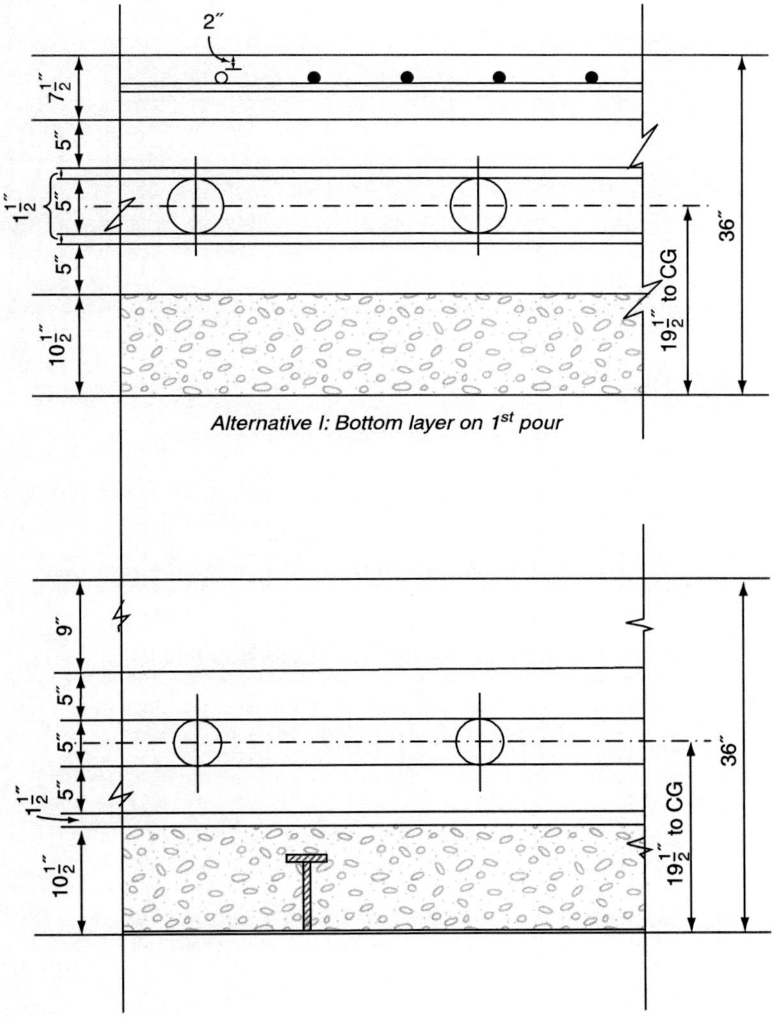

Fig. 6.7 Alternative II: layers touching in middle

Preaent Situation
key: 1″ = 25.4 mm

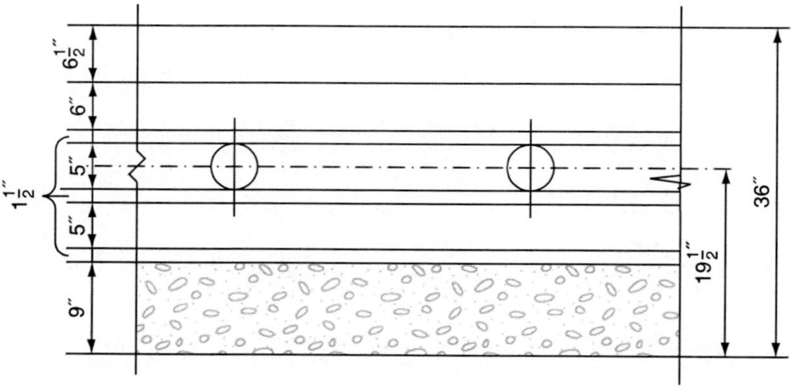

Fig. 6.8 Present case

Refer to Sketch on sheet 1 dated 6-6-72

$$\sin\theta_1 = 40.00' = .39089221$$
$$\theta_1 = 23°00'36'$$

$$Ao = \sqrt{(102.33')^2 - (40)^2} = 94.18826 \; or \; 94.19$$
$$\cos\theta_2 = \frac{60'}{102.33'} = .58633831$$
$$\theta_2 = 54° \, 06'09'$$
$$CD = \sqrt{(102.33)^2 - (60)^2} = 82.89408 = 82.89$$
$$\theta_3 = 90° - (\theta_1 + \theta_2)$$
$$\theta_3 = 12°53'15'$$

*E*l of Transition from elliptical to cir. cure $-AO$

$$882.74 - 94.191 = 788.55$$

*E*l of centre of spherical curve $+CD$

$$788.55 + 82.89 = 871.44$$

Dome Tendon Layout 3 – Layer
key: 1″ = 25.4 mm

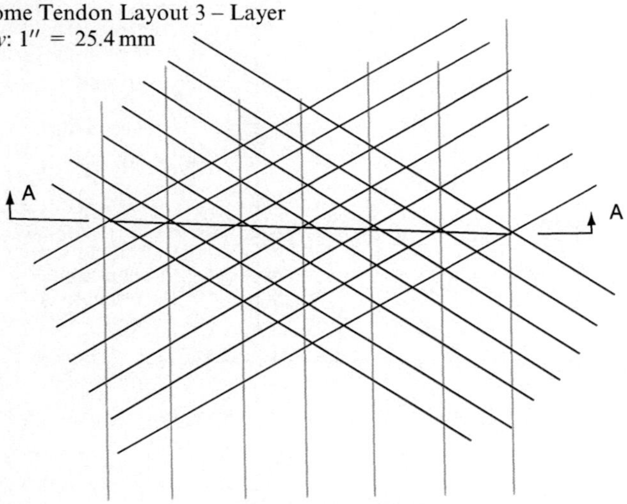

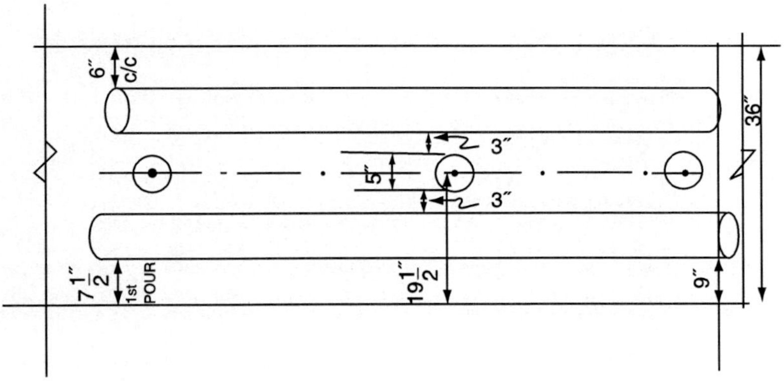

Fig. 6.9 Bar locations and dimensions

Eleve: of Transition from spherical to conical curve= *E*1 871.44

$$\theta_4 = 90^\circ - \theta_2 = 90^\circ - 54^\circ 06' 09'$$
$$\theta_4 = 35^\circ 53' 51'$$

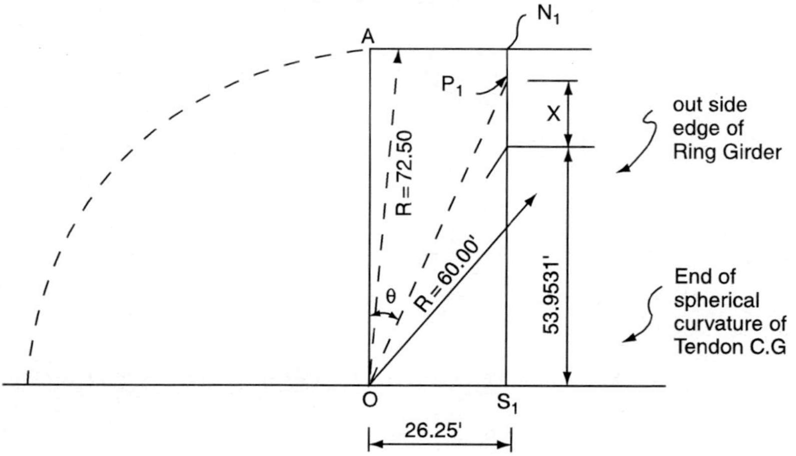

Fig. 6.10 Length of dome

Reactor Bldg. Dome Length of Tendon

$$AN_1 = R\sin\theta_1$$

$$\sin\theta_1 = \frac{26.25'}{72.50} = .36206$$

$$\theta_1 = 21^\circ 13' 36'$$

$$N_1P_1 = R vers\,\theta_1$$

$$vers\,\theta_1 = 1 - \cos\theta$$

$$= 2\sin^2 1/2\theta_1$$

$$\frac{\theta}{2} = 10^\circ 36' 48'$$

$$\sin\frac{\theta}{2} = 0.18418$$

$$N_1P_1 = R\left(2\sin^2 1/2\theta_1\right)$$

$$= 72.50(2)(.18418)^2$$

$$= 4.91872'(1.5m)$$

$$P_1S_1 = 72.50' - 4.91872 = 67.58128'(20.599\,m)$$

$$X = 67.58128' - 53.2531 = 13.6282'(4.15388\,m)$$

$$l_r = \frac{13.63}{\cos\theta_4}$$

6.3.3.6 Reactor Bldg – Dome Length of Tendon

$$Key : 1\,ft = 0.3048\,m$$

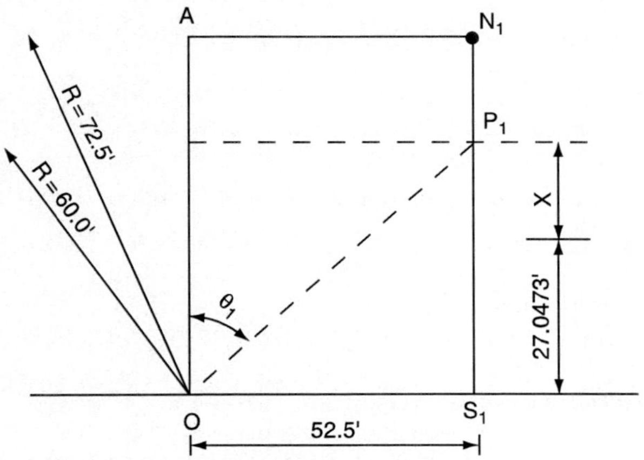

Fig. 6.11 Length of tendon

$$AN_1 = R\sin\theta_1$$

$$\sin\theta = \frac{52.50}{72.50} = 0.72413$$

$$\theta_1 = 46°23'48'$$

$$N_1P_1 = R\left(2\sin^2 1/2\,\theta_1\right)$$

$$1/2\ \theta_1 = 23°11'54''$$

$$N_1P_1 = 72.50(2)(.39391)^2$$

$$\sin 1/2\ \theta = 0.39391$$

$$N_1P_1 = 22.4980$$

$$P_1S_1 = 72.50 - 22.4980 = 50.0011$$

$$X = 50.001 - 29.0473 = 20.9530 = 20.95'$$

6.3.3.7 Dome Layout Horizental Measure

Purpose

To determine the proper spacing of Dome Tendons to obtain Equal Areas of NO Prestress in Ring Girder Region size of Dome from Bulds.

Known Data

41N711

Computations

Assuming the tendon groups divide section in red into 12 equal parts; therefore Each segment is 360°÷12 or 30° from each sketch pg 2 $\theta_2 - \theta_2 = 30^\circ$
And $\theta_2 + \theta_2 = 30^\circ$
Or $2\theta_2 = 30^\circ$

$$\theta_2 = 15^\circ$$

$$\therefore \theta_1 - 15^\circ = 30^\circ \quad \theta_1 = 45^\circ$$

$$\text{from sketch } \cos\theta_1 = X/72.5$$

$$X = 72.5 \times \cos 45^\circ = 51.26524\,ft$$

Say 51'-3" checks value used (ok)

6.3.3.8 Vertical Prestress Calculations

To determine the required vertical prestress force necessary to withstand load cases of design criteria and to comply with BELLEPONTE PSAR.

Assumptions

1- Critical section for vertical Tendons is 2t base of wall (EL620.0)
2- Concrete is not counted on to resist tendon
3- Liner is always in compression.
4- Force in reinforcement balances the force in liner.
5- At accident tendon elongation increases stress in the Tendon.
6- According to latest AEC guidelines earthquakes must be considered to occur in three directions simultaneously the effects of the earthquakes must be computed by taken the square root of the sum of the square of the maximum responses at a particular point caused by each of the three components of motion (two orthogonal horizontal motions and one vertical motion).

Explanation

In order to determine the max. (F) force to be resisted by vertical prestress. The three motions will be combined as

$$E^* = \sqrt{(V)^2 + (H_N - S)^2 + (H_E - W)^2}$$

But when $H_N - E$ is max. $H_E - W$

$$\therefore E^* = \sqrt{(V)^2 + (H)^2 + O^2}$$

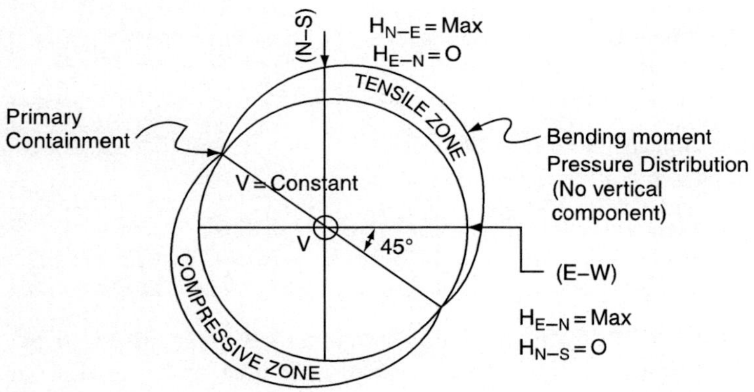

Fig. 6.12 Force resisted by vertical stress

Vertical Prestress Calculations Computations

A-Controlling Load Case.
Vertical free body
Notation:

$F = final\ prestress\ force$
$DL = dead\ = live\ load$
$E = load\ due\ to\ earthquake$

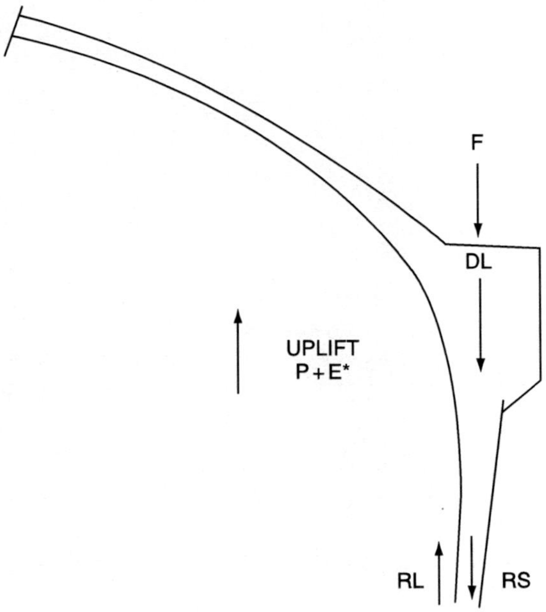

Fig. 6.13 Controlling load

$P = presure\ load$
$RL = force\ in\ liner$
$RS = force\ in\ reinf.\ steel$
Cases or bar number – American Practice

From previous calculations it has been determined that load cases 8 and 9 of design critical control vertical Prestress.

#8 – $1.00 + 1.0L + 1.0F + 1.25P + 1.0R + 1.0Ta + 1.25E$
#9 – $1.00 + 1.0L + 1.0F + 1.0P + 1.0Ta + 1.0Ra + 1.0Rs + 1.0E$

Note:
$1.0Ro$, $1.0R$, $1.0Ra$, are small loads that will be considered to act only locally. These loads are not axists metrical and most be handled separately. Therefore, they will be ignored for prestress determination $1.0T_2$ will 20 be ?

1- LOAD CASE # 8
From free body sheet 2 to balance vertical forces (see assumption 4)

$$1.0F + 1.0D = 1.25P + 1.25E$$

∴ Dead load, Pressure and Earthquake must be determined.

2- DEAD LOAD (1.0D)
$1.0D$ is made of wall and Dome weight
$1.0wall : 229.4m$
$weight = (229.4)(3.54)(0.150\,k/ft) = 120.23\,k/ft = 1754.64KN/m$

2 Dome. Surface of Dome Forms ∂one-half Oblate spheroid with $\partial i = 67.5\, bi = 39.75$ from Ref.#6 pg17, formula for surface area.

$$S = 2\pi\partial^2 + \pi\frac{b^2}{\in}\ln\frac{1+\in}{1-\in}$$

$$\text{where } \in \text{ ecentricity } = \sqrt{\frac{a^2 - b^2}{a}}$$

$$\therefore \in = \frac{\sqrt{(67.5)^2 - (39.75)^2}}{67.5} = \frac{\sqrt{2976.19}}{67.5} = 0.8082$$

$$S = 2\pi(67.5)^2 + \pi\frac{(39.75)^2}{0.8082}\ln\frac{1.8082}{0.1918}$$

$$S = 28{,}627.76 + 13{,}778.59 = 42406.35$$

Dome inside surface area = 1/2 S

$$S_{Ai} = 21{,}203.17\,ft^2\,(1969.84\,m^2)$$

Out side sur face $\partial = 70.5'$ $b = 42.75'$

$$\in = \frac{\sqrt{(70.5)^2 - (42.75)^2}}{70.5} = 0.79517$$

$$S = 2\pi(70.5)^2 + \pi\frac{(42.75)^2}{0.79517}\ln = \frac{1.79517}{0.2048}$$

$$S = 31,229.0 + 15,673.35 = 46902.35$$

$$SA_o = 23,451.176\,ft^2\left(2178.69\,\mathrm{m}^2\right)$$

6.3.3.9 Vertical Prestress Calculations

$\therefore$ Average Surface ∂rea

$$\frac{SAi + SAo}{2} = \frac{21,203.17 + 23,451.18}{2} = 22,327.175\,ft^2\left(2072.36\,\mathrm{m}^2\right)$$

$$Weight\ of\ dome = \left(22,327.175\,ft^2\right)(3.0)\left(0.15\,k/\mathrm{ft}^3\right)$$

$$W_D = 10,047.23\,k(44.690\,MN)$$

$$\text{Weight per foot} = 10,047.23/_{2\pi(69.25)} = 23.09\,k/ft(336.975\,KN/m)$$

3 Ring Girder. Volume of ring girder will be determined by sealing from sketch sh.6

Using plain meter the Un- shaded area of sheet #6 has been established as 14.8 ft^2

$$\therefore W_{RG} = 148ft^2 \times o.15\,k/ft^3 = 22.2\,k(98.75KN)$$

$$a - 1.0D = W_W + W_D + W_{RG}$$

$$= 120.23 + 2.09 + 22.2 = 165.52\,k/ft(2415.60\,\mathrm{KN/m})$$

$$b - \text{pressure}(1.0P)$$

$$\text{Design accident pressure} = 50.0\,\text{psi}\left(0.345\,MN/\mathrm{m}^2\right)$$

Uplift from pressure

$$p = 50^{\#}/\boxdot'' \times \frac{144\boxdot''}{ft^2} \times \frac{1^k}{1000^{\#}} \times \pi(67.5)^2.(\pi)(138.5)^2$$

$$p = 236.86\,\mathrm{k/ft}\ \text{say}\ 237\mathrm{k/ft}\ (50.4773\mathrm{MN/m})$$

6.3.3.10 New Earthquake Response

For *SSE*

$$P_{AE} = 204.0 \times 10^2\,k$$

$$ME = 44.8 \times 10^5\,k - 4$$

$$E = \sqrt{\left[\frac{204.0 \times 10^2}{1523}\right]^2 + \left[\frac{44.3 \times 10^5\,k - 4 \times 69.35}{3,653.332}\right]^2}$$

$$= \sqrt{179.4157 + 7209.1912} = \sqrt{7388.6089}$$

$$E = 85.9570\,\mathrm{k/ft} \times 3.5 = 300.25\,\mathrm{k/ft}\,(4.382\,\mathrm{MN/m})$$

$$1.25E = 1.25 \times 30025 = 376.06\,\mathrm{k/ft}\,(5.4884\,\mathrm{MN/m})$$

For *SSE*

$$P_{AE} = 313.0 \times 10^2\mathrm{kips}$$

$$ME = 67.2 \times 10^2\mathrm{kips} - \mathrm{ft}$$

$$TE = \sqrt{\left(\frac{313.0 \times 10^2\mathrm{kips}}{1523\,\mathrm{ft}^2}\right)^2 + \left(\frac{67.2 \times 10^5\mathrm{kips} - \mathrm{ft} \times 69.25}{3653.832}\right)^2}$$

$$= \sqrt{422.37 + 16222.068} = \sqrt{16643.05}$$

$$= 129.01\ \mathrm{k/ft^2}\,(6.167\,\mathrm{MN/m^2})$$

$$E' = 129.01\,\mathrm{k/ft^2} \times 3.5\,\mathrm{ft} = 45153\,\mathrm{k/ft}$$

$$1.0E = 451.53\,\mathrm{k/ft}$$

Case#8

$$1.0F + 1.0D = 1.25P + 1.25E$$

$$1.0F = 296.52 + 376.06 - 165.52$$

$$= 506.79\,\mathrm{k/ft}$$

Case#9

$$1.0F = 1.0P + 1.oE' - 1.0D$$

$$= 236.86 + 451.53 - 165.52$$

$$1.0F = 522.87\,\mathrm{k/ft}$$

∴ Load case#9 controls design conclusion.

Present prestress force of 530 k/ft (7734.82 KN/m) is more than 523 k/ft reqd. by controlling load CASE.

$$\therefore \text{From pressure } 1.0P = 236.86 \approx 237\,\text{k/ft}\,(3.459\,\text{MN/m})$$

$$1.25P = 296.25\,\text{k/ft}(4.234\,\text{MN/m})$$

$$1.50P = 355.5\,\text{k/ft}(5.2\,\text{MN/m})$$

6.3.3.11 Earthquake Forces $-1.0E$ and $1.0E'$

The effect of three component earthquake must be considered according to AEC. This will be done by taking the square root of the sum of the squares of each component. From Bending Moment pressure distribution may uplift occurs over a 90° quadrant and is constant @ H_{max}.

$$\sigma_E = \sqrt{\left(\frac{P_{AE}}{A_W}\right)^2 + \left(\frac{M_E \cdot C}{I}\right)^2 + (0)^2}$$

where

$$P_{AE} = \text{vertical load from earthquake.}$$

$$M_E = \text{bending moment from horizontal component.}$$

$$A_W = \text{moment of inertia of cylinder.}$$

$$C = \text{radius top of wall}$$

Thus,

$$A_W = \pi\left(r_0^2 - r_i^2\right) = \pi\left((71)^2 - (67.5)^2\right) = \pi(484.75)$$

$$A_W = 1522.89 \text{ say } 1523\,\text{ft}^2$$

$$I = \pi/64\left[d_o^4 - d_i^4\right] = \pi/64\left[142^4 - (135)^4\right]$$

$$= 3,653.882\,\text{ft}^4$$

$$c = 67.5 + 3.5/2 = 9.25\,ft$$

1. 1/2 Safe- Shutdown Earthquake

   From R&D spectrum see sheet # 9 and 10

$$P_{AE} = 2.215 \times 10^4\text{kip}$$

$$M_E = 47.5 \times 10^5\text{kip/ft}$$

$$\sigma_E = \sqrt{\left(\frac{2.215 \times 10^4}{15.23}\right)^2 + \left(\frac{4.75 \times 10^6 \times 69.25'}{3653882}\right)^2}$$

$$\sigma_E = \sqrt{8315.0} = 91.19\,\text{k/ft}^2$$

$$E = (91.19\,k/\text{ft}^2)(3.5ft) = 319.15\,\text{k/ft}$$

$$1.25E = \ 398.94\,\text{k}/ft(5.8354\,\text{MN/m})$$

2. Safe Shutdown Earthquake

From

$$P_{AE} = 3.13 \times 10^4 kip$$

$$M_E = 66.0 \times 10^5 kipft$$

$$\sigma_E = \sqrt{\left(\frac{3.13 \times 10^4\text{k}}{15.23\text{ft}^2}\right)^2 + \left(\frac{66.0 \times 10^5\text{k.ft.}69.25}{3.653882\text{ft}^2}\right) + 0}$$

$$= \sqrt{(20.55\text{k/ft})^2 + (125.09)^2}$$

$$= \sqrt{1606.81}$$

$$\sigma_E = 126.77\text{k}/ft^2$$

$$E' = 126.77\text{k/ft}^2 \times 3.5ft = 443.68\text{k/ft}$$

$$1.0E' = 443.68\text{k/ft}(6.5\text{MN/m})$$

LOAD CASE 8 Required Prestress

$$1.0F + 1.0D = 1.25P + 1.25E$$

$$1.0F + 165.52 = 296.25 + 398.94$$

$$1.0F = 529.67\,k/ft\,(7.76MN/m)$$

II LOAD CASE 9

From free body $5\lambda.2$ statics gives $1.0F + 1.0D = 1.0P + 1.0E$
Substituting values from proceeding pages

$$1.0F + 165.52 = 236.86\,\text{k}/ft + 443.68\text{k}/ft$$

$$1.0F = 515.2\,\text{k/ft}(7.546\,\text{MN/m})$$

$$1.0F_3 > 1.0F_9 \ \therefore \text{load case eight controls}$$

$$\therefore \text{Reqd vertical prestress} \ = \ 529.67\,k/ft$$

$$\text{Say } 530\text{k/ft } (7.765\text{MN/m})$$

Note: This force is reduced by 64 k/ft(0.9rMN/m) over original estimated values. Calculations must be made to optimize the amount of prestress provided to reduce reinf steel necessary to withstand sheer. Fin al prestress @Accident = 502k/ft. manufacturers supply 564k/ft the reserve prestress F_{TR} = 62k/ft@accident use have

$$62 \times \frac{151.3}{144.3} = 65\text{k/ft}(0.9523\,\text{MN/m})$$

$$\therefore F_{\text{TR}} = 65\text{k/ft}\ (0.9523\,\text{MN/m})$$

B. Required Final Prestress

From sheet # 13 the required force is 530 k/ft however the prestress force may be reduced because of Tendon Elongation. Since tendon elongation depends of final stress in tendon, an estimate of this will be made and the reserved @later time.

Based on preliminary Cale by EGB & NP (see reference $\phi = 1$) and calculation by sergeant & Lundy the final stress is 144.9 ksi and @accident fs = 151.3

The estimated final prestress

$$F_F = 530\,\text{k/ft} \times \frac{144.9}{151.3} = 506\,\text{k/ft}(7.4134\,\text{MN/m})$$

C. INITIAL PRESTRESS

In order to estimate losses an initial prestress force must be determined. However initial prestress and losses are interdependent.

From previous calculations the primary containment under different loading it has been determined that the initial stress in tendons is approx $161.8ksi\,(1115.6MN/m^2)$

and final stress is approx $144.9ksi\,(999MN/m^2)$

$$F_t = \frac{506\,k}{\text{ft}} \times \frac{161.8\text{ksi}}{144.9\text{ksi}} = 564.9\text{k/ft}$$

$$565\,\text{k/ft}(8.28\text{MN/m})$$

Check of Initial concrete stress.

$$f_{CI} = \frac{F_I + DL}{AW} = \frac{565 + 165.32}{(12)(42)} = \frac{730.52\,k/ft}{504\,k/ft}$$

$$F_{CI} = 1440\,\text{psi} <= 0.3fc = (0.3)(5500)$$

$$F_{CI} = 1440\,\text{psi} < 1650\,\text{psi} \therefore ok$$

$$9.93\,\text{MN/m}^2 < 11.38\,\text{MN/m}^2$$

Summary

The final prestress force @ Accident = 530 k/ft (7.765 MN/m).
The estimated final prestress before accident = 506 kft (7.4134 MN/m).

6.3.3.12 Vertical Prestress Losses Purpose Of Computations

To determine the prestress losses for vertical Tendons in order to compute the required prestress force for design.

Data

$$E_c = 5.0 \times 10^{\epsilon}, E_c = 29 \times 10^6 psi, n = \frac{29}{5} 5.8$$

$$f_c' = 5500\,\mathrm{psi}, E_{px} = 240\,\mathrm{ksi}(1659 MN/m^2), f_{\mathrm{py}} = 192\,\mathrm{ksi}(1327\,\mathrm{MN/m^2})$$

Assumptions

Final prestress and losses are interdependent therefore ∂ value for final prestress will be estimated and losses will be calculated. Then the process will be repeated to see if changes are required for final prestress.
Assume final prestress of 506 k/ft.

Computations

1-CREEP LOSSES
From PSAR, Creep strain of concrete = 0.20×10^6
Assuming $F_{TF} = 506\mathrm{k/ft}$
Initial prestress $\approx 506\left(\frac{161.8}{144.9}\right) = 565.02\mathrm{k/ft} = (8.28\mathrm{MN/m})$
Concrete stress due to Fi

$$f_a = \frac{565\mathrm{k/ft}}{12 \times 41.75} = 1128\,\mathrm{psi}(7.78\,\mathrm{MN/m^2})$$

Strain change in concrete

$$\varDelta E_c = (1.128)(0.20 \times 10^{6\%}) = 2.26 \times 10^{-4}$$

Steel stress change

$$\varDelta f_s = (29 \times 10^3) \times (2.26 \times 10^{-4}) = 6.54\,\mathrm{ksi}(45.1\mathrm{MN/m^2})$$

5. Losses Due to Elastic Shortening
   Elastic shortening will be calculated according to procedure

$$\Delta fs = \frac{n \times F_0}{Ac \times 2}$$

Where: F_0/A_C = initial concrete stress from prestress

$$n = E_S/E_C$$

Assuming $fo = 1128\,\text{psi}(fo = 565\text{k/ft})$

$$\Delta fs = \tfrac{(5.8)(1.128)}{2} = 3.27\,ksi(22.547\,MN/m^2)$$

∴ In order for tendon stress @anchorage to equal 168 ksi the tendon must be ANCHORED

$$F_S = 168 + 3.27 = 171.27\,ksi(1181\,MN/m^2)$$

6. Friction Loss
   From Ref# 5 the loss due to friction is 11.2%

∴ An increase in stress of 11.2% in necessary to overcome friction loss.

$$11.2\%\ of\ 171.27\,\text{ksi} = 19.18 + 171.27 = 190.45 \approx 191\,\text{ksi}(13.7\,\text{MN/m}^2)$$

$$151 < 0.8 f_{pg} = 192\,\text{ksi}(1324\,\text{MN/m}^2)(\text{OK})$$

Therefore friction and elastic shortening may be overcome by over stressing.

Purpose of Computations

To recalculate the friction loss sustained by vertical Tendons from Tendon Gallery Base Slab to Ring Girder Anchorage. This loss is necessary for calculation of required vertical prestress.

Data

Dimensions from Design
   (SEE sheet 3 for copy portion of this DRG)

$$\left.\begin{array}{ll} K = 0.0003 & \textit{wobble coefficient} \\ \mu = 0.13 & \textit{curvature coefficient} \end{array}\right\} \text{PSAR 3.8.1.4.1.6}$$

Assumptions

Friction ignore for average Tendon will be calculated and used. Deflected Tendons will be considered separate due to insure the additional losses do not alter design. For straight portion of Tendon have wobble friction loss and curved portion has both wobble curvature.

Primary Containment Structure Computations

1. Equation of Friction Loss
   The ring girder has curvature in it.
   Equation from *Ael EQ friction loss*

$$P_S = P_x e^{(KL+\mu x)}$$

Where:

$P_S =$ force@ Jacx $G\ END$

$P_Y =$ Force @ paint of question

$e =$ natural log base

$k =$ wobble friction loss coefficient.

$L =$ length of strand being considered.

$\mu =$ curvature friction coefficient.

$\alpha =$ angle change for strand.

and $k = 0.0003\ \mu = 0.13$

2. Loss Calculations

See sheet 3 for sketch and segment numbering. Segment $+$BC is only segment with curvature $\alpha = 11^{\circ}30' = 0.2007, L = R.\alpha = (30')(0.2007) = 6.021\,\text{ft}$
Then $11^0 30' = \Delta EL/30$

$$\Delta EL = 6.104\,\text{ft}855 = 861.10\,\text{ft}(262.463\,\text{m})$$

3. Table Losses

Segment	L	KL	α	$\mu.\alpha$	$KL + \mu\ \alpha$	$e^{-(KL+\mu x)}$	Tx
A	0	0	0	0	0	1.0	1.0
AB	12.52’	0.003756	0	0	0.003756	0.99625	0.99625
BC	6.021’	0.001806	0.2007	0.0261	0.027897	0.972488	0.968841
CD	248.5	0.07455	0	0	0.07455	0.928161	0.89924

$$P_x = 0.89924' P_S$$

$$P_S = \frac{1}{0.89924} \cdot P_X = 1.11205 P_X$$

∴ To obtain Required Jacking Force (or stress) the desired force must be increased 11.2%.

Conclusions

1. Vertical Friction Loss for Average Tendon is 11.2%

Summary of Forces

	Condition	Prestressing Force
Initial	Jacking	498 k/ft(7.2962 MN/m^2)
	Anchorage	436 k/ft(6.388 MN/m^2)
	Design $F_S = (161.8\,ksi)$	420 k/ft(6.1536 MN/m^2)
Final	Before Tendon elongation @ accident ($f_S = 143.9\,ksi$)	373 k/ft(5.465 MN/m^2)
	After tendon elongation after accident ($f_S = 152.7\,ksi$)	396 k/ft(5.802 MN/m^2)

6.3.4 Dome Prestressing: Description of Tendon Arrangement

Dome tendons are arranged in three groups, each group related 60° apart. Tendon in each group run in vertical planes parallel with vertical plane thru ℄ of tendon group. Spacing of tendons in each group is approximately 42" and spacing is begun at approx: 52'-6" from of roof.

There will be three layers of Tendons, are in each of the three directions.

Each group of Tendons will take 1/1.5 times the required prestressing force. For check on this see roof plane on sheet 2. Red cross hatching indicates areas where 2 groups of tendons are anchored. Single red lines indicates where only one group of tendons are anchored.

Each anchor zone is divided into 12 equal areas. ½ of these areas have one tendons group anchored where each group will take 1/12 of the prestressing force. Also ½ the anchor areas where 2 groups of tending area will take 1/6 of the prestressing force with each direction contributing 1/12.

Summing 12 increments around circumference of dome.

$$6 \times 1/6 = 1.0$$
$$6 \times 1/12 = 0.5$$
$$1.0 + 0.5 = 1.5$$

1.5 times required prestressing of force is furnished by all the tendons, therefore the computed prestress force will be divided 1.5 to act read prestress Assumption and Criteria for design.

(1) All computations for prestressing requirements will be based upon the centre of gravity of the dome tendons.
(2) The tensile stress in the prestressing tendons exert a normal force on dome which depends on the radius of curvature of the tendons. Tendons will have a different curvature which will result in Normal forces that vary through out the dome.
(3) Since the Tendons are placed in three different directions, the normal forces will have to be computed in each direction for certain critical points in dome. Each critical point will have three "normal" forces acting. These forces will have to be summed vertically to obtain a resultant force. Liberal sketches will be provided to locate these critical points.
(4) Final prestressing force in Dome will be based on the resultant effect of the above mentioned "Normal" forces.
(5) Dome Tendons are anchored at different angles (in relation to a horizontal plane). At least every other tendon will be anchored on a different angle at ring due to the fact that:

    (a) The curvature of the dome is constantly changing at location of each Tendon.
    (b) With 3 layers of tendons, each layer has to emerge from ring girder at different elevation. Therefore, an average angle of emergence thru the ring girder will be used to determine the horizontal and vertical components of the prestressing force at the ring girder.

(6) When all prestressing forces have been determined, they will used in a computer analysis of the containment wall and dome to arrive at stresses and deformations in the structure.

6.3.5 *Dome Vertical Components*

6.3.5.1 Statics

Total Uplift Force on Dome Must be Balanced by Vertical Component of Dome Prestressing Force.

$$\left[\sum F_V = 0\right] P = F \times \text{circum} + W_{\text{DOME}}$$

$$W_{\text{DOME}} = (\text{vol.}\, of\ \text{Dome}) \times \left(150\,\text{ibs/ft}^3\right)$$

$$\text{Volume} = \text{Area} \times 3.0'$$

$$\text{Area} \cong (80.6)(2\pi)(35) = 17,700$$

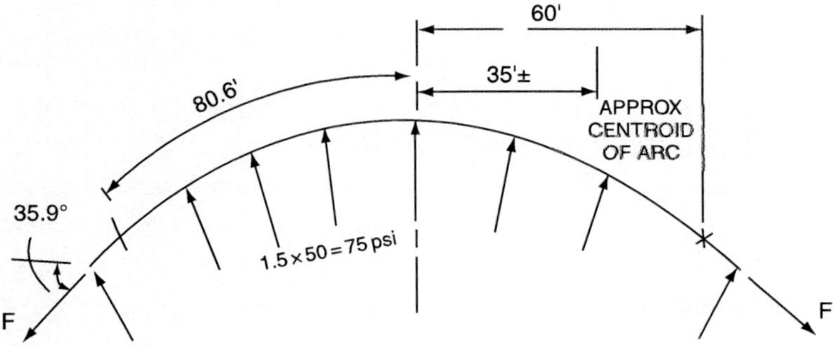

Fig. 6.14 Uplift force on Dome

$$\text{Volume} = 3 \times 17,700 = 53100\,ft^3$$

$$\text{WGT} = 53100 \times 0.150 = 7960 - 10,047\,\text{KN}\,(actual)$$

$$P \cong (0.075\,k)(144)(\pi)(67.5)^2 = 154,500\,\text{k}$$

$$F(k/1)(\pi)(135)\sin 35.9^\circ = 154.500 - 7960 = 146,540$$

$$249F = 146,540$$

$$F = 588\,k/1 \pm (8.615 MN/m)\pm$$

Prestressing provided 593 k/z (8.69 MN/m).

6.3.6 Dome Prestressing Elliptical Dome

For design accident pressure $P = 50\,psi\big(344.75\,KN/\text{m}^2\big)$

$$P = 50\,psi \times 144^{in^2}/ft^2 = 7.20\,k/ft(0.1055\,\text{MN/m})$$

$$\text{Dead load} = 0.15\,k/ft^3(3.0ft) = 0.45\,k/ft$$

$$1.5p = 1.5 \times 7.20 = 10.80ksf(48.04KN)$$

Critical Design Case for Dome $= 1.5P + 1.0TA + 1.0D$
Dead load subtracts from pressure load since pressure is an un lift force.

$$\begin{aligned} \therefore\ \textit{Net load on dome} &= 1.5P - D \\ &= 10.80 - 0.45 \\ &= 10.35ksf(46.04\,\text{KN}) \end{aligned}$$

$$Max.meridional force\ at\ \boldsymbol{\tau}\ \boldsymbol{dome}\ N\phi = \boldsymbol{pa2}$$
$$N\phi = 10.35\,k/ft = (67.5)2 = 593\,k/ft$$
$$2(39.75')$$

$\therefore$ Prestress force $= \frac{593^{K1}}{1.5} = 396\,k/\text{ft}(5.802\,\text{MN/m})$

*Assume that compressive force in liner due to thermal loads in resisted by tendon.

6.3.6.1 Elliptical Dome

Meridional force @ 55 from ∮ dome (this is where dome begins transition to thickened section and where one layer of Tendons begin)

$$N\phi@55' = \frac{p.\gamma_2}{2}$$

$$\gamma_2 = \frac{\alpha^2}{\left(\alpha^2\sin^2\phi + b^2\cos^2\phi\right)^{1/2}}$$

$$\tan\phi = \frac{55'}{110} = 0.5 \quad \phi = 26.5^\circ$$

$$\sin 26.5^\circ = 0.4446 \quad \sin^2 = 0.199$$

$$\cos 26.5^\circ = 0.895 \quad \cos^2 = 0.800$$

$$\gamma_2 = \frac{(67.5)^2}{[67.5^2(0.199 + 39.75^2(0.800)]^{\frac{1}{2}}}$$

$$\gamma_2 = \frac{4560}{(905 + 1265)^{\frac{1}{2}}} = 97.8\,ft(29.81\,\text{m})$$

Meri: force $N\phi = \frac{P_{\gamma_2}}{2} = \frac{10.35\times 97.85\text{ft}}{2} = 506\,\text{k/m}(7.413\,\text{MN/m})$

$$\text{For}\ \phi = 13^\circ$$

$$\sin 13^\circ = 0.225 \quad \sin^2 = 0.0507$$

$$\cos 13^\circ = 0.974 \quad \cos^2 = 0.948$$

$$\gamma_2 = \frac{4560}{(231 + 1500)^{\frac{1}{2}}} = 109.2\text{ft}\ (32.285\,\text{m})$$

Meridional: force will be; $N_{\phi O} = \frac{\text{p}.\gamma_2}{2} = \frac{10.35(109.2)}{2} = 567\,k^2/ft(8.6433\,\text{MN/m})$

From the preceding computation, we see that the meridional force increases. The dome is all reached for check, approx: radius of curvature for upper part of dome = 102'-3" and consider as spherical dome.

$$* \; Min\, N\phi \; @ \text{ top of dome} \frac{ag}{2} 1/2(10.35)102.25') = 530\,\text{k/ft}(.765\,\text{MN/m})$$

$$\sin\phi = 55\text{ft}/102.25' \;=\; 0.557 \quad \phi = 32.4^{\circ}$$

$$N\phi = @55' = \frac{ag}{1 = \cos\phi} = \frac{102.25ft(10.35\text{k/ft})}{1 + \cos 32.4}$$

$$N\phi @ 55' \;=\; 575k/ft(8.4243MN/m)\,\text{max}\,.$$

Tendons become tangent to curvature of dome at this point (see sheet 7)
$\therefore Prestress\, force = \frac{575\,k/ft}{1.5} = 383\,k/i \approx 396$
Therefore, we will continue to use elliptical dome for our analysis.
Note:
The final prestress can be reduced Since @ accident condition tendon elongates and stress is increasing.
$Final\, prestress\, force = \frac{593\,k/ft}{1.5} = 396\,k/ft(5.80MN/m)$
For spherical dome minimum meridional stress occurs at Top while for an elliptical dome, may meridional stress may be occur at Top.
Based on this assumption of Tendon elongation at accident; the final force before accident.

$$\left(\frac{593\,k}{N}\right)\left(\frac{\text{final stress befor eaccident}}{\text{finl stress before accident} \times (\Delta f_{z})\text{aacident}}\right)$$

At accident condition, concrete compressing stress diminishes to zero

$$\text{Before accident} = f_c = \frac{539}{(12)(35.75)} = 1.385\,\text{ksi}\left(4.55\,\text{KN/m}^2\right)$$

and

$$f_c = \frac{f_i}{\in_c} = \frac{1.385}{4.55 \times 10^6} = 0.000304$$

At accident $\in_c -0$
$\Delta f_s = 29 \times 10^6 \times 0.000304 = 8.8\,\text{ksi}.$
Final tendon stress before accident = 140.35;
final tendon strees after accident = 140.35 + 8.8 = 148.35
Thus

$$\text{Final force before @ accident} = \frac{593\,k}{ft}\frac{40.35}{148.35}$$

$$= 593\text{k/ft}\left(\frac{40.36}{148.35}\right)\left(\frac{143.9}{152.7}\right)$$

$$= 560\text{k/ft}(8.205\,\text{MN/m})$$

and

$$\text{Final stress force} = \frac{560}{1.5} = 373\text{k/ft}\,(4.484\text{MN/m})$$

Friction Losses
Assmptions

(1) Tendons stressed from both ends
(2) From sketch of sheet 2 of Dome prestressing

Calculation dated 2-24-72, will take three critical locations in dome for computing normal force acting on dome from prestress. This analysis will require computation of friction losses for the following Tendon locations.

(A) Tendon passing through £ dome
(B) Tendon @ 26′3″ from £dome
(C) Tendon @ 52′6″ from £ dome.

(3) Use C.G of Tendon for all computations.

(A) Tendon Passing through £ dome Ac1 Code equation from Friction Loss

$T_0 = T_x e^{(KL+\mu x)}$
$T_0 =$ Force at jacking and.
$T_x =$ Force at any point
$K =$ Wobble coefficient.
$L =$ Length of prestressing strand being considered.
$\alpha =$ angle change for strand being considered.

From PSAR, pg. 40:

$$K = 0.0003$$
$$\mu = 0.13$$

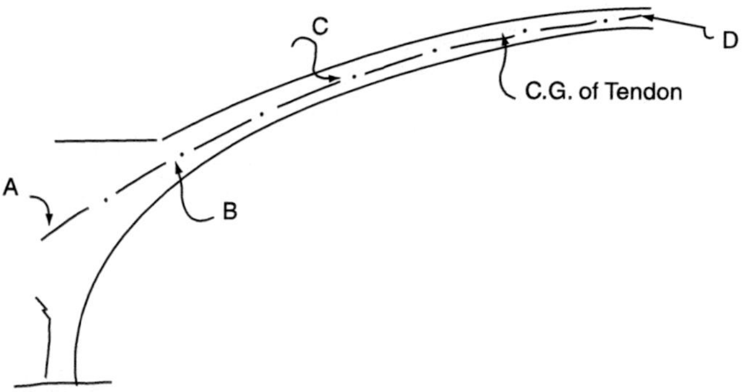

Fig. 6.15 Frictional losses in dome

Segment	L	$K.L$	α	$\mu.\alpha$	$KL+\mu\alpha$	$e^{-(KL-\mu\alpha)}$	T_x
A	0	0	0	0	0	1	1.0
AB	15.43′	.00463	0	0	.00463	0.995	0.995
BC	23.02′	.00690	0.224	.291	.0360	0.965	0.960
CD	42.10′	.01262	0.387	.0503	.0629	0.939	0.900

$$T_X = \frac{T_0}{e^{(KL+\mu\alpha)}} \quad \alpha_{BC} = \frac{L_{BC}}{R} = \frac{23.02}{102.33}$$

$$T_X = T_0 e^{-(KL+\mu\alpha)} \quad \alpha_{BC} = 0.224 \text{ radians}$$

$$T_X = T_0(0.900) \quad \alpha_{CD} = \frac{L_{CD}}{R_{Avg}} = \frac{42.10}{\frac{115.49' + 102.3}{2}}$$

$$T_0 = \frac{T_X}{0.900} = 1.11 T_X \quad \alpha_{CD} = 0.387 \text{ radians}$$

$\therefore$ Friction loss $= 11.1\%$

Elastic Shortening Loss

Final prestress force $=$ 560 kip/ft(8.205 MN/m)
Initial prestress;

$$F_{ct} \cong 560\,\text{k/ft} \times (160.1\text{kbi}) \cong 630\,\text{k/ft}(9.2306 \text{ MN/m}) \qquad (142.2)$$

$$\therefore F_{ci} = \frac{630}{(13'')(36.75'')} = 1470 psi$$

$\therefore$ Assume initial prestress $= 1500\,\text{psi}(10.3425 \text{ MN/m}^2)$

Avg. elastic shortening $(\Delta fs)_{\text{Avg}} = \frac{nfci}{2} = \frac{6.38(1500)}{2} = 4.81\text{ksi/ft}$ (0.705 MN/m)

Creep Loss

$$\Delta \in_c = (1500 \text{ psi})(0.25 \times 10^{-6}\text{in./in./psi}) = 0.000375$$

$$\Delta f_s = \Delta \in_c \times E_s = 0.000375(29 \times 10^6)$$

$$\Delta f_{s_{\text{creep}}} = 10.90\ ksi\,(75.16 \text{ MN/m}^2)$$

Tendon Relayation Loss

$\Delta F_S = 12.50\ ksi\,(86.2\text{MN/m}^2)$

Shrinkage Loss

$$\Delta f_s = \in_{sh} E_s$$

$$\Delta f_s = (100 \times 10^{-6})(29 \times 10^6) = 2.9 \text{ ksi}(20 \text{ MN/m}^2)$$

Summary of losses

Avg elastic shortening = 4.8 ksi(0.705 MN/m)
Creep = 10.90
Shrinkage = 2.90
Tendon relavation = 12.50
AVG friction loss = $\frac{1}{2(11.1\% + 6.0\%)} = 85\%$
Friction loss at 26′3 from the dome = 10.3%

} 26.3 ksi (181.34 MN/m^2

∴ use 10.3% for all friction losses since at least ½ of all tendons will have 1.3 11.1% friction losses. (Note: friction losses in domes are small enough to be compensated for by overstressing to $0.80\,f_5'(0.8 \times 240$ ksi)
Maximum jacking stress $= 0.80\%\ \ fsa = 0.80(280\,ksi) = 192\,ksi(2.813MN/m^2)$

Then max. initial stress available at while overstressing $= \frac{192}{1.103} = 174.07$ ksi <180
$\therefore O.K$
or $= \frac{0.813}{1.103} = 2.55\ MN/m^2 < 2.6372$
Average initial tendon stress at achoring $= 0.70 f_s' = 168$ ksi(2.4614 MN/m^2)
Max initial allow tendon stress at enchoring $= 0.75 f_s' = 180\,ksi(2.6372MN/m^2)$
Avg elastic shortening = 4.8ksi

Therefore initial force stress in first stressed tendon at enchoring
168 + 4.8 = 172.8ksi < 174 *O.K*
Total elastic shortening = − 9.6
Stress in first stressed tendon $= 163.2\,ksi(2.391MN/m^2)$
Stress in last stressed tendon $= 172.8\,ksi(2.52173MN/m^2)$
∴ Avg initial stress in tendons an anchoring (at completion of stressing)
= 168 ksi(2.4614 MN/m^2)
Avg final prestress = 168 − 26.3 = 143.7(2.1054 MN/m^2)

6.3.6.2 Initial Stress for Design

Final stress = 141.7 ksi(2.0761)
Creep loss = 10.90 ksi(0.160)
Shrinkage loss = 2.90 ksi(0.0426)
1/2 Relax loss = 6.30 ksi(0.0425)
Total = 161.8 ksi(2.3705)

Summary of Dome Prestressing Forces

(A) STRESS LEVELS

(1) Maximum jacking stress = 192 ksi(2.813 MN/m^2)
(2) Initial tendon stress at anchorage = 172.8 ksi(2.53173 MN/m^2)

(3) Average initial stress in tendon at anchorage = 168 ksi(2.4614 MN/m^2)
(4) Final tendon stress = 143.70 ksi(2.1054 MN/m^2)

(B) FORCES

(1) Final tendon force = $373\,k/ft(5.4648\,MN/m)$
(2) Initial tendon force at anchorage $= 373 \times \frac{172.8}{141.70} = 454$ k/ft(6.65155MN/m)
(3) Average initial force at anchorage $= 373 \times \frac{168}{147.7} = 436\,k$/ft(6.388 MN/m)
(4) Jacking Force $= 373 \times \frac{192}{147.7} = 506\,k$/ft (7.4134 MN/m)

NOTE: *Effecting* meridional force in dome considering three layers of tendon $S = 1.5$ times.

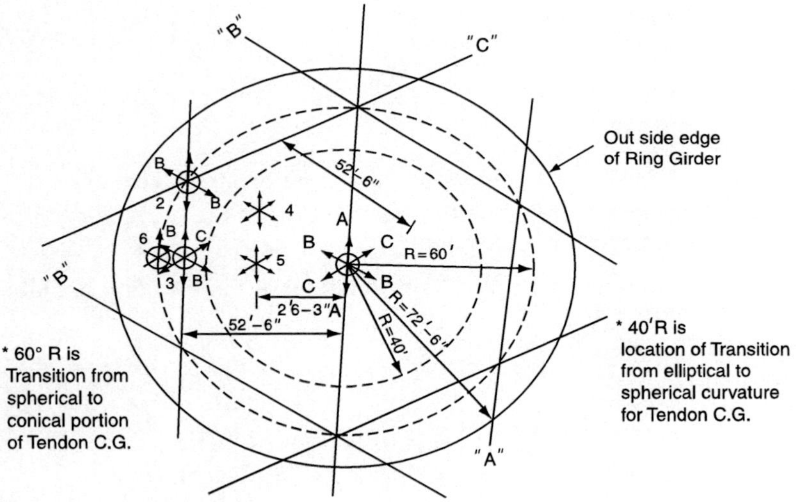

Fig. 6.16 Plan of dome

Dome Prestressing Tendon Normal Forces.

*Scale*1 = 40′0

Compute approx. spacing of tendons based on 186′ − 1/4 wires per tendon.

$$Spacing = \frac{186\ wires \times 0.0471^{B}/_{\text{wires}} \times 143.90\,\text{k}/B''}{373\text{k/ft}} = 42''(1067\text{ mm})$$

For 170 wires Tendon.

$$Speg = \frac{170}{186} \times 42 = 38.6C/C(980.6\text{ mm})$$

Assume 42″(1067 mm) spacing between Tendons measured in horizontal plane.

(1007 mm)42″ spacing will equire approx 31 Tendons in each group,

$$3 \text{ groups } = \frac{\text{tendons table}}{93}$$

Due to the fact that the curvature in the Tendons gets sharper (i.e. radius of curvature) as the tendons become further from £ dome and the normal forces that the Tendon experts on the dome increases as the Tendons becomes more remote form the £ of dome, then we will design on the following basis:

(1) Complete forces on dome assuming constant tending spacing in horizontal plane.
(2) Adjust spacing as necessary to achieve uniform prestress.

Point 1 on Dome (£ of dome at Apex)

$$N = \frac{F}{R}$$

where

$N =$ normal force exerted in a direction from prestress Tendon.
$F =$ force in prestressing tendon.
$R =$ radius of curvature of tendon.

From sketch on sheet. The three groups of tendons in the dome are labelled directions “A”,“B”&“C”.

Critical points are labelled 1, 2, 3, etc.

For point 1

$$NA = NB = NC = \tfrac{373}{115.49} = 3.23 \text{ kip/ft}$$
$$\therefore N_i \text{ total} = 3 \times 3.23 = 9.69 \text{ kip}/ft(0.1242 \text{ MN/m})$$

Check normal force at 1.5P accident condition

$$N_T = 9.69 \text{ kip/ft} \times \tfrac{396 \text{ kip/ft}}{373 \text{ kip/ft}} = 10.30 \text{ ksf}(13\ .955\text{KNm})$$
$$\text{Dead load roof} = \tfrac{0.45}{10.75}$$
$$1.5p \text{ load} = 10.80 \text{ ksfi}(14.63 \text{ KNm})$$
$$\approx 10.75 \text{ ksf prestress load}(14.56 \text{ KNm})$$

Check Average Initial Prestress Force

Average initial prestress = 436 kip/ft

$$N_T = \tfrac{436}{115.49} \times 3 = 11.33 \text{ kip/ft}$$
$$D.L = 0.45 \text{ kip/ft}$$

$$\text{Total load} = 11.78 \text{ kip/ft}$$

Meridional force in dome

$$N_{\phi o} = \frac{P_{ver}}{2} = \frac{11.78 \times 115.5}{2}$$
$$N_{\phi o} = 687\,k/ft$$

Initial conc. stress;

$f_{ct} = \frac{687.000}{12 \times 35 - 75} = 1602$ psi(11.0538 MN/m^2) $<$ 1650 psi(11.2585 MN/m^2)

$\therefore$ ok

Maximum allowable initial stress $= 0.3fc' = 0.3(5500) = 1602$ psi($11.2585\ MN/m^2$)

Point 2 on Dome

$(1\,k/ft = 12.81734KN/m)$

Consider forces in vertical plane along a radial line from £ dome. (This will give $X\&Y$ Components of forces only Z Components will cancel out).

$''A''$ direction $\rightarrow NA = \frac{373}{87.8'} = 4.25$ kip/ft
$''B''$ direction $\rightarrow NB = \frac{373}{102.33} = 3.64$ kip/ft
$''C''$ direction $\rightarrow NC = \frac{373}{87.8} = 4.25$ kip/ft

Computer Vector Components in $X\&Y$ *Plane*

$''A''$ direction $F_{AY} = 4.25 \cos 19.31^\circ = -4.01^K$
$F_{AX} = 4.25 \sin 1931^\circ (\cos 60^\circ) = 0.71^K$
$''B''$ direction $F_{BY} = 3.64. \cos 35.90^\circ = -2.96^K$
$F_{BX} = 3,64 \sin 35.90^\circ = 2.14^K$
$''C''$ direction $F_{CY} = F_{AY} = -4.01^k$
$F_{CX} = F_{AX} = 0.71^k$

Resultants:

$$F_{RY} = 4.01^{k/1} + 2.96^{k/1} + 4.01^{k/1} = 10.98\text{k/ft}$$
$$F_{RX} = 0.71 + 2.14 + 0.71 = 3.56\,k/ft$$
$$F_R = \sqrt{(10.98)^2 + (3.56)^2} = 22.56\,k/ft$$
$$\tan\,\theta(\text{angle from vertical}) = \frac{3.56}{10.96} = 0.324$$

$$N_A = \frac{373\overset{k}{1}}{87.8'} = 2.25^{k/ft}$$
$$N_B = \frac{373\overset{k}{1}}{98.9'} = 3.78^{k/ft}$$
$$N_C = \frac{373}{98.9} = 3.78^{k/ft}$$

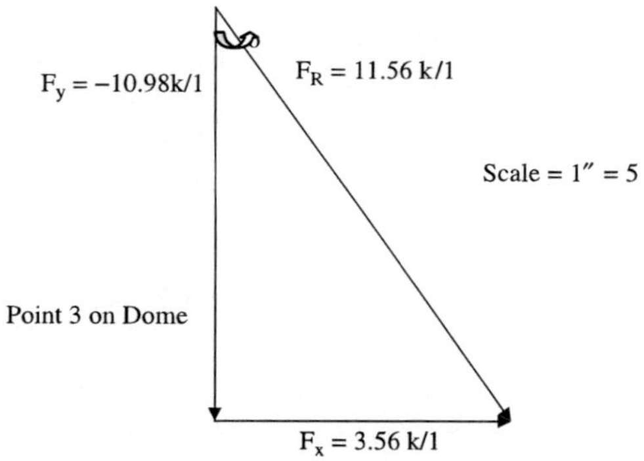

Fig. 6.17 Resolution of forces

key1k/ft = 2.8173 KN/m

Compute vector components in $Y–Y$ plane along radial line

$$''A'' \text{ direction } = \mathrm{F_{AY}} = -4.25\,\mathrm{k/ft}$$
$$F_{AX} = 0$$
$$''B'' \text{ direction } = F_{SY} = 3.78\cos 27^\circ = -3.36\,k/ft$$
$$F_{BX} = 3.78\sin 27^\circ(\cos 30^\circ) = 1.49\,k/ft$$
$$C \text{ direction } = F_{CY} = F_{BY} = -3.36\,k/ft$$
$$F_{CX} = F_{BX} = 1.49\,k/ft$$

Dome Prestressing Tendon Normal Forces
Resultants

$$F_{RY} = 4.25 + 3.36 + 3.36 = 10.97^{k/ft}$$
$$F_{RX} = 0 + 1.49 + 1.49 = 2.98^{k/ft}$$
$$F_R = \sqrt{(10.97)^2(2.98)^2}$$
$$= 11.36\,k/ft$$

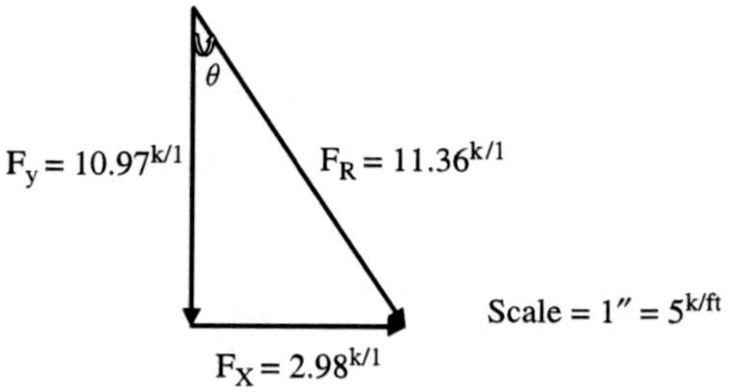

Fig. 6.17(a) Resolution of forces

$$\tan\phi o = \frac{2.98}{10.97} = 0.272$$
$$\phi o = 15.2^\circ$$

Fig. 6.17(b) A check on forces on dome at point 4

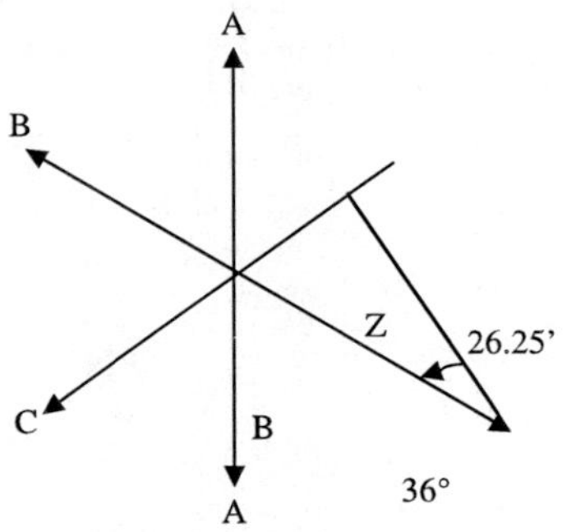

$$\tan 30^0 = \frac{X}{26.25'}$$
$$X = 15.17'$$
$$Z = \frac{26.25}{\cos 30^0} = 30.35'$$

For "A" and "C" Directions @26.25' from ∮ Dome
Properties of ellipse;

$$a = 63.95$$
$$b = 38.28$$
$$x = 115.17$$
$$Y \text{ ISOPAR EEO} = 37.1'$$
$$Y(\text{from olliveti program}) = 37.2'$$
$$\text{ISOPAR EEO} = 163.978$$
$$\gamma_z(\text{from olliveti program}) = 164.92'$$
$$\sin\theta_1 = \frac{15.17}{104.92} = 0.45\ \theta_1 = 8.3^0$$

For "B" directions @ 30.35' from ∮ dome properties of ellipse

$$a = 69.125'$$
$$b = 41.375'$$
$$x = 30.35$$
$$\therefore Y = 37.2'$$
$$\sin\theta_1 = \frac{30.35}{108.18} = 0.279\ \theta_1 = 16.2^0$$

"A" and "C" direction

$$N_A = N_C = \frac{373\frac{k}{ft}}{104.92'} = 3.56^{k/ft}$$
$$N_B = \frac{373\frac{k}{ft}}{108.18'} = 3.45^{k/ft}$$

Compute Vector Components in X-Y Plane Along Radial Line
"A" direction

$$F_{AY} = 3.56^{k}(\cos 8.3^{0}) = 3.52^{k/ft}$$
$$F_{AX} = 3.56(\cos 8.3^{0})(\cos 60^{0}) = 0.26^{k/ft}$$

"B" direction

$$F_{BY} = 3.45(\cos 16.2^{0}) = 3.31^{k/ft}$$
$$F_{BX} = 3.45(\sin 16.2^{0}) = 0.96^{k/ft}$$

"C" direction

$$F_{CY} = 3.52^{k/1}$$
$$F_{CX} = 0.26^{k/1}$$

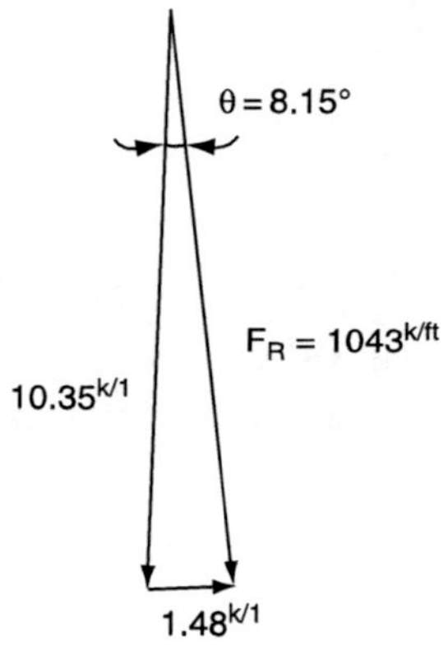

Fig. 6.17(c) Resolution of forces

$$F_{RY} = 3.52 + 3.31 + 3.52$$
$$F_{RY} = 10.35^{k/1}$$
$$F_{RX} = 0.26 + 0.96 + 0.26$$
$$F_{RY} = 1.48^{k/1}$$
$$F_R = \sqrt{(10.35)^2 + (1.48)^2}$$
$$= 10.43^{k/1}$$
$$\tan\theta = \frac{1.48}{10.35} 1.43$$
$$\theta = 8.15°$$

Vector Diagram
Check Point 5 on Done

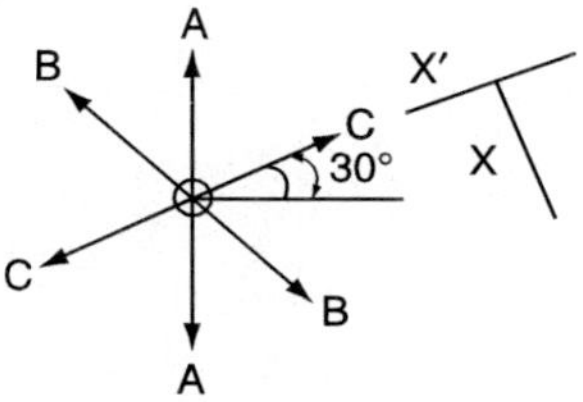

Fig. 6.17(d) X value X′ value in A, B, C Directions.

$$\sin 30^0 = \frac{X}{26.25'}$$
$$X = 13.125$$
$$X' = 26.25 \cos 30^0 = 22.7'$$

Properties of Ellipse in "B" and "C" Directions

$Major\ Axis\ a = \sqrt{(69.125)^2 - (13.125)^2}$
$a = 67.9'$
$b = 40.5'$
$x = 22.7'$
$y = 38.28'$
r_z(from ollivetti Prog) $= 109.97'$
$\sin\theta_1 = \frac{22.7'}{109.97} = 0.2065$
$\theta_1 = 11.9^\circ$

Properties of Ellipse in "A" Direction

$a = 63.95'$
$b = 38.28'$
$x = 0$
$y = 38.28$
r_z(from ollivetti Prog.) $= 106.83'$
Normal forces $= N_A = \frac{373\,k/1}{106.83'} = 3.49\,k/ft$
$N_B = N_C = \frac{373\,k/1}{109.95} = 3.40\,k/ft$

Vector components along Radial Line

"A" Direction

$$F_{RY} = 3.49\,k/ft$$
$$F_{RX} = 0$$

"B" Direction

$$F_{BY} = 3.40\,k/1(\cos\ 11.9^\circ) = 3.33\,k/ft$$

$$F_{BX} = 3.40 \sin 11.9^\circ 1(\cos\ 30^\circ) = 0.61\,k/ft$$

"C" Direction

$$F_{CY} = 3.33\,k/ft$$
$$F_{CX} = 0.61\,k/ft$$

Vector Diagram

Fig. 6.17(e) A check on forces

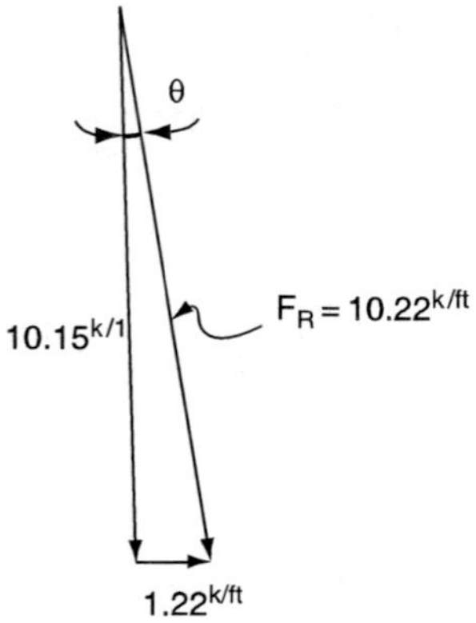

$$F_{RY} = 3.49 + 3.33 + 3.33 = 10.15\,k/ft$$
$$F_{RY} = 0.61 + 0.61 = 1.22\,k/ft$$
$$\tan\,\theta = \tfrac{1.22}{10.15} = 0.1202$$
$$\theta = 6.85^{\circ}$$

$$F_R = \sqrt{(10.15)^2 + (1.22)^2}$$
$$F_R = 10.22\,k/ft$$

Point 6 on Dome

For estimation purposes the curvature in "B" & "C" will be assumed to be the same as for point 3

$$N_A = 0$$
$$N_B = \tfrac{373}{98.9} = 3.78\,k/ft$$
$$N_C = \tfrac{373}{98.9} = 3.78\,k/ft$$

Vector Components in $X - Y$ Plane

"B" Direction

$$F_{BY} = 3.78\cos 27^{\circ} = -3.36\text{k/ft}$$
$$F_{BX} = 3.78\sin 27^{\circ} = +1.49\text{k/ft}$$

"C" Direction

$$F_{CY} = -3.36$$
$$F_{CX} = 1.49$$
$$F_{RY} = 6.72\,k/ft$$
$$F_{RX} = 2.98$$
$$F_r = \sqrt{(6.72)^2 + (2.98)^2}$$
$$= 7.35\,k/ft$$
$$\tan\,\theta = \tfrac{2.98}{6.72} = 23.9150^{\circ}$$

Dome Prestressing Tendon Normal Forces
Key 1 k/ft = 12.81734 KN/m 1ft = 0.3048

Summary of tendon normal forces

POINT	Fry (k/ft)	Frx (k/ft)	Angle from vert. to resultant F_e	Fr (k/ft^2)	Fr and accident 396/373X Fry(k/ft)
1	9.69	0	0	9.69	10.29
2	10.98	3.56	18°	11.56	12.27
3	10.97	2.98	15.2°	11.36	12.06
4	10.35	1.48	8.15°	10.43	11.07
5	10.15	1.22	6.85°	10.22	10.85
6	6.72	2.98	23.92°	7.35	7.80

Summary of forces

Initial	Condition	Prestressing Force
	Jacking	973' kip/ft
	Anchorage	875 kip/ft
	Design ($fs = 154.2^{ksi}$)	781 kip/ft
Final	Before Tendon elongation @ accident (fs 135.7 ^{ksi})	687 kip/ft
	After Tendon elongation due to accident (fs = 144.2 ^{ksi})	729 kip/ft

Prestress Losses
Sources Of Design Information

Reference No.	Reference	Page No.
1	Bellefonte Nuclear Plant, Preliminary Safety Report	
2	Design of Prestressed Concrete Structures, Second Edition T.Y. Lin	
3	Building Code Requirements for Reinforced Concrete (ACI 313 – 71)	
4a	Preliminary Calculations "Hoop Tendon – Prestress Loss" EGB	
4b	"Reactor Containment – Friction Loss in Horz: Tendon RWH	
5	Bellefonte N.P Design Criteria	

6.4 Primary Containment Hoop Prestress

6.4.1 1- Prestress Losses

6.4.1.1 1- Friction Losses

PC1 Code equation for friction loss.

$$T_o = T_x e^{(kl+\mu\alpha)}$$
$$T_x = T_o e^{(kl+\mu\alpha)}$$

Notation

T_o = steel stress (force) at jacking end
T_x = steel stress (force) at any point x.
k = friction wobble coefficient per foot of prestressing steel.
L' = length of prestressing *steel being considered.*
μ = friction curvature coefficient.
α = *Total angular change of prestressingsteel profile in radians from jacking end to any*

Given

$k = 0.0003$
$\mu = 0.13$

For straight portion of Tendon use wobble only. For curved portion of Tendon use wobble and curve friction. When post tensioning from both sides, only half of the Tendon length is substituted for L'.

6.4.2 Prestress Losses

key 1ft = 0.3048 m 1 k = 6.895 MN/m^2
Let $T_0 = 1^{\text{kip}}$
Assume jacking from both ends

Segment	$L(ft)$	$\alpha(rad)$	$kl + \mu\alpha$	$e^{-(kl-\mu\alpha)}$	$T_x(kips)$
D	0.0	0	0.0	1	1.0
DC	15.25	0	0.004575	0.9954	0.9954
BC	101.04	1.44	0.218	0.804	0.800

@ mid point of Tendon point B

$$T_x = 0.800\, T_0$$
$$T_x = T_x/0.800 = 1.25\, T_x$$

Friction loss = 25.07

@ point D $T_x = T_0 = 0.8\, f_{su}$
$= 0.8\,(240\text{ ksi}) = 192\text{ ksi}$

@ point B mid point of Tendon.

$$T_x = 0.8\,(T_0) = 0.8\left(192^{\text{ksi}}\right) = 153.6^{\text{ksi}}$$

6.4.3 Prestress Losses

6.4.3.1 2- Elastic Shortening

$(\Delta ts)_{AVG} = \frac{f_{ci}}{2E_C}$

$\Delta f_{sAVG} = E_S(\Delta ts)_{AVG}(\Delta ts)_{AVG} =$ average value of strain loss in tendons due to elastic shortening.

$f_{ci} =$ initial concrete prestress.

$E_C =$ modulus of elasticity of concrete.

$E_S =$ modulus of elasticity of steel

$\Delta f_{sAVG} =$ average value of steel prestress loss due to elastic shoetening.

$f_{ei} = 1.600\,\text{ksi}$

$(\Delta ts)_{AVG} = \frac{1.600}{2(5{,}000)} = 0.00016\,\text{in./in.}$

$\Delta f_{sAVG} = (29{,}000)(0.00016) = 4.64\,ksi$

6.4.4 Prestress Losses

6.4.5 3- Creep Loss

$\Delta tc = tcfci$

$\Delta ts = \Delta tc$

$\Delta fs = \Delta tsEs$

Notation

$\Delta tc =$ strain in concrete due to creep $(in./in.)$

$tc =$ unit strain in concrete due to creep ($in.$/in./psi)

$fci =$ initial concrete prestress (psi).

$\Delta ts =$ strain in steel due to creep (in./in.)

$Es =$ modulus of elasticity of steel(ksi)

$\Delta fs =$ loss of prestress in steel due to creep (ksi)

$tc = 0.20\Delta \times 10^{-6} in./in./psi$

$fci = 1600\,psi$(estimated)Actual $= 1.558$ psi

$\Delta tc = (0.0000002)(1600) = 0.00032\,in./in.$

$\Delta ts = \Delta tc = 0.00032\,in./in.$

$\Delta fs = (0.00032)(29000\,ksi)$

$\Delta fs = 9.280ksi$

6.4.6 Prestress Losses

6.4.6.1 4- Shrinkage Loss

$\Delta ts = \Delta ki$

$\Delta fs = Es\Delta ts$

Notation

Δtc = strain in concrete due to concrete shrinkage (in./in.)
Δts = loss of strain in steel due to concrete shrinkage (*in./in.*)
Es = modulus of Elasticity of steel (ksi)
Δfs = loss of prestress in steel Tendon due to concrete shrinkage (ksi)
$\Delta tc = 100 \times 10^{-6}$(in./in.)
$\Delta fs = (29.000 \times 0.0001) = 2.9$ ksi
$\Delta fs = 2.9$ ksi

6.4.6.2 5- Tendon Relaxation Loss

$\Delta fs = 0.08(0.65\text{f su})$
or
$\Delta fs = 12.5$ ksi

Notation

Δfs = loss of prestress in tendon due to tendon relaxation (ksi)
fsu = ultimate strength (ksi)
$\Delta fs = 0.08(0.65)(240\,\text{ksi})$
$= 12.48$ ksi
$\therefore$ Use $\Delta fs = 12.5$ ksi

6.4.7 Prestress Loss

6.4.7.1 Summary of Losses

1. Friction losses = 25.0%
2. Average Elastic Shortening =4.64 ksi
3. Creep Loss = 9.28 ksi
4. Shrinkage loss = 2.90 ksi
5. Tendon Relaxation = 12.50 ksi

Total losses due to Creep, Shrinkage and tendon relaxation = 9.28 + 2.90 + 12.50 = 24.68ksi

6.4.7.2 Primary Containment

Hoop Prestress

Prestress Losses
Tendon 1

$t_x = 0.9954$
$\Delta fs = 0.0046(192\,ksi)$
$= 0.883ksi\,.0941$
$\therefore\ T_X \approx 191$ ksi

Average initialtendon stress @ anchoring = $70\% f_{sci}$
= 0.7/2nd
= 168 ksi

For Tendon 1 use anchorage stress equal to 168 ksi plus average value of Elastic Shortening @ anchorage f_s Tendon 1 = 168 ksi + 4.64 ksi = 172.64 ksi @ critical section use f_s tendon 1 = 173 ksi.
Tendon 2
Friction Losses in Tendon @ critical Section using PC1 code Equation for Friction Loss

$T_o = T_X e^{(KL'+\mu\alpha)}$
$T_X = T_o e^{(KL+\mu\alpha)}$
Angle between $P + B$ and critical section = 74°52′ = 131 radians = 1.27
$L' = 70.17 \times 1.31 = 92'$
$KL' = \mu\ \alpha = (0.0003)(92) + (0.13)(1.31) = 0.1976$
$e^{-(KL'+\mu\alpha)=0.821^\circ}$
$T_X = (0.821)(0.995)T_o = 0.817T_o$

∴ For Tendon 2
$f_s = (0.817)(192) = 156.9ksi$ atcritical section.
Average Initial Prestress

Average prestress (elastic shortening) = $\frac{tendon1+tendon2}{2}$
= $\frac{173+156.9}{2} = 164.95ksi$
Average Elastic Shortening Loss = 4.64 ksi
∴Average initial prestress = 164.95 − 4.64 = 160.31 ksi

Average Final Prestress @ Critical Section

To obtain avg. final prestress @ critical section, subtract the losses due to creep shrinkage and tendon relaxation from avg. initial prestress.

Loss due to creep, shrinkage and tendon relaxation = 24.68^{ksi}
Average initial prestress = 160.36 ksi
Losses = 24.68 ksi
Avg = Final Prestress = 135.68 ksi

Stress At 1/4 Point (0°) of Tendon

$$f_{s_{\text{tandon\# 2}}} = f_{s_{\text{tandon\# 2}}} \Rightarrow average\, f_s = f_{s_{\text{tandon\# 1}}}$$

Tendon#1

$\gamma = 70.17'$
$\alpha = 37°.26' = 0.6533\ RADIONS$
$l = 45.84'$
$k = 0.0003$
$\mu = 0.13$

$kl + \mu\,\alpha = 0.0003(45.84') + 1.13(0.6533) = 0.09865$
$e^{-(kl+\mu\alpha)} = 0.90606$
$T_x = 0.906\,T_o(0.998)$
$f_s = 0.906(197) = 173.95\,\text{ksi}$

Average f_s@ critical section: $\frac{191+156.75}{2} = 173.87\,ksi$
Conclusion
Average stress at 1/4 point of Tendon is only slightly larger than average stress at critical section.
Initial Tendon Stresses
Jacking Prestress

Jacking prestress $= 0.8fsu$
Jacking prestress $= 0.8fsu$
$= 192^{ksi}$

Anchorage prestress

$Av.\,initial\,tendon = 0.70fsu$
$stress@anchoring = 0.70(240) = 168^{ksi}$

Achoring Prestress = Avg:initial tendon stress @ anchoring plus avg
= value of elastic shortenig
$= 168' + 4.64'\text{ksi}$
$= 172.64'\text{ksi}$

Initial Stress for Design
To obtain initial stress for design add the losses due to Creep, shrinkage, and 1/2 tendon relaxation loss due to the Avg: Final Prestress

Creep Loss $= 9.28\,\text{ksi}$
Shrinkage Loss $= 2.90\,\text{ksi}$
$\frac{1}{2}$Tendon relaxation loss $= 6.25\,\text{ksi}$
Av. final prestress $\frac{135.68}{154.11\,\text{ksi}}$
Initial stress for Design $= 154.11\,\text{ksi}$

Final Prestress Force
Sketch

Notes:

(1) Controlling load case is Equation T in PSAR pg.
(2) Earthquake load has no effect on required hoop prestress.
(3) NO principal stress check is required for this case.
(4) Temp: analysis calculations by NP 2-1-72 are ok. For Hoop direction.

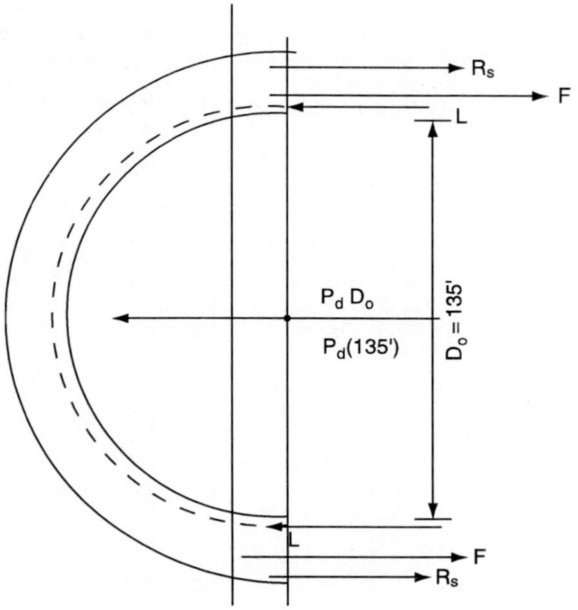

Fig. 6.18 Controlling load final prestressing force

6.4.7.3 Final Prestress Force

Design Assumptions

(1) Concrete is not counted on to resist tension.
(2) Liner load (L) and reinforced steel force (R_s) exactly balance; $R_s = L$

This is reasonable and conservative. As the cylinder expands due to pressure, the value of R_s increases; while the value of L decreases. At some value of cylinder deformation R_s and L are equal and opposite. Further deformation leads to Rs becoming greater than L, As $L \to 0$.

Loading:

$$p = 50\text{psi} \quad P_{\text{Design}} = P_d = 1.5P = 1.5(50)$$
$$P_d = 75\,\text{psi}$$
$$P_d.D_o = (0.75\,k/in.^2)(144\text{in}^2/\text{ft}^2)(135\,\text{ft})$$
$$P_d.D_o = 1459\,K/ft$$
$$\text{from} \sum F_{11} = 0 \quad 2F = P_d.D_o = 1458\,K/\text{ft}$$
$$F = 729\,k/\text{ft}$$

from $\Sigma F_{11} = 0\ 2F = P_d \cdot D_o = 1458\backslash\text{ft}\ F = 729\ k/\text{ft}$
This is the final prestressing force neglecting tendon elongation @accident.
The 'final' force @ accident + (As sargent & Lundy specify it) would be:

$F_{final} = F(\text{final stress before Ave/final stress before accident} + (\Delta f_s)\ \text{accident})$

Before accident $f_e = \frac{F}{Area} = \frac{729}{12(41.75)} \approx 1.455^{ksi}$
Close approximation

$\in_c \cong$ concrete strain (in./in.)
$f_c \cong$ concrete stress (ksi)
$E \cong$ modulus of elasticity (ksi)
$\in_c = \frac{f_c}{E_c} = \frac{1.455^{\text{ksi}}}{5000 ksi} = 0.00291$ in./in.

At accident $\in_c$ goes to zero

Δf_s = increase in steel stress(ksi)
E_s = modulus of elasticity of steel (ksi)
$\Delta f_s = E_s$(increase in steel strain which is equal to the loss of conc : strain)
$\Delta f_s = E_s t_c = 29000^{ksi}(0.000291)$
$\Delta f_s = 8.439^{ksi}$
Final tendon stress before accident = 135.68ksi
Final tendon stress after accident
= Final before accident + Δf_s = 135.68 + 8.349 = 144.12ksi
Final prestress Force = 729, (135.68) = 686.31k/ft

Initial Forces

$$\text{Jaking force} = \frac{\text{Jacking tendon stress}}{\text{Avg : final prestress}}(\text{final prestress force})$$
$$= \frac{192^{ksi}}{135.68}(678\ k/ft)$$
$$= 972.17\ k/ft$$

For Jacking Force use 973 kip/ft
Anchorage Force

$$\text{Anchorage forec}$$
$$= \frac{\text{Avg : initial tendon stress@anchoring}}{\text{Avg final prestres}}(\text{final prestress force})$$
$$= \frac{172.64}{135.68}$$
$$687\ k/ft$$
$$874.14\ k/ft$$

For anchorage force use 875 kip/ft
Initial Forces for Design

$$\text{Initial force design}$$
$$= (\text{finalforce @ accident})\frac{(\text{initial stress for design})}{\text{Avg : final prestress}}$$
$$= (678)\frac{154.11}{135.68}$$
$$= \frac{780.32\,k}{ft}$$

∴ For initial force for design use 781.0k/ft

$$f_{sci} = \text{concrete stress} = \frac{P}{\text{area}} = \frac{780.5}{12\times 41.75}$$
$$= 1.558\,\text{ksi} < f_c\,\text{allowable} = 0.30 f_i = 1650\,\text{ksi}$$

Estimated min spacing for stoop tendons

$$S_{\min} \cong \text{min spacing}$$
$$S_m = \frac{(\text{area of tendon})(\text{Av : final prestress})}{\text{finalprestress force @ acci}}$$
$$= \frac{(170\,\text{wries})(0.00491\,B''/\text{wire})(135.68\,\text{ksi})}{687\text{k/ft}}$$
$$S_m = 1.6485 ft = 19.78 in$$

Equipment Hatch
Tendon layout (horizontal)

Tendon $1(T_1)$
Radius $= 20'$
$a_1' = x_1' = 20\sin 20^\circ = 6.84'$
Avc length $= \frac{\pi(20)(20)}{180} = 6.98'$
$\dot{b}_1 = \dot{y}_1 = 20(20^2 x_1') = 1.21'$
$b_2 = y_2 = 11.375' - 2\dot{y}_1 = 8.96'$
$a_2 = x_2 = \frac{42'}{\tan 20^\circ} = 24.62'$
$2x_1 + x_2 = 38.30$

The straight portion of the tendon is an ellipse but for simplicity the length is assumed to be $(a_2^2 + b_2^2)^{1/2}$
a_2 is an arc length OK
Tendon 2 (T_2)

$$2a_1 + a_2 = 38.30''$$
$$2b_1 + b_2 = 11.375' - 2.625 = 8.750$$

Let $\theta = 14.806^\circ$

$a_1 = 20\sin\theta = 5.111$ radius $= 20'$
Arc length $= = \pi(20)\theta = 5.168$

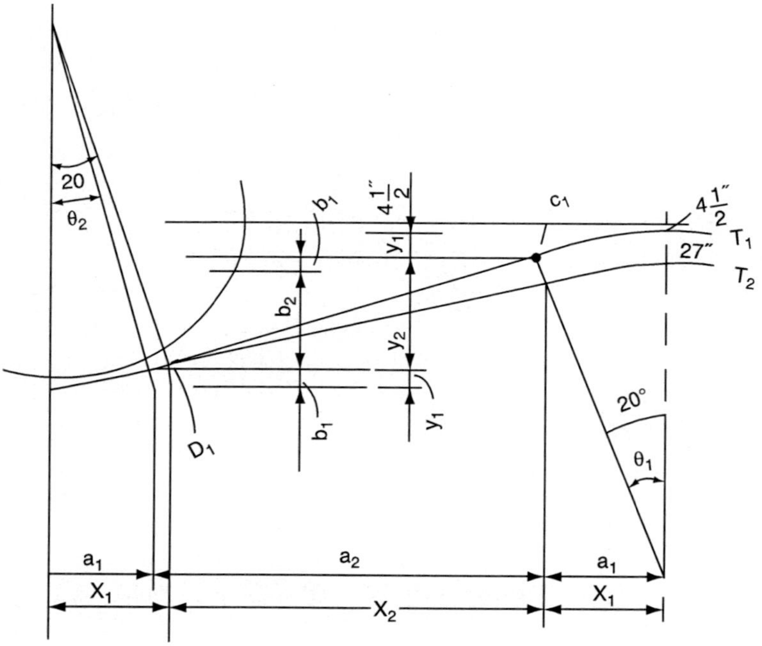

Fig. 6.19 Horizontal tendon layout for equipment hatch

$b_1 = 20 - (2\theta^2 - a_1^2)^{\frac{1}{2}} = 0.644$
$b_1 = 8.75 - 2b_1 = 7.422$
$a_2 = \frac{b_2}{\tan\theta} = 28.078$
$2a_1 + a_2 = 38 - 30$
$2c_1 + c_2 = x - y_1$
$a_1 = R\sin\phi \quad b_1 = R\sin\theta_2$
$Arc_1 = \frac{\pi R\theta_1}{180} \quad Arc_2 = \frac{\pi R\theta_2}{180}$
$c_1 = R - (R^2 - a_1^2)^{\frac{1}{2}} \quad d_1 = (R^2 - b_1^2)^{\frac{1}{2}}$
$c_2 = x - y - 2c_1 \quad d_2 = x - y_2 - 2d_1$
$a_2 = \frac{c_2}{\tan\theta_1} \quad b_2 = \frac{d_2}{\tan\theta_2}$

Solve for θ_1 by trial and error.

$2R; = 38.3\sin\theta_2 + (2R_i - x_i - y_i)a_2\theta_2$

let $\phi_1 = 16.16° \quad R = 21(T_3)$ let $\phi_2 = 11.857°(T_4)$
$2a_1 + a_2 = 39.30 \quad 2b_1 + b_2 = 38.30$
$2c_1 + c_2 = 9.375' \quad 2d_1 + d_2 = 7.125'$
$a_1 = 21\sin\theta_1 = 5.845 \quad b_1 = 21\sin\theta_2 = 4.315$
$\text{Arc} = \frac{\pi 2\phi\theta_1}{180} = 5.923 \quad \text{Arc} = \frac{\pi 21\theta}{180} = 4.346$
$c_1 = 21 - (21^2 + a_1^2)^{\frac{1}{2}} = 8.30 \quad d_1 = 21 - (21^2 + b_1^2)^{\frac{1}{2}} = 0.448$
$c_2 = 9.735' - 2c_1 = 7.716 \quad d_2 = 7.125 - 2d_1 = 6.229$
$a_2 = \frac{c}{\tan\phi_1} = 24.426 \quad b_2 = \frac{d_2}{\tan\phi_2}$

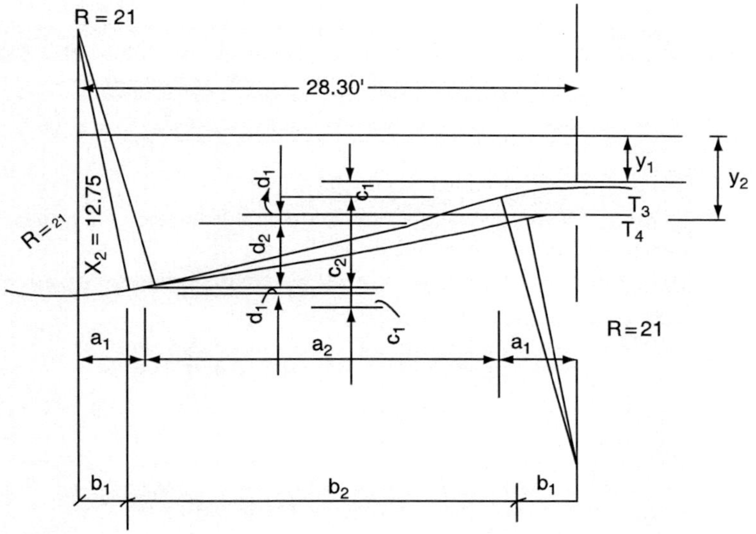

Fig. 6.20 Parameters for tendon layout

$$2b_2 + b_2 = 38.289\,Ok$$
$$2a_2 + a_2 = 34.316\,Ok$$

6.4.7.4 Equipment Hatch

Tendon Layout (Horizental)
Deflected Tendon

Tendon	R	θ_i	a_{1i}	a_{2i}	c_{1i}	C_{2i}	A_{rc}	A	X	y
T1	20'	20°	6.84	24.62	1.21	8.96	6.98	26.200	11.75	0.375
T2	20'	14.806	5.111	22.078	0.661	7.422	5.163	29.042	11.75	2.625
T3	21'	16.16	5.845	26.226	0.830	7.716	5.923	27.721	12.75	3.375
T4	21'	11.875	4.315	29.668	0.448	6.229	4.346	30.315	12.75	5.625
T5	22'	12.41	4.728	28.844	0.514	4.097	4.765	29.134	13.75	6.375
T6	22'	8.306	3.178	33.523	0.231	4.894	3.189	33.878	13.75	8.625
T7	23'	8.789	3.514	31.271	0.270	4.835	3.528	31.643	14.75	9.375
T8	23'	4.917	1.971	34.357	.085	2.956	1.974	34.484	·14.75	11.625
T9	24'	53.47	2.236	33.928	o.104	3.166	2.240	33.976	15.75	12.375
T10	24'	1.715	0.718	36.855	.011	1.103	0.718	36.872	15.75	14.625
T11	25'	2.107	0.919	36.455	.017	1.341	0.919	36.480	16.75	15.375

Purpose

To determine the curvature and radii of the deflected tendons occurred the equipment hatch.

Data

See drawing 41N711-A for size of equipment hatch and tendon size and spacing.

Assumptions

1- Tendons deflected along the circular curve on the developed elevation
2- Maximum angle of deflection of 70°
3- Centreline of tendon assumed 1'-0" from the edge of hatch sleeve.

References

1- Dwgs 41N711-1 thru-4
2- Calculation by HRT "Equipment"
 Hatch Tendon layout 11-9.72
3- Olivetti program by Rw Hanorton "Tendon Deflection"

Discussion

Due to cross found in HRTS calculations (Equp: Hatch Tendon Layout) the following will be a revised set of calculations.

Calculations

Tendon 1 Radius $= 20'$, $x_1 = 20 \sin 20° = 20(0.342°02) = 6.24'$
 Using Pyth's theorem to find y_1 ∴

$$y_1 = 20 - (20^2 - x_1^2)^{1/2} = 20 - 18.79 = 1.21'$$
$$y_2 = 12' - 4\tfrac{1''}{2} - 2y_1 = 12' - 0.375' - 2.42' = 9.21'$$

From ΔAEC

$$X_2 = \frac{y_2}{\tan 20°} = \frac{9.21}{0.363} = 25.3'$$
$$2x_1 + x_2 = 13.68 + 25.3 = 38.98' \text{say } 39'$$
$$Arc\,\text{Length} = \frac{\pi(20)(20)}{180} = 6.98'$$

Table 6.4 Membrane forces in parabolic domes

Loading	N'_ϕ	N'_θ	$N'_{\phi\theta}$
w/unit area of surface per m^2 of surface Normal 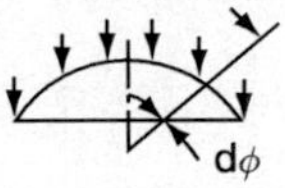 $d\phi$	$-\dfrac{wr_0 1-\cos^3 d\phi}{3\sin^2 d\phi \cos^2 d\phi}$	$-\dfrac{wr_0 2-3\cos^2 \partial\phi+\cos^2 \partial\phi}{3\sin^2 \partial\phi}$	0
p/unit area of projection 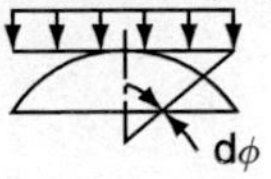 $d\phi$	$-\dfrac{pr_0}{2\cos d\phi}$	$-\dfrac{pr_0}{2}\cos \partial\phi$	0

6.5 Computer Program NIZAM

A typical computer program, NAZAM, is developed for the spherical dome shape made in reinforced concrete. With small changes incorporated, the same dome can be analysed in any material, provided correct data can be included. The NAZAM listing is given below.

Computer program: NAZAM listings

```
C     *    TITLE: COMPUTER AID ANALYSIS OF THIN SHELLS              *
C     *          : Dr. Y. BANGASH                                    *
C     **********************************************************
C     * The original program was modified by NIZAM of Malaysia,
C     * under the supervision of Y. Bangash.
C     **********************************************************
C     *           COMPUTER PROGRAM FOR CASE STUDIES
C     *           ==================================
C     * THIS IS A PROGRAM TO FIND OUT THE MERIDIAN                 *
C     * FORCES, THE HOOP FORCES AND RING TENSION ACTING             *
C     * ON EACH SECTION OF THE SPHERICAL DOME DUE TO ITS            *
C     * DEAD LOAD AND IMPOSED LOAD.                                 *
C     ******************************************************** C
C
C     ==========================================================
C     =                    * NOTATIONS *                         =
C     =                                                          =
C     =           QK  UNIFORMLY DISTRIBUTED LIVE LOAD            =
C     =           GK  UNIFORMLY DISTRIBUTED DEAD LOAD            =
C     =           TY  THICKNESS OF THE DOME                      =
C     =           HI  RISE AT POINT 1                            =
C     =           H0  RISE AT POINT 0                            =
C     =           A   LENGTH OF LATITUDE AT POINT 1              =
C     =           B   LENGTH OF LATITUDE AT POINT 0              =
C     =                                                          =
C     =                                                          =
C     ==========================================================
```

```
C
C
      REAL LY, LX, X, Y, AS, ASR, BAR, SP
C
3000  WRITE (6, 10)
      FORMAT (1X, 'INPUT THE SPAN UNI FORMLY
      DISTRIBUTED LIVE LOAD' )
      READ (5, *) QK
      WRITE (6, 15)
15    FORMAT (/, 1X, 'INPUT THE DEAD LOAD' )
      READ (5, *) GK
      WRITE (6, 20)
20    FORMAT (/, 1X, 'INPUT THE THICKNESS OF THE DOME' )
      READ (5, *) T
      WRITE (6, 30)
30    FORMAT (/, 1X, 'INPUT THE SIN0' )
      READ (5, *) SIN0
      WRITE (6, 40)
40    FORMAT (/, 1X, 'INPUT SIN1' )
      READ (5, *) SIN1
      PRINT*, 'SP=' , SP
      IF (ASR.GE.141.AND.ASR.LE.226.AND.SP.GE.125.AND.SP.LE.200)
      *      GO TO 25
      IF (ASR.GE.226.AND.ASR.LE.355.AND.SP.GE.150.AND.SP.LE.200)
      *      GO TO 27
      IF (ASR.GE.392.AND.ASR.LE.532.AND.SP.GE.150.AND.SP.LE.200)
      *      GO TO 29
      IF (ASR.GE.565.AND.ASR.LE.904.AND.SP.GE.125.AND.SP.LE.200)
      *      GO TO 31
      IF (ASR.GE.1005.AND.ASR.LE.1340.AND.SP.GE.150.AND.SP.LE.200)
      *      GO TO 33
      IF (ASR.GE.1571.AND.ASR.LE.2094.AND.SP.GE.150.AND.SP.LE.200)
      *      GO TO 35
      IF (ASR.GE.2454.AND.ASR.LE.2805.AND.SP.GE.175.AND.SP.LE.200)
      *      GO TO 3740
      WRITE(6,500)
500   FORMAT(/,16X,'HALTED IN REINFORCEMENT' )
      STOP
25    PRINT*, 'BAR=6'
      RETURN
27    PRINT*, 'BAR=8'
      RETURN
29    PRINT*, 'BAR=10'
      RETURN
31    PRINT*, 'BAR=12'
      RETURN
33    PRINT*, 'BAR=16'
      RETURN
35    PRINT*, 'BAR=22'
      RETURN
37    PRINT*, 'BAR=25'
      RETURN
      END
```

```
C
C     **************************************************************
C     *        SUBROUTINE TO FIND MOMENT COEFFICIENT               *
C     **************************************************************
C
      SUBROUTINE TABL12(LY,LX,X,Y)
      IF(RATIO.GT.3.0) GOTO 160
      IF(RATIO.EQ.3.0) GOTO 81
      IF(RATIO.GE.2.5) GOTO 82
      IF(RATIO.GE.2.0) GOTO 83
      IF(RATIO.GE.1.8) GOTO 84
      IF(RATIO.GE.1.5) GOTO 85
      IF(RATIO.GE.1.4) GOTO 86
      IF(RATIO.GE.1.3) GOTO 87
      IF(RATIO.GE.1.2) GOTO 88
      IF(RATIO.GE.1.1) GOTO 89
      IF(RATIO.GE.1.0) GOTO 90
81    X=0.124
      Y=0.014
      RETURN
82    X=0.122
      Y=0.020
      RETURN
83    X=0.118
      Y=0.029
      RETURN
84    X=0.113
      Y=0.037
      RETURN
85    X=0.104
      Y=0.046
      RETURN
86    X=0.099
      Y=0.051
      RETURN
87    X=0.093
      Y=0.055
      RETURN
88    X=0.084
      Y=0.059
      RETURN
89    X=0.0774
      Y=0.061
      RETURN
90    X=0.062
      Y=0.062
      RETURN
160   WRITE(6,170)
170   FORMAT(//,16X,'SLAB TOO NARROW - DESIGN AS ONE WAY SPANNING')
      RETURN
      END
```

6.6 Conoidal dome's surfaces

A conoidal dome is analysed in this section. The dome can be of any material but the geometry of this surface is different from the spherical surface. Using Figure 6.1 the radius of the circle of latitude for the conoidal dome is,$r \sin\phi - r'$.

Consider the conoidal dome shown in Figure 6.17, say, between a plane of latitute through points 0 and 1.

Length of element $= r \times d\phi$

Area of element rotated about axis $= r\, d\phi \times 2\pi(r \sin\phi - r')$

Total surface area A, of dome is:

$$
\begin{aligned}
A &= \int_{\phi_0}^{\phi_1} r\, d\phi \times 2\pi(r \sin\phi - r') \\
&= 2\pi r^2(\cos\phi_0 - \cos\phi_1) - 2\pi r r'(\phi_1 - \phi_0)
\end{aligned}
\tag{6.33}
$$

Assuming a uniform distributed load (M, W) for all elements

Total load between points 1 and 0 is:

$$W_u = 2\pi r^2 W(\cos\phi_0 - \cos\phi_1) - 2\pi r r' W(\phi_1 - \phi_0) \tag{6.34}$$

To find the meridional thrust T

Substitute$r \sin\phi_1 - r'$ for$r \sin\phi$. Then

$$
\begin{aligned}
W &= 2\pi(r \sin\phi_1 - r') \\
T &= W/(2\pi(r \sin\phi_1 - r') \sin\phi_1
\end{aligned}
\tag{6.35}
$$

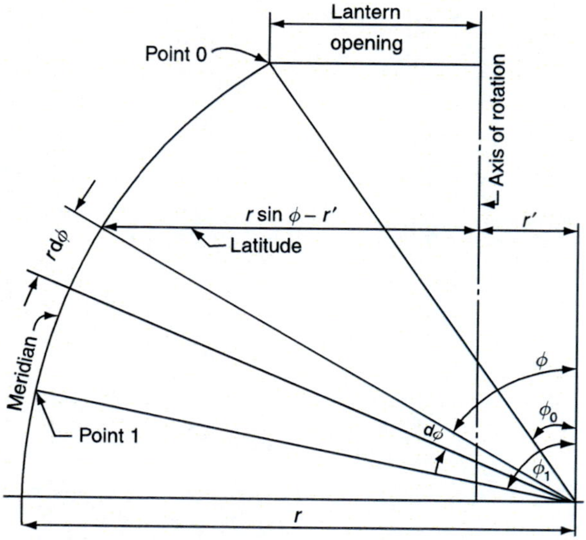

Fig. 6.17(f) Conoidal dome

To find the hoop force H

$$T/r + H/(r\sin\phi_1 - r') \times \sin\phi_1 - [W + W'(\phi_1 - \phi_0)]\cos\phi_1 = 0$$

$$H = \frac{r\sin\phi_1 - r'}{r\sin\phi_1}[-T + [W + W'(\phi_1 - \phi_0)]r\cos\phi_1$$

If the dome is continuous along the circle of latitude through 1, a circular ring through that point is subjected to a unit radial force of $T \times \cos\phi_1$. Therefore

$$S\text{ (ring tension)} = \frac{W_u \times \cos\phi_1}{2\pi \times \sin\phi_1} \tag{6.36}$$

Three examples are included to analyse a conical surface. They are based on:

Example 6.2 –Conical tent under surface or dead weight
Example 6.3 –Conical tent under uniform load W
Example 6.4 –Conical umbrella roof supported on a central column.

Another example (Example 6.5) is given on a truncated conical dome type tent. General solutions are incorporated. By introducing specific data the same shape can be incorporated for any material such as plastic, glass, fabric, concrete and steel/aluminum. Table 6.5 gives other cases under different loads for conical and truncated conical dome shapes.

Example 6 Conical dome tent

The conical dome tent (Fig. 6.18) is generally used for roofs and closures for storage structures and temporary dwellings.

Conical tent under selfweight of deadweight
For the analysis the value of r_1 isα and

$$r_0 = r_e\sin\phi = r_2\cos\alpha$$

where α is the semi-angle at the apex C.

$$W = \text{Load} = \int_0^y 2\pi r_0 w\,\mathrm{d}y = \pi w y^2 \sin\alpha \tag{6.37}$$

Table 6.5 Conical and truncated conical dome shapes

Type	Loading	Resultant components	N_ϕ	N_θ	T_θ
1	p_s	$w\,\mathrm{d}s\sin\alpha = p_s$ $w\,\mathrm{d}t\cos\alpha = p_t$	$\frac{\omega_\alpha}{2S}\frac{L^2-S^2}{\sin\alpha}$	$-\omega\,\mathrm{d}s\cos\alpha\cot\alpha$	0
2	p_t	$p_s = w_L\sin\alpha\cos\alpha$ $p_t = -\omega\cos^2\alpha$	$\omega_L\frac{L^2-S^2}{2S}\cot\alpha$	$-\omega_L S\cos^2\alpha\cot\alpha$	0
3	p_t	$p_t = -p\sin\alpha\cos\theta$	$p[L^2 - S^2]\frac{(1/3)S^2-(1/2S)\sin^2\alpha}{\cos\alpha/\cos\theta}$	$p\frac{L^2-S^2}{3S^2}\sin\theta$	

Fig. 6.18 Conical dome tent

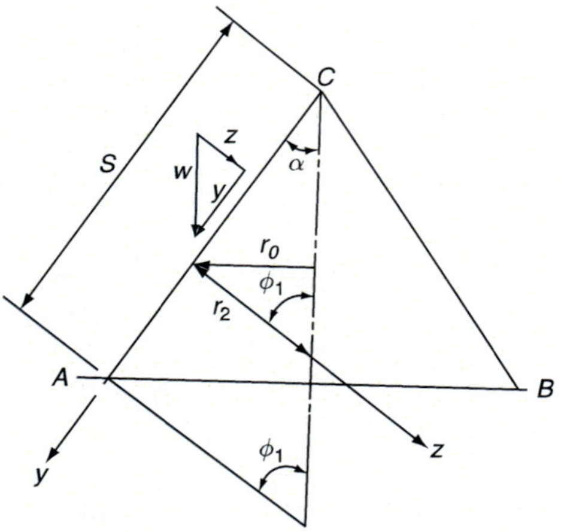

$$N_\phi = -\frac{w_y}{2\cos\alpha} \tag{6.38}$$

$$N_\theta = -w_y \sin\alpha\cos\alpha \tag{6.39}$$

***Example 6 Conical dome tent under uniform load* w**

(Figs. 6.19 and 6.20)

Let the vertical distance be represented by 'z'. The total load w_z and stress resultants N_ϕ and N_θ are computed as given below:

$$\begin{aligned} A &= \text{Surface Area} \\ &= 2\pi\int_0^h z\tan\alpha\sec\alpha\,\mathrm{d}z \\ &= \pi h^2\tan\alpha\sec\alpha \end{aligned} \tag{6.40}$$

$$V = \text{Volume}$$

$$V = \pi\int_0^h z^2\tan^2\alpha\mathrm{d}z \tag{6.41a}$$

$$= \frac{\pi}{3}h^2\tan^2\alpha \tag{6.41b}$$

$$\begin{aligned} w_z &= w\pi z^2\tan\alpha\sec\alpha \\ \phi &= (90^\circ - \alpha) \end{aligned} \tag{6.41c}$$

Fig. 6.19

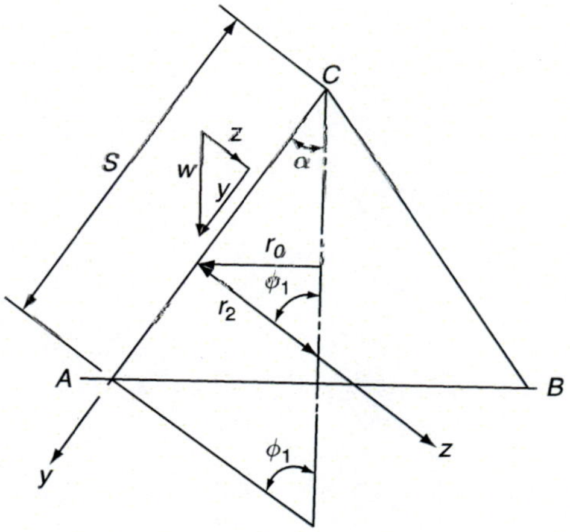

$$N_\phi = -\frac{w_z}{2\pi r_2 \sin^2\phi} = -\frac{w\pi z^2 \tan\alpha \sec\alpha}{2\pi r \cos\alpha}$$

$$= -\frac{w_z^2 \tan\alpha \sec\alpha}{2z \tan\alpha \cos\alpha} = -\frac{1}{2} w_z \sec^2\alpha \qquad (6.41\text{d})$$

$$N_\theta = -r_2 w \sin\alpha = -w_z \tan^2\alpha \qquad (6.41\text{e})$$

Example 6.4 Conical umbrella roof supported on a central column

The analysis of the conical umbrella roof is shown in Figure 6.21. The roof is subjected to a uniform load w/unit area. The stress resultants are computed together with load between parallel circles.

The load on the surface between parallel circles A–A and B–B, i.e. on surface below z''

$$W_{AB} = w\pi(h^2 - z^2)\tan\alpha \sec\alpha \qquad (6.41\text{f})$$

Equilibrium of vertical forces at level A–A gives

$$N_\phi \cos\alpha 2\pi z \tan\alpha = W_{AB} \qquad (6.41\text{g})$$

$$N_\phi = \frac{w(h^2 - z^2)\sec^2\alpha}{2z} \text{ (tensile)} \qquad (6.41\text{h})$$

with$R_2 = Z \tan\alpha \sec\alpha$

$$N_\theta = -w_Z \tan^2\alpha; \quad N_\phi \to \alpha \text{ as} Z \text{ tends to zero.} \qquad (6.41\text{i})$$

This indicates that membrane theory is inadequate since bending moments a brought into play.

Fig. 6.20

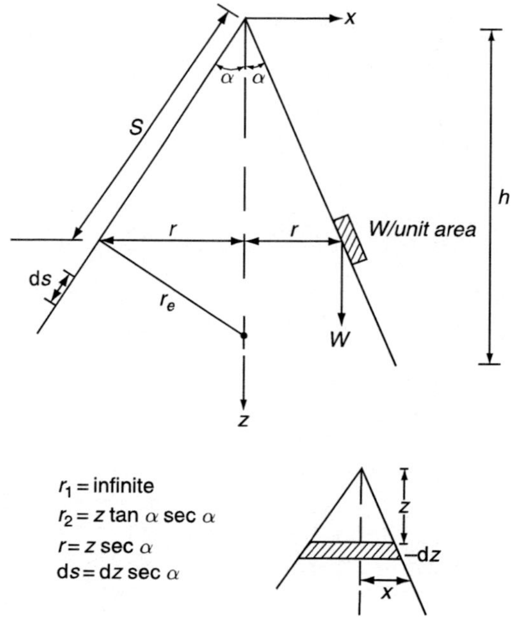

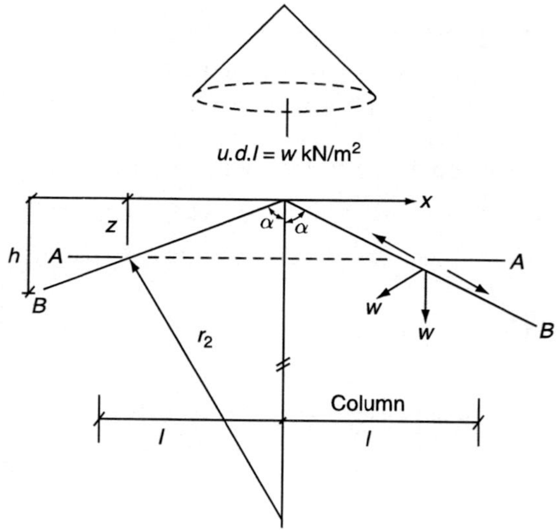

Fig. 6.21 Conical umbrella roof supported on a central column

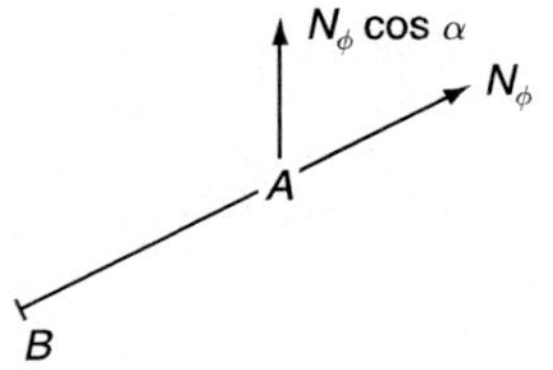

(6.42)

$N_\phi \cos\alpha =$ the value of the vertical force in ϕ direction.

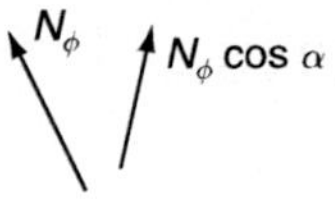

(6.43)

N_ϕ and$N_\phi \cos\alpha =$ two components in ϕ direction.

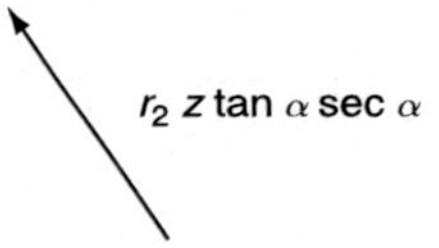

(6.44)

$r_2 z \tan\alpha \sec\alpha =$ the inclined component.

Truncated conical dome

The conical dome is truncated at a length S_0, and obviously the values of N_ϕ and N_θ will be affected. The generalized equation for N_ϕ and N_θ for a full length is given by

Table 6.6 Truncated conical domes

Type	Loading	Resultant components	N_ϕ	N_θ	T
1	w/unit area	$w\,\mathrm{d}s \sin\alpha = p_S$ $-w\,\mathrm{d}\cos\alpha = p_r$	$-w_\mathrm{d}\dfrac{S^2 - S_0^2}{2R}\dfrac{L}{\sin\alpha}$	$-w_{\mathrm{d}S}\cos\alpha\cot\alpha$	0
2	p/unit area	$p_S = w_L \sin\alpha\cos\alpha$ $p_1 = -w_L \cos^2\alpha$	$-w_L\dfrac{\sin^2 - S_0^2}{2S}\cot\alpha$	$-w_{LS}\cos^2\alpha\cot\alpha$	–
3	p_r/radius	$p_r = p\sin\alpha\cos\theta$	$-\dfrac{p_S S}{3}\left[\cos\alpha - \dfrac{1}{3\cos\alpha} - \dfrac{S_0^2}{S^2}\left(\cos\alpha - \dfrac{1}{\cos\alpha}\right) - \left(\dfrac{S_0^3}{S_0}\dfrac{2}{3\cos\alpha}\right)\right]\cos\theta$	$-p_S\cos\alpha\cos\theta$	$-\dfrac{p_S^2 - S_0^2}{3S^2}\sin\theta$
4	p_w/unit area	$S_0 = 0$	$-p_w\dfrac{S_0}{S}\dfrac{1}{\sin\alpha}$	–	0
			$\dfrac{p_w}{2\pi S\sin\alpha\cos\alpha}$	–	$\dfrac{pL^2 - S^2}{2S^2}\sin\theta$

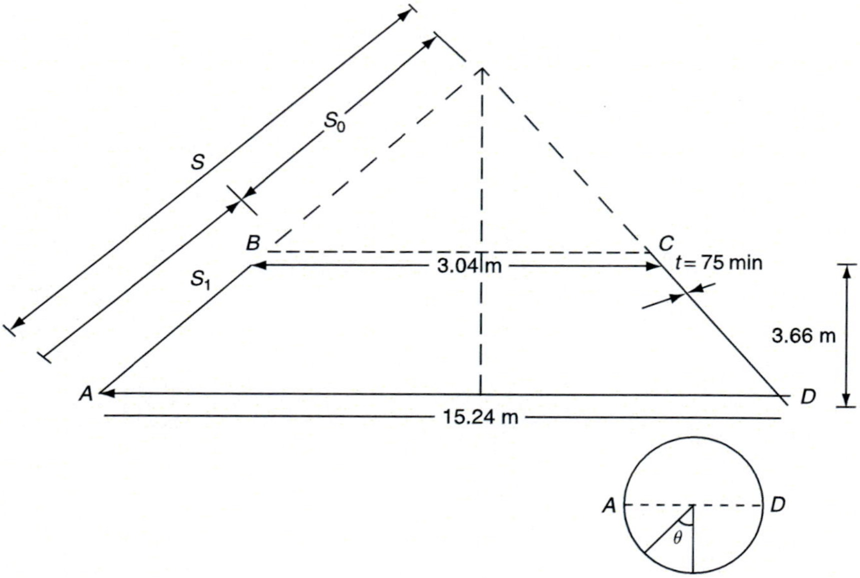

Fig. 6.22 Truncated shapes

$$N_\phi = -\frac{1}{S}\left[\frac{wS^2}{2\sin\alpha} + \text{const } C\right] \tag{6.45}$$

$$N_\theta = -wS\cos\alpha\cot\alpha \tag{6.46}$$

Table 6.5 gives a good description of a truncated conical dome. Table 6.6 shows other loading cases for conical roofs and truncated domes.

Example 6 Truncated conical dome type tent

Data: Uniform load W (excluding at the top of the opening= 2.394 kN/m^2.

Uniform load (daylight lantern = 0.035 kN/m)

Determine the stress resultants and the compressive stress for the truncated conical dome.

$$N_\phi = -\frac{1}{S}\left[\frac{wS^2}{2\sin\alpha} + C\right]$$

For a truncated conical shell for a sloping length of S_0

$$N_S = 0 \text{ at } S_0 0 = -\frac{1}{S_0}\left[\frac{wS_0^2}{2\sin\alpha} + C\right]$$

or

$$C = -\frac{wS_0^2}{2\sin\alpha}$$

$$N_\phi = \frac{w}{2}\frac{(S^2 - S_0^2)}{2S\sin\alpha}; \quad N_\theta = -wS\cos\alpha\cot\alpha$$

Determination of α and S_0

$$\tan\alpha = \frac{3.66}{\left(\frac{15.24}{2} - \frac{3.04}{2}\right)} = 0.6 \quad \therefore \alpha = 30^{\circ}58'$$

Sloping length$S = \sqrt{(3.66^2 + 6.1^2)} = 7.114\,\text{m}$

$$\therefore \frac{7.114}{7.114 + S_0} = \frac{6.1}{7.62}$$

$$S_0 = 1.773\,\text{m}$$

Forces due to dead weight $N_\theta = -wS\cos\alpha\cot\alpha = -1.368S$
For $S = S_0 = 1.773\,\text{m}$ $N_\theta = -2.426\,\text{kN/m}$
$S = 7.114\,\text{m}$ $N_\theta = -9.732\,\text{kN/m}$

$$N_\phi = -2.394\frac{S^2 - (1.773)^2}{2S}\left(\frac{1}{0.5145}\right) = \frac{2.33(S^2 - 3.144)}{S}$$

For $S = S_0 = 1.773\,\text{m};$ $N_\phi = 0$
$S = 7.114\,\text{m};$ $N_\phi = -2.33\frac{(7.104)^2 - 3.144}{3.114}$

$$= -15.546\,\text{kN/m}$$

Forces due to uniform vertical load at top edge

$$N_\phi = -w' \cdot \frac{1}{\sin\alpha} \cdot \frac{1.773}{S}$$

$$= -0.0035 \times \frac{1.773}{S} \times \frac{1}{0.5145}$$

$$= \frac{-0.0121}{S}$$

For $S = S_0 = 1.773;$ $N_\phi = -0.0068\,\text{kN/m}$
$S = 7.114;$ $N_\phi = -0.0027\,\text{kN/m}$

Total $N_{\phi(\max)} = (-0.0027) + (-15.546) = 15.5487\,\text{kN/m}^2$

$$\text{Max. compressive stress} = \frac{15.5487}{0.075 \times 1} = 207.316\,\text{kN/m}^2$$

$$= 0.207316\,\text{N/mm}^2$$

Computer program: Conoidal domes

```
C      **************************************************
C      *    TITLE: COMPUTER AID ANALYSIS OF THIN SHELLS      *
C      *         : Dr. Y. BANGASH                              *
C      **************************************************
C      * Assisted by NIZAM of Malaysia
C
C      **************************************************
C      * THIS IS A PROGRAM TO FIND OUT THE MERIDIAN           *
C      * FORCES, THE HOOP FORCES AND RING TENSION ACTING      *
C      * ON EACH SECTION OF THE CONOIDAL DOME DUE TO ITS      *
C      * DEAD LOAD AND IMPOSED LOAD.                          *
C      **************************************************
C
C
C      ==================================================
C      =                    * NOTATIONS *                   =
C      =                                                    =
C      =    GK UNIFORMLY DISTRIBUTED LIVE LOAD               =
C      =    QK  UNIFORMLY DISTRIBUTED DEAD LOAD              =
C      =     TX  THICKNESS OF THE DOME                       =
C      =      R  THE RADIUS OF CURVATURE                     =
C      =     R1  DISTANCE FROM AXIS OF ROTATION              =
C      =   X(I)  LENGTH OF LATITUDE AT POINT I               =
C      =  AE(I)  AREA BETWEEN POINT 1 AND I                  =
C      =     WC  COLLAR LOAD                                 =
C      =  WU(I)  TOTAL LOAD BETWEEN POINT 1 AND I            =
C      =   T(I)  MERIDIONAL THRUST AT POINT I                =
C      =  TS(I)  STRESS DUE TO MERIDIONAL THRUST AT POINT I  =
C      =   H(I)  HOOP FORCE AT POINT I                       =
C      =  HS(I)  STRESS DUE TO HOOP FORCE AT POINT I         =
C      =   S(I)  RING TENSION AT POINT I                     =
C      =     SC  COMPRESSION IN EDGE MEMBER AT THE OPENING   =
C      =  SN(I)  SIN ANGLE AT POINT 0                        =
C      =  CS(I)  COSINE ANGLE AT POINT 0                     =
C      = PIE(I)  ANGLE AT POINT 0 IN RADIAN                  =
C      =  SN(I)  SIN ANGLE AT POINT 1                        =
C      =  CS(I)  COSINE ANGLE AT POINT 1                     =
C      = PIE(I)  ANGLE AT 1 IN RADIAN                        =
C      ==================================================
C
C
       REAL QK, GK, WC, TX, R, R1, Q1, PI, SC
       REAL X(1:6), Y(1:6), Z(1:6), SN(1:6), CS(1:6), PIE(1:6),
       REAL S(1:6), AE(1:6), WU(1:6), T(1:6), TS(1:6), H(1:6)
       REAL HS(1:6), XB(1:6)
3000   WRITE(6, 10)
10     FORMAT (1X, 'INPUT THE UNIFORMLY DISTRIBUTED LIVE LOAD')
       READ(5, *) QK
       WRITE(6, 15)
15     FORMAT(/, 1X, 'INPUT THE DEAD LOAD')
       READ(5, *) GK
```

```
      WRITE(6, 17)
17    FORMAT(/, 1X, 'INPUT COLLAR LOAD')
      READ(5, *) WC
      WRITE(6, 20)
20    FORMAT(/, 1X, 'INPUT THE THICKNESS OF THE DOME')
      READ(5, *) TX
      WRITE(6, 50)
50    FORMAT (/, 1X, 'INPUT THE RADIUS OF THE CURVATURE')
      READ(5, *) R
      WRITE(6, 60)
60    FORMAT (/, 1X, 'INPUT DISTANCE FROM THE AXIS OF ROTATION')
      READ(5, *) R1
      WRITE(6, 70)
70    FORMAT(/, 1X, 'INPUT THE LENGTH ALONG THE X-AXIS')
      READ(5, *) X(1), X(2), X(3), X(4), X(5), X(6)
      DO 80 I=1,6
      XB(I)=R1+X(I)
80    CONTINUEDO 81 I=1,6
      Y(I)=SQRT((R**2)-(X(I)**2))
81    CONTINUE
      Q1=2400
      PI=3.142
      DO 90 I=1,6
      Z(I)=SQRT((Y(I)**2)+((R1+X(I))**2))
90    CONTINUE
      DO 100 I=1,6
      SN(I)=(R1+X(I))/Z(I)
100   CONTINUE
      DO 100 I=1,6
      CS(I)=Y(I)/Z(I)
110   CONTINUE
      DO 120 I=1,6
      PIE(I)=ASIN(SN(I))
120   CONTINUE
      DO 130 I=1,6
      AE(I)=2*PI*R**2*(CS(1)-CS(I))-2*PI*R*R1*(PIE(I)-PIE(1))
130   CONTINUE
      DO 140 I=1,6
      WU(I)=AE(I)*(GK+QK)+WC*2*PI*X(I)+(QK+GK)*(PIE(I)-PIE(1))
140   CONTINUE
C     SC=COMPRESSION IN EDGE MEMBER AT OPENING
      SC=(WC*2*PI*(R1+X(1))*CS(1))/(2*PI*SN(1))
C     T1=MERIDIONAL THRUST AT POINT 1
      DO 150 I=1,6
      T(I)=WU(I)/((2*PI)*(R*SN(I)-R1)*SN(I))
150   CONTINUE
C
C     TS1=STRESS AT POINT 1 DUE TO MERIDIONAL THRUST
      DO 160 I=1,6
      TS(I)=T(I)/TX
160   CONTINUE
C
C     TO FIND THE HOOP FORCE
```

```
C       H1=HOOP FORCE AT POINT 1
        DO 170 I=1,6
        H(I)=((R*SN(I)-R1)/(R*SN(I)))*(-T(I)+(GK+QK)*R*CS(I))
170     CONTINUE
C
C       HS1=STRESS DUE TO HOOP FORCE
        DO 180 I=1,6
        HS(I)=H(I)/TX
180     CONTINUE
C
C       S1=RING TENSION IN EDGE MEMBER
        DO 190 I=1,6
        S(I)=(WU(I)*CS(I))/(2*PI*SN(I))
190     CONTINUE
C
        PRINT 191
        PRINT 192
191     FORMAT(///,31X,'OUTPUT RESULTS')
192     FORMAT(31X, '**************')
        WRITE(6,200) XB(1),XB(2),XB(3),XB(4),XB(5),XB(6)
200     FORMAT(//,1X,'POINTS',1X,F7.3,4X,F7.2,4X,F7.2,4X,
        * F7.2,4X,F7.2,4X,F7.2)
        WRITE(6,210) AE(1),AE(2),AE(3),AE(4),AE(5),AE(6)
210     FORMAT(/,1X,'AREA',5X,F7.2,4X,F7.2,4X,F7.2,4X,F7.2,4X,
        * F7.2,3X,F8.2)
        WRITE(6,220) WU(1),WU(2),WU(3),WU(4),WU(5),WU(6)
220     FORMAT(/,1X,'WEIGHT',2X,F8.2,3X,F8.2,3X,F8.2,3X,F8.2,3X,
        * F8.2,3X,F8.2)
        WRITE(6,230) T(1),T(2),T(3),T(4),T(5),T(6)
230     FORMAT(/, 1X, 'THRUST',2X,F8.2,3X,F8.2,3X,F8.2,3X,F8.2,3X,
        * F8.2,3X,F8.2)
        WRITE(6,240) TS(1),TS(2),TS(3),TS(4),TS(5),TS(6)
240     FORMAT(/,1X,'TSTRESS',1X,F8.2,3X,F8.2,3X,F8.2,3X,F8.2,3X,
        * F8.2,3X,F8.2)
        WRITE(6,250) H(1),H(2),H(3),H(4),H(5),H(6)
250     FORMAT(/,1X,'HOOP',4X,F8.2,3X,F8.2,3X,F8.2,3X,F8.2,3X,
        * F8.2,3X,F8.2)
        WRITE(6,260) HS(1),HS(2),HS(3),HS(4),HS(5),HS(6)
260     FORMAT(/,1X,'HSTRESS',1X,F8.2,3X,F8.2,3X,F8.2,3X,F8.2,3X,
        * F8.2,3X,F8.2)
        WRITE(6,270) S(1),S(2),S(3),S(4),S(5),S(6)
270     FORMAT(/,1X,'RING',4X,F8.2,3X,F8.2,3X,F8.2,3X,F8.2,3X,
        * F8.2,3X,F8.2)
        WRITE(6,280) SC
280     FORMAT(/,1X,'COMPRESSION IN EDGE MEMBER AT
        * OPENING =',1X,F10.2)
        END
```

Elliptical dome-shaped surfaces

The following analysis is carried out for the elliptical dome (Fig. 6.23). Equation of ellipse:

$$\frac{X^2}{a^2} + \frac{Y^2}{b^2} = 1$$

where $X = a \sin\phi$ and $Y = b\cos\phi$. Consider a ring produced by rotating an element ds about the Y axis of the dome.

$$\begin{aligned} \mathrm{d}s &= \sqrt{[(\mathrm{d}x)^2 + (\mathrm{d}y)^2]} \\ &= \sqrt{[(a^2\cos^2\phi + b^2\sin^2\phi)]\mathrm{d}\phi} \end{aligned}$$

Weight of element ring is

$$\begin{aligned} \mathrm{d}w &= 2\pi \times X \times W \times \mathrm{d}s \\ &= 2\pi \times w \times a^2 \sin\phi\sqrt{(1 - [((a^2 - b^2)/a^2)\sin^2\phi]} \times \mathrm{d}\phi \end{aligned} \quad (6.47)$$

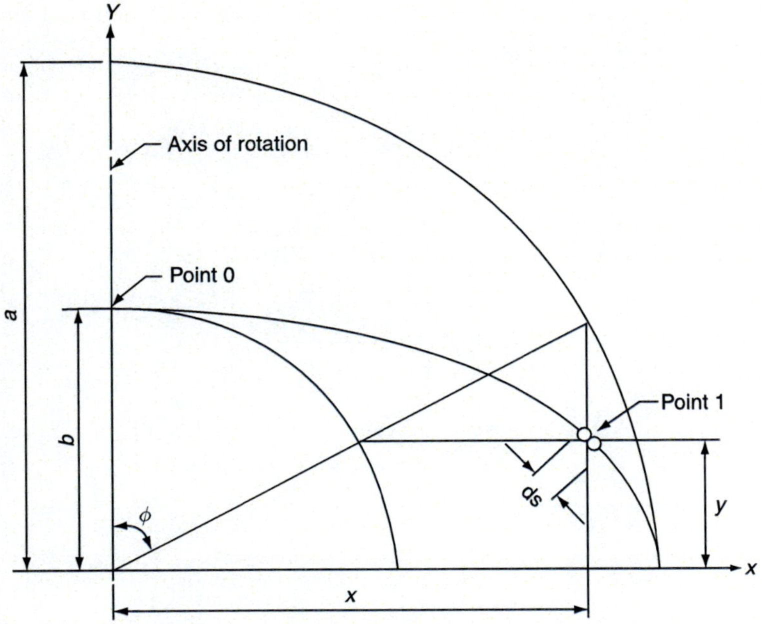

Fig. 6.23 Elliptical dome

Let $K^2 = (a^2 - b^2)/a^2$

$$W_u = \text{Total load between points 0–1}$$

$$W_u = \mathrm{d}w = 2\pi \times a^2 w\left[\cos\phi(\sqrt{(1 - k^2\sin^2\phi)} - \frac{1}{2}k(1 - k^2)\right.$$

$$\left.\times \log k \cos\sqrt{(1 - k^2\sin^2\phi))}\right]$$

Let $\cos\phi = y/b = g$ and$\sin^2\phi = 1 - \cos^2\phi = 1\,g^2$.
Substitute into Eq. (6.47):

$$W = 2\pi \times a^2 w\left[\frac{1}{2} + \frac{1}{2}k(1 - k^2) \times \log(1 + k) - \frac{1}{2}g\left(\sqrt{(1 - k^2(1 - g^2))}\right.\right.$$

$$\left.\left. - \frac{1}{2}k(1 - k^2) \times \log gk + \sqrt{(1 - k^2(1 - g^2))}\right)\right] \tag{6.48}$$

$$= (2\pi a^2 w)C$$

where C is the quantity in the bracket.

To find the meridional thrust

$$T = \frac{W}{2\pi\sin\phi x}$$

$$= \frac{2\pi \times a^2 wC}{2\pi \times a(\sqrt{(1 - g^2\sin\phi)})}$$

$$= wCa^2\sqrt{\frac{1 - k^2(1 - g^2)}{b(1 - g^2)}} \tag{6.49}$$

$$= \frac{wa^2}{b} \times \frac{cQ}{1 - g^2}$$

where $Q = \sqrt{1 - k^2(1 - g^2)}$.

Table 6.7 Elliptical surfaces

System	Loading	N_s	N_ϕ	T
(a) $\eta = a^2\cos^2\phi + b^2\sin^2\phi$ 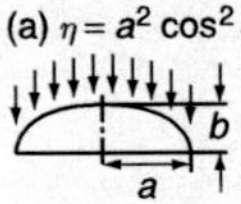	$p_y = -p_E\cos\phi$ $p_z = p_E\sin\phi$	$-p_E\dfrac{s}{2}(l-s)\times\dfrac{3a^2b^2-\eta^2}{a^2b^2\eta^{1/2}}\sin\phi$	$-p_E a^2b^2\dfrac{\sin\phi}{\eta^{3/2}}$	$-p_E\left(\dfrac{1}{2}-s\right)\dfrac{2a^2+(a^2-b^2)\sin^2\phi}{\eta}\cos\phi$
(b) 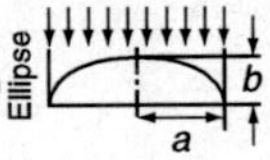	$p_y = -p_S\sin\phi\cos\phi$ $p_z = p_S\sin^2\phi$	$-p_S\dfrac{3s(l-s)}{2a^2b^2\eta^{1/2}}$ $\times[b^2(a^2\cos^2\phi - b^2\sin^2\phi)$ $+2\eta^2(\sin^2\phi-\cos^2\phi)]$	$-p_S a^2b^2\dfrac{\sin^2\phi}{\eta^{3/2}}$	$p_S\left(\dfrac{1}{2}-s\right)\dfrac{3\sin\phi\cos\phi}{\eta}(b^2-2\eta)$
(c) 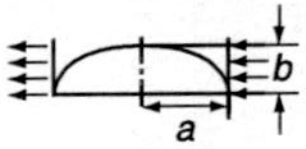	$p_z = p_W\cos\phi$	$-p_W\dfrac{s}{2}\dfrac{(l-s)\cos\phi}{a^2b^2\eta^{1/2}}[\eta^2+3\eta$ $\times(b^2-a^2(1-3\sin^2\phi)$ $-6(b^2-a^2)^2\sin^2\phi\cos^2\phi]$	$-p_W a^2b^2\dfrac{\cos\phi}{\eta^{3/2}}$	$p_W\left(\dfrac{1}{2}-s\right)3\dfrac{b^2-a^2}{\eta}$ $\times(1+\cos^2\phi)\sin\phi$
(d) 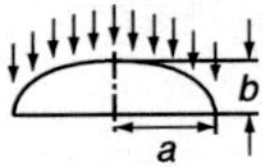 $\dfrac{\sqrt{a^2-b^2}}{a}=\varepsilon$	$p_x = p_E\cos\theta$ $p_z = p_E\cos\theta$	$-\dfrac{p_E}{2}\dfrac{\sqrt{a^2\mathrm{tg}^2\theta+b^2}}{a^2\sin\theta\,\mathrm{tg}\theta}$ $\times\left[a^2-\dfrac{a^2b^2\sqrt{1+\mathrm{tg}^2\theta}}{b^2+a^2\mathrm{tg}^2\theta}\right.$ $\left.+\dfrac{b^2}{\varepsilon}\ln\dfrac{(1+\varepsilon)\sqrt{b^2+a^2\mathrm{tg}^2\theta}}{b(\varepsilon+\sqrt{1+\mathrm{tg}^2\theta})}\right]$	$p_E\left[\dfrac{(b^2+a^2\,\mathrm{tg}^2\theta)^{3/2}}{2\mathrm{tg}^2\theta\sqrt{1+\mathrm{tg}^2\theta}}\right.$ $\times\left(\dfrac{1}{\varepsilon a^2}\ln\dfrac{(1+\varepsilon)\sqrt{b^2+a^2\mathrm{tg}^2\theta}}{b(\varepsilon+\sqrt{1+\mathrm{tg}^2\theta})}\right.$ $\left.\left.+\dfrac{1}{b^2}-\dfrac{\sqrt{1+\mathrm{tg}^2\theta}}{b^2+a^2\mathrm{tg}^2\theta}\right)-\dfrac{a^2}{\sqrt{b^2+a^2\mathrm{tg}^2\theta}}\right]$	0
(e) 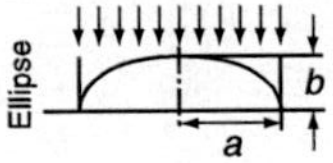	$p_x = p_S\sin\theta\cos\theta$ $p_z = p_S\cos^2\theta$	$-\dfrac{p_S}{2}\dfrac{a^2\sqrt{1+\mathrm{tg}^2\theta}}{\sqrt{b^2+a^2\mathrm{tg}^2\theta}}$	$-\dfrac{p_S}{2}\dfrac{a^2}{b^2}\dfrac{b^2-a^2\mathrm{tg}^2\theta}{\sqrt{b^2+a^2\mathrm{tg}^2\theta}\sqrt{1+\mathrm{tg}^2\theta}}$	0
(f) 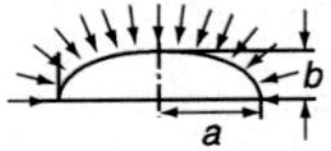	$p_z = p$	$-p\dfrac{a^2}{2}\dfrac{1}{\sqrt{b^2+(a^2-b^2)\sin^2\theta}}$	$-p\dfrac{a^2}{2b^2}\dfrac{b^2-(a^2-b^2)\sin^2\theta}{\sqrt{b^2+(a^2-b^2)\sin^2\theta}}$	0

To find the hoop force

$$H = \frac{-Tx}{R\sin\phi} + \frac{wx\cos^2\phi}{\sin\phi}$$
$$= \frac{-wx^2}{2R\sin^2\phi} + \frac{wx\cos^2\phi}{\sin\phi} \tag{6.50}$$
$$= \left(\frac{wa^2}{2b}\right) \times \frac{2g^2 - 1}{Q}$$

If the dome is continued along the circle of latitude through point 1, an edge memb must be provided along that circle and that member is subjected to ring tension,

$$S = \frac{Wa}{2\pi \times b} \times \frac{g}{\sqrt{(1 - g_1)}} \tag{6.51}$$

Example 6 shows a typical example for elliptical dome surfaces, and Table 6.7 give certain specific cases for elliptical shell surfaces.

Torospherical shell surfaces

The torospherical shell surface is formed when a toroidal segment is inserted between a spherical and a cylindrical cap.

Main analysis
The geometry of this shell is shown in more detail in Figure 6.24.

In terms of the radii of curvature of the sphere a_1, the toroid a_3, and the cylinder a_2, the meridional angle at the sphere–toroid junction is given by

$$\sin\phi_1 = \frac{a_2 - a_3}{a_1 - a_3} \tag{6.52}$$

In practice, the radius of the toroid is selected so as to provide smooth transitions at both ends.

To derive the membrane theory stress resultants, one again uses the overall equilibrium method. At a general section within the toroid,$\phi_1 < \phi < \pi/2$, the resultant axial load consists of two parts: $P_1 =$ the load on the spherical cap, and $P_2(\phi) =$ the load between ϕ_1 and ϕ on the toroid. For a uniform positive pressure p, the magnitude of the first force is the pressure multiplied by the projected base area of the cap.

Toroidal-shaped surfaces

$$P_1 = p\pi(a_1 \sin\phi_1)^2 \tag{6.53}$$

For the second force, first compute the horizontal radius of the toroid

$$R_0(\phi) = (a_1 - a_3)\sin\phi_1 + a_3 \sin\phi \tag{6.53a}$$

where upon the magnitude equals the pressure multiplied by the area of the projected annulus,

$$\begin{aligned} P_2(\phi) &= p\pi[R_0^2 - (a_1 \sin\phi)^2] \\ &= p\pi\left[(a_1 - a_3)^2 \sin^2\phi_1 + a_3^2 \sin^2\phi + 2(a_1 - a_3)a_3 \sin\phi_1 \sin\phi - a_1^2 \sin^2\phi_1\right] \end{aligned}$$

which reduces to

$$P_2(\phi) = p\pi a_3(\sin\phi - \sin\phi_1)[a_3(\sin\phi - \sin\phi_1) + 2a_1 \sin\phi_1] \tag{6.53b}$$

Summing Eqs. (6.53a) and (6.53c)

$$P(\phi) = p\pi\bar{p}(\phi) \tag{6.53c}$$

where,

$$\begin{aligned} \bar{p}(\phi) = (a_1 \sin\phi_1)^2 &+ a_3(\sin\phi - \sin\phi_1) \\ &\times [a_3(\sin\phi - \sin\phi_1) + 2a_1 \sin\phi_1] \end{aligned} \tag{6.54}$$

$$\begin{aligned} N_\phi &= \frac{1}{2\pi R_0 \sin\phi}\bar{p}(\phi) \\ &= \frac{p\bar{p}(\phi)}{2[(a_1 - a_3)\sin\phi_1 + a_3 \sin\phi]\sin\phi} \end{aligned} \tag{6.55}$$

To compute the circumferential stress resultant, the equation for N_θ is

$$N_\theta = R_\theta\left(p - \frac{N_\phi}{R_\phi}\right) \tag{6.56}$$

From Eq. (6.53b), since $R_0 = R_\theta \sin\phi$,

$$R_\theta = \frac{a_2 - a_3}{\sin\phi} + a_3 = (a_1 - a_3)\frac{\sin\phi_1}{\sin\phi} + a_3 \tag{6.57}$$

From Figure 6.24

$$R_\phi = a_3$$

Substituting Eqs. (6.55a) and (6.56) into Eq. (6.55b) gives

$$N_\theta = p\left[\frac{(a_1 - a_3)\sin\phi_1 + a_3\sin\phi}{\sin\phi}\right] \times \left\{1 - \frac{\bar{Q}(\phi)}{2a_3\sin\phi[(a_1 - a_3)\sin\phi_1 + a_3\sin\phi]}\right\} \tag{6.58}$$

Equations (6.5) and (6.7) constitute the membrane theory solution for the called toroidal knuckle portion of the torospherical head.

Toroidal knuckle:

$$Z^2 + R^2 - 95.7574Z - 126.72R + 5234.9519 = 0$$

Cylindrical segment:

$$R - 96.1 = 0 \tag{6.59}$$

Table 6.8 gives some cases of a toroidal shell surface under load.

Example 6.6 Toroidal shell shapes

Referring to Figure 6.24, and allowing for the wall thickness $h = 5\,\text{mm}$, the geometrical properties are

$$a_1 = 4392\,\text{mm}; \quad a_2 = 2441\,\text{mm}; \quad a_3 = 823\,\text{mm}$$

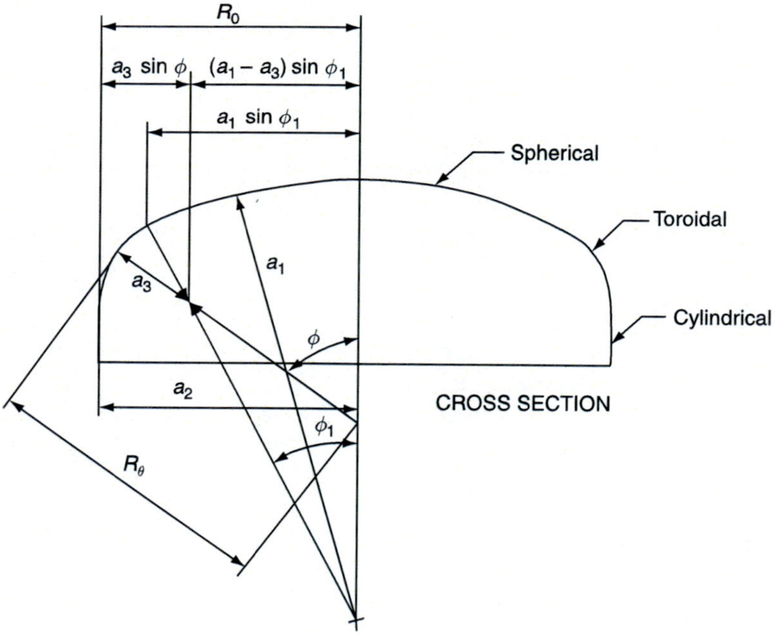

Fig. 6.24 Torospherical shell geometry

Table 6.8 Toroid shell

System	Loading	N_θ	N_ϕ	T
(a) 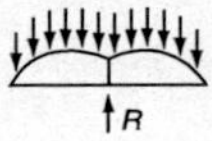	$p_x = p_E \sin\theta$ $p_z = p_E \cos\theta$ $R = 2p_E\pi\left(r\theta_0\sin\theta_0 - 2\sin^2\frac{\theta_0}{2}\right)$	$-p_E r\,\frac{1-\cos\theta+\theta\sin\theta_0}{\sin\theta(\sin\theta+\sin\theta_0)}$	$-p_E r\left[\cos\theta - \frac{1-\cos\theta}{\sin^2\theta} + \sin\theta_0\left(\mathrm{ctg}\,\theta - \frac{\theta}{\sin^2\theta}\right)\right]$	0
(b) 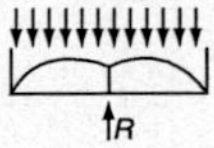	$p_x = p_S \sin\theta\cos\theta$ $p_z = p_S\cos^2\theta$ $R = p_S\pi r^2\sin^2\theta_0$	$-p_S\frac{r}{2}\,\frac{\sin\theta+2\sin\theta_0}{\sin\theta+\sin\theta_0}$	$-p_S\frac{r}{2}(\cos 2\theta - 2\sin\theta\sin\theta_0)$	0
Ring axis does not bisect the cross-section		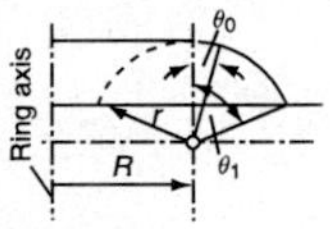	For the direction of the load components p_x, p_z and of the unit normal forces N_θ, N_ϕ	
(c) 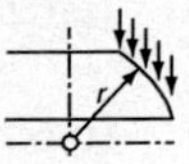	$p_x = p_E\sin\theta$ $p_z = p_E\cos\theta$	$-p_E\,\frac{Rr(\theta-\theta_0)+r^2(\cos\theta_0-\cos\theta)}{(R+r\sin\theta)\sin\theta}$	$-\frac{p_E}{\sin\theta}\left[(R+r\sin\theta)\cos\theta - R(\theta-\theta_0) - r(\cos\theta_0-\cos\theta)\right]$	0
		For $\theta_0 = -\theta_1$ (symmetrical cross-section)		
		$-p_E\,\frac{Rr\theta+r^2(1-\cos\theta)}{(R+r\sin\theta)\sin\theta}$	$-\frac{p_E}{\sin\theta}\left[(R+r\sin\theta)\cos\theta - R\theta - r(1-\cos\theta)\right]$	0
(d) 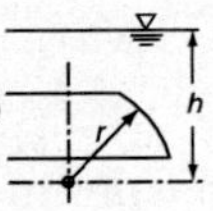	$p_z = \gamma(h - r\cos\theta)$	$-\frac{\gamma r}{(R+r\sin\theta)\sin\theta}\Big[-Rh(\sin\theta_0-\sin\theta) + \frac{rh}{2}(\cos^2\theta_0-\cos^2\theta) + \frac{Rr}{2}(\sin\theta_0\cos\theta_0-\sin\theta\cos\theta-\theta+\theta_0) - \frac{r^2}{3}(\cos^3\theta_0-\cos^3\theta)\Big]$	$-\frac{\gamma}{\sin^2\theta}\Big[(h-r\cos\theta)(R+r\sin\theta)\times\sin\theta + Rh(\sin\theta_0-\sin\theta) - \frac{rh}{2}(\cos^2\theta_0-\cos^2\theta) - \frac{Rr}{2}\times(\sin\theta_0\cos\theta_0-\sin\theta\cos\theta-\theta+\theta_0) + \frac{r^2}{3}(\cos^3\theta_0-\cos^3\theta)\Big]$	0
		For $\theta_0 = -\theta_1$ (symmetrical cross-section)		
		$-\frac{\gamma r}{(R+r\sin\theta)\sin\theta}\Big[Rh\sin\theta + \frac{rh}{2}\sin^2\theta - \frac{Rr}{2}(\sin\theta\cos\theta+\theta) - \frac{r^2}{3}(1-\cos^3\theta)\Big]$	$-\frac{\gamma}{\sin^2\theta}\Big[\frac{rh}{2}\sin^2\theta - \frac{Rr}{2}(\sin\theta\cos\theta-\theta) - r^2\left(\cos\theta\sin^2\theta - \frac{1-\cos^3\theta}{3}\right)\Big]$	0
(e) 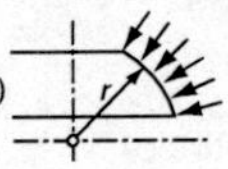	$p_z = p$	$-\frac{p}{2(R+r\sin\theta)\sin\theta}\left[(R+r\sin\theta)^2 - (R+r\sin\theta_0)^2\right]$	$-\frac{p}{2\sin^2\theta}\left[2R\sin\theta_0 + r(\sin^2\theta_0+\sin^2\theta)\right]$	0
		For $\theta_0 = -\theta_1$ (symmetrical cross-section)		
		$-\frac{pr}{2}\,\frac{2R+r\sin\theta}{R+r\sin\theta}$	$-\frac{pr}{2}$	0
(f) 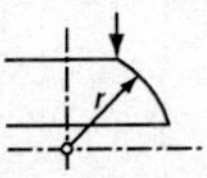	Edge load p_L	$-p_L\,\frac{R+r\sin\theta_0}{(R+r\sin\theta)\sin\theta}$	$\frac{p_L}{r}\,\frac{R+r\sin\theta_0}{\sin^2\theta}$	0

as shown in Figure 6.25, so that from Eq. (6.58)

$$\sin\phi_1 = \frac{2441 - 823}{4392 - 823} = 0.45205 \tag{6.60}$$

and

$$\phi_1 = 26.88^\circ.$$

The equation of the meridian was established with the origin set at the pole, and the resulting equations are as follows:

Spherical cap:

$$Z^2 + R^2 - 345.8Z = 0 \tag{6.61}$$

$$\bar{p}(\phi) = \text{Eq. (6.54b)} \tag{6.62}$$

$$P(\phi) = p\pi\bar{p}(\phi) \tag{6.63}$$

The values of N_ϕ and N_θ can now be computed, as in other domical shapes.

Reference is made to Table 6.8 and Eqs. (6.55a) and (6.57) for evaluation of these values.

Example 6 Elliptical dome

The reinforced concrete dome shown in Figure 6.25 is an ellipsoid of revolution about the minor axis. The dome carries a vertical live load of 2.5 kN/m^2 on plan. Calculate the meridional and hoop forces and stresses for zones $C1$, 2 and 3 and

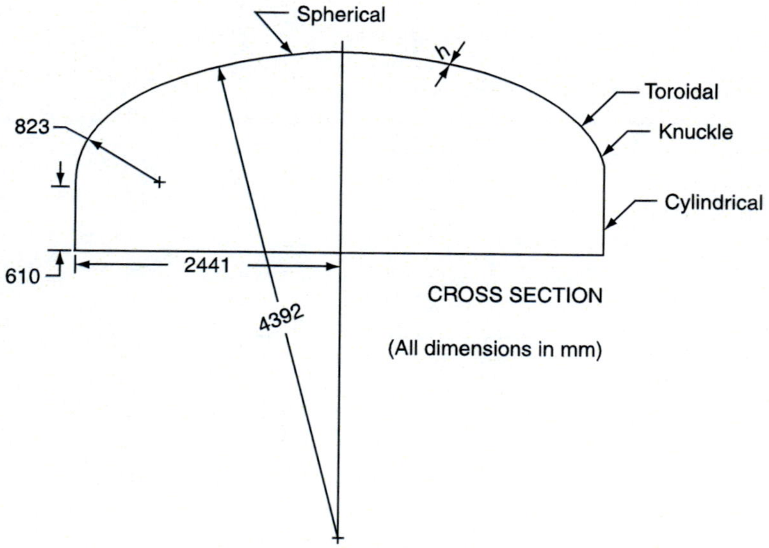

Fig. 6.25 Test specimen middle surface geometry

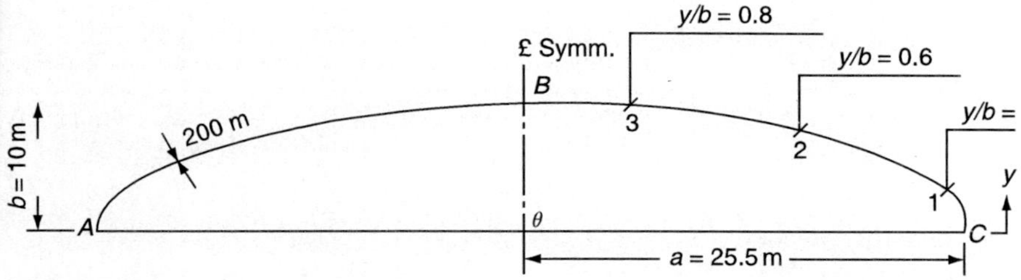

Fig. 6.26 Elliptical dome surface

draw the hoop stress diagram for one-half of the dome. Calculate the areas of reinforcement at points $C1$, 2, 3 and B. Use the following data:
Density of concrete= 2400 kg/m^3

N_ϕ = meridional thrust from the live load $= \frac{2.5a^2}{2b}Q = 1.25\frac{a^2}{b}Q$

N_θ = hoop force $= 1.25\frac{a^2}{b}\left(\frac{2(y/b)^2-1}{Q}\right)$

Total meridional thrust $= N_\phi + N_{\phi,1} = \bar{T}$

Total hoop thrust $= N_\theta + N_{\theta,1} = \bar{H}$

$g(y/b)$	0	0.2	0.6	0.8
$\bar{L}$	0.632	0.552	0.334	0.183
$\bar{M}$	0.392	0.433	0.671	0.833
$\bar{T}$(kN/m)	+107.60	111.30	163.2	197.3
$\bar{H}$(kN/m)	−700	−525.70	−86.0	+85.4
Stresses due to $\bar{T}$(N/mm)^2 $= \bar{T}/0.2 \times 10^{-3}$				
	0.538	0.556	0.816	0.987
Stresses due to T_1(N/mm)^2 $= T_1/0.2 \times 10^{-3}$				
	−3.50	−2.66	−0.43	+0.43

Allowing no tension in the concrete (compressive stress in the reinforcement steel = 125 N/mm^2)

$$A_S = \frac{\bar{H}}{125} = \frac{700}{125} \times 10^3 = 5600\,\text{mm}^2/\text{m}$$

0.2% concrete sectional area for temperature and shrinkage

$$= 5610 + 200 \times 1000 \times \frac{0.2}{100}$$

$$= 6000\,\text{mm}^2/\text{m in the latitude direction}$$

and

$$400\,\text{mm}^2/\text{m in the meridional directions}$$

List of reinforcement for various zones:

g	0	0.2	0.6	0.8	1.0	
N_ϕ	400	400	400	400	400	meridional (mm^2/m)
N_θ	6000	4600	1088	400	400	latitude

Self weight $W_D = 2400 \times 9.81 \times 10^{-3} \times 0.2 = 4.71\,\text{kN/m}^2$
Stresses at the end of minor axis:
$N_\phi = N_\theta = Wa^2/2b$ for both live and dead loads is:

$$4.71 \times (25.5)^2 + \frac{2.5 \times (25.5)^2}{20} = 235\,\text{kN/m}$$

$$\text{Stresses} = \frac{235}{0.2} = 1175\,\text{kN/m}^2$$

$$= 1.175\,\text{N/mm}^2$$

Computer program: analysis of thin shells

```
C        * TITLE: COMPUTER AID ANALYSIS OF THIN SHELLS                  *
C       * TUTOR: Dr. Y. BANGASH                                         *
C       *************************************************************
C
C
C       *************************************************************
C       * THIS IS A PROGRAM TO FIND OUT THE MERIDIAN FORCES,            *
C       * THE HOOP FORCES AND RING TENSION ACTING ON                    *
C       * EACH SECTION OF THE ELLIPTICAL DOME DUE TO ITS                *
C       * DEAD LOAD AND IMPOSED LOAD.                                   *
C       *************************************************************
C
C
C       =============================================================
C       =                    * NOTATION *                               =
C       =                                                               =
C       = QK  UNIFORMLY DISTRIBUTED LIVE LOAD                           =
C       = GK  UNIFORMLY DISTRIBUTED DEAD LOAD                           =
C       = TX  THICKNESS OF THE DOME                                     =
C       = B  HALF HEIGHT OF THE DOME                                    =
C       = AX  HALF LENGTH OF THE DOME                                   =
C       = R  RADIUS OF CURVATURE                                        =
C       = H  HOOP FORCE DUE TO LIVE LOAD                                =
C       = H1  HOOP FORCE DUE TO DEAD LOAD                               =
C       = HBAR TOTAL HOOP FORCE                                         =
C       = DELTH  STRESS DUE TO HOOP FORCE                               =
C       = T1  MERIDIONAL THRUST DUE TO DEAD LOAD                        =
C       = T2  MERIDIONAL THRUST DUE TO LIVE LOAD F                      =
```

```
C      = TBAR  TOTAL MERIDIONAL STRESS                              =
C      = DELTT  STRESS DUE TO MERIDIONAL THRUST                     =
C      ==============================================================
C
       REAL G(1:6), C(1:6), Q(1:6), A(1:6), WU(1:6), TBAR(1:6)
       REAL DELTT(1:6), HBAR(1:6), DELTH(1:6), Y(1:6), H(1:6),
       T(1:6
       REAL E(1:6), F(1:6), O, Z, Q1, PI, K, H1(1:6), T1(1:6)
       REAL QK, GK, TX, B, AX, R
3000   WRITE(6, 10)
10     FORMAT(1X, 'INPUT THE UNIFORMLY DISTRIBUTED LIVE LOAD')
       READ(5, *) QK
       WRITE(6, 14)
14     FORMAT(/, 1X, 'INPUT THE DEAD LOAD')READ(5, *) GK
       WRITE(6, 18)
18     FORMAT(/, 1X, 'INPUT THE THICKNESS OF THE DOME')
       READ(5, *) TX
       WRITE(6, 22)
22     FORMAT(/, 1X, 'INPUT THE HALF HEIGHT OF THE DOME')
       READ(5, *) B
       WRITE(6, 26)
26     FORMAT(/, 1X, 'INPUT THE HALF LENGTH OF THE DOME')
       READ(5, *) AX
       WRITE(6, 29)
29     FORMAT(/, 1X, 'INPUT THE RADIUS OF THE CURVATURE')
       READ(5, *) R
       WRITE(6, 31)
31     FORMAT(/, 1X, 'INPUT THE VALUE OF Y')
       READ(5, *) Y(1), Y(2), Y(3), Y(4), Y(5), Y(6)
       DO 33 I=1,6
       G(I)=Y(I)/B
33     CONTINUE
       Q1=2400
       PI=3.142
       K=SQRT((AX**2-B**2)/(AX**2))
       DO 40 I=1,6
       Q(I)=SQRT(1-(K**2)**(1-G(I)**2))
40     CONTINUE
       DO 41 I=1,6
       E(I)=(-0.5*G(I)*Q(I))
41     CONTINUE
       DO 42 I=1,6
       F(I)=((1-K**2)/(2*K))*LOG(K*G(I)+Q(I))
42     CONTINUE
       O=1
       Z=(-0.5)-(1-K**2)/(2*K)*LOG(K+O)
       DO 43 I=1,6
       C(I)=E(I)-F(I)-Z
43     CONTINUE
       W=QK + GK
       DO 44 I=1,6
       A(I)=(2*PI*AX**2)*C(I)
44     CONTINUE
```

```
      DO 45 I=1,6
      WU(I)=W*A(I)
45    CONTINUE
C
C     TO FIND THE HOOP FORCE
C     H = HOOP FORCE DUE TO THE LIVE LOAD, QK
      DO 46 I=1,6
      H(I) = (QK*AX**2/B)*(G(I)-C(I)/(1-G(I)**2)/Q(I))
46    CONTINUE
C     H1 = HOOP FORCE DUE TO DEAD LOAD, GK
      DO 47 I=1,6
      H1(I) = (GK*AX**2)/(2*B)*(2*G(I)**2-1)/Q(I)
47    CONTINUE
C     HBAR = TOTAL HOOP FORCE
      DO 48 I=1,6
      HBAR(I)=H(I)+H1(I)
48    CONTINUE
C     TO FIND THE STRESS DUE TO HOOP FORCE, DELTH
      DO 49 I=1,6
      DELTH(I) = HBAR(I)/(TX*1000)
49    CONTINUE
C
C
C     TO FIND THE MERIDIONAL THRUST
C     T=MERIDIONAL THRUST DUE TO DEAD LOAD, GK
      DO 50 I=1,6T(I)=(GK*C(I)*Q(I)*AX**2)/((1-G(I)**2)*B)
50    CONTINUE
C     T1=MERIDIONAL THRUST DUE TO LIVE LOAD, QK
      DO 51 I=1,6
      T1(I)=(QK*Q(I)*AX**2)/(B*2)
51    CONTINUE
C     TBAR = TOTAL MERIDIONAL THRUST
      DO 52 I=1,6
      TBAR(I)=T(I)+T1(I)
52    CONTINUE
C     TO FIND STRESSES DUE TO MERIDIONAL THRUST, DELTT
      DO 53 I=1,6
      DELTT(I)=TBAR(I)/(TX*1000)
53    CONTINUE
C
C
      PRINT 55
      PRINT 56
55    FORMAT(////,2X, '*SHELL DIMENSION (m)*')
56    FORMAT(/,2X, 'THICKNESS',8X,'HALF HEIGHT', 6X,
      *'HALF LENGTH', 6X, 'RADIUS')WRITE(6,57) TX,B,AX,R
57    FORMAT(/,3X,F7.2,9X,F7.2,9X,F7.2,9X,F7.2)
      PRINT 58
58    FORMAT(//,31X,'OUTPUT RESULTS')
      PRINT 59
59    FORMAT(31X,'**************')
      WRITE(6,60) G(1),G(2),G(3),G(4),G(5),G(6)
60    FORMAT(//,1X,'G=Y/B',6X,F4.3,7X,F4.3,7X,F4.3,7X,F4.3,
```

```
      *7X,F4.3,7X,F4.3)
      WRITE(6,72) C(1),C(2),C(3),C(4),C(5),C(6)
72    FORMAT(/,1X,'C',7X,F7.2,4X,F7.2,4X,F7.2,4X,F7.2,4X,
      *F7.2,4X,F7.2)
      WRITE(6,84) Q(1),Q(2),Q(3),Q(4),Q(5),Q(6)
84    FORMAT(/,1X,'Q',7X,F7.2,4X,F7.2,4X,F7.2,4X,F7.2,4X,
      *F7.2,4X,F7.2)
      WRITE(6,96) A(1),A(2),A(3),A(4),A(5),A(6)
96    FORMAT(/, 1X, 'AREA', 4X, F7.2, 4X, F7.2, 4X, F7.2, 4X, F7.2,
      *4X,F7.2,4X,F7.2)
      WRITE(6,100) WU(1),WU(2),WU(3),WU(4),WU(5),WU(6)
100   FORMAT(/,1X,'WEIGHT',1X,F8.2,3X,F8.2,3X,F8.2,3X,
      *F8.2,3X,F8.2,4X,F7.2)
      WRITE(6,110) TBAR(1),TBAR(2),TBAR(3),TBAR(4),TBAR(5),
      *TBAR(6)
110   FORMAT(/,1X,'TBAR',4X,F7.2,4X,F7.2,4X,F7.2,4X,F7.2,
      *4X,F7.2,4X,F7.2)
      WRITE(6,120) DELTT(1),DELTT(2),DELTT(3),DELTT(4),
      *DELTT(5),DELTT(6)
120   FORMAT(/,1X,'DELTT',3X,F7.2,4X,F7.2,4X,F7.2,4X,F7.2,
      *4X,F7.2,4X,F7.2)
      WRITE(6,130) HBAR(1),HBAR(2),HBAR(3),HBAR(4),HBAR(5),
      *HBAR(6)
130   FORMAT(/,1X,'HBAR',4X,F7.2,4X,F7.2,4X,F7.2,4X,F7.2,
      *4X,F7.2,4X,F7.2)
      WRITE(6,140) DELTH(1),DELTH(2),DELTH(3),DELTH(4),
      *DELTH(5),DELTH(6)
140   FORMAT(/,1X,'DELTH',3X,F7.2,4X,F7.2,4X,F7.2,4X,F7.2,
      *4X,F7.2,4X,F7.2)END
```

Execution begins...

```
INPUT THE UNIFORMLY DISTRIBUTED LIVE LOAD
?
2.25

INPUT THE DEAD LOAD
?
1.5

INPUT COLLAR LOAD
?
10

INPUT THE THICKNESS OF THE DOME
?
0.2

INPUT THE RADIUS OF THE CURVATURE
?
50
```

```
INPUT DISTANCE FROM THE AXIS OF ROTATION
                                                          MORE...  IBMA

?
10


INPUT THE LENGTH ALONG THE X-AXIS
?
0, 10, 20, 30, 40, 50

OUTPUT RESULTS
*************
POINTS    10.00     20.00     30.00       40.00       50.00      60.00
AREA       0.00    262.68   1059.98     2448.77     4704.97   11089.70
WEIGHT          0.00     1614.16     5233.17      11070.29     20160.35
44733.53
THRUST     0.00     76.38     87.45       98.26      113.81     177.97
TSTRESS    0.00    381.88    437.26      491.29      569.04     889.83

MORE...  IBMA

HOOP      -3.64     45.77     44.07       24.62      -13.30    -142.37
HSTRESS  -18.21    228.87    220.36      123.08      -66.48    -711.86
RING       0.00    629.20   1272.09     1761.66     1924.92       0.00

COMPRESSION IN EDGE MEMBER AT OPENING = 500.00

Execution begins....
INPUT THE SPAN UNIFORMLY DISTRIBUTED LIVE LOAD
?
291.0
INPUT THE DEAD LOAD
?
0
INPUT THE THICKNESS OF THE DOME
?
1.2
INPUT SIN
0
?
0
INPUT SIN
1
?
1INPUT THE RADIUS OF CURVATURE
?
55.5
```

```
INPUT COS0
?
1
INPUT COS1
?
0
MERIDIONAL THRUST = 5631950.00 KN/M
HOOP FORCE = 8075.00 KN/M
```

Execution begins. . ..

```
INPUT THE SPAN UNIFORMLY DISTRIBUTED LIVE LOAD
?
344.75
INPUT THE DEAD LOAD
?
0
INPUT THE THICKNESS OF THE DOME
?
1.37
INPUT SIN0
?
0
INPUT SIN1
?
1
INPUT THE RADIUS OF CURVATURE
?
22.71
INPUT COS0
?
1
INPUT COS1
?
0
MERIDIONAL THRUST = 7829.3 KN/M
HOOP FORCE = 3924.63 KN/M
ASR = 15048.8 MM**2/M
BAR = 25
**************************************************************
```

Chapter 7
Concrete Containment Subject to Aircraft Crashes and Seismic Effects and Over Pressurisation

7.1 Introduction

A comprehensive analysis of the containment using numerical techniques has been given in this text. More examples are given in the authors' following book: '*Shock, Impact and Explosion*', Springer, Heidelberg (2009). First, the Belle Fonte vessel given in chapter 6 is taken with respective design parameter.

Secondly, this time Size well B containment vessel (Fig. 7.1) is taken as an example. Figures 7.2, 7.3, 7.4 and 7.5 show typical reinforcement system. Figure 7.6 shows a secondary dome provided to protect the primary containment from the external hazards and public from the internal hazards.

The same containment is taken as an example to assess its behaviours in the seismic environments and an overpressurisation. Results obtained are clearly established in the form of acceleration. Time historys, displacement and stress-states and acceleration spectra.

7.2 Aircraft Impact/Crashes on Containment

The design and analysis of containment vessel to withstand the effect of an aircraft crash or impact involves many complex parameters, including loadings and durations. Bangash has produced modified force–time impact functions for various aircrafts, and these are shown in Fig 7.6. First, the Belle Fonte vessel parameters discussed earlier were chosen for the analysis. The impact point is chosen to be the apex of the vessel and at the junction of the dome and the cylinder. Results are produced in the form of the acceleration–time histories, displacement and stress states for selected times and acceleration response spectra. Newmark Wilson θ direct integration methods are used for the analysis. Crack propagation and linear and non-linear displacements for four aircrafts are shown in Figs. 7.7 and 7.8a. The impact zones are shown in Fig. 7.9. Table 7.1 shows comparative results for penetrations and perforations based on both empirical and finite element analysis for a number of aircraft impact loads. The endochronic model is used throughout the analysis. A similar analysis was carried out for the Size well B vessel using load-time functions for the F-16

DOI 10.1007/978-3-642-12560-7_7, © M.Y.H. Bangash 2011

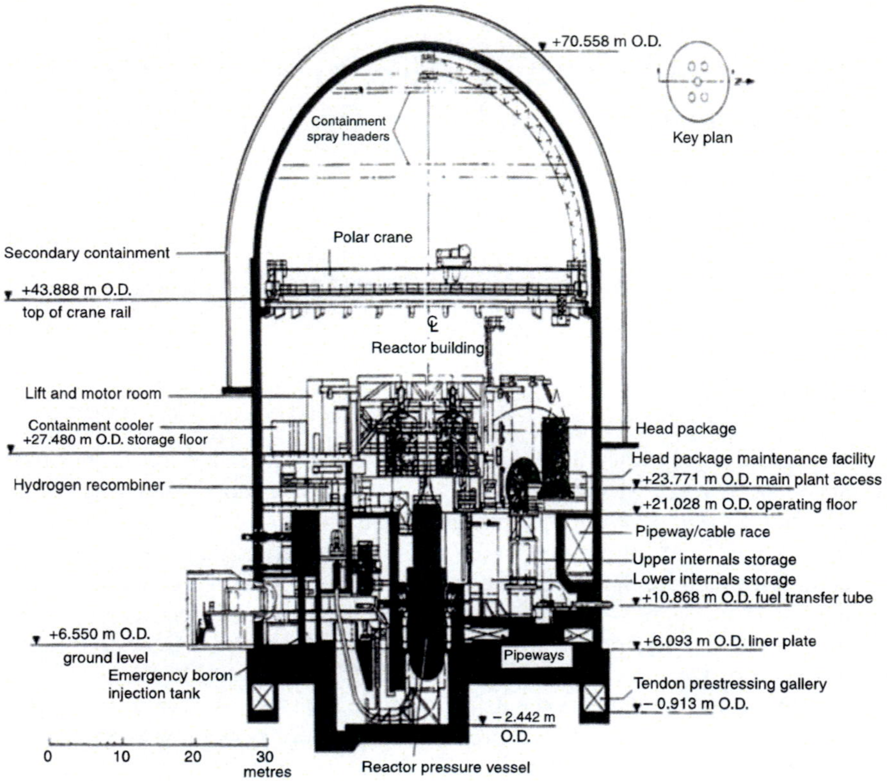

Fig. 7.1 Sizewell B power station reactor building – sectional elevation on north/south centre line (with compliments of CEGB, UK)

aircraft. 'Post-mortem' of the vessel is shown in Fig. 7.10. The load-displacement as a function of time is given in Fig. 7.8. For this analysis the vessel foundations are assumed to be consisting of strings at different directions with parameters given in general by Table 7.2. The springs represent the soil behaviour when aircraft crashes occur as part of the integrated system. In reality a proper soil structure interaction analysis would have given accurate assessment of vessel behaviour. This analysis is avoided due to costs. The application of the soil-structure interaction analysis is more appropriate and is cost-effective in case seismic analysis of such vessels is carried out.

7.2.1 Aircraft Impact Using Finite Element and Ultimate Limit State Analysis

A reference is made to the finite element and ultimate state analysis given in this text and the author books stated in section 7.2. These analyses are not repeated

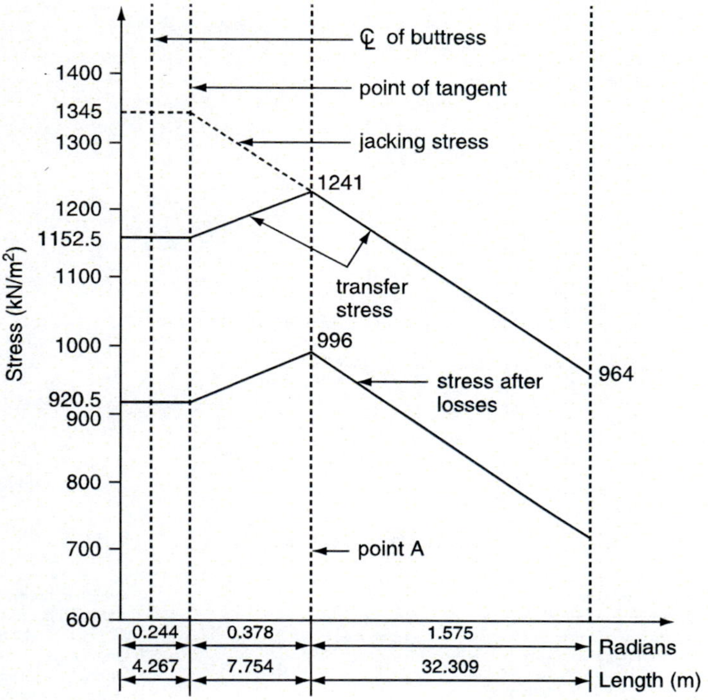

Fig. 7.2 Post-tensioning loads — circumferential tendons

and are avoided due to space limitation. A non-linear finite element model is given in Fig. 7.11 for the Sizewell B containment vessel parameters given in Fig. 7.3. Both finite element and ultimate limit state analysis stated earlier are compared and plotted in Figs. 7.12 and 7.13 while assuming different boundary conditions. For convenience the results are plotted on the containment outside face. The containment is also analysed when it is integrated with other ancillary buildings. The impact zone is assumed on the visible part of the containment outside the ancillary building shown in Fig. 7.10. Throughout the analytical work the impacting aircraft was assumed as F-16 multirole combat aircraft for which the parameters are given. The post-mortem of the containment vessel is given in Fig. 7.14 for the final disaster scenario. Throughout this analysis the following data were maintained and used:

- No. of vessel solid elements 1005 isoparametic type, 20 noded.
- Vessel material properties are stated earlier with bonded steel and tendon indicated in Figs. 7.3, and 7.4 placed either on solid nodes or in the body of the element. Total 5000 elemtents are taken as measured and compared.
- The aircraft elements/variable with the total element numbering 500,000 solid elements.

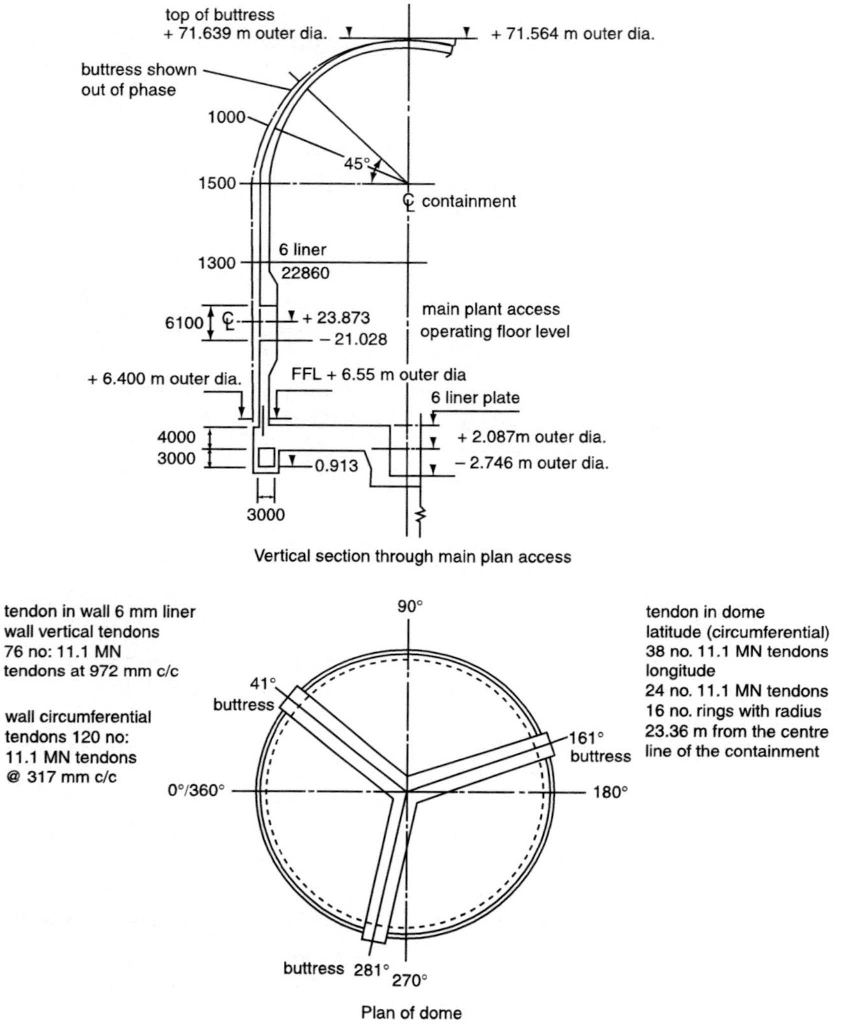

Fig. 7.3 Sizewell B dimensional layout

- The gap or contact elements were provided for the interaction and integrated analysis adopted at suitable places to be 9500 elements.
- Program ISOPAR conveniently checked by ANSYS took the total time $3\frac{1}{4}$ h to produce the disaster scenario. Stresses, displacements and strains were obtained. Cracks were produced together with plasticity indices and zones.
- A four-parameter crack theory was considered for the concrete failure.
- Mostly tendons ruptured or deformed with visible plastic zone.
- Similar time dependent displacements have been produced.

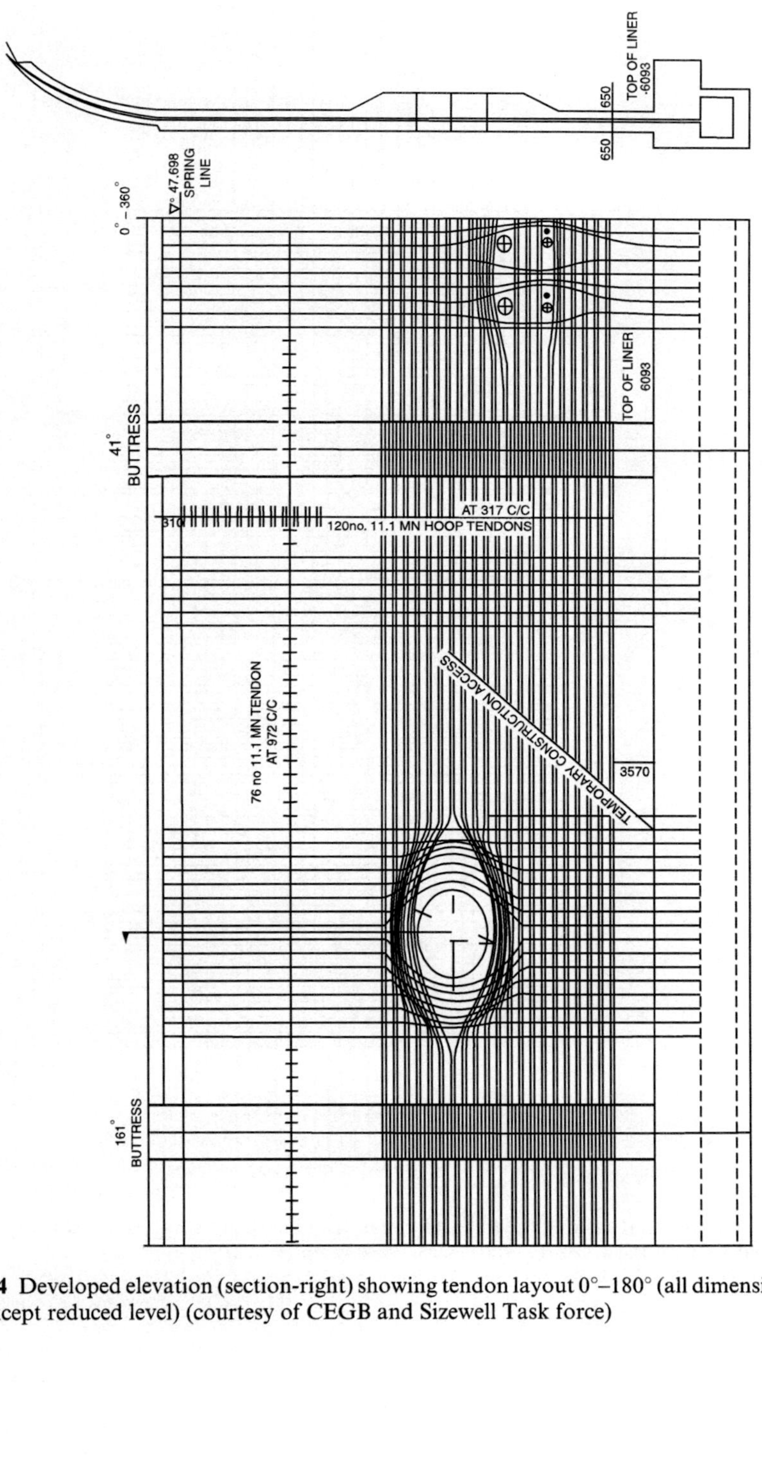

Fig. 7.4 Developed elevation (section-right) showing tendon layout 0°–180° (all dimensions in mm except reduced level) (courtesy of CEGB and Sizewell Task force)

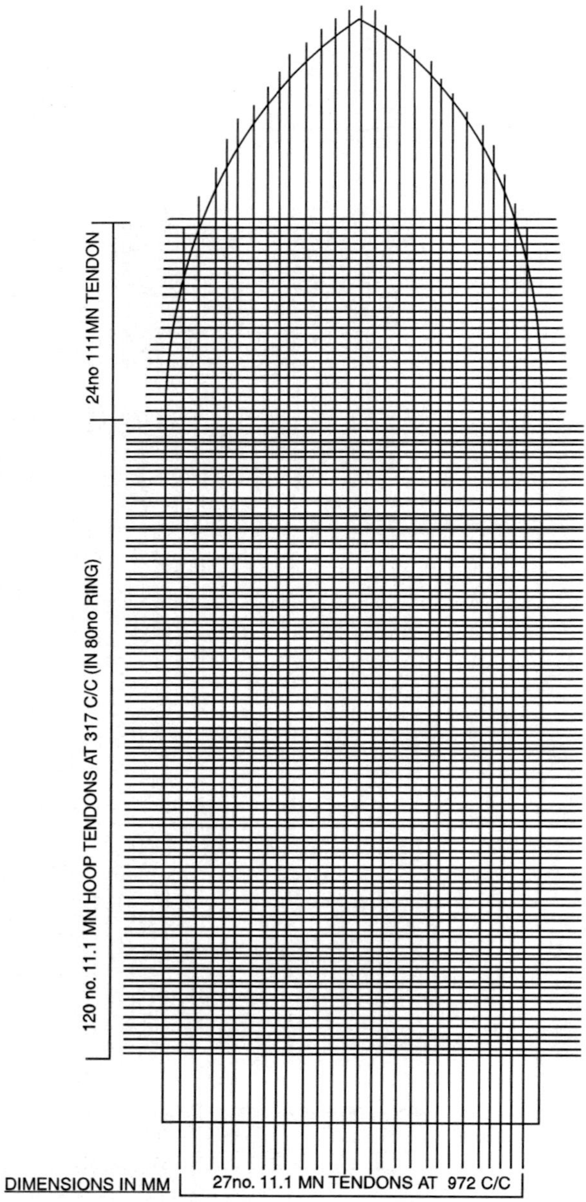

Fig. 7.5 Arrangement of longitudinal and circumferential tendons (all dimensions in mm) (EGB and Sizewell B Task force, 2001) with compliments)

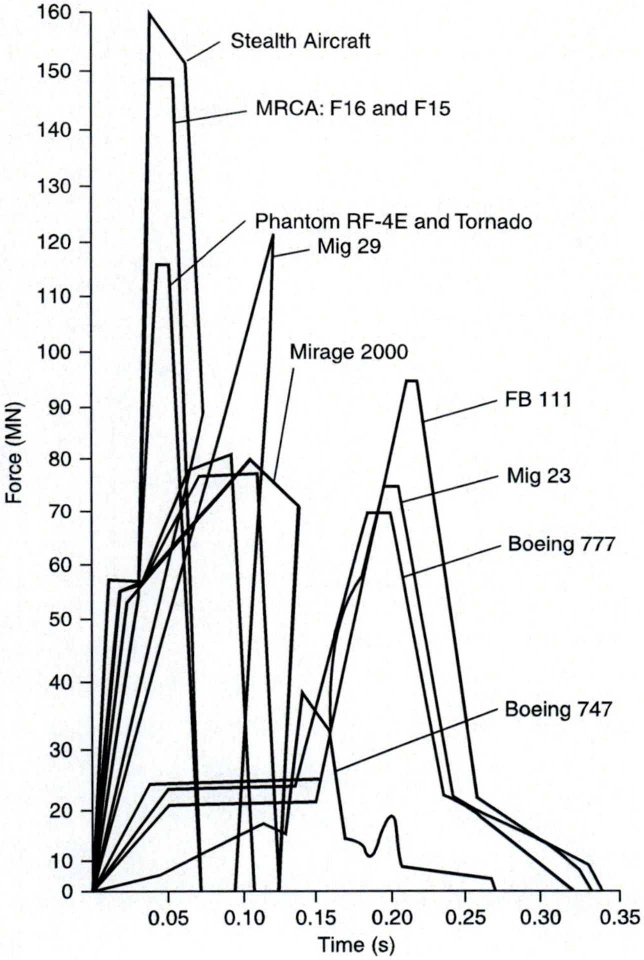

Fig. 7.6 Force–Time function

- A comparative study of the finite element analysis and the limit state approach of this containment has been carried out. Figures 7.7 and 7.8 and the results are in good agreement.

Note: For more detailed analysis the reader is referred to the author's available books on

1. *Explosion Resistant Buildings*
2. *Shock, Impact & Explosions*
   Springer Verlag 2001, 2005 respectively
   Hiedelberg, Germany

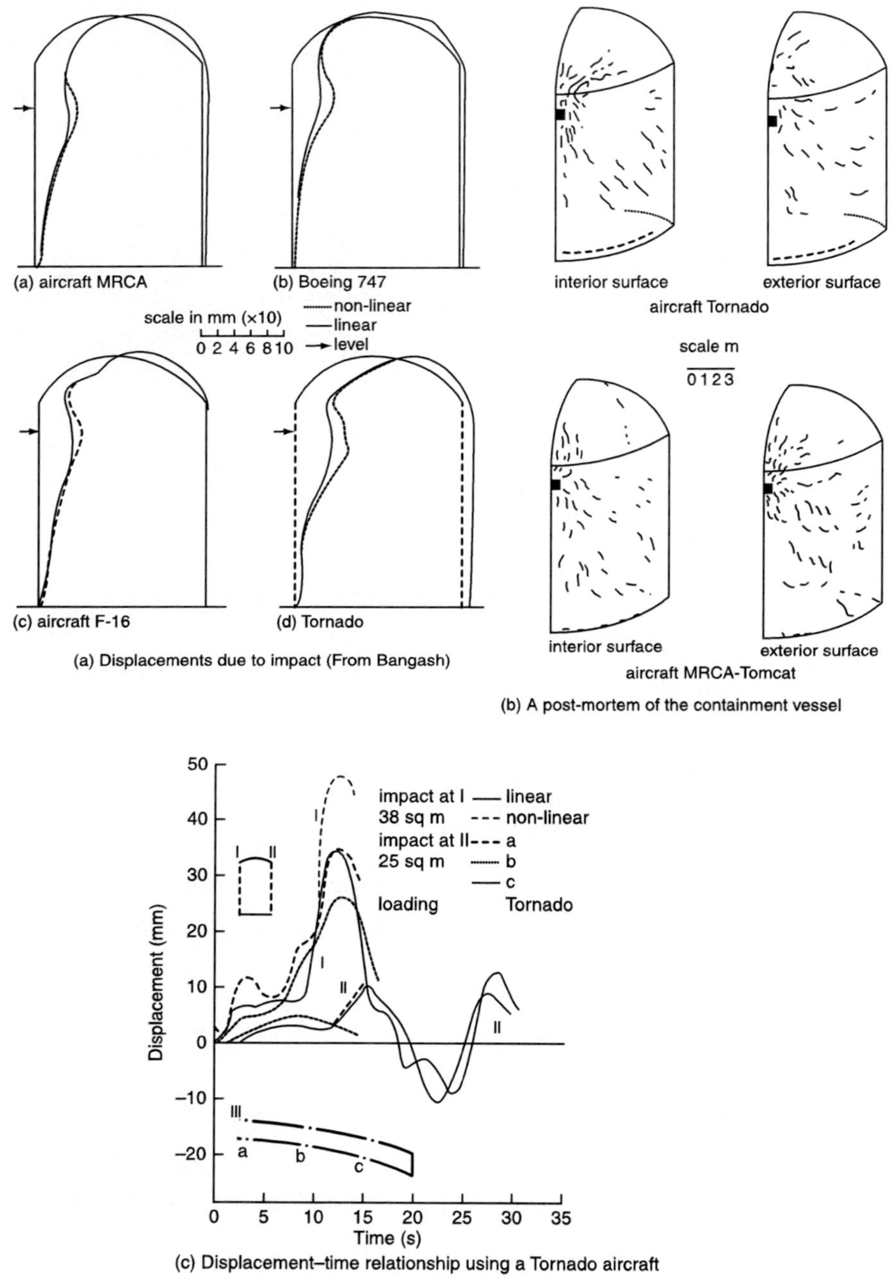

Fig. 7.7 A comparative study of displacement–time relations and post-mortem of containment vessel

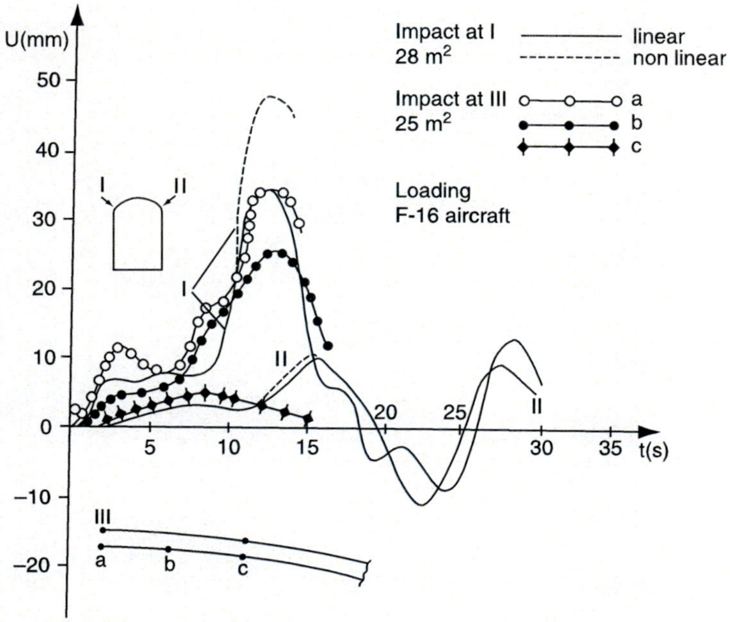

Fig. 7.8 Linear and non–linear displacement as a function of time

7.3 Seismic Analysis for PWR/BWR

A non-linear dynamic analysis was carried out for the containment vessel. A direct approach was adopted in which the whole system (containment/soil/bedrock) was modelled and analysed in the time domain using finite elements. Inertia forces were reproduced by distributed masses by involving consistent matrices. In general the equation of motion contain a mass matrix (M), damping matrix (C) and a stiffness matrix (K). The excitation is taken in the form of a base motion. As shown in Figs. 7.15 and 7.16, 20-noded isoparametric elements together with line and panel elements representing concrete, prestressing tendons and reinforcement and steel liner are used to model the structure and the supporting medium. This method allows treatment of non-linear behaviour and material properties which are adjusted at each step. Special care is exercised in using the numerical integration procedure together with time steps in order to achieve stability in the solution. In the stiffness matrix of the evaluation of the vessel, the four-parmeter model is considered. The cracking cases are evaluated using the endochronic cracking model. Table 7.3 shows the foundation parameter for impact and seismic problems. Figure 7.16a and b indicates damage scenarios of the PWR containment of BelleFonte NPS with and without seismic devices.

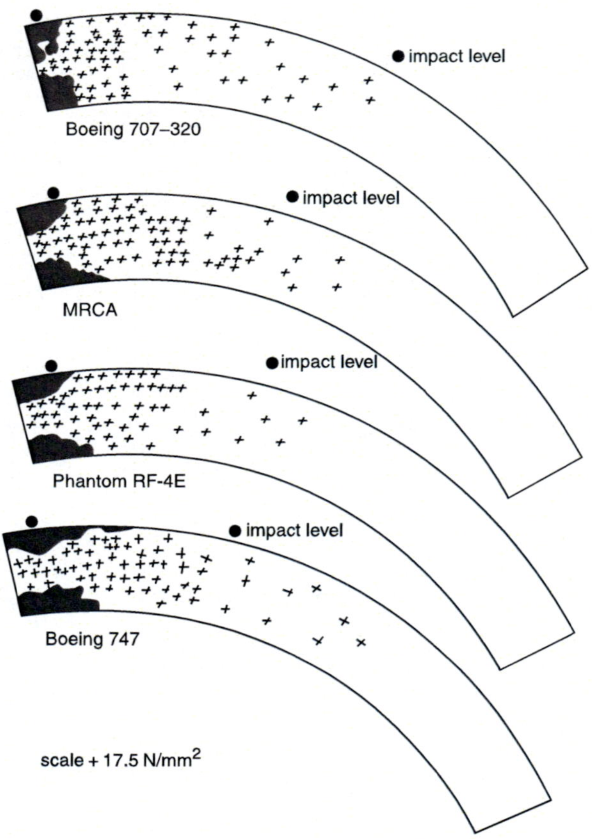

Fig. 7.8 (a) Impact on the dome for various aircraft

The Program ISOPAR has been written to perform seismic analysis incorporating the effects of circumferential cracking. The program uses numerical integration rather than the method of normal modes. The cracks were obtained when the overall stiffness of the vessel was no longer constant but was a function of the shearing stresses in the vessel.

The equation of motion was solved using the Wilson and Newmark θ methods. The convergence criterion for ending the iterations was 0.0001.

The size of the time step (t) is critical for obtaining accuracy and rapid iteration convergence (optimum $t = 0.0025$s).

Stiffness changes will occur while giving some time steps. A typical subroutine that keeps track of where each crack is on the hysteresis loop is shown in Figs. 7.17 and 7.18 and makes in the crack stiffness when necessary. A time step is repeated only when stiffness changes from line to line. This is because the high velocities which occur during the unloading from certain lines can cause the

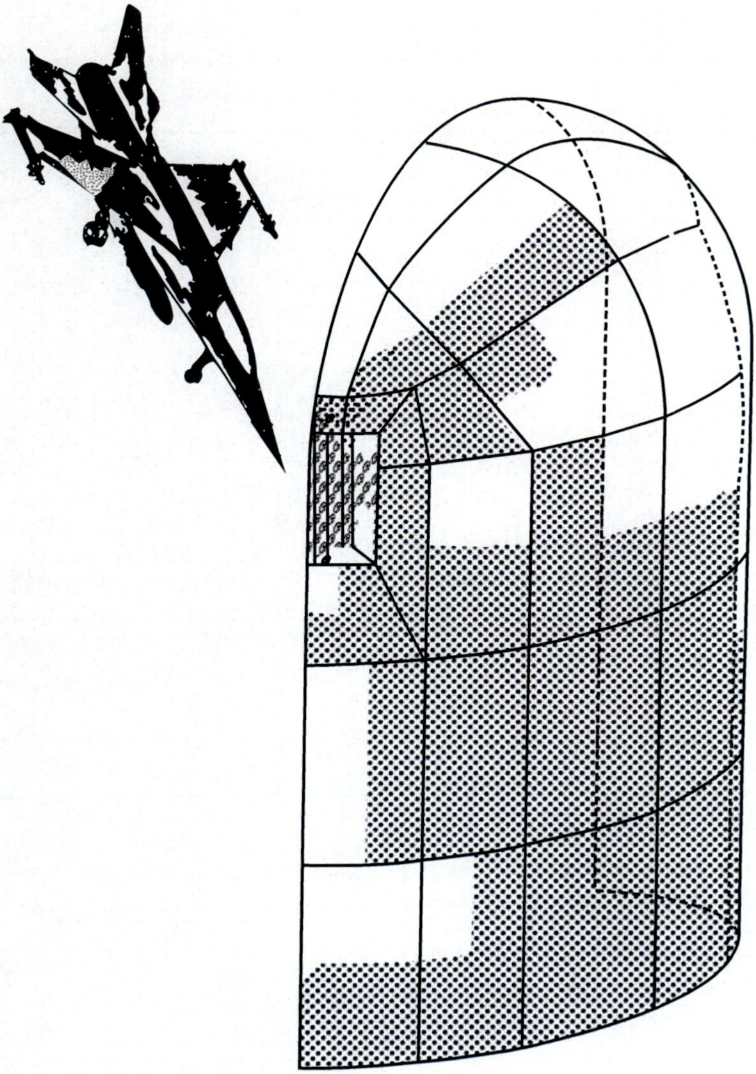

Fig. 7.9 Post-mortem of the containment vessel

crack tip to go far below these specified slips at certain points, which causes the loop to grow much wider than originally specified. Shear crack relation for a typical containment is given in Fig. 7.18. These techniques are given in this text with exclusive details.

At time t, one of the cracks is at the positions marked on a line by the hysteresis loop. At time $(t + \Delta t)$, a certain point can be missed by a significant

Table 7.1 Aircraft impact load versus comparative vessel thickness

	Depth of penetration (m)						Finite element	
	Thickness for no perforation (m)						Thickness[a]/displacement (mm)	
Aircraft [speed (mph)]	IRS	HN-NDRC	ACE	DF-BRL	BRL	DF-ACE	Depth of penetration (m)	Thickness[b]
MRCA	2.0	3.1	2.5	1.57	1.59	2.5	1.80	3.0
	[2.59]	[3.75]	[2.85]	[2.30]	[2.75]	[2.95]	(55 max)	–
Phantom RF-4E	1.85	2.90	2.25	1.25	1.35	2.10	1.55	2.35
[482]	[2.50]	[3.70]	[3.00]	[2.00]	[1.75]	[2.75]	(55 max)	–
Phantom (C)	1.75	2.80	2.15	1.10	1.35	2.0	1.55	2.35
[F482]	[2.50]	[3.50]	[2.85]	[1.75]	[1.75]	[2.75]	(60 max)	–
Boeing 707-320 (E)	1.80	2.75	2.00	1.20	1.50	2.0	1.10	2.00
[730]	[2.50]	[3.25]	[2.75]	[1.75]	[2.0]	[2.75]	(55 max)	–
Boeing 747	2.0	3.0	2.30	1.55	1.85	2.30	1.90	2.75
[250]	[2.50]	[3.75]	[2.85]	[2.30]	[2.50]	[2.85]	(65 max)	–
FB-11 (CH)	2.0	2.50	2.00	1.05	1.35	1.75	1.25	2.30
[200]	[2.50]	[3.75]	[2.75]	[2.00]	[2.00]	[2.25]	(45 max)	–
Mirage	1.75	2.85	2.20	1.25	1.50	2.50	1.45	2.35
[470]	[2.75]	[3.75]	[3.00]	[2.00]	[2.20]	[3.00]	(40 max)	–
Mig-23	1.2	1.8	2.2	1.0	1.0	2.1	1.80	2.85
[480]	[1.6]	[5.0]	[6.0]	[3.0]	[1.15]	[3.10]	(45 max)	–

[a]All thicknesses mentioned under finite element are computed thicknesses
[b]For wide perforation

Table 7.2 Foundation parameters (major data)

Foundation radius $R = 24.86$ m
Foundation thickness $t_f = 3.76$ m
Seismic shear force = 1822.52 kN/m
Uplift from vertical earthquake component = 470 kN/m
Tangent shear from earthquake = 1272.96 kN/m
K_u = translational spring constant $= \frac{32(1-\nu)GR_o}{(7-8\nu)}$
K_z = vertical constant $= \frac{4GR}{(1-\nu)}$
K_O = rotational spring constant $= \frac{8GR_0^3}{3(1-\nu)}$
K_t = torsional spring constant $= \frac{16GR^3}{3}$
$\nu = 0.14$
$\beta' = 0.5$
Material damping 8% for the soil and 2% for the vessel

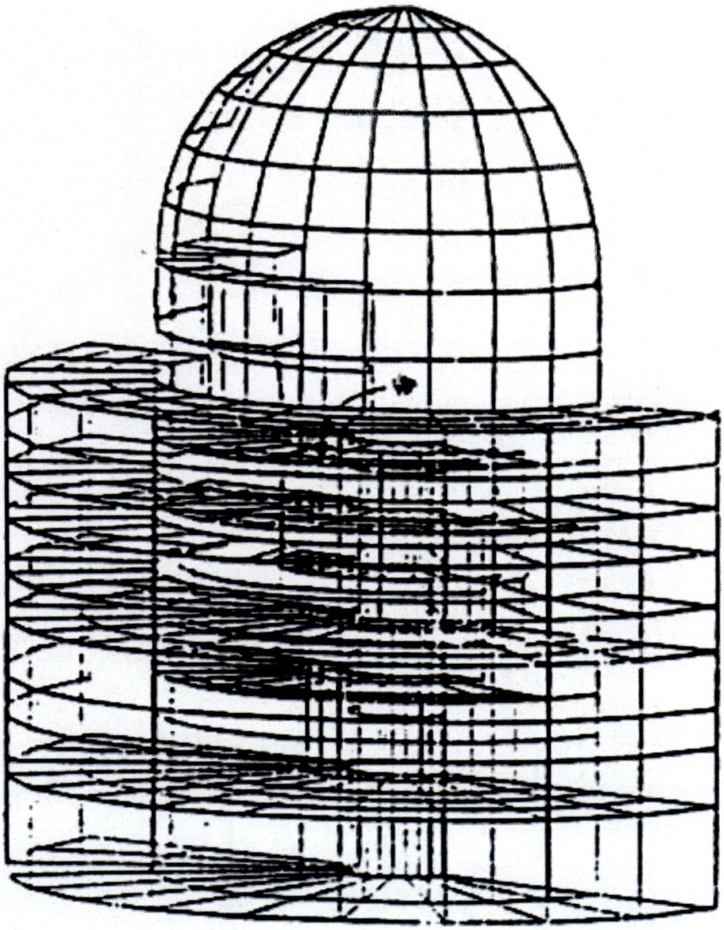

Fig. 7.10 Finite element mesh scheme for PWR reactor vessel integrated with other buildings

amount. The program at this point goes back to time t and refines the time step and computes the shear stress and crack displacement with the new refined time step. If the shear stress is not within a specified limit then the time step is refined further and the process is repeated until the specified limits are met. The same process is performed for the stiffness change from time to time.

The changing of hysteresis loops due to cycling is shown in Figs. 7.16 and 7.17. All cracks start on the cycle1 line. Once a shear stress is exceeded, unloading starts and proceeds along a line parallel to the dotted line A–B. The second cycle is reached when A–B intercepts a line. This process continues.

The circumferential (horizontal) and longitudinal (vertical) cracks in the vessel have a significant effect on the dynamic response of the vessel due to

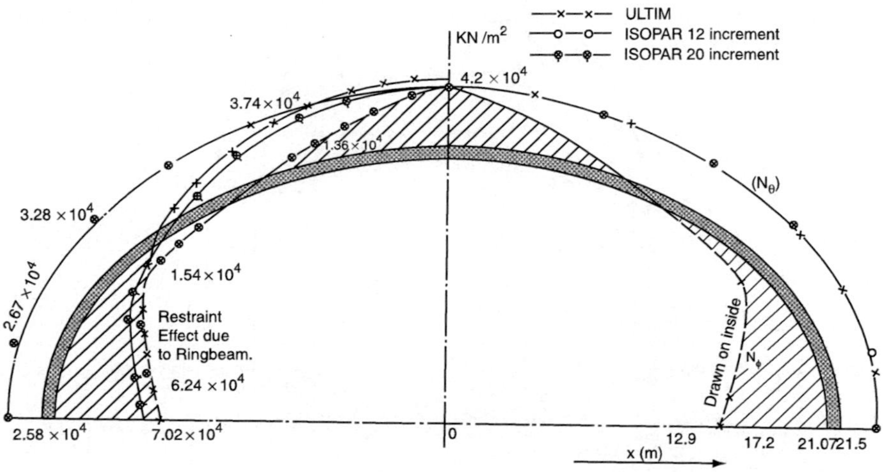

Fig. 7.11 Stress distribution on dome

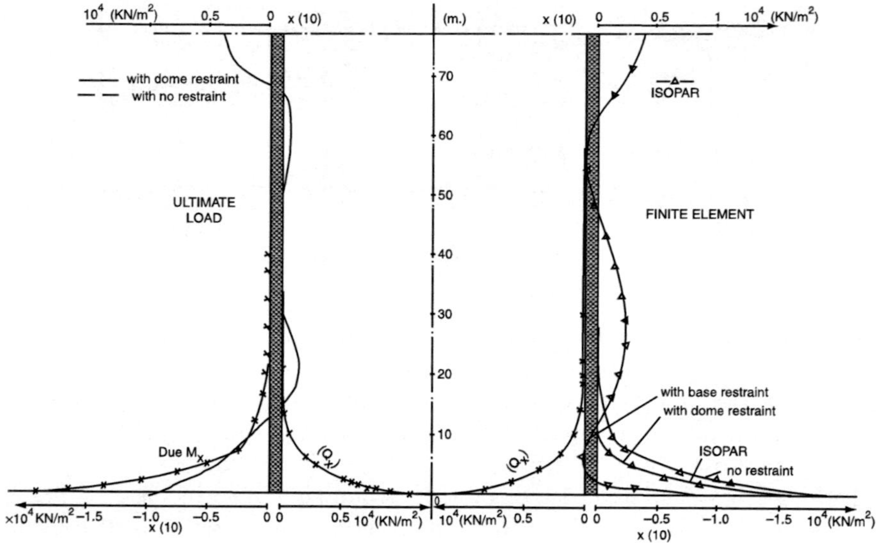

Fig. 7.12 Stress on the barrel wall

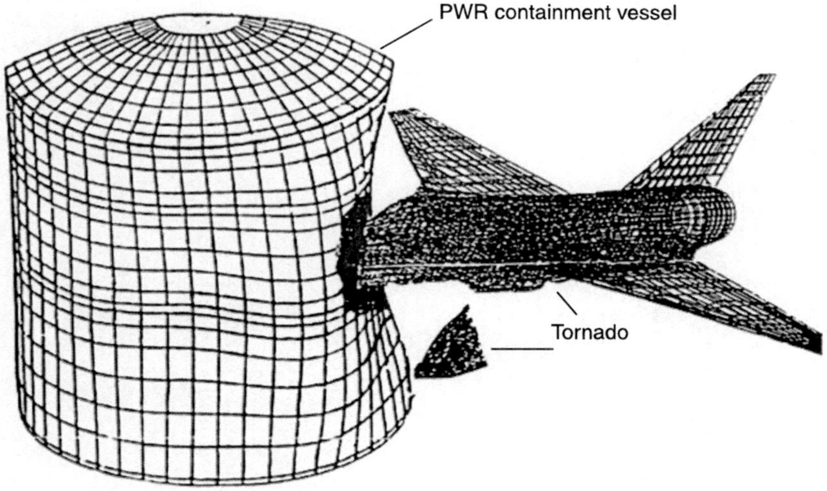

Fig. 7.13 Final disaster scenario

the SSE. The important design parameters which are affected are the liner distortion and the maximum shear stress in the concrete vessel.

Current design criteria specify that the nuclear containment vessel must be able to withstand the simultaneous occurrence of a LOCA with strong (SSE) earthquake motions. The internal pressure creates tensile stresses in both the longitudinal and circumferential directions, while the earthquake causes inertia forces which in turn cause shearing stresses and bearing stresses in the vessels. These stresses cause forces which must be transferred across horizontal and vertical cracks. The crack patterns, crack width and spacing are computed. Crack widths vary between 0.25 and 0.38 mm.

To predict failure mechanisms and to establish the seismic safety margin of a structure, the most severe earthquake which possibly could occur at the site is applied. For this extreme loading case pounding will often occur. Extensive non-linear analyses have to be performed. Pounding introduces impact loads which have to be superimposed on those caused by the ground acceleration itself. While, in general, the structure can withstand the latter, if sufficient ductility is provided, the former loads often introduce stress and displacement distributions not envisaged in the original design. When these impact loads from pounding are too high, the structural system has to be modified to reduce the response [6]. This can be achieved, e.g. by 'tuning' the two involved structures by introducing a spring-dashpot system between them or using Hallquist contact elements.

The shearing stiffness of the cracks due to aggregate interlock affects the shear stress distribution in the containment vessel after immediate short-term internal pressurisation.

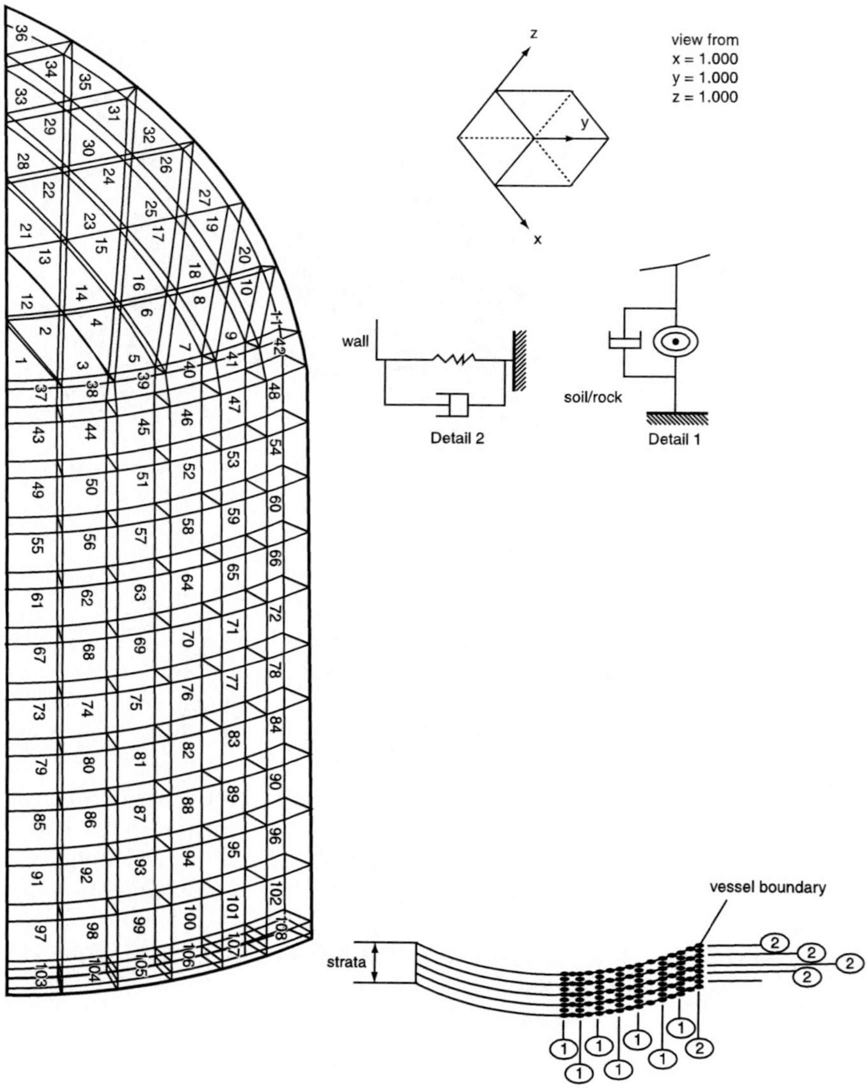

Fig. 7.14 Non-linear model

During an earthquake, the stresses caused by inertia forces do cause the crack width to change. Where the crack was too narrow over the whole of some portion of its circumference, it results in alteration in the shear stress distributions. Computer results show that for a total unbonded length at the crack of 64 mm the change in crack width is small compared to the initial crack width. Therefore the shear stress distribution is not significantly altered from the

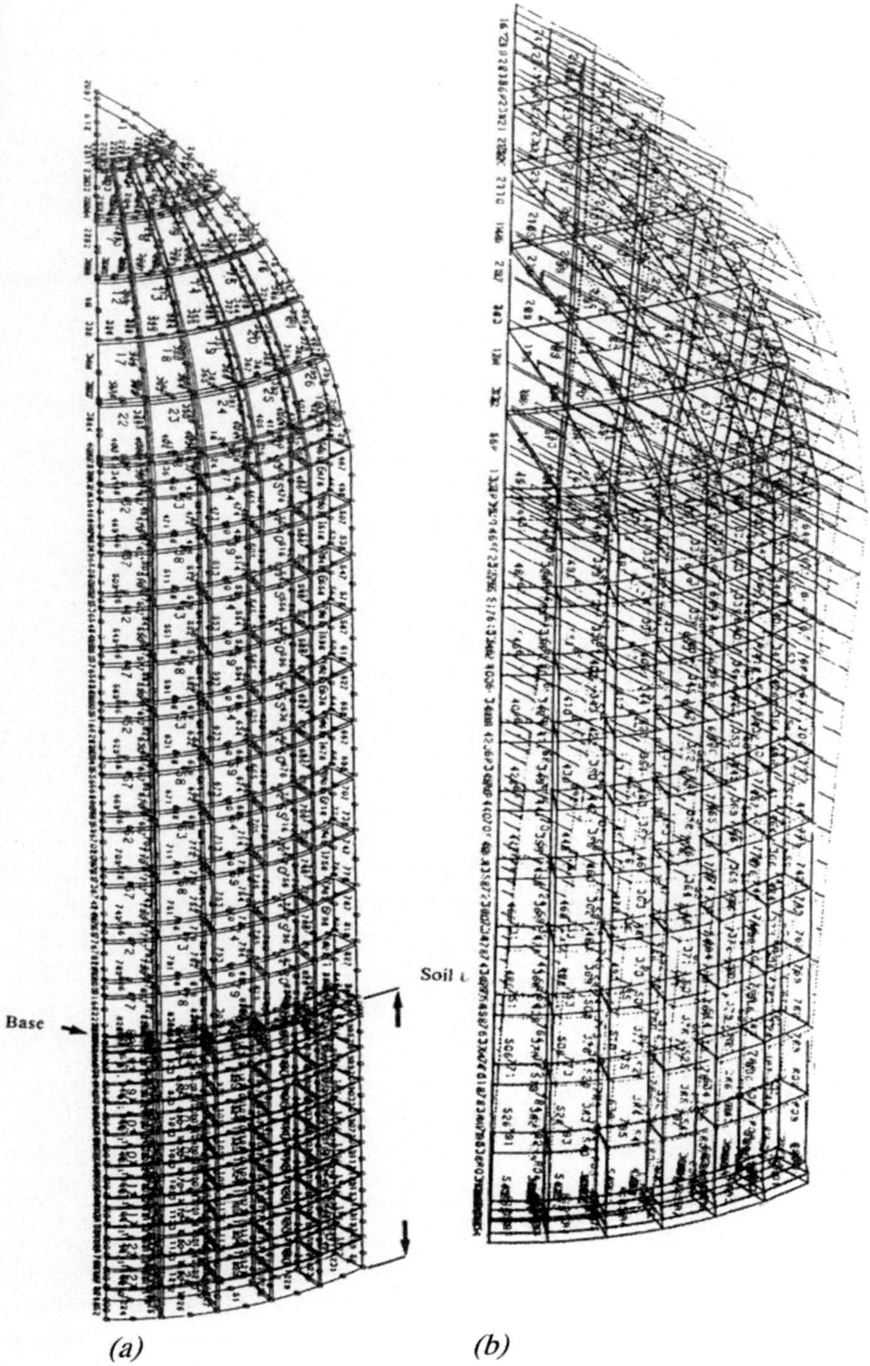

Fig. 7.15 Finite elements for the Bellefonte vessel: **(a)** with seismic devices and **(b)** without seismic devices

Table 7.3 Parameters, design loads and stresses

Conventional steel	
σ_y = yield strength	451.6 MN/m^2
$E_p = 0.1E$	
Liner 6–12 mm thickness	
Coefficient of linear expansion	10 $\mu M/m^\circ C$
Thermal conductivity	41.6 $W/m^\circ C$
ε_{cu}–ultimate strain for concrete 0.0035	$200 \times 10^3\ MN/m^2$
E (steel)	
E (concrete)–short term	$38 \times 10^3\ MN/m^2; 34 \times 10^3\ MN/m^2$
Long term	$20.7 \times 10^3\ MN/m^2$
(soft zones)	$0.74E$
E_p–plastic modulus	$0.476E$
Poisson's ratio	0.15 (concrete)
	0.3 (steel)
Coefficient of thermal expansion of concrete	$10 \times 10^{-6}/^\circ C$; $12 \times 10^{-6}\ /^\circ C(*)$
Short-term specific creep	$1830 \times 10^{-9}\ MN/m^{2^\circ} C$
Long-term specific creep	$2407 \times 10^{-9}\ MN/m^{2^\circ} C$
Minimum crushing strength at 28 days	41.7 MN/m^2
Rigid target	
Steel and alloys	
$m_a = 127.5$ Mg	
$m_a = 38.6$Mg included in m_a	
$\varepsilon_y = 2 \times 10^{-3}; \varepsilon_r = 5 \times 10^{-2}$	
Deformable target	
Impact area = 37.2 m^2; 38 $m^2$and 25 m^2	
Sizewell B parameter	
R_1	22.5 m
d (wall)	2.5 m
R (outer dome)	22.860 m
d (dome)	1500 mm spherical shape
$D + F + 1.5P + T_a$	
$D + F + 1.25P_a + T_a + 1.25E$	
$D + F + P_a + T_a + E$	
D–dead load; F–prestressing loads;	P_a–local pressure;
T_a–local temperature load; E–seismic (OBE) load;	E'–siesmic (SSE)
P_a = 50 psig (345 kN/m^2);	design accident temperature 271°F
$E = 0.05\,g$; $E = 0.11\,g$ and 0.17; →	Time 100
E'' (SSE) uplift 6.9 MN/m;	$f_{cu} = 41.3\ MN/m^2$
E(OBE) = uplift 4.7 MN/m	
$\mu = 0.14$; $K = 0.0003$ for friction;	tendon losses 21 %

Note: For detailed 3D hybrid analysis with and without seismic devices using finite element, the reader is referred to the author main text on "Earthquake Resistant Buildings" published by Springer Verlag, Heidelberg in 2010.

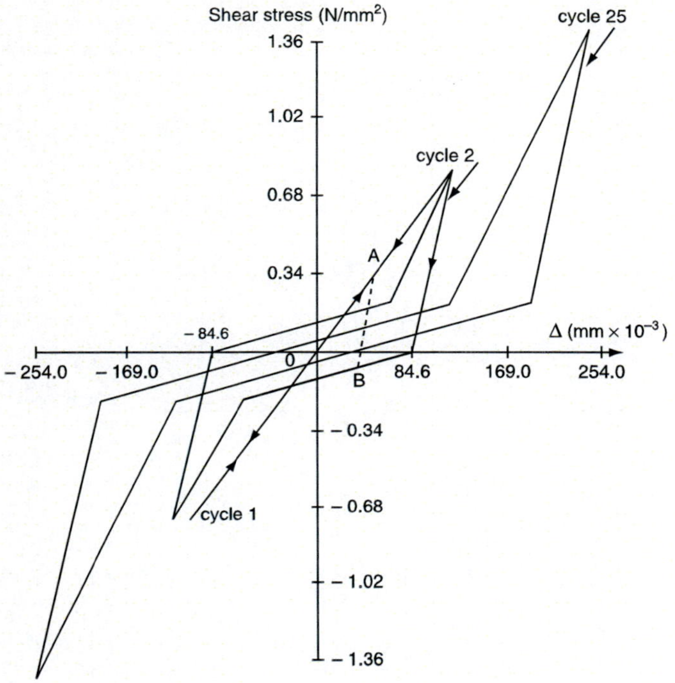

Fig. 7.16 Shear stress versus slip

sinusoidal distribution. The vertical cracks present may effectively decrease the flexural and shear stiffnesses of the containment vessel.

Figure 7.20 shows a mass displacement–time relation for the containment. Figure 7.21 shows response spectra under various conditions and at the top and bottom of the vessel. The control motion is specified at the free surface of the soil deposit.

The curves show the motion specified at bedrock and at the foundation level, respectively. Figure 7.16 shows a 'post-mortem' of the Bellefonte containment at X, Y = 200 gals and Z = 50 gals. Throughout the vessel excitation the damping for the vessel was 2% and that of the soil was 8%.

7.3.1 Mutual Pounding of Containment Building with Auxiliary Building–The Pushover Analysis

7.3.1.1 Introduction

The seismic response of a structure which, in addition to earthquake loading, is subjected to the impact force resulting from pounding by an adjacent containment

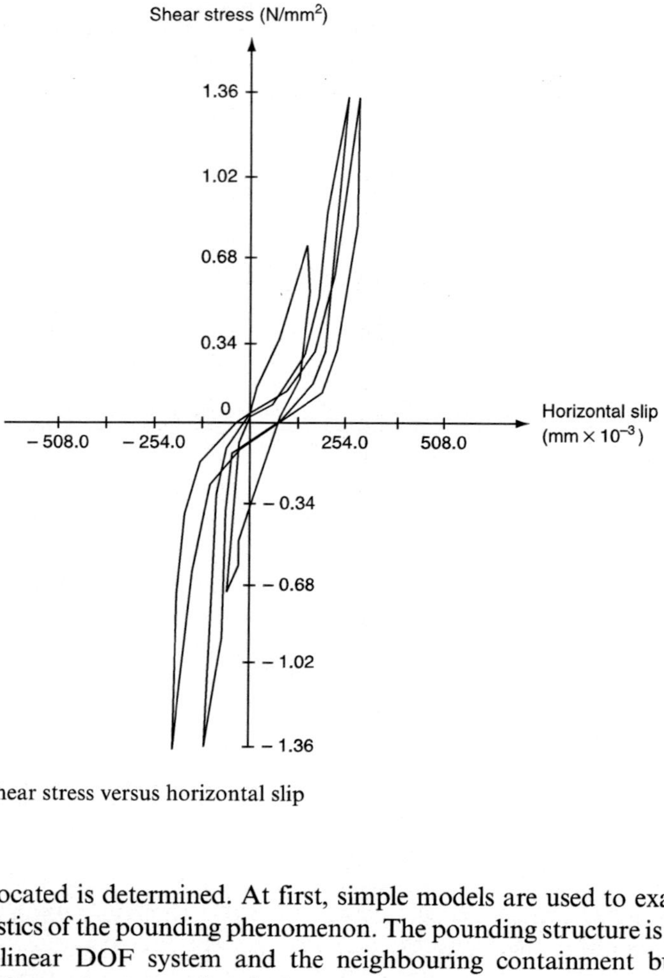

Fig. 7.17 Shear stress versus horizontal slip

Building located is determined. At first, simple models are used to examine the characteristics of the pounding phenomenon. The pounding structure is modelled as a non-linear DOF system and the neighbouring containment building is represented by an impact-spring-dashpot mechanism. Parametric studies, varying the structural parameters, are performed both for steady-state and transient excitations. Finally, the pounding of a typical reactor building by an adjacent auxiliary building during an earthquake is analysed using a simple DOF system and a very detailed dynamic model. The stresses and in-structure response spectra are compared to those for the case in which no pounding occurs.

The gap size between adjacent structures or between adjoining structural units can turn out to be too small in an actual earthquake in which plastic displacements of the soil and of the structure develop, as it is customary to design the structure, and thus also the gap size, for a moderate earthquake, assuming elastic behaviour. The problem of insufficient gap size can also arise when retrofitting nuclear power plants for increased seismic requirements can occur.

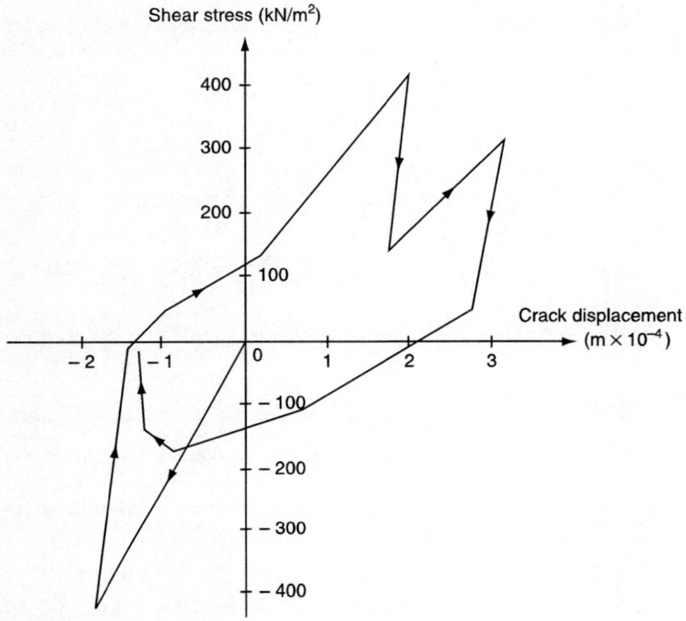

Fig. 7.18 Shear – crack relation

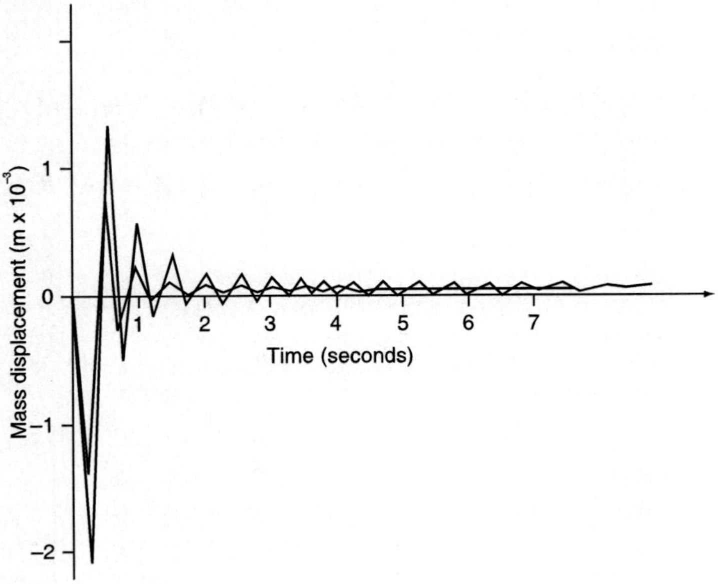

Fig. 7.19 Mass displacement – time relation for typical containment structure

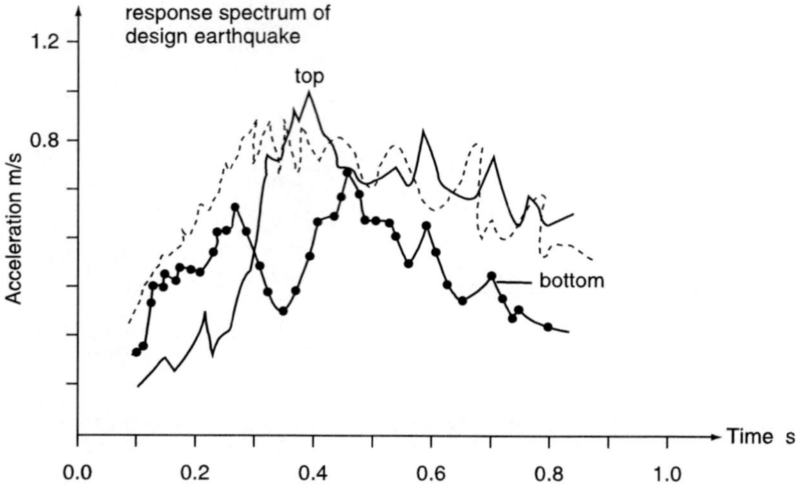

Fig. 7.20 Containment vessel response spectrum (- - - , measured; —, non-linear; -•-, linear)

$$|K|_{\text{Element}}=\frac{1}{\det}\begin{bmatrix} \left(\frac{L}{EI}+C\right) & -\left(\frac{L^2}{2EI}+B\right) & -\left(\frac{L}{EI}+C\right) & \left(B-\frac{L^2}{2EI}-CL\right) \\ -\left(\frac{L^2}{2EI}+B\right) & \left(\frac{L^3}{3EI}+\frac{L}{GA_S}+\frac{N}{K_C}+A\right) & \left(\frac{L^2}{2EI}+B\right) & \left(\frac{L^3}{6EI}+BL-\frac{L}{GA_S}-\frac{N}{K_C}-A\right) \\ -\left(\frac{L}{LI}+C\right) & \left(\frac{L^2}{2EI}+B\right) & \left(\frac{L}{EI}+B\right) & \left(\frac{L^2}{2EI}+B\right) \\ \left(B-\frac{L^2}{2EI}-CL\right) & \left(\frac{L^3}{6EI}+BL-\frac{L}{GA_S}-\frac{N}{K_C}-A\right) & \left(\frac{L^2}{2EI}+B\right) & \left(\frac{L^3}{3EI}+\frac{L}{GA_S}+\frac{N}{K_C}+A\right) \end{bmatrix}$$

$$A=\left(\frac{1}{I}-\frac{1}{I}\right)\frac{L^2L_u}{E}\sum_{i=1}^{N}\left(\frac{2i-1}{2N}\right)^2 \qquad B=\left(\frac{1}{I_C}-\frac{1}{I}\right)\frac{LL_u}{E}\sum_{i=1}^{N}\left(\frac{2i-1}{2N}\right) \qquad C=\left(\frac{1}{I_C}-\frac{1}{I}\right)\frac{NL_u}{E}$$

$$\det=\frac{L^4}{12E^2I^2}+\frac{L}{EI}\left(\frac{L}{GA_S}+\frac{N}{K_C}\right)+\frac{AL}{EI}+C\left(\frac{L^3}{3EI}+\frac{I}{GA_S}+\frac{N}{K_C}+A\right)-\frac{BL^2}{EI}-B^2$$

GLOBAL STIFFNESS MATRIX

$$\begin{matrix} u_1\theta_1 & u_2\theta_2 & u_3\theta_3 & u_4\theta_4 & u_5\theta_5 \end{matrix}$$

$$\begin{bmatrix} K_{11} & -K_{12} & 0 & 0 & 0 \\ -K_{21} & K'_{11}+K_{22} & K_{23} & 0 & 0 \\ 0 & -K_{32} & K'_{22}+K_{33} & -K_{34} & 0 \\ 0 & 0 & -K_{43} & K'_{33}+K_{44} & -K_{45} \\ 0 & 0 & 0 & -K_{54} & K'_{44}+K_{55} \end{bmatrix}$$

$$K_{ii} = \frac{EI_i}{1+4\lambda_i}\begin{bmatrix} \frac{-12}{L_i^3} & -\frac{6}{L_i^2} \\ -\frac{6}{L_i^2} & \frac{4}{L_i}(1+\lambda_i) \end{bmatrix} \quad K'_{ii} = \frac{EI_i}{1+4\lambda_i}\begin{bmatrix} \frac{12}{L_i^3} & \frac{6}{L_i^2} \\ \frac{6}{L_i^2} & \frac{4}{L_i}(1+\lambda_i) \end{bmatrix}$$

$$K_{ij} = \frac{EI_i}{1+4\lambda_i}\begin{bmatrix} -\frac{12}{L_i^3} & -\frac{6}{L_i^2} \\ \frac{6}{L_i^2} & \frac{2}{L_i}(1-2\lambda_i) \end{bmatrix} \quad M_{ij} = K_{ij}^{T''}$$

$$[K_E] = \frac{EI}{1+4\lambda}\begin{bmatrix} \frac{12}{L^3} & -\frac{6}{L^2} & -\frac{12}{L^3} & -\frac{6}{L^2} \\ -\frac{6}{L^2} & \frac{4}{L}(1+\lambda) & \frac{6}{L^2} & \frac{2}{L}(1-2\lambda) \\ -\frac{12}{L^3} & \frac{6}{L^2} & \frac{12}{L^3} & \frac{6}{L^2} \\ -\frac{6}{L^2} & \frac{2}{L}(1-2\lambda) & \frac{6}{L^2} & \frac{4}{L}(1+\lambda) \end{bmatrix}$$

$$\lambda = \frac{3EI}{L^3}\frac{L}{A_S G}$$

$$\begin{matrix} u_1\theta_1 & u_2\theta_2 & u_3\theta_3 & u_4\theta_4 & u_5\theta_5 & u_f\phi_f \end{matrix}$$

$$\begin{bmatrix} K_{11} & K_{12} & 0 & 0 & 0 & 0 \\ K_{21} & K'_{11}+K_{22} & K_{23} & 0 & 0 & 0 \\ 0 & K_{32} & K'_{22}+K_{33} & K_{34} & 0 & 0 \\ 0 & 0 & K_{34} & K'_{33}+K_{44} & K_{45} & 0 \\ 0 & 0 & 0 & K_{54} & K'_{44}+K_{55} & K_{5f} \\ 0 & 0 & 0 & 0 & K_{f5} & K'_{55}+K_{ff} \end{bmatrix}$$

$$K_{ff} = \begin{bmatrix} K_u & 0 \\ 0 & K_\phi \end{bmatrix} \quad K_{5f} = \frac{EI}{1+4\lambda_5}\begin{bmatrix} -\frac{12}{h_5^2} & -\frac{6}{h_5^2} \\ \frac{6}{h_5^2} & \frac{2}{h_5}(1-2\lambda_5) \end{bmatrix}$$

$$K_{f5} = K_{5f}^{T''}$$

$$F_i(t_k) = M_{1i},[u_{1i},(t_k) - u_{ni}] = M_{\mathrm{II}},i[u_{ni}, -u_{\mathrm{II}i},(t_k)] \tag{7.1}$$

$$F_i(t'_k) = M_{1i},[u_{ni}, -u_{Ii}(t'_k)] = M_{\mathrm{II}i},[u_{\mathrm{II}i},(t'_k) - u_{ni},] \tag{7.2}$$

Hence

$$\bar{e} = [\dot{u}_{\mathrm{II}i},(t'_k) - u_{\mathrm{I}i},(t'_k)]/[u_{\mathrm{I}i},(t_k) - u_{\mathrm{II}i},(t_k)] = F_1(t'_k)/F(t_k) \tag{7.3}$$

In terms of this coefficient, the final velocities are as follows:

$$\dot{u}_{\mathrm{II},1}(t'_k) = \dot{u}_{\mathrm{I},\iota}(t_k) - (1+\bar{e})[M_{\mathrm{II}i}\dot{u}_{\mathrm{II}i}(t_k) - M_{\mathrm{II}i}\dot{u}_{\mathrm{II}i}(t_k)]/(M_{\mathrm{I}i} + M_{\mathrm{II}i}) \tag{7.4}$$

$$\dot{u}_{IIi}(t'_k) = \dot{u}_{IIi}(t_k) + (1 + \bar{e})[M_{IIi}\dot{u}_{Ii}(t_k) - M_{Ii}\dot{u}_{IIi}(t_k)]/(M_{Ii} + M_{IIi}) \quad (7.5)$$

The loss of kinetic energy is given as

$$\text{KE(loss)} = \frac{1}{2}\left[\frac{(M_{Ii}M_{IIi})}{(M_{Ii} + M_{II_i})}\right](1 - 2^2)(\dot{u}_{Ii}(t_k) - \dot{u}_{II_i}(t_k))^2 \quad (7.6)$$

The energy loss vanishes when the impact is on elastic build-up.
It means $\bar{e} = 1$, the kinetic energy loss will become

$$\text{KE(loss)} = \frac{1}{2}\left[\frac{(M_{Ii}M_{IIi})}{(M_{Ii} + M_{IIi})}\right](\dot{u}_{Ii}(t_k) - \dot{u}_{IIi}(t_k))^2 \quad (7.7)$$

$$(\bar{e} = 0)$$

For completely plastic ($\bar{e} = 0$), since there is no coupling existing between pair of the floors of the build-up. Equations (7.1), (7.2), (7.3), (7.4), (7.5), (7.6), (7.7), (7.12), (7.13), (7.14), (7.15), and (7.16) can be applied to any pair of floors in the two or with one conventional type building and one reactor vessel buildings whenever they come into comtact. Equations (7.11), (7.12), (7.13), and (7.14) and can be summed up as Eqs. (7.11) and (7.12) without the value of $[F]$ on the right-hand side. Equation (7.13) stays as it is. However

$$\{\bar{e}\dot{u}_I(t_k)\} = \{\dot{u}_I(t'_k) - \{\dot{u}_I(t_k)\}\} \quad (7.8)$$

$$\{\bar{e}\dot{u}_{II}(t_k)\} = \{\dot{u}_{II}(t'_k) - \{\dot{u}_{II}(t_k)\}\} \quad (7.9)$$

$$(t_k \leq t < t_{k+1;}k = 1, 2, 3, \ldots, e_1 = 0) \quad (7.10)$$

where (t_k) and (t_{k+1}) are instants corresponding to two successive different states of contacts and Eqs. (7.3) and (7.9) represent the incremental velocities of buildings I and II at time (t_k), respectively. The velocities $\dot{u}_I(t'_k)$ and $\dot{u}_{IIi}(t'_k)$ after the collision cannot be evaluated from Eqs. (7.1), (7.2), (7.3), (7.4), (7.5), and (7.6). The response of the entire system can be useful between the limits given in Eq. (7.10).

7.3.1.2 Problem Formulation for Publisher Analysis

A containment building and nuclear auxiliary building exist and they are shown in Plate 8.1. Both structures are named as follows:

Nuclear Containment Building → I
Auxiliary Lab Building → II

Both structures are modelled by discrete MDOF systems with concentrated masses. The adjacent masses at the same height are considered as *Contact Impact Case*. One can imagine each pair of masses vibrating initially along the line joining their centres and finally colliding head-on and moving along the same straight line, joining their centres, without rotation after collision which can be treated as inelastic such that the local energy absorption methodologies during impact will be taken into account. Two separate dynamic equations of equilibrium are written for buildings I and II, taking into consideration the geometric compatibility conditions: looking at Plate 7.1, the equations are written as

$$[M_{\mathrm{I}}]\{\ddot{u}\}_{\mathrm{I}} + [C_{\mathrm{I}}]\{\dot{u}\}_{\mathrm{I}} + [k_{\mathrm{I}}]\{u\}_{\mathrm{I}} = -[M_I][I]\ddot{x}_g - \{F\} \tag{7.11}$$

$$[M_{\mathrm{II}}]\{\ddot{u}\}_{\mathrm{II}} + [C_{\mathrm{II}}]\{\dot{u}\}_{\mathrm{II}} + [k_{\mathrm{II}}]\{u\}_{\mathrm{II}} = -[M_{II}][I]\ddot{x}_g - \{F\} \tag{7.12}$$

$$\{u\}_{\mathrm{I}} - \{u\}_{\mathrm{II}} \leq \{\bar{e}\} \tag{7.13}$$

$$\{F\} \geq \{0\} \tag{7.14}$$

in which the displacement, velocity and acceleration column vector, respectively. are relative to the structures base; $[M]$ = structure diagonal mass matrix; $[C]$=damping matrix; $[K]$ = the structure stiffness matrix; $\ddot{x}_g$ represents the acceleration of support motion and $\{I\}$ is the influence coefficient column vector and = the initial distance between each pair of spheres and $\{F\}$= the impulsive forces developed between them during collision; $\{0\}$= zero column vector. The subscripts I and II represent buildings $_{\mathrm{I}}$ and $_{\mathrm{II}}$, respectively. Note since the whole system has now turned to be a series of zones, the problem of determining the motion of restrained buildings after collision from the motion before collision can be solved independently between each pair of zones, having masses. For those not in touch, all kinematical terms remain constant. For those which are in contact the liner impulse momentum law to the normal velocities of colliding zones yields in terms of common normal velocity $u_{n,i}$ at the end of approach, called the Goldsmith approach:

$$M_{\mathrm{I},i}u_{\mathrm{I},i}(t_k) + M_{\mathrm{II},i}u_{\mathrm{II},i}(t_k) = M_{\mathrm{I},i}u_{\mathrm{I},i}(t'_k) + M_{\mathrm{II},i}u_{\mathrm{II},i}(t'_k) = u_{n,i} \tag{7.15}$$

or

$$u_{n,i} = [M_{\mathrm{I},i}u_{\mathrm{I},i}(t_k) + M_{\mathrm{II},i}u_{\mathrm{II},i}(t_k)]/(M_{\mathrm{I},i} + M_{\mathrm{I},i})(M_{\mathrm{I},i} + M_{\mathrm{II},i}) \tag{7.16}$$

where t_k = instant time when an impact occurs,
t'_k = instant time when an impact ends.

The normal impact force is $F_{\mathrm{i}}(t_k)$ and $F_{\mathrm{i}}(t'_k)$ for the approach and restitution periods.

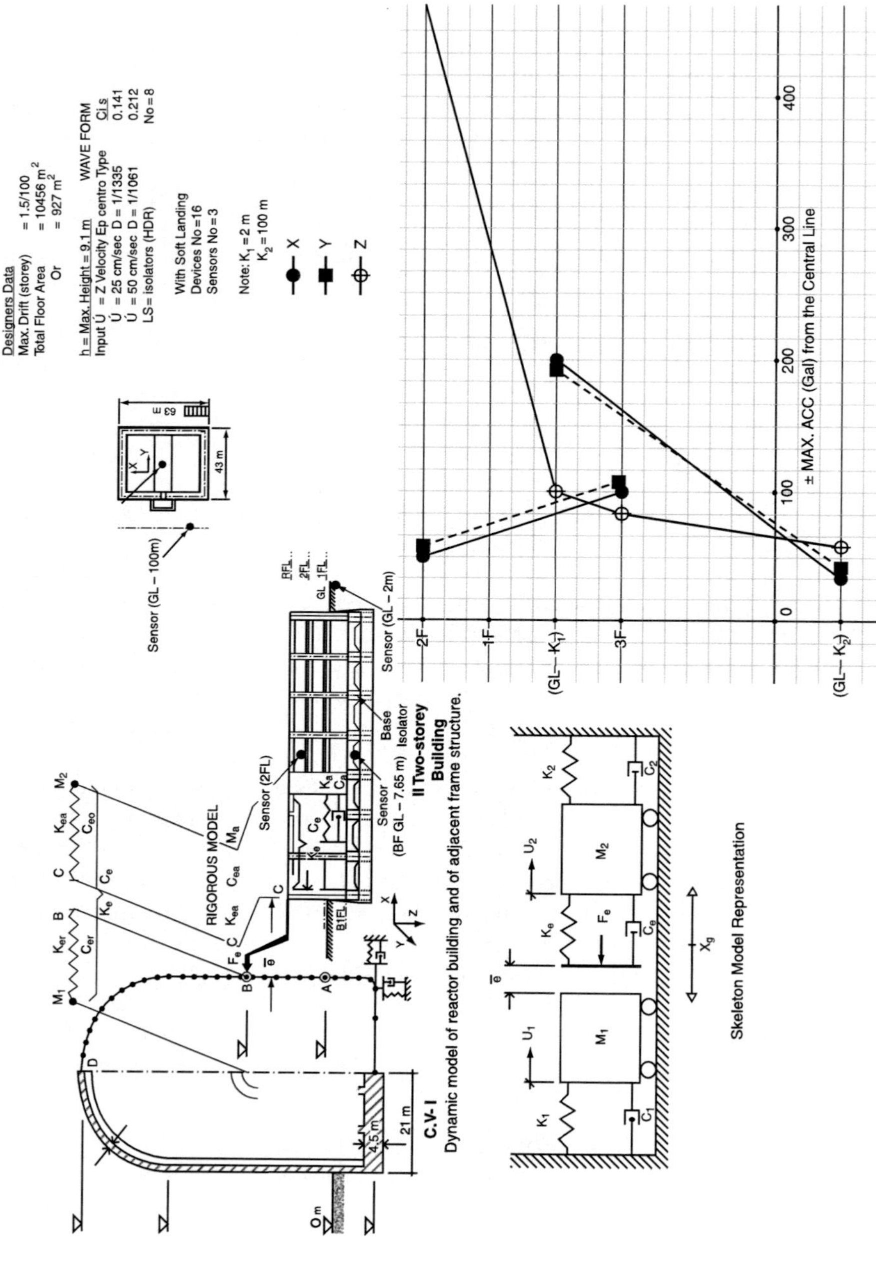

Plate 7.1 Mutual pounding of adjacent structures during earthquakes

Auxiliary Buildings

The adjacent building associated with the containment structure is shown in Plate 7.1. Various stiffness matrices are given in Plates 7.2, 7.3 and 7.4 which have been used in the program ISOPAR with input shown in Plate 7.1.

Analysis of Results

Pounding occurs between the two adjacent structures when the displacement of containment 1 (relative to the ground $\ddot{X}_g$) exceeds that of the frame at c of the auxiliary building by the value of the gap size $\bar{e}$. The impact spring stiffness K_e becomes equal to the force required for a unit rotation of half of the length of

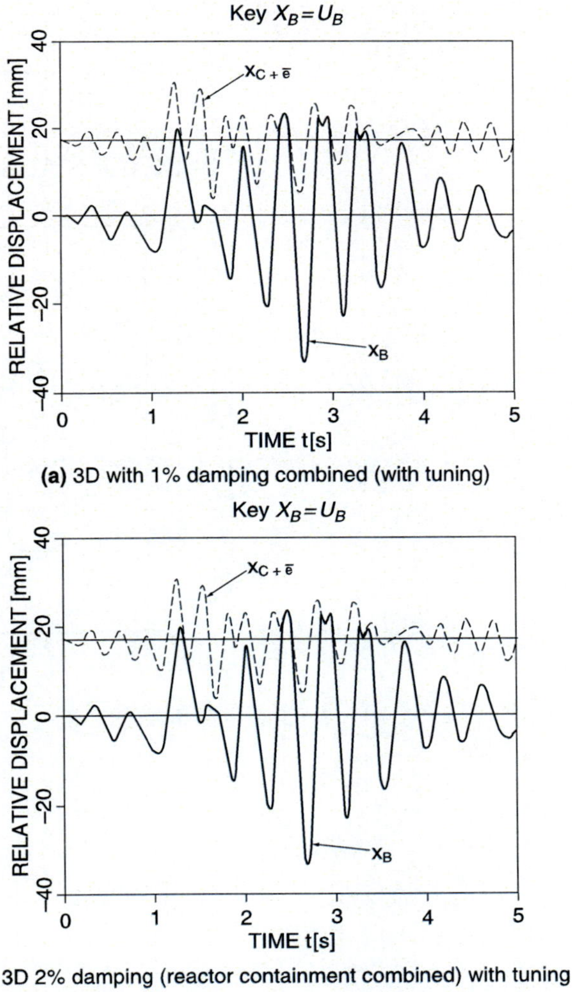

Plate 7.2 Seismic displacement–time history: $\bar{e} = 17\,\mathrm{m}$

Plate 7.3 Seismic impact force–time history

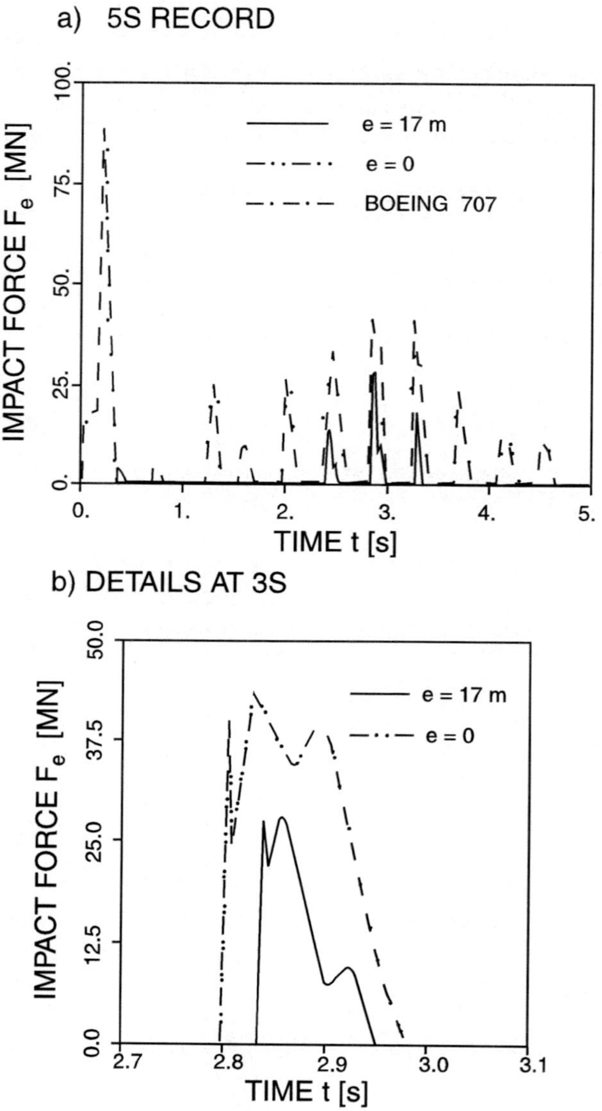

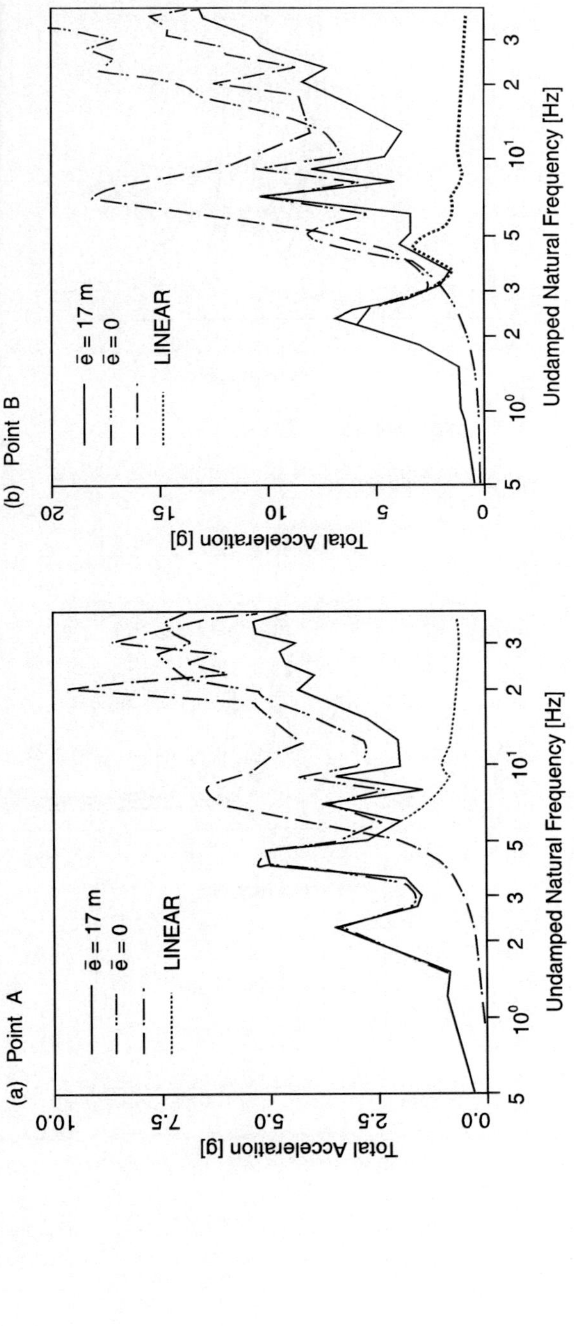

Plate 7.4 Frequency acceleration time relations: **(a)** $\bar{e} = 0, \bar{e} = 17$ m and **(b)** when $\bar{e} > 17$m

Plate 7.5 Response spectra for floor 1 and 2 for different relaxation time parameter values containment: 1, vessel; 2, building

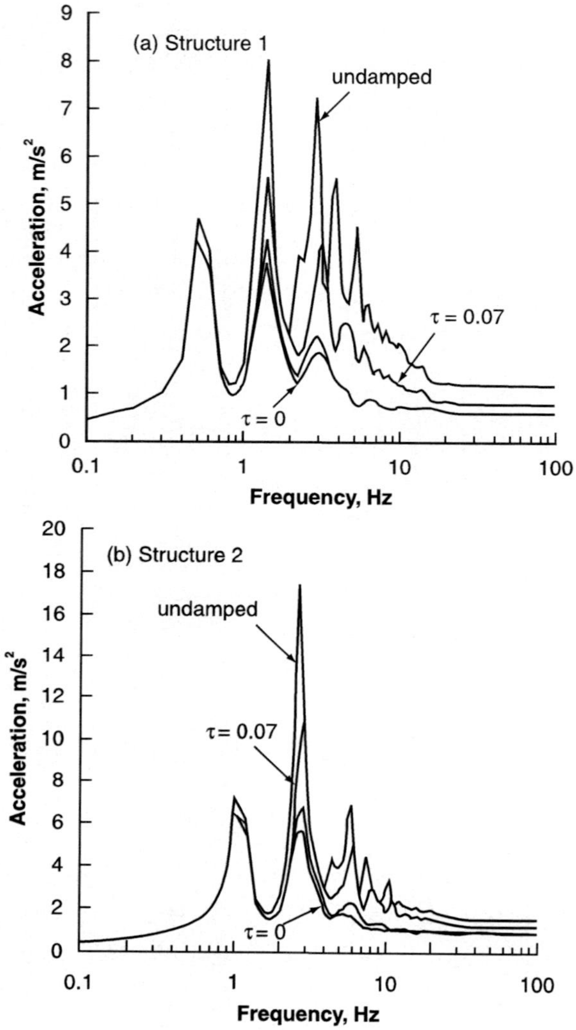

the frame ($K_{ea} = 306\,N/m$). The impact damping coefficient C_{ea} can be determined at point *C*. $C_{ea} = 61.8\,MN/m$ was found. The spring stiffness K_{er} was found to be 2.75 GN/m structure. K_e was evaluated as 2.51 GN/m. The value of C_e was 7.12 MN/m. Two modes are taken as typical example and they are 2.4 and 4.78 HZ with respective damping ratios of the associated module motion as 0.0693 and 0.0710.

The maximum relative displacement of point B and point C based on non-linear calculations for $\bar{e} = 17$m between the two structures is given in graphical form as shown in Plate 7.2. They give relative displacement time history. The impact force time history is shown in Plate 7.3. The total acceleration time and frequency relations are summarised in Plate 7.5.

For greater understanding of the analysis a reference is made to the authors' following book.

Earthquake Resistant Buildings – Analysis and Design
(*With and Without Seismic Devices*)
Springer Verlag 2009.

7.4 Data for Program ISOPAR–NLFEA

Table 7.4 gives data for the two structures (structure 1 and structure 2) for program ISOPAR–NLFEA. The theory behind the seismic analysis is clearly established in the text.

7.5 Containment Overpressurisation and Blowdown

7.5.1 General Introduction

This section deals with the dynamic loadings of overpressurisation and pressure suspension in containment due to pressure fluctuations which are induced during blowdown as a result of steam condensation in the suspension pod. In the later cases regarding blowdown two experiments have been conducted: one on Brunsbuttle Power Plant (Germany in 1974) and the other one at Marviken Power Plant (Sweden in 1972–1973). A comprehensive paper on blowdown has been published:

"*Dynamic Loading of the Containment During Blow Down: Reviews of Experimental Data from Marviken and Brunsbuttle* "

J. Kadlec and R.A. Muller. *Nuclear Engineering and Design*, vol. 38P, 143–158(1976) published by North-Holland Publishing Company, Amsterdam, Holland.

The reader is referred to this paper. The details are not mentioned in this text.

The containment overpressurisation analysis has been carried out using program ISOPAR and results obtained are fully collaborated with the 1/6

Table 7.4 Data from program ISOPAR–NLFEA

1. Vessel parameters and auxiliary building dimensions as stated in Plate 8.1	
2. Containment vessel data	
Concrete solid elements 20 noded	2500/sector of $22\frac{1}{2}^{\circ}$ –2 deep
Isoparametric type: →	30,000 nodes
Plastic elements:	
8-noded isoparametric 1000 elements/sector of $22\frac{1}{2}^{\circ}$ –1 depth	
Representing liner or membrane	8000 nodes
Prestressing elements 4 noded	5000
Isoparametric	20,000 nodes/sector of $22\frac{1}{2}^{\circ}$
Total embedded or on surface	2000/sector elements
Bond elements	20,000
Bar elements: 4 noded	
Isoparametric	60,000 nodes/sector of $22\frac{1}{2}^{\circ}$
P_g = containment vessel internal pressure	$\not> 344.75\ \mathrm{KN/m^2}$
Material parameters are given in Chapter 2	
The failure of concrete and cracking criteria there in	
ω_g = natural frequency = 15.7 rad/s	
3. Building data	
Line elements	
Nodes 4–noded ISOPAR	
The rest of data visible in Plate .1	
Each storey mass $\not> 5.52 \times 10^8$ kg	
Storey stiffness calculated = 5.404×10^8 KN/m	
When VED is considered	100 total used
Total number	= 30
γ = scale factors	= 100
η = loss factor	= 1.12
ξ_g = soil damping ratio	= 0.61
5_w = white noise input data	= 1.0 g
k = each brace	$\not>$ storey stiffness
α weighting ratio	= 4.120
$E_s = 200\ \mathrm{GN/m^2}$; $\sigma_y = f_{yt} = 460\ \mathrm{N/mm^2}$ considered	
Plate elements	= 500/floor, edge nodes are linked
10 bond elements	
4. Seismic data	
$\dot{u}$ velocity: 50 cm/s;	
$\ddot{u}$ acceleration	
X- direction	200 gal
Y- direction	200 gal
Z- direction	50 gal
D = maximum drift = 2%	
$\bar{e}$ = gap maintained = 17 m = d	
$\ddot{x}_g$ = ground acceleration as shown	

Table 7.4 (continued)

Plate 8.1
For undamped vibration $c = c_e = 0$
$\frac{\bar{e}}{u_{g0}} \ngtr 5$ for $\xi = 0$
$\omega_g/\omega \ngtr 1.8$ computer program limitation
Test case
I. One-sided impact
$K=4$, $\xi = 0.05$
e – coefficient of restriction = 0
e – pounding and impact distance 0–17 m

scale R C model containment test results by Lawrence Livermore laboratory and many others.

7.5.2 *Containment Overpressurisation*

A 1/6 th scale R C model (Figs. 7.20 and 7.21) has been designed for a design pressure of 317 KN/m^2. The reinforcement layouts are shown in Figs. 7.22 and 7.23. A pre-test analysis has been carried out using the program ISOPAR. The endochronic model representing the failure of concrete along with cracking criteria indicated in the text has been adopted.

The results are compared with others in Figs. 7.24, 7.25, 7.26 and 7.27 using the 3D finite element analysis with non-linear dynamic solutions. The mesh scheme for this technique for a model is shown in Fig. 7.28. The final model is shown in Fig. 7.29 and some cracks and their zones are indicated in a portion of wall in Fig. 7.30.

The Sizewell B model provided by then CEGB was overpressurised. Figure 7.33(a) gives the displacement pattern. The vessel model failed at 3.15 times the design pressure by rupturing reinforcements crossing the developed cracks. The model is of the type indicated in Figs. 7.30 and 7.31. The post-mortem of the Sizewell B containment model with crack patterns is shown in Fig. 7.31.

For the Lawrence Livermore model, the pressure -displacements are collaborated well with the experimental and theoretical models. They are summarised in Figs. 7.24, 7.25, 7.26, 7.27, 7.28, 7.29, 7.30, 7.31. Thus, these results validate the analytical work and computer program ISOPAR-NLFEA.

Concrete structures are equal to 30 GPa, 0.2, 2.5 Mg/m^3 and 0.07, respectively. The horizontal earthquake excitation is represented by the input–time history of Fig. 16, scaled to a maximum acceleration of 0.3 g. No vertical or rotational input components are considered. The structures of the plant are constructed on a deep firm soil layer whose (uniform) material properties, compatible with the strain level of the prescribed earthquake, are as follows:

Fig. 7.21 Schematic of the 1/6th scale reinforced concrete containment model elevation view (Courtesy of Lawrence Livermore, USA)

Dynamic shear modulus = 0.6 GPa, Poisson's ratio = 0.4, density = 2.4 Mg/ m^3 and coefficient of hysteretic damping = 0.05.

The dynamic model is also shown schematically in Fig. 7.21. The auxiliary building is indicated on the right-hand side of this figure. It is assumed that only the roof slab of the structure, which is hardened for aircraft impact, hits the reactor building during the earthquake. To extend the scope of this investigation, the gap size e is varied. For the purpose of this analysis, where primarily the response of the reactor building is of concern, the auxiliary building is adequately represented by a simple frame with one (horizontal) degree of freedom. The dimensions of the roof slab (length 72 m, width = 30 m, thickness 1.2 m) result in a mass ma = 6.48 Gg of this 1-DOF system. Assuming a representative natural frequency of 3.50 Hz, a stiffness coefficient k_a is calculated as 3.13 GN/m. A damping ratio of 0.07 determines the damping coefficient C_a = 20.0 MN s/m. The axisymmetric reactor-shield building is modelled with 32 curved higher-order isoparametric frusta in the meridional direction. Inasmuch as (in this investigation) the results are calculated on the outer shield building only, the

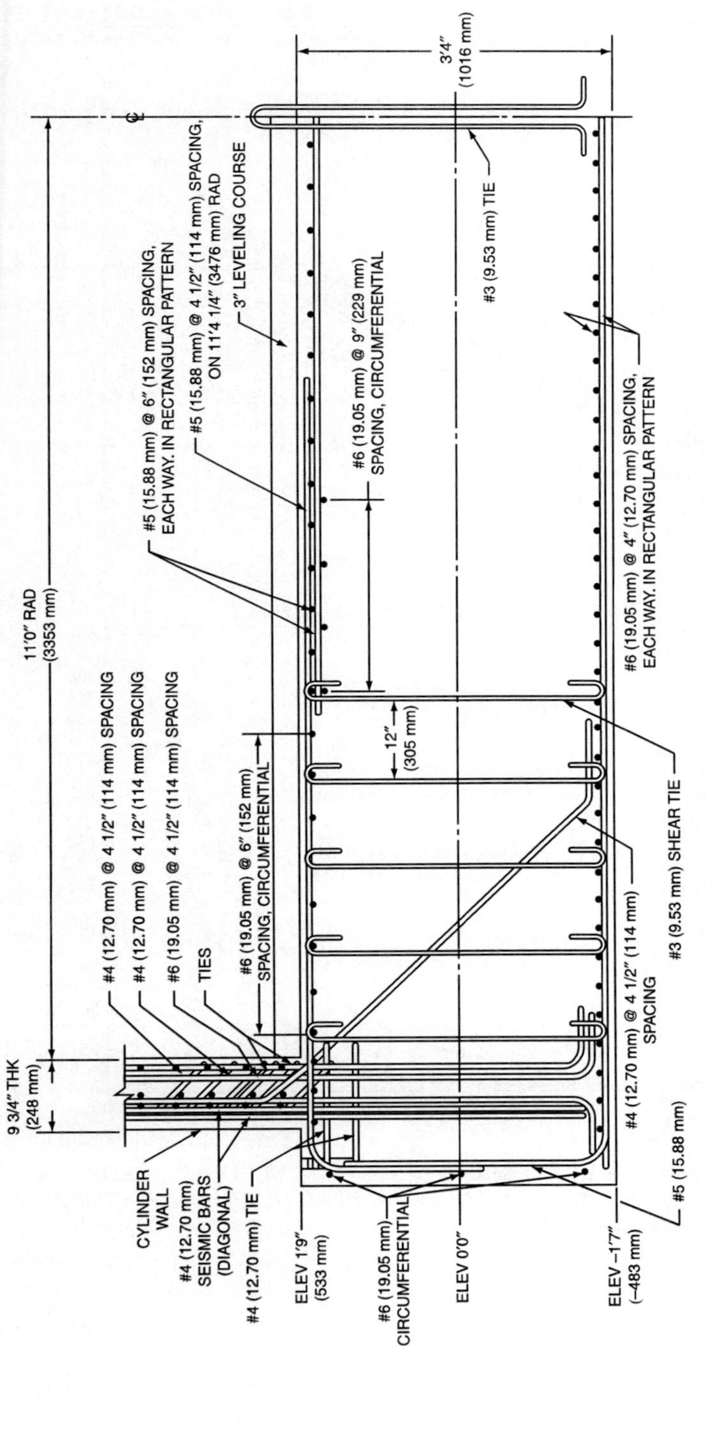

Fig. 7.22 Reinforcement in the basement and cylinder basement junction (Courtesy of Lawrence Livermore, USA)

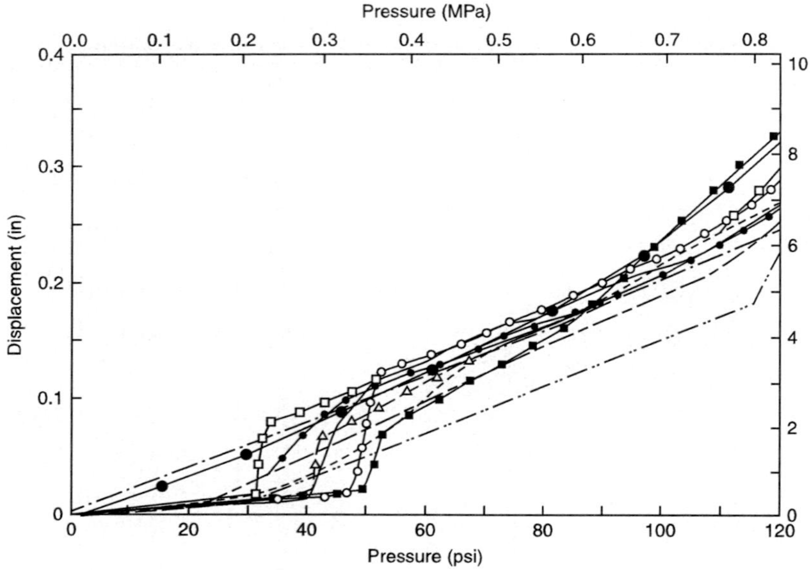

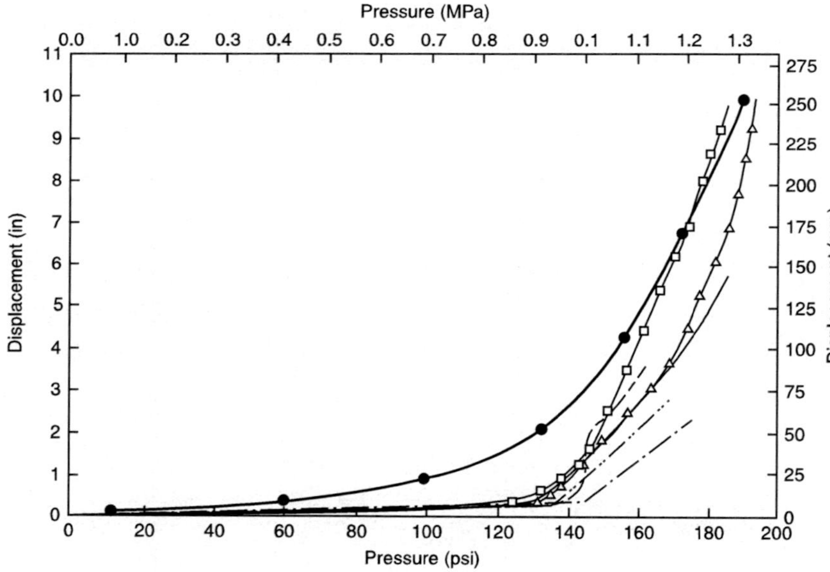

Fig. 7.23 Comparative study – (– · ·–, SNL; - -, NIL; – – –, SERD; –□–, CEGB; –○–, EPRI; – – –, BNL; –□–, ENEA; – – –, ANL; • – •, CEA; –△–, GRS; –●–, Bangash) (Courtesy of Lawrence Livermore, USA)

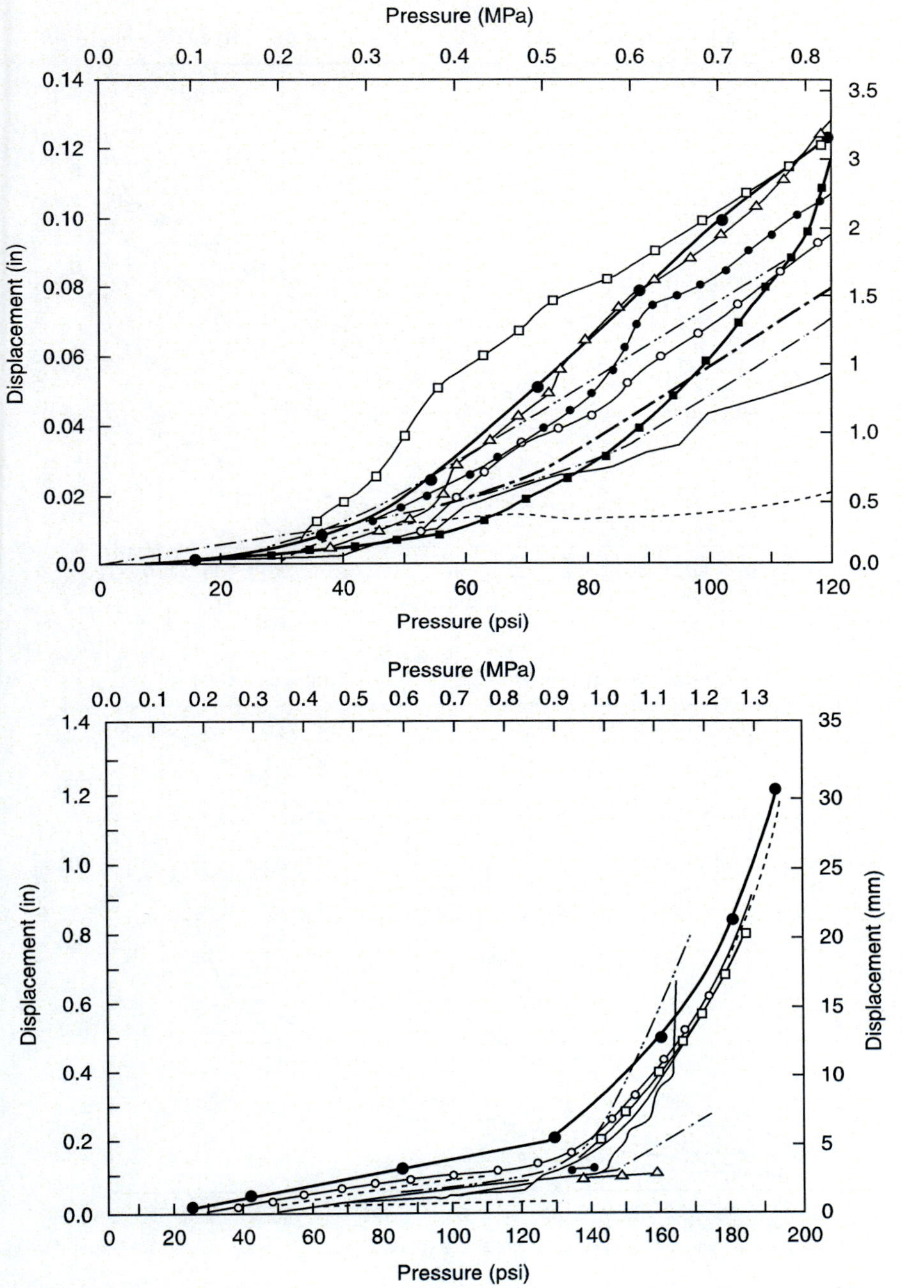

Fig. 7.24 Comparative study – 2 (see Fig. 7.26 for key) (Courtesy of Lawrence Livermore, USA)

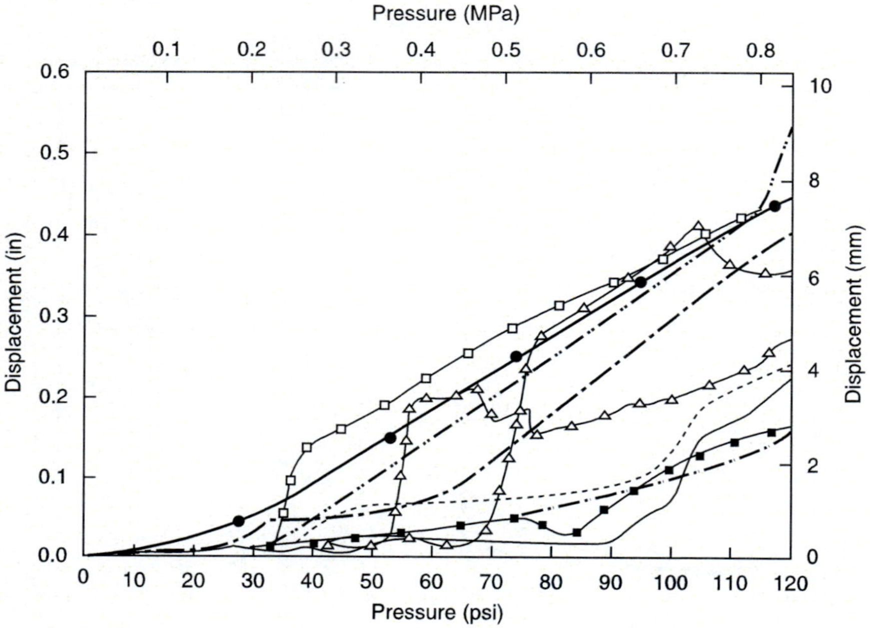

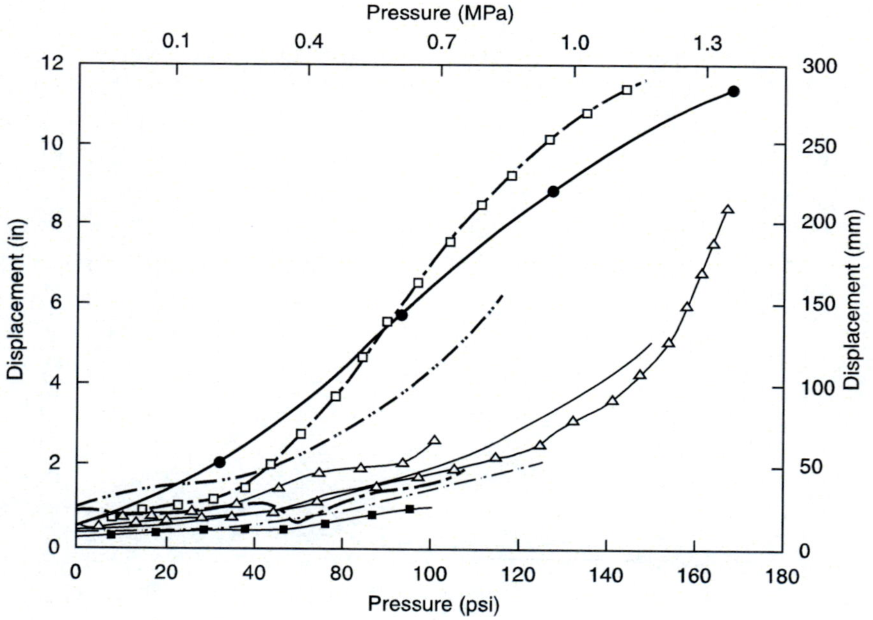

Fig. 7.25 Comparative study – 3 (see Fig. 7.26 for key) (Courtesy of Lawrence Livermore, USA)

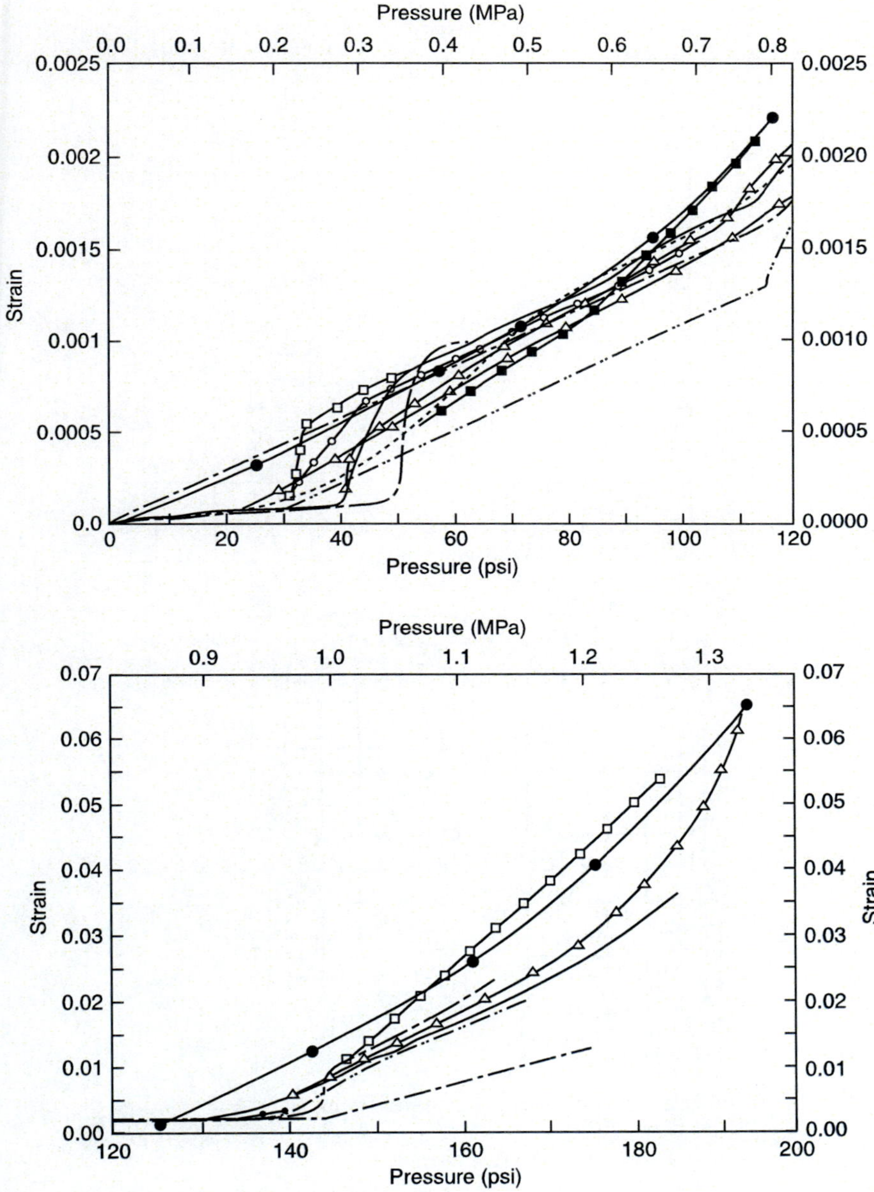

Fig. 7.26 Comparative study – 4 (see Fig. 7.26 for key)

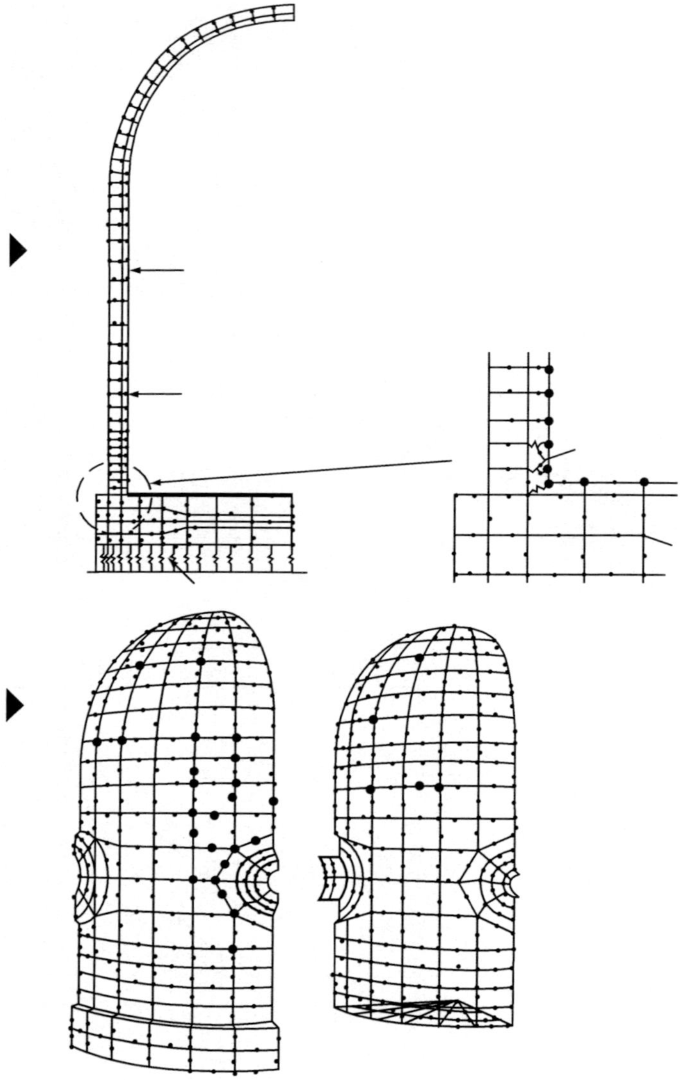

Fig. 7.27 One-sixth scale RC containment – finite element mesh

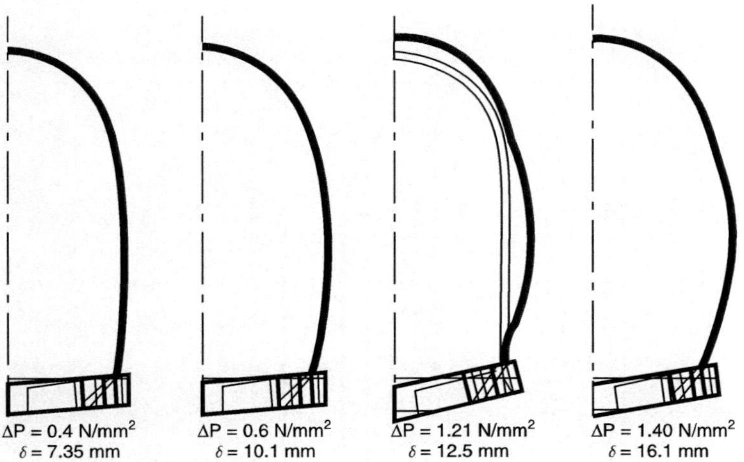

Fig. 7.28 Containment vessel final mode [1006; ISOPAR]

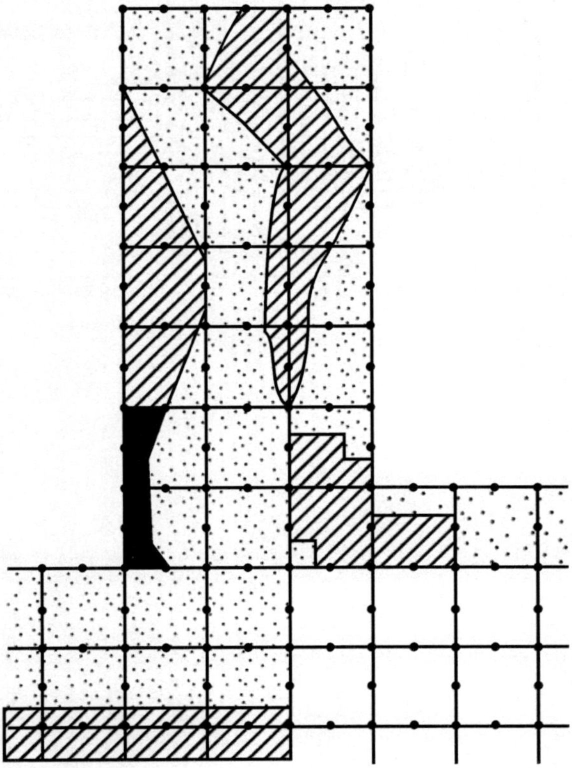

Fig. 7.29 Major crack zones [1006](denotes zero tensile stress; denotes crushing; denotes radial vertical cracking)

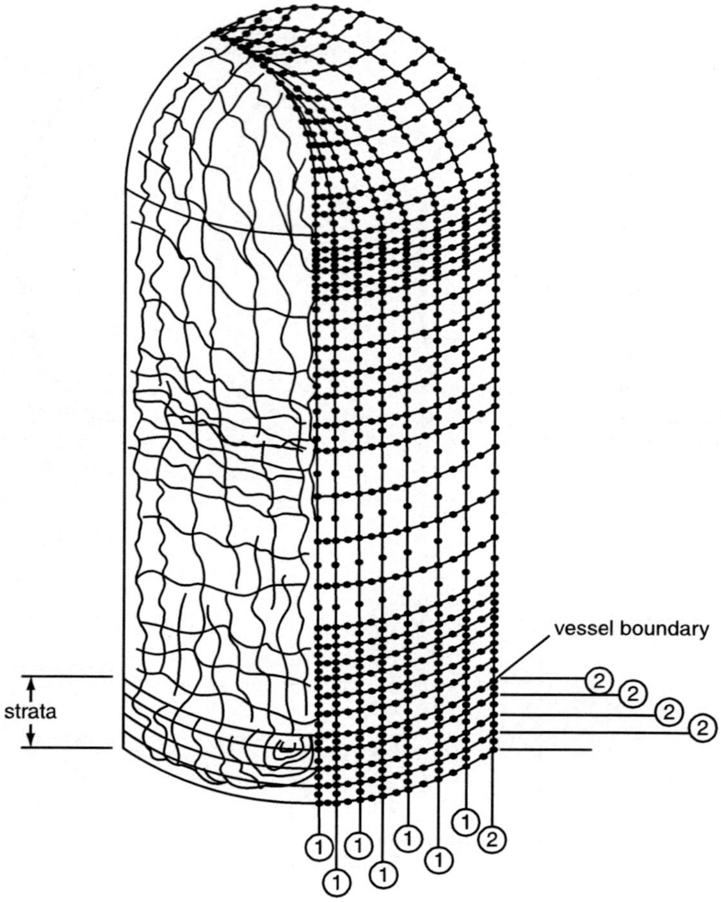

Fig. 7.30 Cracked containment

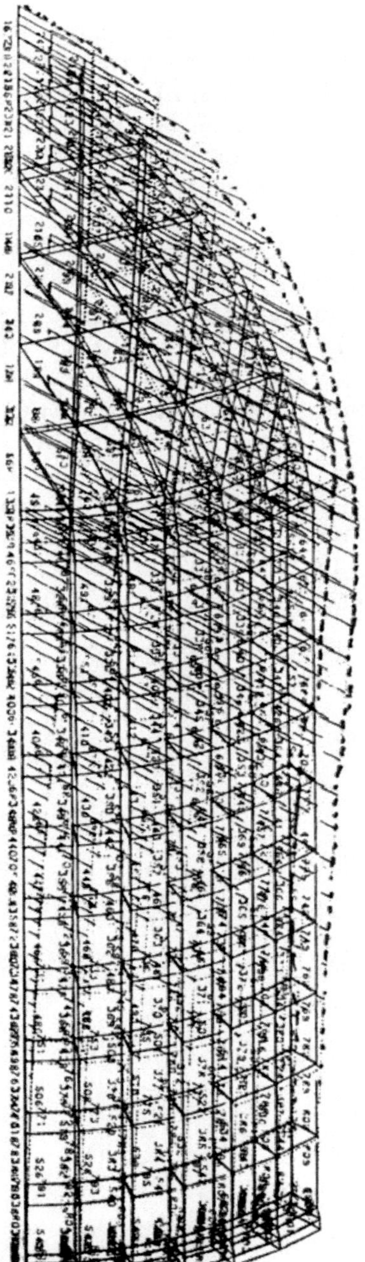

Fig. 7.31 Sizewell B overpressurisation analysis – final mode

internal structures and the base are modelled as a rigid body. In the circumferential direction, the standard Fourier expansion method is used. It is assumed that the impact force from the frame acts horizontally on an area on the middle surface of the shell equal to 3 m in the meridional times 12 m in the circumferential direction at point B. This evenly distributed load is decomposed circumferentially into 16 Fourier terms. For each harmonic, 297 dynamic degrees of freedom result for the reactor building. Structural damping of 0.07 times the critical value is selected. The impedance function of the soil, modelled as a visco-elastic half space, is approximated as being frequently independent. The corresponding springs and dashpot are shown schematically in Fig. 7.20.

In addition to this quite elaborate analysis, referred to as the 'rigorous' solution further on, the pounding structures are also represented as a two-degree-of-freedom (2-DOF) system (Fig. 7.21). The properties of the first structure are associated with the response of the reactor building in the first mode of the first harmonic (fundamental mode). The corresponding frequency equals 2.30 Hz. Using this mode shape as the shape function, the generalised mass $m_1 = 53.1$ Gg and the generalised stiffness $k_1 = 11.1$ GN/m are calculated. The corresponding damping coefficient $C_1 = 107$ MNs/m. The modelling of the second structure in Fig. 7.22 is identical to that of the frame in Fig. 7.20 ($m_2 = m_a$, $k_2 = k_a$, $C_2 = C_a$).

Chapter 8
Bonded Reinforcement in the Concrete Reactor Pressure and Containment Vessels

8.1 Introduction to the Main Philosophy

The quantity and disposition of main bonded reinforcement in the Hunterston—'B' P C P V is taken as a typical case study. In this chapter covering only main and local bonded reinforcement are covered. An attempt has been made to discover the need and function of main bonded steel in the prestressed concrete cylindrical pressure vessel. The same is adhered to the containment vessel. At present there is no established method available to determine the amount of such steel needed in the vessel and to adjudge its performance during working and overload conditions of the vessel.

The elastic analysis carried out for the vessel hardly suggests the need of such reinforcement apart from where maximum stress concentrations occur. The present ultimate load analysis for the vessel needs very little help (about 10%) to develop along with prestressing tendons and the liner steel a full assumed ultimate strength of the vessel. In both these stages, namely elastic and ultimate load, certain assumptions are adopted, the truth of which is yet to be discovered. Moreover, many other factors are unknown about the vessel behaviour including those from creep, shrinkage, construction and general accidents. All of them individually as well as in combination will create infinite number of cracks in the body of the vessel, thereby causing a premature failure of the vessel before ultimate load is reached. Greater emphasis is therefore laid on these factors in this report and it is maintained that cracks developing due to above causes must be controlled by bonded steel and they should also be sufficient in quantity so that a rise of pressure would lead a progressive increase in depth of cracks incipient or already present under working conditions. More research work is recommended for the accurate assessment of the bonded steel quantities. Local areas with and without penetrations obviously require special analysis such as finite element to assess whether or not unacceptable stresses and cracks are present. Those areas need to offset them. These analyses obviously act as a scanning processes.

M.Y.H. Bangash, *Structures for Nuclear Facilities*,
DOI 10.1007/978-3-642-12560-7_8, © M.Y.H. Bangash 2011

8.1.1 General Rules for the Disposition of the Main Bonded Reinforcement in the Pressure or Containment Vessels

The following general rules are taken into account for the disposition of main bonded reinforcement.

8.1.1.1 Disposition Type-1

The behaviour of the vessel must be thoroughly studied under working and overload conditions prior to the assessment of bonded steel. Sufficient bonded steel is provided in areas of maximum stress concentration under working conditions and is gradually reduced for economic reasons to areas of low tensile or high compressive stresses. Where cracks are predicted at pressure above design value, bonded steel provided there must be sufficient to effectively control the initiation and propagation of these cracks.

8.1.1.2 Disposition Type-2

Sufficient strain must be developed in the vessel due to cracking to enable the prestressing tendons, bonded steel and liner to develop their full assumed ultimate strength, i.e., the number of cracks times the average crack width must be equal to the additional elongation necessary to bring tendons up to 90% GUTS.

8.1.1.3 Disposition Type-3

Bonded reinforcement in all areas must be sufficient to accept, without failure, the tensile load transferred to it during vessel cracking. In general, a minimum percentage of steel based on the area of cross-section of the vessel (in this case not more than 0.3%) should be investigated for ultimate load reasons and then distributed in proportion on the outside and inside faces of the vessel.

8.1.1.4 Disposition Type-4

In assessing the contribution of bonded reinforcement to the ultimate strength of the vessel due consideration shall be given to the stress–strain curve of the steel concerned. The stress allowed in the bonded steel shall not exceed that corresponding to a strain of 1%. The minimum average strain in bonded steel shall be given by dividing crack width by crack spacing from a well-distributed crack pattern formed in the vessel.

8.1.2 Choice of Main Bonded Reinforcement

The choice of main bonded reinforcement is based on its high tensile strength and interaction between it and concrete, resulting in higher bond length and an improved degree of crack control. A considerable experimental work has been carried out by research on steel bars such as GK 60 deformed bars in close cooperation with the manufacturers. Experience of design companies with this material in the past suggests to use GK + 60 $1\frac{1}{4}''$dia. (31 mm) bar as main bonded reinforcement steel. It develops a stress of 60,000 psi at 0.2% strain and has an ultimate strength of about 100,000 psi (735 MN/m^2) at minimum guaranteed elongation of 14%. The low carbon content (comparable to mild steel) of GK 60 reinforcing bar simplifies welding in local areas and can expect to carry the full tensile strength of the bar. Bond tests carried out on G.K 60 according to BS8110 of EC-2 indicate that where an end slip of 0.001$''$ occurs its bond strength exceeds a plain bar by more than twice the 40% suggested by B8110 of EC-2 and with initial end slip, the anchorage value increases continuously up to the maximum value. This suggests that by using this type of reinforcement, no hooks are required in the vessel concrete.

All main bonded reinforcement shown on vessel drawings cannot be less than $1\frac{1}{4}''$dia. (31 mm) high tensile GK + 60 bar. Minimum requirement of bonded steel in different areas of the vessel is, therefore, based on the properties of GK + 60 bar or similar. For other types approved by authorities, special tests shall be carried out.

8.1.2.1 Choice of the Minimum Percentage of Main Bonded Reinforcement

The phrase 'Percentage of Bonded Reinforcement' is ambiguous in many existing vessels; nevertheless, minimum percentage of steel based on cross-sectional area of the vessel is still regarded as the basis of design quantity in the vessel with some variations depending upon different requirements, discussed elsewhere. The minimum percentage of bonded GK 60 reinforcement steel for pure tensile case is shown in Fig. 8.1. This figure is obtained on a simple basis of assuming concrete failure stress of 300 psi (2.07 MN/m^2) with a steel stress of 90,000 psi (620 MN/m^2). The figure becomes 0.46% if corresponding steel stress is assumed to be 60,000 psi. (441 MN/m^2). The research on other containments inclusive WYLFA vessel indicated that the steel is likely to rupture following the formation of primary cracks if the amount is greater than 0.2% of the cross-sectional area of concrete. Therefore 0.2% steel is considered as minimum steel to be placed in any cross-section of the vessel concrete.

In Hunterston 'B', a desirable assumption is made that the bonded reinforcement is capable of maintaining a strain of 100 micro-strain throughout the vessel in order to assist in the formation of a well-disposed crack pattern, hence a strain of 100×10^{-6} can be propagated if the percentage reinforcement is of the order

$$\frac{250}{90,000} \times 100 = 0.277\% \text{ of concrete area}$$

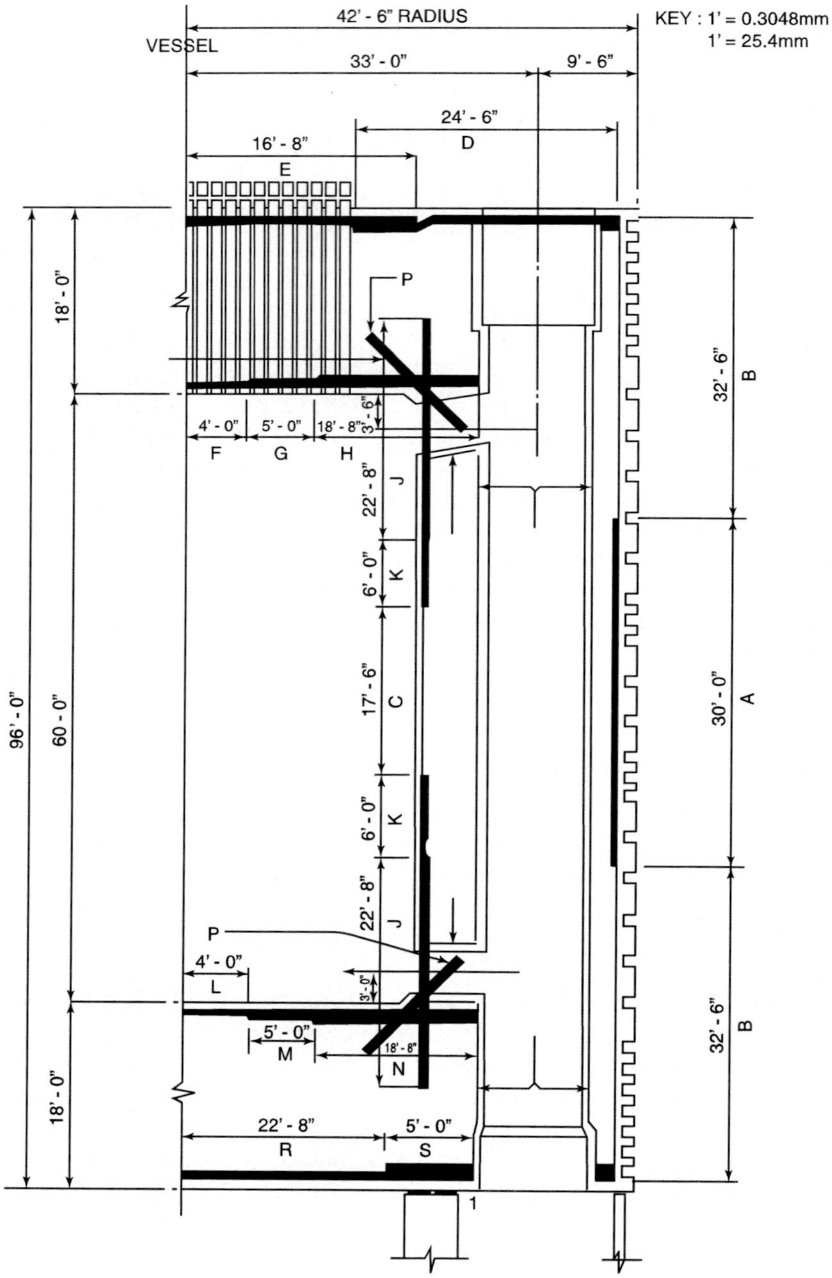

Fig. 8.1 Disposition of main bonded reinforcement in the vessel

This value is found experimentally in case of WYLFA. A conservative value of 0.3% of the area of cross-section is allowed for ultimate load purpose. As indicated in the general rules the obvious choice is that this should be distributed in a proportion 0.2% on outer face of the vessel and 0.1% at the liner. Being a conservative choice, this selection is kept in areas of the vessel where maximum hinges or cracks occur at ultimate load. The reason of not putting the whole amount of 0.3% on the outside face comes from the fact that it would be difficult in practice to fix steel and pour concrete in the congested outer face area of the vessel. Areas outside the predicted maximum crack zones are reinforced with 0.15% (approx.) of the area of cross-section of the vessel (refer to Fig. 8.1c, d). To limit cracking, should any occur under pressure test conditions, 0.05% of bonded steel is provided near the inner face in a similar manner to the outer layer. Therefore away from the critical areas of crack formation at ultimate load, the choice of minimum bonded steel is always 0.15% on outside face and 0.05% near the inner face. It is of interest to note the following additional points regarding the requirements of minimum steel.

8.1.2.2 Disposition Requirement Type-1

Bonded reinforcement is considered up until now to control tensile stresses in the concrete of the vessel. In fact, concrete is subjected to both tensile and flexure effects. The flexural stresses arise from the temperature cross fall through the vessel barrel. As a pure flexural member (refer to Fig. 8.1), the estimate is that only one-half of steel areas mentioned above would be necessary. Since the conditions in the vessel are partly between these two cases, it is justifiable to take 0.2% bonded steel, made up of 0.15 and 0.05% on the outer and inner faces, respectively, as discussed earlier.

8.1.2.3 Disposition Requirement Type-2

In actual vessel tensile stresses up to 500 or 600 psi (3.45–9.14 MN/m^2) will exist at the outer face of vessel concrete during working conditions for 15°C temperature cross fall.

This is based on a fact that smaller cracks have already been initiated at construction joints or any other areas of low tensile strength and at areas of maximum stress concentrations. These cases are very difficult to analyse, hence engineering judgement is required to assess the amount of steel in areas under these worse effects on top of above requirements.

8.1.2.4 Disposition Requirement Type-3

'Anti-crack steel' is necessary to be placed as splay bars in the haunch area of the vessel to control major cracks predicted at ultimate load. Preferential cracking at this position is possible due to stress concentration and will already be existing in order to bring about premature failure before the development of a major crack. 'Anti-crack steel' helps to a greater extent, if placed as close to the liner as possible, in contributing along with horizontal and vertical bars towards controlling these cracks.

8.1.2.5 Design of Main Bonded Steel

General Vessel Analysis

As explained earlier, it is expected that concrete adjacent to the outer face of the vessel will lose moisture at a greater rate than in the general body of concrete. This moisture gradient causes differential shrinkage and tensile stresses are introduced in the outer face of the vessel. In addition to these, tensile stresses due to differential cooling strains are introduced in the vessel together with stresses due to stress concentration, constructional stresses and thermal gradient stresses. The overall tensile stress value between 500 and 600 psi is taken on the outside face of the vessel. It is also noted that a tensile stress of 500 psi (3.45 MN/m^2) is permitted on the outside face due to temperature cross fall of 15°C at working conditions. For preliminary design, areas outside haunch and equator, carrying bonded steel, are designed on the basis of Fig. 8.2 (a, b). At equator and at the haunch Fig. 8.2 (c, d), has been considered in order to arrive at the quantity of bonded steel. Figure.8.2 shown in calculation I is maintained for preliminary design and is selected after careful study of the vessel behaviour during working conditions.

A special case around the haunch due to local effects with and without fillet is taken to check the steel in that area. Figure. 8.3 in calculation I represents the principal tensile stresses around haunch both with and without fillet. These values are taken from computer program for prestress plus proof pressure case. Four different cases are analysed both across the barrel wall and the cap with and without principal compressive stresses.

As shown in Section 8.2.1.2, a barrel wall is analysed as a in infinite cylinder for combined loading of hoop prestress, temperature and design pressure with the following limitation:

$$+\left[f_{\mathrm{to}} + f_{p10} + f_{p20}\right] \bar{<} f_{\mathrm{s}} = +500\,\text{psi (outside tension)} \tag{8.1}$$

$$-\left[f_{\mathrm{ti}} + f_{p2i}\right] \bar{<} f_{\mathrm{c}} = -2200\,\text{psi (inside compression)} \tag{8.2}$$

Refer to Fig. 8.4 where

f_{ti} = hoop stress on inside due to temperature;

f_{p10} = hoop stress on inside due to internal pressure;

f_{p20} = hoop stress on inside due to external pressure;

f_{p2i} = hoop stress on inside due to external pressure.

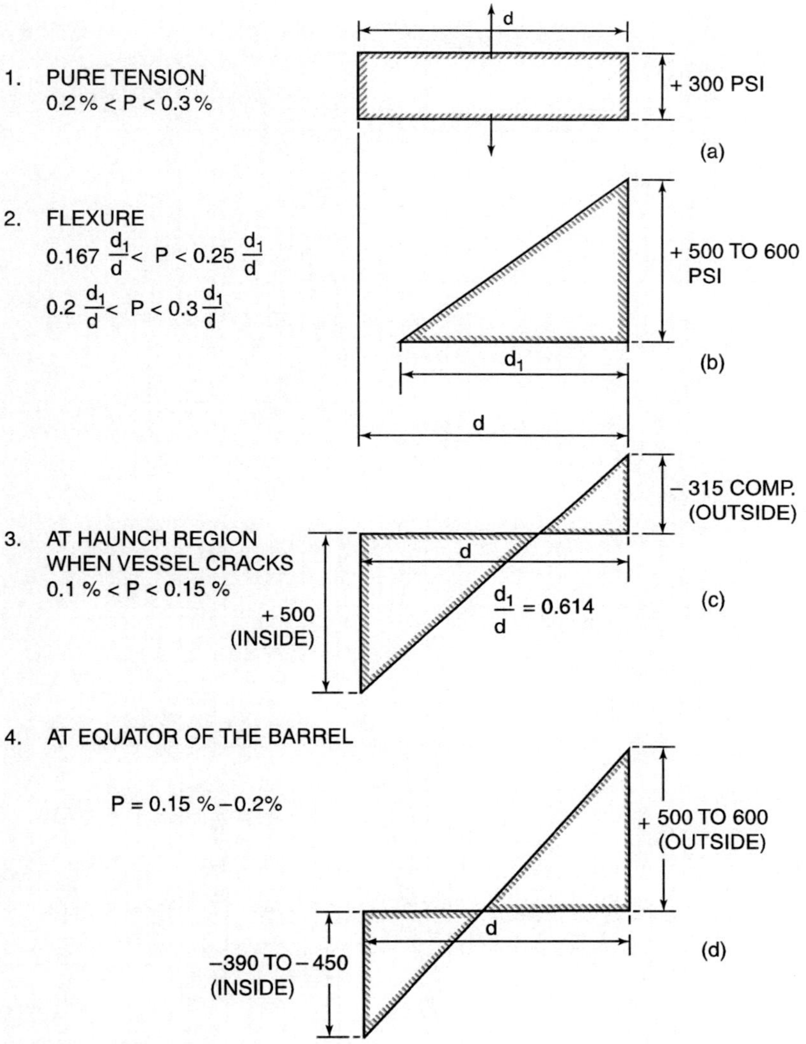

Fig. 8.2 General criteria for nominal main bonded reinforcement

The results tabulated indicate that the stresses are well within the limiting cases. Hence reinforcement design based on 500 psi tension allowed for 15°C temperature cross fall on the outer face is safe. It is of interest to note that compressive stress on the inside face of the vessel due to above combination of loads is well within the allowable compressive stress, i.e. 2200 psi for Hunterston 'B' concrete mix of 5400 psi cube strength at 28 days.

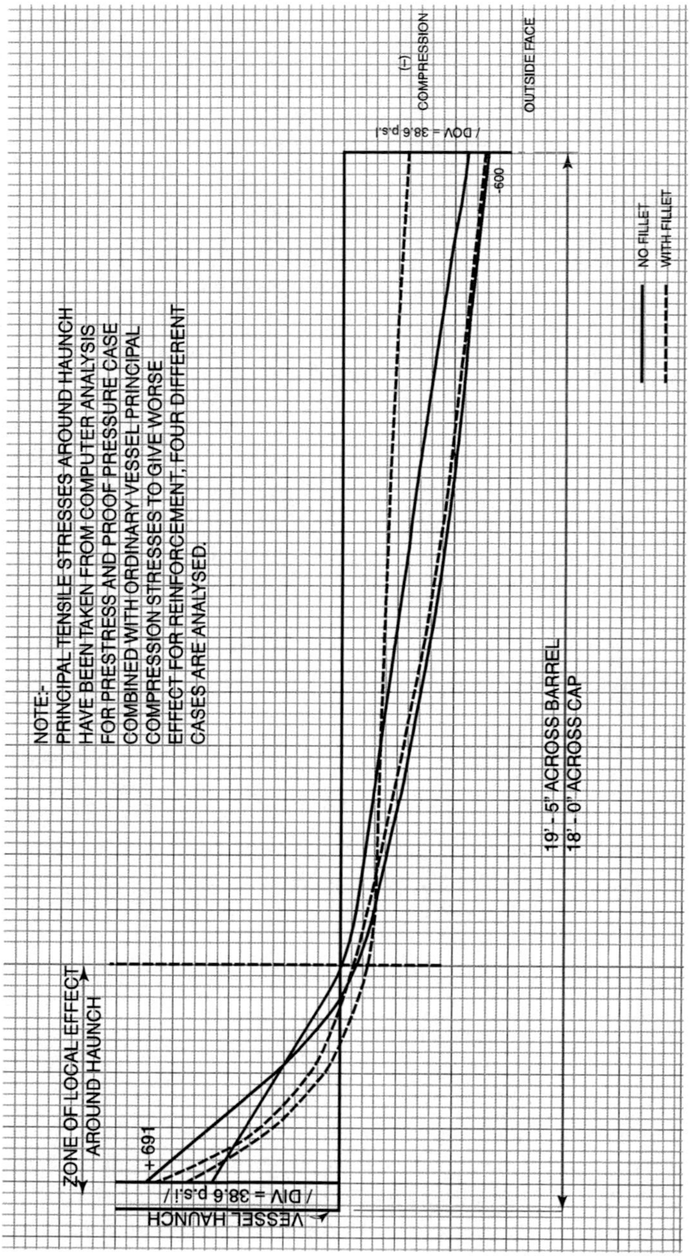

Fig. 8.3 Reinforcement check for principal stresses at haunch region

Fig. 8.4 Cylindrical part

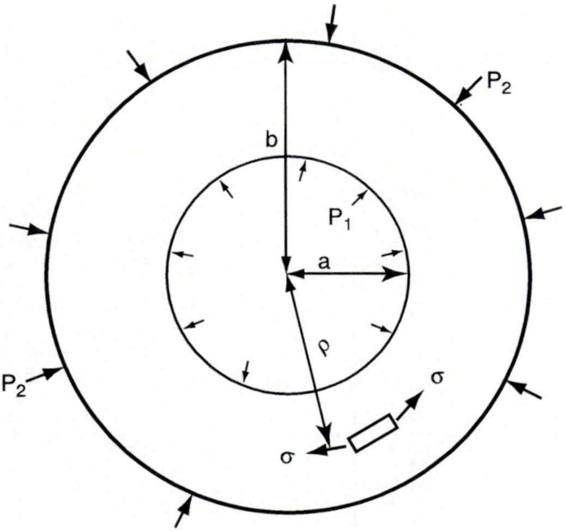

Figure 8.5 shows from zero tensile stress to 500 psi tension for various internal pressures and ratios of inside and outside radii, i.e. wall thicknesses. This is based on a simple formula:

$$P_1 = \left(f_t + 0.45f_c'\right)\frac{\frac{1-a^2}{b^2}}{\frac{1+a^2}{b^2}} \tag{8.3}$$

All these notations are defined in Fig. 8.5. It is interesting to note that tension in concrete becomes more significant for internal pressure only when it is reaching ultimate load somewhere halfway; 500 psi (3.45 MN/m^2) tension is hardly reached for a ratio of approximately 1 between internal and external radii. Thus the choice of 500 psi is conservative if no temperature is considered as in the case of ultimate load. This graph covers various barrel wall thicknesses and will be useful for future reports on bonded reinforcement. In Section 8.2.1.3, a good attempt has been made to analyse the cap for the design pressure plus temperature. By assuming the cap fixed all round circumferentially with certain elastic fixity ($n = 0.5$) in vertical direction due to scattering positions of the vertical tendons, it is assumed that the steel does not reach yield point under design pressure and that the principal problem of the plastic design of two-way reinforced cap is represented by the determination of their redistribution factor values. The cap is transformed into a mechanism shown in Fig. 8.6. The 'minimum re-distribution criterion' is established for a given design pressure and cap dimensions.

Bending stress in the cap is found to be insignificant and the caps arc too thick to develop bending under design pressure. However, on top of it, direct stresses due to temperature (339 psi (3.34 MN/m^2) tension on

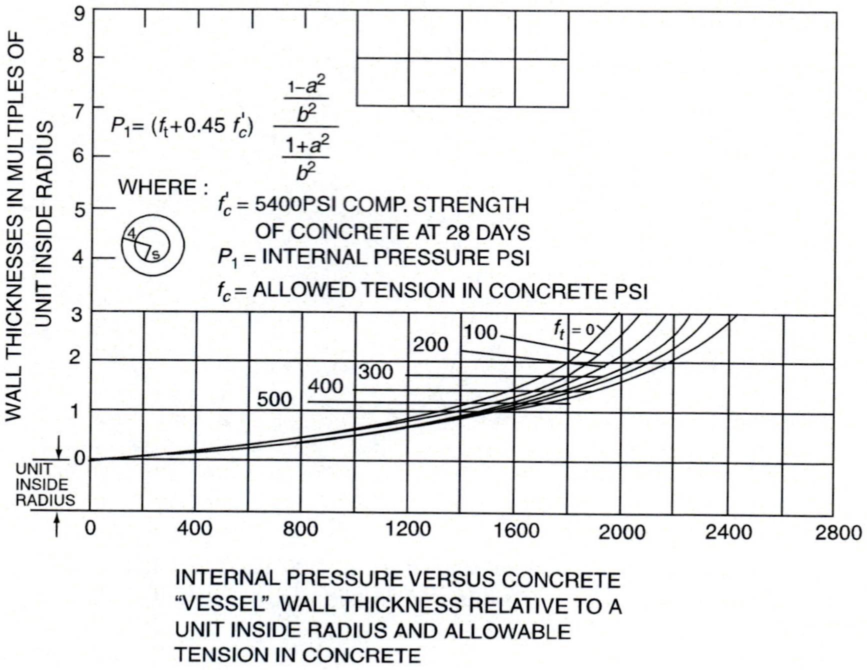

Fig. 8.5 Graphs showing variations of internal pressure and concrete tensile stress for a given strength of concrete

outside and 131 psi (0.904 MN/m^2) compression on inside) taken from computer program as a worse case gave the total area of steel necessary to be 1.1565 in^2. (10.04 cm^2) based on 16.3″ to standpipes, 21″ dia. GK + 60 bars are required on outside face of the cap, which is less than what is required from conditions established in Fig. 8.5 and for the 'Ultimate Load Analysis'. In general the bonded steel 5-layers of $1\frac{1}{4}''$ (30 mm) dia. bars on the outside face is more than sufficient. This is justified later on under 'Quantity and Spacing of Bonded Steel'.

Taking a segment of the cap between the centres of two boilers and by means of virtual work method, the cap reinforcement is analysed at the edge of standpipe area and at the inside of the barrel wall, i.e. at the junction between the cap and the wall, subject to external hoop prestress and ultimate gas pressure. The amount of steel thus required is

$$3.47\text{in.}^2 \text{ or } 3 - 1\frac{1''}{4} \text{ dia bars at } 16.3'' \text{c/c}$$

In SI system (3T30 @ 415 mm) centres or 3T30 – 415 centres

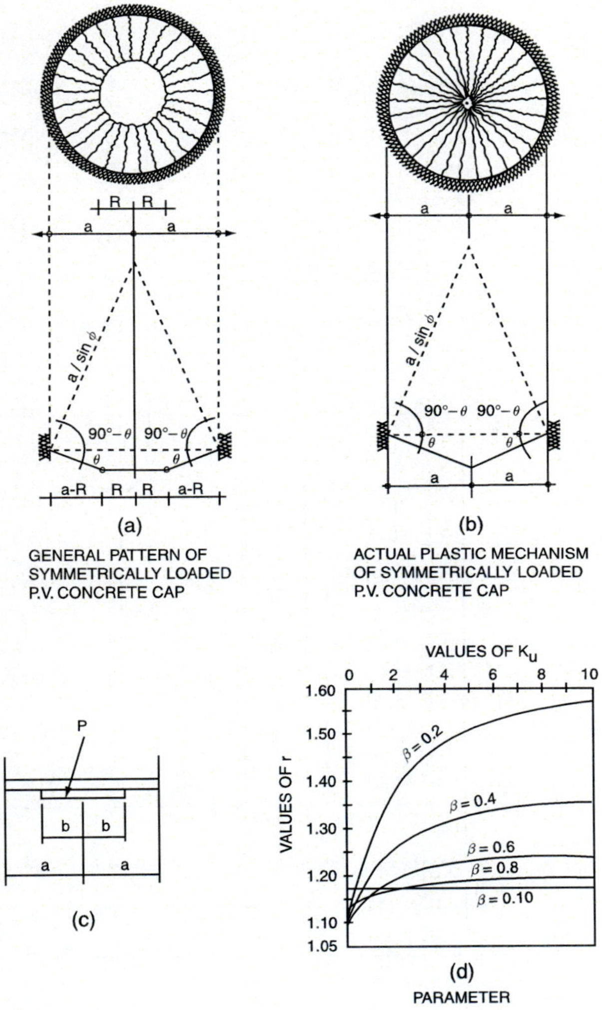

Fig. 8.6 Plastic design of caps under symmetrical loads (Special yield Line Analysis)

As mentioned earlier, the minimum steel is between 0.25 and 0.3% 5-layers of $1\frac{1}{4}''$ dia. (30mm) bars on the outside are given for the 'Ultimate Load Analysis' and 3D Finite Element Analysis.

In Section 8.2.1.5, with a given cap reinforcement shown in Figs. 8.1 and Table 8.1 computer results from 'Finite Element Analysis' are taken to check the cap for shear and principal stresses. In Fig. 8.8 the loads are shown (combined loading from vertical and hoop prestress and internal ultimate pressure). Random values arc taken from the program ISOPAR and stresses

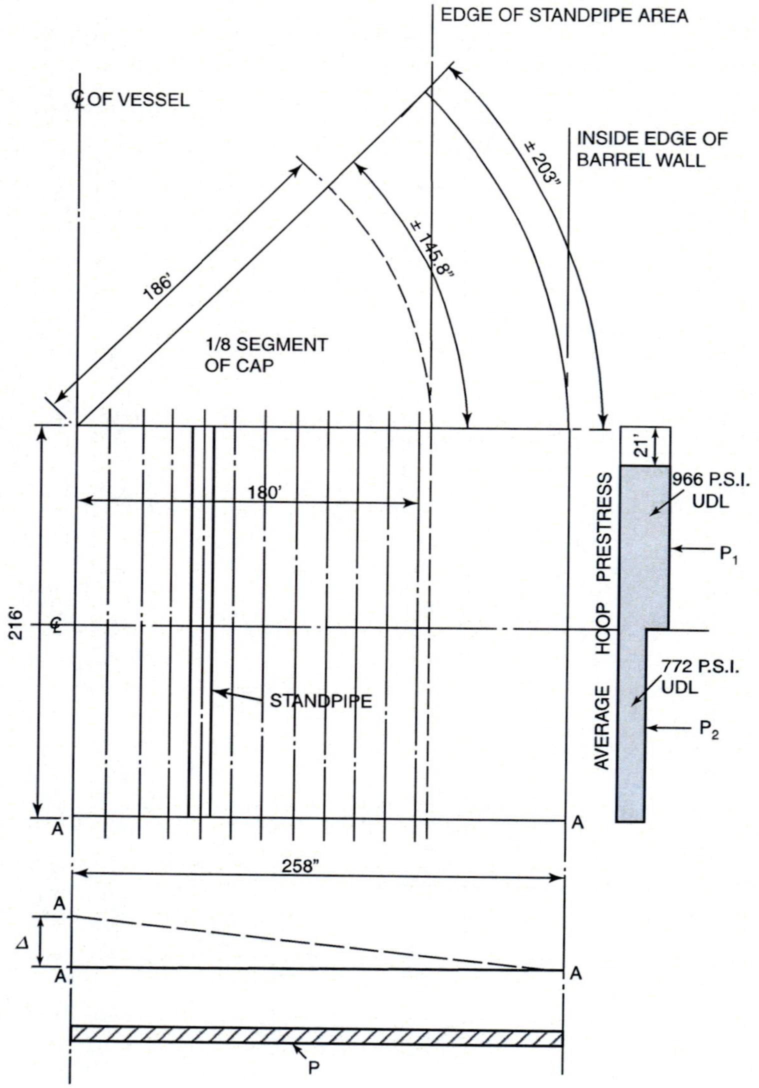

Fig. 8.7 Cap reinforcement at ultimate load

are checked at the edge and at the predicted failure plane. Principal tension is 135 psi (0.923 MN/m^2) at the failure plane which is close to allowable value and the cap is not necessary to be reinforced in that area. On the other hand at the edge nominal barrel shear is 21 psi. Reinforcement in this area is sufficient to look after any effect due to shear or principal stresses.

Table 8.1 Location of Main Bonded Reinforcement (key: 1 ft = 0.3098 m, 1 in = 25.4 mm)

Area	Description of reinforcement bars
A	$1\frac{1}{4}''$ dia. bars in 3 layers at 10″crs. both ways
B	$1\frac{1}{4}''$ dia. bars in 2 layers at 10″crs. both ways
C	$1\frac{1}{4}''$ dia. bars in single layer at 10″crs. both ways
D	$1\frac{1}{4}''$ dia. bars in 2 layers between boilers (circumferential)
	$1\frac{1}{4}''$ dia. bars in 3 layers at 10″crs. (radial)
E	$1\frac{1}{4}''$ dia. bars in 5 layers at $16\frac{1}{4}''$ crs. both ways
	$1\frac{1}{4}''$ dia. bars in 5 layers at $16\frac{1}{4}''$ crs. (orthogonal grid)
F	$1\frac{1}{4}''$ dia. bars in 2 layers at $16\frac{1}{4}''$ crs. both ways
	$1\frac{1}{4}''$ dia. bars in 5 layers at $16\frac{1}{4}''$ crs. (orthogonal grid)
G	$1\frac{1}{4}''$ dia. bars in 3 layers at $16\frac{1}{4}''$ crs. both ways
	$1\frac{1}{4}''$ dia. bars in 3 layers at $16\frac{1}{4}''$ crs. (orthogonal grid)
H	$1\frac{1}{4}''$ dia. bars in 4 layers at $16\frac{1}{4}''$ crs. both ways
	$1\frac{1}{4}''$ dia. bars in 4 layers at $16\frac{1}{4}''$ crs. (orthogonal grid)
J	$1\frac{1}{4}''$ dia. bars in 3 layers at 10″ crs. (vertical)
	$1\frac{1}{4}''$ dia. bars in 2 layers at 20″ crs. (horizontal)
K	$1\frac{1}{4}''$ dia. bars in 2 layers at 10″ crs. (vertical)
	$1\frac{1}{4}''$ dia. bars in 2 layers at 20″ crs. (horizontal)
L	$1\frac{1}{4}''$ dia. bars in single layer at 11″ crs.
	$1\frac{1}{4}''$ dia. bars in single layer at 11″ crs. (orthogonal grid)
M	$1\frac{1}{4}''$ dia. bars in 2 layers at 11″ crs. both ways
	$1\frac{1}{4}''$ dia. bars in 2 layers at 11″ crs. (orthogonal grid)
N	$1\frac{1}{4}''$ dia. bars in 3 layers at 11″ crs. both ways
	$1\frac{1}{4}''$ dia. bars in 3 layers at 11″ crs. (orthogonal grid)
P	$1\frac{1}{4}''$ dia. splay bars in 3 layers with $1\frac{1}{4}''$ dia spacer.
	bars in 2 layers
R	$1\frac{1}{4}''$ dia. bars in 4 layers at 12″ crs. both ways
	$1\frac{1}{4}''$ dia. bars in 4 layers at 12″ crs. (orthogonal grid)
S	$1\frac{1}{4}''$ dia. bars in 6 layers at 12″ crs. (circumferential)
	$1\frac{1}{4}''$ dia. bars in 6 12″ layers (radial)

Note: This should be read in conjunction with Fig. 8.1
Note: $1\frac{1}{4}''$ in metric T-30m.s. All GK. 60 bar

8.1.3 Crack Distribution and Control Analyses of the Vessel

Various crack distributions shall be summarised in a separate section and indicates distribution of stresses between cracks of a well-disposed crack pattern of the wall. Figure 8.9 can be drawn in which the embedment and tension areas in the barrel wall are defined. Main bonded steel in these areas is defined on the basis of a single bar embedded in vessel concrete which would span a certain crack width and also between two cracks of a well-disposed crack pattern as indicated earlier. As the strain in main bonded steel varies from maximum at the crack to a reduced value in between the two cracks, it becomes necessary to know the minimum average strain in the bonded steel.

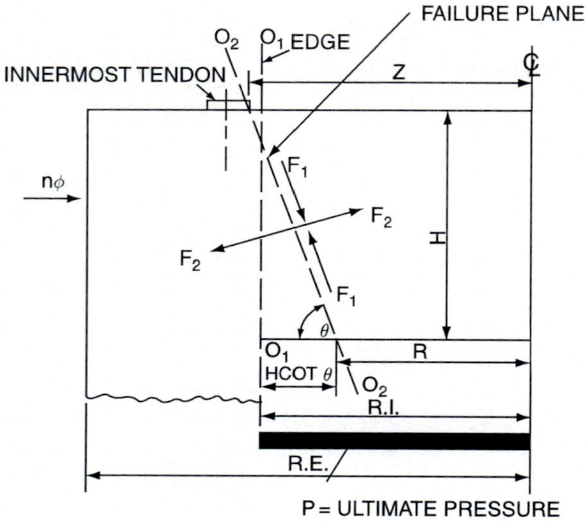

Fig. 8.8 Shear and principal stresses in cap at ultimate load

The minimum value is defined as

$$\frac{W}{L} \tag{8.4}$$

where W is the width of crack; L the average crack spacings. 'W' and 'L' both depend on the shape of the vessel, bonded steel and the way the internal forces are distributed in the vessel. Moreover, construction joints, local stress concentrations and thermal stresses will be additional factors to assist in the formation of these cracks.

In the uncracked part of the vessel concrete, tie strain assumed is 100 microstrain. Therefore, the average strain in the bonded steel is

$$100 \text{ mirco} \quad - \quad \text{strain} + \frac{W}{L} \tag{8.5}$$

Various theories gave different values of 'W' and 'L' and they are tabulated in Section 8.2.1.2.

The average values are taken as $W = 0.05''$ (1.27″ mm);

$$L = 30''(762\,\text{mm}).$$

As the pressure increases more and more cracks are formed and the value of 'L' will be reduced. It is necessary therefore to limit the size of the cracks by providing sufficient bonded reinforcement. The size of the crack is limited to 0.3 mm (0.0118″) and the number of G.K +60 bars of $1\frac{1}{4}''$ dia. required to do so

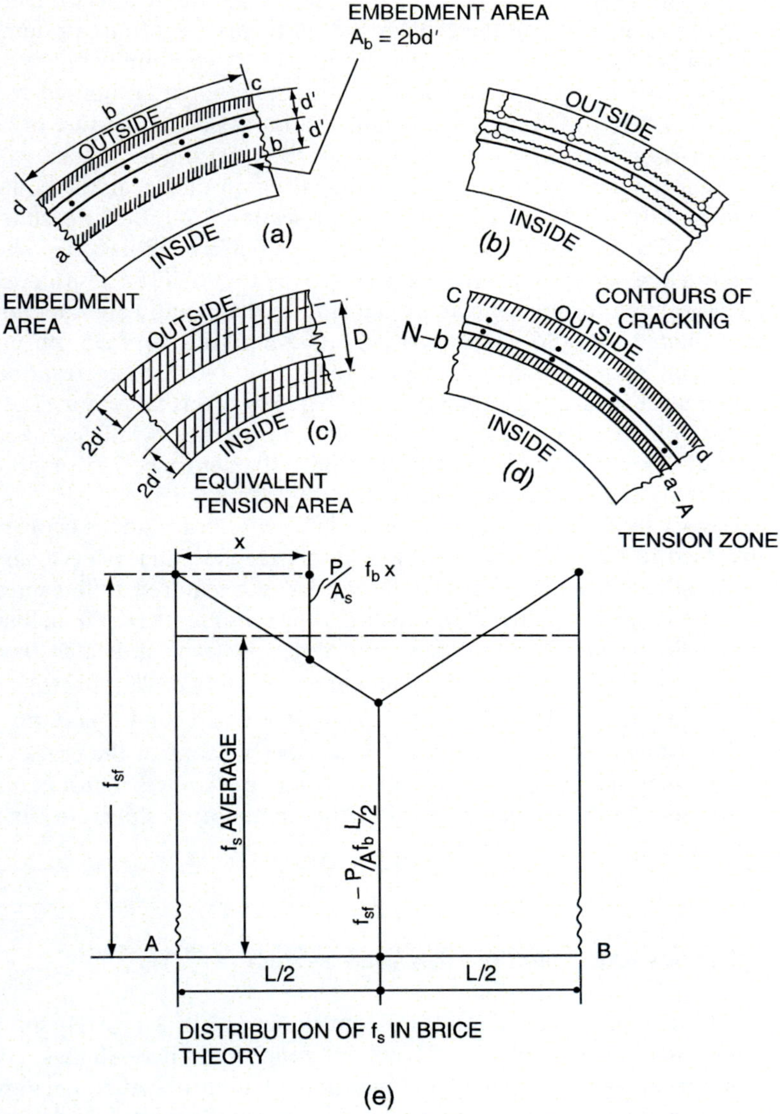

Fig. 8.9 Embedment and tension areas location in the barrel

is 3. This condition is applied only near the equator and at the haunch. In all other areas it is taken for granted that at a stress of 60,000 psi in a bar, a crack width of 0.05″ (1.27 mm) can be expected. Thus in the whole vessel the minimum number of bars necessary to be placed for obvious reasons of these cracks is to be 2– 1 $\frac{1}{4}$″ dia. G.K. + 60 type. They are shown in Figs. 8.1 and Table 8.1 of this chapter. This matches very well with the reasons discussed earlier.

In an axisymmetric vessel the cracks in the longitudinal direction in the barrel and radial direction in the top cap will in theory be infinite in number. In nonaxisymmetric vessel such as in the Hunterston 'B' containing boiler pods, the number of these cracks is assumed at ultimate load to be limited by the number of weak sections. Since peak bending moments occur because of loading and inter-restraint, between cap and wall, the meridional cracks will be bounded by a few major cracks in the haunch region and near the equator. These cracks at ultimate load are tabulated in Section 8.2.1.3. The results are taken from 'Ultimate Load Analysis' and the crack pattern in this case is shown in Fig. 8.11. These cracks vary from 1.8″ (38 mm) to 3.6″ (91.5 mm). Also in Section 8.2.1.4 crack formulation at ultimate load corresponding to liner failure strain is evaluated for stud spacings of $8\frac{1}{4}''$ (205 mm) and 9″ (235 mm). The crack width that a liner can span comes out to be 1.57″ (40 mm), corresponding to 28% liner steel strain and 14% GK +60 bar. However, the crack sizes are compatible. Thus a common figure of 2″ (50 mm) crack width is taken at ultimate load. This is checked by the 3D Finite Element whether or not the bars lie matching the solid element nodes.

It is assumed that smaller crack developed in major crack zones is progressively increased to 2 inch (50 mm) width at ultimate load and it is necessary to check the bursting stress in the bar and the bond length required at that stage, in Section 8.2.1.5, analysis is carried out to check the minimum GK-60 inch bond length of steel provided in the vessel. Previous research indicates that at ultimate load $1\frac{1}{2}''$ (40 mm) dia. GK + 60 bar can span over 2″ (50 mm) crack for a stress differential of 40,000 psi. Thus, at ultimate load conditions 2 no. $1\frac{1}{4}''$ (30 mm) dia bars will still be a reasonable number to control the crack. The number of bars from Brice theory in this area comes out to be 3. Thus, in major crack zones, this number is maintained in either direction as shown in Fig. 8.1 and Table 8.1.

8.1.4 Quantity and Spacings of Bonded Steel

The disposition of main bonded reinforcement shown in Fig. 8.1 clearly shows a change of the amount of bonded steel from one zone to another. All these zones are named and are given reasonable dimensions. Quantities and spacings of bonded steel in these zones are based on design requirements, as discussed earlier.

As a general rule, outer skin and inner skin bonded steel are taken to be 2-layers at 10″ (250 mm) c/c and single layers at 10″ c/c both ways, respectively. Departure from this general rule in some areas is based on certain requirements to be met at working and overload conditions.

In the top and bottom caps, the number of layers and spacings vary but the quantity of steel necessary is kept more or less the same. In top cap standpipe area, the bar spacings are rearranged to fit between standpipes while at the inside face of the bottom cap, the spacings are limited by the grout holes. These bars are placed in orthogonal grid. All radial and circumferential bars in top and bottom caps are fitted in between the tendons and hence their spacings are not constant. In order to adjust the minimum requirement of 3 layers at 16.3″ (415 mm) c/c or 2 layers at 10″ (250 mm) c/c, the number of bars is increased. Additional reinforcement is shown in the caps around boiler pods in order to meet ultimate load requirements.

All bars have been given full bond lengths. Various lap lengths given in the vessel have been worked out from the construction programme of lifts and bays.

After the final arrangement of bonded steel, each zone is further checked on the basis of the following limitation:

$$\mu \times f_{sf} \geq t_{ca} \tag{8.6}$$

where

μ is the bond steel ratio;
f_{SF} the rupture stress of bond steel;
t_{ca} the tensile stress of concrete.

8.2 Calculation 1

8.2.1 General Criteria for Nominal Main Bonded Reinforcement

8.2.1.1 General Problems

It may be expected that concrete adjacent to the outer surface of the 'vessel' will lose moisture at a greater rate than in the general body of the vessel. This moisture gradient through the thickness of the vessel will cause differential drying shrinkage and the tensile stresses, hence, are introduced in the outer surface of the vessel. These stresses are effectively controlled by means of bonded reinforcement.

In Additional to these tensile stresses, some random tensile stresses are introduced due to differential cooling strains which cannot be covered by analysis; constructional stresses locked in, local stress concentration and thermal gradient stresses are some more to be considered while assessing the amount of reinforcement.

For preliminary disposition of main bonded reinforcement, it is necessary to study carefully the 'vessel' behaviour during elastic and ultimate load conditions and a proposed construction technique. After careful study of the 'vessel', general criterion for main bonded reinforcement is established. This criterion is indicated in Fig. 8.2. It covers preliminary requirements for general 'vessel' areas and specific haunch and equatorial regions. A tensile stress of 1500 lb/m^2 (3.45 MN/m^2) at working conditions, worse case of 600 psi (4.15 MN/m^2) at ultimate load is also considered.

Check on reinforcement around haunch based on stress concentration figures (principal tensile stresses) with and without fillets is also made, taking zones across the barrel wall and the cap. Four different cases shown in Fig. 8.3 have been taken for analysis, with and without principal compression. These are the stresses for proof pressure case from elastic analysis carried out on computer using program ISOPAR or equivalent.

8.2.1.2 Prestress, Temperature and Design Pressure

After selecting and locating reinforcement in the barrel, various areas are checked for combined loading of prestress, temperature and pressure.

ρ = radius
a = internal radius = 21 − 6′(6.5532 m)
b = external radius = 42′ − 16″ (excluding channel shutters)
p_1 = internal pressure = 644 psi design pressure
p_2 = external pressure due to circumferential prestress = (700 psi average)
T_x = temperature cross fall = 15°C
α = thermal coefficient = 9.2×10^{-6}
$E_c = 5 \times 10^6 \text{psi} \left(34{,}500\,\text{MN/m}^2\right)$
f_c = Permissible working compressive stress = 200 psi (p/s + pressure)
f_s = Permissible working tensile stress = 500 psi
f_t = hoop stress due to temperature cross fall
f_{p1} = hoop stress due to internal pressure
f_{p2} = hoop stress due to external pressure
υ = Poisson's ratio = 0.15
f_{ti} = hoop stress on inside due to temperature
f_{to} = hoop stress on outside due to temperature
$f_{p_1^i}$ = hoop stress on inside due to internal pressure
$f_{p_2^o}$ = hoop stress on outside due to external pressure

All other suffixes 'i' and 'o' are for inside and outside, respectively.

General Cases

$$f(\text{hoop}) = \frac{p_1 a^2 - p_2 b^2 + \left(\frac{a^2 b^2}{\rho^2}\right)(p_1 - p_2)}{(b^2 + a^2)} + \text{compressive tensile compression} \quad (8.7)$$

$$f(\text{radial}) = \frac{p_2 b^2 - p_1 a^2 + \left(\frac{a^2 b^2}{\rho^2}\right)(p_1 - p_2)}{(b^2 - a^2)} \quad (8.8)$$

(a) Internal pressure only

$$f_{p_1} = \frac{a^2 p_1}{b^2 - a^2}\left(1 + \frac{b^2}{\rho^2}\right) \quad (8.9)$$

(b) External pressure only

$$f_{p_2} = \frac{-b^2 p_2}{b^2 - a^2}\left(1 + \frac{a^2}{\rho^2}\right) \tag{8.10}$$

(c) Temperature cross fall

$$f_t = \frac{E_c \alpha T_x}{2(1 - v)\log_c b/a}\left[1 - \log_c \frac{b}{\rho} - \frac{a^2}{b^2 - a^2}\left(1 + \frac{b^2}{\rho^2}\right)\log_c b/a\right] \tag{8.11}$$

Limiting Cases

$$+\left|f_{to} + f_{p_1 o} + f_{p_2 o}\right| \bar{<} +f_s = +500\,\text{psi}\left(\text{MN/m}^2\right) \tag{8.12}$$

$$-\left|f_{ti} + f_{p_2 i}\right| \bar{<} -f_c = -2200\,\text{psi}\left(-15.18\,\text{MN/m}^2\right) \tag{8.13}$$

	p_1 f_{p_1}i	f_{p_1}o	p_2 f_{p_2}i	f_{p_2}o
psi	+ 11.15	+ 473	− 1918	− 1215
(MN/m^2)	(+7.94)	(+3.28)	(−13.235)	(−8.384)

	f_{ti}	f_{to}
psi	−219	+1090
(MN/m^2)	(−1.511)	(+7.5215)

Applying Limiting Cases

$$1090 + 473 + (-1215) = 348 < 500 = 3.45\,\text{MN/m}^2$$

$$-1918 - 219 = -2137 < -2200 = -15.18\,\text{MN/m}^2$$

OK

Minimum reinforcement $2 - 21\frac{1}{4}\phi$ bars at 10 c/c is required. In some areas such as near the haunch, additional reinforcement is introduced for the reason indicated elsewhere in this chapter.

8.2.1.3 Plastic Design of Caps Under Symmetrical Working Loads

Notation

$$a = \text{radius of cap periphery}$$

$$b = \text{radius of a load surface}$$

$$g_s = \text{permanent load uniformly distributed on a unit area}$$

$$G_{s'}G = \text{over load and design lond}$$

$$G_{s'}G\ \gamma_g = 1$$

L_M, L_G, L_P = virtual works of plastic moments, design permanent load and design gas load, respectively

M_{1a} = absolute value of the negative unit elastic radial moment on the cap periphery

M_{p_1}, M_{p_2} = positive unit plastic radial and annual moments, respectively

M_{p_1a}, M_{p_2a} = absolute values of the negative unit plastic radial and annular moments in the slab periphery

p = gas load distribution on a unit area or on a unit length of a certain circle of the cap

$P_{s'}p_1$ = stand pipe and design loads respectively

$$(P_u = \gamma_p p_s)$$

P_1 = total gas load corresponding to the attainment of the design yield point is only one point of the steel reinforcement

$$\gamma = \text{redistribution factor} = \frac{G + P_U}{G + P_1}$$

γ_a = redistribution factor for $P_1 = P_s$

U_{10}, U_{20}, U_{1a} = coefficient of unit elastic moments produced by total permanent load

γ_S, γ_P = overload factor of permanent and gas loads, respectively

η = elastic fixity of the cap periphery = 0.5 in this case

= $M_{1a}/$ absolute value of unit elastic moment on the peiphery of a perfectly fixed ended slab

$$K_s = \frac{P_s}{G}; \quad K_1 = \frac{P_1}{G}; \quad K_u = \frac{P_u}{G}; \quad \beta = b/a$$

$\rho_1 g$ = radius corresponding to the circle defined by all the zero points of the diagram of the elastic moments produced by the permanent load, exclusively, assuming rigid fixity of the cap periphery

$\rho_1 p$ = same as for $\rho_1 g$ except the load is gas load

ρ_1 = approximate true radius of the circle defined by all zero points of the diagram of the elastic negative radial moments produced by the loads, assuming an elastic fixity of the slab periphery.

8.2.1.4 The General Plastic Equilibrium Equation for Reinforcement

Let θ be the virtual angular displacement of the plastic mechanism (Fig. 8.6). In this case the virtual work principle gives the following plastic equilibrium equation:

$$L_{\mathrm{M}} + L_{\mathrm{G}} + L_{\mathrm{P}} = 0 \quad (8.14)$$

The virtual work of the plastic moments acting along all the yield lines after rearranging becomes

$$L_{\mathrm{M}} = 2\pi\left[RMP_1 + aM_{p1\mathrm{a}} + (a - R)M_{p2}\right]\theta = L_{\mathrm{G}} + L_{\mathrm{P}} \quad (8.15)$$

For obtaining a unique solution of the plastic moments $MP_1, M_{P\mathrm{a}}$ and M_{P2} satisfying the above equation, supplementary equations are necessary.

The criterion for solving the problem of the plastic design of two-way reinforced cap is the ‘ minimum redistribution criterion’. In the case of a fixed ended circular cap subjected to a symmetrical pressure load, the most stressed points are its centre and the unit elastic radial and annular moments are maxima. Let

$$M_{10} = U_{10}G + \upsilon_{10}P \quad (8.16)$$

$$M_{20} = U_{20}G + \upsilon_{20}P \quad (8.17)$$

$$M_{1a} = U_{1a}G + \upsilon_{1a}P \quad (8.18)$$

be the values of the unit elastic moments at these points. Taking into account both the definition of the plastic moment and the redistribution criterion

$$M_{10} = M_{p1}$$
$$M_{20} = M_{p1} \text{ and } p = p_1$$
$$M_{1\mathrm{a}} = M_{p1\mathrm{a}}$$

On the other hand, the elastic analysis of symmetrically loaded reinforced concrete cap shows

$$M_{10} = M_{20} \quad \text{(a)}$$

Therefore

$$v_{10} = v_{20} \quad \text{(b)}$$
$$u_{10} = v_{20} \quad \text{(c)}$$

Hence

$$M_{p1} = M_{p2} \quad \text{(d)}$$

Therefore

$$2\pi a\left(M_{p2} + M_{p1a}\right)\theta = L_{G} + L_{P} \tag{8.19}$$

The sum

$$L_{G} + L_{p} = \text{maximum if R} = 0 \tag{8.20}$$

Hence the actual plastic mechanism of the cap subjected to symmetrical pressure loads

$$L_{G} = 2\pi U_l G\theta \tag{8.21a}$$
$$L_{P} = 2\pi v_l P_{u}\theta \tag{8.21b}$$

From the above consideration, the maximum value of ρ_1 is

$$P_1 = \frac{(U_l - U_{20} - U_{1a})G + v_1 P_{u}}{v_{20} + v_{1a}} \tag{8.22}$$

Therefore,
γ = Expression for redistribution factor becomes

$$\frac{G + P_u}{G + P_1} = \frac{(v_{20} + v_{1a})(1 + K_u)}{U_L - U_{20} - U_{1a} + v_{20} + v_{1a} + v_l K_u} \tag{8.23}$$

$$G = \pi a^2 g;\ L_G \frac{1}{3} aG\phi;\ U_e = \frac{1}{6\pi} \tag{8.24}$$

$$P_{v} = \pi\beta^2 a^2 p_u;\ v_e = \frac{3 - 2\beta}{6\pi} \tag{8.25}$$

$$u_{20} = \frac{19 - 2\eta}{96\pi};\ v_{1a} = \frac{\eta}{8\pi} \tag{8.26}$$

$$v_{20} = \frac{24 - 5\beta^2 - 28\ln\beta - 12\eta(2 - \beta^2)}{96\pi} \tag{8.27}$$

$$v_{10} = \frac{\eta(2 - \beta^2)}{8\pi} \tag{8.28}$$

Final value for 'γ'

$$\gamma = \frac{(24 - 5\beta^2 - 28 \ln \beta)(1 + K_u)}{21 - 5\beta^2 - 28 \ln \beta + 16(3 - 2\beta)K_u} \tag{8.29}$$

Taking the expression p_1 by using the general expression of the redistribution factor

$$p_1 = -\left(1 - \frac{1}{\gamma}\right)G + \frac{P_u}{\gamma} \tag{8.30}$$

$$M_{p1} = M_{p2} = \left[u_{20} - \upsilon_{20}\left(1 - \frac{1}{\gamma}\right)\right]G + \frac{v_{20}}{\gamma}P_u \tag{8.31}$$

$$M_{p1a} = \left[u_{1a} - \upsilon_{1a}\left(1 - \frac{1}{\gamma}\right)\right]G + \frac{\upsilon_{1a}}{\gamma}P_U \tag{8.32}$$

The above plastic moments become that the actual redistribution factor corresponding to the loads 'G' and 'P_U' are equal to γ_a. In this case a partial redistribution of internal forces takes place under the action of loads G and P_U and also $P_1 = P_s$

Therefore

$$M_{p1} = M_{p2} = u_{20}G + v_{20}P_s \tag{8.33}$$

$$M_{P_{2a}} = u_{1a}G + v_{1a}P_s \tag{8.34}$$

Position of the Zero Points

$$P_1 = (1 - \eta)a + \eta\frac{\rho_1 g + K_1\rho_{1p}}{1 + K_1} \tag{8.35}$$

(a) When $\gamma < \gamma_a$

$$K_1 = \frac{1 + ku}{\gamma} - 1$$

(b) When $\gamma \geq \gamma_a$

$$K_1 = K_s$$

The absolute values of the negative radial moments decrease from the periphery to the centre of the cap. Obviously, the maximum absolute value of these moments corresponds to the cap periphery. For particular purposes it is first necessary to obtain the position of the zero point of the elastic negative radial moment. Subsequently, the fulfilment of the condition is obtained when radial steel reinforcement is placed along the length of the above-mentioned moment diagram so that the intensity of this reinforcement is the same at all points and equal to the corresponding radial plastic moment $M_{P_{2a}}$. It is clear that this is on the safe side.

$$\text{Total effective dia} = (43' - 0) + 40' = 83' - 0 \quad (25.3\,\text{m})$$

Ignoring channel shutters approximate C.G of tendon from outside = 150″ (3.810 m)

$$\text{effective } 2a = 58' - 0 \text{ and } a = 29' - 0 = 883.82\,\text{cm}$$

$$2b = 43' - 0 \text{ and } b = 21' - 6'' = 655.32\,\text{cm}$$

$$\eta = \text{elastic fixity degree of the cap periphery} = 0.5(\text{worse condition})$$

Value of g_s

The value of g_s is the built-in stresses, i.e. construction stresses. They vary from 100 psi to a very negligible amount. These stresses are usually ignored. The amount of 10 psi (0.069 MN/m^2) is taken for *a* circle of the cap.

$$g_s = 0.703\,\text{kg/cm}^2 = 10\,\text{Psi}\left(0.069\ \text{MN/m}^2\right)$$

$$p_s = 45\,\text{kg/cm}^2 = 644\,\text{Psi}\left(4.443\text{MN/m}^2\right)$$

$$\gamma_8 = 1.0 - \text{no overload factor}$$

$$\gamma_p = 1.15 - \text{overload factor for proof pressure}$$

$$\text{design value of } G_s = \pi a^2 g_s = 2.207(883.92)^2$$

$$= 172 \times 10^4\text{kg}\left(15.627\,\text{MN/m}^2\right) = G$$

$$P_s = \pi b^2 p_s = 6 \times 10^7\,\text{kg}(545.13\,\text{MN})$$
$$P_U = \gamma_p P_s = 6.9 \times 10^7\,\text{kg}(629.91\,\text{MN})$$

$$K_s = \frac{P_s}{G} = \frac{6 \times 10^7}{1.72 \times 10^6} = 39.4$$

$$K_U = \frac{P_U}{G} = \frac{6.9 \times 10^7}{1.72 \times 10^6} = 40.1$$

$$\gamma_a = \frac{1 + k_u}{1 + k_s} = \frac{41.1}{40.4} = 1.02$$

$$\beta = \frac{b}{a} = \frac{21.5}{29} = 0.741(\text{from Fig. 8.6d})$$

$$\gamma > \gamma_a, \quad \gamma = 1.22$$

$$U_{20} = \frac{19 - 2\eta}{96\pi} = 0.0425$$

$$v_{1a} = \frac{\eta}{8\pi} = 0.0199$$

$$v_{20} = \frac{24 - 5\beta^2 - 28 \ln \beta - 12\eta(2 - \beta^2)}{96\pi}$$

$$= 0.04$$

$$v_{1a} = \frac{\eta(2 - \beta^2)}{8\pi} = 0.0289$$

For $\gamma > \gamma_a$

$$\begin{aligned}
M_{p1} = M_{p2} &= \left[U_{20} - \upsilon_{20}\left(1 - \frac{1}{\gamma}\right)\right]G + \frac{\upsilon_{20}}{\gamma}\rho_u \\
&= \left[0.0425 - 0.04\left(1 - \frac{1}{1.22}\right)\right]1.72 \times 10^6 + \left[0.0328 \times 6.9 \times 10^7\right] \\
&= 0.0072 \times 1.72 \times 10^6 + 0.22632 \times 10^7 \\
&= 0.12384 \times 10^5 + 0.22632 \times 10^7 \\
&= 2,275,584\,\text{kg cm}\ (20.675\,\text{MN cm})
\end{aligned}$$

$$\begin{aligned}
M_{p1a} &= \left[U_{1a} - \upsilon_{1a}\left(1 - \frac{1}{\gamma}\right)\right]G + \frac{\upsilon_{1a}}{\gamma}\rho_u \\
&= \left[0.0199 - 0.0289\left(1 - \frac{1}{1.22}\right)\right]1.72 \times 10^6 + \left[\frac{0.0199}{1.22} \times 6.9 \times 10^7\right] \\
&= 0.01468 \times 1.72 \times 10^6 + 0.11247 \times 10^7 \\
&= 0.25250 \times 10^5 + 0.11247 \times 10^7 \\
&= 1,149,950\ \text{kg cm}\ (10.448\ \text{MN cm})
\end{aligned}$$

$$f_c = \frac{1}{3} \times 5400 = 1800\ \text{psi} = 126.54\,\text{kg cm}^2$$

$$f_t = \frac{1900\,\text{kg}}{\text{cm}^2}; \frac{f_t}{f_c} = \frac{1900}{126.54} = 15 = \gamma_1$$

$$\eta_1 = \text{neutral axis factor} = \frac{m}{m+\gamma_1} = 0.5$$

$$a = 1 - \frac{\eta_1}{3} = 1 - \frac{0.5}{3} = 0.833$$

$$Q = \frac{1}{2} f_c \eta_1 a = \frac{1}{2} \times 126.54 \times 0.5 \times 0.833 = 26.4$$

$$M = Qbd^2 = 26.4 \times 1 \times (648.64)^2$$

$$= 26.4 \times 1 \times 42 \times 10^4$$

$$= 1108.8 \times 10^4 \text{ kg cm}$$

$$11.088 \times 10^6 \text{ kg cm}$$

$$> 2.275584 \times 10^6 \text{ kg cm (20.875MNcm)}$$

Also $> 1.149950 \times 10^6$ kg cm (10.448MNcm)
Thus at working load the bending stress is not very significant:

$$\alpha = 0.01 \times af_t = 16.112$$

$$\alpha d = 1.0368 \times 10^4$$

$$A_t = \frac{M(\text{kg m})}{\alpha d} = 2.19\,\text{cm}^2/m = 0.107\,\text{m}^2/\text{ft}$$

Less than minimum

Maximum tensile stress in the cap on outside surface taken from 'Dynamic Relaxation Program' closed to edge = 339 psi with corresponding compression stress of 131 psi. Therefore combinated of these with 'Design Pressure' shall give the required minimum reinforcement in top and bottom caps

$$f_s = 131; \frac{f_t}{\text{m}} = 339; f_t = 5085$$

$$\eta_1 = 0.315; l_a = 0.895$$

$$Q = \frac{1}{2}(131)(0.315)(0.895) = 18.5$$

$$Qbd^2 = 18.5 \times 12 \times (210)^2 = (222)(44,100)$$

$$\alpha = l_a f_t = 4560$$

$$A_s = \frac{222 \times 44,100}{4560 \times 210} = 1.06\,\text{in.}^2/\text{ft}$$

Since the steel owing to proof pressure is negligible, the following amount of steel by direct ratio of proof pressure and design pressure will be close to reality:

$$A_t = 0.0965\,\text{in}^2/\text{ft}$$
$$A_T = \text{total steel} = 1.1565$$

Area based on 16.3″ c/c (415 mm c/c) spacings

No. bars $= \frac{16.3}{12} \times 1.565 \times \frac{1}{1.227} = \frac{1.57}{1.227} = 1.28$ bars

Minimum steel

2 no.$1\frac{1}{4}''$ Q bars (2 – 30 mm HTS)

<5 no.$1\frac{1}{4}''$ Q bars (5 – 30 mm HTS)

Based on 600 psi (tension).

Note: Check this for ultimate load by combining ultimate pressure and hoop prestressing.

8.2.1.5 Cap Reinforcement at Ultimate Load with Penetration

Yield Line at Edge of Standpipe Area

Refer to Fig. 8.7

Approx. no. of standpipes crossing a radius = 9

O.D of standpipes = 12″ (300 mm)

(1/2″ thickness)

Ligament $1\frac{1}{2}''$ and spacing c/c of stand pipes = 16.3″ (415 mm)

$$\text{replacing } 12'' \text{ concete by1 steel compensation} = \frac{1\times 35{,}000}{12\times 4000}$$
$$= 0.73 \text{ take } 0.5 \text{ for safety reasons}$$
$$\text{equivalant concrete on radius } = 186 - \tfrac{1}{2} \times 9 \times 12 = 132$$

Let m be the resisting moment/inch around circumference

$$\text{Area} = \tfrac{1}{2} \times 186 \times 145.8 = 13{,}600$$
$$\text{Total gas pressure} = 13{,}600 pF_s$$

For gas cooled reactor p = pressure 644 psi, F_s = safety factor 3 for cap

$$\text{external work} = 13{,}600 pF_s \tfrac{\Delta}{3}$$
$$\text{internal work} = \tfrac{145.8}{186} m\Delta$$

Thus equating work

$$m = 11.1 \times 10^6 \text{ Ib in./in.}$$

Average pressure over top 9′ – 0 cap = 966 psi $\left(5.665\,\text{MN/m}^2\right)$ (based on 77 t/in.^{2}). Similarly for containment $p = 50$ psi. similar calculations are carried out.

Average pressure over bottom $9' - 0\,\text{cap} = 722\,\text{psi}\ \left(4.2341\ \text{MN/m}^2\right)$

$$d = 118''$$

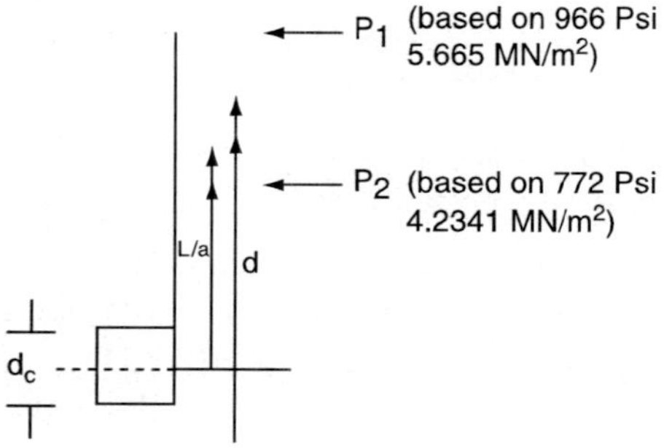

$$l_a = 118 - \frac{d_1}{2}$$

$$f_c\, d_1\, l_a = 11.1 \times 10^6$$

Concrete stress $4000 \times \frac{132}{186} = 2800\,\text{psi}$

$$2800 d_1 \left(118 - \frac{d_1}{2}\right) = 11.1 \times 10^6$$

Therefore

$$d_1 = 39.75''$$

Lever arm for $P_1$is $l_{a1} = (216 - 19.875 + 21) = 175.125''$

$$M_1 = 966(108 - 21) \times 175.125 = 14.72 \times 10^6 (\text{lbin})$$

Lever arm for P_2 is $l_{a2} = 108 - 19.875 = 88.125''$

$$M_2 = 772 \times 108 \times 88.125 = 7.44 \times 10^6 (\text{lbin})$$

$$M_1 + M_2 = 25.16 \times 10^6 > 11.1 \times 10^6$$

No reinforcement is required in this area.
Yield Line at the Inside of Barrel Wall

$$\text{equivalent area} = 258 - \tfrac{1}{2} \times 9 \times 12 = 204\ in^2$$
$$\text{area of the element} = \tfrac{1}{2} \times 258 \times 203 = 26187\, in^2.$$

Similarly equating external and internal works

$$26187 pF_s \frac{\Delta}{3} = \frac{203}{258} m\Delta$$

$m = 11,100;\ pF_s = 11,100 \times 1932 = 21.4452 \times 10^6\,\text{lbin/in.}$

$d = 118''$, concrete stress $= 205/258 \times 4000 = 3180\,\text{psi}$

Similarly

$$3180d_1\left(118 - \frac{d_1}{2}\right) = 21.4452 \times 10^6$$

$$d_1 = 97.39''$$

$$M_1 = \frac{146.31}{175.125} \times 14.72 \times 10^6 = 12.3 \times 10^6\,\text{lbin/in.}$$

$$M_2 = \frac{108 - 48.698}{88.125} \times 7.44 \times 10^6 = 5 \times 10^6\,\text{lbin/in.}$$

$$M_1 + M_2 = 17.3 \times 10^6 < 21.4452 \times 10^6\ \text{lbin/in.}$$

Bonded reinforcement is required.

Assume bonded reinforcement on top and bottom faces

$$\text{External work} = 8729 \times 1932\Delta = 16.85 \times 10^6\Delta$$

$$\text{Internal work} = 0.786m\Delta + 2 \times 0.786m'\Delta$$

$$\text{m due to prestress} = 17.3 \times 10^6\ \text{psi}$$

(assuming load can be transferred to ring beam – cracked but not separated)

$$16.85 \times 10^6 = 0.786 \times 17.3 \times 10^6 + 1.572m'$$

Therefore $m' = 20.6 \times 10^6$

$$A_{st} = \frac{20.6 \times 10^6}{\left(210 - \frac{97.39}{2}\right) \times 6000} = 0.213\ \text{in.}^2/\text{in.}$$

$$\text{total no bars} = \frac{2.56}{1.227} \approx 2$$

In standpipe area from this total effect

$$= \frac{16.3}{12} \times 2.556 = 3.47\,\text{in.}^2\ or\ 2.84\ \text{bars}$$

$$\text{say } 3 - 1\frac{1''}{4}\ \phi\text{bars G. K. 60} + \text{(minimum requirement)}$$

Elsewhere in the cap –2 no. $1\frac{1}{4}\phi$bars G.K. 60+placed at 12″c/c (305 mm *c/c*) Same as before.

Note: Area of steel in particular zones should be increased if spacings required are more than 12" c/c or of varied nature.

8.2.1.6 Shear and Principal Stresses in Cap at Ultimate Load

Data

$$\mathrm{SF} = \text{shear force at the hinge}$$
$$H = \text{cap thickness} = 216''$$
$$A = \text{area of concrete } t_c = \frac{\theta}{\pi}\left[\pi R_E^2 - \pi RI^2 - \pi RD^2\right] = 26841 \text{ in.}^2$$

Where

$$\theta = \text{ half symmetrical angle} \tag{8.36}$$

$$\mathrm{RE},\ \mathrm{RI} = \text{ external and internal radii}$$
$$D = \text{boiler diameter}$$

$$n_\phi = \frac{\text{radial force on the element of cap}}{(2\mathrm{R_E} - \mathrm{DP})\sin\theta} \tag{8.37}$$

$$t_\phi = \text{yield force of cap reinforcement per inch ACT} \times \text{FYS} \tag{8.38}$$

$$\tau_b = \text{shear stress in the barrel} = \frac{SF}{A}$$
$$\tau_{xy} = \text{nominal shear at } \theta_1 - \theta_1 \text{ edge plane}$$
$$= \frac{P \times 2\pi \mathrm{RI}^2}{H \times 2\pi \mathrm{R_I}} = \frac{\mathrm{RI}}{2H} P = 0.597P \text{ psi} \tag{8.39}$$

$$\sigma_x = \text{nominal compression} = \frac{n_\phi + t_\phi}{H}$$
$$\sigma = \text{principal stresses (psi)} \tag{8.40}$$

The following table shows various values at $\theta_1 - \theta_1$ of shear and principal stress derived based on data available from ultimate load program IBM140:

AT Plane $\theta_1 - \theta_1$

EEH	ρ_{cap}	ρ_{barrel}	SF at X	τ	n_ϕ	t_ϕ	σ_x
0.016	1885	1610	−1,374,839	−21	138,394	76,663	996 psi 6.8724 MN/m^2

	τ_{xy}	$\frac{\sigma_x}{2}$	$\sqrt{\left(\frac{\sigma_x}{2}\right)^2 (\tau_{xy})^2}$	σ_1	σ_2
psi	961	498	±1082	+1530	−534
MN/m^2	6.631	3.436	±7.466	10.56	−3.685

At plane $\theta_2 - \theta_2$

$$z = R + H\cot\theta$$

$$\text{where } \cot\theta = \frac{Z - R}{H} \tag{8.41}$$

$$\tan 2\theta = \frac{2\tan\theta}{1-\tan^2\theta} = \frac{2H(Z-R)}{(Z-R)^2 - H^2}$$

$$\alpha = \theta \tan 2\alpha = \frac{2\tau}{\sigma} = \frac{PR}{2H} = \frac{PR}{n_\phi} \tag{8.42}$$

$$\frac{2H(Z-R)}{(Z-R)^2 - H^2} = \frac{P_s R}{n_\phi}$$

Rearranging, the cubic equation becomes

$$PR^3 - 2\rho Z R^2 + R\left[\rho\left(Z^2 - H^2\right) + 2Hn_\phi\right] - 2HZn_\phi = 0$$

Taking random value from computer program

$$H = 216;\ Z = 250.5;\ n_\phi - 163,676.96$$
$$P = -1885.5532$$

The cubic equation reduces to

$$R^3 - 501R^2 - 2.14 \times 10^4 R + 0.84 \times 10^7 = 0$$

Transforming into the form

$$x^3 - px \pm q = 0$$
$$\text{Putting}\, x = y - \frac{b}{3a} = y - 167$$

cubic equation becomes

$$y^3 - y\left(10.5 \times 10^4\right) - 34.9 \times 10^4 = 0$$

The only applicable root $+\sqrt{(4p/3)}\cos(\theta + 240°)$ where $\cos 3\theta = \sqrt{27q^2/4p^2}$

$$\text{when } 4p^3 > 27q^2$$

$$4p^3 = +4630.5 \times 10^{12}$$

$$27q^2 = +32,886 \times 10^8$$

$$\cos 3\theta \approx o$$

$$\theta \approx 30°$$

$$R = 167 +$$

$$\tan\alpha = \frac{216}{83.5} = 2.584,\ \alpha \approx 60^\circ$$

$$\sigma_x = 845$$

$$F = \frac{845}{2} \pm \frac{1}{2}\sqrt{(845)^2 + 4(730)^2} = 422.5 \pm 55.75$$

$$F_1 = +980\,\text{psi}$$

$$F_2 = -135\,\text{psi not critical}$$

The vale of θ as 60° is established from computer program and lab tests.

Ratios, Coefficients

$$\mu = \frac{A_s}{bd}(\text{bending}); \frac{A_S}{A_b}(\text{tension})$$

$$m = \text{modular ratio}$$

$$\upsilon = \frac{p}{ACT}$$

$$K_B, K_R, K_W = \text{coefficients}$$

$$\alpha_s, \alpha_c = \text{coefficient of thermal expansion of steel and concrete respectively}$$

$$K_1, C_1, C_2, C_3 = \text{coefficients}$$

$$a = \sqrt{\frac{A_b}{n}}$$

$$\lambda = \frac{a}{\phi}$$

8.2.1.7 Bending Area (Circular)

(a) Two consecutive cracks and a section in between having a constant bending moment
(b) At each crack where $t_c = 0$ the maximum stress is attained in the reinforcement and its magnitude is

$$f_{SF} = \frac{M}{A_s J} \tag{8.43}$$

Moving away from cracks, the bond between concrete and steel is brought into play; the reinforcement transfers an increasing proportion of the total tensile force to the concrete. The maximum stress in concrete is reached at $x = \frac{L}{2}$ and cracking takes place if this value $t_c = t_{ca}$. Thus two extreme values are achieved for the distance between the cracks

$$t_c < t_{ca} - L\,\text{maximum distance}$$

$$t_c = t_{ca} \text{ at } \frac{L}{2}\text{maximum distance } x = \frac{L}{2}$$

$$\text{Average mean spacing } L_a = 0.75L \tag{8.44}$$

At Section 8.2.1.10, total tensile force is carried by partly:

(a) Reinforcement $f_s A_s$
(b) Bond between steel and concrete is

$$p \int_0^x f_b\left(F\left(\frac{x}{L}\right)\right) dx$$

where

$$f_{bx} = f_b F\left(\frac{x}{L}\right) \tag{8.45}$$

Denoting the distribution bond stress along distance between cracks as shown in Fig. 10.10

$$f_{SF} A_s = f_s A_s p \int_0^x f_b F\left(\frac{x}{l}\right) dx \tag{8.45a}$$

or

$$f_s = f_{SF} - \frac{p}{A_s} \int_0^x F\left(\frac{x}{l}\right) dx \tag{8.45b}$$

For constant moment domain

$$M_A = M_B = f_{SF} A_s y = M_x = f_s A_s J + t_c z_c$$

Together with Eq. (8.45b)

$$t_c = \frac{p f_b J}{z_c} \int_0^x F\left(\frac{x}{l}\right) dx \tag{8.45c}$$

$$\text{at } x = 0.5L \; t_c = t_{ca}$$

$$t_{ca} = \frac{pf_b JL}{z_c} \int_0^{\frac{1}{2}} F\left(\frac{x}{l}\right) d\left(\frac{x}{l}\right) \tag{8.45d}$$

$$L = \frac{1}{\int_0^{\frac{1}{2}} F\left(\frac{x}{l}\right) d\left(\frac{x}{l}\right)} \frac{1}{p} \frac{z_c}{J} \frac{t_{ca}}{f_b} \tag{8.45e}$$

$$= C_1 \frac{1}{p} \frac{z_c}{J} \frac{t_{ca}}{f_b} \tag{8.45f}$$

where

$$\frac{1}{C_1} = \int_0^{1/2} F\left(\frac{x}{l}\right) d\left(\frac{x}{l}\right) \tag{8.45g}$$

Referring to the assumption regarding the functions $F(x/l)$ shown in Fig. 8.10

$$\left.\begin{array}{l} C_1 = 2 \text{ for } F\left(\frac{x}{l}\right) = 1 \text{ in Brice theory} \\ C_1 = \pi \text{ for } F\left(\frac{x}{l}\right) = 1 \text{ in Saligers theory} \\ C_1 = 4 \text{ for } F\left(\frac{x}{l}\right) = 1 \text{ in Wastlund theory} \end{array}\right\} \tag{8.45h}$$

Crack Width

Width of cracks denotes elongation in net value between steel and concrete, i.e. the total slip between them, or

$$w = 2 \int_0^{L/2} s dx \tag{8.46a}$$

where $s = (e_s - e_c)$ per unit length

$$= \frac{f_s}{E_s} - \left(\frac{t_c}{E_s} - e_{sh}\right) \pm T(\alpha_s - \alpha_c) \tag{8.46b}$$

taking into consideration the influences of shrinkage and temperature variation of $\pm$T. Introducing the modular m

$$s = \frac{f_s - mtc}{E_s} + e_{sh} \pm T(\alpha_s - \alpha_c) \tag{8.46c}$$

Using Eq. (8.46a)

$$w = 2\int_0^{L/2} \frac{1}{E}(f_s - mtc)dx + 2\int_0^{L/2} \{e_{sh} \pm T(\alpha_s - \alpha_c)\}dx \qquad (8.46d)$$

Substituting for f_s and t_c from Eqs. (8.45b) and (8.45c) respectively, and after rearranging terms:

$$w = \frac{2}{E_s}\int_0^{L/2} \left[f_{SF} - 2\frac{pLf_bJ}{z_c}\left\{m + \frac{z_c}{JA_s}\right\}\int_0^x F\left(\frac{x}{l}\right)dx\right]dx + L\{e_{sh} \pm T(\alpha_s - \alpha_c)\} \qquad (8.47)$$

$$= \frac{L}{E_s}\left[f_{SF} - 2\frac{pLf_bJ}{z_c}\left(m + \frac{z_c}{JA_s}\right)\int\int_0^{1/2} F\left(\frac{x}{L}\right)d\left(\frac{x}{L}\right)d\left(\frac{x}{L}\right)\right] + L\{e_{sh} \pm T(\alpha_s - \alpha_c)\} \qquad (8.48)$$

With the aid of Eqs. (8.45d) and (8.45)

$$w = \frac{L}{E_s}\left[f_{SF} - t_{ca}\left(m + \frac{z_c}{JA_s}\right)2C_1\int\int_0^{1/2} F\left(\frac{x}{L}\right)d\left(\frac{x}{L}\right)d\left(\frac{x}{L}\right)\right] + L\{e_{sh} \pm T(\alpha_s - \alpha_c)\} \qquad (8.49)$$

or more compactly

$$w = \frac{L}{E_s}\left[f_{SF} - C_2 t_{ca}\left(m + \frac{z_c}{JA_s}\right)\right] + L\{e_{sh} \pm T(\alpha_s - \alpha_c)\} \qquad (8.50)$$

where

$$C_2 = 2C_1\int_0^{1/2}\int_0^{1/2} F\left(\frac{x}{L}\right)d\left(\frac{x}{L}\right)d\left(\frac{x}{L}\right) \qquad (8.51)$$

For bond distribution shown in Fig. 8.10

$$\left.\begin{aligned} C_2 &= \tfrac{1}{2} \text{ in Brice theory} \\ C_2 &= \tfrac{1}{2} \text{ in Saligers theory} \\ C_2 &= \tfrac{1}{3} \text{ in Wastlund theory} \end{aligned}\right\} \qquad (8.52)$$

Neglecting the elastic deformation in concrete (i.e. $m = 0$) and the shrinkage and temperature effects

$$w = \frac{L}{E_s}\left[f_{SF} - C_2 t_{ca}\left(m + \frac{z_c}{JA_s}\right)\right] \tag{8.53}$$

where L is given by Eqs. (8.45e) and (8.45f).

Direct Tension Area

These equations are taken by Watstien and Parsons. In Fig. 8.9a the embedment area $Ab = abcd = 2d'b$ when under direct tension having a large depth D, consisting of two flanges each of area $Ab = 2d'b$ as shown in Fig. 8.9c:

$$z_c = 2Ab\frac{D^2}{4}\frac{2}{D} = A_b D \tag{8.54a}$$

$$J = Ab\frac{D^2}{2}\Big/ A_b\frac{D}{2} = D \tag{8.54b}$$

or

$$f_c = JA_b \tag{8.54c}$$

Substituting these values in Eqs.(8.45c), (8.45d), (8.45e) and (8.45f) and expressing

$$\frac{A_s}{A_b} = \mu \text{ and } \frac{p}{A_s} = \frac{4}{\phi}$$

$$t_c = \frac{pf_b J}{z_c}\int_0^x F\left(\frac{x}{l}\right)dx = p\frac{f_b}{A_b}\int_0^x F\left(\frac{x}{l}\right)d\left(\frac{x}{L}\right) \tag{8.55}$$

$$\begin{aligned} t_{ca} &= \frac{pf_b JL}{z_c}\int_0^{1/2} F\left(\frac{x}{l}\right)d\left(\frac{x}{L}\right) = \frac{p}{A_s}\frac{A_s}{A_b}f_b L\int_0^{1/2} F\left(\frac{x}{l}\right)d\left(\frac{x}{L}\right) \\ &= 4\frac{\mu}{\phi}f_b\int_0^{1/2} F\left(\frac{x}{l}\right)d\left(\frac{x}{L}\right) \end{aligned} \tag{8.56}$$

$$\begin{aligned} L &= C_1\frac{1}{p}\frac{Z_c}{J}\frac{t_{ca}}{f_b} = C_1\frac{1}{p}\frac{A_s}{\mathrm{A}_s}A_b\frac{t_{ca}}{f_b} \\ &= \frac{C_1}{4}\frac{\phi}{\mu}\frac{t_{ca}}{f_b} = C_3\frac{\phi}{\mu}\frac{t_{ca}}{f_b} \end{aligned} \tag{8.57}$$

in which

$$C_3 = \frac{C_1}{4} \tag{8.57a}$$

In a similar manner Eq. (8.50) transforms into

$$W = \frac{L}{E_s}\left[f_{SF} - C_2 t_{ca}\left(m + \frac{1}{\mu}\right)\right] + L\{e_{sh} \pm T(\alpha_s - \alpha_c)\} \tag{8.58}$$

The similar expression comparable to Eq. (8.58) is

$$w = \frac{L}{E_s} f_{SF} - C_2 \frac{t_{ca}}{\mu} \tag{8.59}$$

For practical reasons, a considerable amount of simplification is achieved by neglecting the term $C_2 \frac{t_{ca}}{\mu}$ in Eq. (8.59). The assumption is on the same side:

$$w = L\frac{f_{sp}}{E_s} = C_3 \frac{f_{SF}}{E_s} \frac{t_{ca}}{f_b} \frac{\phi}{\mu} \tag{8.60a}$$

[It seems that the limitation of stresses f_{sf} and t_{ca} may sometimes be necessary to restrict the width of cracking. Reduction in crack.]

Width is also based on f_b which in turn is dependent on bar surface. GK 60 is more effective in this respect.

Writing in different ways

$$w = c\frac{t_{ca}}{f_b} \frac{\phi}{\mu} f_{SF} \tag{8.61a}$$

Equation (8.53a) can be written as

$$w = A\frac{\phi}{\mu} f_{SF} \tag{8.61b}$$

where

$$A = C_3 \frac{t_{ca}}{E_s f_b}$$

The following are various theories taking into consideration various formulae for f_b or f_{bx}

8.2.1.8 Brice Theory

Bond between steel and concrete is essentially a frictional force and the elastic plastic strains in concrete are negligible compared to those in reinforcement. This force is uniform:

$$f_{bx} = f_b = \frac{2\,KB}{1 + 3\frac{l_s}{l_c}} \tag{8.62a}$$

KB for GK 60 = 1.6

l_s, l_c = the sum of bar diameters and the sum of thickness of concrete traversed by a line of possible cracking whose last two non-coincident sides are normal to the face of concrete.

The line of possible cracking should be so chosen that it encounters the largest number of bars and minimum of concrete, i.e l_s/l_c is as large as possible. In Fig. 8.9b the contour of cracking may be abcde or abfg or gfhi.

When the bars are fairly uniformly distributed over the embedment area. Brice suggests that the area Ab may be assumed to be made up of as many square areas as the number of bars with the bar embedded in the centre of each of these elemental areas.

In such a case

$$a^2 = \frac{Ab}{n} \tag{8.63}$$

$$l = n\phi$$

$$l = na - n\phi$$

$$\lambda = \frac{a}{\phi}$$

Therefore

$$1 + 3\frac{l_s}{l_c} = 1 + \frac{3p}{a - \phi} = \frac{\lambda + 2}{\lambda - 1}$$

Expression (8.47a) reduces to

$$f_{bx} = f_b = 2KBt_{ca}\frac{\lambda - 1}{\lambda + 2} \tag{8.63a}$$

$$\text{i.e. } F(x/L) = 1; C_1 = 2.0; C_2 = \frac{1}{2} \text{and } C_3 = \frac{1}{2}$$

As given earlier by Eqs. (8.61b), (8.52) and (8.50) substitution of C_1, C_2 and C_3 in the appropriate expression of the above general theory gives results in Table 8.2.

Theoretically, flexural cracks will not occur so long as the applied moment $M<M_c$ but accidental cracks can nevertheless take place. Brice assumes that in the circumstance the bars behave as if they were fixed in concrete on either side of the crack and proposes a simple formula for limiting the stresses in steel in relation to the width of cracks.

The variation of f_s in this theory is shown in Fig. 8.9a from which the average stress in steel f_{sa} can be computed as:

$$f_{sa} = f_{SF} - \frac{p}{4A_s} L f_b = f_{SF} - \frac{f_b L}{\phi} \tag{8.63b}$$

Therefore, the crack width on $L_a(0.75L)$ is

$$w_a = \frac{0.75L}{E_s}(f_{SF} - 0.75\frac{L}{\phi}f_b) \tag{8.63c}$$

Substituting L from Eq. (8.8b)

$$w_a = \frac{0.75L}{E_s}\left(f_{SF} - 0.75\frac{L}{\phi}f_b\right) \tag{8.63d}$$

reduces to

$$0.75\frac{L}{\phi}f_b = 0.75C_1\frac{1}{p}\frac{z_c}{J}t_{ca} \tag{8.63e}$$

With $c = 2, \frac{A_s}{p} = \frac{\phi}{4}; M_c = t_{ca}z_c; M_f = f_{SF}A_sJ$
Therefore

$$0.75\frac{L}{\phi}f_b = 0.75 \times 2 \times \frac{r}{p\phi}\frac{A_s f_{SF}}{A_s f_{SF}}\frac{z_c}{J}t_{ca} = \frac{3}{8}f_{SF}\frac{M_c}{M_f} \tag{8.63f}$$

and

$$0.75L = \frac{\phi}{f_b}f_{SF}\left\{\frac{3}{8}\frac{M_c}{M_f}\right\}$$

Table 8.2 Parameters in general theory

Quantity	General theory: Flexure $C_1 = 2; C_2 = 0.5$	General theory: Tension $C_2 = 0.5; C_3 = 0.5$
f_{bx}		$f_{bx} = f_b = 2KBt_{ca}\frac{1}{1+3\frac{l_s}{l_c}} = 2KBt_{ca}\frac{\lambda-1}{\lambda+2}$
t_c	$\frac{pf_bJx}{Z_c}$	$\frac{pf_{bx}}{Ab} = \frac{4\mu}{\phi}f_{bx}$
f_s	$f_{SF} - \frac{p}{A_s}f_{bx} = f_{SF} = \frac{4\mu}{\phi}f_{bx}$ from $0 - \frac{1}{2}$ $f_{SF} - \frac{p}{A_s}f_{bx} = f_{SF} = \frac{4\mu}{\phi}f_{bx}$ from $0 - L$	
L	$\frac{\phi}{4KB}\frac{z_c}{J}\left(1 + 3\frac{l_s}{l_c}\right)$	$\frac{\phi}{4KB}\frac{1}{\mu}\left(1 + 3\frac{l_s}{l_c}\right)$
w	$L/E_s\left\{f_{SF} - \frac{t_{ca}}{2}\frac{z_c}{JA_s}\right\}$	$L/E_s\left\{f_{SF} - \frac{t_{ca}}{2\mu}\right\}$
	Simplified solution	
f_{bx}	As above	
w	$\frac{1}{2}\frac{\phi}{\mu}\frac{t_{ca}}{f_b}\frac{f_{SF}}{E_s} = \frac{\phi}{4\mu}\frac{1}{KB}\frac{\lambda+2}{\lambda-1}\frac{f_{SF}}{E_s}$	

$$E_s W_a = \frac{\phi}{f_b} f_{SF}^2 \left[\left\{ \frac{3}{8} \frac{M_c}{M_f} \right\} \left\{ 1 - \frac{3}{8} \frac{M_c}{M_f} \right\} \right] = \frac{\phi}{f_b} f_{SF}^2 P \tag{8.64a}$$

in which

$$P = \frac{3}{8} \frac{M_c}{M_f} \left(1 - \frac{M_c}{M_f} \right) \tag{8.65}$$

When $w_a = 0$ there will be no cracking when either $M_c = 0$, i.e. $t_{ca} = 0$ or $1 - \frac{3}{8} \frac{M_c}{M_f} = 0$

$$\text{i.e. } M_c = 2.67 M_f \tag{8.65a}$$

Since $t_{ca} > 0 M_c = 2.67 M_f$ can be attained if reinforcement is provided.

On the other hand w_a attains the maximum magnitude when p is the largest which is easily determined. Thus putting

$$\begin{gathered} \frac{3}{8} \frac{M_c}{M_f} = y \\ P = Y(1 - Y) = Y - Y^2 \end{gathered} \tag{8.66}$$

8.2.1.9 Summary of Crack Distribution Theories

Notation

(A) Stresses

- t_c =tensile stress in concrete at any section
- t_{ca} =tensile strength of concrete
- f_{su}, f_{pr} =crushing strength of cubes and prisms, respectively
- f_s =tensile stress in steel at any section
- f_{sp} =tensile stress in steel across a crack
- f_b =maximum value of bond stress between concrete and steel
- f_{ba} =average bond stress
- f_{bx} =bond stress at any section at a distance X from a crack
- E_s =modulus of elasticity of steel

(B) Strains

- $e_c e_s$ =strain in concrete and steel, respectively
- e_{sh} =shrinkage of concrete/unit length
- s =slip of steel per unit length

(C) Dimensions and section properties

- b =width of tension zone
- d, h =effective and overall depth, respectively, in a section
- d' =cover to the centre of tensile reinforcement $d = h - d'$
- z_c =sectional modulus of concrete section

J =lever arm
Ab =embedment area of concrete
A_{ct} =area of concrete in tension zone
A_d =area of tension reinforcement in one direction per foot length (width)
A_t =active stretched area of concrete $= 12\ (2\ d' + Q)$
ϕ =diameter of reinforcing bar
A_s =area of reinforcement $= \frac{n\pi\phi^2}{4}$ for 'n' bars
p =perimeter of reinforcing bars $= = n\pi\phi$ for 'n' bars
l_s, l_c =explained
L =maximum distance between cracks
L_a =average (mean) distance between cracks
w =maximum width of cracks
w_a =average width of cracks
T =temperature
S =spacing of bars

(D) Moments
M =moment
M_f =moment at a crack
M_c =moment of resistance of concrete above $t_{ca}z_c$

Setting dp/dy $= 0$

$$T - \frac{1}{2} \cdot P = 0.25$$

Then

$$E_s w_a = \frac{\phi f_{SF}^2}{4 f_b} \tag{8.67}$$

Limiting value for stress in steel is given by

$$f_{SF}^2 \leq 4 \frac{E_s w_a f_b}{\phi}$$
$$\leq 8 E_s \frac{\lambda - 1}{\lambda + 2} K_S \frac{w_a}{\phi} t_{ca}$$

8.2.1.10 Saliger's Theory

In this theory the bond stress distribution in the form of sine curve (Fig. 8.10) is defined by

$$f_{bx} = f_b \sin \frac{2\pi x}{L} \tag{8.68}$$

Saliger takes into consideration elastic strain in concrete and uses the following value for stresses and modular ratio

$$t_{ca} - \frac{f_{pr}}{4.4}\,\mathrm{kg/cm^2}\ \text{flexure} \quad E_s = 29{,}000\,\mathrm{kg/cm^2}$$

$$t_{ca} - \frac{f_{pr}}{12}\,\mathrm{kg/cm^2}\ \text{tension} \quad m = 8.8\ \text{flexure}$$

$$\frac{f_{pr}}{3}\, to\, \frac{f_{pr}}{4}\ \text{for deformed bars} \quad m = 24\ \text{tension}$$

with $C_3 = \frac{\pi}{4}$ as given by Eq (8.9b).

Then from Eq. (8.44a)

$$w = \frac{\pi}{4}\frac{f_{SF}}{E_s}\frac{t_{ca}}{f_b}\frac{\phi}{\mu} \tag{8.69}$$

8.2.1.11 Rüsh's Theory

In this theory the crack width is expressed as

$$w = \frac{K_R}{\upsilon} f_{SF} \tag{8.70a}$$

$$K_R - \text{coefficient}$$

The value of w is easily transformed to the general form of Eq. (8.44b) by relating v to u

$$v = \frac{p}{A_{ct}} = \frac{p}{A_{ct}}\frac{A_s}{A_s}\frac{A_b}{A_s}$$

$$= \frac{4\mu}{\phi}\frac{A_b}{A_{ct}}$$

$v = 0$ percent age bond between concrete and steel denoted by the ratio of the bond area to the area in the tension zone abcd in Fig. 8.9d i.e.

$$v = \frac{p}{A_{ct}}$$

and

$$w = \frac{K_R}{4}\frac{A_o}{A_{ct}}\frac{\phi}{\mu} f_{SF} = \frac{K_R}{2}\frac{\phi}{\mu} f_{SF} \tag{8.70b}$$

$K_R = 0.10 - 0.14 \times 10^{-6}$ for the type similar to GK 60

8.2.1.12 Efsen and Krenchel Theory

A_b = gross area of concrete

Net area $= \left(A_b = n\frac{\pi}{4}b^2\right) = AN$

Part of the load transferred to the concrete by bond stress $= \pi\phi A_t = AN(f_s) = A_b f_s$

$f_b$maximum bond stress

$$A_b = Cf_b\frac{L}{2}$$
$$f_m = \frac{1}{2}\pi\phi = A_b f_{SF} \tag{8.70c}$$
$$L = \frac{2Eb}{\pi\phi} \times \frac{f_{SF}}{f_m} = 2\beta\frac{f_{SF}}{f_m}$$

W from this work was found out by Henshuholt to be 0.05 mm for total elongation of 120 cm prism.

Ratio between mean values w_m/L_a is independent of bar type and of the concrete cross-section and practically only dependent on the reinforcing bar stress

$\frac{w_m}{L_a}$ and f_{SF} are plotted as shown in Fig. 8.10a;

f_{SF} between 2000 and 4000 kg/cm^2.

$$w_m = L_d\frac{f_{SF} - 600}{2 \times 10^6} \tag{8.70d}$$
$$\tan\theta = \frac{\text{stress}}{\text{strain}} = 2 \times 10^6 \text{ kg/cm}^2$$

For

$$\frac{w_m}{L_a} = (f_{SF} - 600)\cot\theta$$

$f_{SF} > \frac{2500\,\text{kg}}{\text{cm}^2}$ L_a practically is independent of the steel stress

Crack spacing for $f_{SF} = 3000$ kg/cm^2 $\left(42,600\,\text{lb/m}^2\right)$

$$\beta = \frac{A_b}{\pi\phi}$$
$$L_a = K_1 + K_1 p \tag{8.70e}$$

Variation K_1, K_2 with f_{SF} shown in Fig. 8.10b

$$L_a = 3 + 0.63\beta \tag{8.70f}$$

Crack width L can also be computed directly from

$$w_m = 0.03 + 3.6\beta\frac{f_{SF} - 800}{10^6}$$

$$w = 1.1\beta\frac{f_{SF}}{10^6}\ (\text{mm})\ \text{Refer to Fig. 8.12c}$$

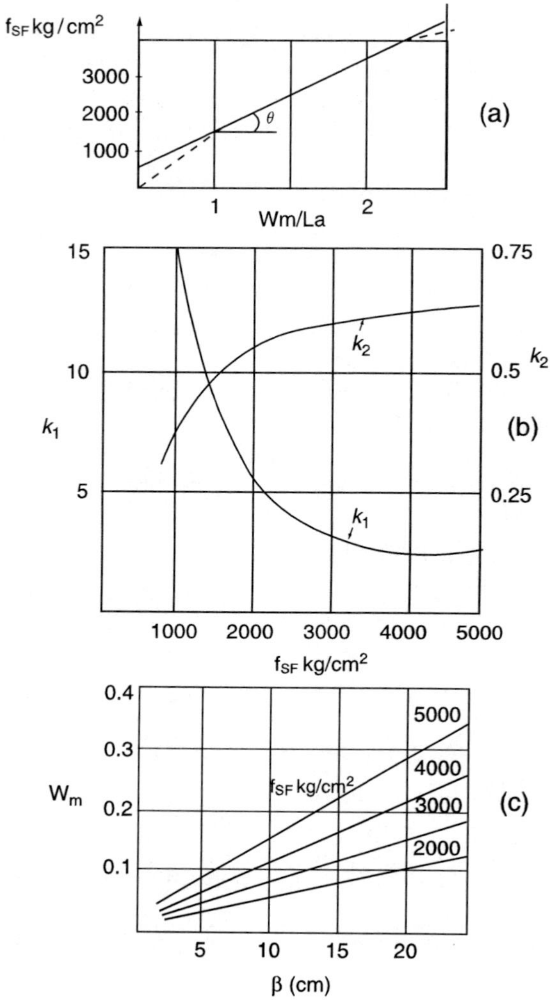

Fig. 8.10 Graphs for various parameters in Efsen and Krenchel theory

8.2.1.13 Wastlund, Jonsson and Osterman Theory

The investigation deals with flexure only. The maximum value of the bond stresses assumed to occur at the edges of the cracks is expressed as

$$f_b = kt_{ca}\sqrt{\frac{w}{\phi}} \tag{8.71}$$

$$t_{ca} = \frac{f_{cu}}{10};\ K\,\text{coefficient}$$

The distribution of bond stress gives

$$f_{bx} = \left(1 - \frac{2x}{L}\right) K t_{ca} \sqrt{\frac{w}{\phi}} \tag{8.72}$$

Elastic strains in concrete is taken for $m = 10$

$$w(\text{flexure only}) = k_w \phi \left[\frac{z_c f_{SF}}{A_s J E_s}\right]^{2/3} \tag{8.73}$$

$$K_w = 0.1$$

For transformation of Eq. (8.73) to the case of direct tension, using Eq. (8.54c)

$$\frac{Z_c}{A_s J} = \frac{A_b}{A_s} = 100u \tag{8.74}$$

$$\text{Therefore } w = K_w \phi \left[\frac{100 f_{SF}}{\mu \; E_s}\right]^{\frac{2}{3}}$$

$$= 100 \frac{K_R}{E_s} \sqrt{K_R} \sqrt{\frac{w}{\phi}} f_{SF} \frac{\phi}{\mu}$$

Generalised expression

$$A_w = 100 \frac{K_R}{E_s} \sqrt{\frac{w}{\phi}} K_R \tag{8.74a}$$

For GK 60 bars $\sqrt{w/\phi = 14}$

Summary of Results

Table 8.3 summarises various crack width and spacings based on the above theories.

To limit the crack to 0.3 mm 3 no. GK60+ bars $1\frac{1}{4}\phi$ are required in the embedded area on the tension side. Usually in RC structures, hair crack size permissible with ordinary reinforcement having tensile strength of 20,000 psi is 0.15 mm. With the type of reinforcement provided in p.v and with the size of structure like pressure vessel, crack size at working permissible is assumed to be 0.3 mm. Therefore where greater tension in concrete is to be expected on the outside surface or haunches, the minimum number of bars <3.

Lower average strain in

$$\text{Bonded steel} = 100 \times 10^{-6} + \frac{w}{L} \tag{8.75}$$

Table 8.3 Summary of crack width and crack spacing

Method of analysis	Crack width (w) (mm)	Crack spacing (L) (mm)	GK60 + Minimum reinforcement embedded	Stress assumed in GK60 bar	Tensile stress developed in concrete
Brice	1.59	1130			
Saliger	1.85	990			
Rüsch	1.96	930			
Efsen and Krenchel	1.815	1005	Single bar $1\frac{1}{4}''$ ϕ 30mm ϕ or T30 only	60,000 psi (415) MN/m^2	500--600 psi (345) MN/m^2
Wastlund, Jonsson, Osterman	1.90	710			

Maximum crack spacing = 30″ (750 mm)
Maximum crack width = 0.05″ (1.25 mm)

8.2.1.14 Crack Formation at Ultimate Load Corresponding to Liner Failure Strain

Maximum crack width the liner can span is estimated for a failure strain of 25%. A stress–strain curve for the AGR liner is not yet available after aging and irradiation. In this analysis, a typical mild stress–strain curve is used and is assumed that the stress–strain curve in bending tension is similar to that in direct tension.

Studs spacing are fixed from buckling analysis and neglecting elastic strains and friction, a crack width is determined for ultimate pressure. The results are summarised in the following pages. The ultimate capacity of studs is investigated according to C P 117 and results established by 'Nelson Studs' of USA.

8.2.1.15 Crack Widths at Ultimate Load Case

Combination of vertical prestress, hoop prestress and ultimate pressure.

Data

HL = internal height from equator to the underside of the cap
H = cap thickness
2θ = angle of segment
RE = external dia
RI = internal dia
X = position of hinge for equator
EEH = hoop strain at equator
ECH = hoop strain at cap
EOH = hoop strain at hinge
l = length of crack
D = boiler dia
γ = depth of compression block

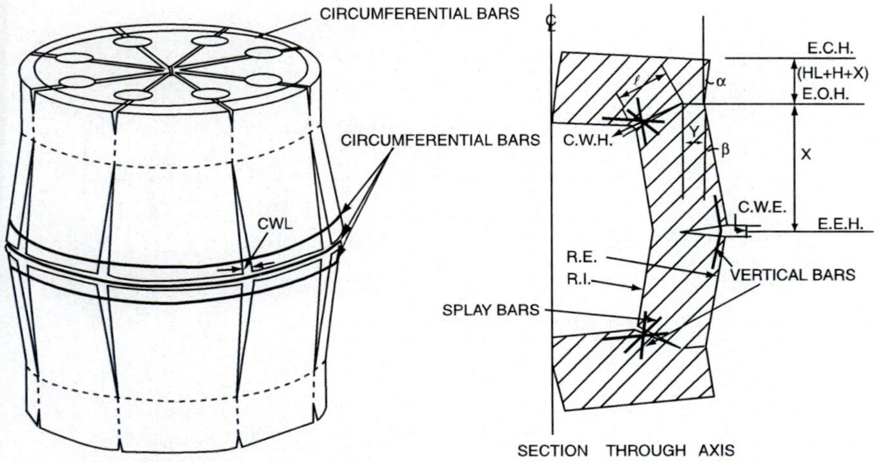

Fig. 8.11 Deformation pattern of vessel

The general patterns are shown for various cracks in Fig. 8.11.

CWL = longitudinal crack width at equator

$$= 2\sin\theta \mathrm{RI}\ (\mathrm{EEH} - \mathrm{EOH}) \tag{8.76}$$

CWH = horizontal crack at haunch

$$= \left[(\mathrm{EEH} - \mathrm{EOH})\frac{\mathrm{RE}}{X} + \frac{(\mathrm{ESH} - \mathrm{EOH})\mathrm{RE}}{(\mathrm{HL} + H - X)}\right] \times L \tag{8.77}$$

where

$$l = \sqrt{(\mathrm{RE} - \mathrm{RI} - Y)^2 + (X - \mathrm{HL})^2}$$

CWE horizontal crack at equator

$$= \frac{2\mathrm{RE}(\mathrm{EEH} - \mathrm{EOH})}{X}\left[\mathrm{RE} - \mathrm{RI} - Y\frac{\mathrm{RE}}{\mathrm{RI}}\right] \tag{8.78}$$

For 1610 psi ultimate pressure, Table 8.4 shows the estimated crack width.

Table 8.4 A comparative study of cracks

Type	IBM 1401 programcalculated values	IBM 7094 programcomputer values from ISOPAR
CWL	1.8 in. (46 mm)	2.1 in. (54 mm)
CWH	2.8 in. (70 mm)	2.3 in. (58 mm)
CWE		3.6 in. (92 mm)

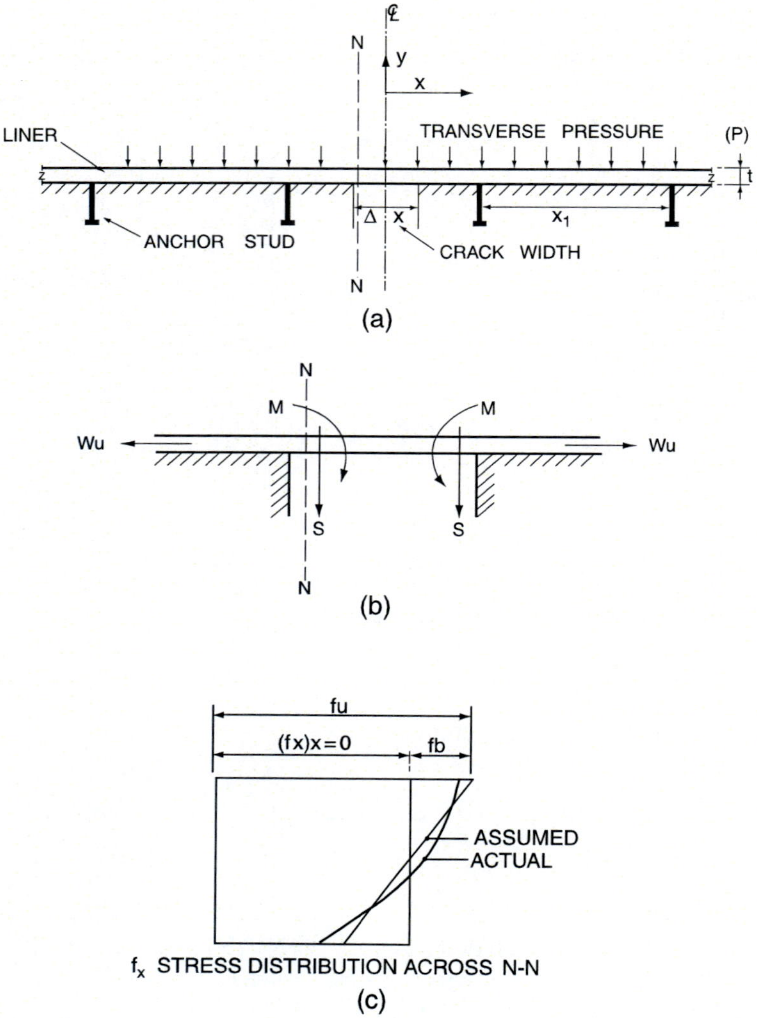

Fig. 8.12 Crack formation at ultimate load corresponding to linear failure strain

Crack formation ultimate load corresponding to liner failure strain is shown in Fig. 8.12.

Data

x_1 = square pitch of studs (assumed) = 9″ and 1/8″
t = liner thickness = 3/4″
μ = coefficient of friction = 0.4 (assumed)

ρ_s = shear loading on studs (between 12.75 and 14.3 t)
P = transverse pressure = 1610 lb/in^2,
n = number of pitches
$(f_x)x$ = stress in the liner at distance X from crack

$$= \frac{(f_x)_{x=0}x_1 t - \mu x_1 p x - [np_s]^{(2n-1)\frac{x_1}{2}<x<(2n+1)\frac{x_1}{2}}}{tx_1} \tag{8.79}$$

$$\text{for } (2n-1)\frac{x_1}{2} < x < (2n+1)\frac{x_1}{2} \tag{8.79a}$$

$$\begin{aligned} (f_x)_{x=0} \text{ in tension} &= 55,000 \tfrac{\text{lb}}{\text{in}^2} (380 \text{ MN/m}^2) \\ (\in_x)_{x=0} &= 12\% \text{ corresponding strain} \\ f_b = \tfrac{M}{Z} &= \tfrac{P\Delta^2 x_1}{12} \tfrac{6}{x_1 t^2} = \tfrac{p\Delta^2 x}{2t^2} \end{aligned} \tag{8.80}$$

Assuming this, since stress–strain curve is not linear

$$f_u = \text{limiting stress} = 60,000 \text{ psi } (415 \text{ MN/m}^2)$$

Failure strain 25–28% (25% taken)

$$(f_x)_x = 55,000 - 855x - n(4220)\text{lb/in}^2, \tag{8.81}$$

$$\in_x = \left[\frac{f_x - 36,000}{58,000}\right]^{1.633} \tag{8.82}$$

$$\begin{aligned} \Delta x &= 2\int_0^{4.5} t_x dx + 2\int_{4.5}^{13.5} \in_x dx = 1.8''(\text{crack width}) \\ f_b &= \tfrac{1610(1.8)^2}{2\times(0.75)^2} = 4650 \text{ lb/in}^2. \end{aligned} \tag{8.83}$$

From Fig 8.12c

$$\begin{gathered} (f_x)_{x=0} + f_b < 6000 \\ 55,000 + 4650 = 59,650 < 60,000 \end{gathered}$$

The liner in this case takes $\Delta x = 1.8''$ (47 mm), take as 50 mm

$$\text{shear} \frac{1610}{0.75} = 2280 \text{ lb/in}^2. < (f_x)_{x=0} \textit{ and } f_b$$

Considering 25% strain

$$\in_x > \left(25 - \frac{13x}{4.5}\right)$$

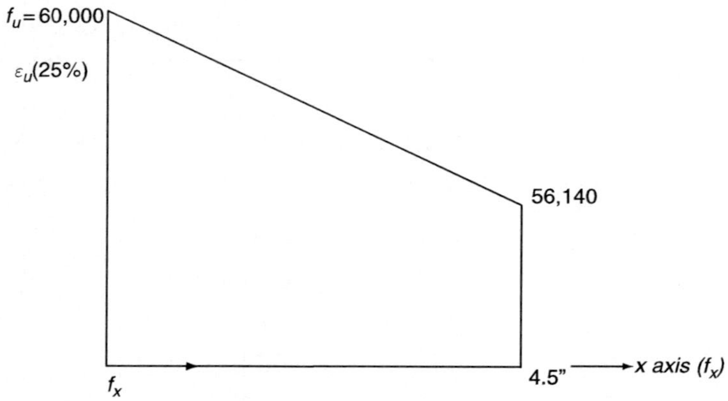

Ratio proportion

$$\Delta x > \left[25x - \frac{2.89x^2}{2}\right]_0^{4.5} \left\{\frac{1}{100} \times 2\right\}$$

$$\Delta x > 0.847 \times 2$$

i.e. $\Delta x > 1.694$ (hence 1.8 is equivalent reasonable)
With 14.3 value of ρ_s for $x_1 = 8\frac{1}{8}''$ (207mm)

$$\Delta x = 2\int_0^{4,0625} \in_x dx + 2\int_{4.0625}^{12.1875} \in_x dx = 1.35''(4.29 \text{ mm})(\text{crack width})$$

Note: The value of the 14.3 t from standard tests on anchors by Nelson company on 5/8″ size studs. If 12.7 tons anchor value is taken, the value of Δx will reach about 2″.

If idealised stress–strain curve is used with friction stress in x-direction $= f_x$

$$w_u \rightarrow f_y t \text{ or } f_u t \qquad x_1 = 8\frac{1}{8}'' \ (207 \text{ mm})$$

$$f_x = [w_u - \mu P x] \times \frac{1}{t}$$

$$x_{1/2} = p_1 = \text{Half spacing} = 4.0625$$

$$\in_x = \frac{f_x}{E}$$

$$\Delta x = 2\int_0^{p_1} \frac{1}{Et}(w_u - \mu Px)dx \qquad E = 27.9 \times 10^6\ lb/in.^2$$

$$= \frac{2}{Et}\left[w_u P_1 - \frac{\mu P P_1^2}{2}\right] \qquad w_u = 60,000 \times 0.75 = 45,000(311\text{MN/m}^2)$$

$$= 0.17''(\text{crack width})$$

The most valuable consideration is friction factor. However, idealised stress–strain curve is definitely meaningless and also in this case exact value of E cannot be correctly determined.

The liner is more ductile than either prestressing strand or GK60 bar, having elongation twice that of GK60 and therefore an average crack width of 1.57 in. over which the liner can span before failure against 25% strain is a reasonable assessment. To prove this, it is proposed that further tests on liner should be carried out.

Crack width = 1.57 in. (40. mm)

8.2.1.16 Analysis of a Single Bar Spanning a Crack at Ultimate Load

Major cracks are formed due to the stress in concrete exceeding the permissible stress in tension. Concrete on the edge of crack and to a distance x on either side of it will become progressively ineffective in transmitting the force from the reinforcement to the body of concrete with the increase of the force in steel up to the ultimate load.

It is assumed that the concrete adjoining the crack will fail in the manner shown in Fig. 8.13a, c.

Area of truncated cone on which spalling is assumed to take place:

$$A = \pi sa = \frac{\pi x}{\cos\theta}(\gamma + \phi) \qquad (8.84)$$

where $a = (\gamma + \phi)$

Tensile force in concrete developed on the plane of weakness just before cracking = force of difference in the bar.

Let concrete fail in tension when the tensile stress along the plane of failure is 'f_t' ie. stress in concrete in the direction of reinforcement $= f\cos\theta$

Tensile force in concrete on the area of conical plane of weakness just before spalling occurs

$$= T = Af_t \sec\theta \rightarrow \text{comonent of the force parallel to the bar } Af_t \qquad (8.85)$$

For 'x' bar length change of force

$$f_b = \frac{\pi}{4}\phi^2(f_{SF} = f_y) \qquad (8.86)$$

For maximum crack width say 2″ with $\gamma = x\tan\theta$

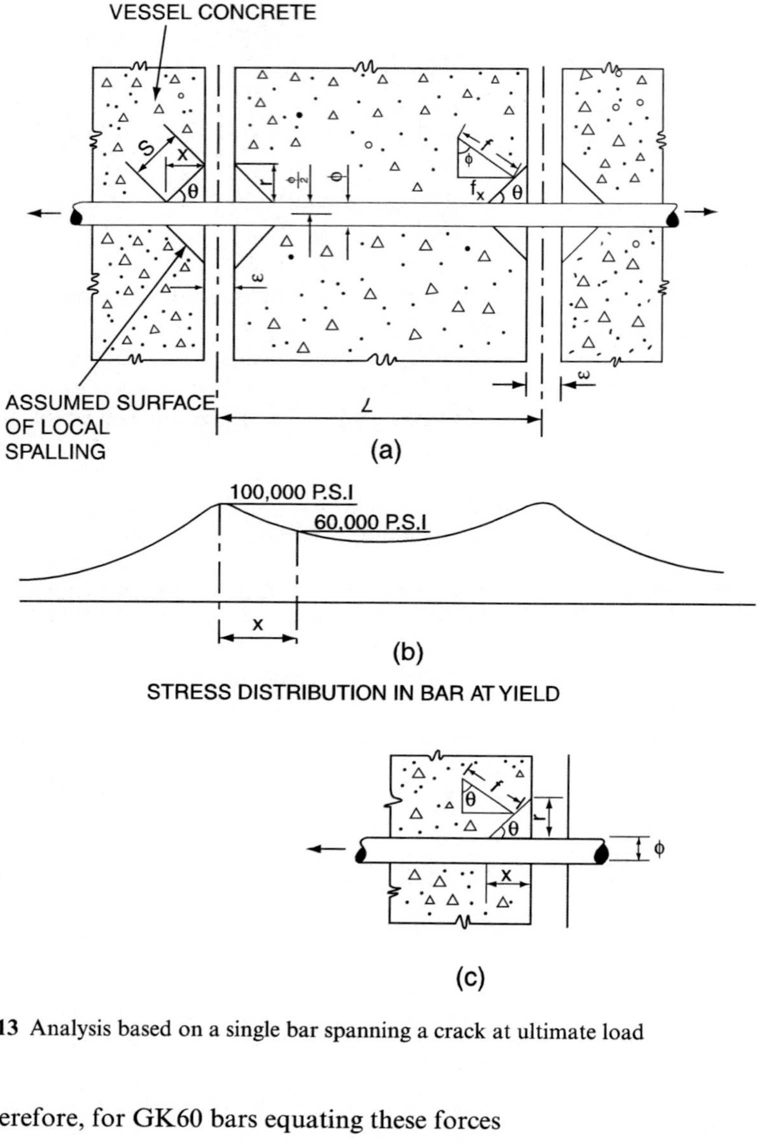

Fig. 8.13 Analysis based on a single bar spanning a crack at ultimate load

Therefore, for GK60 bars equating these forces

$$\frac{\pi x}{\cos\theta}(x\tan\theta+\phi)f_t=\frac{\pi}{4}\phi^2(100{,}000-60{,}000)\text{ from Fig. 8.15b} \qquad (8.86a)$$

$$x^2\tan\theta+x\phi-10{,}000\frac{\phi^2}{f_t}\cos\theta=0$$

$$f_t=500\,\text{psi at ultimate}$$

$$\phi = 1\frac{1''}{4}\, size\, 6 - k \; - 60\,\text{bar}$$

$$x^2 \tan\theta + 1.25x - 31.25 \cos\theta = 0$$

$$\text{for } \theta = 45^\circ$$

$$x^2 + 1.25x - 22.1 = 0$$

$$x = 4.2''$$

$$\text{For } \theta = 0 \text{ along the length of bar}$$

$$x = \frac{31.25}{1.25} 25''$$

The assumptions are as follows:

(a) No external compression at right angle to the bar.
(b) Effect of mechanical anchorage is ignored.
(c) Failure occurs in a single bar, i.e. all bars fail over maximum crack width.

The value of $x = 2.5''$ bond length on both sides is just about all right. Assume still the stress distribution in bar at ultimate load is as shown in Fig. 8.13 having a bond length of 60″ (1500 mm) minimum on both sides of the crack, i.e.

$$\text{minimum bond length} = 5' - 0(1.524\,\text{m})$$

The final bond stress is

$$\frac{60}{\cos\theta}(60 tan\theta + \phi) f_t = \phi^2/4(f_{sf} - 60{,}000) \tag{8.87}$$

$$\theta = 0\,,\phi = 1.25$$

$$60 f_t = \phi/4[40{,}000]$$

$$f_t = 0.00521 \times 40{,}000$$

$$= 208.4\,\text{psi} > \text{allowable bond stress } 180 \text{ psi}$$

Conclusion

$$\text{minimum bond length} \;=\; 5' - 0\,(1.52\,\text{hm})$$

$$\text{Maximum crack width} \;=\; 2''\,(50\,\text{mm})$$

$1\frac{1}{2}$GK 60+ can span a crack of 2″(50 mm) and hence one bar $1\frac{1'}{4}$(30 mm)ϕGK 60 is considered a minimum requirement for a crack of 2″ size anywhere in vessel:

$$\frac{\text{stress at bond distance}}{\text{A}_\text{s} = x - \text{sectional area}} x = 60$$

The force in the tensile reinforcement at a distance $x = 60''$ from the crack

$$F = f_{\text{SF}} - p/A_s \int_0^{x=60} f_{\text{bdx}} \qquad p - \text{perimeter} \tag{8.88}$$

Bending moment is constant and before cracking starts, it is taken up partly by the steel and partly by concrete.

$$f = 100{,}000 - 4/\phi f_b.x \tag{8.89}$$

$$= 100{,}000 - 4/1.25 \times 208 \times 4 \times 60$$

$$\approx 60{,}000 \text{ psi o.k.}$$

Take an average crack width at ultimate load = 2″ = 50 mm.

Since all theories are giving identical results, for example, Brice theory is taken into consideration:

$$L - \text{between boilers} = 391.179''$$

$$W\text{max} = A\frac{\phi}{\mu} f_{SF} \text{ where } A = \frac{C_3 \times t_{ca}}{E_s f_b} \tag{8.90}$$

$$C_3 = \frac{1}{2}; E_s(\text{assumed}) = 30 \times 10^6 \text{ psi} = 2.12 \times 10^4 \text{ kg/mm}^2$$

$$t_{\text{ca}} = 42.2 \text{ kg/cm}^2 \text{ (max)}$$

$$f_b = \text{bond stress (max)}$$

$$14.65 = 2 \times 1.6 \times 42.2\left(\frac{\lambda - 1}{\lambda + 2}\right)$$

$$\left(\frac{\lambda - 1}{\lambda + 2}\right) = \frac{14.65}{135} = 0.1085$$

$$\lambda = 1.365$$

$$f_{b(\text{max})} = 208.4 \text{ psi} \approx 14.65 \text{ kg/cm}^2$$

taking 60″ bond length as discussed above.

Notation

$$w, \phi - mm; \mu - \%; f_b, t_{ca} - kg/cm^2; f_{sf}, E_s - kg/mm^2$$

$$A = \frac{1}{2} \times \frac{42.2}{2.12 \times 10^4 \times 14.65} = 0.68 \times 10^{-4}$$

Therefore,

$$50.8 = w_{\max} = 0.68 \times 10^{-4} \frac{\phi}{\mu} f_{SF}$$

$$= 0.68 \times 10^{-4} \times \frac{\phi}{\mu} 42.2$$

$$\frac{\phi}{\mu} = \frac{50.8 \times 10^{-2}}{0.287}$$

$$\phi = 31.8\,\text{mm}$$

$$\mu = 0.00018\%$$

$$\text{assumed embedded area} = 1.365 \times 31.8 = 43.3$$

Therefore

$$\%\ \text{area of steel} = 0.0018 \times 43.3 = 0.08\%\ \text{(minimum through out)}$$

Assuming if 3 bars are used

$$\text{area of steel} = 0.25\%$$

Taking 12×240 embedded area across the barrel thickness (excluding shutters)

$$\text{steel area} = 2880 \times 0.0018 = 5.184\ \text{in}^2$$

$$\text{using } 1\frac{1''}{4}\ \phi\text{GK60 bars, area} = 1.227\ \text{in}^2$$

$$\text{total bars} = \frac{5.184}{1.227} = 4.23\ \text{bars (based on 60,000 psi yield strength)}$$

Since crack size of approx. 2″(50 mm) is developed at ultimate load plastic range is considered for steel.

$$\text{No.bars} = \frac{60,000}{90,000} \times \frac{5.184}{1.227} \approx 3\,\text{bars GK601}\frac{1''}{4}\phi \text{ in metric 3T30}$$

Therefore, where crack sizes are 2″(50 mm) width, 3 no.1 $\frac{1''}{4}$ ϕGK60 bars should be embedded, giving minimum 50″(1270 mm) length on either side.

Chapter 9
Concrete Nuclear Shelters

9.1 Introduction

There is increasing current concern about safety from nuclear hazards, including nuclear blasts and radiation. There will be greater involvement in protecting people against such hazards. A nuclear shelter is just one of many ideas to protect and shield a person from the effects of nuclear explosions. These structures can range from a deep buried rigid structure to a concrete framed box covered with soil.

The vast quantities of nuclear energy released by detonation are distributed approximately in the following way:

Blast and shock wave: 45%
Light and heat radiation: 35%
Initial nuclear radiation: 5%
Residual radiation: 15%

Secondary effects such as flooding and soil liquefaction are also considered in the overall safety analysis of shelters. Factors which affect the response of shelters to air blast are the following:

(a) strength and mass
(b) structural layout
(c) ductility

The airburst from an atom bomb propagates through the atmosphere to great distance with minimum energy losses. On the other hand, in an underground burst much of the energy is absorbed in cratering and melting of the ground. A burst occurring above 30,000 m is known as an airburst provided the fireball at maximum brilliance does not touch the surface of the earth. If it touches, it becomes a 'surface burst'. These cases together with their loading conditions are described in this section. A typical concrete shelter design for a six-person family is chosen for finite element analysis using the five-parameter constitutive law. Analytical results show agreement with those provided by the various codes.

M.Y.H. Bangash, *Structures for Nuclear Facilities*,
DOI 10.1007/978-3-642-12560-7_9, © M.Y.H. Bangash 2011

9.2 Characteristics of the Blast Wave in Air

Most of the material damage from an airburst (nuclear weapon) is caused mainly by the shock (or blast) wave which accompanies the explosion. The majority of structures will suffer some damage from air blast when the overpressure in the blast wave, i.e. the excess over atmospheric pressure (101.3 kN/m^2 at sea level), is about 3.5 kN/m^2 or more. The distance to which this overpressure level will extend depends on the yield of the explosion and on the height of the burst.

A difference in air pressure acting on separate surfaces of a structure produces a force on that structure; the size of these forces is dependent on the difference between ambient air pressure and the overpressure. The maximum value of overpressure, known as the peak overpressure, occurs at the front of the blast wave.

As the blast wave travels in the air away from its source, the overpressure at the front steadily decreases, and the pressure behind the front falls off in a regular manner. After a short time, when the shock front has travelled a certain distance from the fireball, the pressure behind the front drops below that of the surrounding atmosphere and a so-called 'negative phase' of the blast wave forms, as given in Fig. 9.1.

For the curves marked t_1 to t_4 the pressure in the blast wave has not fallen below atmospheric pressure, but on the curve t_5 it is seen that at some distance behind the shock front the overpressure is below that of the original atmosphere so that an 'underpressure' exists.

During the negative (rarefaction or Suction) phase a partial vacuum is produced and the air is sucked in, instead of being pushed away, as it is when the overpressure is positive.

The peak values of the underpressure are generally small compared with the peak overpressures, the former having a maximum value of 27.5 kN/m^2 below the

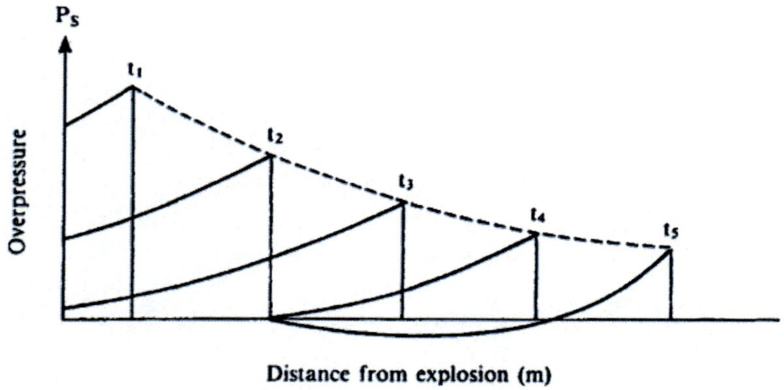

Fig. 9.1 Variation of overpressure with distance at successive times

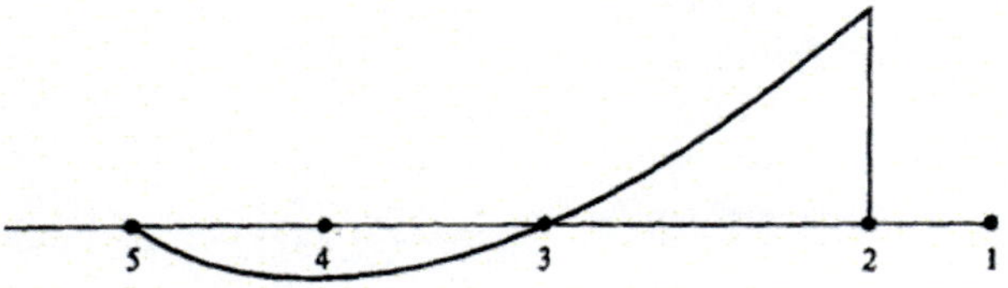

Fig. 9.2 Variation of pressure with time at a fixed location

ambient pressure. With increasing distance from the explosion, both peak values decrease, the positive more rapidly than the negative, but they do not approach equality until the peak pressures have decayed to a very low level (Fig. 9.2).

For a short interval after the detonation, point 1, there will be no increase in pressure since it takes the blast wave some time to travel the distance from the point of explosion to the given location. Point 2 indicates the time of arrival of the shock front; a strong wind commences to blow away from the explosion. This is often referred to as a 'transient' wind, as its velocity decreases fairly rapidly with time.

Following the arrival of the shock front the pressure falls rapidly, and at the time corresponding to point 3 it is the same as that of the original atmosphere. Although the overpressure is now zero, the wind will continue in the same direction for a short time. The interval from points 2 to 3 (roughly 2–4 s for a 1 megaton explosion) represents the passage of the positive (or compression) phase of the blast wave. It is during this interval that most of the destructive action of the airburst will be experienced.

As the pressure in the blast wave continues to decrease, it sinks below that of the surrounding atmosphere. In the time interval from point 3 to point 5, which may be several seconds, the negative (or suction) phase of the blast wave passes the given location. For most of this period the transient wind blows towards the explosion. As the negative phase passes, the pressure at first decreases below ambient and then increases towards that of the ambient atmosphere which is reached at the time represented by point 5. The blast wind has then effectively ceased and thedirect destructive action of the airblast is over.

9.2.1 Dynamic Pressure

Although the destructive effects of the blast wave are usually related to values of the peak overpressure, there is another quantity of equivalent importance called the 'dynamic pressure'.

The dynamic pressure is proportional to the square of the wind velocity and to the density of the air behind the shock front. For very strong shocks the dynamic pressure is larger than the overpressure but below 480 kN/m^2 (4.7 atmospheres) overpressure at sea level; the dynamic pressure is smaller. The peak overpressure dynamic pressure decreases with increasing distance from the explosion centre, although at a different rate.

Some indication of the corresponding values of peak overpressure, peak dynamic pressure and maximum blast wind velocity for an ideal shock front in air at sea level are given in Table 9.1.

Table 9.1 Overpressure, dynamic pressure, and wind velocities, in air at sea level for an ideal shock front

Peak overpressure, KN/M^2 (atoms)	Peak dynamic pressure, KN/M^2 (atoms)	Maximum wind velocity, Km/h
1378 (13.78)	2274 (22.74)	3347
1034 (10.34)	1537 (15.37)	2860
689 (6.89)	848 (8.48)	2275
496 (4.96)	551 (5.51)	1883
345 (3.45)	275 (2.75)	1512
207 (2.07)	110 (1.10)	1078
138 (1.38)	55 (0.55)	756
69 (0.69)	15 (0.14)	467
34 (0.34)	5 (0.05)	257
14 (0.14)	0.7 (0.007)	97

When the shock front reaches a given point, both the overpressure and the dynamic pressure increase almost immediately from zero to their maximum values and then decrease. The dynamic pressure (and wind velocity) will fall to zero some what later than the overpressure because of the momentum of the air in motion behind the shock front, but for the purposes of estimating damage the difference is not significant. During the negative (suction) phase of the blast wave the dynamic pressure is very small and acts in the opposite direction.

9.2.2 *Arrival Time and Duration*

There is a finite time interval required for the blast wave to move out from the explosion centre to any particular location. This time interval is dependent upon the energy yield of the explosion and the distance involved. At 1.6 km from a 1 megaton burst, the arrival time would be about 4 s. Initially, the velocity of the shock front is quite high, many times the speed of sound, but as the blast wave progresses outwards, so it slows down as the pressure at the front weakens. Finally, at long ranges, the blast wave becomes essentially a sound wave and its velocity approaches ambient sound velocity.

The positive phase duration is shortest at close ranges and increases as the blast wave moves outwards. At 1.6 km from a 1 megaton explosion, the duration of the positive phase of the blast is about 2 s. The period of time over which the positive dynamic pressure is effective may be taken as essentially the positive phase of the overpressure.

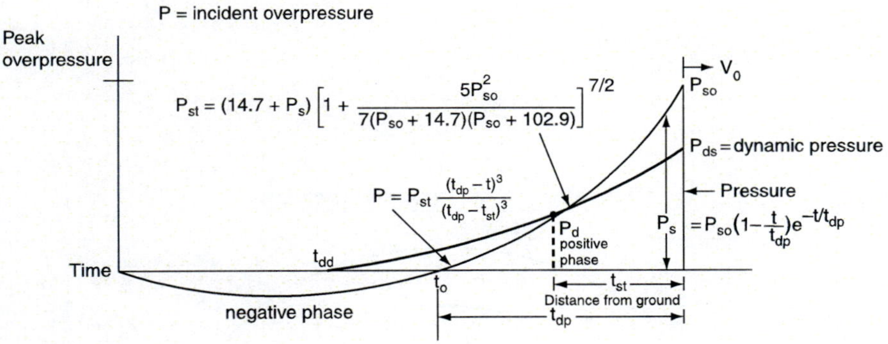

Fig. 9.3 Variation of overpressure with time at fixed location

9.2.3 Reflection of the Blast Wave at a Surface

When the incident blast wave from an explosion in air (Fig. 9.3) strikes a more dense medium, such as the Earth's surface, either land or water, it is reflected.

When such reflection occurs, an object precisely at the surface will experience a pressure increase, since the reflected wave is formed instantaneously the value of the overpressure thus experienced at the surface is generally considered to be entirely a reflected pressure. In the region near ground zero, this total reflected overpressure will be more than twice the value of the peak overpressure of incident blast wave.

There are two important destructive aspects of the blast wave involved reflection:

(i) Only a single pressure increase is experienced in the Mach region below the triple point as compared to the separate incident and reflected waves region of regular reflection.

(ii) (a) Since the Mach stem is nearly vertical, the accompanying blast wave travelling in a horizontal direction at the surface.

(b) The transient winds are approximately parallel to the ground, thus, the Mach region, the blast forces on aboveground structures other objects are directed nearly horizontally, so that vertical surfaces are loaded more intensely than horizontal surfaces.

9.2.4 Blast from a Surface Burst

In a surface explosion, the front of the blast wave in air is hemispherical in (Fig. 9.4), there is no region of regular reflection, and all objects and on the

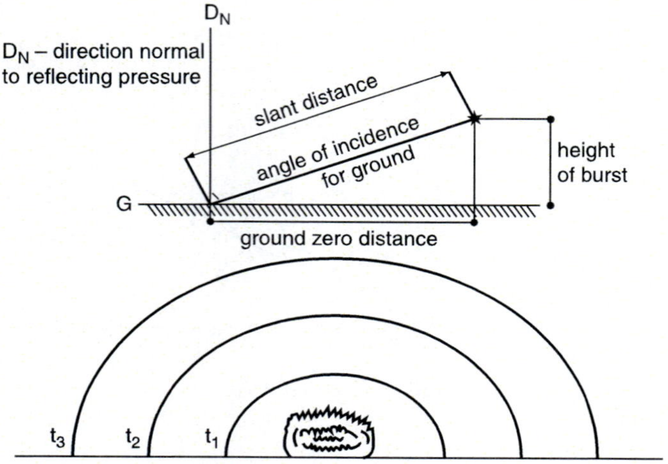

Fig. 9.4 Blast wave from a contact burst

structures on the surface, even close to ground zero, are subjected to air blast similar to the Mach region below the triple point of an airburst.

The diameter of the rupture zone D_r is 1 .5D_a; the overall diameter, including the D_D is 2D_a; and the height of the lip H_t is 0.25H_a.

The dependence of crater radius and crater depth upon the depth of burst for a 1 kiloton explosion in dry soil is shown in Fig. 9.5. Also shown are the range of... dimensions possible from a surface burst to the approximate maximum for an

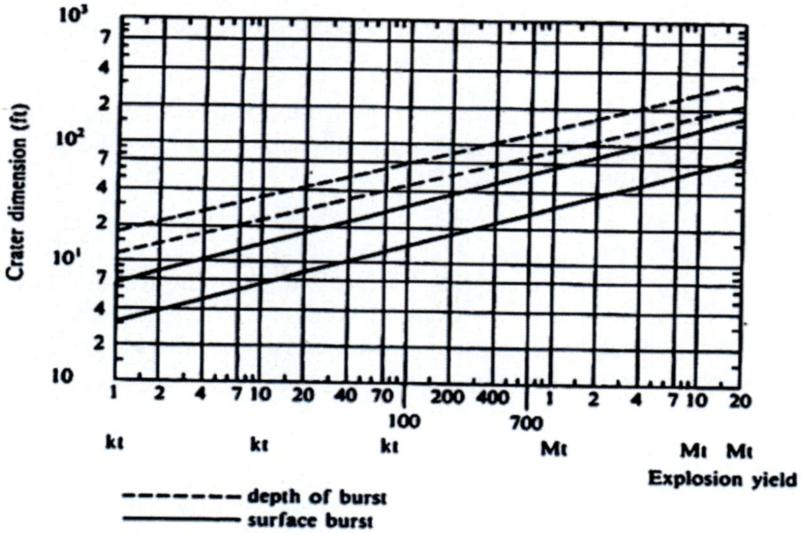

Fig. 9.5 Crater versus yield (courtesy of US Airforce)

underground burst, for any explosion energy yield from 1 kiloton to 10 megatons. Therefore a 1 megaton blast wave front may be assumed to be vertical for most structures, with both overpressure and dynamic pressure decaying at different rates behind the blast wave as previously stated.

All the above descriptions are based on an ideal blast wave in ideal atmospheric conditions: unfortunately, nothing is perfect, especially meteorological conditions, which can greatly alter the expected performances of explosions, either increasing them or decreasing them.

9.2.5 Ground Shock from Air Blast

Another aspect of the blast wave problem is the possible effect of an airburst on underground structures as a result of the transfer of some of the blast wave energy into the ground. A minor oscillation of the surface is experienced and a ground shock is produced. The strength of the shock at any point is determined by the overpressure in the blast wave immediately above it. For large overpressures with long-positive phase duration, the shock will penetrate some distance into ground; but blast waves which are weaker and of shorter duration are attenuated more rapidly. The major principal stress in the soil will be nearly vertical and about equal in magnitude to the air blast overpressure.

For high airbursts where relatively large blast pressures are not expected at ground level, the effects of ground shock induced by the air blast will be negligible, but if the overpressure at the surface is large there may be damage to buried structures. However, even if the structure is strong enough to withstand the effect of the ground shock, the sharp jolt resulting from the impact of the shock front can cause injury to occupants and damage to loose equipment.

9.2.6 Technical Aspects of a Blast Wave

The basic relationships among the properties of a blast wave having a sharp front at which there is a sudden pressure discontinuity, i.e. a true shock front, are given in Tables 9.2–9.5 and Fig. 9.5 respectively.

9.2.7 Air Blast Loading and Target Response

The behaviour of an object or structure exposed to the blast wave from a nuclear explosion may be considered under two main headings.

1. Loading, the forces which result from the action of that pressure.
2. Response or distortion of the structure due to the particular loading.

Table 9.2 Blast loadings[a]

p_{do} = peak dynamic pressure
U_o = velocity of the shock front
p_s = overpressure at time
t = time after detonation of yield 'y'
t_{dd} = delayed time (dynamic)
t_{dp} = duration of dynamic overpressure
$U_o[U_s = 1117 \times \sqrt{1 + 6p_{so}/7p_o}$
p_o = atmospheric pressure = 14.7
p_r = reflected pressure = $2p_o\left(\frac{102.9+4p_{so}}{102.9+p_{so}}\right)$
t_c = clearing time of the reflection effect
= $3x$ height of the reflecting surface above ground/U_o

Scaling laws (for two bombs)
$R_2 = R_1(y_2/y_1)^{1/3}$ R, R_1, R_2 = radii or distances from explosions
$t_{dd} = t_{dp}(y_2/y_1)^{1/3}$
Reference should be made to Fig. IX. 163 for other details
Additional loadings
$\frac{p_{so}}{p_o} = 3.2 \times 10^6 R^{-3}\sqrt{1 - \left(\frac{R}{87}\right)^2 \times \left(\frac{800+R}{800}\right)}$
time in seconds $= t_{dp} = (y)^{1/3}\left\{\frac{180[1+(0.01R)^3]^{1/2}}{1+(R/40)\sqrt[6]{1+(R/285)^5}\sqrt[6]{1+0.0002R}}\right\}$

$$IMt : Pso = 20\text{lbf/in}^2 \rightarrow 7100ft$$

$$\frac{1}{2}Mt : \frac{R_2}{7100} = \left(\frac{0.5}{1}\right)^2$$

Therefore

$$R_1 = 5650\,\text{ft}$$

(Note: 1ft = 0.3048 m.)

As a general rule, response may be taken to be synonymous with damage, since permanent distortion of a sufficient amount will impair the usefulness of a structure.

Direct damage to structures attributable to air blast can take various forms; the blast may deflect structural steel frames, collapse roofs, dish-in walls, shatter panels and break windows.

9.2.7.1 Loading

(a) Diffraction Loading

When the front of an air blast wave strikes the face of a building, reflection occurs. As a result the overpressure builds up rapidly to at least twice that in the incident wave front; the actual pressure attained is determined by various factors, such as the peak overpressure of the incident wave and the angle between the direction of motion of the wave and the face of the building. As the wave front moves forward, the reflected overpressure on the face drops

Table 9.3 Characteristics of blast waves from 1- and 10-megaton nuclear burt optimum height air blasts (UK Home Office Data)

	Peak overpressure kpa (psi)		Windspeed, m/s (mph)		Arrival time (s)		Duration (s)	
Distance from ground zero kilometer miles	1 Mt	10 Mt	1 Mt	10 Mt	1 Mt	10 Mt	1 Mt	10 Mt
1 (0.6)	980 (140)	4200 (600)	980 (2200)		0.9	—	0.9	1
2 (1.25)	200 (29)	1050 (150)	400 (900)	980 (2200)	4	1.8	1.8	2
3 (1.9)	130 (18)	380 (54)	180 (400)	805 (1800)	6	4.7	2.3	2.4
4 (2.5)	63 (12)	220 (31)	128 (285)	490 (1100)	9	8	2.6	3.7
5 (3.1)	56 (8)	170 (24)	106 (238)	277 (620)	12	11	2.8	4.2
6 (3.8)	42 (6)	135 (19)	81(182)	200 (450)	14	12.5	3	4.7
7 (4.4)	33 (4.7)	110 (16)	69 (155)	170 (380)	18	16	3.2	5
8 (5)	27 (3.8)	90 (13)	56 (125)	143 (320)	22	18	3.4	5.3
9 (5.6)	24 (3.4)	77 (11)	48 (108)	125 (280)	26	20	3.5	5.5
10 (6.2)	21(3)	63 (9)	40 (90)	114 (255)	28	22	3.6	5.9
12 (7.5)	15 (2.2)	46 (6.5)	31 (70)	96 (215)	31	30	3.7	6.3
15 (9.4)	10 (1.5)	34 (4.8)	24 (53)	72 (160)	45	38	3.8	6.7
20 (12.5)	7 (1)	22 (3.3)	17 (37)	46 (103)	5.9	53	4	7.5
25 (16)	5 (0.75)	15 (2.2)	9 (20)	32 (72)	70	67	4.2	7.8

Table 9.4 Blast loads on shallow buried surface shelters

	Shallow buried		
Member	In dry ground	High water level	Surface
Roof and floors	P_{so}	P_{so}	P_{so}
Walls	$0.5P_{so}$	P_{so}	$2.3P_{so}$

P_{so} = overpressure
Dead loads, soil and water loads should be added to the blast loads given above.

Table 9.5 Generalised data in relation to loads

1-megatonburst at a distance of 1.6 km (1 mile) from ground zero
Velocity of shock front 0.500 m
Ductility ratio $\mu = 5$
Maximum allowable time =0.15 s
Where applicable, drag coefficients: roof = −0.4, wall = +0.9
Yield strength of steel reinforcement =410 or 425 N/mm^2
f_{cu} (dynamic) = 1.25 f_{cu} (static)
P_y (dynamic) = 1.10 P_y (static)
Main reinforcement $\nless$ 0.25% bd (b = width of section, d = effective depth of section)
Secondary reinforcement $\nless$ 0.1 5 bd
The ultimate shear stress $\ngtr$ 0.04 f_{cu}
The dynamic shear stress for mild steel $\ngtr$ 1.172 N/mm^2
Bolts should be black bolts to BS 4190
Allowable dynamic stresses:
Bolts: tension 275 N/mm^2
shear 170 N/mm^2
bearing 410
Welds: tension/compression 274 N/m^2
Shear 170 N/mm^2

rapidly to that produced by the blast wave without reflection, plus an added drag force due to the wind (dynamic) pressure. At the same time the air pressure wave bends or ‘diffracts’ around the structure so that the structure is eventually engulfed by the blast (Fig. 9.7) and approximately the same pressure is exerted on the side walls and the roof.

The front wall, however, is still subject to wind pressure, although the back wall is shielded from it.

The pressure differential between the front and back faces will have its maximum value when the blast wave has not yet completely surrounded the structure. In this case, such a pressure differential will produce a lateral (or translational) force, tending to cause the structure to deflect and thus move bodily, usually in the same direction as the blast wave. This force is known as the ‘diffraction loading’ because it operates while the blast wave is being diffracted around the structure. The extent and nature of the actual motion will depend upon the size, shape and weight of the structure, and how firmly it is attached to the ground.

When the blast wave has engulfed the structure, the pressure differential is small, and the loading is almost entirely due to the drag pressure exerted on the front face.

Fig. 9.6 Megaton weapon curves (US Department of Defence and Atomic Energy Commission). For all other weapon yields Table 9.1 should be used. For example

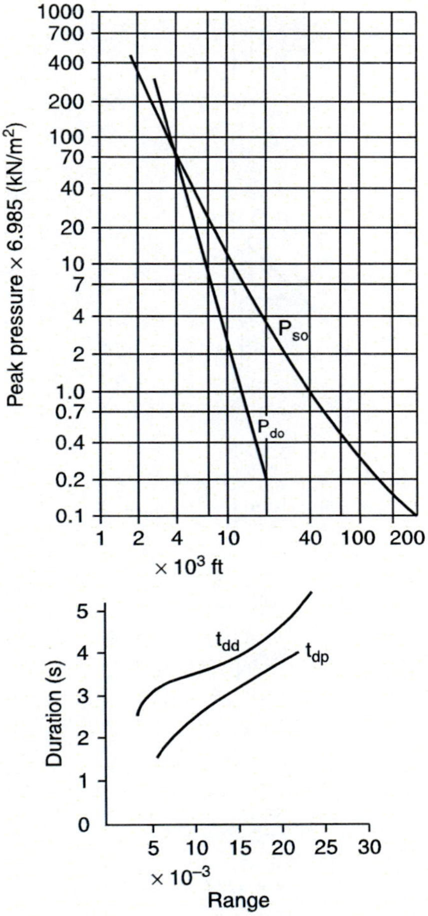

The actual pressures on all faces of the structure are in excess of the ambient atmospheric pressure and will remain so, although decreasing steadily, until the positive phase of the blast wave has ended, thus the diffraction loading on a structure without openings is eventually replaced by an inwardly directed pressure.

The damage caused during the diffraction stage will be determined by the magnitude of the loading and by its duration. The loading is related to the peak overpressure in the blast wave and this consequently is an important factor. If the structure under consideration has no openings, as has been assumed for this discussion, the duration of the diffraction loading will be very roughly the time required by the wave front to move from the front to the back of the building, although wind loading will continue for a longer period.

If the building exposed to the blast wave has openings, or if it has Windows, panels, light sidings or doors which fail in a very short space of time, there will

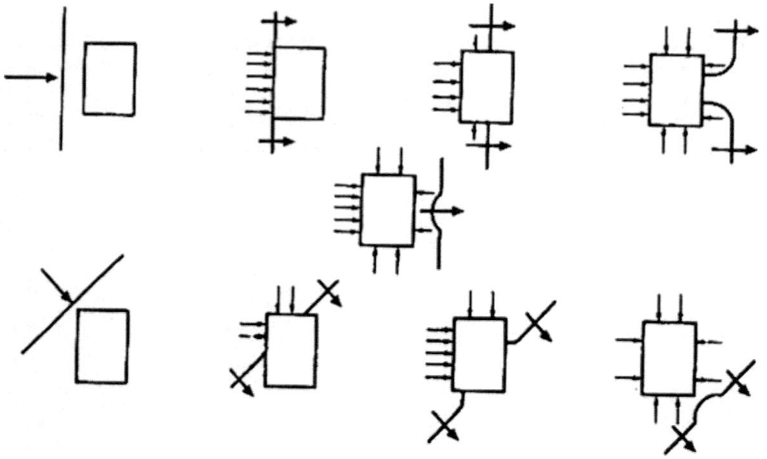

Fig. 9.7 Stages in diffraction (plan view) without openings (Courtesy of US Department of Defence and Atomic Energy Commission)

be a rapid equalisation of pressure between the inside and the outside of the structure. This will tend to reduce the pressure differential while diffraction is taking place. The diffraction loading on the structure as a whole will thus be decreased, although the loading on interior walls and partitions will be greater than for essentially closed structures. Furthermore, if the building has many Openings after the diffraction stage, the subsequent squeezing (crushing) action, due to the pressure being higher outside than inside, will not occur.

9.2.7.2 Loading Due to Drag

During the whole of the overpressure positive phase, a structure will be subjected to the dynamic pressure loading or drag loading caused by transient winds behind the blast wave front. Like the diffraction loading, the drag loading, especially in the Mach region, is equivalent to a lateral (air translational) force acting upon the structure.

9.2.7.3 Loading Versus Structural Characteristics

There are basically two types of building when considering blast wave loading:

(1) Diffraction type
(2) Drag type

As these names imply, in a nuclear explosion, the former would be affected mainly by diffraction loading and the latter by drag loading. While it is true that some structures will respond mainly to diffraction forces and others mainly to drag forces, all such buildings will respond to both types of loading.

A diffraction type building is one that is primarily sensitive to the peak overpressure in the shock wave, e.g. reinforced concrete buildings with small window area and large wall-bearing structures.

When the pressures on different areas of a structure are quickly equalised, because of its small size, the characteristics of the structure or the rapid formation of numerous openings by the action of blast, the diffraction forces operate for a very short time. The response of the structure is then mainly due to the dynamic pressures (or drag forces) of the blast wind, e.g. for telephone poles, radio and television transmitter towers, and tall chimneys.

9.3 Introduction to Codified Design

This section is devoted to the analysis and design of reinforced concrete nuclear shelters. Calculations are given for a particular study using both the British and American codes. Details are also given in this section regarding the Swedish Civil Defence Administration Code.

9.3.1 US Code Ultimate Strength Theory: General Formulae

Figure 9.8 shows cracking, crushing and disengagement cases recommended in successive ACI building codes.

9.3.1.1 General Equation: Ultimate Static Moment Capacity

Cross-Section Type I

(1) The ultimate unit resisting moment M_u of a rectangular section of width b, with tension reinforcement only, is given by

$$M_u = (A_s f_s/b)(d - a/2) \tag{9.1}$$

where A_s = area of tension reinforcement within the width b
f_s = static design stress for reinforcement
d = distance from the extreme compression fibre to the centroid of tension reinforcement
a = depth of equivalent rectangular block = $A_s f_s/0.85bf_e'$
b = width of compression face
f_e' = static ultimate compressive strength of concrete.

The reinforcement ratio p is defined as

$$p = A_s/bd \tag{9.2}$$

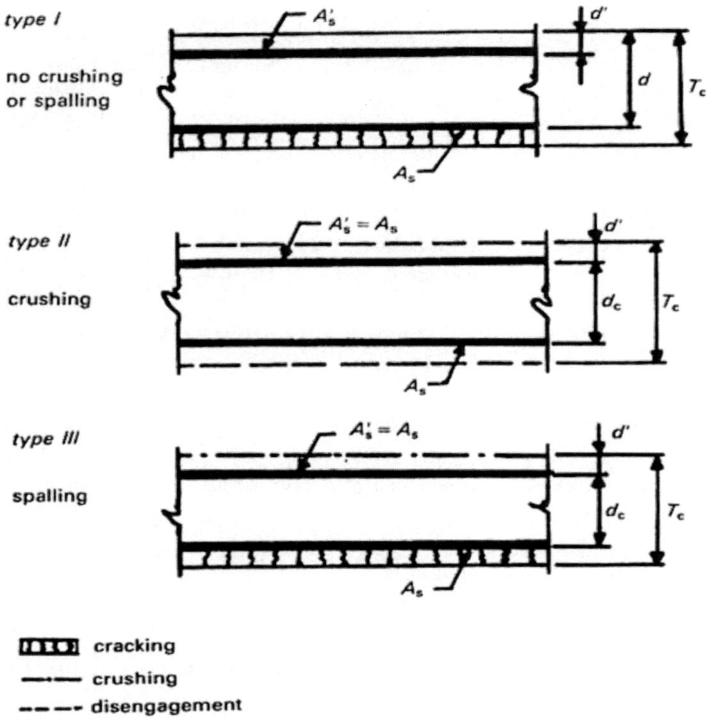

Fig. 9.8 Reinforced concrete cross-section

(2) To ensure against sudden compression failures, p must not exceed 0.75 of the ratio P_b, which produces balanced conditions at ultimate strength and is given by

$$P_b = (0.85K_1\, f_o'/f_s)[87\,000/(87\,000 + f_s)] \tag{9.3}$$

where $K_1 = 0.85$ for ft up to 4000 psi and is reduced by 0.05 for each 1000 psi in excess of 4000 psi.

(3) For a rectangular section of width b with compression reinforcement, the ultimate unit resisting moment is

$$M_u = \left[\left(A_s - A_s'\right)f_s/b\right](d - a/2) + \left(A_s' f_s/b\right)\left(d - d'\right) \tag{9.4}$$

where A_s' = area of compression reinforcement with in the width b,

d' = distance from the extreme compression fibre to the centroid of compression reinforcement,

Table 9.6 Reinforcement for one-and two-way elements

Pressure design range	Reinforcement	Two-way elements	One-way elements
Intermediate	Main	$A_s = 0.0025bd$	$A_s = 0.0025bd$
and low	Other	$A_s = 0.0018bd$	$A_s + A_s^{'} = 0.0020bT_c$
High	Main	$A_s = A_s^{'} = 0.0025bd_c$	$A_s = A_s^{'} = 0.0025bd_c$
	Other	$A_s = A_s^{'} = 0.0018bd_c^{*}$	$A_s = A_s^{'} = 0.0018bd_c^{*}$

*But not less than $A_s/4$ used in the main direction (see Fig. 9.9 for coefficients)

a = depth of the equivalent rectangular stress block:

$$\left(A_s - A_s^{'}\right)f_s/0.85bf_c^{'} \qquad (9.4a)$$

The minimum area flexural reinforcement is given in Table 9.6.

9.3.1.2 Ultimate Static Shear Capacity

Diagonal Tension

(1) The ultimate shear stress υ_u as a measure of diagonal tension, is computed for type I sections from

$$\upsilon_u = V_u/bd \qquad (9.5)$$

and for type II and III sections from

$$\upsilon_u = V_u/bd_c \qquad (9.6)$$

where V_u is the total shear on a width b at the section a distance d (type I) or d_c (type II and III) from the face of the support. The shear at sections between the face of the support and the section d or d_c therefor need not be considered critical.

(2) The shear stress permitted on an unreinforced web is limited to

$$\upsilon_c = \phi\left[1.9\sqrt{f_c^{'}} + 2500p\right] \leq 2.28\phi\sqrt{f_c^{'}} \qquad (9.7)$$

where ϕ is the capacity reduction factor and is equal to 0.85 for all sections.

(3) When the ultimate shear capacity $\upsilon_u > \upsilon_c$ shear reinforcement must be provided. When stirrups are used, they should be provided for a distance d beyond the point theoretically required, and between the face of the support and the cross-section at a distance d. The required area for stirrups for type I cross-sections is calculated using

$$A_v = [(\upsilon_u - \upsilon_c)b_s s_s]/[\phi f_s(\sin\alpha + \cos\alpha)] \qquad (9.8)$$

while for cross-sections conforming to types I, II and III, the required area of lacing reinforcement is calculated and shown in Fig. 9.10.

$$A_v = [(\upsilon_u - \upsilon_c)b_l s_l]/[\phi f_s(\sin\alpha + \cos\alpha)] \qquad (9.9)$$

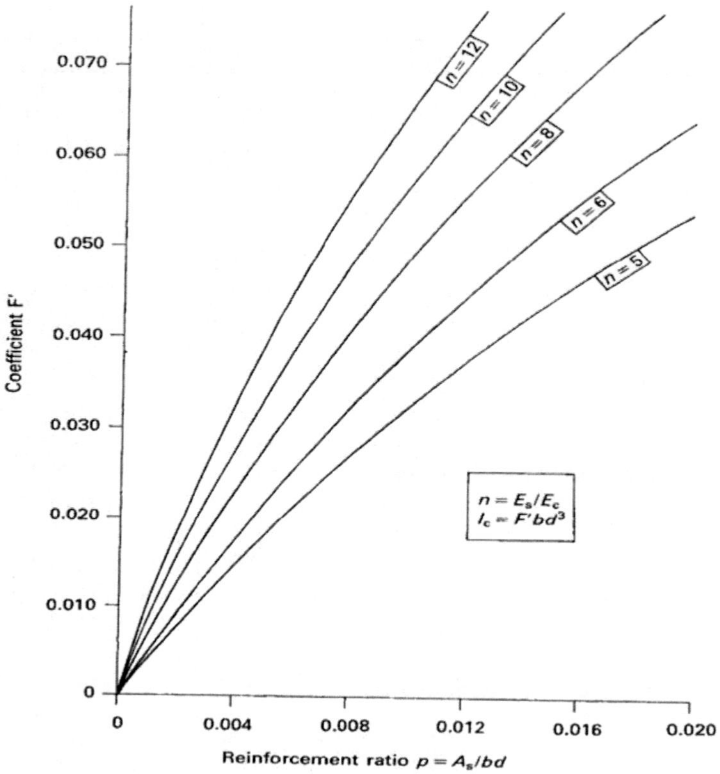

Fig. 9.9 Coefficient for the moments of inertia of cracked sections with tension reinforcement only (courtesy of ACI)

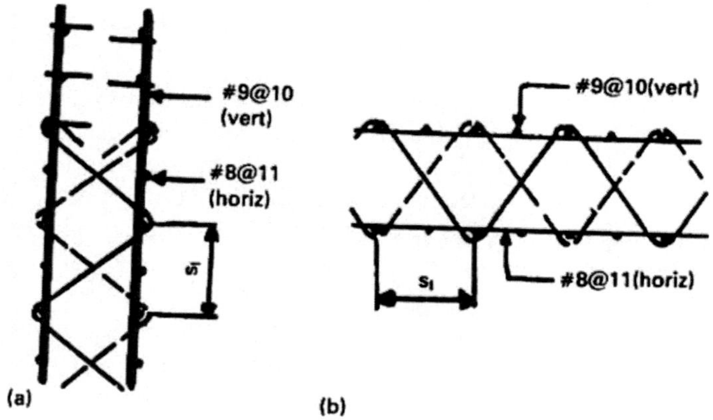

Fig. 9.10 Designs of lacings: (**a**) vertical and (**b**) horizontal

where A_{υ} = total area of stirrups or lacing reinforcement in tension within a width b_s, b_l and distance s_s or s_l

$(\upsilon_u - \upsilon_c)$ excess shear stress

b_s = width of concrete strip in which the diagonal tension stresses are resisted by stirrups of area A_v

b_l = width of concrete strip in which the diagonal tension stresses are resisted by lacing of area

s_s = spacing of stirrups in the direction parallel to the longitudinal reinforcement

s_l = spacing of lacing in the direction parallel to the longitudinal reinforcement

α = angle formed by the plane of the stirrups or lacing and the plane of the longitudinal reinforcement

The excess shear stress $\upsilon_u - \upsilon_c$ is as follows.

Limits	Excess shear stress $\upsilon_u - \upsilon_c$ stirrups	Lacing
$\upsilon_u \leq \upsilon_c$	0	υ_c
$\upsilon_c < \upsilon_u \leq 2\upsilon_c$	$\upsilon_u - \upsilon_c$	υ_c
$\upsilon_u > 2\upsilon_c$	$\upsilon_u - \upsilon_c$	$\upsilon_u - \upsilon_c$

The ultimate shear stress υ_u must not exceed $10\phi\sqrt{f_c'}$ in sections using stirrups. In sections using lacing there is no restriction on υ_u because of the continuity provided by this type of shear reinforcement.

Wherever stirrups are required $(\upsilon_u > \upsilon_s)$, the area A_v should not be less than $0.0015bs_s$ and for type III rectangular sections of width b:

$$M_u = A_s f_s d_c / b \tag{9.10}$$

where A_s = area of tension or compression reinforcement within the width b;
d_c = distance between the centroids of the compression and the tension reinforcement

The reinforcement ratios p and p' are given by

$$p = p' = A_s / bd \tag{9.11}$$

The reinforcement ratio p' is given by

$$p' = A_s' / bd \tag{9.12}$$

Equation (9.11) is valid only when the compression steel reaches the value f_s at ultimate stress, and this condition is satisfied when

$$p - p' \geq 0.85K_1 \frac{f_c' d'}{f_s d}\left(\frac{87000}{87000 - f_s}\right) \quad (9.13)$$

Cross-Section Types II and III.

(1) The ultimate unit resisting moment of type II.

9.3.1.3 Modulus of Elasticity

Concrete.

The modulus of elasticity of concrete, E_c, is given by

$$E_c = w^{15} 33\sqrt{f_c'}\text{psi} \quad (9.14)$$

The value of w, the unit weight of concrete, lies between 90 and 155 lb/ft^2.

Reinforcing Steel.

The modulus of elasticity of reinforcing steel, E_s, is

$$E_s = 30 \times 10^6\text{psi} \quad (9.15)$$

Modular Ratio.

The modular ratio, n, is given by

$$n = E_s/E_c \quad (9.16)$$

and may be taken as the nearest whole number.

9.3.1.4 Moment of Inertia

The average moment of inertia, 'I_a', to be used in calculating the deflection is

$$l_a = (l_g + l_c)/2 \quad (9.17)$$

where 'I_g' is the moment of inertia of the gross concrete cross-section of width b about its centroid (neglecting steel areas) and is equal to

$$l_g = bT_c^3/12 \quad (9.18)$$

and I_c is the moment of inertia of the cracked concrete section of width b considering the compression concrete area and steel areas transformed into equivalent concrete areas and computed about the centroid of the transformed section. I_c is calculated from

$$I_c = Fbd^3 \tag{9.19}$$

The coefficient F varies as the modular ratio n and the amount of reinforcement used. For sections with tension reinforcement only, F is given in Fig. 9.9.

9.4 Design of a Concrete Nuclear Shelter Against Explosion and Other Loads Based on the Home Office Manual

Figure 9.11 shows a typical layout of a domestic nuclear shelter for a family of six.

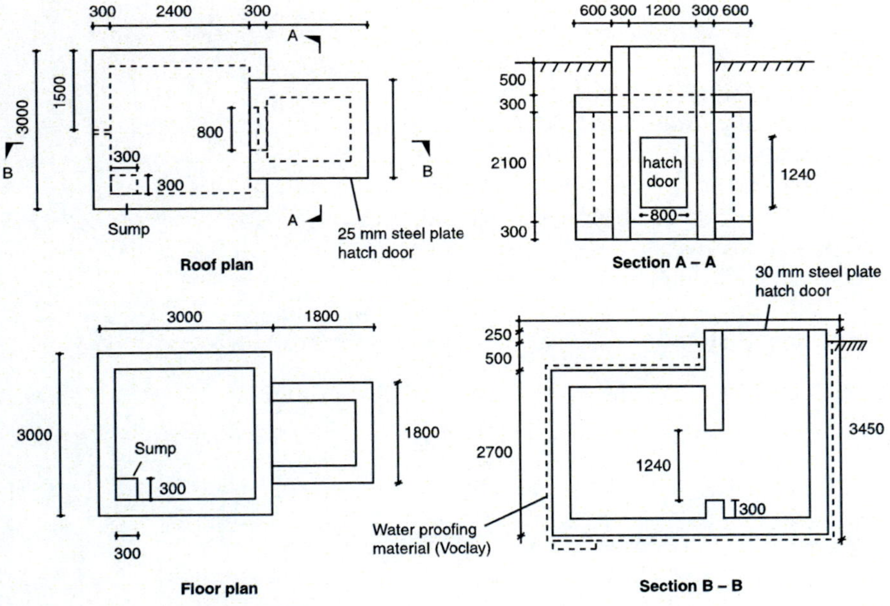

Fig. 9.11 Domestic nuclear shelter: general arrangement

9.4.1 Basic Data (Home Office Code)

For a 1 Mt ground burst at a distance of 1.6 km from ground zero:

Ductility ratio, μ : 5
Main reinforcement $\not> 0.25\% bd$

Secondary reinforcement $\not> 0.15\% bd$
Ultimate shear stress $\not> 0.04 f_{cu}$
Dynamic shear stress (mild steel) $\not> 172\,\text{N /mm}^2$
Protective factor: 4000
Concrete f_{cu} (static): 30 N/mm^2 (grade 30)
Concrete f_{cu} (dynamic): 1.5f f_{cu} = 37.5 N/mm^2
Reinforcement f_{cu} (static): 420 N/mm^2
Reinforcement f_{cu} (dynamic): 1.10= 462 N/mm^2
Young's modulus, E_c: 20 GN/m^2
Young's modulus, E_s: 200 GN/m^2
Clear span: 3 m
Slab thickness: 300 mm (with minimum cover 50)
Blast load: 0.17 N/mm^2, $F_1(t) = P_{do}$

9.4.2 Additional Data for Designs Based on US Codes

Dynamic increase factors (DIF)

Concrete		
compression	1.25	
diagonal tension	1.00	
direct shear	1.10	
Reinforcement: bending		1.10
shear		1.00

Dynamic stresses:
concrete f_c' (cylindrical strength) $= 0.87 f f_{cu}$
$= 3000\ \text{lb/in}^2$ (psi)
concrete f_y (static) $= 60000\ \text{lb/in}^2$ (psi)

$$R_m = r_u = \left(\frac{1}{1-\frac{1}{2\mu}}\right) F_1(t) = 1.1 F_1(t) = 0.187\,\text{N/mm}^2 \tag{9.20}$$

Deadload of concrete plus soil $= 0.014\ \text{N/mm}^2$

$$r_u = 0.187 + 0.014 = 0.201\,\text{N/mm}^2$$

For a two-way slab

$$M_u = r_u - L^2/16 = 0.201(3000)^2/16 = 11,3062.5\,N\,mm/mm$$

300 mm thick slab
T16—200 bars; $A_s = 1005\frac{mm^2}{m} : d = 300 - 50 - 8 = 242$

$$z = d - (0.84 f_{yd} A_s / f_{cu}(\text{dyn})$$
$$= 242 - (0.84(462)(1005)10^{-3}/37.5)$$
$$= 231.58\,\text{mm}$$

Note: Later on, based on finite-element analysis, the T20-200 bars adopted were checked

Area of the roof $= 9\ \text{m}^2 = A_t; \sqrt{A_t} = 3m$

$$H - x = 2.7 - 0.3 = 24\ or\ 3.4 - 0.3 = 3.1$$
$$\sqrt{(A\_t)}(H - X) = 1.25\ \text{and}\ 0.97$$

Weight of overhead material $= 1340\ \text{kg/m}^2$
$R = 0.025\%$ (roof contribution)

$$\text{PF} = 100/(R + G_{\text{T}})$$
$$= 100/(0.025 + 0)$$
$$= 4000\ (\text{safe})$$

where G_{T} is the percentage wall contribution, ignored in the worst case. Figure 9.12 gives structural details of the reinforced concrete shelter.

Steel blast doors
Clear opening 800 mm × 1200 mm.

$F_1(\text{t}) = P_{\text{do}} = 2.3P_{\text{so}} = 2.3(0.17) = 0.39\ \text{N/mm}^2$
$r_{\text{u}} = 1.1\ F_1(\text{t}) = 0.43\ \text{N/mm}^2$
M_{u} (simply supported) $0.43(800)^2/8 = 34{,}400\ \text{N/mm}$

20 mm thick steel door

$$z = bd^2/4 = 1(20^2)/4 = 100\ \text{mm}^2$$
$$\text{Also}\ z_{\text{p}} = M_{\text{u}}/1.1(265) = 118\ \text{mm}^2$$

Calculated thickness of steel doors $= (118/100)20 = 23.6\ 25\ \text{mm}$
A 25 mm thick door was adopted.

The thickness of the glass door may have to be increased for protection against radiation fall-out. One possibility is a steel-concrete sandwich construction. One possible steel door design is given in Fig. 9.13 or $z = 0.95(242) = 229.9\,mm \approx 230\,mm$

Walls: 300 mm thick

Blast load on walls $= p_{do} \times 05 = 0.085\,\text{N/mm}^2$
$r_u = 1.1\ F_1(t) = 1.1(p_{\text{do}}) = 0.0935\ \text{N/mm}^2$
Total (including soil) $= 0.0935 + 0.08 = 0.1735\ \text{N/mm}^2$

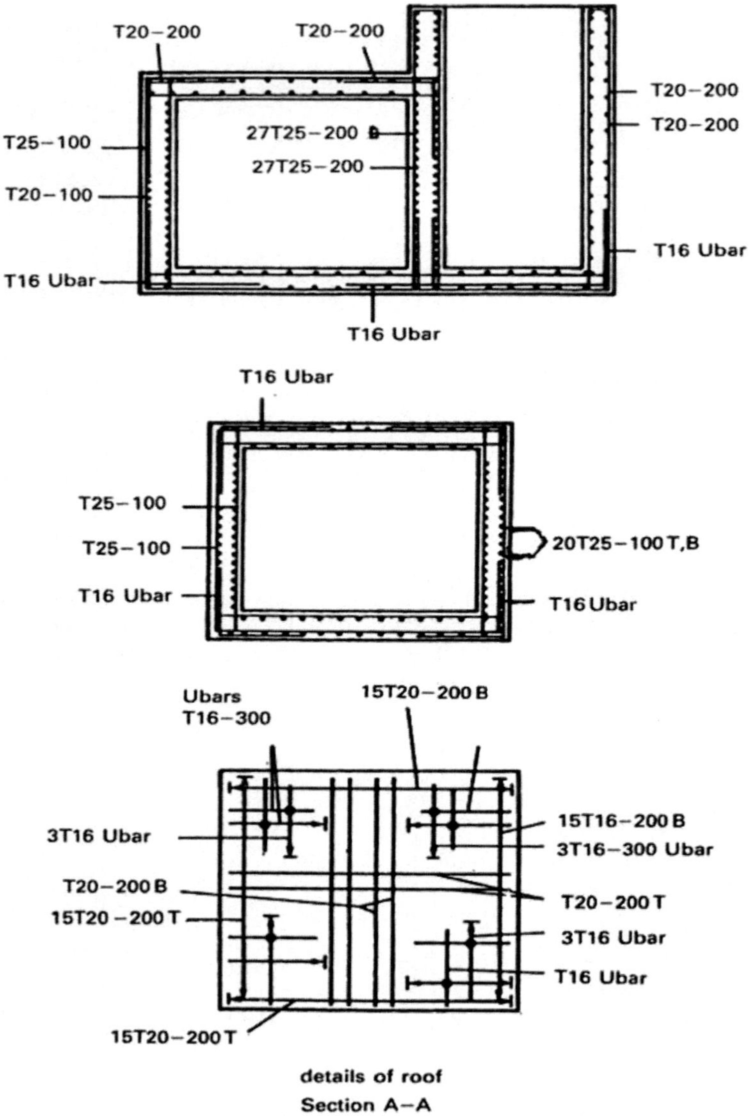

Fig. 9.12 Domestic nuclear shelter (reinforcement concrete): detail

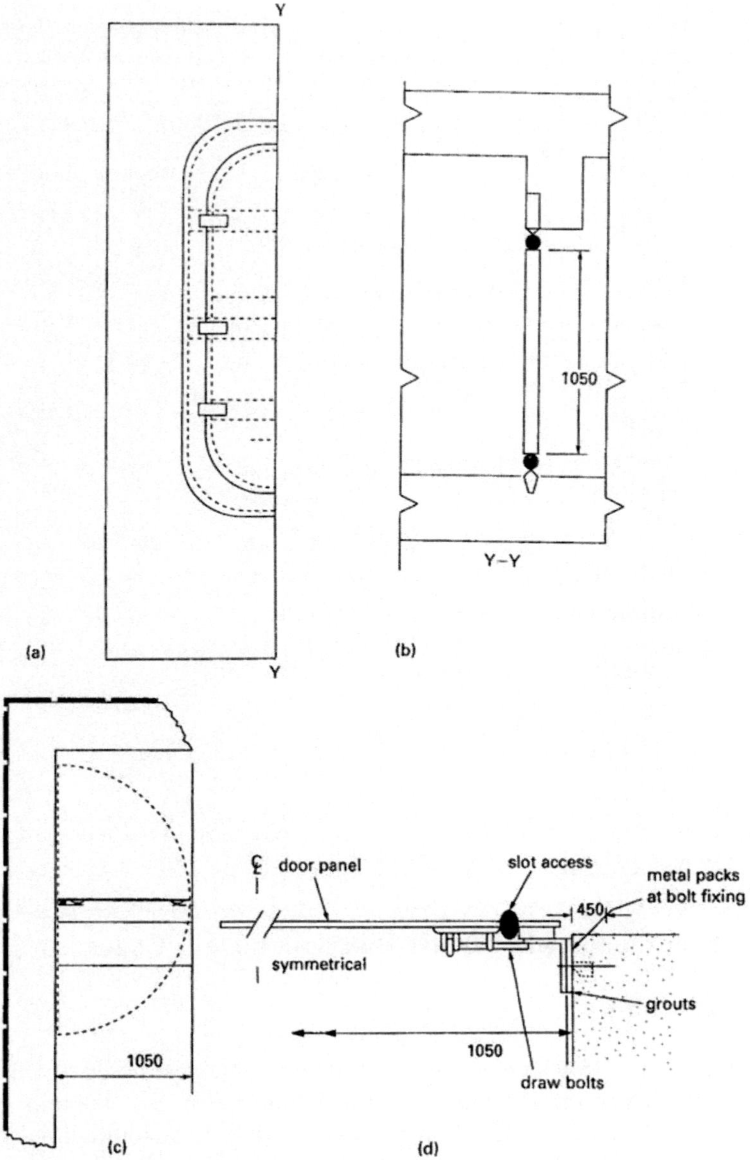

Fig. 9.13 Design of steel blast doors. (**a**) Elevation; (**b**) vertical section; (**c**) door location; (**d**) horizontal section: structural details

Two-way slab

$$M_u = r_u L^2/16 = 0.1735(2700)^2/16 = 79\,050.941$$

Both walls (2700 mm and 3400 mm)

$$\left.\right\}\quad \text{also } M_u = \left[(3400)^2/(2700)^2\right](79\,050.94)$$

$$= 125\,353.75\ \text{N mm/mm (adopted)}$$

$$M_u = 125\,353.75 = A_s(230)(462)$$

$$A_s = 1.18\ \text{mm}^2/\text{mm} = 1180\,\text{mm}^2/m$$

adopted T20–200 (in some critical areas T20–100 and T25–100)
Shaft wall bars: T12–200 links T16-300 U-bars
Minimum steel:

Main → $0.25\% \times 1 \times 242 = 0.605\ \text{mm}^2/\text{mm}\ (605\,\text{mm}^2/\text{m})$
$1005\,\text{mm}^2/\text{m} > 605(\text{T}16-200)$ adopted

Secondary → $0.15\% \times 1 \times 242 = 0.363\,\text{mm}^2/\text{mm}\ (363\,\text{mm}^2/\text{m})$
(T16–200 or 300) adopted

Shear: allowable shear $= 0.04 f_{cu} = 1.2\ \text{N/mm}^2$

$$\begin{aligned}\text{shear} &= r_u[(L/2 - d)/d]\\ &= (2700/2 - 242)/2700\\ &= 0.41 < 1.2\,N/\text{mm}^2(\text{safe})\end{aligned}$$

$$\begin{aligned}\text{or} &= \left(\tfrac{3400}{2} - 242\right)/3400\\ &= 0.43 < 1.2\,\text{N/mm}^2\end{aligned}$$

Protective factor (PF) in the middle of the shelter and at 0.25–0.30 m above the floor level.

9.5 Design of a Nuclear Shelter Based on the US Codes

9.5.1 Introduction

Many codes in the USA have empirical equations which are based on imperial units. The reader is given the conversions in SI units. However, the bulk of the calculations given here are based on imperial units (conversion factors shown below).

Conversion factors

$1\,\text{ft} = 0.3048\,\text{m};\ 1\,\text{lb/ft}^2 = 47.88\text{N/mm}^2$
$1\,\text{lb}f = 4.448\,\text{N};\ 1\,\text{lb/ft}^2 = 16.02\ \text{kg/mm}^2$
$1\,\text{lbin} = 0.113\ \text{Nm};\ 1\,\text{kg} = 9.806\,\text{N};$
$1\,\text{lb/in}^2 = 6895\ \text{N/mm}^2 1\,\text{in} = 25.4\,\text{mm}$

Dynamic stresses
Concrete:
Comp — 1.25(3000) = 3750 psi
Diagonal tension — 1.00(3000) = 3000 psi
Direct shear — 1.10(0.18) (3000) = 600 psi

Reinforcement:
Bending — 1.10(60,000) = 66,000 psi
Shear — 1.10(60,000) = 60,000 psi
since $f_c' = 3000$ psi and f_y(static) = 60,000 psi.

9.5.2 Wall Design

Figure 9.14 shows a one-way slab fully restrained at the supports.

Wall thickness (T_c) = 300 mm (12 in) (see Fig. 9.14).

The US recommended covers are 0.75 in. and 1.5 in. (37 mm) rather than 50 mm (adopted by the Home Office, in the UK).

For a negative moment, $d = 12 - 1.5 - 0.3125 = 10.1875$ in. (assuming # 5 bars). For a positive moment, $d = 12 - 0.75 0.3125 = 10.935$ in.

$$A_s = 0.0025 \times 12 \times 10.935 = 0.328\, in.^2/\text{ft}$$

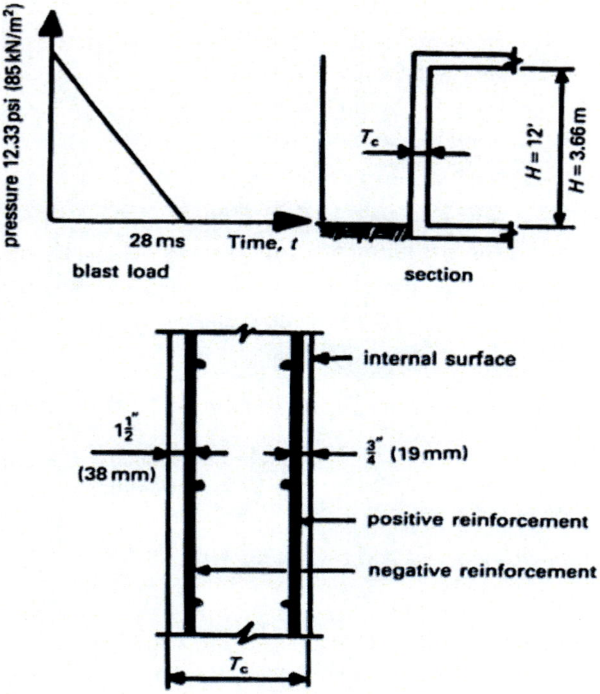

Fig. 9.14 Wall analysis and design

#5 bars at 11 in (275 mm), $A_s = 0.34 > 0.328\,\text{in.}^2$ the wall blast load $= 0.085\ \text{N/mm}^2 = 12.33\,\text{lb/in}^2$. The ultimate moment is given by

$$M_u = (A_s\, f_{yd}/b)(d - a/2)$$

where $a = A_s\, f_{yd}/0.85\, bft(\text{dyn}) = 0.586\,\text{in.}$

$$b = 12\,\text{in.}$$

$$M_u(\text{positive}) = M'_p = 19900\,\text{in.} - \text{lb/in.}$$
$$M_u(\text{negative}) = M'_N = 18500\,\text{in.} - \text{lb/in.}$$

$$\begin{aligned} \text{E}_c \text{ for concrete} &= D^{1.5}\{33\sqrt{f'_c}\} \\ &= (150\,\text{lb/in}^3)^{1.5} \times 33(3000)^2 \\ &= 3.32 \times 106 \text{ psi} \end{aligned}$$

$$\rho = D = \text{density of concrete} = 150\ \text{lb/in.}^3 (23.6\ \text{kN/m}^3)$$

$$E_s \text{ for steel} = 30 \times 10^6 \text{ psi } (200\ \text{GN/m}^2)$$
$$n = E_s/E_c = 9.03$$

Average moment of inertia for a 1 in. strip.

l_g (gross) $= bT_c^3/12 = 144$ in.^{4}
T_c (thickness of the wall) $= d$
d (average) $= 10.5625$ in.
p (average) $= A_s/bd = 0.00268 = \rho_s$
l (cracked section) using Fig. 2.1
$F' = 0.0175$; $I_{\text{cracked}} = I_c = F'bd^3 = 20.6\,\text{in.}^4$
I_a = average moment of inerita $= \frac{I_g + I_c}{2} = 82.3\,\text{in.}^4$
Elastic (K_e) and elasto – plasto (K_{ep}) stiffness
$K_e = (384E_cI_a)/bL^4 = 244\,\text{lb/in.}^3$; $b = 1$
$K_{ep} = 384E_cI_a)/5bL^4 = 48.8\,\text{lb/in.}^3$

Elastic and elasto-plastic deflection

$\delta_e = X_e = r_e/K_e = 10.71/244 = 0.0439\,\text{in.}$
$\delta_{ep} = X_p - X_c = (r_u - r_e)/K_{ep} = 0.084\,\text{in.}$
$X_p = 0.1279\,\text{in.}$

Equivalent elastic deflection and stiffness

$$\begin{aligned} X_E &= X_e + X_p(1 - r_e/r_u) \\ &= 0.0793\,\text{in} \end{aligned}$$
$$K_E = r_u/X_E = 1.86.8\,\text{lb/in}^3$$

Load-mass factors and effective mass
Fig. 9.14 gives

K_{LM}	Range
0.77	elastic
0.78	Elasto-plastic
0.66	Plastic

K_{LM} (elastic and elasto-plastic) = 0.78 (average)
K_{LM} (elastic and plastic) = 0.72 (average)

$$M = \rho T_c/g = 150 \times 1 \times 10^6/32.3(1728)$$
$$= 2700\,\text{lb} - \text{ms}^2/\text{in.}^3$$

$$M_{\text{effetive}} = K_{LM} \times M = 1944\text{lb} - \text{ms}^2/\text{in.}^3$$
Natural period $= T_N = 2\pi\sqrt{(M_e/K_e)} = 20.3\,ms$

Where $g = 32.2\text{ft/s}^2$; $k_E = 186.8$

Response chart parameters
Reference is made to Fig. 9.15

Peak pressure B = 12.33 psi
Peak resistance r = 14.81 psi
The chart $B/r_u = 0.8325 \rightarrow T/T_N = 28/20.3 = 1.38$
$X_m/X_E = 1.50$ as this is < 3 the section is safe

The corresponding $t_m/T_N = 0.50 \rightarrow t_m/t_o = t_m/T$
$$= (t_m/T_N)/(T/T_N) = 0.50/1.38$$
$$= 0.3623$$

This lies within the range $3.0 > t_m/t_o > 0.1$, hence the response is satisfactory.
Diagonal tension at a distance d from the support

$$v_u = r_u[(L/2) - d_e]/d_e = 14.81(72 - 10.1875)/10.1875$$
$$= 89.9\,\text{psi}$$

The allowable shear stress, v_c, is given by

$$v_c = \phi[1.9\sqrt{f_c'} + 2500p] \leq 2.28\phi\sqrt{f_c'}$$

(Where ϕ = 0.85)

v_c = 94.4 psi, as this is > 89.9 psi, OK with no stirrups

Ultimate shear

$$V_s = r_u L/2.0 = 14.81 \times 144/2.0 = 1066\,\text{lb/in}$$

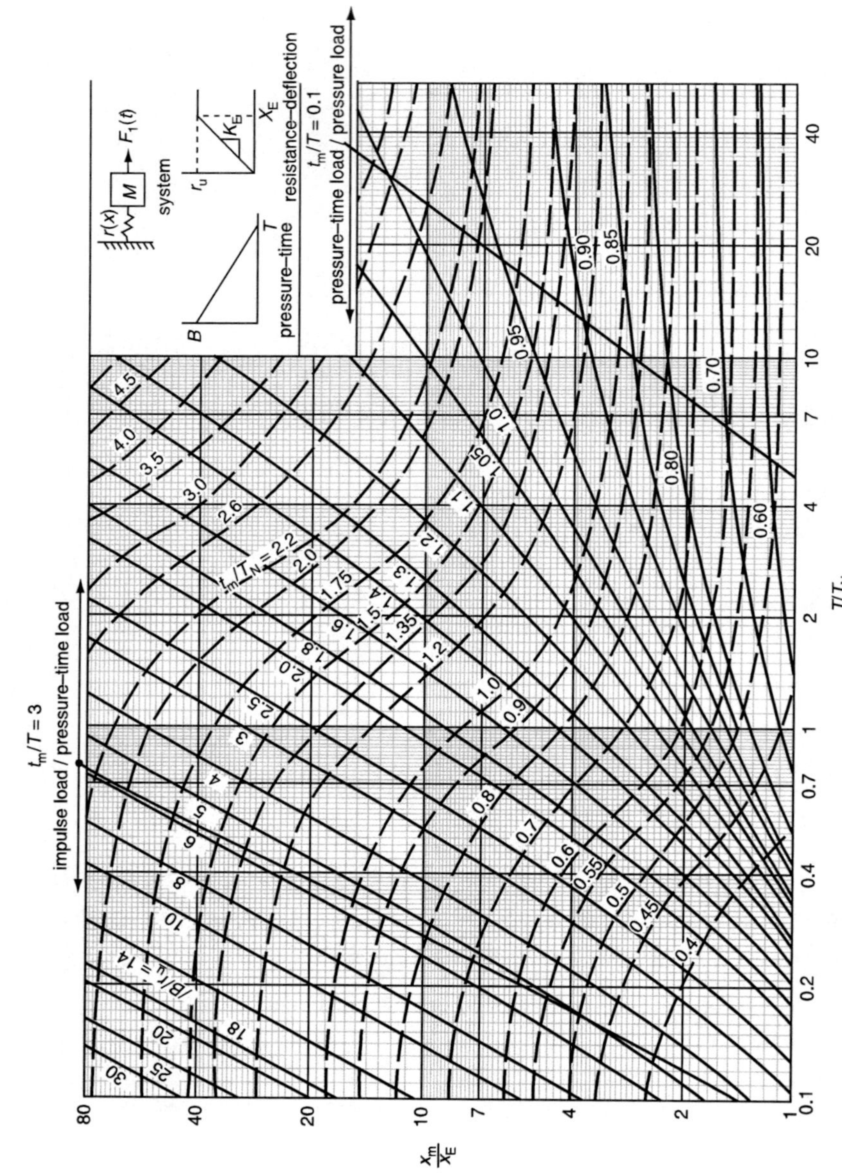

Fig. 9.15 Maximum response of simple spring-mass system (LU.S Corpe of Engineer, Washington DC)

Allowable shear

$$V_d = 0.18 f_c'(dyn)\text{bd} = 6050\text{lb/in} > 1066\,\text{lb/in}$$

Hence the 300 mm (12 in.) wall designed against the same blast load in both codes (British and US) is safe. The roof slab can be checked in the same way as for the gas explosion, described earlier in the text.

9.6 Lacing Bars

Design of lacing bars

Where lacing bars are needed, the following calculations will help in the design of nuclear shelters. The lacings can be in both the vertical and horizontal directions.

Vertical lacing bars

The wall thickness is kept the same. Data: $d_l = 10$ in. $s_l = 22$ in. no of bars = 6; $D_o = 0.75\,in.$

$$d_l = 21 + 1.13 + 2 + 0.75 = 24\,88\,\text{in.}$$

$$Re = 3D_o$$

$$\text{for } S_l/d_l = 0.884$$

$$(2R_e + D_o)/d_l = 7D_o/d_l = 0.211$$

L_l is measured along the centre line of the lacing bar between points *a* and *b*.

$$L_l = \frac{s_l - (2R_l + D_o)\sin\alpha}{\cos\alpha} + \pi\,(2R_l + D_o)\left(\frac{\alpha}{180}\right) \tag{9.21}$$

$$\cos\alpha = \frac{-2\text{B}(1-\text{B}) \pm \sqrt{\{2\text{B}(1-\text{B})^2 - 4(1-\text{B})^2 + \text{A}^2](\text{B}^2 - \text{A}^2)}}{[2(1-\text{B})^2 + \text{A}^2]}$$

$$A = \frac{s_l}{d_l} \text{ and } B = \frac{2R + D_o}{d_l}$$

$$\alpha = 51.5°$$

$$A_v = (v_{uv} - v_c)b_l S_l/\Phi f_s(\sin\alpha + \cos\alpha) = 0.378\,\text{in}^2. \tag{9.22}$$

$$A_{v\min} = 0.0015\; b_l s_l = 0.330\,\text{in}^2.;\ no.\text{ of bars} = 6;\ A_s = 0.44\,\text{in}^2.\,OK$$

Horizontal lacing bars

No. of bars 6; $D_o = 0.75$ in.

$$d_l = 21.0 + 1.13 + 0.7522.8\text{ in.}$$

$$R_{l\min} = 3D_o$$

for $s_l d_l = 20/22.88 = 0.874$

$$(2R_e + D_o)/d_l = 7D_o/d_l = 0.229$$

$$A_v = 0.339\ in^2.$$

$A_{v\,min} = 0.0015\ b_l s_l = 0.0015 \times 11 \times 20 = 0.330\,\text{in}^2$ no of bars $= 6$; $A_s = 0.44\ in^2$.

Still OK.

Additional reinforcement details from the British code are given in Figs. 9.18 and 9.24.

When a ring forced concrete element is subject to a blast load, the element deflects far beyond the stage of well-defined cracking until:

(1) The strain energy of the element is developed sufficiently to balance the kinetic energy created by the applied load when it comes to rest.
(2) Fragmentation of the concrete element results in either its partial or total collapse.

For the development of the available energy of the concrete elements, it is necessary to make changes in the reinforcement layouts and details. Each

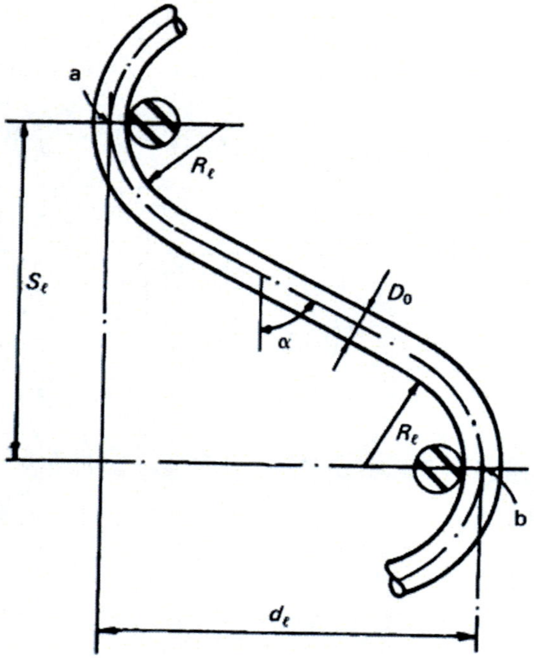

Fig. 9.16 Length of lacing bar

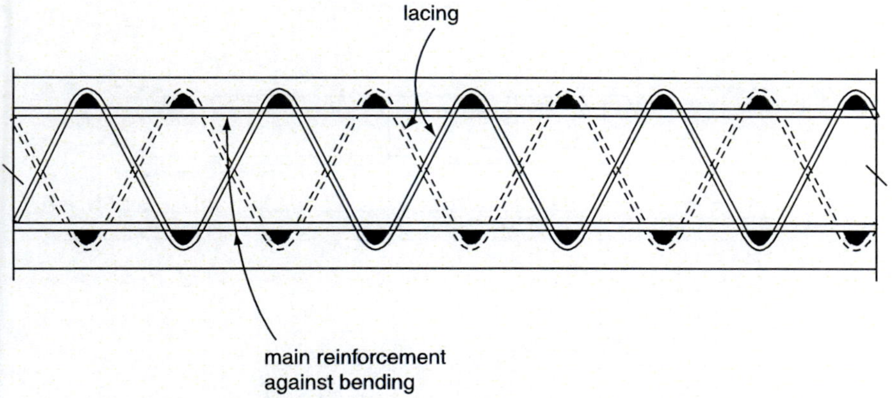

Fig. 9.17 Lacing reinforcement

element is reinforced symmetrically. The reinforcement and the intervening concrete are laced together, as shown in Figs. 9.16 find 9.17 with continuous bent diagonal bars. This system offers forces which will contribute to the integrity of the protective element. Where structural elements are located outside the immediate high blast intensity, they should be designed without lacing. All other types are given in Figs. 9.18, 9.19, 9.20, 9.21 and 9.22.

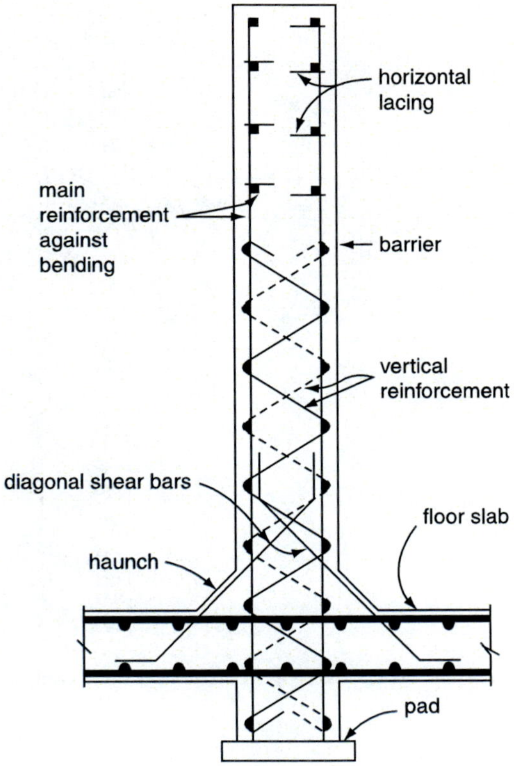

Fig. 9.18 Typical laced wall

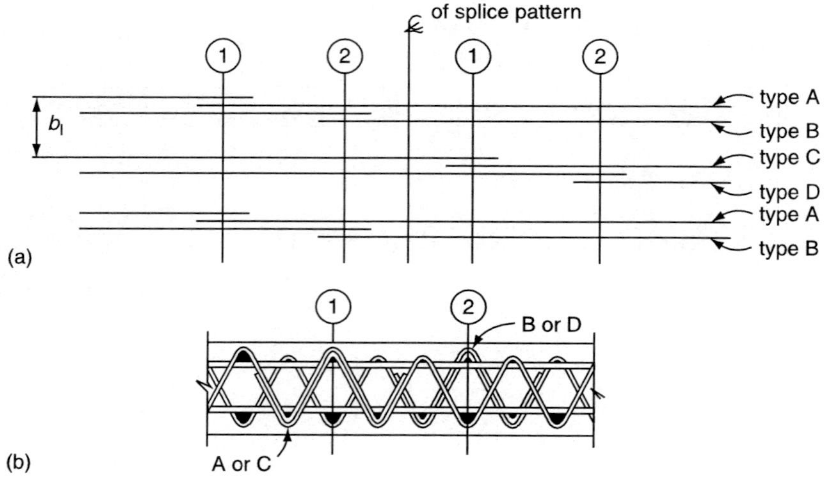

Fig. 9.19 Typical details for splicing of lacing bars: (**a**) splice pattern; (**b**) lacing splice

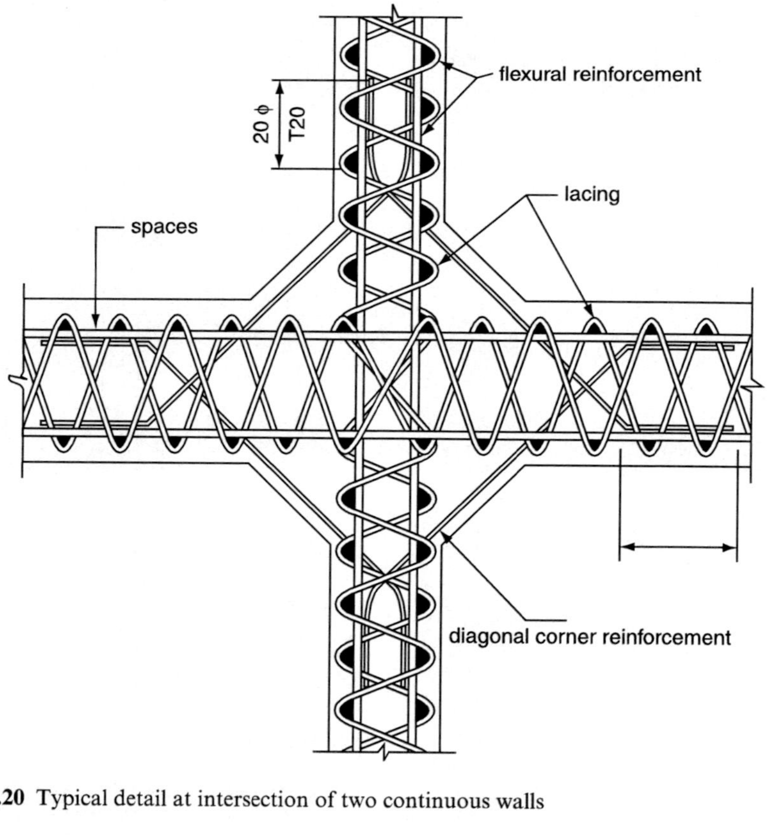

Fig. 9.20 Typical detail at intersection of two continuous walls

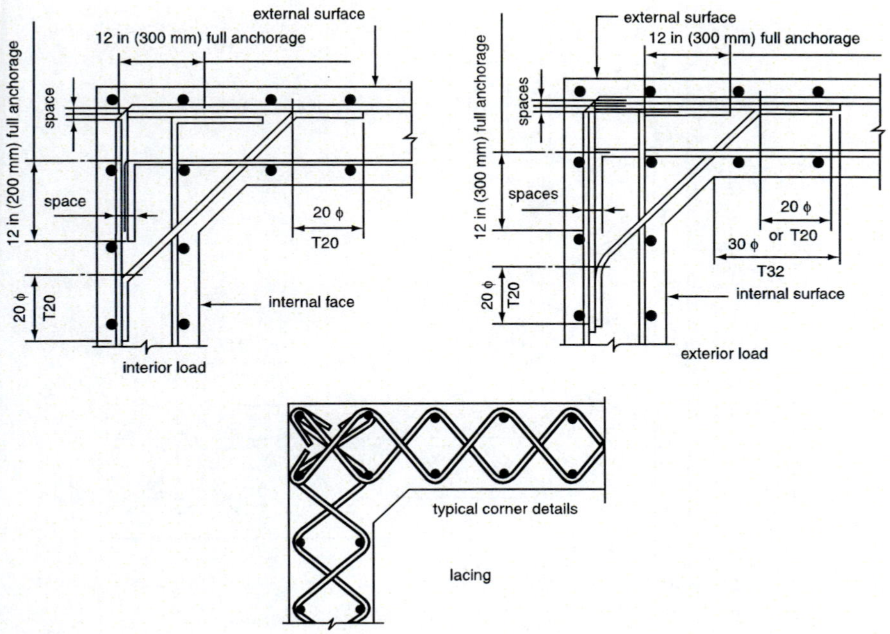

Fig. 9.21 Flextural reinforcement in lacing

9.7 Finite Element Analysis

A 3D isoparametric, finite-element analysis has been carried out by Bangash. Figure 9.25 shows the finite-element mesh scheme for a dynamic model for a nuclear shelter. Figure 9.26 gives the relationships between pressure and time. The results are given in Fig. 9.27.

9.7.1 The Swedish Design and Details

The Swedish code TB78E provides novel details of the nuclear shelter. They are presented here by courtesy of the Civil Defence Administration of Sweden. Figures 9.28 and 9.29 show structural details of a roof slab and sectional details illustrating various reinforcements.

9.8 Damage Classification

Damage to structures or objects above ground can be divided into three categories:

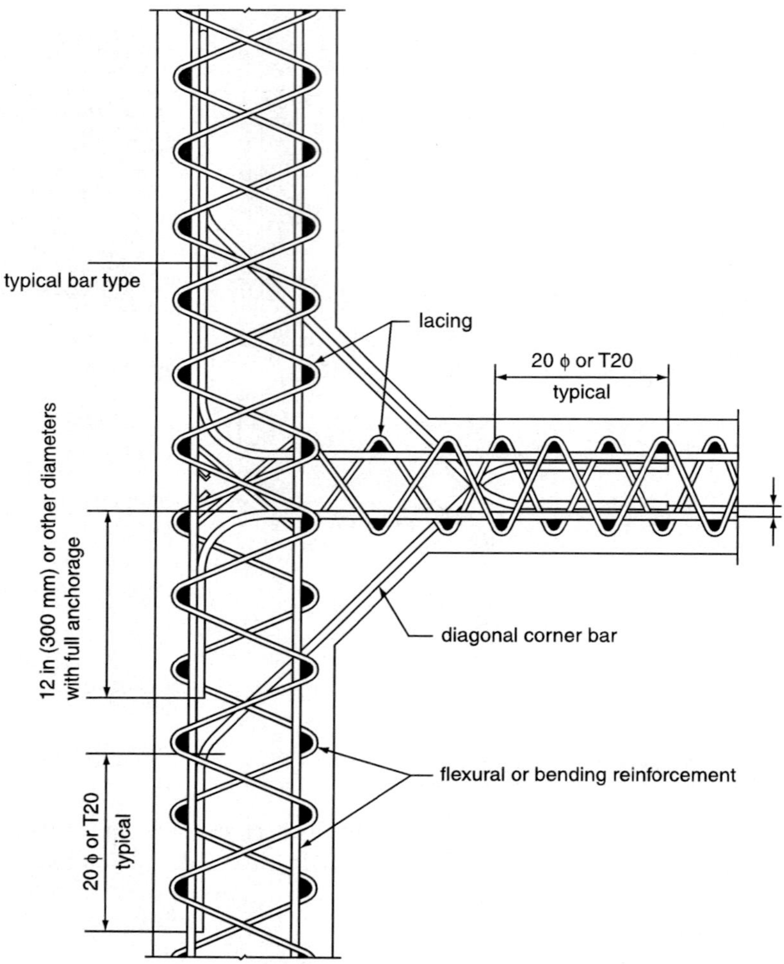

Fig. 9.22 Typical detail at intersection of continuous and discontinuous walls

- *Severe damage*: A degree of damage that precludes further ijse of the structure or object for its intended purpose, without essentially complete reconstruction. For a structure or building, collapse is generally implied.
- *Moderate damage*: A degree of damage to principal members that precludes effective use of the structure or object for its intended purpose, unless major repairs are made.
- *Slight damage*: A degree of damage to buildings resulting in broken windows, slight damage to roofing and cladding, blowing down of light interior partitions and slight cracking of certain walls in buildings.

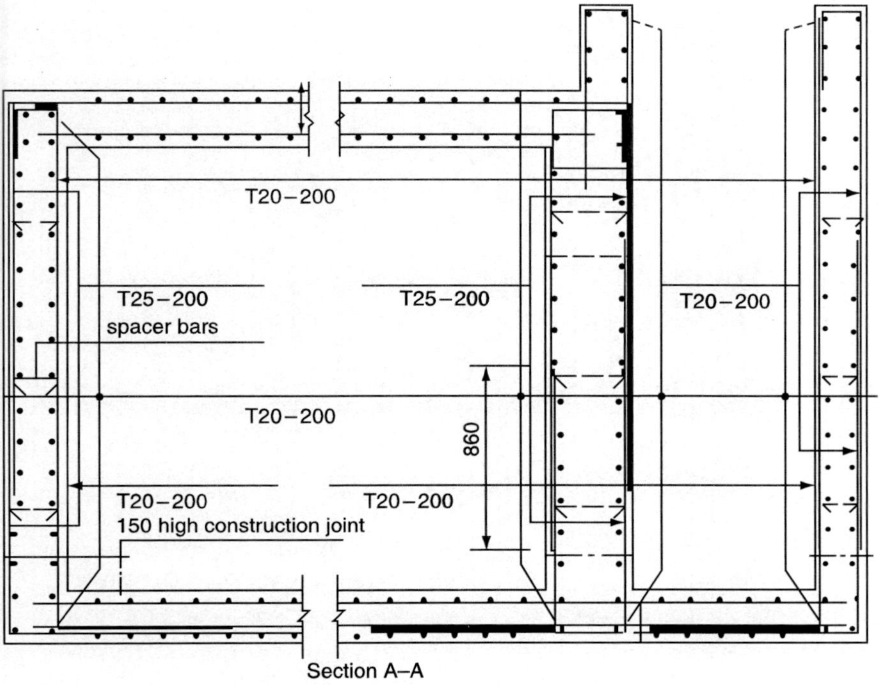

Fig. 9.23 Reinforcement through section

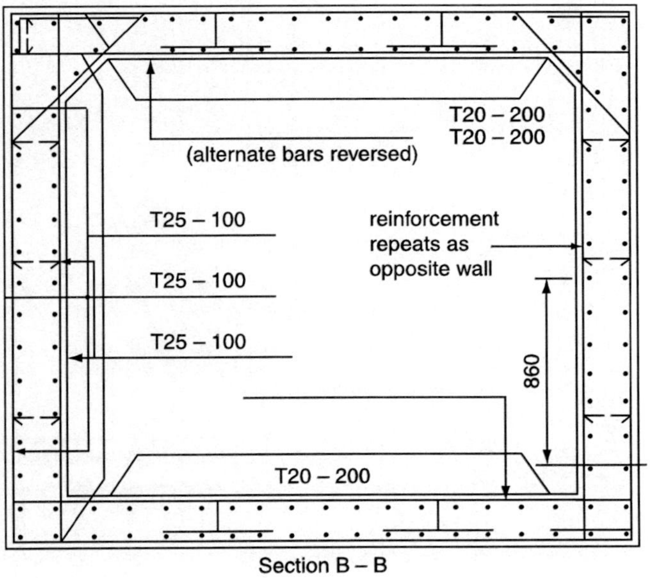

Fig. 9.24 Section B–B through shelter

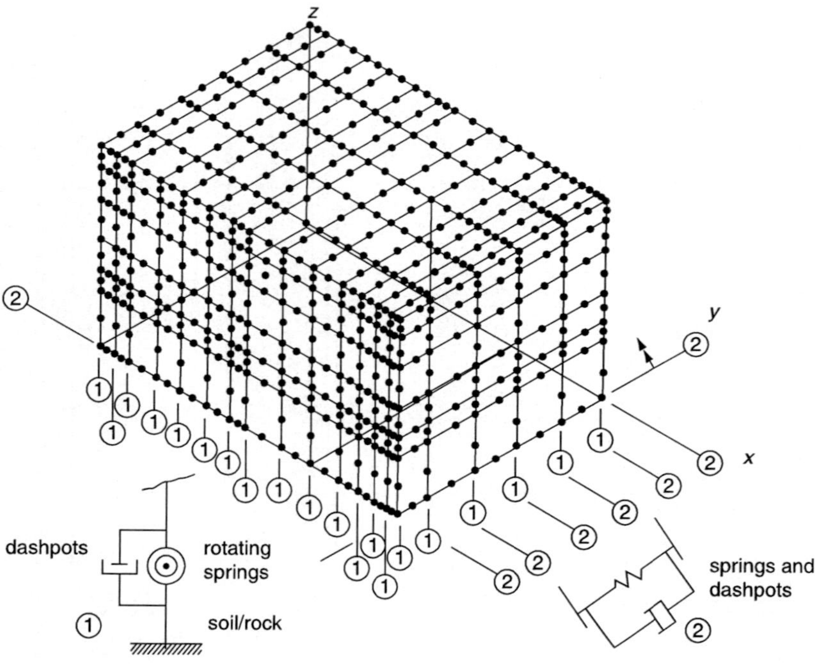

Fig. 9.25 Dynamic model for a nuclear shelter

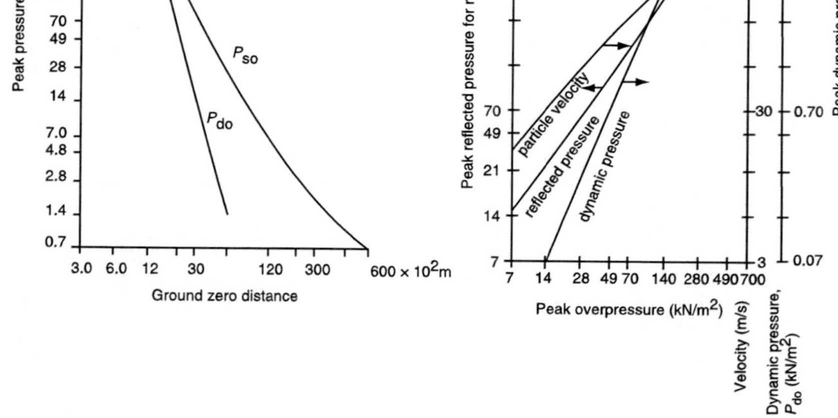

Fig. 9.26 Relationships between pressures and time

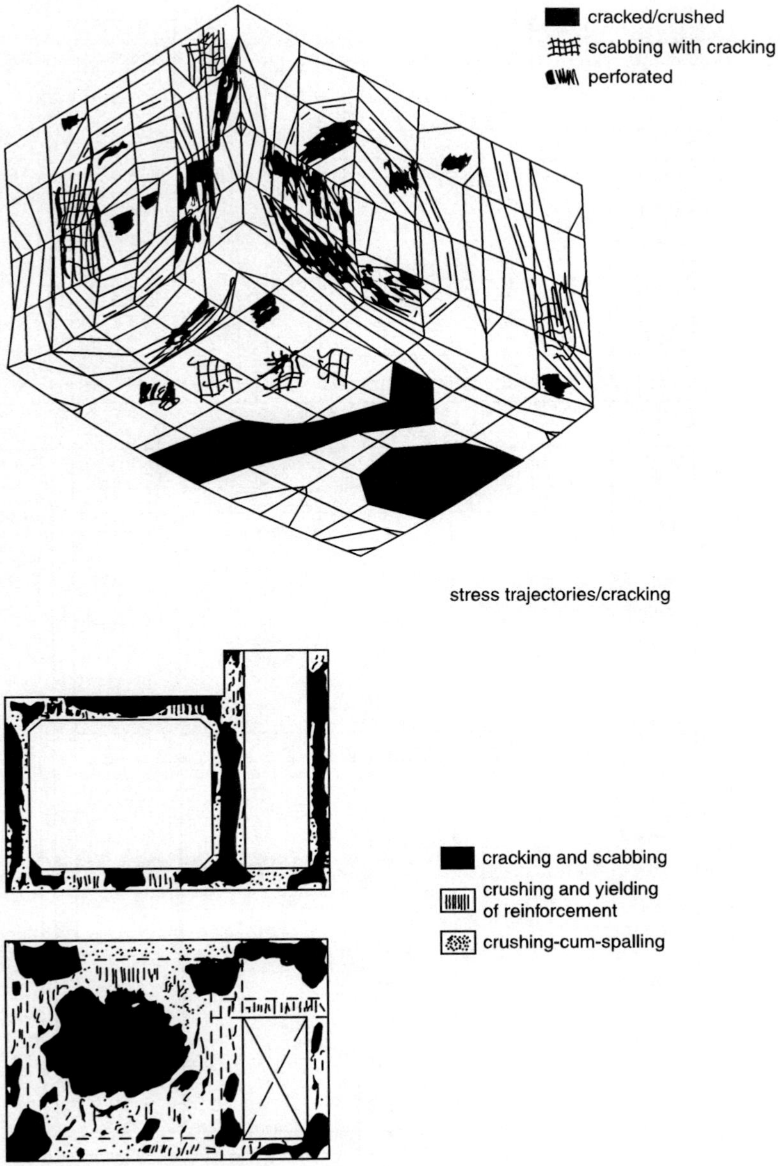

Fig. 9.27 Typical results from the finite-element analysis

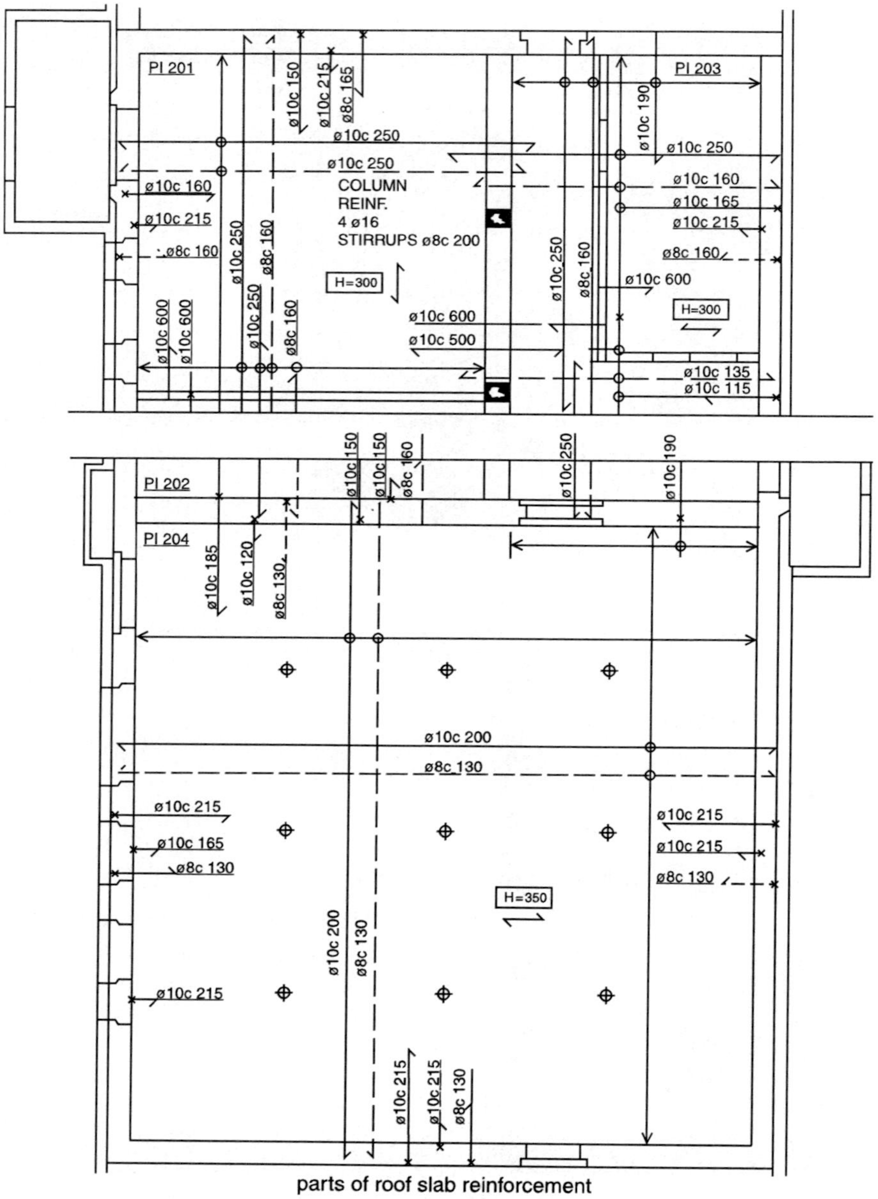

Fig. 9.28 Structural details – I Swedish code TB78E

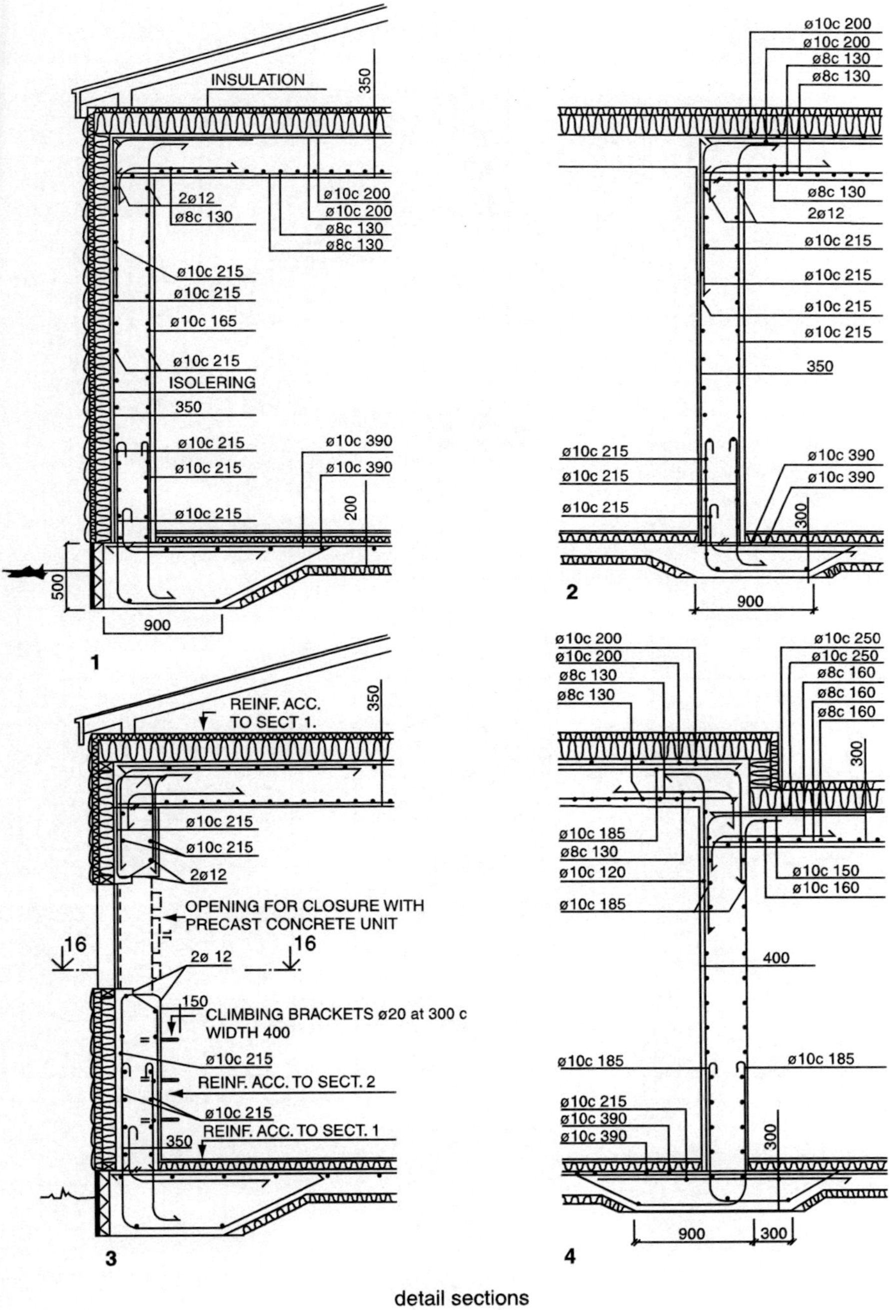

Fig. 9.29 Structural details – II Swedish code TB78E

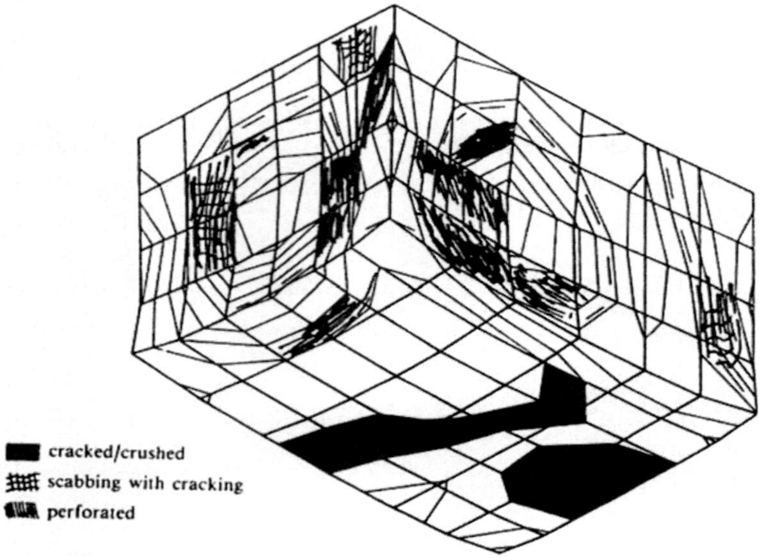

Fig. 9.30a Stress trajectories/cracking

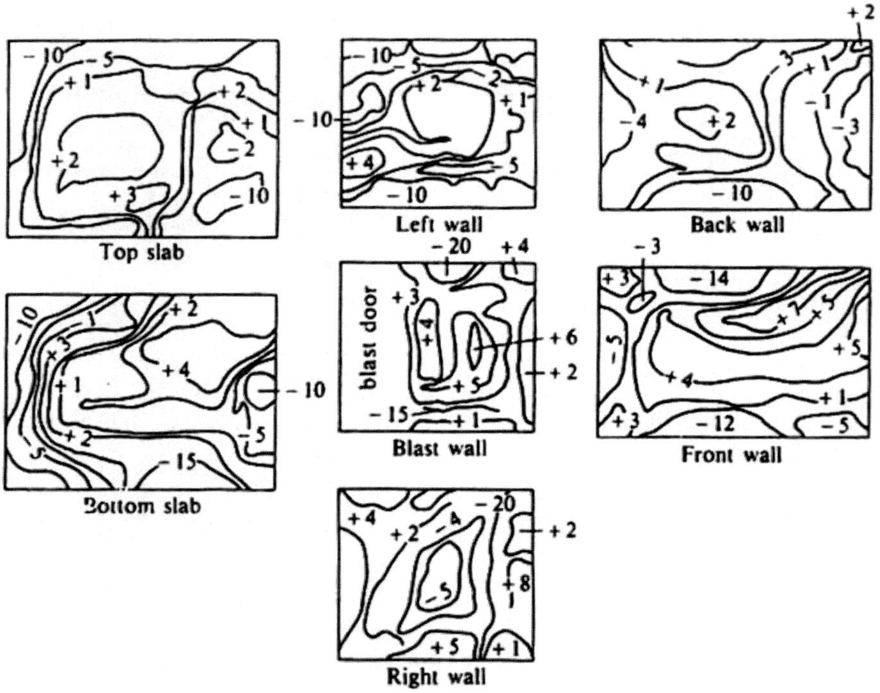

Fig. 9.30b Principal stresses for a nuclear shelter (all stresses are maximum principal stresses)

9.9 Blast Loads and Stesses

Table 9.2 and Fig. 9.10 give a summary of the blast loads generated from a bomb burst. Table 9.5 gives generalised data, including allowable stresses used later on in the analysis.

Based on HMSO reports, American Defence Agency carried out limit state analysis for a six-person domestic shelter, as given in Figs. 9.11a and 9.12b. This was designed for a 1 megaton weapon yield, for which the data are given in Table 9.3. For shallow shelters Table 9.4 indicates the blast loads.

9.10 Finite Element Analysis of a Domestic Nuclear Shelter

Figure 9.30 shows a typical finite element mesh for the dynamic analysis of this shelter. Again 20-noded elements representing concrete in the walls and slabs with smeared reinforcements are used to analyse the shelter resistance against a 1 megaton weapon yield. The dynamic pressure p_{do} is applied in the form of nodal loads. Time duration is included corresponding to the specific values of p_{do}. The dynamic finite-element analysis together with cracking is given in Chapter 6, and the numerical model for the material law developed for the five-parameter case has been adopted. The program ISOPAR is modified to include the blast loads. The total number of increments considered is 12. The structure is yielded, cracked and crushed in vulnerable areas when assumed to be placed above ground. The same structure is placed underground with 0.5 m overburden. The stress trajectories are plotted. Figures 9.30a shows the cracking of the structure when placed above ground. Figure 9.30b shows the stresses and partial cracking when placed underground. The safety factor for the structure is 3.1 when placed underground and the protection factor based on the HMSO reports [639 – 640] is 4000. The shelter received no damage above ground for $\frac{1}{2}$ Mt yield at a ground zero distance of 20,000 m.

Chapter 10
Elemental Design Analysis for Auxillary Structures Associated with Nuclear Facilities

10.1 General Introduction

In this chapter design calculations are given for auxiliary structures associated with nuclear facilities. They are divided into three parts. Part A deals with the design analysis of steel elements using steel Euorcode EC-3. Part B deals with the design analysis of concrete elements based on Eurocode EC-2. Parts C is developed entirely based on the design of a nuclear laboratory. It is assumed that the dimensions and thicknesses of various elements are in accordance with the nuclear requirements and have satisfied the radioactive and the radiation parameters normally adopted by the respective experts. Any necessary changes and alterations needed must be adhered to the demands imposed by those experts. Part C can only claim on the structural design analysis of the structural elements of the nuclear laboratory, using British Codes such as BS 8110, BS 5950 and other codes.

PART A DESIGN ANALYSIS OF STEEL ELEMENTS FOR AUXILLARY STRUCTURES OF NUCLEAR FACILITIES BASED ON EC-3

10.2 Background to Eurocode EC-3

NV1993-1-1 Eurocode-3. Design of Steel Structures: Part 1.1. General Rules and Rules for Buildings (EC-3) gives general rules for design of all types steel structures. In principle this code covers both ultimate and serviceability limit states. Safety factors are applied to loading (actions) and resistances. The National Application Documents (NAD) have been prepared by the authorities in various CEN member countries including United Kingdom. The U.K version which is followed adds rules for lying forces or actions etc. from BS5950 Part 1. In order to minimise any difficulty that those engineers who are already familiar with BS5950 Part 1 may find in switching to EC-3, many of the EC-3 provisions have been

M.Y.H. Bangash, *Structures for Nuclear Facilities*,
DOI 10.1007/978-3-642-12560-7_10, © M.Y.H. Bangash 2011

recast into tables and charts to improve their useability. The system of symbols used in EC-3 is at level of sophestication above that used in BS5950 Part 1. Majority of symbols are found quite familiar. The parametric comparisons are as follows:

Parameter	Symbols (BS5950)	Symbols (EC-3)
Axial force	F	N
Elastic section modulus	Z	w_{el}
Plastic section modulus	S	w_{pl}
Radius of gyration	r	i
Conventions of Member Axes		
(a) Definition of longitudinal Axis of a member	ZZ	XX
(b) Major Axis of cross-section	XX	YY
(c) Minor Axis of cross-section	YY	ZZ
Values of γ_m		
(a) When failure is by Plastic yielding		γ_{MO}
(b) When failure is by any type of buckling		γ_{MI}
(c) When failure occurs by any net section at bolt holes		γ_{M2}

On the analysis side both elastic and plastic methods are allowed. In case of plastic global analysis either rigid plastic or elasto-plastic methods are used in EC-3.

Imperfections

The adverse effects of imperfections are allowed in EC-3. Frame imperfections are included in the global or individual member's analysis.

Classification of cross-section

The main purpose is to identify the extent to which its resistance is modified by the propensity of its compression elements for local buckling which depends on width-to-thickness b/t ratio of the steel elements and upon the patterns of stresses and levels of strains imposed on them.

Various classes

Class 1:- For plastic global analysis the usage of cross-sections with large rotation capacity, at least at the plastic hinge location. The b/t ratios are efficiently stocky to permit the necessary substantial deformations *without local buckling*.

Class 2:- Most rolled sections have b/t values of sufficiently stocky for the cross-section to reach M_{pl} and they are termed Class-2 or compact sections. Here some strain hardening occurs and some limited rotation is possible before falling part of the curve drops (rotation capacity) below M_{pl}.

Class 3:- Here semi-compact section is envisaged for the section. The values b/t are less. The maximum moment reaches the elastic resistance moment M_{el} but falls short of the M_{pl}, the plastic resistance moment.

Class 4:- In this class, the slender cross-section, the maximum moment is governed by local buckling and occurs at a value less than the elastic moment of resistance. In this case the stress at the extreme fibres reach the yield strength. $f_y' < f_y$

In BS5950 classes 1, 2, 3, 4 as comparison representing plastic, compact, semi-compact and slender classifications, are replaced by the above classes. The principles are the same for classifications.

10.2.1 Axially Loaded Members: Definitions and Formulae

(a) Section properties

Shear lag effects when for outstanding flange elements

$$c \leq Lo/20 \tag{10.1}$$

$$b \leq Lo/10 \tag{10.2}$$

If they are exceeded, the effective breadth specified in EC-3 should be used.

10.2.2 Tension Members

(a) The design tension resistance is smaller than $N_{pl.Rd}$, the design plastic resistance of the gross area. In this case f_y = yield strength of with γ_{M0} = 1.05 shall be used. The design ultimate resistance of the net area = 0.9 × ultimate tensile strength f_u with γ_{M2} = 1.2.

$N_{u \cdot Rd} \not< N_{Pl \cdot Rd}$ and can be achieved inductile behaviour as

$$\frac{0.9A_{net}}{A} \geq \left\{\frac{f_y}{f_u}\right\}\left\{\frac{\gamma_{M2}}{\gamma_{M0}}\right\} \tag{10.3}$$

10.2.3 Compression Members

They shall be designed both for cross-section resistance and buckling resistance. All classes 1, 2, 3 shall be considered. For class 4 sections, the additional moment due to the eccentricity of the centroidal axes of the effective section should be allowed for.

10.2.4 Buckling Resistance

Calculated values of the compressive strength, similar to BS5950 Part 1, in EC-3 are calculated according to various slenderness ratio $\lambda = \frac{\text{buckling length}}{\text{radius of gyration}}$. The

compressive strengths are tabulated against values of $\bar{\lambda}\sqrt{\beta_A}$ in Table 5.14 (a) and (b) for Fe 430 and Fe 510 steels, respectively.
Where

$$\beta_A = 1 \text{ for classes } 1, 2, 3 \text{ cross sections}$$

$$\beta_A = A_{\text{eff}}/_{\text{A}} \text{ for class 4 cross sections}$$

10.3 Beams

The yield strength in shear $= f_y/\sqrt{3}$ which is a plastic shear resistance. Comparing with BS5950 which is $0.6 f_y$, the EC-3 takes sometime a lower value of $0.577 f_y$ for a rolled section W_{pl}, the plastic modulus, is used for class 1 or 2 while W_{el}, the elastic modulus is used for class 3, each with γ_{M0} for class 4 cross-section W_{el} and W_{eff}, the effective cross-section with γ_{M1} are both used.

The beams should be checked for the following:

(a) Moment of resistance for cross-section with high shear
(b) Effect of transverse forces on moment of resistance
(c) Lateral torsional buckling

The methods are explained in the examples associated with this part.

10.4 Shear Buckling

Tables and clauses related to shear buckling are given in Section 5.5.6 of the code. This is needed for which the shear failure mode is a consequence of buckling rather than yielding. "Simple post critical method" is a prominant one. Interaction of moment resistance and shear buckling is an important phenomenon which needs full checking. The following relations are to be noted

$$m_{\text{sd}} = \text{max . value of moment} \leq M_{f\cdot\text{Rd}}, \text{ the plastic moment resitance due to two flanges alone} \tag{10.4}$$

$$v_{sd} = \text{max .value of shear} \leq V_{ba\cdot Rd}, \text{ the design shear buckling.} \tag{10.5}$$

10.5 Axially Loaded Member

The following processes are to be noted:

(a) Axial force and bending with low shear
(b) Axial force and bending with high shear] Clause 5.6.1.2
Clause 5.6.1.3
Clause 5.6.1.4 to 5.6.1.5(3)

(c) Buckling resistance with combined bending and axial tension and with combined bending and axial compression tacking into consideration major axis bending and minor axis bending.] Clauses 5.6.3.2 to 5.6.3.4(4)

10.6 Resistance of Webs to Transverse Compression Forces

(a) Crushing resistance
(b) Crippling resistance
(c) Buckling resistance] Clauses 5.7 to 5.7.6(10)

(d) The choice of transverse stiffness
(e) Flange induced buckling] Clauses 5.7.7

Note. All these are thoroughly explained in the design examples using various formulae in the EC-3.

10.6.1 WE10.1

A steel beam of Fig. 10.1 has an effective span of 6.5 m fully restrained along length. At one end its simply supported and at the other end is supported by a roller. The beam shall be designed in F_e430 steel (S275) based on EC-3. The followings are unfactored loads:- loads (udl) $g_{kl} + G_{kl} = 15.0\text{KN} + 41\text{ KN}$, $q_{kl} + Q_{kl} = 35\text{KN} + 45\text{KN}$

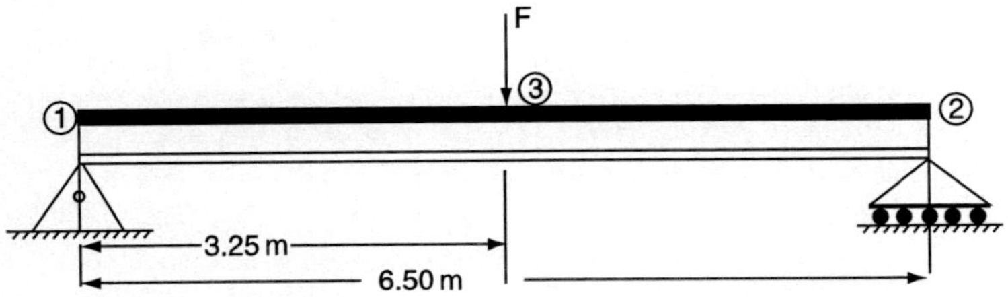

Fig. 10.1 A simply supported steel beam (with permanent and variable loads)

Solution:
Factored loads Stiffness bearing length at supports = 50 and 75 mm at concentrated load

(a) Permanent $g_{dl} = \gamma_G\, g_{kl} = 20.25$ KN/m - distributed
concentrated load $= \gamma_Q\, G_{kl} = 55.35$ KN/m - distributed
g_{dl} (total) $= 73$ KN/m
G_{dl} (total) $= 122.85$
(b) Variable $q_d = \gamma_Q\, q_{dl} = 52.55$ KN/m
conentrated load $Q_{dl} = \gamma_Q\, Q_{kl} = 67.5$ KN

Trial section and material properties

$533 \times 210 \times 92UB$
$t_1 = 533.1$ mm; $b = 209.3$ mm
$t_w = 10.2$ mm; $t_f = 15.6$ mm
$d = 476.5$ mm; $A_v = 58.1\ \text{cm}^2$
$A = 118\ \text{cm}^3$; $W_{ely} = 2076\ \text{cm}^3$
$W_{\text{Pl.y}} = 2366\ \text{cm}^3$; $c/t_f = 6.71$
$I_y = 55390\ cm^4$; $d/t_w = 46.7$
$t \leq 40$ mm Grade F_e 430 ($S275$)
$F_y = 275\text{N/mm}^2$; $F_u = 430\ \text{N/mm}^2$

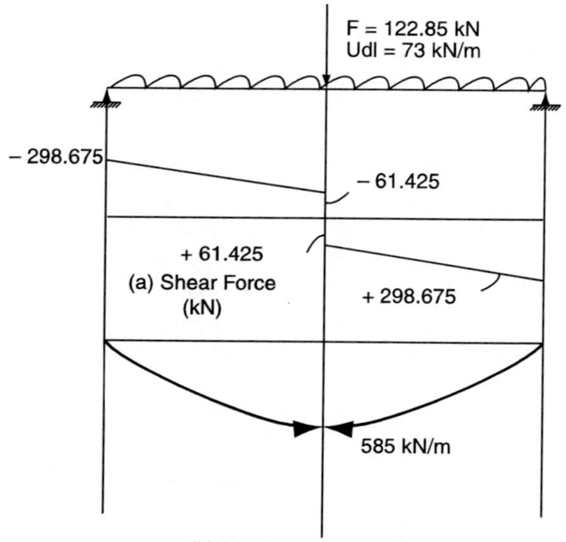

(b) Bending moment diagram

Classification:-

I-section-rolled section $= \frac{c}{t_f} \leq 9.2$
Flange :- class I type
Web subject to bending when NA @ mid depth

For selected item $\frac{d}{t_w} = 46.7$
For I class web $\frac{d}{t_w} \leq 66.6$
Section is class I

(i) Shear resistance of crossection
$V_{\text{sd}} = $ shear force at x - *section* $\not> V_{pl,Rd}$

$$V_{pl,Rd} = \frac{AV}{\gamma_{M0}}\left[\frac{f_y}{\sqrt{3}}\right]$$

$$= \frac{58.1 \times 10^2}{1.05}\left[\frac{275}{\sqrt{3}}\right] \times 10^{-3} = 879\text{KN}$$

$V_{sd}\,(298.075KN) < V_{pl,Rd} = 879$ KN OK

Hence 533 × 210 × 92UB is $O.K.$ for shear resistance.

(ii) Moment resistance at x-section

Coexisting shear force $V_{sd} \leq 0.5V_{pl,Rd}$, then the moment of resistance cannot be reduced by shear

$M_{c.Rd} = \frac{W_{pl.y.fy}}{\gamma_{mo}}$ for class 1 or 2 section $= 620\ KNm$

Maximum moment is calculated as $585 < 620\ KNm$

The section 533 × 210 × 92UB is sufficient.

(iii) Beam for lateral torsional buckling (LTB)

Since it is fully restrained in lateral direction it will not be affected by lateral torsional buckling.

(iv) Shear buckling

For steel grade

Fe 430 (s = 275)

$$\frac{d}{t_w} > 63.8$$

For the section $\frac{d}{t_w} = 46.7 < 63.8$

No check is needed.

(v) Resistance of web to transverse forces

The unstiffened web to forces applied to a flange is governed by crushing resistance or by crippling resistance or by buckling resistance. Here stiff bearing length can be taken into consideration.

S_s = bearing at supports = 50 mm

S_{ms} = bearing at midspan = 75 mm

Crushing resistance $R_{y.Rd}$ on the web.

At support

$$R_{y.Rd} = (S_s + S_y)\frac{t_w f_{yw}}{\gamma_{m1}}$$

$$S_y = t_f\left[\frac{b_f}{t_w}\right]^{0.5}\left[\frac{f_{yf}}{f_{yw}}\right]^{0.5}\left[1 - \left\{\frac{\gamma_{mo}\sigma_{fE.d}}{f_{yf}}\right\}^2\right]^{0.5}$$

$$= 15.6\left[\frac{20.93}{10.35}\right]^{0.5}\left[\frac{275}{275}\right]^{0.5}[1, 0]$$

70.67 mm

$$R_{y.Rd} = (50 + 70.67)\frac{10.2 \times 275}{1.05 \times 10^3} = 322KN$$

But

$$V_{\mathrm{sd}} = 298.675 < R_{y.Rd} = 322KN$$

The section is OK

(1) At mid span
Concentrated load = 122.85KN
(2) Crippling resistance

At supports

$$R_{a.Rd} = 0.5t_w^2(Ef_{yw})^{0.5}\left[\left\{\frac{t_f}{t_w}\right\}^{0.5} + 3\left\{\frac{t_u}{t_f}\right\}\left\{\frac{s_s}{d}\right\}\right]\frac{1}{\gamma_{m1}}$$

But $\left\{\frac{s_s}{d}\right\} \leq 0.2$
At support $s_s = 50\,mm \quad \frac{s_s}{d} = \frac{50}{476.5} = 0.105 < 0.2$

$$\gamma_{m1} = 1.05 \quad E = 210 \times 10^3\,N/mm^2$$

$$R_{a.Rd} = 543KN$$

Maximum Design shear force $V_{sd} = 298.675KN < R_{a.Rd} = 543KN$
The section is O.K
At mis span
For $S_{sm} = 75mm$ $R_{a.Rd} = 583KN$
And $V_{sd} = 61.425KN < R_{a.Rd} = 583KN$
The web has sufficient crippling resistance
When the beam is subject to bending moment, then

$$\frac{F_{sd}}{R_{a.Rd}} + \frac{M_{sd}}{M_{c.Rd}} \leq 1.5$$

$$\frac{122.85}{583} + \frac{585}{620} = 1.154 < 1.5\,O.K$$

Buckling Resistance
At supports:

$$h = 533.1\mathrm{mm}$$

$$a = 0\,\mathrm{mm}$$

$$b_{\mathrm{eff}} = \frac{1}{2}[h^2 + s_s^2]^{0.5} + a + \frac{s_s}{2}$$

$$= 292.7\,\mathrm{mm} < 535.4\,\mathrm{mm}$$

The buckling resistance of the web $R_{b.Rd} = \frac{B_A f_{cA}}{\gamma m_2}$

$$\beta_A = 1,\ A = b_{\text{eff}} \times t_w = 2986\,\text{mm}^2$$

$$\gamma_{M1} = 1.05,\ \lambda = \frac{2.58}{t_w} = 116.8$$

$$f_c = 104\,\text{N/mm}^2$$

$$R_{b.Rd} = 300\,\text{KN} > V_{\text{sd}} = 298.675\,O.K$$

The buckling resistance is sufficient laterally.
Serviceability limit state

$$\gamma_G = 1.0, \gamma_Q = 1.0$$

$$g_d = 1.0 \times 15 = 15\,\text{KN/m} \quad \text{(dead)}$$

$$q_d = 1 \times 35 = 35\,\text{KN/m} \quad \text{(imposed)}$$

Concentrated loads

$$G_d = 1.0 \times 41 = 41KN; Q_d = 1 \times 45 = 45KN$$

δ_2 – deflection due to variable load

$$= \frac{5}{384}\left[\frac{q_d L^4}{EI}\right]\left[\frac{1}{48}\left(\frac{Q_d L^3}{EI}\right)\right]$$

$$I = I_y = 553.3 \times 10^6\,\text{mm}^4;\ E = 210 \times \frac{10^3\,N}{\text{mm}^2}$$

$$\delta_2 = 9.21\,\text{mm}$$

$$\delta_2(\text{limiting}) = \frac{L}{20} = 26\,\text{mm}$$

$$\delta_2 < \delta_2(\text{limiting})\ \text{deflection}\ O.K$$

Hence the section

$$533 \times 210 \times 92UB, Fe(430)S = 275\,O.K$$

10.6.2 WE10.2

In the Section 10.6.1, the beam is laterally restrained at the ends only and as a result the span is increased to 90 m.

g_d = characteristic load 24.2 KN/m (permanent) inclusive $\gamma_G = 1.35$
q_d = charateristic variable load = 30 KN/m inclusive $\gamma_Q = 1.5$

No concentrated load exists. Design basis EC3
Solution

Total u.d.l = 54.2 KN/m material grade Fe 430
Shear force = 243 KN $\gamma_{mo} = 1.05$

BM diagram parabolic with maximum ordinate = 550 KNm
Section $610 \times 305 \times 149UB$ (Fe430) and s275
Class I section; $h = 609.6$ mm; $b = 304.8$ mm, $t_w = 11.9$ mm

$t_f = 19.7$ mm, $d = 537.2$ mm; $A_v = 79\ \text{cm}^2$
$A = 190\ \text{cm}^2$, $c/t_f = 7.74$; $d/t_w = 45.1$
$w_{pl.y} = 4575\ \text{cm}^3$, $w_{el.y} = 4093\ \text{cm}^3$
$I_y = 124700\ \text{cm}^4$; $i_{LT} = 7.75$ cm; $a_{LT} = 201$ cm
$t \leq 40$ mm; $f_y = 275\ \text{N/mm}^2$; $f_u = 430\ \text{N/mm}^2$

Shear resistance plastic

$$V_{pl.Rd} = \frac{A_V}{\gamma_{M0}}\left\{\frac{f_y}{\sqrt{3}}\right\} = 1195\,\text{KN}$$

$$V_{sd} = 243\,\text{KN} < V_{pl.Rd} = 1195\,\text{KN. O.K section sufficient}$$

Moment resistance applied shear force

$$M_{c.Rd} = \frac{W_{pl.yfy}}{\gamma_{mo}} = 1198\,\text{KNm} \quad 0.5V_{pl.Rd} = 59.7 > 243$$

$$> M_{sd} = 550\,\text{KNm}\ O.K.\ \text{Section sufficient}$$

Lateral Torsional Buckling (LTB)
Class I rolled I section
For loading support:-

$$\left[\frac{k}{c_1}\right]^{0.5}.L = 0.94 = 8460 > 0.35 i_{LT} = 2713\ O.K$$

$$M_{y.sd} \leq M_{b.Rd} = \frac{B_w f_b w_{pl.y}}{\gamma_{M1}}$$

$$f_b = \text{bending strength} = \lambda_{LT}\sqrt{B_w}$$

where λ_{LT} = spenderness Equivalent

$$= \frac{\left[\frac{k}{c_1}\right]^{0.5} L}{i_{LT}} \Bigg/ \left\{ 1 + \frac{\left(\frac{L}{a_{LT}}\right)^2}{25.66} \right\}^{0.25} = 94.5$$

$$M_{b.Rd} = \text{buckling resistance moment}$$

$$= \frac{B_w f_b w_{pl.y}}{\gamma_{m1}} = 723.3KNm > M_{sd} 550KNm$$

OK for sufficient resistance.
Shear Buckling

$$\frac{d}{t_w} > 63.8 \; and \; \frac{d}{t_w} \text{(chosen section)}$$

$$= 45.1 < 63.8$$

Shear buckling not needed.

Example 10.1 gives a method for checking resistance of the web to transverse forces.

Section 610 × 305 × 149UB Fe430,(S275) O.K

10.6.3 WE10.3

The ground floor steel column assumed pinned at top and bottom has total length 6.0 m. From the top floors the column receives a total load of 2500KN axially only. Design is steel column based on EC3 of material properties Fe430 (s275, f_y = 275N/mm^2; f_u = 430 N/mm^2. Show that the column 365 × 368 × 129UC class 3 section is adequate for load 2300KN (factor load)

Solution

$h = 355.6$; $b = 368.3$ mm; $t_w = 10.7$ mm; $t_f = 17.5$ mm; $d = 290.2$ mm

$A_v = 43.5\,\text{cm}^2$; $A = 165\,\text{cm}^2$; $i_y = 15.6\,\text{cm}^2$; $i_z = 9.39$ cm; $c/t_f = 10.5\; d/t_w = 27.1$

$\gamma_{M0} = 1.05$; $\gamma_{M1} = 1.05$

Compressive Resistance

$N_{sd} \le N_{c.Rd} = \frac{Af_y}{\gamma_{Mo}}$ class 3 section $N_{c.Rd}$ = design compressive resistance $= 4321KN > N_{sd} = 2300$ KN

Section OK against compression.

Buckling Resistance

$$N_{sd} \leq N_{c.Rd} = \frac{\beta_A f_y A}{\gamma_{M1}}$$

Buckling about YY axis $l_y = 1 \times 6000 = 6000$ mm
Buckling about ZZ axis $l_z = 1 \times 6000 = 6000$ mm

Hence

$$\lambda_y = \frac{l_y}{i_y} = 38.5 < 180$$

$$\lambda_z = \frac{l_z}{i_z} = 63.9 < 180$$

The buckling about YY axis and $\frac{h}{b} = 0.966 < 1.2$ curve 'b' Table 5.14

$\beta_A = 1.0\, \lambda_y \sqrt{\beta_A} = 38.5$ and $t \leq 40$ mm
$f_c = 249.5\, N/\text{mm}^2$
$N_{b.y.Rd} = 3921\,\text{KN} > 2300\,\text{KN}$

Buckling about ZZ axis

$$\frac{h}{b} = 0.966$$

$$\beta_A = 1.0\, t \leq 40\,\text{mm}$$

$$\lambda_z \sqrt{\beta_A} = 63.9,\ f_c = 193\text{N/mm}^2$$

The design buckling resistance @ ZZ axis

$$N_{b.z.Rd} = 3033\,\text{KN} > N_{sd} = 2700\,\text{KN}$$

Section $356 \times 368 \times 129UC\ Fe430\ (S275)$ O.K

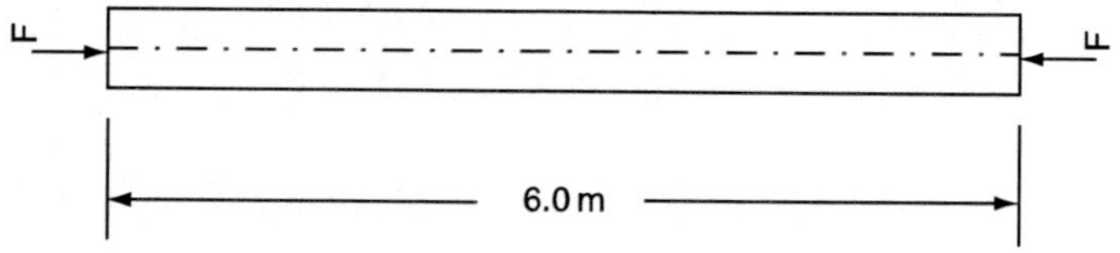

Fig. 10.2 Axially load column with pinned end

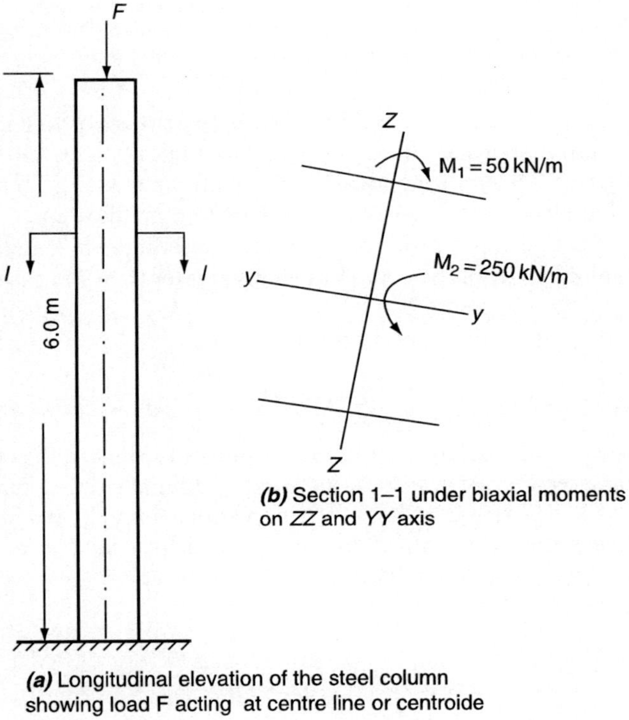

Fig. 10.3 Steel column under axial load and biaxial moment

PART B DESIGN ANALYSIS OF CONCRETE ELEMENTS FOR AUXILARY STRUCTURES OF NUCLEAR FACILITIES BASED ON EC-2

10.7 A Brief of Systematic EC-2 Design

Suitable dimensions for b and d can be decided by a few trial calculations as follows

1. For no compression reinforcement

$$K = M/bd^2 f_{ck} < K_{bal} \tag{10.6}$$

where

$$K_{bal} = 0.167\ for\ f_{ck} \le C50 \tag{10.7}$$

With compression reinforcement it can be shown that

$$M/bd^2 f_{ck} < 8/f_{ck} \tag{10.8}$$

approximately, if the area of bending reinforcement is not to be excessive.

2. The maximum design shear force $V_{Ed.\max}$ should not be greater than 0.1 $8b_w d(1\text{-}f_{ck}/250)f_{ck}$. To avoid congested shear reinforcement, $V_{Ed.\max}$ preferably be somewhat closer to half (or less) of the maximum allowed.
3. The span-effective depth ratio for spans not exceeding 7 m should be within the basic values given of the code. For spans greater than 7 m the basic ratios are multiplied by 7/span.
4. The overall depth of the beam is given by

$$h = d + cover + t \tag{10.9}$$

where t = estimated distance from the outside of the link to the centre of the tension bars, see Fig. 10.4. For example with nominal sized 12 mm links and one layer of 32 mm tension bars, t = 28 mm approximately. It will, in fact, be slightly larger than this with deformed bars as they have a larger overall dimension than the nominal bar size.

5. Calculate

$$z = d\left[0.5 + \sqrt{0.25 - \frac{K}{1.134}}\right] \tag{10.10}$$

6. Calculate

$$A_s = \frac{M}{0.87 f_{yk} z} \tag{10.11}$$

7. Check that the area of steel provided is within the maximum and minimum limit required.

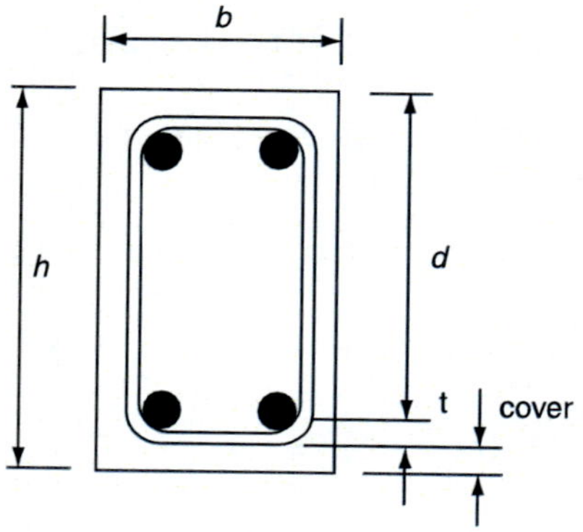

Fig. 10.4 Beam dimensions

10.7.1 Rectangular Sections with Tension and Compression Reinforcement with Moment Redistribution Applied (Based on the UK Annex to EC2)

The steps in the design are

8. Calculate

$$X_{bal} \leq (\delta - 0.4)d \tag{10.12}$$

9. Calculate

$$K = M/bd^2 f_{ck} \tag{10.13}$$

10. Take K_{bal} from Table 7.1 or alternatively calculate

$$K_{bal} = 0.454(\delta - 0.4) - 0.182(\delta - 0.4)^2 \; for \; f_{ck} \leq C50 \tag{10.14}$$

If K> K_{bal}, compression steel is required.

11. Calculate the area of compression steel from

$$A'_s = \frac{(K - K_{bal}) f_{ck} b d^2}{f_{sc}(d - d')} \tag{10.15}$$

where f_{sc} is the stress in the compression steel

If $d'/x \leq 0.38$ the compression steel has yielded and $f_{sc} = 0.87 f_{yk}$
If $\frac{d'}{x} > 0.38$ then the strain ε_{sc} in the compressive steel must be calculated from the proportions of the strain diagram and $f_{sc} = E_s \varepsilon_{sc} = 200 \times 10^3 \varepsilon_{sc}$

12. Calculate the area of tension steel from

$$A_s = \frac{K_{bal} f_{ck} b d^2}{0.87 f_{ykz}} + A'_s \frac{f_{sc}}{0.87 f_{yk}} \tag{10.16}$$

where $z = d - 0.8 X_{bal}/2$

13. Check Eq. 7.5 for the areas of steel required and the areas provided that

$$\left(A'_{s.prov} - A'_{s.req}\right) \geq \left(A_{s.prov} - A_{s.req}\right) \tag{10.17}$$

This is to ensure that the depth of the neutral axis has not exceeded the maximum value of X_{bal} by providing an over-excess of tensile reinforcement.

10.7.2 Design of Shear

The theory and design requirements for shear have been covered in EC3 and the relevant design equations were derived based on the requirements of EC2 using the variable strut inclination method.

The shear reinforcement will usually take the form of vertical links or a combination of links and bent-up bars. Shear reinforcement may not be required in very minor beams such as door or window lintels with short spans of less than say 1.5 m and light loads.

The following notation is used in the equations for the shear design

A_{sw} = the cross-sectional area of the two legs of the stirrup
s = the spacing of the stirrups
z = the lever arm between the upper and lower chord members of the analogous truss
f_{ywd} = the design yield strength of the stirrup reinforcement
f_{yk} = the characteristic strength of the stirrup reinforcement
V_{Ed} = the shear force due to the actions at the ultimate limit state
V_{wd} = the shear force in the stirrup
$V_{rd.s}$ = the shear resistance of the stirrups
$V_{Rd.max}$ = the maximum design value of the shear which can be resisted by the concrete strut

10.7.3 Vertical Stirrups or Links

The procedure for designing the shear links is as follows

14. Calculate the ultimate design shear forces V_{Ed} along the beam's span.
15. Check the crushing strength $V_{Rd.max}$ of the concrete diagonal strut at the section of maximum shear, usually at the face of the beam support.

For most cases the angle of inclination of the strut is $\theta = 22°$, with $\cot\theta = 2.5$ and $\tan\theta = 0.4$ so that from Eq. 5.6:

$$V_{Rd.\max} = 0.1\,24 b_w d\,(1 - f_{ck}/250)\,f_{ck} \tag{10.18}$$

and if $V_{Rd.\max} \geq V_{Ed}$ then go to step (16) with $\theta = 22°$ and $\cot\theta = 2.5$, but if $V_{Rd.max} < V_{Ed}$ then $\theta > 22°$ and θ must be calculated from Eq. 10.18 as VEd 1

$$\theta = 0.5\,\sin^{-1}\left\{\frac{V_{Ed}}{0.18 b_w d f_{ck}(1 - f_{ck}/250)}\right\} \leq 45°$$

16. The shear links required can be calculated from Eq. 10.19

$$\frac{A_{sw}}{s} = \frac{V_{Ed}}{0.78 d f_{yk} \cot \theta} \tag{10.19}$$

where A_{sw} is the cross-sectional area of the legs of the stirrups ($2 \times \pi\phi^2/4$ for single stirrups)
For a predominately uniformly distributed load the shear V_{Ed} should be taken at a distance d from the face of the support and the shear reinforcement should continue to the face of the support.

17. Calculate the minimum links required by EC2 from

$$\frac{A_{sw,min}}{s} = \frac{0.08 f_{ck}^{0.5} b_w}{f_{yk}} \tag{10.20}$$

and the shear resistance for the links actually specified

$$V_{min} = \frac{A_{sw}}{s} \times 0.78 d\, f_{yk} cot\theta \tag{10.21}$$

This value should be marked on the shear force envelope to show the extent of these links.

18. Calculate the additional longitudinal tensile force caused by the shear force

$$\Delta F_{td} = 0.5 V_{Ed} cot\theta \tag{10.22}$$

This additional tensile force increases the curtailment length of the tension bars. The minimum spacing of the links is governed by the requirements of placing and compacting the concrete and should not normally be less than about 80 mm. EC2 gives the following guidance on the maximum link spacing:

(a) Maximum longitudinal spacing between shear links in a series of links

$$S_{1,max} 0.75 d(1 + \cot \alpha)$$

where α is the inclination of the shear reinforcement to the longitudinal axis of the beam.

(b) Maximum transverse spacing between legs in a series of shear links

$$S_{b,max} = 0.75 d \ (\leq 600\, mm)$$

Types of links or stirrups are shown in Fig. 7.15. The open links are usually used in the span of the beam with longitudinal steel consisting of top hanger bars and bottom tensile reinforcement. The closed links are used to enclose top and bottom reinforcement such as that near to the supports. Multiple links are used when there are high shear forces to be resisted.

10.7.4 Floor Slab with Different and Boundary Conditions

After evaluating moments, the rest of the procedures given in the EC3 shall be followed.

10.7.5 Short and Slender Columns in Reinforced Concrete

10.7.5.1 Slenderness Ratio of a Column

The slenderness ratio of a column bent by an axis is given by

$$\lambda = \frac{l_o}{L^\circ} = \frac{l_o}{\sqrt{I}/A^\circ} \tag{10.23}$$

where l_o = effective height of the column
i = radius of gyration about an axis
I = second momment of inertia
A = the cross - sectionalarea of the

l_o the effective height , based on EC2 gives two formulae:

$$l_o = 0.5\,l\sqrt{\left(1+\frac{k_1}{0.45+k_1}\right)\left(1+\frac{k_2}{0.45+k_2}\right)}\text{(braced members)} \tag{10.24}$$

$$l_o = l\sqrt{1+10\frac{k_1k_2}{k_1+k_2}} \tag{10.25}$$

and

$$l_o = l\left(1+\frac{k_1}{1+k_1}\right)\left(1+\frac{k_2}{1+k_2}\right) \tag{10.26}$$

The values k_1 and k_2 are respectively relative flexibilities and rotations. The restraints at ends 1 and 2 of the columns respectively do exist.

At each end k_1 and k_2 can be taken as

$$k = \frac{\text{column stiffness}}{\sum \text{beamstiffness}} = \frac{\left(\frac{EI}{l}\right)_{\text{column}}}{\sum 2\left(\frac{EI}{l}\right)_{\text{beam}}} = \frac{\left(\frac{I}{l}\right)_{\text{column}}}{\sum 2\left(\frac{I}{l}\right)_{\text{beam}}} \tag{10.27}$$

A typical column is a symmetrical frame with spans of approximately equal length, k_1 and k_2 can be calculated as

$$k_1 = k_2 = k\frac{\left(\frac{I}{l}\right)_{\text{column}}}{\sum 2\left(\frac{I}{l}\right)_{\text{beam}}} = \frac{\left(\frac{I}{l}\right)_{\text{column}}}{2 \times 2\left(\frac{I}{l}\right)_{\text{beam}}} = \frac{I\left(\frac{I}{l}\right)_{\text{column}}}{4\left(\frac{I}{l}\right)_{\text{beam}}} \quad (10.28)$$

The EC2 places an upper limit on a slenderness ratio of a single member below which second-order effects can be ignored.

$$\lambda_{\text{limit}} = 20 \times A \times B \times C/\sqrt{n} \quad (10.29)$$

$$A = 1/(1 + 0.2\phi_{\text{ef}});\ B = \sqrt{1 + 2w};\ \ c = 1.7 - \gamma_m \quad (10.30)$$

ϕ_{ef} effective creep ratio (if not known $A = 0.7$)
$w = A_s/f_{yd}(A_c f_{ed})$ (if not nown $B = 1.1$)
f_{yd} = reinforcement yield strength
f_{cd} = design compressive strengthof concrete
A_s = Total area of longitudenal reinforcement
$n = N_{ED}/(A_s f_{cd})$; N_{ED} = Design ultimate load
$\gamma_M = \frac{M_{01}}{M_{02}}$ (if not know $c = 0.7$)
= *moments at the end of the column* $|M_{02}| \geq |M_{01}|$

The following conditions apply to the value of C:

(a) If the end moments M_{01} and M_{02} give rise to tension on the same side of the column r_m should be taken as positive from which it follows that $C \leq 1.7$.
(b) If the converse to (a) is true, i.e the column is in a state of double curvature, then r_m should be taken as negative from which it follows that $C > 1.7$.
(c) For braced members in which the first-order moments arise only from transverse loads or imperfections; C can be taken as 0.7.
(d) For unbraced members, C can be taken as 0.7.

For an unbraced column an approximation to the limiting value of λ will be given by

$$\begin{aligned}\lambda_{\text{limit}} &= 20 \times A \times B \times C/\sqrt{n} = 20 \times 0.7 \times 1.1 \times 0.7/\sqrt{N_{ED}/(A_s f_{cd})} \\ &= 10.8/\sqrt{N_{ED}/(A_s f_{cd})}\end{aligned} \quad (10.31)$$

The limiting value of λ for a braced column will depend on the relative value of the column's end moments that will normally act in the same clockwise or anti-clockwise direction as in case (b) above. If these moments are of approximately equal value then, $r_m = -1, C = 1.7 + 1 = 2.7$ and a typical approximate limit on λ will be given by

$$\lambda_{\text{limit}} = 20 \times A \times B \times C/\sqrt{n} = 20 \times 0.7 \times 1.1 \times 0.7/\sqrt{N_{ED}/(A_s f_{cd})}$$
$$= 41.6/\sqrt{N_{ED}/(A_s f_{cd})} \tag{10.32}$$

Alternatively for a braced column the minimum limiting value of λ will be given by taking C 1.7. Hence

$$\lambda_{\text{limit}} = 20 \times A \times B \times C/\sqrt{n} = 20 \times 0.7 \times 1.1 \times 0.7/\sqrt{N_{ED}/(A_s f_{cd})}$$
$$= 26.2/\sqrt{N_{ED}/(A_s f_{cd})} \tag{10.33}$$

If the actual slenderness ratio is less than the calculated value of λ_{limit} then the column can be treated as short. Otherwise the column must be treated as a slender one and the second-order effects must be accounted for in the design of the column.

10.7.6 WE.10.5

A simply supported beam of an effective span of 6.0 m is subject to permanent load $g_k = 60KN/m$ inclusive self-weight and variable load q_k of $18KN/m$. Design this beam taking into consideration $f_{ck} = 30\ N/mm^2; f_{yk} = 500\ N/mm^2$

Solution

Ultimate loading and maximum moment

Ultimate load $w_u = = (1.35g_k + 1.5q_k)kN/m$

$$= (1.35 \times 60 + 1.5 \times 18) = 108KN/m$$

Maximum design moment $M = \frac{w_u L^2}{8} = \frac{108 \times 6.0^2}{8} = 486\,\text{kNm}$

Bending reinforcement

$$K = \frac{M}{bd^2 f_{ck}} = \frac{486 \times 10^6}{300 \times 540^2 \times 30} = 0.185 > k_{bal} = 0.167$$

Therefore compression reinforcement, A'_s is required.

$$d'/d = \frac{50}{540} = 0.092 < 0.171 \text{ therefore } f_{sc} = 0.87 f_{yk}$$

Compression steel

$$A'_s = \frac{(k - k_{bal})bd^2 f_{ck}}{f_{sc}(d' - d)}$$
$$= \frac{(0.185 - 0.167) \times 30 \times 300 \times 540^2}{0.87 \times 500(540 - 50)} = 222\text{mm}^2$$

Provide H16 bars, $A'_z = 402\,\text{mm}^2$

Tension steel, $A_s = \frac{0.167bd^2 f_{ck}}{0.87 f_{ykz}} + A'_s$

$I_a = 0.82$ therefore of steel as

$$A_s = \frac{0.167 \times 30 \times 300 \times 540^2}{0.87 \times 500 \times (0.82 \times 540)} + 222$$
$$= 2275 + 222 = 2497\,\text{mm}^2$$

Provide 2NOH32 bars and 2NOH25 bars, area 2592 mm^2, 100 A_s/bd = 1.6 > 0.15.

Two bars are curtailed near the support. Tension bars shall be anchored over the support.

Span-effective depth ratio

$$\rho = 100A_{sreq}/bd = (100 \times 2497)/(300 \times 540) = 1.54\%$$
$$A_s = \text{2NOH32 and 2NOH25 steel } A'_s = \text{2NOH16}$$

From Table 6.10 or Fig. 6.3 basic span-effective depth ratio = 14

Modification for steel area is provided as

Modified ratio = $14.0 \times \frac{2592}{2497} = 14.5$

Span-effective depth ratio provided = $\frac{6000}{540} = 11.1 <$ allowable upper limit O.K

10.7.7 *WE10.6*

A simply supported T-beam 600 mm (Fig. 10.4(b)) flange width and overall height 580 mm with a 6 m span is subjected to the design ultimately distributed load of 40 KN/m. As shown in Fig. 10.5, the beam has flange thickness of 150 mm and web thickness is 250 mm. Using $f_{ck} = 25\ N/mm^2$ and $f_{yk} = 500\ N/mm^2$ design the beam based on EC2

Solution

$$B.M_{\max} = \frac{40 \times 6^2}{8}$$
$$= 180\,\text{KNm}$$

Longitudinal reinforcement

$$\frac{M}{b_f d^2 f_{cx}} = \frac{180 \times 6^2}{600 \times 530^2 \times 25} = 0.0273$$

$$l_a = 0.95$$

$$z = \text{lever arms} = l_a d = 0.95 \times 530 = 503\,\text{mm}$$

$$s = \text{depth of stress block} = 2(d - z) = 54\,\text{mm} < b_f = 150\,\text{mm}$$

$$A_s = \frac{M}{0.87 f_{ykz}} = \frac{180 \times 6^2}{0.87 \times 500 \times 503} = 823\,\text{mm}^2$$

Provision:
2 No. H25 bars; $A_s = 982\,\text{mm}^2$
For these bars

$$\frac{100 A_s}{b_w d} = \frac{100 \times 982}{250 \times 530} = 0.74\% > 0.13\%$$

Transverse steel in the flange

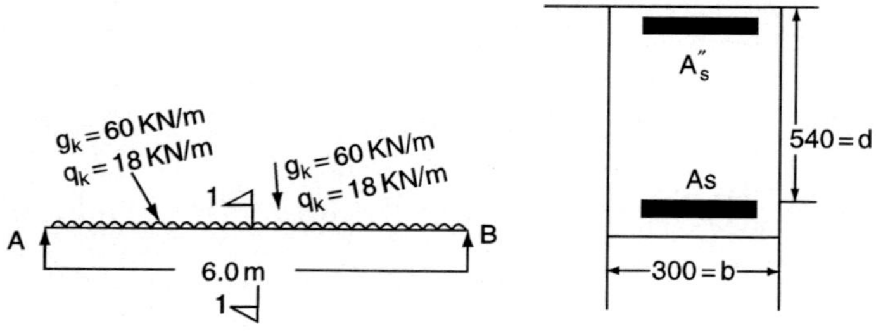

Fig. 10.4(a) Simply supported T beam

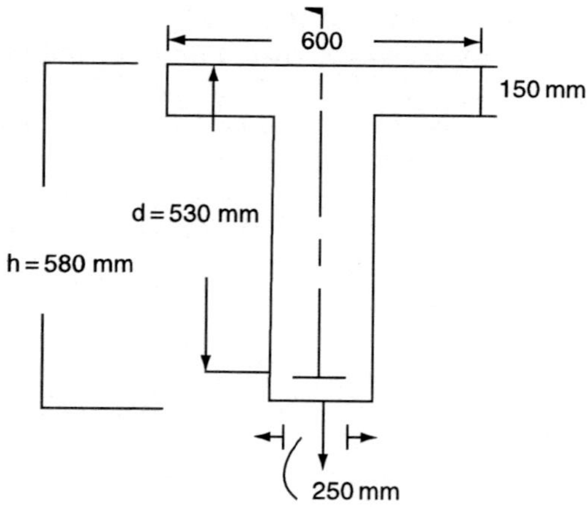

Fig. 10.4(b) T-section

Fig. 10.5 Type of shear link

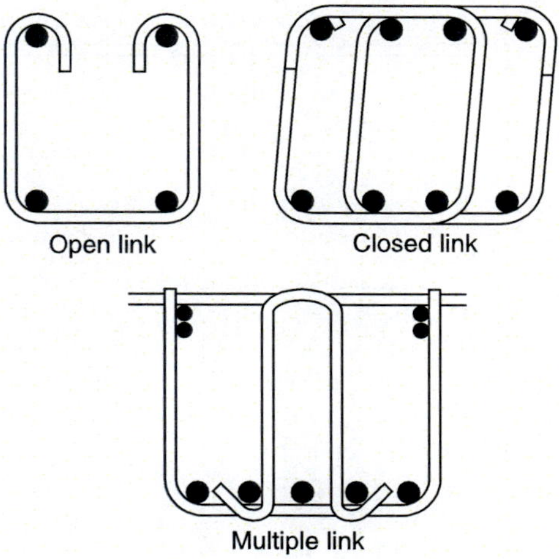

Using the code Section 5.14

$$\Delta x = 0.5\frac{L}{2} = \frac{6000}{4} = 1500\,\text{mm}$$

$$\Delta M = \text{change in moments over distance } \Delta x = \frac{L}{4} \text{ from moment}$$

$$= \frac{w_u L}{2} \times \frac{L}{4} - \frac{w_u L}{8} \cdot \frac{L}{8} = \frac{3 \times 40 \times 6^2}{32} = 135\,\text{KNm}$$

The change in the longitudinal force ΔF at the flange web interface is

$$= \frac{\Delta M}{(d - h_f/2)} \left[\frac{(b_f - b_w)/2}{b_f}\right] = 87\text{KN}$$

V_{Ed} = Longitudinal shear

$$Stress = \frac{\Delta F_d}{h_f \times \Delta x} = \frac{87 \times 10^3}{150 \times 1500} = 0.4\,\text{N/mm}^2$$

Check the concrete steel strength
To prevent crushing of the compressive steel in the flange

$$V_{Ed} \leq \frac{0.6((1 - f_{ck})/250)f_{ck}}{1.5(\cot\theta_f + \tan\theta_f)}$$

Note:- moments are sagging-compression steel in the flange

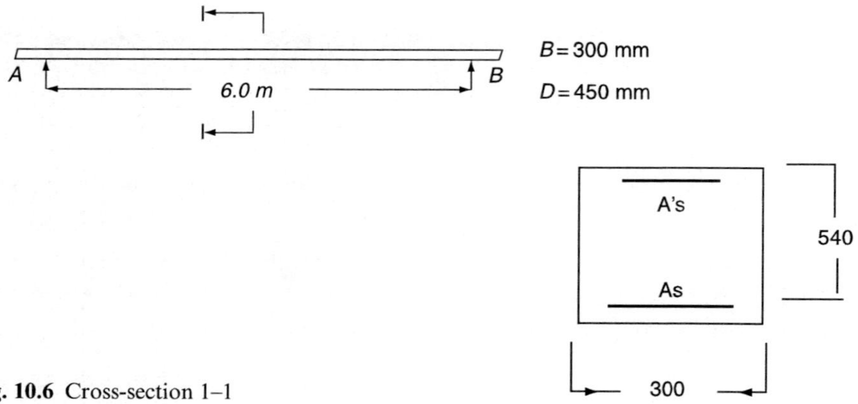

Fig. 10.6 Cross-section 1–1

$\leq 5.4N/\text{mm}^2 > 0.43N/\text{mm}^2$

And the concrete strength has sufficient strength for flange in tension θ is $45° \geq 38.6°$.

Limits for $\theta f\, 26.5° \leq \theta f \leq 45°$ take $\theta f = 26.5°$ mini.

Transverse steel

Transverse shear reinforcement required if

$$v_{Ed} \geq 0.27 f_{ctk} = 1.85\text{N/mm}^2\, for\, 25 \text{ grade of concrete}$$

$$0.27 f_{ck} = 0.49\text{N/mm}^2 > 0.43\,\text{N/mm}^2$$

Transverse reinforcement *Not Needed*

A minimum are 0.13%

$$A_{sf} = 0.1bh_f/100 = 0.13 \times 1000 \times 150/100 = 195\,\text{mm}^2/\text{m}$$

$$\text{Provide } H10@300 = 262\,\text{mm}^2/\text{m}$$

Reinforcement

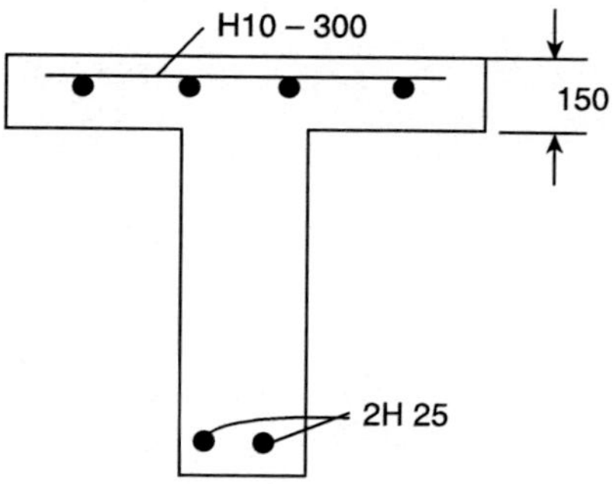

Fig. 10.7

(c) Reinforcement details

10.7.8 WE10.7

A simply supported slab, spanning 5 m as an effective span is to be designed using EC2. The following data are taken:

1. Floor finishes + ceiling loads = 1KN/m^2, basic span/depth = 0.19
2. Variable load = 3KN/m^2
   Characteristic material properties: $f_{ck} = 25N/mm^2; f_{yk} = 500N/m^2$
   Class × $c1$ exposure cover = 25 mm (c25/30) l = 25%

Solution

Slab depth = span/depth = 27 take 40% more
Minimum depth = span/27 × correction factor $c.f$

$$= \frac{5000}{27} \times c.f = 185.6/c.f$$

Since the span < 7 m; $c.f = 1.0$
Overall depth of the slab = 185.6 + 25 + 5 = 215.6 mm
Take initially the slab 200 mm on trial basis with d = 170 mm
Self-weight = 200 × 25 × 10^{-3} = 5KN/m^2
Total permanent load = 1 + 5 = 6KN/m^2
Ultimate load parameter width of the slab = $(1.35g_k + 1.5q_k)$ (5.0)
= (1.35 × 6 + 1.5 × 3) (5.0) = 63.0 KN
M = ultimate moment = 63 × 5/8 = 39.375 KNm
Bending reinforcement

$$\frac{M}{bd^2 f_{ck}} = \frac{39.375}{1000} \times (170)^2 \times 25 = 0.054$$

Code:- l_a = 0.96 adopt upper limit = 0.95; z = lever arm l_ad = 0.95 × 170 = 161 mm
A_s = area of steel = $M/0.87f_{ykz}$ = 562 mm^2
Provide H12-150; A_s = 754 mm^2
Check span–depth ratio = $\frac{100(562)}{1000(170)} = 0.33 > 0.13\%$ min
Code: span/depth = 32 while actual span/depth = $\frac{5000}{170} = 29.41$
Hence the chosen depth is OK.
Shear at the face of the support v_{Ed}

shear $v_{Ed} = \frac{61.7}{2}\left[\frac{2.25-0.5\times0.3}{2.25}\right] = 28.79\,\text{KN}$

$$P_1 = \frac{100 \times 754}{1000 \times 170} = 0.44$$

$$v_{R.dc} = v_{R,dc}bd, \quad V_{R,dc} = 0.55 \text{ around}, \quad \rho_1 = 0.4 \text{ no adjustment}$$

$$= 0.55 \times 1000 \times 170 = 0.935\,\text{KN}$$

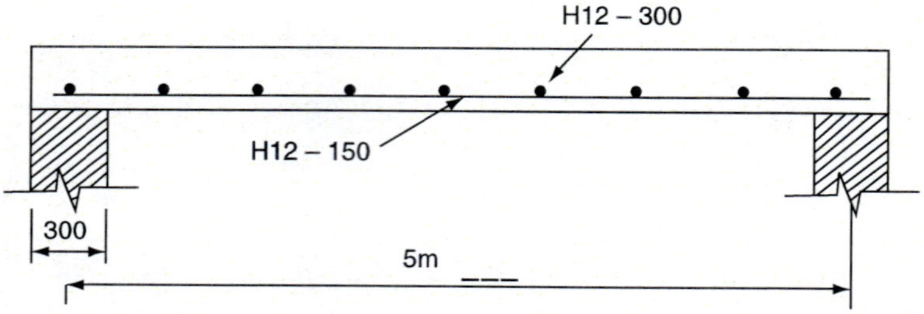

Fig. 10.8 A simply supported slab

Enchorage length $= 40\phi = 40 \times 12 = 480$ mm
Distributed steel $= 0.13bd = 0.13 \times 1000 \times 170 = 221\ \text{mm}^2/\text{m}$
Provided H12–300:- $A_s' = 377\text{mm}^2/\text{m} > 22\ \text{mm}^2/\text{m}$
The slab reinforcements are shown in Fig. 10.8.

10.7.9 WE10.8

Assuming the column is in a braced frame using EC2 and the following data:
Ultimate axial load $= 1279$KN; column $= 400 \times 300$

λ_{yy} = heighest slenderness ratio; beam of the frame 300×500
$f_{ex} = 25\ \text{N/mm}^2$; hcolumn $= 3$ m; $l_{\text{beam}} = 4$ m; $l_o = 0.59 \times 3.0 = 1.77\ m$

Solution

$$I_{col} = 400 \times 300^2/12 = 9 \times 10^8\text{mm}^4$$

$$l_{beam} = 300 \times 500^3/12 = 31.25 \times 10^8\text{mm}^4$$

$$k_1 = k_2 = \frac{\frac{I_{col}}{l_{col}}}{\sum\left(\frac{2I_{\text{beam}}}{l_{\text{beam}}}\right)} = \frac{\frac{40\times10^8}{3} \times 10^8}{\frac{2(2\times3.12\times10^8)}{4} \times 10^3} = 0.96$$

$$\lambda = \frac{l_o}{i} = 20.4$$

$$i = \text{radius of gyration} = \sqrt{\frac{bh^3/12}{bh}}$$

$$= 86.6\,\text{mm}$$

$$\lambda_{Lrm} = \textit{limited slenderness ratio} = \sqrt{\frac{26.2}{\{N_{Ed}/A_c f_{cd}\}}}$$

$$f_{cd} = \frac{25}{1.5} = 16.67$$

Hence $\lambda_{Lim} = 26.2\sqrt{\left\{\frac{1279}{400\times300\times16.67}\right\}} = 32.8 > 20.4 = \lambda$
The column is short since $\lambda_{Lim} > \lambda$.

10.7.10 WE10.9

A slender column 450 × 350 mm cross-section is subject to an ultimate load of 1650 KN and end moments at $A = 65$ KNm and at $B = 10$ KNm as shown in Fig. 10.9. The double curvature exist about y–y axis. Use the following data and EC2 design requirements, check the sufficient reinforcement are catered for $l_{ey} = 6.75\text{m}$; $l_{ez} = 7.5\text{m}$; $f_{yk} = 500\text{N}/mm^2$; $\phi_{\text{eff}} = \text{creep ratio} = 0.87$.

Solution

$$e_{ob} = \frac{MB}{N_{Ed}} == 6.06\,\text{mm};$$

$$e_{oA} = \frac{MA}{N_{Ed}} = \frac{-65 \times 10^3}{1650} = -40.0\,\text{mm (column bent in double base curvature)}$$

Limiting slenderness ratio

$A =^1 /_{(1+0.2\phi_{eff})} = 0.85$

B = the default value 1.1

$C = 1.7 - \frac{M_{oB}}{M_{oA}} = 1.7 - \frac{10}{65} = 1.854$

$\lambda_{Lim} = [20 \times A \times B \times C]/\sqrt{n}$

$[20 \times A \times B \times C]/\sqrt{\{N_{Ed}/A_c f_{cd}\}}$

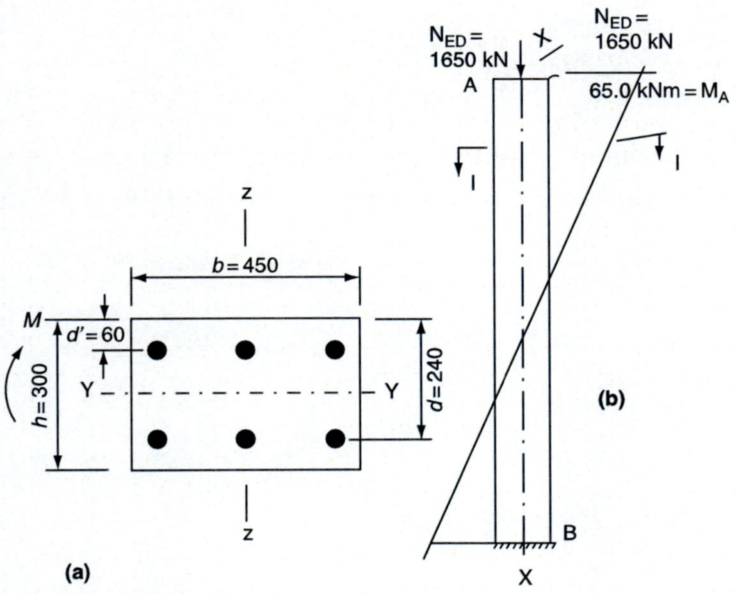

Fig. 10.9 Axial load with two acting moments

$\frac{20(0.85)(1.1)\times 1.854}{\sqrt{\frac{1650\times 10^3}{300\times 450\times 0.567\times 25}}}$

$\lambda_{\mathrm{Lim}} = 43.71;$

Actual slenderness ratio

$\lambda_y = \frac{l_{ey}}{l'_z} = \frac{6.75}{0.3} \times 3.46 = 77.85 > 43.71$

$\lambda_z = \frac{l_{ez}}{l'_z} = \frac{7.5}{0.45} \times 3.46 = 61.35 > 31.19$

The column is parallel to be slender and λ_y is critical.

Equivalent eccentricities $= e$

$$0.6e_{oA} + 0.4e_{oB} = 0.6(40) + 0.4(-6.06)$$

$$= 24.0 - 2.424 = 21.576\,\mathrm{mm}$$

$$0.4e_{oA} + 0.6 = 0.4(40) = 16\,\mathrm{mm}$$

Hence v = accidental eccentricity $= \frac{1}{200}$

$$e_{eq} = v\frac{l_{ey}}{2} = 16.88\,\mathrm{mm}$$

$$e_{so2} = \text{second} - \text{order eccentricity}$$

$$= \frac{k_1 k_2 l_o^2 f_{yk}}{\pi^2(103500d)}$$

Since $k_1 = \left(1 + 0.354 + \frac{f_{ck}}{200} - \frac{\lambda}{150}\right)\phi_{eff} = 0.96 \geq 1$

Hence $e_{so2} = 92.92$ mm

Initial value of $k_2 = 1$ is taken prior to iteration

First iteration

$$e_{tot} = e_{total} = e_o + e_{eq} + e_{so2} = 21.756 + 16.88 + 92.92 = 131.556\,\mathrm{mm}$$

$$M_{total} = M_{tot} = N_{Ed} \cdot e_{tot} = 1650 \times 131.556 \times 10^{-3}$$

$$\frac{N_{Ed}}{bhf_{ck}} = {}^{1650\times 10^{-3}}/_{450\times 300\times 25} = 0.489$$

$$\frac{M_{tot}}{bh^2 f_{ck}} = {}^{217\times 10^{6}}/_{450\times 300\times 25} = 0.214$$

From standard design chart on EC2

$$\frac{A_s f_{yk}}{bhf_{ck}} = 0.8 \text{ looking at the chart } k_2 = 0.78$$

Second iteration

Similarly when $k_2 = 0.78$; $e_{so2} = 72.48$; $e_{tot} = 111.116$ mm

$$M_{tot} = 183\,\text{KNm}; \frac{A_s f_{yk}}{bhf_{ck}} = 0.8; \frac{M_{tot}}{bh^2 f_{ck}} = 0.158$$

$$A_s = \frac{0.6bhf_{ck}}{f_{yk}} = 4050\,\text{mm}^2 \; k_2 = 0.74$$

$$N_{\text{balance}} = 0.29 f_{ck} A_c = 0.29 \times 25 \times 450 \times 300 \times 10^{-3} = 978\,\text{KN}$$
$$N_{Ed} = 0.567 f_{ck} A_c = 0.87 f_{yk} A_s = 3675\,\text{KN}$$

$$k_2 = \frac{3675 - 1650}{3675 - 978} = 0.75\,\text{is almost}\,k = 0.74\,\text{above}$$

PART C A NUCLEAR LABORATORY CASE STUDY BASED ON BRITISH CODES

10.8 A Nuclear Laboratory: A Case Study

10.8.1 Proposal and Preliminary Data

The provision is for an open laboratory involving only British Codes such as BS 8110, BS 5950 and BS 6399. Here a general design is given. The clear height is given to be 408 m from the nuclear concerns with a clear structural layout necessary for a number of classified testing and storage facilities. The nuclear concerns allocated an area of 50m $\times$ 50m. Many options regarding the structural layout can be available. Figures 10.10 and 10.11 are the structural schemes proposed for this text. A composite steel frame is chosen. The roof and cladding are of precast concrete units. It is proposed to have a concrete slab for the roof to withstand the blast load. The vertical frame is of trust type matching the roof truss supporting the concrete slab with the vertical stanchions made up of trusses. The stanchions or column are designed to withstand extreme wind loads and blast effects. Cladding play a significant part to resist such loads. Since the building is to be clayed in precast concrete units with insulation board combgated steel cover. The main door is to be top up hung when it is open to give a clear opening of 30 m $\times$ 40 m. Several doors are be located for access to personnel. They should not be more than 3.

10.8.2 Loadings

(a) Roof = 3 KN/m^2
(b) Point load applied 500 KN + 10% impact to roof loading at the centre

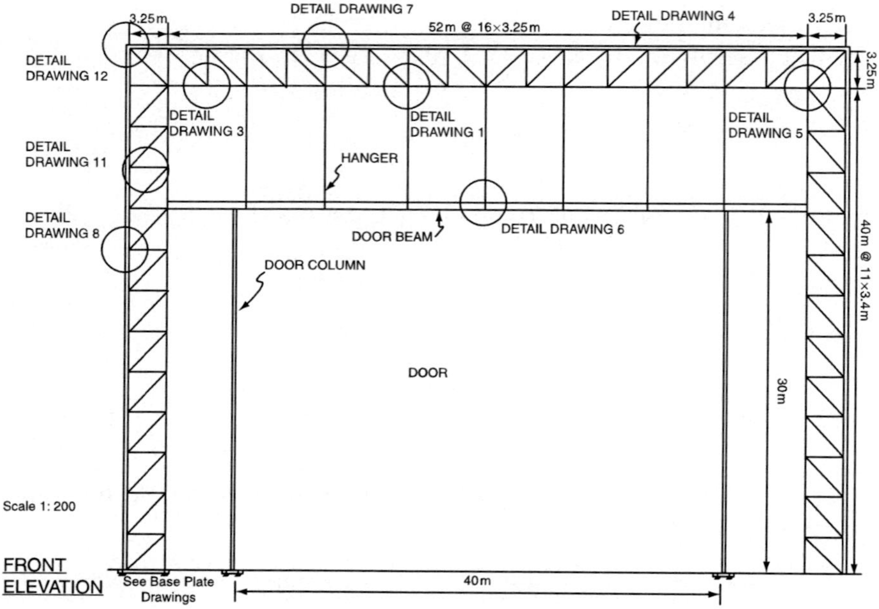

Fig. 10.10 Structural elevational details

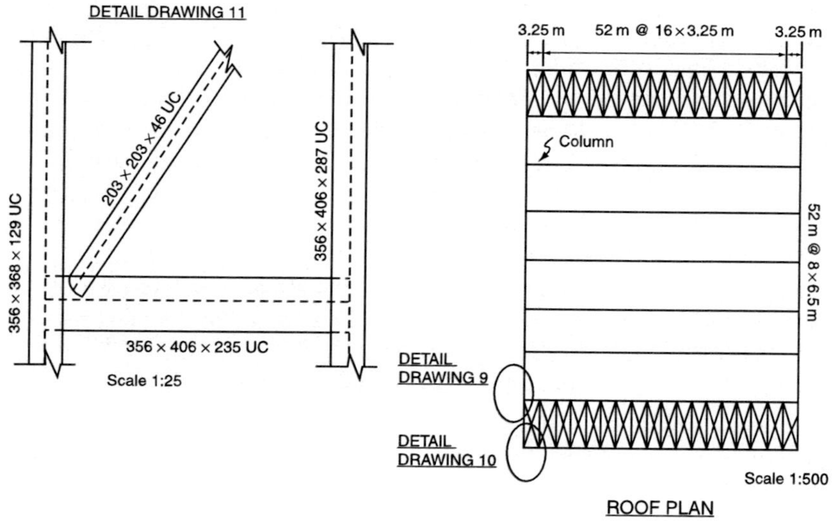

Fig. 10.11

(c) Internal blast pressure $= 5\ \text{KN/m}^2$
(d) basic design wind speed $= 40$ m/sec

10.8.3 Site Condition

0.0–0.31m fill
0.31–2.5m clayeysilt
0.25–18.0m solid chalk

Ground water not allowed.
The bearding capacity for the soil type is evaluated as $75\ \text{KN/m}^2$
Piles and pile caps can be a possibility.

10.8.4 Loading on the Roof

1. Downward loading

$$\text{Imposed load} = 3\ \text{KN/m}^2\ \text{(adverse)}$$
$$\text{Dead load} = \text{snow load} + \text{finishes} + \text{dead load (adverse)}$$
$$= 0.75 + 1.5 + \text{self-weight}$$

Assuming 200 mm slab

$$\text{Self-weight} = 0.2 \times 24 = 4.8\ \text{KN/m}^2$$
$$\therefore \text{Deadload} = 0.75 + 1.5 + 4.8 = 7.05\ \text{KN/m}^2$$
$$\text{Design load} = 1.6 \times LL + 1.4 \times DL$$
$$= 1.6 \times 3 + 1.4 \times 4.8$$
$$= 14.67\ \text{KN/m}^2$$

2. Upward loading

$$\text{Imposed load} = 2\ \text{KN/m}^2\ \text{(benificial)}$$
$$\text{Imposed load} = 5\ \text{KN/m}^2\ \text{blast load (adverse)}$$
$$\text{Dead load} = \text{finishes} + \text{self-weight (benificial)}$$
$$= 1.5 + 0.2 \times 24$$
$$= 6.3\ \text{KN/m}^2$$
$$\text{Design load} = -6.3 \times 1.4 - 3 \times 1.6 + 5 \times 1.6$$
$$= -5.62\ \text{KN ie downward}$$

The downward loading is chosen for this is the worse case. The negative value in the second calculation shows when blasting occurs. The load is balanced by the downward imposed load, dead load and self-weight of the roof slab.

$$\text{Design load} = 14.67\ \text{KN/m}^2$$

10.8.5 Design of Roof Slab

The roof slab will be connected to the top beam flange by bolting, it is assumed that the slab is simply supported at four sides. For in reality it is actually partially fixed, reinforcement is provided at the top of the slab to overcome the fixed end moment.

In general this type of slab could deflect about both axis under load and the corner lift and curl up from supports causing torsional moments then the moment coefficient is used and the maximum moments are given by

$$M_{sx} = \alpha\ S_x\, nL^2x$$

And

$$M_{sy} = \alpha\ S_y\, nL^2x$$

Where M_{sx} = moment at the mid span on the strips of unit width in Lx direction;

M_{sy} = moment at the mid span on the strips of unit width in Ly direction;

$n=$ total ultimate load per unit area;
$Lx=$ the length of the short side;
$Ly=$ length of the long side;
$\alpha\, S_x$ and $= \alpha\, S_y$ are the moment coefficients.

Design for bending and short span of slab

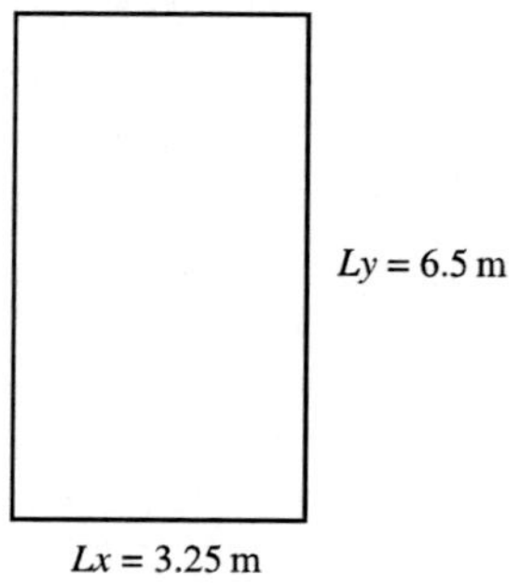

Fig. 10.12 Loading area of the roof slab. **(a)** Slab depth

Table cover = 40 mm

$$d = 200 - 40 = 160$$

$$M_{sx} = \alpha S_x nL^2x$$

$$= 0.118 \times 14.67 \times (3.25)^2 = 18.28\,\text{KNm}$$

$$\frac{M_{sx}}{bd^2 f_w} = \frac{18.28 \times 10^6}{1000 \times (160)^2 \times 30} = 0.023$$

(a)

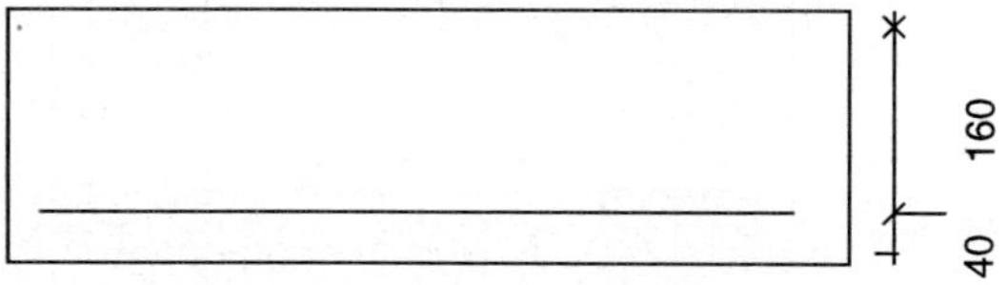

Lever arm, $z = 0.94d = 0.94 \times 160 = 150\,\text{mm}$

$$A_s = \frac{M_{sx}}{0.87 f_{yz}} = \frac{18.28 \times 10^6}{0.87 \times 460 \times 150} = 305\,\text{mm} \;=\; \text{per m width}$$

Use T12 @ 300 centre-to-centre checking for maximum span

$$\frac{M_{sx}}{bd^2} = \frac{18.28 \times 10^6}{1000 \times (160)^2} = 0.71$$

$$\text{Service stress} = \frac{5}{8} \times 460 \times \frac{305}{377} = 233$$

$$\text{maximum} \frac{\text{span}}{\text{effective depth}} = 20 \times 1.8 = 36m$$

$$\text{Reactual} \frac{\text{span}}{\text{effective depth}} = \frac{3250}{160}$$

$= 20.3\,\text{m} < 36\,\text{m}$ $O.K$ 200 mm thick slab is adequate

10.8.6 Design for Bending and Long Span of Slab

$$M_{sy} = \alpha S_y n L^2 x$$
$$= 0.029 \times 18.28 \times (3.25)^2 = 5.6\,\text{KNm}$$

As this reinforcement is at inner depth after the previous one the effective depth is reduced to

$$Z = 160 - 12 = 148\,\text{mm}$$

$$\therefore \text{A}_\text{z} = \frac{M_{sy}}{0.87 f_{yz}} = \frac{5.6 \times 10^6}{1000 \times 200} = 95\,\text{mm/m width}$$

$$\frac{100\text{A}_\text{s}}{bh} = \frac{100 \times 95}{1000 \times 200} = 0.05 < 0.13$$

Minimum transverse steel ratio $= 0.13$

$$\frac{100A_s}{bh} = 0.13$$

$$A_s = \frac{0.13 \times 1000 \times 200}{100} = 260\,\text{mm}^2/\text{m width}$$

Use T10 @ 300 centre to centre. (T10–300)

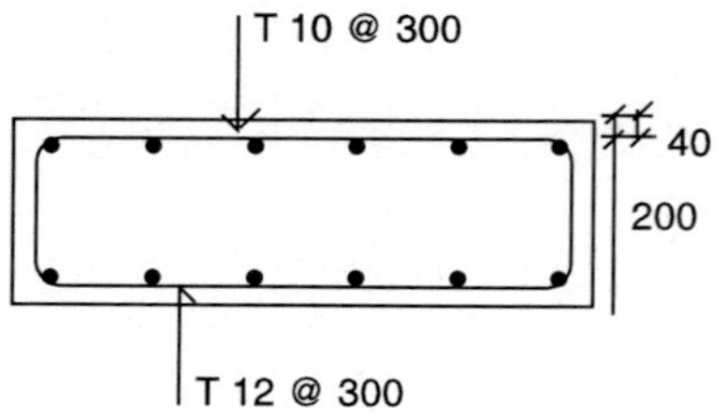

10.8.7 Design of Roof Frame

Calculation of forces in the chords

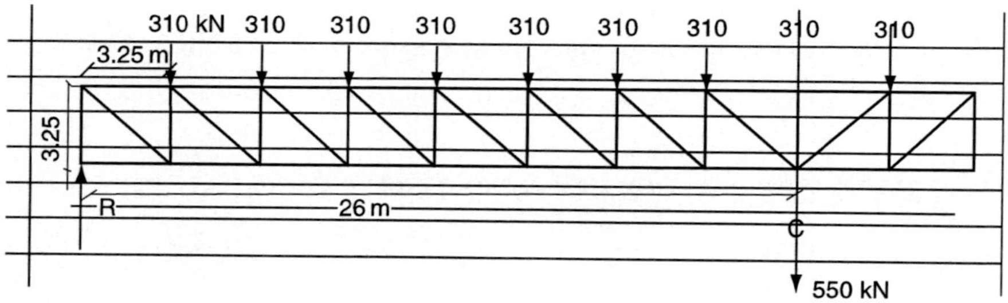

Fig. 10.13 Roof frame

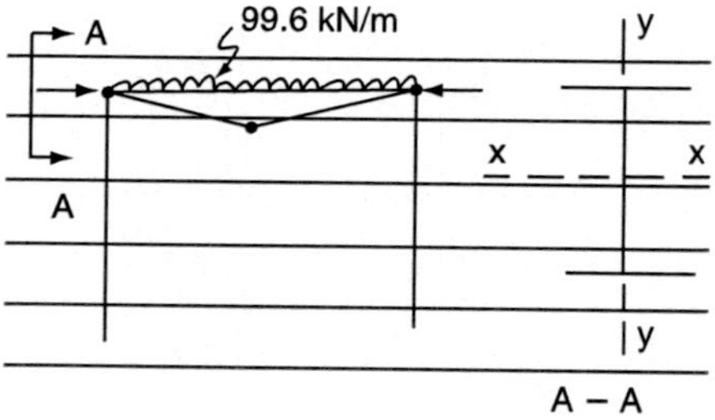

Fig. 10.14 Top chord

From previous calculations, roof load = 14.67 KN/m^2
Requirement, point load = $14.2 \times 3.25 \times 6.5 = 310$ KN
Self-weight of the girder, assume = $3\ KN/m \times 1.4 = 4.2$ KN/m
BM @ centre = $2710 \times 26 - 310(22.75 + 19.5 + 16.25 + 13 + 9.75 + 6.5 + 3.25 - \frac{55}{2}(26) = 35,100$ KNm

Force in chords = $\frac{35100}{3.25} = 10,860$ KN i.e top chord10,860 compression bottom chord 10,860 tension

Design of top chord
udl on top chord = $14.67 \times 6.5 + 3 \times 1.4 = 99.6$ KNm
3 KN/m = self-weight of girder

Buckling about Y–Y axis

With the precast slab top flange is fully restrained

$$\Delta g = \frac{F}{P_y} \text{ take } P_y = 410\ \text{N/mm}^2$$

$$\Delta g = \frac{10800 \times 10^3}{410 \times 10^3} = 264\ \text{cm}^2$$

Try $\Delta g = 300\ \text{cm}^2$; 356 × 406 × 235 *UB*

$$M = \frac{wL^2}{16} = \frac{99.6 \times (3.25)^2}{16} = 65.75\ \text{KNm}$$

$$P_y \Delta g = \frac{410 \times 300 \times 10^2}{10^3} = 1923\ \text{KNm}$$

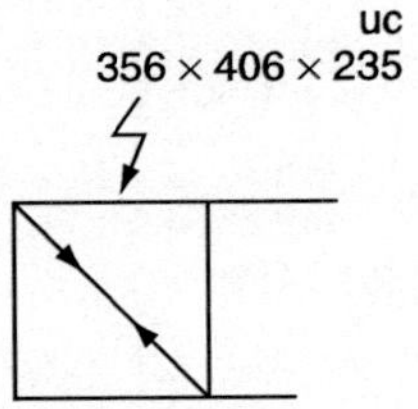

$$\frac{F}{P_y \Delta g} + \frac{M}{pbs_x} = \frac{10800}{12300} + \frac{65.75}{1923} = 0.91 < 1\ O.K$$

Design of bottom of chord
Direct load = 10,800 tension

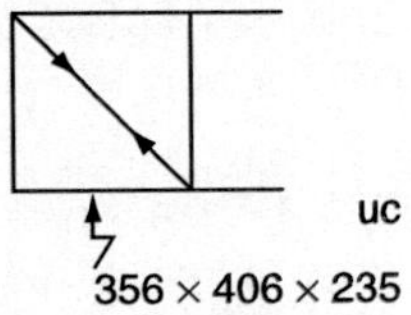

$$\Delta g = \frac{f}{P_y} = \frac{10800 \times 10^3}{410 \times 10^2} = 264\,\text{cm}^2$$

Use $356 \times 406 \times 235\,UB\,(\Delta g = 300\,\text{cm}^2)$
Design of vertical member

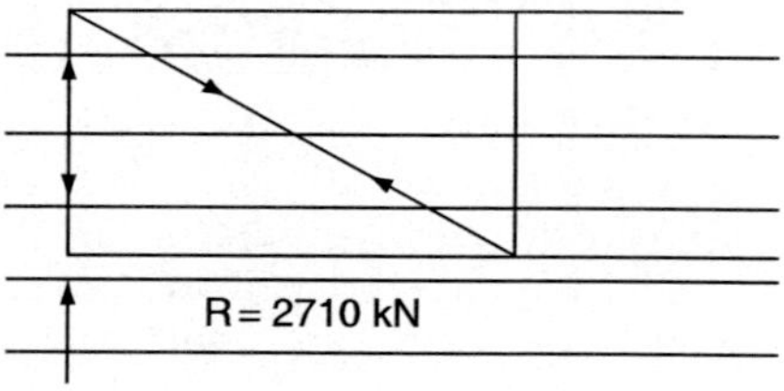

$$l_{ey} = 0.85 \times 3.25 = 2763\,\text{mm}$$

$$l_{ex} = 0.85 \times 3.25 = 2763\,\text{mm}$$

Taking $533 \times 210 \times 82\,UB$

$r_x = 21.3\,\text{cm}\ \Delta\text{g} = 104\,\text{cm}^3$
$r_y = 4.38\,\text{cm}$
$\lambda_y = \frac{2763}{4.38 \times 10} = 63.08$
$\lambda_x = \frac{2763}{21.3 \times 10} = 12.97$
$P_c = 327\text{N/mm}^2$

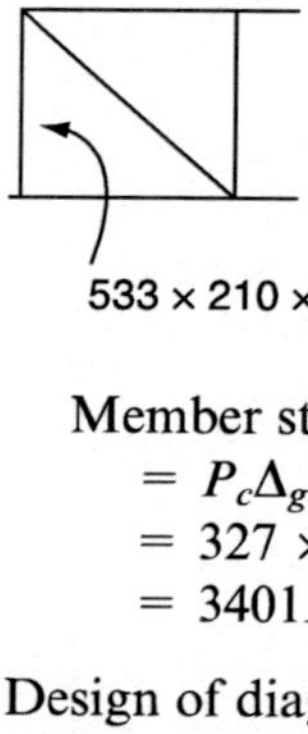

Member strength
$= P_c \Delta_g$
$= 327 \times 104 \times 10^2$
$= 3401KN > 2710KN\ O.K$

Design of diagonal member

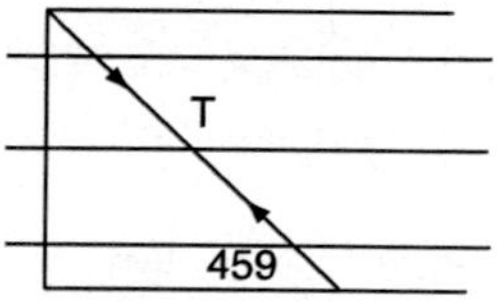

$$T = \frac{R}{\sin 45}$$
$$= 2710\sqrt{2} = 3833KN$$

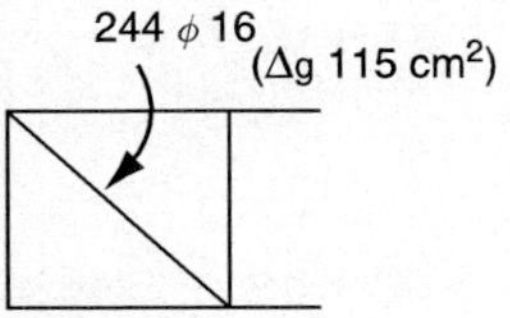

Use circular hollow section

$$\Delta g = \frac{3833 \times 10^3}{410 \times 10^2} = 93.4\,cm^2, \text{Use } 244ø16(\Delta g15\,cm^2)$$

10.8.8 Design of Braced Column

Structural analysis to evaluate the moment in outer column.

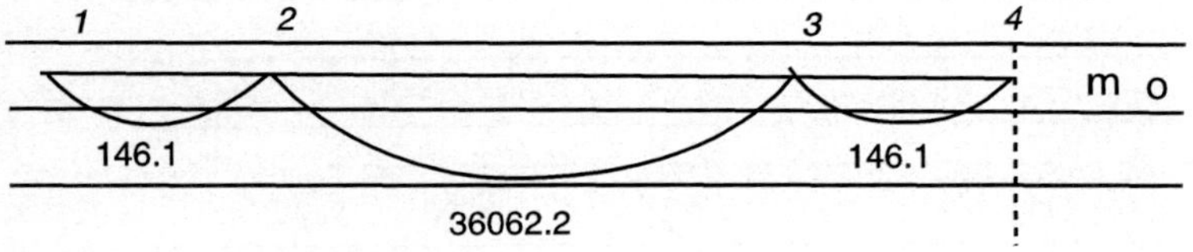

Uniformly distributed liad roofing truss beam = 95.4*KN*

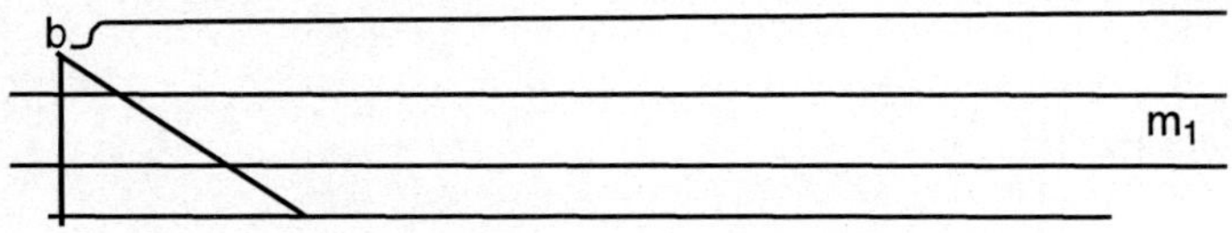

Point load specified $= 500KN + \frac{10\times51}{100} = 550KN$

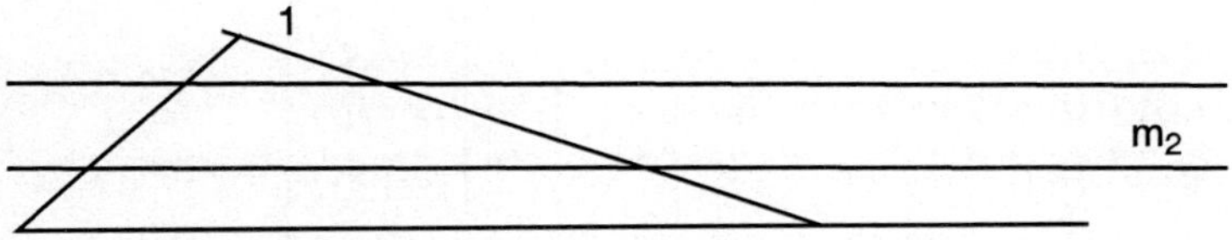

$$f_{11} = f_{44} = 1/3(3.25)(1)(1)EI$$

$1.083/EI$

$$f_{22} = f_{33} = 1/3(3.25)(1)(1)EI$$

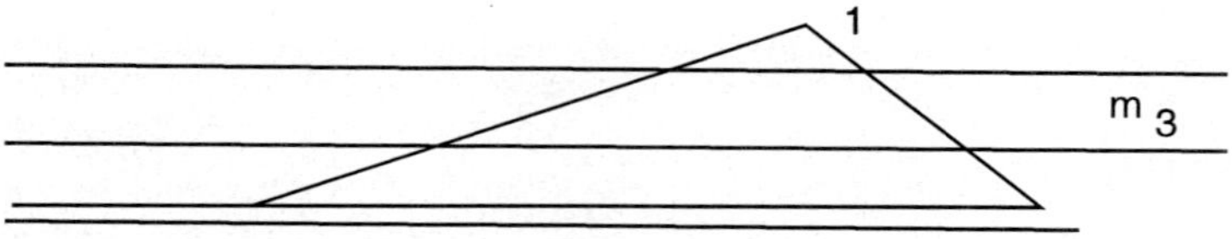

$17.75/EI$

$$f_{12} = f_{34} = 1/3(1)(1)(3.25)/EI$$

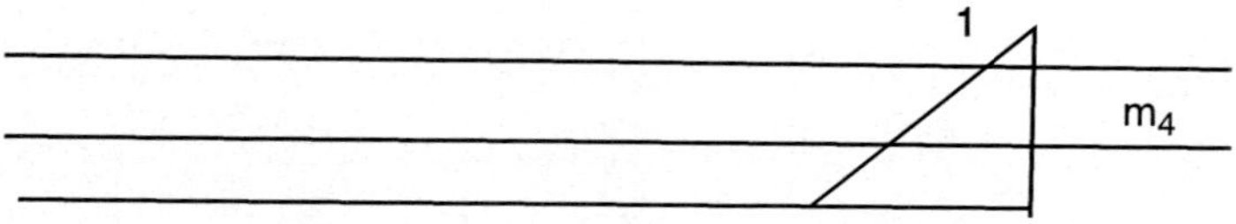

1.083/EI

$F_{43} = f$

$f_{41} = f_{14} = 0$

$f_{31} = f_{13} = 0$

$f_{32} = f_{412} = 1/3(50)(1)(1)/EI$

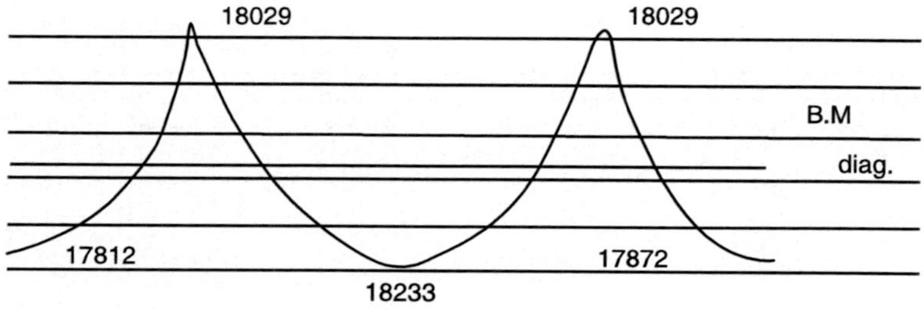

Fig. 10.15 Braced column using flexibilities analysis

$= 16.67/EI$

$D_o = D_o = 1/3(3.2)(146.1)(1)/EI$

$= 170.45/EI$

$D_{20} = D_{20} = \frac{1/3(50)(36062.2)(1)}{EI} + 17.75/EI$

$= \frac{601207}{EI}$

$$\begin{bmatrix} 1.083 & 1.083 & 0 & 0 \\ 1.083 & 17.75 & 16.67 & 0 \\ 0 & 16.67 & 17.75 & 1.083 \\ 0 & 0 & 1.085 & 1.083 \end{bmatrix} \begin{bmatrix} x_1 \\ x_2 \\ x_3 \\ x_4 \end{bmatrix} = \begin{bmatrix} 170.45 \\ 601207 \\ 601207 \\ 170.45 \end{bmatrix}$$

$$\begin{bmatrix} x_1 \\ x_2 \\ x_3 \\ x_4 \end{bmatrix} \begin{bmatrix} -16.57 & 166.7 & -166.7 & 166.68 \\ 166.6 & -166.7 & 166.7 & -166.7 \\ -166.7 & 16.67 & -166.7 & 166.7 \\ 166.7 & -166.7 & 166.7 & -166.7 \end{bmatrix} = \begin{bmatrix} 170.45 \\ 601207 \\ 601207 \\ 170.45 \end{bmatrix}$$

$$\text{or} = \begin{bmatrix} -17872 \\ 18.29 \\ 18029 \\ -17871 \end{bmatrix}$$

$M = M_0 + M_1x_1 + M_2x_2 + M_3x_3 + M_4x_4$
$M_1 = -17871.7KNm$
Δt mid 2 and 3

$$M = 36062.2 - 18029.1$$
$$= 18233.2\text{KNm}$$

From the calculation it can be seen that if the outer column is totally fixed and the inner column behave as a simply supported column, the outer column will take the rotation moment and the inner column will take the maximum load.

$$\text{Load supported by the inner column} = 1/2(550) + 95.4(50/2) + 3.25(95.4/2)$$
$$= 2815KN$$

Load from blasting UDL $= 5 \times 6.5 = 176.8\ KN$
By assuming there is no lateral displacement at the top of column

$$R = 176.8 \times 5.5 = 972.4\ \text{KNm}$$
$$BM\ at\ ø = 97.24 \times 20.4 - 176.8(17 + 13.6 + 10.2 + 6.8$$
$$= 10820\ \text{KNm}$$

$$\text{Force in chords due to blast loading} = \frac{10820}{3.25} = 3329\ \text{KN}$$

Inner Column C Compression.

Total axial load $= 2815 + 3329 = 6144$ KN
$UDL = \frac{176.8}{3.4} = 52\ \text{KN/m}$
$m = \frac{52 \times 3.4^2}{16} = 37.57$

Use UC
Assume $P_c = 200\text{N/mm}^2$

$$\Delta g\,\text{required} = \frac{6144 \times 10^3}{200 \times 10^2} = 307.2\,\text{cm}^3$$

Try UC 356 × 406 × 287

$r_x = 16.5$ $r_y = 10.3$ $S_x = 5820$

$\lambda_x = \dfrac{289}{16.5} = 17.5$ $\lambda_y = \dfrac{340}{10.3} = 33$

$P_c = 242\,\text{N/mm}^2$

$$P_b = 265\,\text{N/mm}^2$$

$$P_c\Delta g = \frac{242 \times 366 \times 10^2}{10^3} = 8857.2$$

$$P_c S_x = \frac{265 \times 5820}{10^3} = 1542.3$$

Iteration

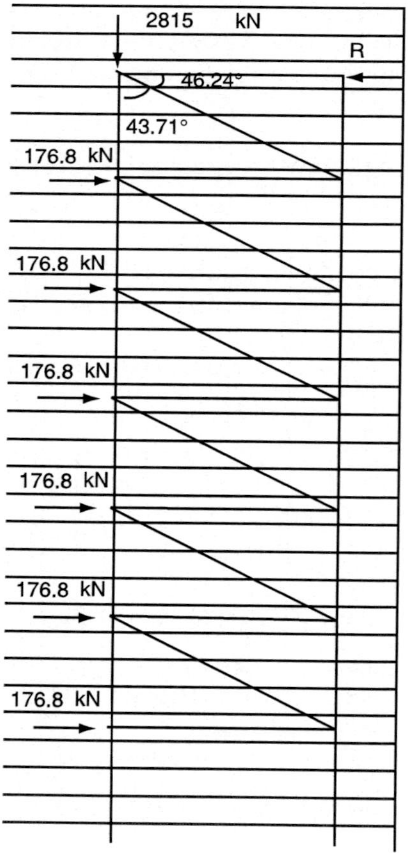

Fig. 10.15 Braced columns of a frame

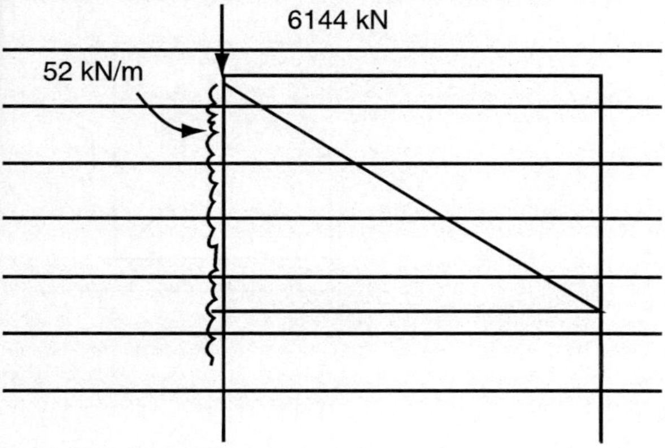

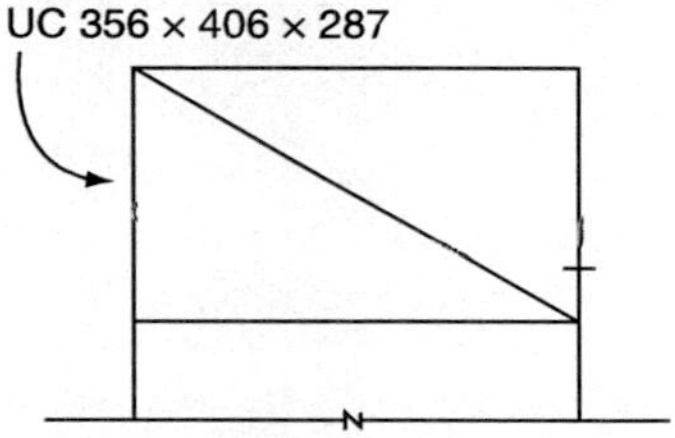

$$\frac{6144}{8857.2}+\frac{37.52}{1542}=0.718<1\ O.K$$

Design of outer chord
Bending moment due to roof loading = 17,872 KNm
Therefore tension in the chord = $\frac{17872}{40.8}$ = 438 KN

Force in chord due toblasting = 33.92 KN
Total = 3767 *KN*
Asuume $P_c = 0.9 \times 265 = 240\,\text{KN/mm}^2$
$\Delta g = \frac{3767\times10^3}{240\times10^2} = 156.9\,\text{cm}^2$

Try

$UC = 356 \times 368 \times 129(165)$
Tension resistance = 265 × 165
= 4372.5 *KN*> 3767 *KN is O.K*

10.8.8.1 Design of Diagonal Ties

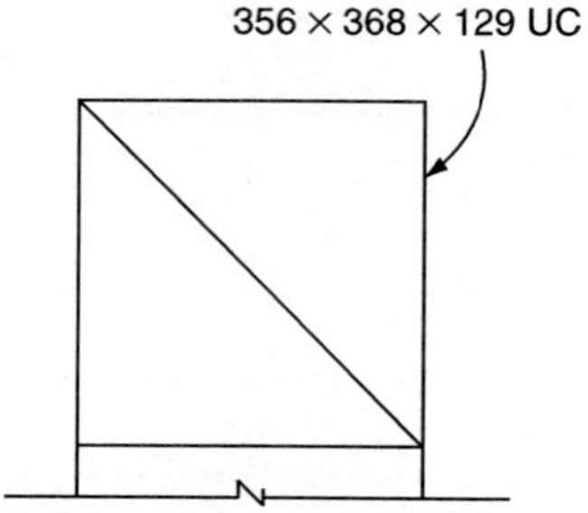

$F_1 = \frac{972}{\cos 46.29} = 1406.6\text{KN}$
$P_y = 265\text{N/mm}^2$
$\Delta g = \frac{1407 \times 10^3}{265 \times 10^2} = 53.1\,\text{cm}^2$

Try $203 \times 203 \times 46(58.8)UC$
Strength $\frac{265 \times 58.5 \times 10^2}{10^3} = 1588KN > 1406KN\,O.K$

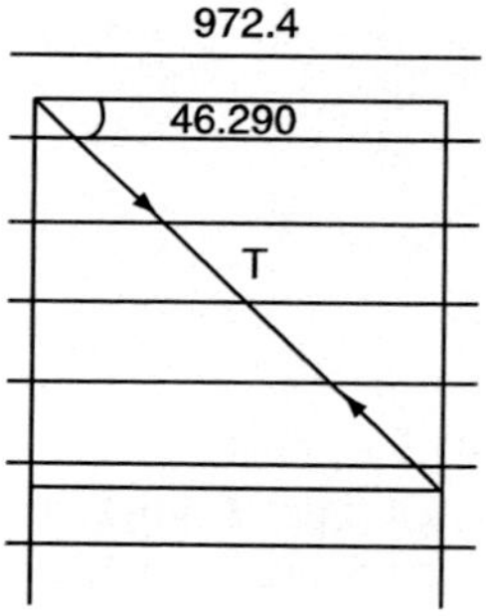

Fig. 10.16 Diagonal ties

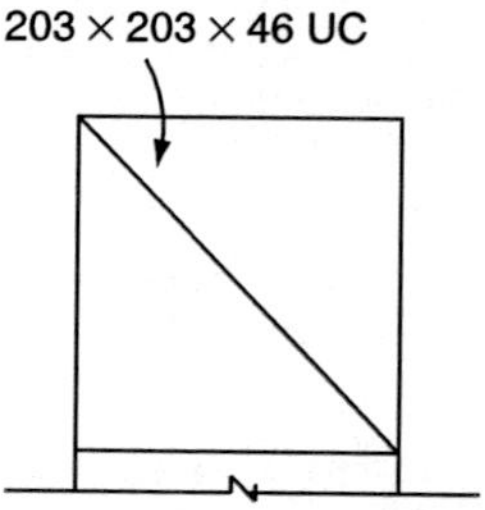

10.8.9 Design Base Plate for Inner Column

Design load = 6144 KN

Chosen design strength of the concrete foundation = 7.5 N/mm^2

As required = $\frac{6144\times10^3}{7.5} = 819,200\,mm^2 = 905 \times 905\,\mathrm{mm}^2$

Use 910 square base plate

Actual pressure under the plate = $\frac{6144\times10^3}{905\times905} = 7.42\,\mathrm{N/mm}^2$

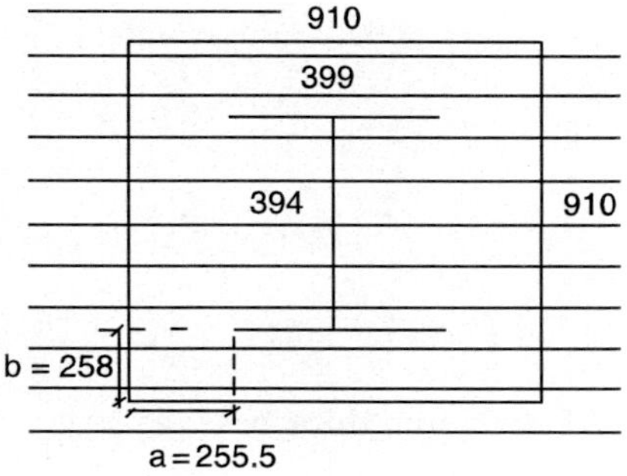

Fig. 10.17 Inner column base plate

Thicknes of base plate $t = \sqrt{\frac{2.5\times7.42}{265}\left[255.5^2 - 0.3\,(258)^2\right]} = 56.3\,\mathrm{mm}$

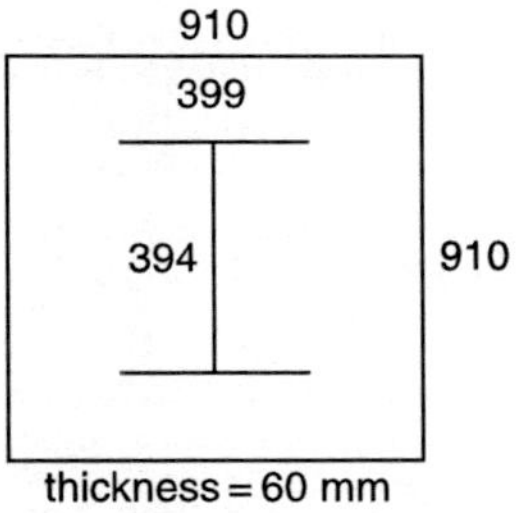

User 60 mm thick plate.

10.8.10 Design of Base Plate for Outer Column

Design load = $95.4 \times \frac{3.25}{2} = 155\,\mathrm{KN}$

For concrete foundation = 7.5 N/mm^2

As required $= \frac{155\times10^3}{7.5} = 20,670\,\mathrm{mm}^2 = 144\,\mathrm{mm}^2$

700 small for the column

Chosen 500 × 500 square base plate

Actual pressure under the plate = $\frac{155\times10^3}{500\times500} = 0.62\,\mathrm{N/mm}^2$

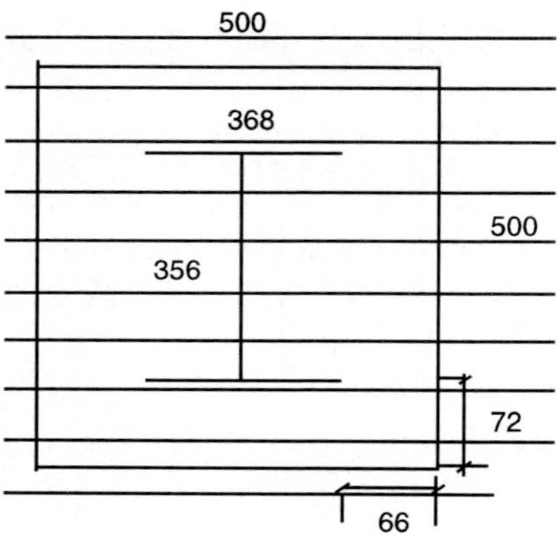

Fig. 10.18 Outer column base plate

Thickness of plate t $= \sqrt{\frac{2.5 \times 0.62[66^2 - 0.3(72)^2]}{265}}$

$t = 4$ mm

Use 25 mm thick base plate.

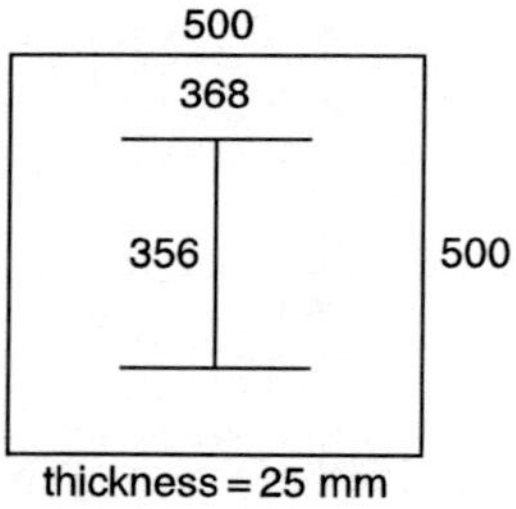

10.8.11 Design for Wind-Loading Resistance

Basic design wind speed $= 40\ m/s$

Class 3B

Total weight $= 40.8 + 3.25 = 44.05$

$s_2 = 1.02$

$v_z = v \times s_1 \times s_2 \times s_3$

$s_1 = s_3 = 1$

$v_z = 40 \times 1 \times 1.02 \times 1 = 40.8\,\mathrm{m/sec}$

$q = kv_z^2$

$k = 0.613$

$q = 0.613 \times 40.8 = 1020\,\mathrm{N/m^2} = 1.02\,\mathrm{KN/m^2} < 5\,\mathrm{KN/m^2}$

As the loading is less than the blasting load which was the design load for the column frame in direction parallel to the main frame no bracing required in this direction.

Design of bracing in direction perpendicular to the main frame.

$\frac{l}{w} = \frac{58.5}{52} = 1.125$

$\frac{h}{b} = \frac{44.05}{52} = 0.847 \approx 1$

$cf = 0.95$

$q = 0.969 < 5\,\mathrm{KN/m^2}$

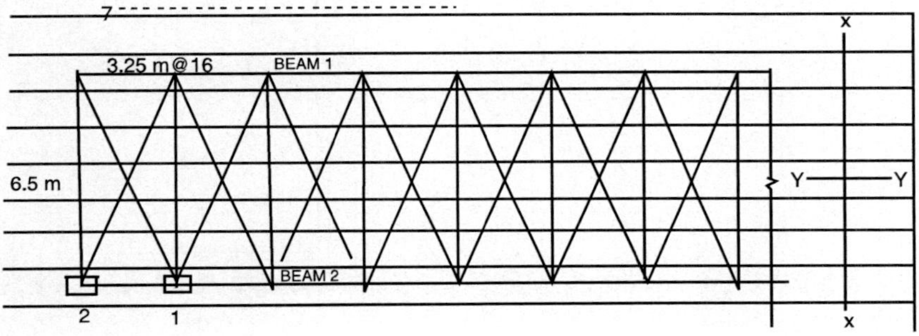

Fig. 10.19 Plan of T100 beam at the edge braced for wind

Pressure in x direction

$q = 0.969 \times 1.4 = 1.3566\,KN/mm^2$

Horizontal load on column $x - x$ (1)

$$= \left(\frac{1}{3}(3.25) + \frac{1}{2}(50)1.3566\right)$$

$= 26.625 \times 1.3566 = 36.12KN/m$

Therefore horizontal load on topof column= $36.12 \times \frac{40.8 \times 3.25}{2} = 795.53KN$

But

Horizontal load on the front joint of thewind girder $= 3.25 \times 1.3566 = 4.4\,KN$

$$R = \frac{4.4 \times 16}{2} = 35.27\,KN$$

BM at the centre $= 35.27\left(\frac{52}{2}\right) - 4.4(22.75 + 19.57 + 16.25 + 13 + 9.75 + 6.5 + 3.25) = 516.66\,KNm$

$$\text{Force in chord} = \frac{216.66}{6.5} = 79.49\,KN$$

$$F = \frac{79.49}{\cos 63.4} = 177.52\,KN$$

Use M 24 MSFG

Slip resistance $= 207\,KN > 177.52\,KN$

For the size and no. of bolt

Use $200 \times 100 \times 10$ unequal angle

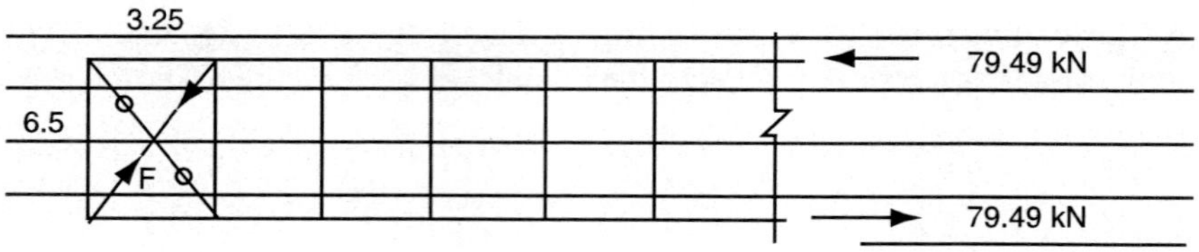

Fig. 10.20 Analysis for diagonal ties

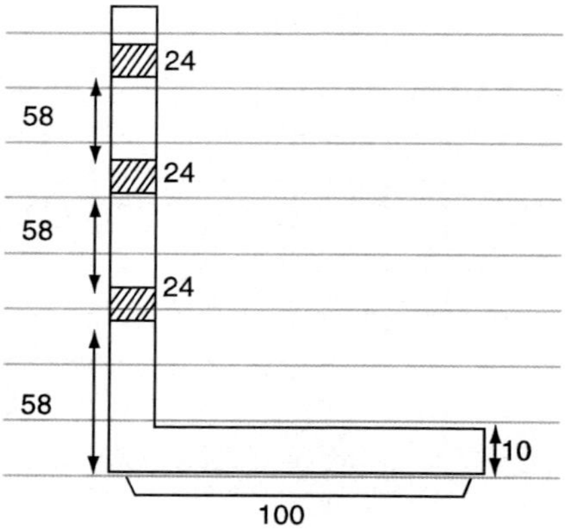

Fig. 10.20a Diagonal ties

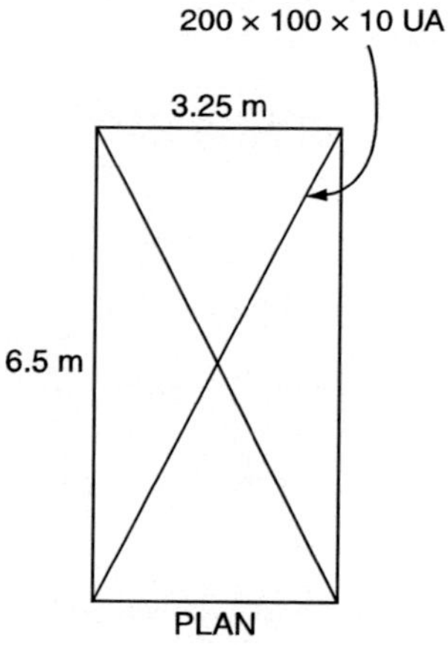

$$a_e = a_1 + \left(\frac{3a_1}{3a_1 + a_2}\right)a_2$$

$$a_2 = (200 - 48)10 = 1520$$

$$a_2 = 1520 + \left(\frac{3(1520)}{3(1520 + 1000)}\right)1000 = 2340\,\text{mm}^2$$

$$Strength = p_y \Delta e = \frac{1240 \times 265}{10^3}$$

$$= 620\,\text{KN} > 177.52\,\text{KN O.K}$$

10.8.12 Design of Diagonal Ties of Two Edge Column

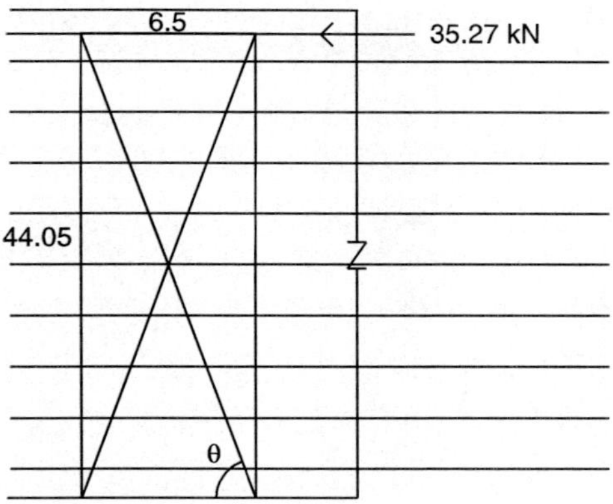

$\tan\theta = \frac{44.05}{6.8}$
$\theta = 61.61^\circ$
$F = 35.27$
$\cos 82.59 = 241.6KN$

Use 3 M 24 HSFG

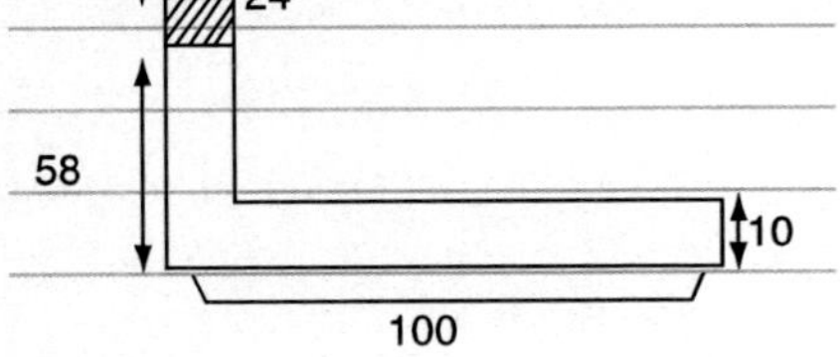

Use 200 × 100 × 10

$a_e = a_1 + \left(\frac{3a_1}{3a_1+a_2}\right)a_2$
$a_1 = (200 - 3(24)10) = 1280\,\text{mm}^2$
$a_e = 1280 + \left(\frac{3(1280)}{3(1280)+1000}\right)1000$
$207\ mm^2$
Strength $=p_y\Delta e = \frac{2073\times265}{10^3} = 549.4KN > 241.6KN$ ∴ O.K

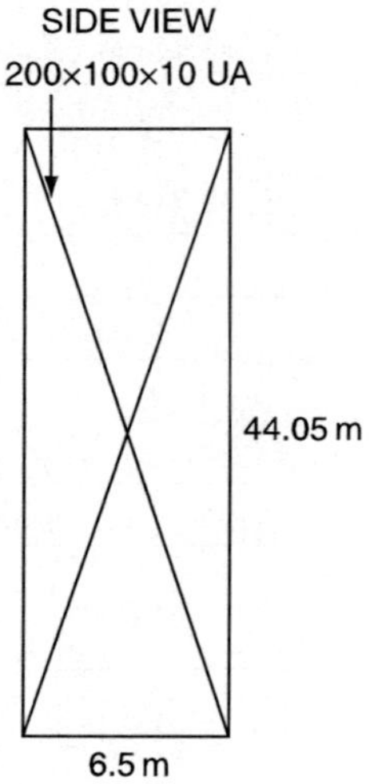

10.8.13 Foundation Design

$N = 6144\,KN$
$M = 37.57\,KN$
$N = 6144\,KN$
Pile load $= \frac{6144}{4} = 1536\,KN$

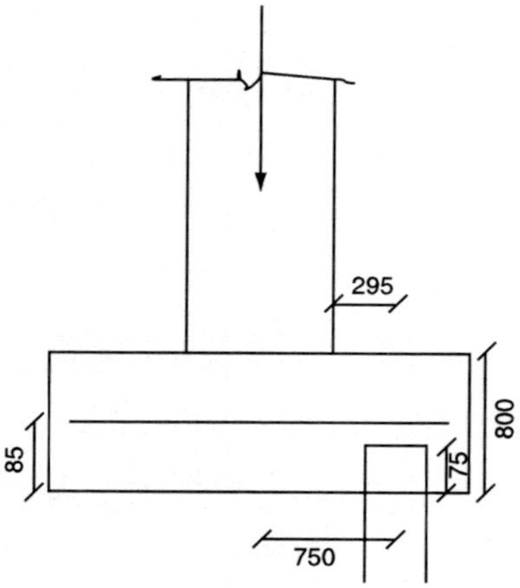

$M = 2 \times 1536 \times 0.295 = 906.2\,\text{KNm}$

$\frac{M}{bd^2 f_{cu}} = \frac{906 \times 10^6}{30 \times 2300 \times 715} = 0.03$

Lever armz $=$ $0.94d$

$$A_s = \frac{906 \times 10^6}{0.87 \times 460 \times 0.94 \times 715} = 3368\,mm^2$$

Use 8v 25 (3930 mm^2)

(b) Shear checkingPunching around the base plate

$$v = \frac{v}{u_c d} = \frac{6144 \times 10^3}{4 \times 910 \times 715} = 2.36\,N/\text{mm}^2$$
$$< 0.8\sqrt{f_{cu}}\ \ Shear\ O.K$$
$$1.5d = 1.5 \times 715 = 997.5$$
$$> 950$$

Distance from column centre to outer diameter of pile. No checking for shear perimeter is required. Enhanced Shear

Thickness of cap = 800 *mm*
Pilediameter = 500 *mm*

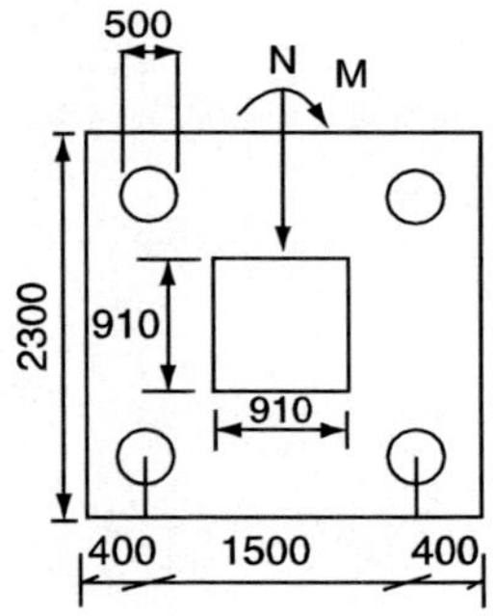

Fig. 10.21 Foundation design

(a) Reinforced design Enhanced *shear*

Adverse	Beneficial
$N = \frac{6144}{2} = 3070$	$N = \frac{6144}{2(1.5)} = 20.48$
$\frac{M}{d} = \frac{37.57}{1.5} = 25.05$	$\frac{M}{d} = \frac{37.57}{1.5} = 25.05$
Total P = 3097	Total P = 2073
Load per pile = 1548 *KN*	Load per pile = 1037 *KN*

$$v = 15.48\,\text{KN}$$

$$v = \frac{2 \times 1548 \times 10^3}{2300 \times 715} = 1.88\,\text{N/mm}^2$$

$\frac{100A_s}{bd} = \frac{100\times3900}{2300\times715} = 0.239$
$v_c = 0.42\,\mathrm{N/mm^2}$
$v_c' = v_c \times \frac{2d}{a_c}$
$= 0.42 \times 2 \times \frac{715}{295}$
$= 2.04\,\mathrm{N/mm^2}$
$> 0.42\,N/mm^2\; v_c'\; O.K$

No link required.

10.8.14 Wall Design

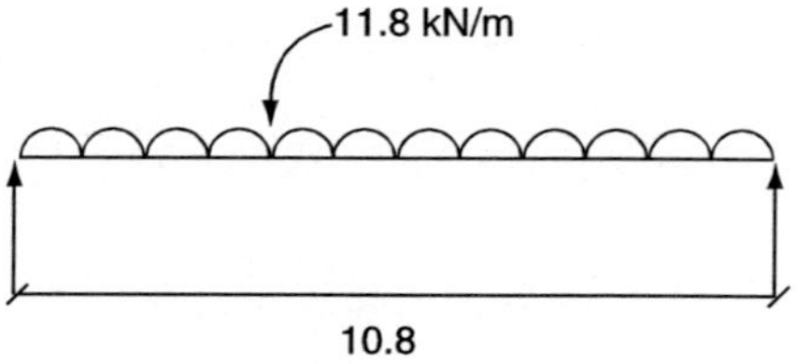

Wall loading

$$w = 11.8\,\mathrm{KN}/m$$
$$\text{Bending moment } M = \frac{wL^2}{8} = \frac{11.8\times10.8^{\circ}}{8}$$
$$= 172.0\,\mathrm{KNm}$$

$$k = \frac{M}{bd^2 f_{cu}} = \frac{172\times10^6}{1000\times200^2\times30} = 0.14$$
$$Lever\,armz = 0.8d$$

$$A_s = \frac{172\times10^6}{0.87\times30\times0.8\times200}$$
$$= 2686.16\,mm^2$$

Use 725–175 (28100 mm^2)

10.8.15 Beam Design

Assume each beam is of 6.5 m span and simply supported
Area of slab precast unit = 50×10.8 = 540 m^2

Self-weight of the precast unit = 0.2×24 = 4.8 KN/m^2
Design loading = $1.4 \times 4.8 \times 10.8$ = 72.58 KN/m
Bending moment M $= \frac{72.58\times6.5^2}{8} = 383.4\mathrm{KNm}$

Fig. 10.22 Beam design

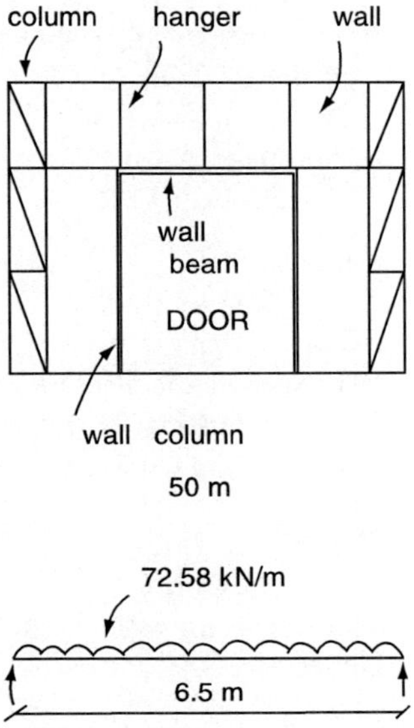

Assume use of compact section

S_x required $= S = 5\,M$

$= 5 \times 383.4$

$= 1917\ \text{cm}^3$

Consider 533 × 210 × 92 UB

Section properties

$D = 533.1$ mm

$B = 209.3$ mm

$t = 10.2$ mm

$T = 15.6$ mm

$r = 12.7$ mm

$d = 476.5$ mm

$r_y = 4.51$ cm

$S_x = 2370\ \text{cm}^2$

$x = 36.4$

$p_y = 265\ \text{N/mm}^2$

Compact check

$$b = \frac{B-t}{2} - r = \frac{209.3 - 10.2}{2} - 12.7$$
$$= 86.85\,mm$$

$\therefore \text{Flange}, \frac{\beta}{T} = \frac{86.85}{15.6} = 5.57$

$E = \left(\frac{275}{265}\right)^{\frac{1}{2}} = 1.02$
$9.5E = 9.5 \times 1.02 = 9.69$
$> \frac{b}{T}$ flange check O.K
$\therefore Web, \frac{d}{t} = \frac{476.5}{10.2} = 46.7$
$98E = 98 \times 1.02$
$> \frac{d}{t}$ web check O.K

Shear
$R = \frac{72.6 \times 6.5}{2} = 234KN$
$Shear\ capacity = 0.6 \times 265 \times 10.2 \times 533.1 = 864.6KN$
$>$ R Shear capacity O.K
Bending capacity $= 265 \times 10.2 \times [20 + (15.6 + 12.7)] = 245.3$KN
$>$ R Bending capacity O.K
Buckling
Slenderness ratio $\lambda = \frac{2.5 \times 476.5}{10.2} = 116.8$
$p_c = 98\,\text{N/mm}^2$
Web buckling $= 98 \times 10.2 \times \left(20 + \frac{533.1}{2}\right) = 286.4KN$
Deflection

$$d = \frac{5}{384}\frac{wL^4}{EI}$$

$$= \frac{5 \times 72.6 \times 6.5^4}{384 \times 205 \times 10^6 (5.54 \times 10^{-4})} = 0.015\,\text{m}$$

$$\frac{L}{200} = \frac{6.5}{200}$$

$$= 0.032\,\text{m} > 0.015\,\text{m Deflection O.K.}$$

10.8.16 Column Design

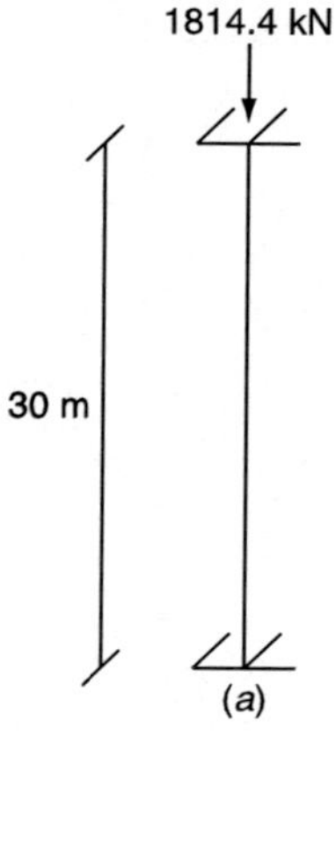

Fig. 10.23 Column design

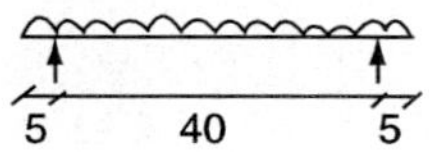

Height of column = 30 m

Working load $= 1.4 \times 4.8 \times \frac{540}{2} = 1814.4\,KN$

For grade 50 steel

Design strength $p_y = 265\ \text{N/mm}^2$

Effective length $L_E = 0.7L$

$= 0.7 \times 30 = 21\,m$

Design load $= 1814.4\,KN$

Column strength $= p_c A_g$

Assume

$P_c = 50\,\text{N/mm}^2$

$A_g = \frac{1814.4 \times 10^3}{50} = 36,288\,mm^2 = 363\,cm^2$

try $356 \times 406 \times 393\,UC$ ($A_g = 505\ \text{cm}^2$)

$r_y = 10.5\ \text{cm}$

$\lambda_y = \frac{2100}{10.5} = 200$

$P_c = 41\,\text{N/mm}^2$

$$\text{Strength} = \frac{41 \times 505 \times 10^2}{10^2} = 2070.5KN$$

$$> 1814.4\,\text{KN}\ \ \text{Column strength O.K}$$

$$\frac{F}{P_c A_g} + \frac{M}{M_b} = 1$$

$$\frac{1814.4}{2070.5} + \frac{44.6}{636.5} = 0.95 < 1$$

10.8.16.1 Design of Base Plate

Design load = 1814 KN

Given design strengthof concrete $= 7.5\ \text{N/mm}^2$

Area $= \frac{1814 \times 10^3}{7.5} = 241866.7\,\text{mm}^2 = 491.8 \times 491.8$

Use 500 square base plate

Actual pressure under plate $= \frac{1814 \times 10^3}{500 \times 500} = 7.26\,\text{N/mm}^2$

Plate thickness

$$t - \sqrt{\frac{2.5 \times 7.26}{265}[(47 \times 72) - (0.5 \times 47 \times 72)]} = 10.77\,\text{mm}$$

$\not< T$

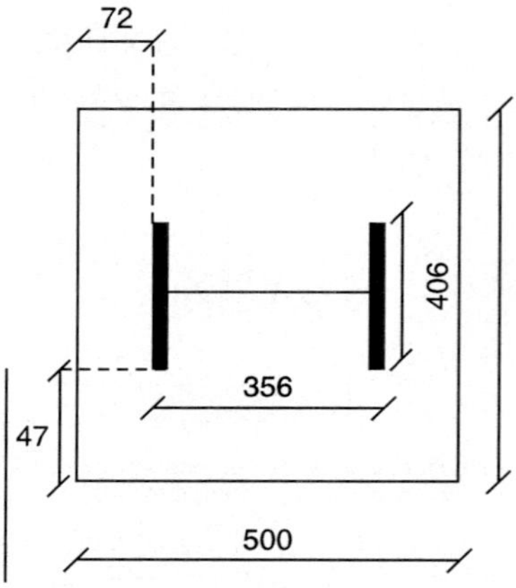

(a)

∴ Use 50 mm thick plate

$P_c A_g = \frac{41\times505\times10^2}{10^3} = 2070.5KN$
$S_x = 4160\,\text{cm}^2$
$x = 7.86$
$\lambda_y = 200$
$P_b = 153\,\text{N/mm}^2$
$P_b s_x = 153 \times 4160$
636.3 KNm

Hanger Design
Hangers Chosen *203 × 203 × 46 UC*

$P_y = 460\,\text{N/mm}^2$
$E = \left(\frac{275}{460}\right)^{\frac{1}{2}} = 0.59$

Fig. 10.24 Column design

$\frac{b}{T} = \frac{101.6}{11} = 9.23$
$15\varepsilon = 11.72 > \frac{b}{T}$

Effective area = 53.1 cm^2
Minimum value of elastic modulus Z = 365 cm^2
Moment capacity of major axis
$M_c x = 363 \times \frac{450}{10^3} = 163.35$ KNm
The applied axial load
$f = (2 \times 120 + 590) \times 0.0088 = 7.30$ KNm
Substitute

$$\frac{F}{A_e P_y} + \frac{M_x}{M_{cx}} = \frac{830 \times 10}{53.1 \times 450} + \frac{7.3}{163.35} = 0.40$$

$< 1 \therefore$ O.K

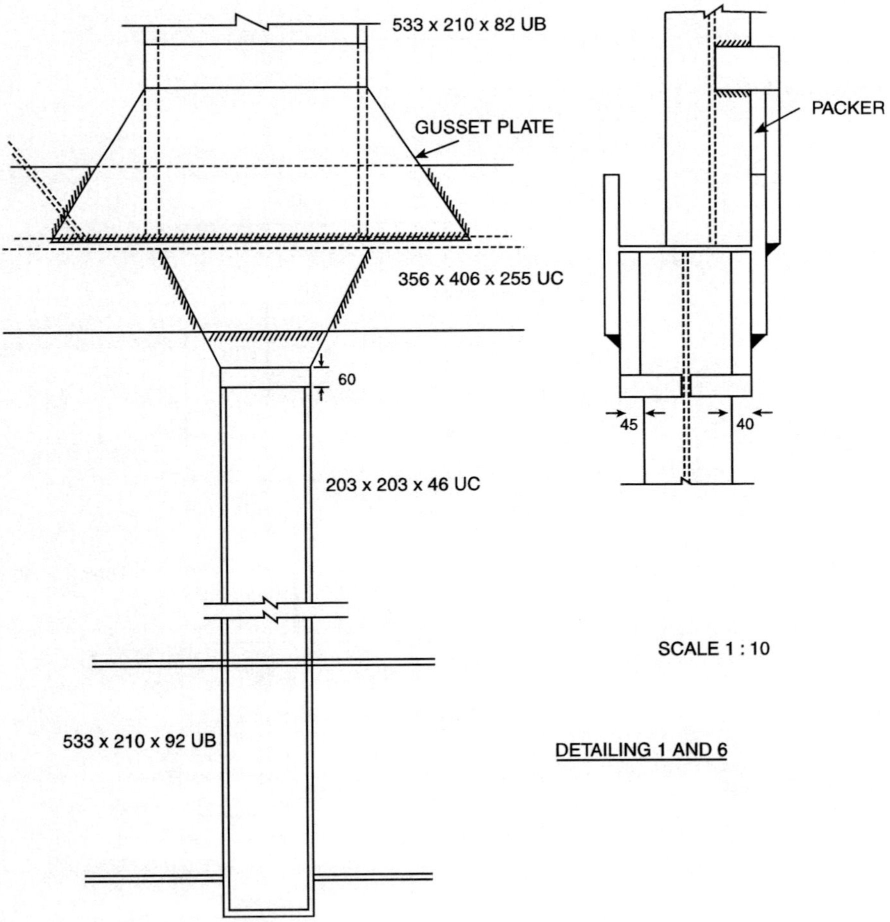

Fig. 10.25 Connections details

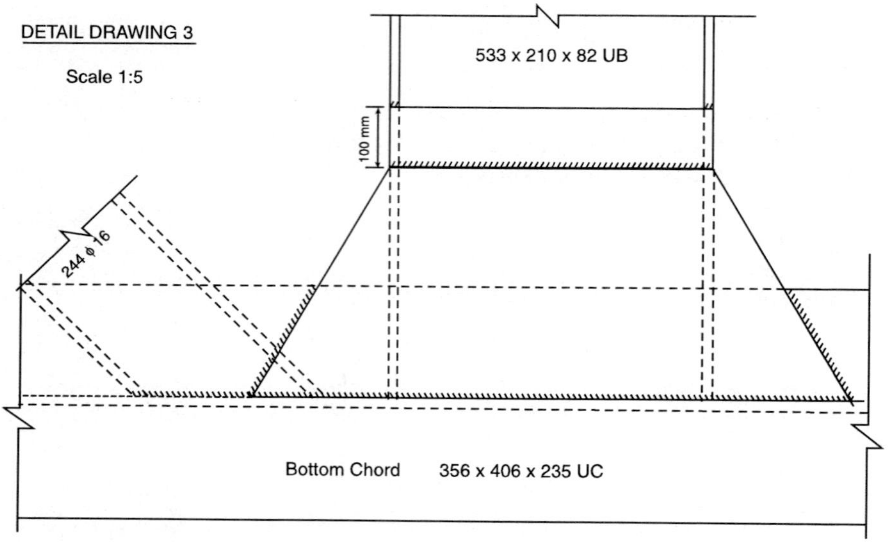

Fig. 10.25a Structural detailing for the bottom chord

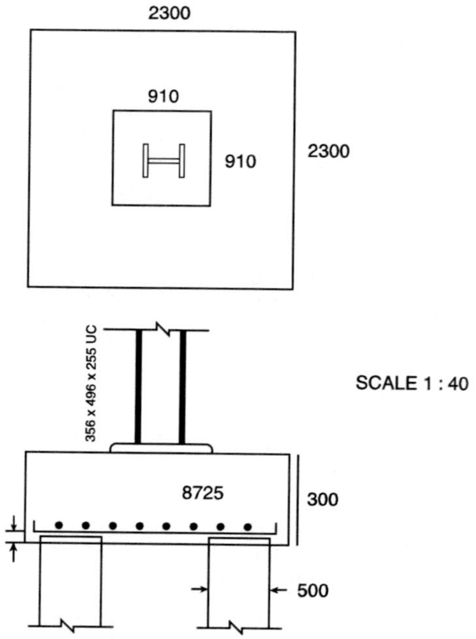

Fig. 10.26 Pile cap arangement

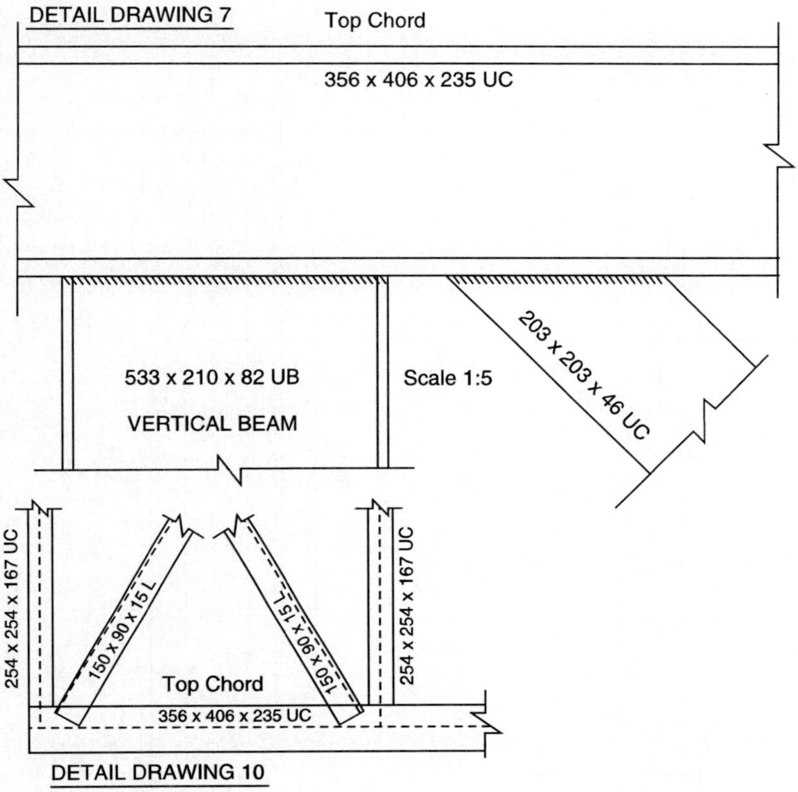

Fig. 10.27 Detailing top chord and vertical beams

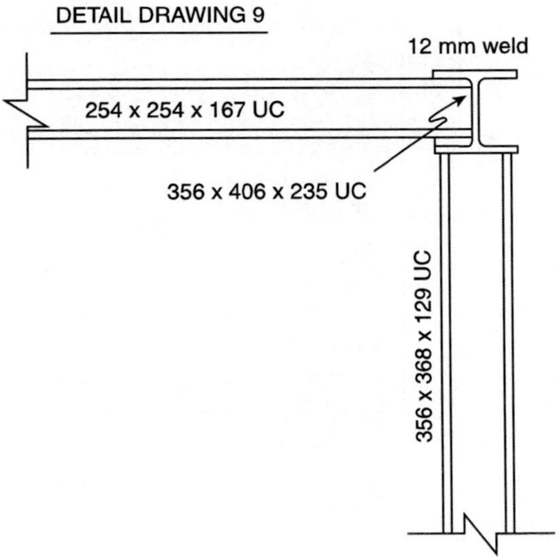

Fig. 10.28 Beam – column connectors

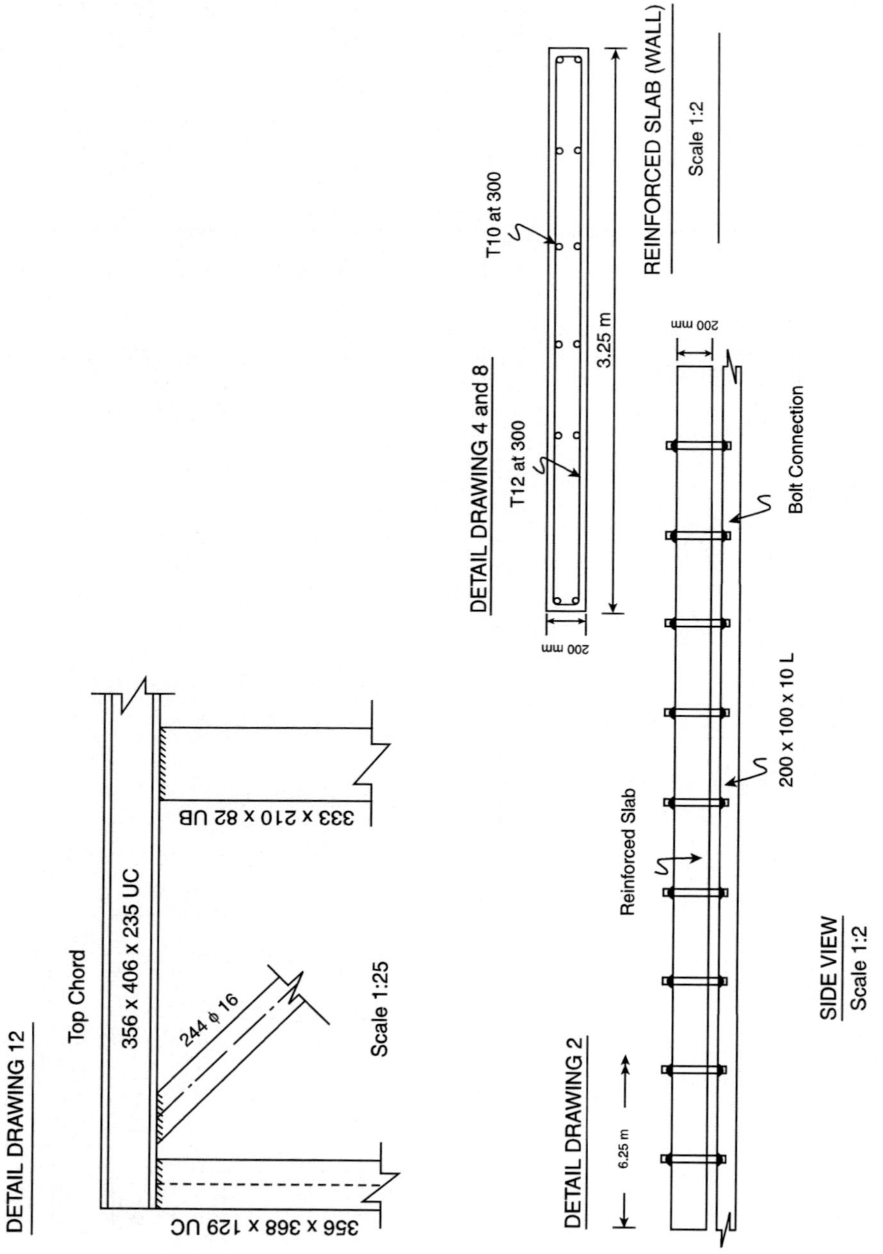

Fig. 10.29 Roof slab

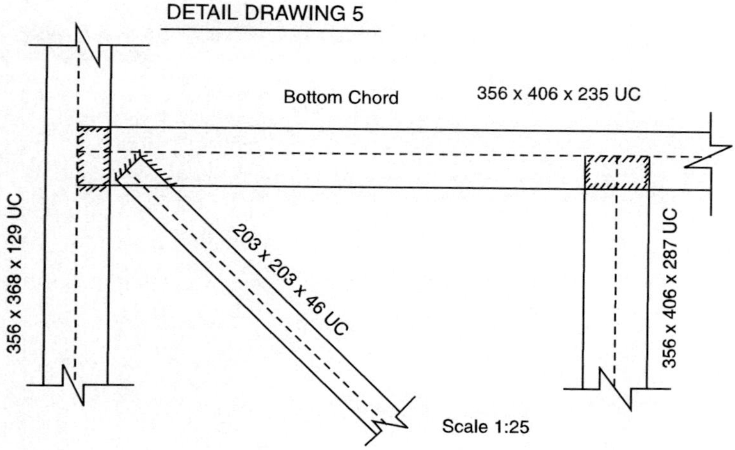

Fig. 10.30 Columns top and bottoms with bracings

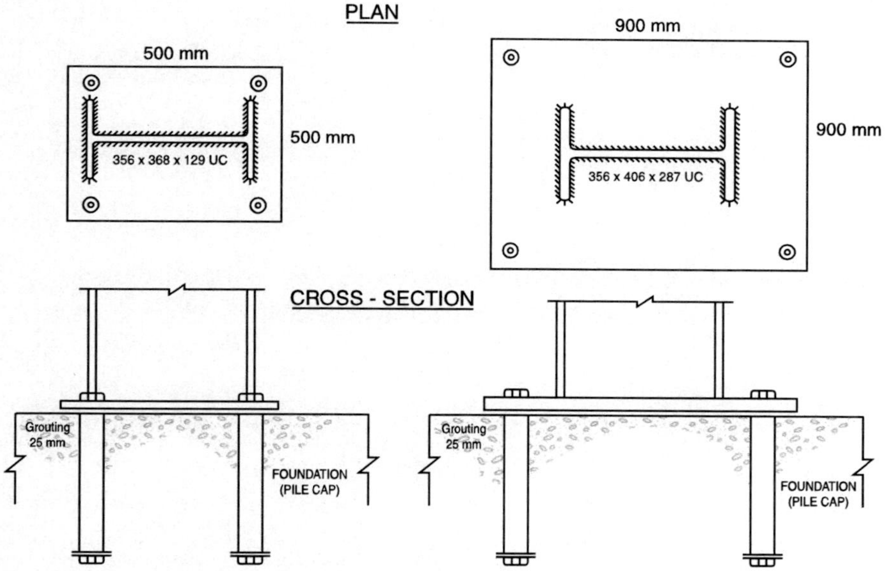

Fig. 10.31 Columns to foundations

Appendix A
Prestressing Systems and Anchorage Design

General Introduction

Prestressed concrete has attained worldwide recognition in the development of industrialised construction and design. Prestressing consists of introducing imposed deformations by tensioning prestressing wires, cables or strands and tendons to a high stress which decreases with time due to various losses such as shrinkage, creep, steel relaxation, friction and wobbling effects. The word *prestress* is associated with the following:

(a) *pretensioned concrete*
(b) *post-tensioned concrete*

In the case of pretensioned concrete structures, the tensioning of the tendon is carried out before concreting. The prestressing force is held temporarily either by a specially constructed prestressing bed or by a mould or form. When the concrete strength reaches the specified transfer strength, detensioning and stress transfer to such structures can be performed. In practice these structures are prefabricated.

In the case of post-tensioned concrete structures, the tensioning of the tendon is carried out after casting and hardening of the concrete. This method is more effective in the design and construction of high-rise and long-span concrete structures. The design and detailing of such structures are influenced by the serviceability classification, which includes the amount of flexural tensile stresses allowed while carrying out the design/detailing of such structures. They are then classified into individual classes which are given below:

Class 1: no flexural tensile stresses.
Class 2: flexural tensile stresses but no visible cracking.
Class 3: flexural tensile stresses but surface width of cracks not exceeding 0.1 mm for members in very severe environments and not exceeding 0.2 mm for all other members.

The structural detailing of prestressed concrete members must take into consideration durability, fire resistance and stability. The relevant codes EC2

M.Y.H. Bangash, *Structures for Nuclear Facilities*,
DOI 10.1007/978-3-642-12560-7, © M.Y.H. Bangash 2011

and BS 8110 which should strictly be followed for the correct evaluation of design ultimate loads and the characteristic strength of concrete and steel.

Generally, high-strength concrete is used for prestressed concrete work. The steel used in prestressed concrete is generally of a much higher strength than mild steel. This aspect is discussed later on in the choice and evaluation of prestressing systems.

Material data and prestressing systems are given on Plates AIA.1 to AIA.7 provided by the manufacturers. In such the prestressing tendons can be bonded and unbonded. The structural detailing is affected when the prestressed concrete structure is designed with bonded and unbonded tendons. Before the prestressing load is transmitted into various zones of concrete with bonded or unbonded tendons, it is necessary to protect the areas immediately under the anchorages against bursting effects caused by large loads generated by prestressing tendons. In such cases, the areas below anchorages must be designed by providing anticrack or burst steel.

'FREYSSI' MONOGROUP – 15/15 mm NORMAL STRAND SYSTEM

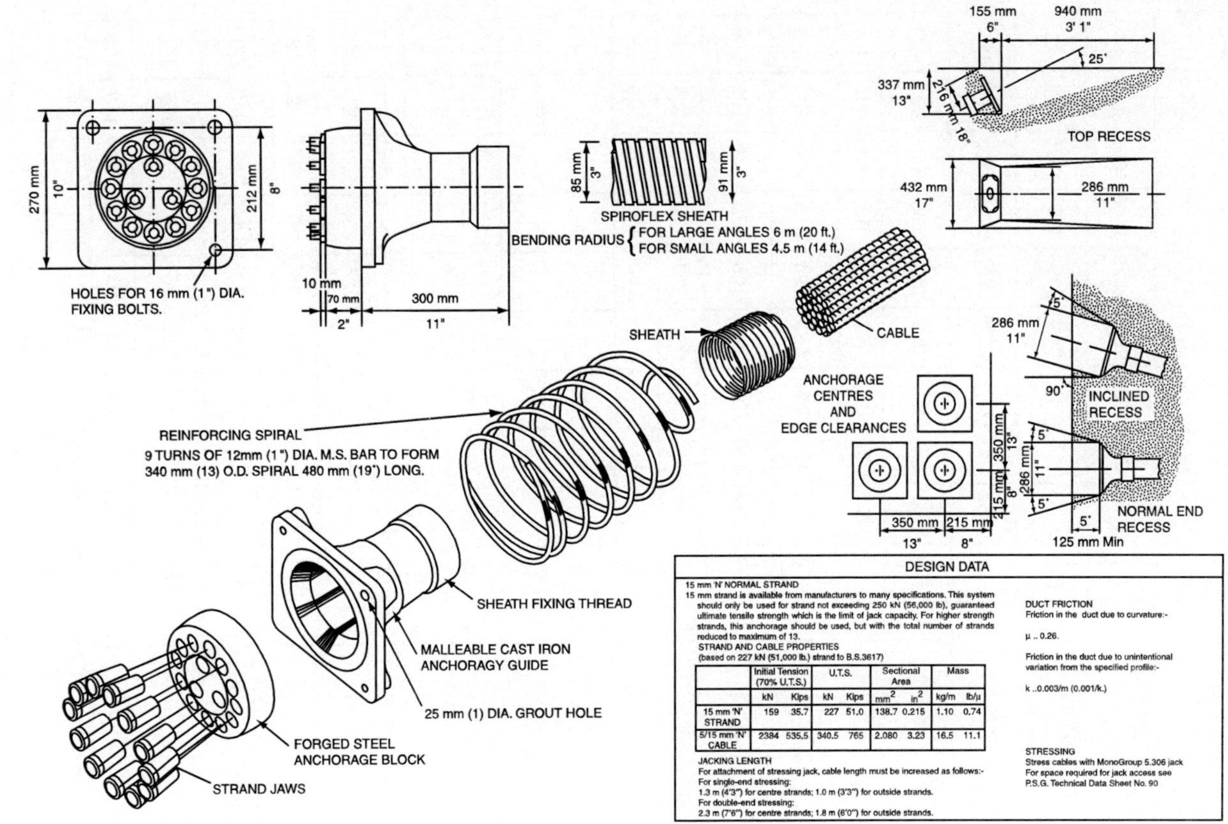

DESIGN DATA

15 mm 'N' NORMAL STRAND

15 mm strand is available from manufacturers to many specifications. This system should only be used for strand not exceeding 250 kN (56,000 lb), guaranteed ultimate tensile strength which is the limit of jack capacity. For higher strength strands, this anchorage should be used, but with the total number of strands reduced to maximum of 13.

STRAND AND CABLE PROPERTIES
(based on 227 kN (51,000 lb.) strand to B.S.3617)

	Initial Tension (70% U.T.S.)		U.T.S.		Sectional Area		Mass	
	kN	Kips	kN	Kips	mm^2	in^2	kg/m	lb/μ
15 mm 'N' STRAND	159	35.7	227	51.0	138.7	0.215	1.10	0.74
5/15 mm 'N' CABLE	2384	535.5	340.5	765	2.080	3.23	16.5	11.1

JACKING LENGTH
For attachment of stressing jack, cable length must be increased as follows:-
For single-end stressing:
1.3 m (4′3″) for centre strands; 1.0 m (3′3″) for outside strands.
For double-end stressing:
2.3 m (7′6″) for centre strands; 1.8 m (6′0″) for outside strands.

DUCT FRICTION
Friction in the duct due to curvature:-

μ .. 0.26.

Friction in the duct due to unintentional variation from the specified profile:-

k ..0.003/m (0.001/k.)

STRESSING
Stress cables with MonoGroup 5.306 jack
For space required for jack access see
P.S.G. Technical Data Sheet No. 90

Plate AIA.1 'Fryssi' Monogroup–15/15 mm Normal Strand System

BBRV TENDON SIZES AND THEIR PROPERTIES

Tendon reference		B16	B24	B34	C42	C55	L73	L85	L97	L109	L121
No. of wires (7 mm dia.)		16	24	34	42	55	73	85	97	109	121
Characteristic strength using 1570 N/mm^2 wire		967	1450	2054	2538	3323	4411	5136	5861	6586	7311
Jacking force (kN) at 80% of C.S.		773	1160	1643	2030	2659	3529	4109	4689	5269	5849
Jacking force (kN) at 75% of C.S.		725	1088	1540	1903	2492	3308	3852	4396	4939	5483
Jacking force (kN) at 70% of C.S.		677	1015	1438	1776	2326	3087	3595	4103	4610	5118
Bearing plate Side lenght (sq.)	A	178	220	250	280	300	335	360	385	405	425
Thickness	B	15	20	25	35	35	45	50	50	60	60
Trumpet Outside diameter	C	120	133	154	154	165	194	219	229	229	245
Anchor Head Thread diameter	D	100	115	130	90	98	125	130	145	150	155
Standard length	E	60	80	90	53	63	110	120	126	134	146
Overall diameter	F	—	—	—	—	—	160	165	185	190	200
Pull Sleeve Diameter	G	—	—	—	130	144	—	—	—	—	—
Standard length	H	—	—	—	118	138	—	—	—	—	—
Lock Nut Diameter	J	135	155	180	178	198	—	—	—	—	—
Thickness	K	30	40	50	45	53	—	—	—	—	—
Chocks Diameter	L	—	—	—	—	—	228	250	262	270	288
Min. thickness	M	—	—	—	—	—	48	52	55	59	62
Max. anchorage projection (Stnd. comps.)	N	67	87	97	118	138	205	219	228	240	255
Sheathing Internal diameter	O	40	50	60	65	75	85	95	100	105	110
External diameter		48	58	68	73	83	93	103	108	113	118

Lenght of trumpet 'p' ≏ Extension + 105 mm (B type anchors)
Extension + 120 mm (C type anchors)
(L type anchors)

All dimensions are in millimetres.

Plate AIA.2

SMALL CAPACITY TENDONS UP TO 2658 kN

Characteristic tensile strength f_u	Maximum working load (80% f_u)	Number of 7 mm dia. wires	Anchorages: Stressing	Anchorages: Fixed	Anchorages: Nominal reference	Standard single bearing plate	Combined bearing plate: Minimum centres of anchorages	Combined bearing plate: Minimum concrete edge distance	SIM-TUBE sheathing internal diameter	Stressing equipment clearance dimensions with dynamometer	Stressing equipment clearance dimensions without dynamometer
kN	kN	No.	Type	Type	Ref. No.	mm	mm	mm	mm	mm	mm
						a × b	c × d	e		f × g dia.	h × g dia.
			B			150 × 150	138 × 138	76			
			J			150 × 150					
483	387	8		F	32	120 × 120			30	1524 × 310	1344 × 310
				SR		120 × 150					
				SL		60 × 300					
			B			175 × 175	171 × 171	89			
			J			175 × 175					
966	773	16		F	64	160 × 160			40	1524 × 310	1344 × 310
				SR		150 × 220					
				SL		80 × 400					
			B			200 × 200	197 × 197	108			
			J			200 × 200					
1450	1160	24		F	100	220 × 220			50	1524 × 310	1344 × 310
				SS		220 × 220					
				SR		160 × 300					
				SL		80 × 560					
			C			250 × 250	235 × 235	127			
				E		235 dia.					
1872	1498	31		SS	130	260 × 260			55	1524 × 340	1324 × 340
				SR		180 × 360					
				SL		120 × 560					
			B			250 × 250	235 × 235	127			
			J			250 × 250					
2054	1643	34		F	138	260 × 260			55	1524 × 340	1324 × 340
				SS		260 × 260					
				SR		180 × 360					
				SL		120 × 560					
			C			280 × 280	241 × 267	152			
				E		270 dia.					
2537	2029	42		SS	170	300 × 300			65	1880 × 440	1580 × 440
				SR		200 × 450					
				SL		140 × 650					
			C			300 × 300	267 × 305	152			
				E		300 dia.					
3322	2658	55		SS	220	340 × 340			75	1880 × 440	1580 × 440
				SR		220 × 500					
				SL		160 × 700					

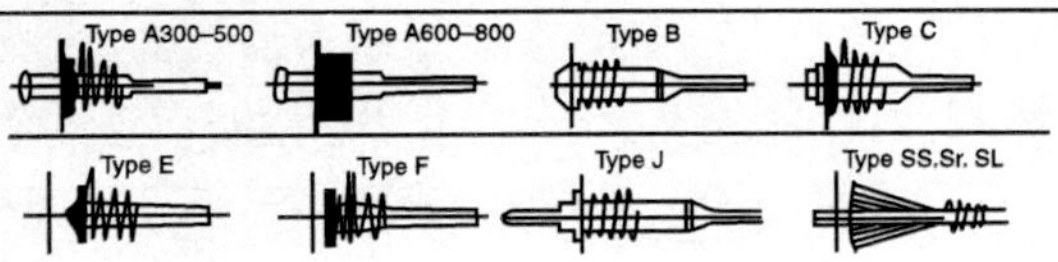

Plate AIA.3

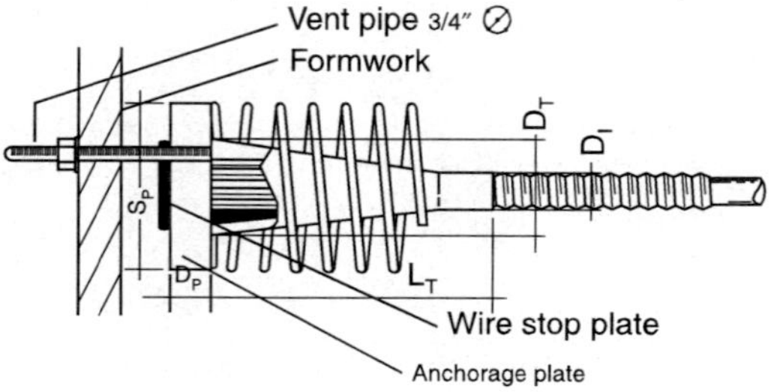

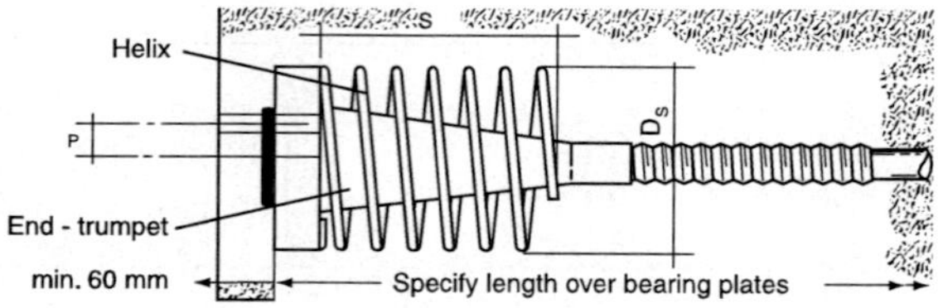

Plate AIA.4

TYPE 'F' FIXED ANCHORAGE

A fixed anchorage-consisting of a rectangular steel plate drilled to receive the individual button-headed wires seated directly on this plate. A thin cover plate retains the button heads during fixing.

With this anchorage, the prestressing force is transferred to the concrete through the plate.

Type 'F' anchorages are normally used in connection with movable anchorages of the Type 'B' series.if there are. several tendons in one part of a structure.it is advisable to put half the Type 'F' anchorages at one end and half at the other.

The anchorage is fastened to the formwork with a pipe which serves at the same time as a vent or as a grout connection.

This pipe passes through the plate and should be well-greased prior to fixing to facilitate removal after grouting.

Type F anchorage	Type designation		F32		F64		F100		F138	
Steel	wires per anchorage, number 7 mm	maximum (0.276") dia.	8		16		24		34	
			mm	in	mm	in	mm	in	mm	in
Anchorage	side length	S_P	120	4.72	160	6.30	220	8.66	260	10.24
	Thickness	D_P	15	0.59	25	0.98	40	1.58	50	1.97
End trumpet	outer diameter	D_T	87	3.43	112	4.41	128	5.04	148	5.83
	Connection, outer dia.	D_I	35	1.38	45	1.77	55	2.17	60	2.36
	Length	L_T	200	7.87	250	9.84	300	11.81	350	13.78
Helix	outer diameter	D_S	120	4.72	160	6.30	220	8.66	260	10.24
	Length (approx.)	L_S	250	9.84	250	9.84	250	9.84	250	9.84
Distance of centre vent pipe to tendon axis		e	0	0	0	0	35	1.38	45	1.77

BBRVF 163 No/7 mm Tendon

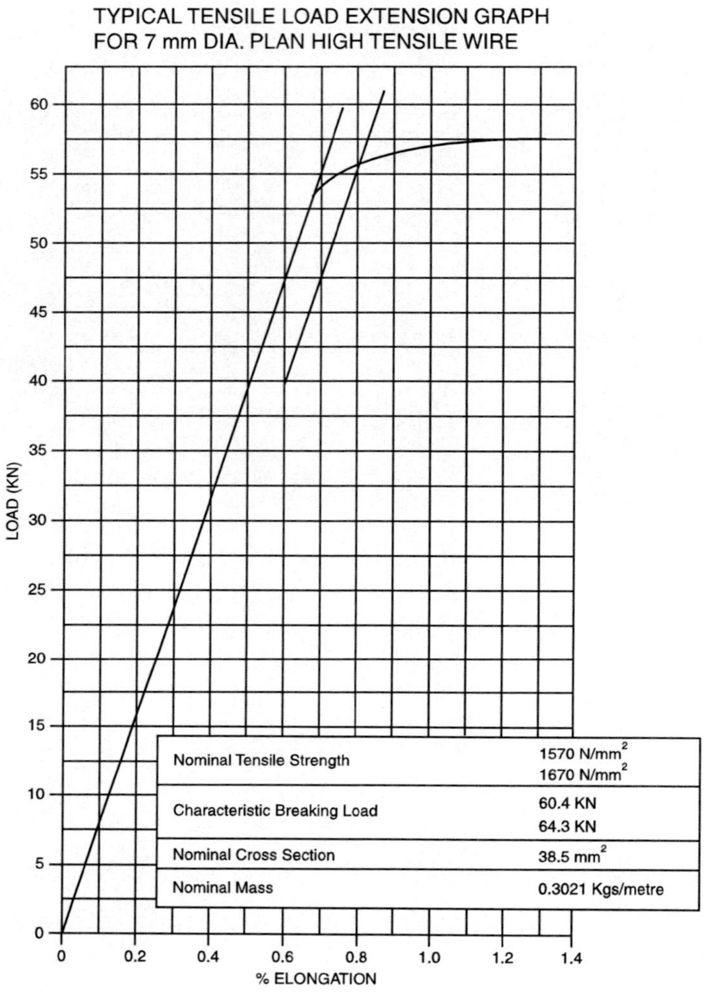

Plate AIA. 5 Load-Elongation diagram

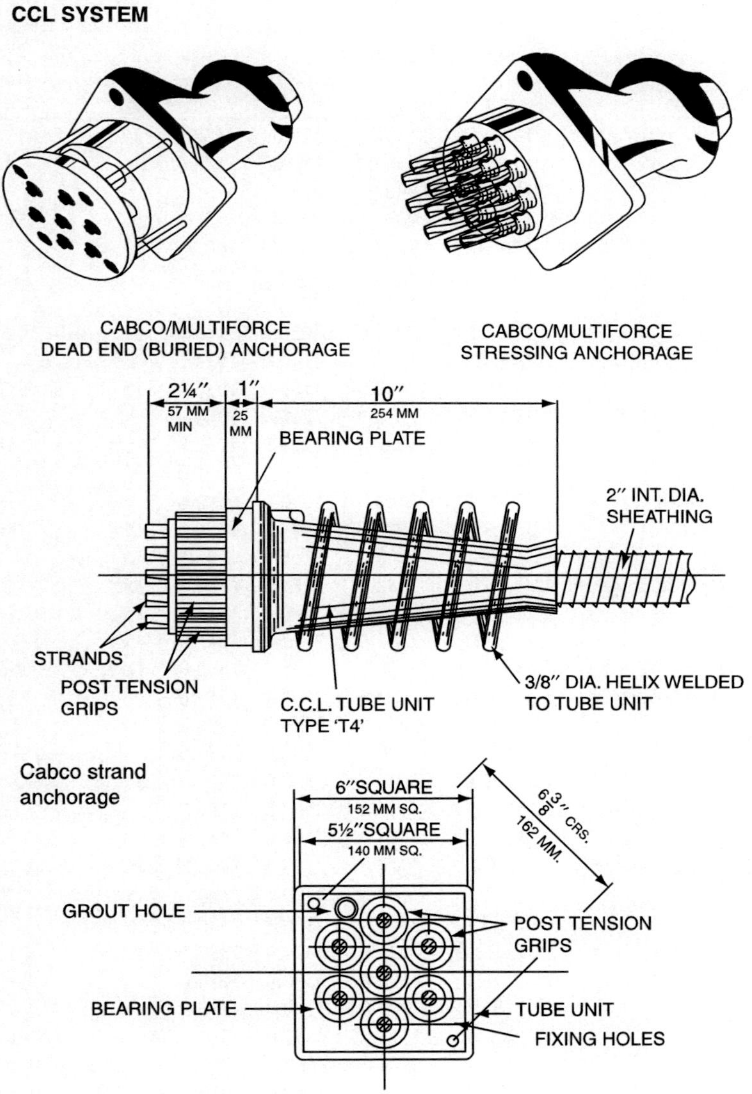

Plate AIA.6 C.C.L. System

CABCO PRESTRESSING TENDON MAIN DATA

Anchorage	System	Tendon*	Tendon Forces (kN) P_k	0.8 P_k	0.7 P_k	Anchorage size mm	Sheath internal dia. mm
		4/13 STD.	660.0	528.0	462.0	130	42
U1	Cabco	4/13 SUP.	736.0	588.8	515.2	"	"
		4/13 DYF.	836.0	668.8	585.2	"	"
		4/15 STD.	908.0	726.4	635.6	175	51
		4/15 SUP.	1000.0	800.0	700.0	"	"
		4/15 DYF.	1200.0	960.0	840.0	"	"
U2	Cabco	7/13 STD.	1155.0	924.0	808.5	"	"
		7/13 SUP.	1288.0	1030.4	901.6	"	"
		7/13 DYF.	1453.0	1170.4	1024.1	"	"
		4/18 STD.	1480.0	1184.0	1036.0	175	51
		4/18 DYF.	1520.0	1216.0	1064.0	"	"
		7/15 STD.	1589.0	1271.2	1112.3	215	75
		7/15 SUP.	1750.0	1400.0	1225.0	"	"
	Cabco	7/15 DYF.	2100.0	1680.0	1470.0	"	63
U3		12/13 STD.	1980.0	1584.0	1386.0	215	81
		12/13 SUP.	2208.0	1766.4	1545.6	215	"
	Multiforce	12/13 STD.	1980.0	1584.0	1386.0	215	"
		12/13 SUP.	2208.0	1766.4	1545.6	215	"
		7/18 STD.	2590.0	2072.0	1813.0	245	81
		7/18 DYF.	2660.0	2128.0	1862.0	"	75
U4	Cabco	12/13 DYF.	2508.0	2006.4	1755.6	"	75
		12/15 STD.	2724.0	2179.2	1906.8	"	81
		12/15 SUP.	3000.0	2400.0	2100.0	"	81
	Cabco	12/15 DYF.	3600.0	2880.0	2520.0	265	81
		19/13 STD.	3135.0	2508.0	2194.5	"	"
U5		19/13 SUP.	3496.0.	2796.8	2447.2	"	"
	Multiforce	12/15 STD.	2724.0	2179.2	1906.8	245	"
		12/15 SUP.	3000.0	2400.0	2100.0	245	"
		12/15 DYF.	3600.0	2880.0	2520.0	265	"
		25/13 STD.	4125.0	3300.0	2887.5	300	90
U6	Multiforce	25/13 SUP.	4600.0	3680.0	3220.0	"	"
		13/15 DYF.	3900.0	3120.0	2730.0	"	84
		31/13 STD.	5115.0	4092.0	3580.5	335	100
		31/13 SUP.	5704.0	4563.2	3992.8	"	"
U7	Multiforce	19/15 STD.	4313.0	3450.0	3019.1	"	"
		19/15 SUP.	4750.0	3800.0	3325.0	"	"
		19/15 DYF.	5700.0	4560.0	3990.0	"	"
U8	Multiforce	19/18 DYF.	7220.0	5776.0	5054.0	"	"
		5/18 STD.	1776.0	1480.0	1295.0	362 × 89	114 × 25
		5/18 DYF.	1824.0	1520.0	1330.0	"	"
	Strand-force	10/18 STD.	3530.0	2960.0	2590.0	362 × 171	" twin
		10/18 DYF.	3648.0	3040.0	2660.0	"	" twin

*STD. Standard
SUP. Super Strand
DYF. Dyform Strand

Plate AIA.7

The Largest Tendon Is the BBRV Tendon and Its Description Is as Under

BBRV—Tendons

General

BBRV tendons consist of any number of parallel wires having diameters of 4–12 mm (0.16–0.47 in.), anchored in a common anchor head by means of cold-upset button-heads. For nuclear pressure, and containment vessels, tendons are used with 121, 163 or 187 wires of 7 mm (0.28 in.) diameter, corresponding to ultimate loads of 770, 1020 or 1200 tons with the quality of steel chosen. Intermediate sizes are likewise possible, as is the use of even larger tendons. Parallel-wire tendons have the following advantages and disadvantages in comparison with strands or wire ropes.

Tendons made from a number of strands

The individual strand is more flexible, but its component wires are subject to additional fiexural and torsional stresses and thus (with the exception of stabilized strands) to greater relaxation losses. A strand is composed of wire of a smaller diameter (maximum approximately 4 mm [0–16 in.]) and is for this reason, as well as on account of the additional stresses, more susceptible to corrosion and to stress corrosion. It also manifests greater losses due to friction.

A coupling is threaded over the outside threading of the anchor's bearing nut. As soon as the required elongation has been reached, the anchor is blocked by placing the two semi-circular chocks under the bearing nut. The anchor plate consists of a concrete block with an external helix that is completely closed and an internal reinforcement of the end of the duct. The surface of the concrete block is covered by a steel plate 25 mm (098 in.) thick in order to obtain a clean working surface, so that the relatively heavy-tensioning equipment can be fastened on and for the purpose of local load distribution. The principal dimensions of this anchor construction are given below, for example, with 163 wires of 7 mm (028 in.) diameter.

Outer diameter of anchor head	294 mm (11.57 in.)
Height of anchor	120 mm (4.72 in.)
Bearing block	610 × 610 × 275 mm (24.0 × 24.0 × 1083 in.).

It was first demonstrated in 1964, at the Cable testing station in Germany, that this anchorage will sustain the full ultimate load of the tendon without undergoing any plastic deformation itself; since then, the proof has been repeated in at least 24 other ultimate tensile tests carried out on tendons. A supplementary test carried out by the Swiss Federal Laboratories for Testing Materials and Research (EMPA), in order to establish the actual reserve bearing capacity of the anchor head, showed that the anchor with 163.7 mm

(0–28 in.) wires had an ultimate load of 1376 t, while that of the tendon which it may anchor is 1020 tons (EMPA Report No. 40879, BORVD964).

Whenever possible, tendons are assembled in the factory. The wires are first measured off to the required length on a fully automatic cutting bench. Then the bundle of wires is passed through a template and the one anchor head, and button-heads are cold-upset onto the wires at this end. In order to make all the wires parallel, the tendon is then pulled through the template by pulling the anchor head. To keep the differences in the lengths of the wires in a curved duct to a minimum, and to preserve the parallelism of the wires even when they are wound onto reels for transport, the tendon is given three to five complete twists and then bound at intervals of approximately 3 m (approx. 10 ft). Once the tendons are reeled the factory work is finished.

On the construction site, once the wire bundle has been unrolled off the reel and pulled directly into its duct, the other anchor head is mounted. Thus the tendon is ready to be stressed from one or both ends.

Anchorage Analysis and Design

Based on research reports 9 and 13 by C & CA (Now Concrete Society London).

General Formulation

A reference is made to Fig. AIA.1 for the design analysis

- f_x general longitudinal stresses
- f_y general transverse ‘bursting’ stresses
- ε_x general longitudinal strains
- ε_x general transverse ‘bursting’ strains
- P load applied to anchorage
- $2a_1$ width of bearing plate
- $2a$ width of end block
- t thickness of anchor bearing plate
- f_c uniform longitudinal compressive stress

$$= \frac{P}{(4a^2 - \text{area of duct})} \tag{AIA.1}$$

- E modulus of elasticity
- γ Poisson’s ratio
- σ_1, σ_2, σ_3 principal stresses, $\sigma_1 > \sigma_2 > \sigma_3$ tension + ve
- ε_1, ε_2, ε_3 principal strains corresponding to γ_1, etc.
- γ_m mean normal stress

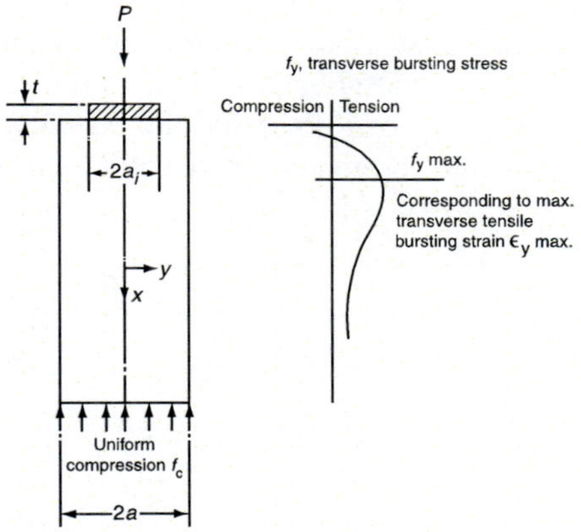

Fig. AIA.1 Idealised end block showing principal notation

$$= \frac{\sigma_1 + \sigma_2 + \sigma_3}{3} = \tag{AIA.2}$$

Octahedral normal stress

The C & CA experimental tests were compared with those of six researchers and the results are plotted in Fig. AIA.2 by Taylor S.G.C. The 3D finite-element analysis given in the text has been implemented in the program ISO-PAR. The analytical results are plotted on Fig. AIA.2.

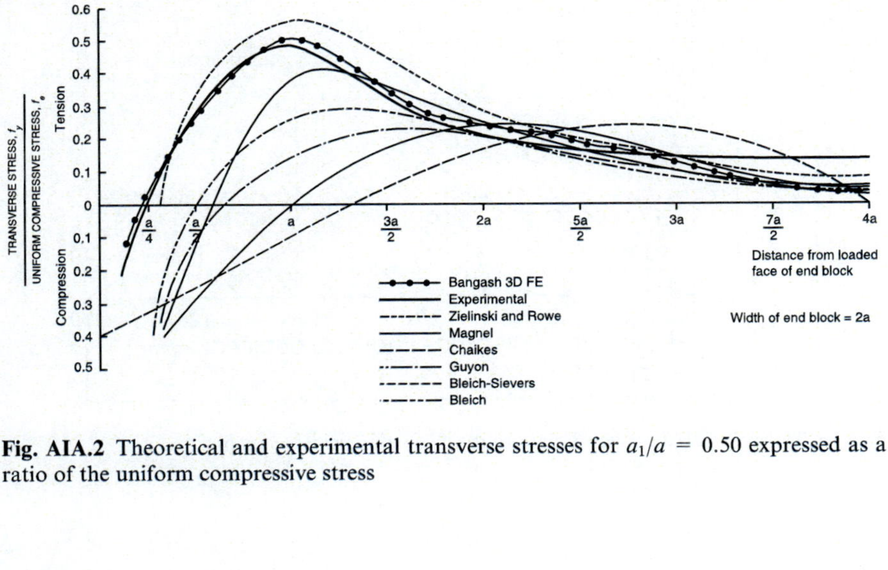

Fig. AIA.2 Theoretical and experimental transverse stresses for $a_1/a = 0.50$ expressed as a ratio of the uniform compressive stress

The plotting of conventional steel bars below the anchorages to take an extra tensile stresses beyond the concrete tensile stresses can be plotted with in the hatched area in Fig. AIA.3. In other words, the tensile forces at a distance from loaded face can be easily plotted with hatched area at suitable spacings. It is important to replace the circular area by an equivalent square area.

The force to be taken by the reinforcement is suggested as

$$T_{\lambda} = T\left[1 - \left(\frac{f_t \text{ permissble}}{f_y \text{ maximum}}\right)^2\right] \tag{AIA.3}$$

where T is the total tensile force.

Example Based on Freyssinet Anchorage System

Freyssinet system has been widely used in the reactor containment vessels
Data:

Cable 12 N 0.5 mm wires
Concrete prism 150 × 150 mm section
Female cone diameter = 100 mm
Total prestressing force = 222.4 KN
a_1 = equivalent square area = 88.9 mm

Solution

$$\frac{a_1}{a} = \frac{88.9}{150} = 0.59$$

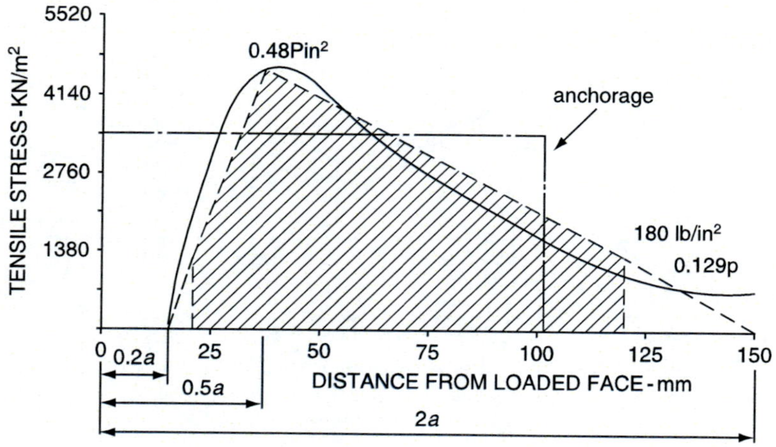

Fig. AIA.3 Tensile force corresponding to the calculated reinforcement

$$p = \text{uniform compressive stress} = \frac{222.4 \times 10^3}{150 \times 150} = 9.88\,\text{N/mm}^2$$

T = maximum tensile force = 0.22 × 222.4 = 48.928 KN
$f_y = f_{t_{\max}}$ = maximum tensile stress = 0.48 × p = 4.74 N/mm^2
$f_1 = f_t$ (permissible) = 0.129 p = 1.273 N/mm^2
Total tensile force (hatched area as a fraction of total traingular area)

$$= T_\lambda = T\left[1 - \left(\frac{f_1\,\text{permissible}}{f_y\,\text{max}}\right)^2\right]$$

=0.928T
The force taken by the reinforcement = 0.928 × 48.928 = 45.4 KN
Mild steel: Area = $\frac{45.4\times10^3}{250}$ = 181.6 mm^2

Select bars and place at 0.2*a* first bar, the next bars at 0.5*a* and pass through 125 mm with cover to 150 mm.

The 3D Finite-Element Analysis

A 3D finite element analysis has been carried out using, respectively, isoparametric element for solid concrete and bar (noded) element for reinforcement. Figure AIA.4 shows a finite element meshscheme for adjacent anchorages. Dimension loading and other parameters are kept the same of example under Section A.2.2.1. the 3D displacements are shown in Fig. AIA.5 from the output opf ratio of stress f_y/f_c are plotted in Fig. AIA.2 against the depths of the immediate blocks. The results are also compared experimental and other theories prominently shown in Fig. AIA.2.

Explanatory Notes

The mode of failure for BBRV tendons has been described earlier in the text. The following summarizes the behaviour for the tendons used in the example A.2.2.1

The mode of failure for an end block utilizing the tendon V anchorage unit as used in the model test example of

First – Uniform tensile load applied to the anchorage zone reduces the load carrying capacity of the anchorage unit and causes a reduction in the maximum tensile strain at cracking.

Second – Uniform compressive stress applied to the anchorage zone increases the load carrying capacity of the anchorage unit, the maximum tensile strain at cracking being increased.

Third – For biaxial compression–tension applied to the anchorage zone and a constant anchorage force, there is a marked reduction in the tensile stress

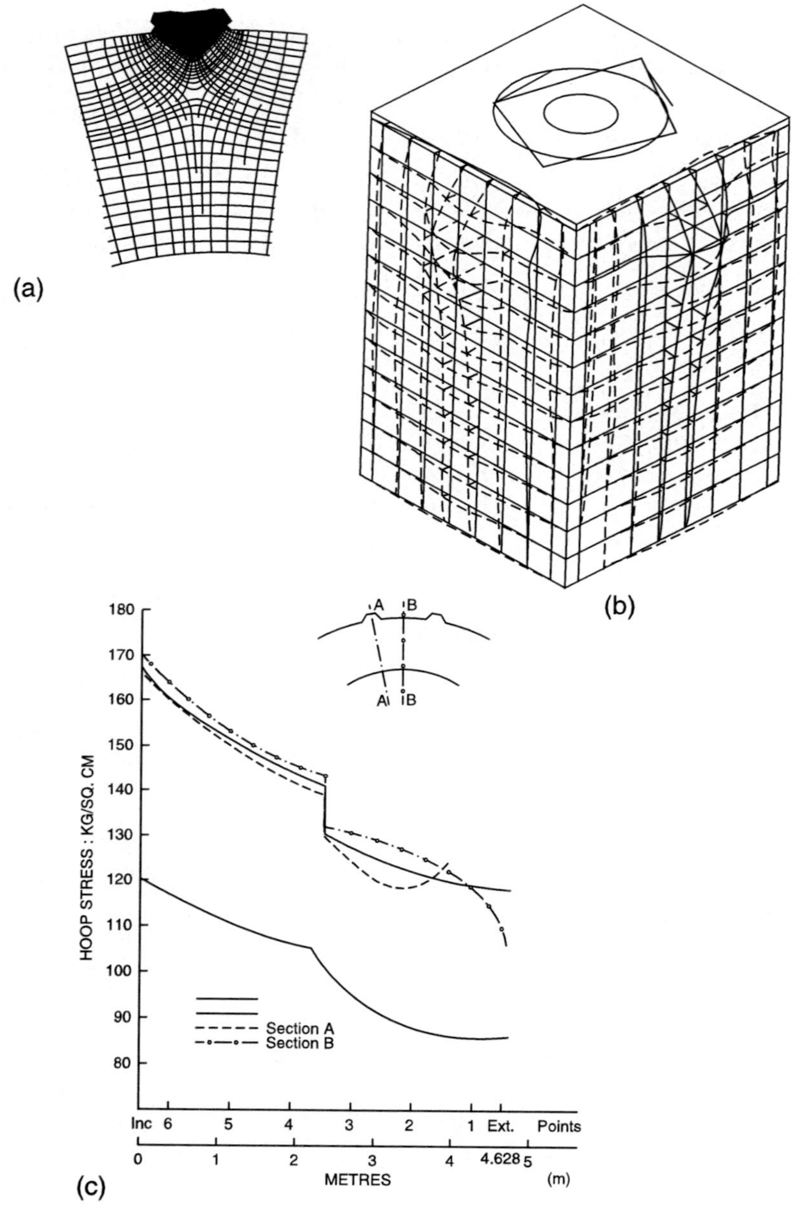

Fig. AIA.4 Analytical Results. (**a**) F.E mash system. (**b**) Displacement. (**c**) Hoopstress on external and internal faces that can be applied to the anchorage zone as compressive stress is increased. An increase in the maximum tensile stress at cracking was noticed.

Fourth – There does not appear to be a limiting tensile stress, or strain criteria of failure for concrete under the influence of combined states of loads

Fifth – The failure envelopes for imposed loads on the anchorage zone are of a similar nature to that of plain concrete.

Sixth – It would appear that an analogy can be drawn between the behaviour of the anchorage zone and that of plain concrete.

Seventh – The distribution of transverse strains in the anchorage zone appears to be uniformly affected by imposed states of stress except for the case of applied biaxial compression–tension, where a movement of the maximum tension towards the loaded face was noticed.

Eighth – The results of the research described in this section show that a fairly simple empirical design method can be US stated under Section A.2.2.1 used for the detailing of end blocks which have uniform stresses imposed on the anchorage zone. This is based on the relation between the maximum tensile strain at cracking and the mean normal stress (octahedral normal stress). Throughout, the uniform stress due to the anchorage load has been used in calculating. However, in order to take into account variations of a_1/a, it may be better to use the stress under the bearing plate in assessing octahedral stress.

Tenth – It must be borne in mind that the complete results are based on fully saturated concrete and were short-term station.

Typical Prestressing Tendons Layout

Plates AIA.5, AIA.6 and AIA.7 give the typical layouts of the prestressing tendon layouts of the pressure and containment vessels.

Wire Winding of the Vessels

The latest vessel are provided with a wire winding system in which the wires are wound under tension into steel-lined channel either formed in the vessel walls or made of precast concrete. The band comprises of layers of wires anchored at both ends. In this way the maximum circumferential pressures are exerted with in a limited area.

Plate AIA.8 shows the general layout of the channel and Plate AIA.9 gives a brief analysis of how to find maximum pressure with in the confined area of the channel subject to wire-winding. Where standards instead of wires are used, the same analysis can simply be adopted. Care must be taken to supplement the analytical results with those of the experiments. A reference is made to the following paper of the author for detailed investigations.

(Prog. Inst. Civil. Engrs, Part 2,1975,59, Mar;195–200 and discussions 7710. Main paper Prog. Inst. Civil Engrs. Part 2, 1974, 57, Sept., 437–450)

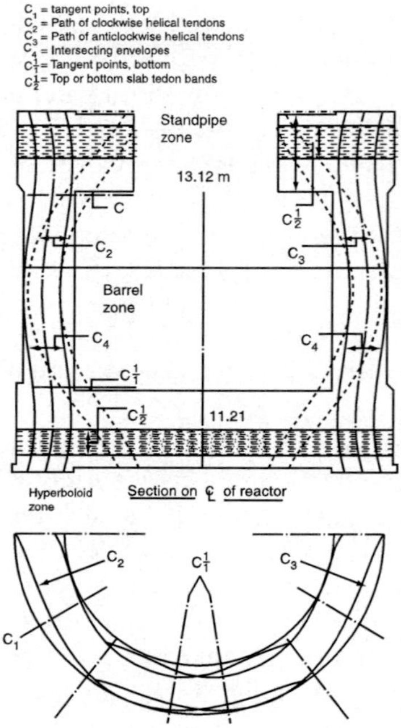

Typical layout of prestressing systems. Oldbury
(With compliments from Sir Robert McAlpine.)

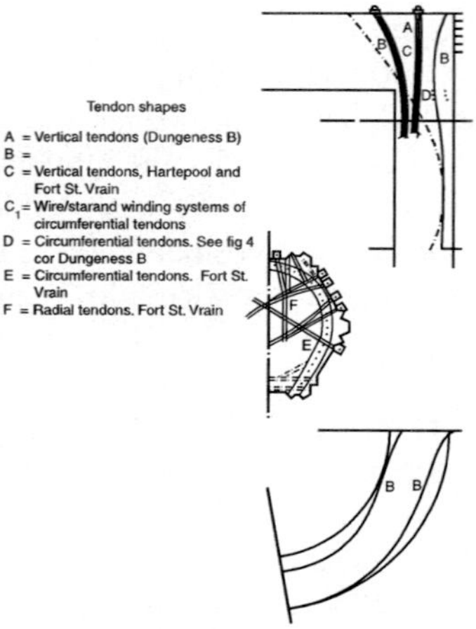

Typical layout of prestressing systems.

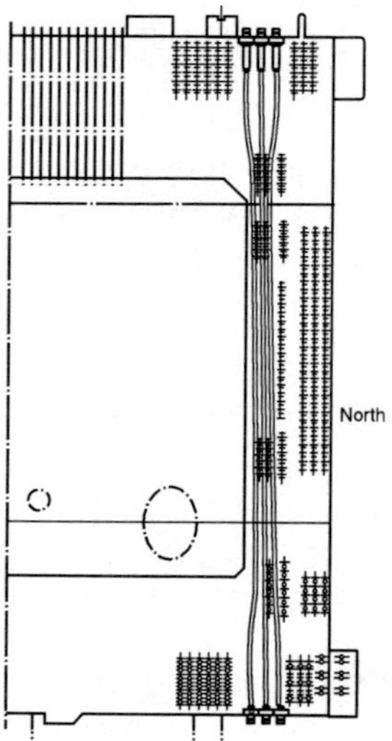

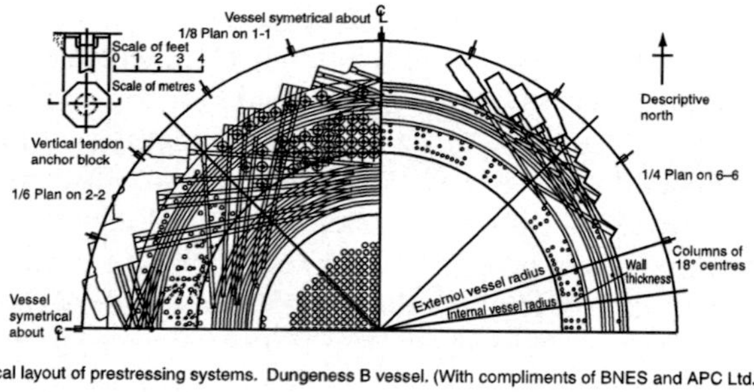

Typical layout of prestressing systems. Dungeness B vessel. (With compliments of BNES and APC Ltd.)

Fig. AIA.5 Prestressing Layouts

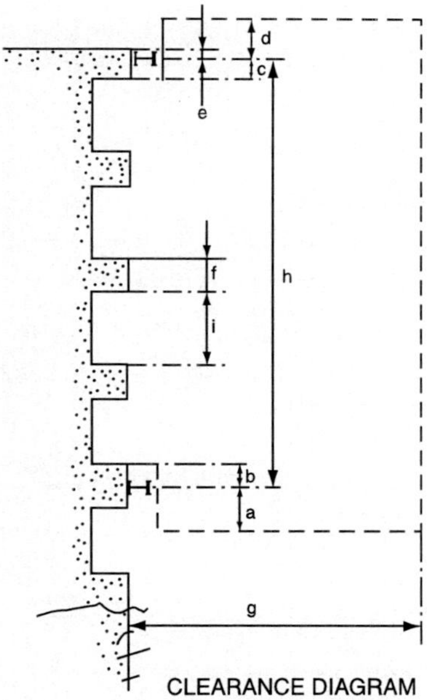

a = 180 mm
b = 210 mm
c = 320 mm
d = 310 mm
e = To suit track fixing detail.
f = 205 mm minimum
g = 2200 mm minimum
h = 4.00 m maximum
2.00 m minimum
i = Dependent on vessel diameter

Fig. AIA. 6 General layout of the channel

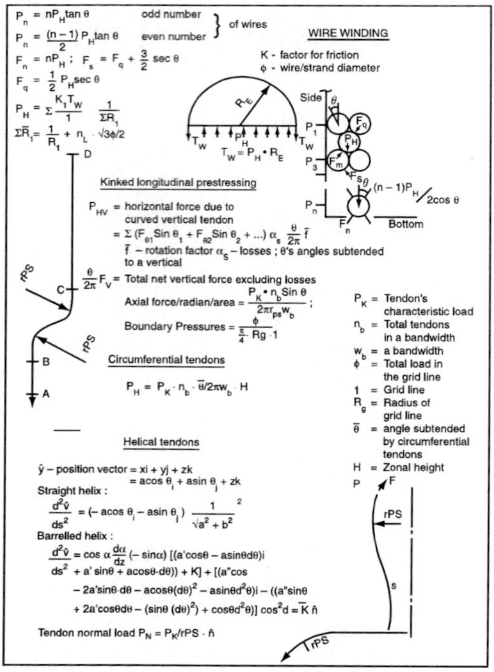

Fig. AIA. 7 Wire/Strand Analysis

Acknowledgements 51. Appendix A is a summary of part of a PhD thesis by the author who is grateful to Professor A. L. L. Baker for supervising the research described, to the United Kingdom Atomic Energy Authority for providing finance for the experimental work, and to Sir Robert McAlpine & Sons, Ltd., for their generous understanding and encouragement. It will not be out of place to mention a great contribution in wire-winding and strand winding by Taylor Woodraw and BBRV in the United Kingdom. The stress distribution under anchorages and its methodology by Newman. K, Robinson G.S., Bleich F and Rusheeduzzafar given below (1–23).

References

1. Rasheeduzzafar. An investigation of the stress distribution in the anchorage zones of post-tensioned concrete beams. Ph.D. thesis, University of London, 1964.
2. Newman, K. The structure and engineering properties of concrete, hit. Symp. Theory Aréh Dams, Southampton, 1965.
3. Robinson G.S. Methods of detecting the formation and propagation of microcracks in concrete. International Conference Structural Concrete, London, Sep. 1965.
4. Bleich, F. Der gerade Stab mit Rechteckquerschnitt als ebenes Problem. Der Bauingenieur, 1923, No. 9, 255–259 and 1923, No. 10, 304–307.

5. Morsch, L. Uber die Berechnung der Geienkquader. Beton und Lisen 1924. No. 12. 151–161.
6. Mafl, G. Prestresseif concrete. 2nd edn. Concrete Publications Ltd, London. 1950.
7. Guvo, Y. Contraintes dans les pièces prisntatiques sournises a des forces appliquées sur leurs bases, au voisinage de ces bases. lnt.Ass Bridge Struct. Engng Publications, 1951, 11, 195–226.
8. Guyor, Y. Prestressed concrete, 1st edo, London Contractors Record & Municipal Engineering. 1953.
9. Silvers, H. Die Berechnung von Auflagerbanken und Auflagerquarden von BrUckenpfeilern. Der Bauingenieur, 1952, 27, No. 6 (Jun) 209–213.
10. St€vnss, H. Uber den Spannungszustand im Bereich der Ankerplatten von Spanngliedern vorgespannter Stahlbetonkonstruktjonen Der Bauingenieur, 1956, 31, No. 4 (Apr) 134–135.
11. Douglas, D.J. and Traiiir, N.S. An examination of the stresses in the anchorage zone of a post tensioned prestressed concrete beam. Magazine Concrete of Research, 1960, 12, No. 34 (Mar) 9–18.
12. Rydzewskj J.R. and Whitbread, F.J. Short end blocks for a prestressed beams. iruct. Engr, 1963, 41, No. 2 (Feb) 41–53.
13. Ziellnski, J. and Row, R.E. An investigation of the stress distribution in the anchorage zones of post-tensioned concrete members. Res. Rep. No. 9. Cement and Concrete Association, September 1960.
14. Lensc HOW, J. and Sozen, M.A. Practical analysis of the anchorage zone problem 'in prestressed beams.Journal of American Concrete Instiute, 1965, 62 (Nov) 1421–1439.
15. Chrjstodoulides, S.P. A two-dimensional investigation of the end anchorages of post-tensioned concrete beams. Structural Engineering, 1955, 33, No. 4 (Apr) 120–133.
16. Christodouljdes, S.P. Three-dimensional investigation of the stresses in the end anchorage blocks of a prestressed concrete gantry beam. Structural Enginerring, 1957, 35, No. 9 (Sep). 349–356.
17. Bortsch, R. Wãlzgelenke und Steizenlager aus Eisenbeton. Beta,, and Eisen, 1938, 37, No. 19 (Oct.) 315–318 and 1938, No. 20 (Oct) 328–332.
18. Jones R. A method of studying the formation of cracks in a material subjected to stress. British Journal of Applied Physics, 1952, 3, No. 7 (Jul) 229–232.
19. Zielinski, J. and Rowe R.E. The stress distribution associated with groups of anchorages in post-tensioned concrete members. Res. Rep. No. 13. Cement and Concrete Association, October 1962.
20. Taylor, S.J. A failure criterion for concrete at cable anchorages. Ph.D. thesis, University of London, 1967.
21. Bortsch, R. Die Spannungen in Wãlzgelenkquardern. Beton and Eisen, 1935, 35, No. 4 (Feb) 6 1–66.
22. Ban S. et al. Anchorage zone stress distributions in posttensioned concrete members. Proceeding World of Conference, Prestress. Concrete, San Francisco, July 1957, 16.1–1 6.14.
23. Taylor, M.A. and Broms, B.B. Shear bond strength between coarse aggregate and cement paste or mortar. Journal of American Concrete Institute, 1964, 61, No. 8 (Aug.) 939–956.

Appendix B
Analytical Formulation for the Liner and Penetration

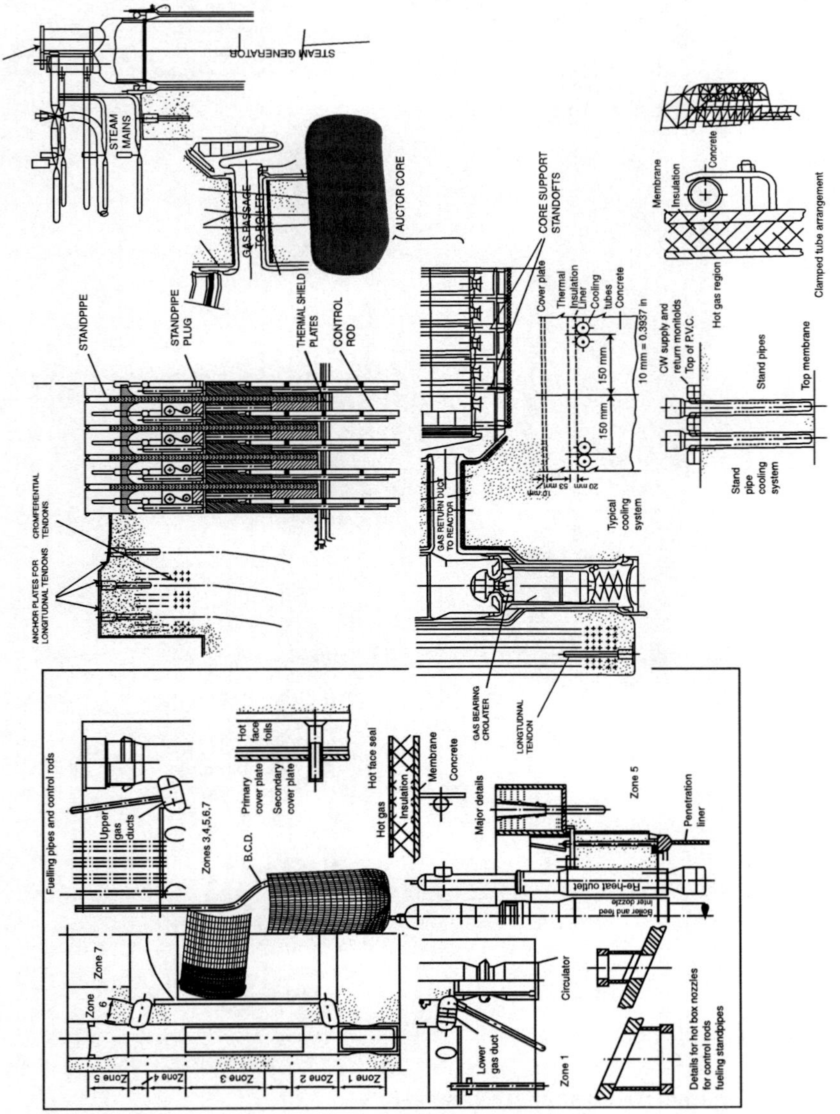

Plate AIB.1 Gas ducts, cooling pipes, control rods and other penetractions with finite element mesh schemes

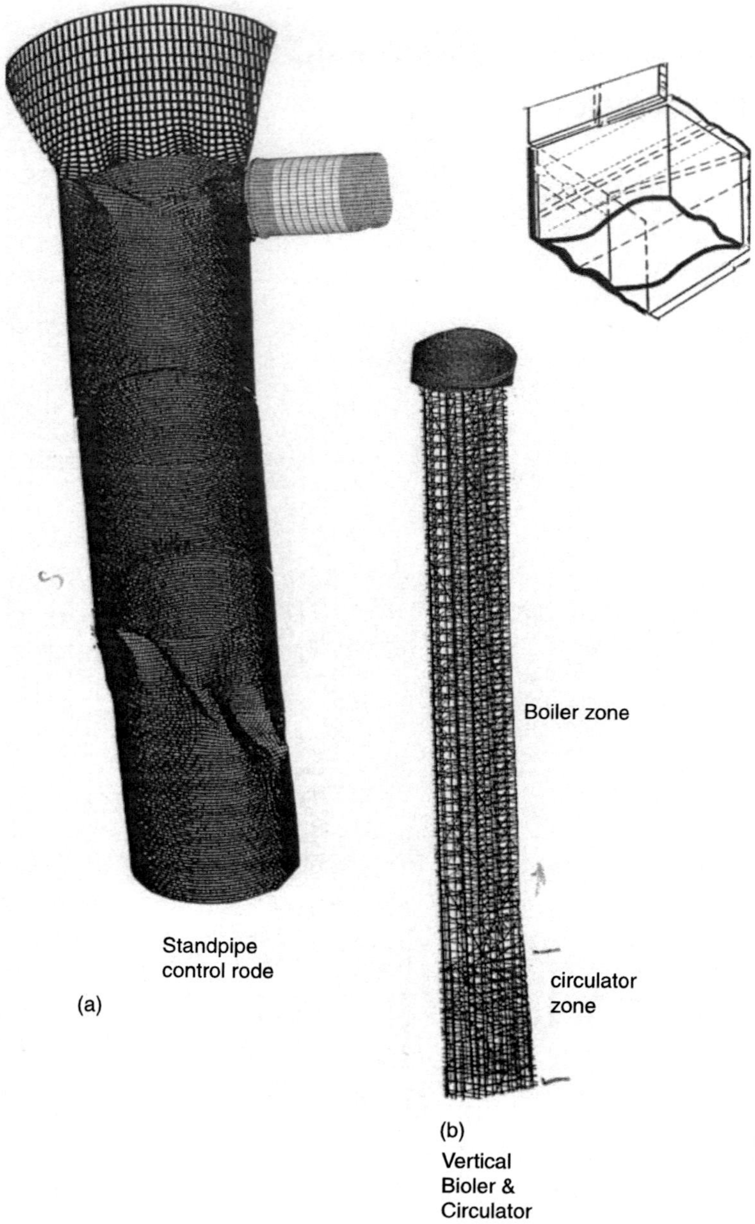

Plate AIB.2 F.E. Meshes

Q_d = Characteristic strength (kips) (KN)

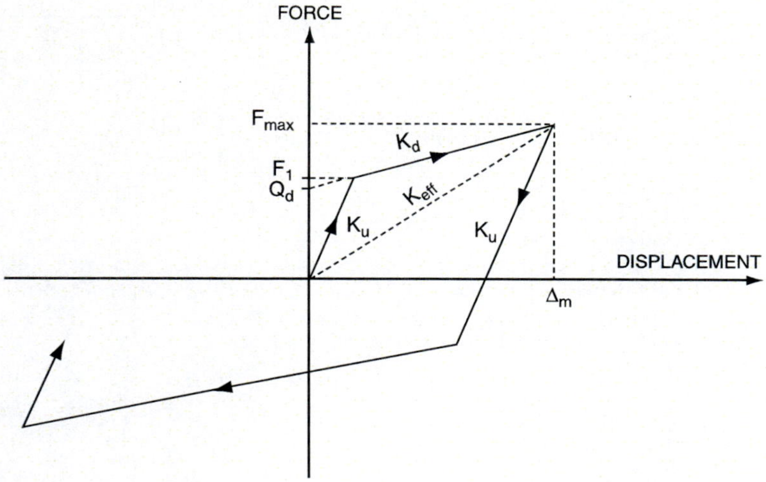

Plate AIB.3 Typical bilinear hysteresis loop

F_1 = Yield force (kips) (KN)
F_{max} = Maximum force (kips) (KN)
K_d = Post-elastic stiffness (kip/inch) (KN/mm^2)
K_u = Elastic (unloading) stiffness (kip/inch)
K_{eff} = Effective stiffness
Δ_m = Maximum bearing displacement

General Formulation for the Liner

In any computer package involving 2- and 3D finite-element technique, Table AIB.1 giving analytical formulation can be simulated to analyse the liner between the studs, standpipes and cooling pipes. For the global analysis of the pressure and containment vessels, these local effects can form the worst case. The vessel can be analysed with and without the contribution of the liner. Under ultimate conditions the liner local areas must be thoroughly assesed prior to the final decision of the factor safety above the elastic conditions under (pressure + prestress + temperature) the combined loading:

Table AIB.1 Analytical formulation of the steel liner

$$[K_{\mathrm{TOT}}]\{\delta\}^* + \{F_T\} - \{R_T\} = 0$$

where

$$[K_{\mathrm{TOT}}] = [K_l] + [K_a]$$

$$\{\delta\}^* = \begin{Bmatrix} \delta_{un} \\ \delta_b \end{Bmatrix}; \{F_T\} = \begin{Bmatrix} F_{un} \\ F_b \end{Bmatrix}; \{R_T\} = \begin{Bmatrix} R_{un} \\ R_b \end{Bmatrix}$$

$[K_{\mathrm{TOT}}]$ = Total stiffness matrix
$[K_l]$ = liner stiffness matrix
$[K_a]$ = A stud stiffness matrix
$\{F_T\}$ = Total initial load vector
$\{R_T\}$ = Total external load vector
$_{un}$ = quantities corresponding to unknown displacement
$_b$ = quantities corresponding to restrained boundaries
$[K_{lr}]\{\delta_{un}\} + \{F_{un}\} = 0$
$\{\varepsilon\}$ = $[B]\{\delta\}$
$\{\sigma\}$ = $[D](\{\varepsilon\}-\{\varepsilon_0\})$
$\{\delta\}$ = anchor shear forces
= $[K_a] \cdot \{\delta_{un}\}$
$\{F_{un}\}$ = $\int_v [B]^T[D]\{\varepsilon_0\}\,dv = \int_v [B]^T[D]\{\varepsilon_0\}\det[J]\,d\xi\,d\eta\,d\zeta.$

The plastic bucking matrix is given by

$$(K + \lambda K_G)F_T = 0$$

where

K= elasto-plastic stiffness matrix as a function of the current state of plastic deformation
K_G = initial stress geometric stiffness matrix
The determinant $|K + \lambda K_G| = 0$.

The essential equation is characteristically triangularized for the ith loading step as

$$(K^i + \lambda_c K_G^i)F_T^i = 0$$

$$\lambda_c = 1 + E_{ps} \quad E_{ps} - \text{accuracy parameter.}$$

Note: Table AIB.1 is a part of Programs, ISOPAR, ANSYS and others.

Penetrations Analysis: Classical and Finite Element Techniques

Penetrations in Reactor and Containment Vessels – General Introduction

Several penetrations, their sizes or locations are related to the reactor systems and their functions and layouts. Some of them have already been discussed in earlier chapters of the text under reactor and containment vessels. Their solutions and results have already been established. This appendix gives some penetration areas which may be treated as added discussions. The finite-element techniques of solving them has been fully discussed. There is no need for a reader to use it again. However, a classical approach is given to evaluate penetrations. The results are fully collaborated from the classical method with these from the finite-element technique. Plates AIB.1 and AIB.2 are giving some penetrations which need detailed discussions.

Classical Method of Analysis

In this method a thin plate theory is assumed with one to several penetrations and it is documented below.

A Thin Circular Plate with a Central Hole

We will compare in our investigations a "thin" circular plate uniformly loaded and fixed at its edges with a similarly loaded plate containing a central hole. This central hole is assumed to be closed by a simply supported plug. These two plates are shown in their radial cross-sections in Fig. 4a, b, respectively.

The solution of the problem for a monolithic circular plate is given in Timoshenkots 'Strength of Material'. Part 11 on page 97.

The moments at a distance x from the centre of the plate are

radially $M\text{rm} = \frac{q}{16}\left[a^2(I+v) - x^2(3+v)\right]$ (AIB.1)

tangentially $M\text{tm} = \frac{q}{16}\left[a^2(I+v) - x^2(1+3v)\right]$ (AIB.2)

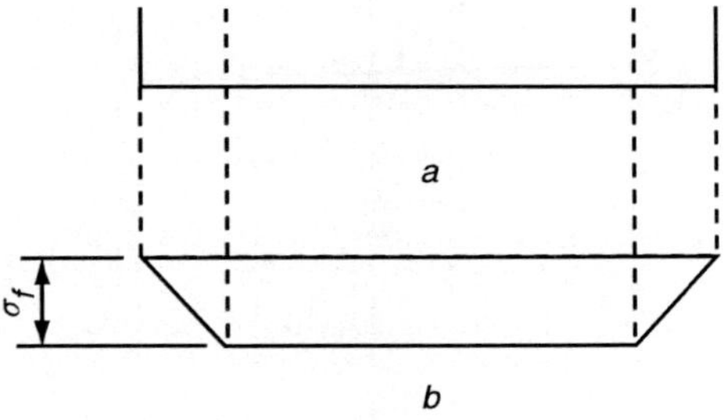

Fig. A1B.1 Moments in radial and tangential directions

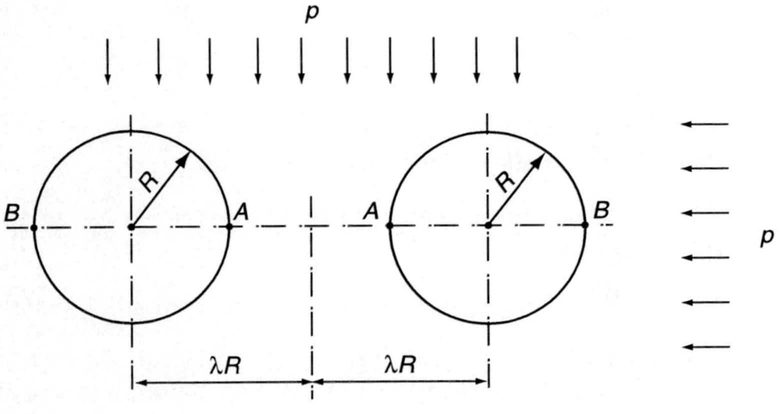

Fig. AIB.2 Circular hole in the circular plate

Fig. AIB.3

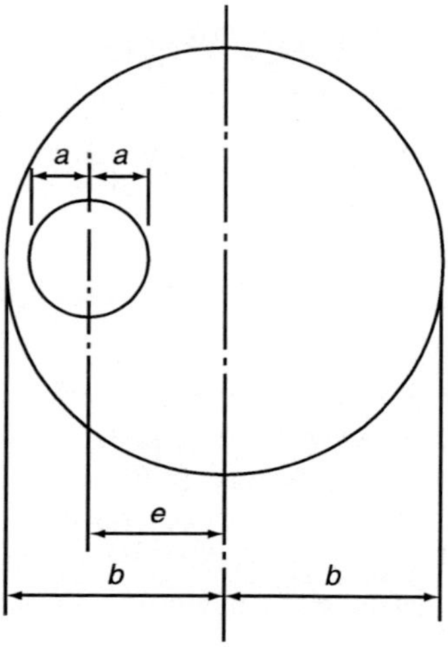

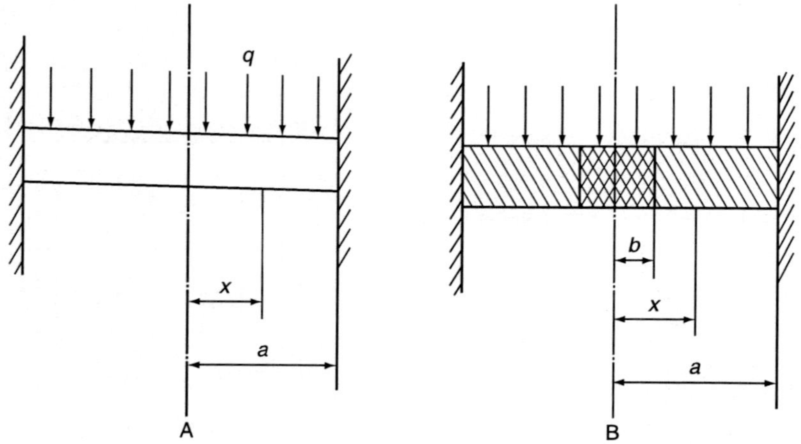

Fig. AIB.4

Where a is the outer radius of the plate.
On assuming Poisson's ratio ν to be equal to zero one obtains

$$M\mathrm{rm} = \frac{q}{16}(a^2 - 3x^2) \ldots\ldots\ldots \quad \text{(AIB.3)}$$
$$M\mathrm{tm} = \frac{q}{16}(a^2 - x^2) \ldots\ldots\ldots \quad \text{(AIB.4)}$$

A plate with a central closed penetration can be considered as a monolithic plate additionally loaded by a radial bending moment M' uniformly spaced along the central hole. Assuming that this plug is fully discontinuous with the plate

$$M = \frac{q}{16}(a^2 - 3b^2) \quad \text{(AIB.4)}$$

For a *'ring plate' loaded uniformly over its inner edge by a moment M'* we have (from equation of Timoshenko's 'Strength of Materials')

The angular deflection

$$\phi = \frac{c_1 x}{2} + \frac{c_2}{x} \quad \text{(AIB.6)}$$

$$M'rx = D\frac{d\phi}{dx} = D\left[\frac{c_1}{2} + \frac{c_2}{x^2}\right] \quad \text{(AIB.7)}$$

$$M'tx = D\frac{d\phi}{x} = D\left[\frac{c_1}{2} + \frac{c_2}{x^2}\right] \quad \text{(AIB.8)}$$

For the boundary conditions we have

$$\phi = 0$$

when $x = 0$

Which gives $c_2 = c_1 - \frac{a^2}{b^2}$ (a) (AIB.9)

We also have $M'rx = M'$ when $x = b$ (b)

And hence $M' = D\left(\frac{c_1}{2} + \frac{c_1}{2}\frac{a^2}{a^2}\right)$ (c)

or $c_1 = \frac{2M'}{D}\frac{b^2}{a^2+b^2}$ (d)

And $c_2 = \frac{-M'}{D}\frac{b^2 a^2}{a^2+b^2}$ (e)

Substituting these values into Eqs. (AIB.7) and (AIB.8), one has

$$M'rx = M'\left[\frac{b^2}{a^2+b^2} + \frac{a^2}{x^2}\frac{b^2}{a^2+b^2}\right] = M'\frac{b^2}{x^2}\frac{a^2+x^2}{a^2+b^2} \quad \text{(AIB.10)}$$

$$M'tx = M'\left[\frac{b^2}{a^2+b^2} - \frac{a^2}{x^2}\frac{b^2}{a^2+b^2}\right] = -M'\frac{b^2}{x^2}\frac{a^2-x^2}{a^2+b^2} \quad \text{(AIB.11)}$$

Substituting the values of *M'* from equation one obtains

$$M'rx = \frac{a}{16}(a^2 - 3b^2)\frac{b^2}{x^2}\frac{a^2+x^2}{a^2+b^2} \quad \text{(AIB.12)}$$

$$M'tx = \frac{a}{16}(a^2 - 3b^2)\frac{b^2}{x^2}\frac{a^2-x^2}{a^2+b^2} \quad \text{(AIB.13)}$$

Adding, respectively, these Equations one obtains the formulae for the perforated plate.

$$Mrp = \frac{a}{16}\left[a^2 - 3x^2 - (a^2 - 3b^2)\frac{b^2}{x^2}\frac{a^2 + x^2}{a^2 + b^2}\right] \; for\; x \geq b \quad \text{(AIB.14)}$$

$$Mtp = \frac{a}{16}(a^2 - x^2)\left[1 + \frac{b^2}{x^2}\frac{a^2 - 3b^2}{a^2 + b^2}\right] \; for\, x \geq b \qquad \text{(AIB.15)}$$

Comparing the Eqn. (AIB.1) and (AIB.2) with Eq. (AIB.30) we can calculate ratios of bending moments and hence of stresses in plates with and without a central hole under one uniform load.

The radial bending stress concentration for factor Cr is assumed to be

$$Cr = \frac{Mrp}{Mrm} = 1 - \frac{a^2 - 3b^2}{a^2 + 3x^2}\frac{b^2}{x^2} = \frac{a^2 + x^2}{a^2 + b^2} \qquad \text{(AIB.16)}$$

If $(x^2 \neq a^2)$

The tangential bending stress concentration factor C_t was assumed to be

$$Ct = \frac{Mtp}{Mtm} = 1 - \frac{b^2}{x^2}\frac{a^2 - 3b^2}{a^2 + b^2} \qquad \text{(AIB.17)}$$

The values of Cr are

$$\begin{aligned} Cr &= 0 \; for\, x = b \\ Cr &= 1 + \tfrac{b^2}{a^2}\tfrac{a^2 - 3b^2}{a^2 + b^2} \; for \;\; x = a \end{aligned} \qquad \text{(AIB.18)}$$

If $a_2 = 9.68b^2$ and similarly it could be any ratio between a and b for 9 ft hole, $Cr9 = 1.0646$ and for $11'.0$ hole central

$$\begin{aligned} a^2 &= 6.48b^2 \\ Cr11 &= 1.0718 \end{aligned} \qquad \text{(AIB.19)}$$

The maximum value Ct if reached at the edge of the hole $(x = b)$, then

$$Ct = \frac{1 + \frac{a^2}{b^2} - 3}{\frac{a^2}{b^2} + 1} \qquad \text{(AIB.20)}$$

Then $Ct9 = 1.627$

$$Ct11 = 1.465 \qquad \text{(AIB.21)}$$

$$\text{Radial stress } Cr = \frac{\text{Bending stress with penetration}}{\text{Bending stress with no penetration}} \qquad \text{(AIB.22)}$$

Let us say $= \frac{624\,\text{psi}\;()}{700\,\text{psi}\;()} = 0.90$

$$Ct = \frac{\text{Bending stress with penetration}}{\text{Bending stress with no penetration}} = \frac{57\,\text{psi}}{250\,\text{psi}} = 2.05 \qquad \text{(AIB.23)}$$

Analysis of the Effect of the Large Penetrations

No reliable general method of analysing 3D stress problems in a body without circular symmetry was available for use in this. For this reason, the analysis of the stress distribution in a vessel with large penetrations consists of the extrapolation and combination of the results obtained by several less sophisticated methods of analysis.

The practical problems of the large penetrations can be divided into the following categories:-

a) The effect of a non-symmetrical arrangement of penetrations on the stress distribution in the whole vessel.
b) The local stress concentrations in the perforated vessel cap or *slab*.

For the reason given above, the latter problem was analysed separately for the biaxial and triaxial loading conditions. Stress concentration factors are derived for the biaxial loading condition. Additional stress concentration factors for the triaxial loading condition are calculated. The use of these factors is explained in many relevant text on stress concentration.

Biaxial Loading Conditions

Method of Analysis

The analysis was performed using the finite-element method. The results obtained by this technique are compared with the results for the same problem solved by a theoretical analysis based on classical method.

Description of the 3D Finite–Element of Computer Programme

Three-dimensional is a method for solving static problems. The solution is obtained using a set of stress–strain relationships and dynamic–equilibrium equations. The set of equations is solved using already given in the text the finite–element forms of the equations. As the equilibrium equations contain damping factors, the required static solution is approached. By an appropriate choice of the damping factors, the structure vibrates with a 'dead beat' form and so a sufficiently accurate static solution is quickly reached.

The Simplified Model of Vessel Cap

The bottom cap or top of the vessel is analysed as a perforated disc under biaxial, loading and triaxial. This simplification gives a good approximation to the conditions of cap prestressed by hoop cables. The effect of a uniform bending moment can also be treated as a special case of the plane stress analysis. However, the effect of a varying bending moment (caused by the curvature of the prestressing cables and internal pressure) cannot be truly estimated by means of such an analysis. Hence 3D finite–element is involved for this kind of work.

The Arrangement of the Loading

The radial stresses in the cap or slab are created either by the hoop tendons or by the interaction between the cap and the cylindrical art of the vessel. For this reason the radial forces loading the disc were assumed to be.

Results of the Finite-Element Analysis

The following types of bottom cap were analysed under different loading conditions. Cases 1–4 relate to an internal vessel diameter of 28 ft. Case 5 considers a vessel internal diameter of 29 ft.

(1) A disc of 44′.0″ dia without any penetration was analysed under the loading arrangements 'A'.

(2) A disc of 44′.0″ dia with two 11′.0″ dia holes was analysed under the loading arrangements 'B' and 'D'.

The biggest stress concentration was found to be loading along the axis of these two penetrations. The resultant stresses together with the field stress are shown for loading 'B' in graph (). The reduction of these stresses under loading 'D' was rather small. The two peak values under 'B' were 2790 and 2400 psi; and the loading 'D' stresses at the same respective points were 2690 and 2100 psi.

The curve giving the ratio of the local stress in the perforated cap over the 'field stress' in the monolithic cap is given on Graph ().

Th comparison of the average radial deformation of the monolithic and perforated cap gives the ratio of 1.78 (the value could be used to simulate presence of the holes in the finite-element program ISOPAR).

(3) A disc 44′.0″ dia with two 11′.0″ hole and two 5′.0″ holes was analysed under loading condition 'B'.

The result of this analysis is given on the graph ().

The ratio of the average radial deformation in this case was equal to 1.97

(4) The ring 44′.0″ outside and 28′.0″ inside diameter was loaded by the forces type 'C' identical in value to loading 'D' of disc (2) (44′.0″ outside dia with two 11′.0″ dia holes.)

The contours of the principal stresses on the ring are shown on graph.

Theses stresses vary from 3700 to 2240 psi on the inside face and from 2550 to 1500 psi on the outside face.

Therefore the variation over average stress is about ±25% on the inside face for the 'worst' (i.e. most unsymmetrical) arrangement of holes.

(5) The disc of 46′.0″ dia with one 9′.0″ central hole and two 5′.0″ and from 4′.0″ dial holes concentrically spaced around it was loaded uniformly spaced in the radial direction over a ring of thickness equal to the thickness of the vertical wall of the vessel.

The distribution of field stresses (σ_1) in unperforated disc follows the diagram shown in Fig. ()

A loading of radial forces uniformly spaced around the vessel parameter represents the loading due to action of hoop stresses.

The loading imposed on the cap by the interaction between the cap and the barrel of the vessel can be uniform only for a circularly symmetrical arrangement of holes in the disc. For the unsymmetrical arrangement of the holes (say two 11′.0″ dia holes) this interaction will not be uniform around the circumference of the vessel. An uneven radial distortion of this cap will produce an uneven distribution of the 'continuity force' between the cap and the barrel. The effect of this variation will reduce the absolute values of the maximum local stresses in the cap, and increase locally, stresses in the barrel. The variation of these continuity forces can be assumed to be proportional to the deflection of the cap.

To simulate the above effects the following loading arrangements were analysed by the finite element.

(A) A disc without penetration was loaded by radial forces uniformly distributed both around its circumference and across the width equal to the vessel's wall thickness. Infact, it is a thick-walled slab.

The "stress field" obtained by this loading is used to calculate stress concentration factors in the next calculation.

(B) A disc with penetration is loaded by the same forces as for case 'A'.

The ratio of stresses obtained by this loading over the 'field stresses' from case (A) gives the values of the stress concentration factors.

(C) A ring representing the vessel's barrel was loaded by radial forces uniformly distributed along the radial direction but not circumferentially. The total value of these factors is the same as for cases 'A' and 'B'. The magnitude of the varies circumferentially in the inverse proportion to the ratio of the actual over the average radial deformation of the disc under loading 'B'.

(D) A perforated disc (as in case B) is then loaded in the same manner as the ring in case 'C'.

By the force type 'B', the results were compared with the unperforated disc of 44′.0″ diameter.

The error introduced by this difference is estimated to be in the order of ± 4.5%. The directions of the maximum principal stresses for perforated disc are shown in graph later on. The values of the maximum principal stresses are given as a ratio of the 'field' stress obtained from the analysis of the unperforated plate. The ratio of the average deflection for the two cases is equal to 1.655 (also with an error of $\pm 4.5\%$).

The Comparison with the Theoretical Formulae

Some of the results of the finite-element calculations referred to above can be compared with results obtained from a simple classical analysis. Two methods of theoretical analysis were investigated.

The problem of an infinite plate with two holes loaded uniformly by a biaxial stress field is analysed in G.N. Savins book 'Stress Concentration Around Holes'. (pp. 134–139)

From the ratio

$$\lambda = \frac{7'6''}{5'6''} = \frac{15}{11} \tag{AIB.24}$$

One has from *Savins table 25* the values of maximum stresses

$$\sigma A = \left(2.887 - \tfrac{0.133}{0.5} \times 0.476\right)p = 2.761p \tag{AIB.25}$$

$$\sigma B = \left(2.255 - \tfrac{0.133}{0.5} \times 0.097\right)p = 2.23p \tag{AIB.26}$$

The formulae given by Timoshenko (Trans. Royal. Society Series A, volume 221, p. 267, 1921) gives the maximum stress (of point B due to an internal pressure in a thick cylinder with an eccentric bore.)

The 'slice' of such a cylinder is shown in Fig. AIB.7. The formulae for the notation

$$\sigma' \max. = pi\left[\frac{2b^2(b^2 + a^2 - 2ae - e^2)}{(b^2 + a^2)(b^2 - a^2 - 2ae - e^2)} - 1\right] \tag{AIB.27}$$

In our case we have

$$a = 5'6''$$
$$b = 22'0''$$
$$c = 7'6''$$

And hence

$$\sigma' \max. = pi \times 1.244 \tag{AIB.28}$$

In the absence of a similar formula for the effect of external pressure, compilation with some other theories is required. It seems a safe implication to assume that an external pressure instead of internal pressure would increase the above coefficient in the same proportion as for a cylinder with a central bore. Accepting this assumption we have F at the edge of the hole

$$\text{Hoop stress}\,\sigma = \frac{a^2p}{b^2 - a^2}\left(1 + \frac{b^2}{a^2}\right) = \frac{2pb^2}{b^2 - a^2} \qquad \text{(AIB.29)}$$

Hence

$$pi = p.\frac{2b^2}{a^2 + b^2} \qquad \text{(AIB.30)}$$

Therefore,

$$\sigma A \approx \sigma\,\text{max}. = p \times 1.244\frac{2b^2}{a^2 + b^2} = 2.34p \qquad \text{(AIB.31)}$$

Comparing the factors and those shown on graph () derived from the finite-element method, one has at point A.

Stress concentration factor obtained by computer analysis = 3.15
Stress concentration factor obtained by mathematical theory = 2.76
Ratio = $\frac{3.15}{2.76}$ = 1.14
Also at point B one has
Stress concentration factor obtained by mathematical theory = 2.34
Ratio
$\frac{2.7}{2.34}$ = 1.15

On can see from these ratios that the finite-element methods give both close and 'safe' results. The larger results obtained by the computer analysis are probably due to the fact that a more realistic boundary condition was used.

Triaxial Loading Conditions

The finite-element analysis described in this text deals adequately with radial loadings applied at the periphery of a disc but does not take into account the variation in radial load over the depth or radius of the disc. These variations arise from the effect of pressure loading, tendon curvature, continuity effects and the pressure of penetrations.

The study which follows attempts to evaluate the effects ignored in the simple biaxial analysis. Sections do examine three alternative approaches from which the most probable bending stress concentration factors are chosen.

Utilisation of the 'Yield Line Method' in the Elastic Analysis

Johansen's yield line method was developed for structures working in a plastic state stresses. For this reason, this method ignores the condition of continuity of deflection and retains only the condition of equilibrium of the internal and external forces. Nevertheless, this method can be safely applied in analysis of elastic problems as long as it is remembered that

(a) The moments along a yield line now represent average, not constant, bending moments.
(b) The ratio of the moments along different edges of any 'failed' panel must be in the same ratio to each other as the actual elastic average moments in an elastic stage of the structure.

This second condition creates great but not insurmountable difficulty. As long as the approximate range of these ratios can be estimated 'safe' results may be obtained by accepting in turn the 'worst' extremes of this range.

Two of the most probable yield line patterns for the basic vessel bottom cap are shown in Figs. AIB.5 and AIB.6. The simplified panels of these pattern are shown in Figs. AIB.7 and AIB.8, respectively, These figures are of an internal diameter of 29'0".

Let us consider the arrangement shown in Fig. AIB.7.

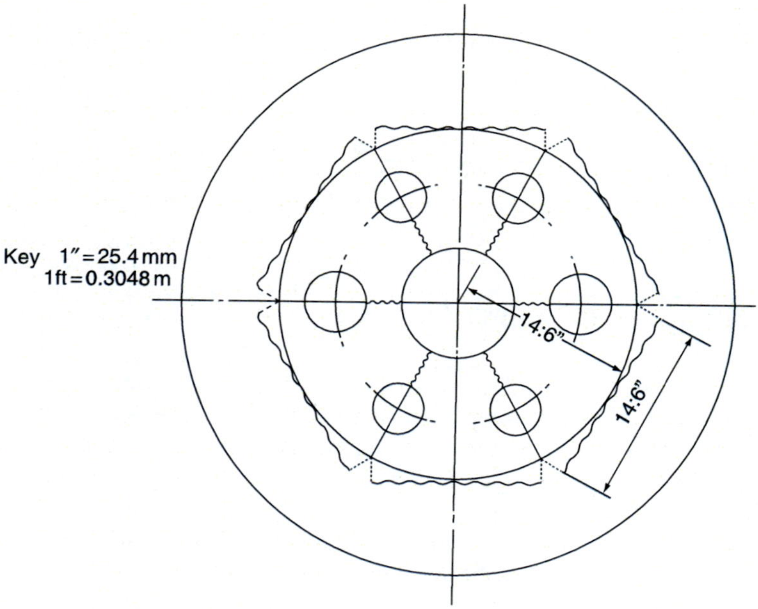

Fig. AIB.5 Vessel capat bottom

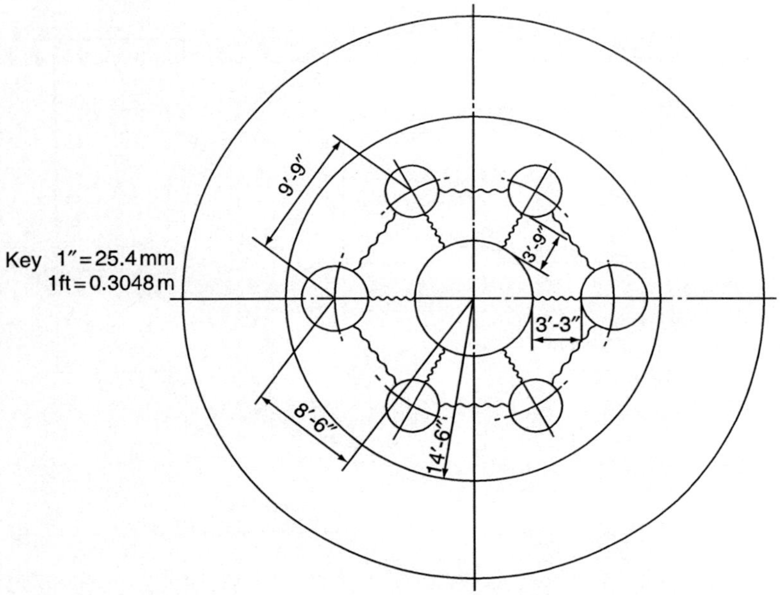

Fig. AIB.6 Vessel with penetrations

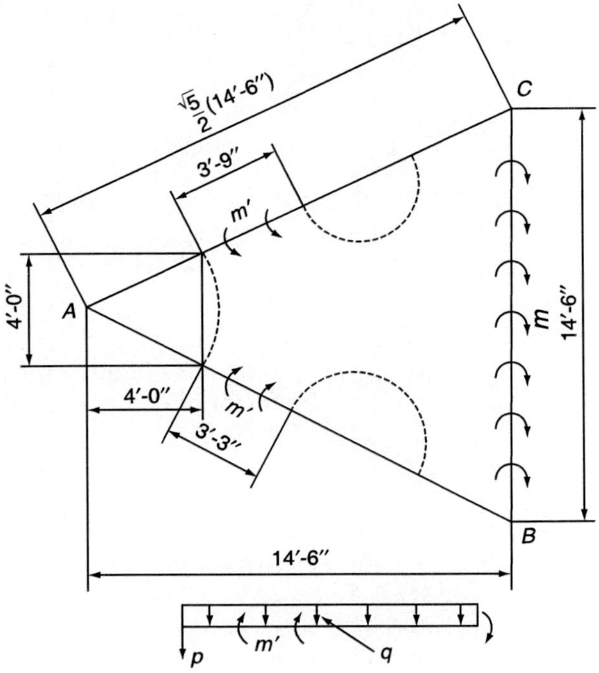

Fig. AIB.7 Sectoral cut zone

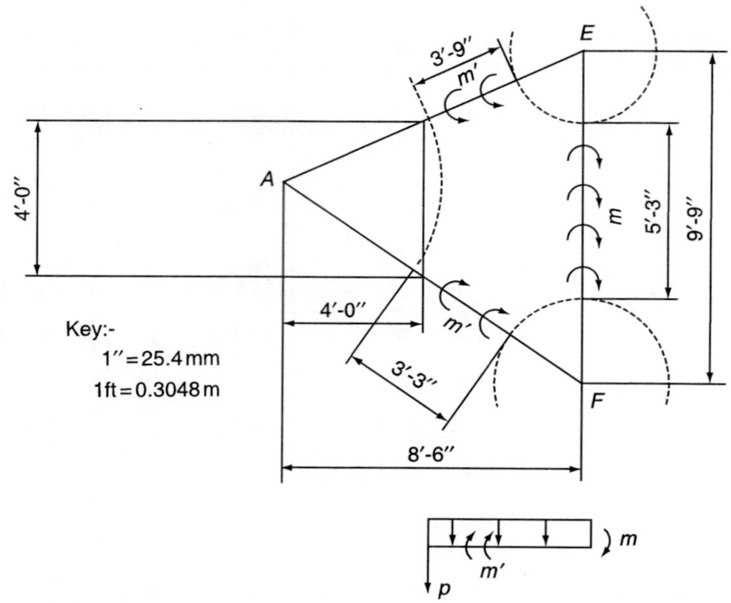

Fig. AIB.8 Sectoral cut with forces or moments

The edge force *P*/*ft* (reaction from the plug) is obtained by considering the equilibrium of the small triangle at the centre.

Hence,

$$p \times 4 = \frac{4}{2}\,4q \tag{AIB.32}$$

or

$$P = 2q \tag{AIB.33}$$

Ignoring the effect (an error on the safe side) of the tangential moments 'm' between the outer ring of penetrations and the barrel position of the vessel, one has the equation of equilibrium for moments about line BC:

$$M_7 \times 14.5 + M_7 \times = \frac{7}{\sqrt{5}} = P \times .4 \times 10.5 + q\left(4 \times \frac{10.5^2}{2} + \frac{10.5}{3} \times \frac{10.5^2}{2}\right) \tag{AIB.34}$$

$$\text{or } M_7 + 0.216\,M_7 = 5.8q + 28.5q = 34.3q \tag{AIB.34a}$$

or for the 28′0″ internal diameter vessel

$$M_7 + 0.216M'_7 = 34.3q \times \frac{14^2}{14.5^2} = 32.0q \qquad \text{(AIB.35)}$$

For the arrangement shown in Fig. AIB.8, the equilibrium condition for moments about the line EF as

$$\begin{aligned} 5.25M_8 + M'_8 \times = \tfrac{7}{\sqrt{5}} = P \times 4 \times 4.5 + q\left(4 \times \tfrac{4.5^2}{2} + \times \tfrac{4.5^2}{6} \times 5.75\right) \\ \text{i.e. } M_8 + 0.596\, M'_8 = 6.86q + 11.42q = 18.28q \end{aligned} \qquad \text{(AIB.36)}$$

or for the 28'0" internal diameter vessel

$$M_8 + 0.596M'_8 = 18.26q \times \frac{14^2}{14.5^2} = 17.05q \qquad \text{(AIB.36a)}$$

For comparison we will analyse the yield line pattern for the cap without any penetrations. The failure yield line pattern of a disc is shown in Fig. AIB.9. The elementary failure panel is shown in Fig. AIB.10.

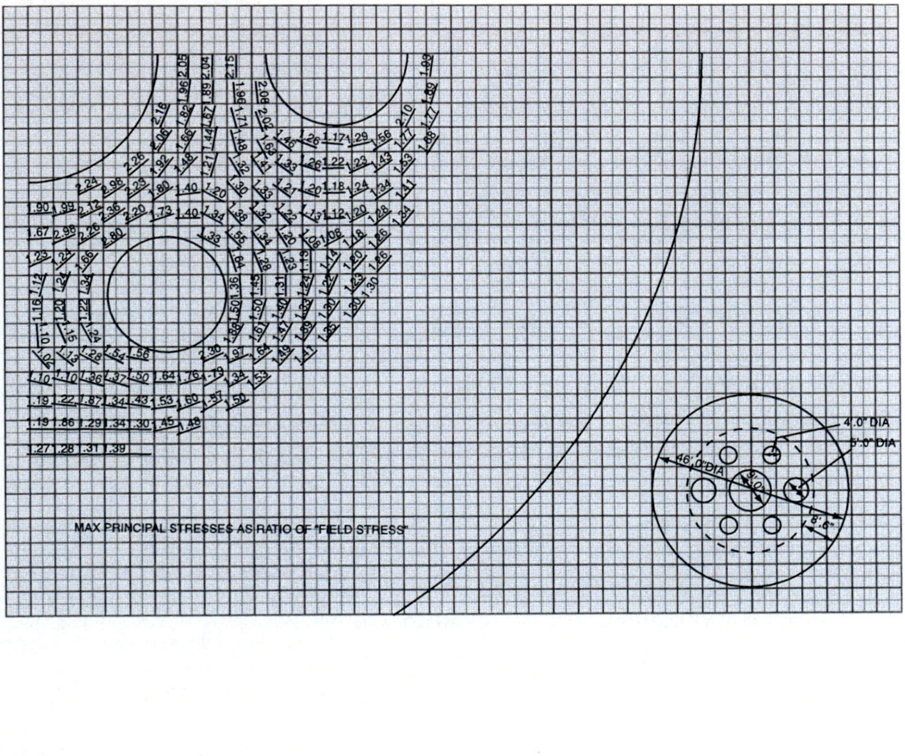

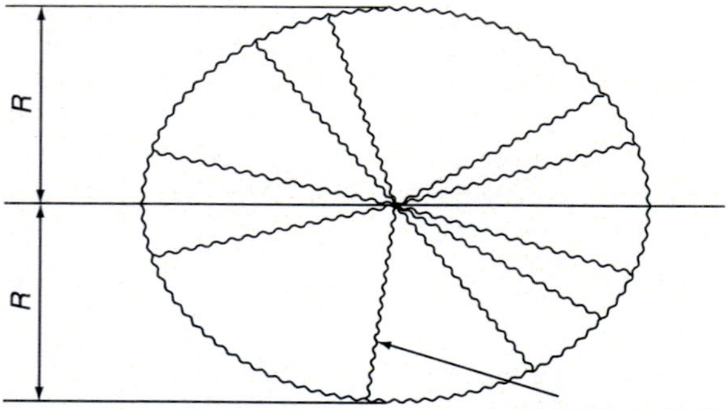

Fig. AIB.9 Yield line arrangement

Taking moments about the outer edge we have

$$(M_0 + M_0')Rd\phi = q\tfrac{Rd\phi}{2} \times \tfrac{R^2}{3}$$
$$\text{Or } M_0 + M_0' = q\tfrac{R^2}{6}$$

Hence for $R = 14'0''$ we have

$$M_{01} + M_{01}' = 32.7q \quad \text{(AIB.37)}$$

And R = 8'6"

$$M_{02} + M_{02}' = 12.03q \quad \text{(AIB.38)}$$

Comparing these values with Eqs. (AIB.35a) and (AIB.36a)

$$M_7 + 0.216M_7' = \frac{32.0}{32.7}(M_{01} + M_{01}') = 0.98(M_{01} + M_{01}') \quad \text{(AIB.39)}$$

$$M_8 + 0.596M_8' = \frac{17.05}{12.03}(M_{02} + M_{02}') = 1.42(M_{02} + M_{02}') \quad \text{(AIB.40)}$$

The ratio of M_0/M_0' is estimated from the Vany analysis for a vessel with no penetrations. The values given below have been taken from sketches PCPVD. The moments are assumed to be proportional the variation in stress through the depth of the cap.

$$\left.\begin{array}{l} M_{01}' \text{ is approximately proprtional to 800 psi} \\ M_{01}' \text{ is approximately proprtional to 150 psi} \end{array}\right\} \text{at 14 ft radius}$$

$$\left.\begin{array}{l} M_{02} \text{ is approximately proprtional to 370 psi} \\ M_{02}' \text{ is approximately proprtional to 200 psi} \end{array}\right\} \text{at } 9'9'' \text{ radius}$$

As stated previously we wish to compare the local stresses (which we have said are proportional to the local moments, M_7, M'_7, M_8 and M'_8 with the Vany results for vessels with simulated penetrations. Hence, the following moment values have been taken from sketches PCPVD.

M_{p1}–Average radial (longitudinal) bending moment at the foot of the vertical walls. This moment is approximately proportional to 800 psi (5.52 MN/m^2)
M_{p2}–Average radial bending moment along the circle through the centres of the outer penetrations. This moment is approximately proportional to 385 psi (2.655 MN/m^2)
M'_{p1}–Average tangential (hoop) bending moment across the ring between the central penetration and the outer ring of penetrations. This moment is approximately proportional to 500 psi (3.448 MN/m^2)

We can now rewrite Equations using these new moments:
One has from Eq. (AIB.39)

$$M_7 = 0.216M'_7 = 0.98M_{pl}(1 + 0.188) = 1.165M_{pl} \tag{AIB.41}$$

Similarly $\frac{M_{02}}{M_{02}} = \frac{370}{200}$
we have from Eq. (AIB.40)

$$M_8 + 0.596M_8 = 1.42M_{02}(1 + 0.54) = 2.22M_{02}$$

Furthermore since,

$$\frac{M_{02}}{M_{p2}} = \frac{370}{385} \; and \; \frac{M_{02}}{M_{p2}} = \frac{370}{500}$$

We have from Eq. (AIB.40) either

$$M_8 + 0.596M'_8 = 1.42M_{p2} \tag{AIB.42}$$

or

$$M_8 + 0.596M'_8 = 1.62M_{p2} \tag{AIB.43}$$

In Eq. (AIB.42) the value of M_8 cannot be smaller than M_{p2}.
Hence one may write

$$0.596M'_8 = 2.14M_{p2} - M_8 < 1.14M_{p2}$$

or

$$M'_8 < 1.92M_{p2}$$

or since $\frac{M_{p2}}{M_{p2}} = \frac{385}{500}$

$$\frac{M'_8}{M_{p2}} < 1.48 \tag{AIB.44}$$

On the other hand, knowing that M'_8 cannot be smaller than M'_{p2} we have from Eq. (AIB.37)

$M_8 = 1.64M'_{p2} - 0.596M'_8 < 1.044M'_{p2}$ and since

$$\frac{M'_{p2}}{M_{p2}} = \frac{500}{385}$$
$$\frac{M_8}{M_{p2}} < 1.36 \tag{AIB.45}$$

Because theoretically we have $M'_7 = M'_8$ we have from Eq. (AIB.37)

$$M_7 = 1.165 \quad M_{pl} - 0.216M'_8$$

or since M'_8 is bigger than M'_{p2}

Also $\frac{M'_{p2}}{M_{pl}} = \frac{500}{800}$

Therefore $\frac{M_7}{M_{pl}} < 1.03$

The formulae (AIB.44), (AIB.45), and (AIB.46) should give 'very safe' coefficient for the stress concentration at 'critical' zones in the bottom cap. Similar formulae could be found for other arrangements of the penetrations.

Large Penetration Under Pressure and Temperature

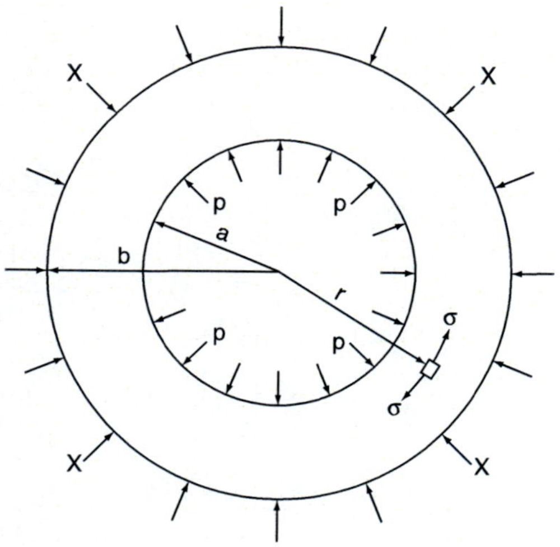

Notation

Giving assumed values where appropriate

R = radius

a = internal radius = 14′

b = external radius = 22′
p = internal pressure = 705 psi
x = external pressure = prestress
t = temperature
ti = temperature crossfall 30°C
α = coefficient of expansion = 13×10^{-6} (c degrees)$^{-1}$
E = module of elasticity = 5.25×10^{6}
γ = poisons ratio = 0.15 or 0 (where neglected)
σ_c = permissible working compressive stress = 2200 psi (for prestressed and internal pressure)
σ_s = permissible working tensile stress = 500 psi
σ_t = hoop stress due to temperature crossfall
σ_p = hoop stress due to internal pressure
σ_x = hoop stress due to prestress
σ_{ti} = hoop stress on inside due to temperature
σ_{to} = hoop stress on outside due to temperature
σ_{pi} = hoop stress on inside due to internal pressure
σ_{po} = hoop stress on outside due to intenal temperature
σ_{xi} = hoop stress on inside due to pressure
σ_{xo} = hoop stress on outside due to prestress

The vessel is assumed to consist of a long thick cylinder.
For this condition the following formulae apply
Internal pressure only:

$$\sigma p = \frac{a^2 p}{b^2 - a^2}\left(1 + \frac{b^2}{r^2}\right) \tag{AIB.46}$$

External pressure (prestress)

$$\sigma x = -\frac{b^2}{b^2 - a^2}\left(1 + \frac{a^2}{r^2}\right) \tag{AIB.47}$$

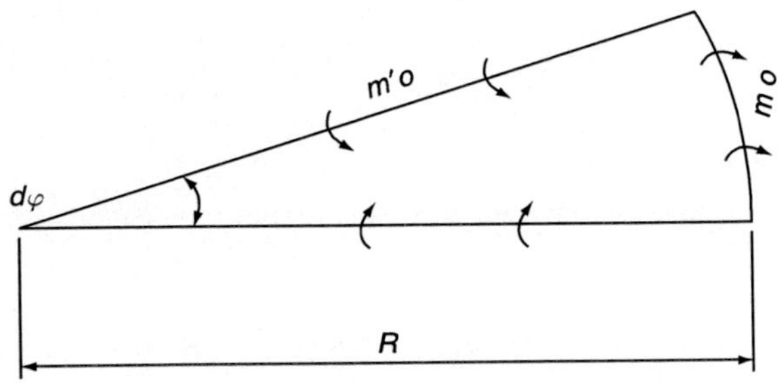

Fig. AIB.10 Sector for yield line analysis

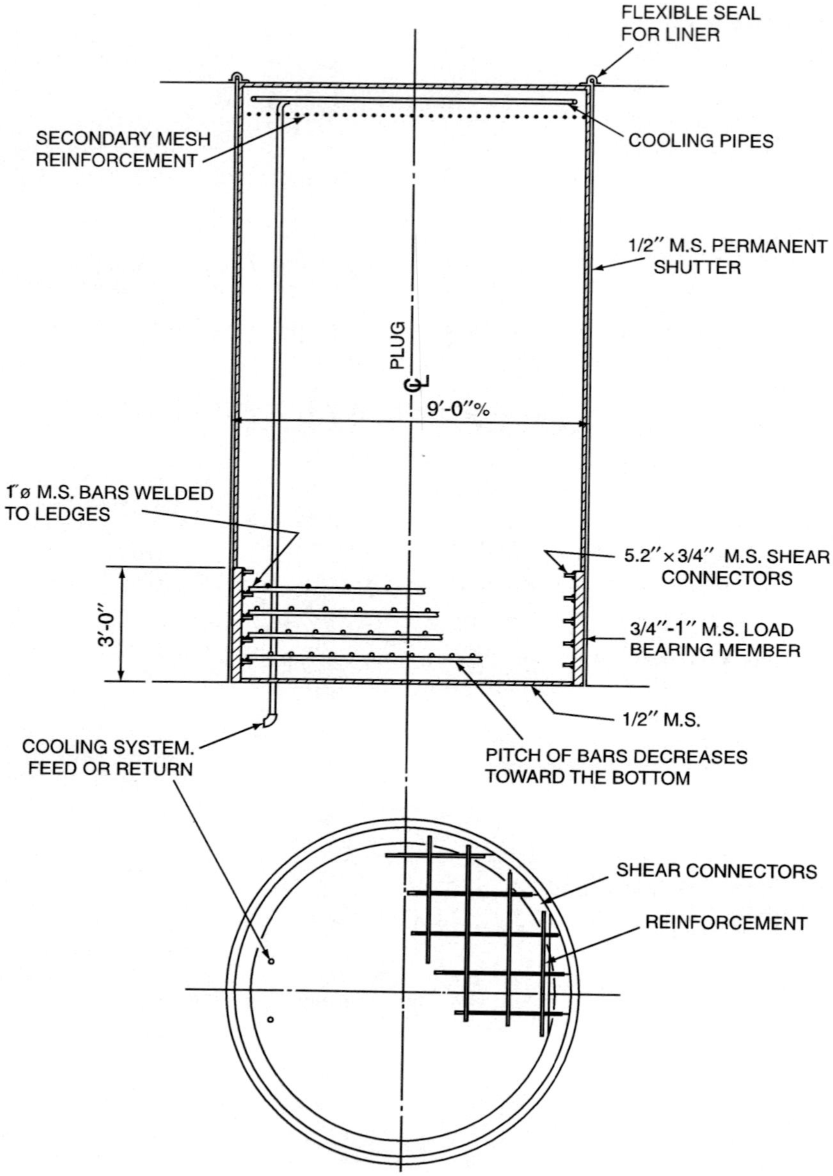
FLEXIBLE SEAL
FOR LINER
SECONDARY MESH
REINFORCEMENT
COOLING PIPES
1/2″ M.S. PERMANENT
SHUTTER
PLUG
9′-0″%
1″ø M.S. BARS WELDED
TO LEDGES
5.2″×3/4″ M.S. SHEAR
CONNECTORS
3′-0″
3/4″-1″ M.S. LOAD
BEARING MEMBER
1/2″ M.S.
COOLING SYSTEM.
FEED OR RETURN
PITCH OF BARS DECREASES
TOWARD THE BOTTOM
SHEAR CONNECTORS
REINFORCEMENT

Fig. AIB.11

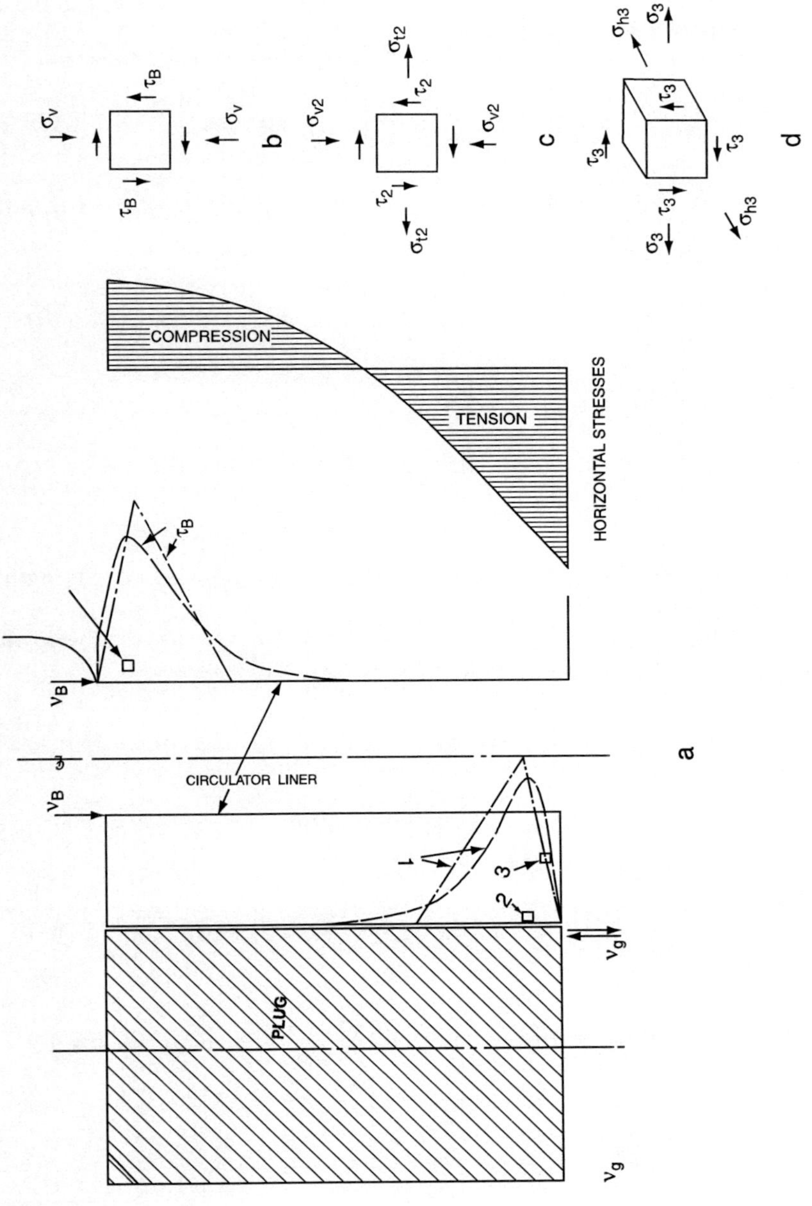
COMPRESSION
TENSION
HORIZONTAL STRESSES
CIRCULATOR LINER
PLUG
a
b
c
d

Fig. AIB.12

For a general temperature distribution (symmetrical with respect to the axis and constant along the axis)

$$\sigma_t = \frac{E}{1-\gamma}\left[\frac{1}{r^2}\int_a^r \alpha\,\mathrm{trdr} + \frac{r^2+a^2}{r^2(b^2-a^2)}\int_a^b \alpha\,\mathrm{trr} - \alpha\, t\right] \quad \text{(AIB.47a)}$$

With a crossfall of magnitude *ti*, the equilibrium temperature distribution may be represented by the function

$$t = \frac{ti\,\log_e b/r}{\log_e b/a} \quad \text{(AIB.47b)}$$

Therefore we have the following:

$$\sigma t = \frac{E\alpha\, ti}{\frac{2(1-\gamma)\log_e b}{a}}\left[1 - \frac{\log_e b}{r} - \frac{\frac{a^2}{(b^2-a^2)}\left(1+\frac{b^2}{r^2}\right)\log_e b}{a}\right] \quad \text{(AIB.48)}$$

The derivation of the above formulae is given in Timoshenko, Strength of Materials, Part II, pp. 228–232.

Substituting in eq. (i), (ii), (iii) the values of a, b, E etc, we have the following:

$$\sigma p = 480\left[1 - \frac{(22)^2}{r}\right] \quad \text{(AIB.49)}$$

$$\sigma_x = -1.68\left[1 - \frac{(14)^2}{r}\right] \quad \text{(AIB.50)}$$

$$\sigma_t = \left[-\frac{396}{r^2} - 2.66\log_e\frac{(22)}{r} + 1.05\right]\times 10^3 \quad \text{(AIB.51)}$$

With $\gamma = 0.15$

The values of the hoop stresses on the inside ($r = 14'$) and outside ($r = 22'$) are tabulated below.

Internal press 705 psi		Prestress X psi		Temp. cross fall 30°C	
σ_{pi}	σ_{po}	σ_{xi}	σ_{xo}	σ_{ti}	σ_{to}
+1670	+960	$-3.36X$	$-2.36X$	($\gamma = 0.15$)	($\gamma = 0.15$)
				−1380	+1030
				($\gamma = 0$)	($\gamma = 0$)
				−1180	+875

The two limiting cases are
On the inside

$$|\sigma ti + \sigma xi| \lesseqgtr |\sigma c| \tag{AIB.52}$$

On the outside
For $\gamma = 0.5$

$$|-1380 - 3.36X| \lesseqgtr |-2200| \tag{AIB.52a}$$

$$\text{Or } 3.36X \lesseqgtr 820$$

$$or\ X \lesseqgtr 244\ \text{psi}$$

For $\gamma = 0$, the condition is

$$|-1180 - 3.36\,X| \lesseqgtr |-2200 \tag{AIB.52b}$$

$$\text{or } 3.36X \lesseqgtr 1020$$

$$i.e.\ X \lesseqgtr 304\ \text{psi}$$

Equation (iii) gives
For $\gamma = 0.15$

$$\begin{aligned} &|+1030 + 960 - 2.36X| \lesseqgtr |+500| \\ &\text{i.e. } 1990 - 2.36X \lesseqgtr 500 \\ &\text{or } 2.36X \gtreqless 1490 \\ &\text{i.e.} X \gtreqless 6.32\ \text{psi} \end{aligned} \tag{AIB.52c}$$

For $\gamma = 0$ the condition is

$$\begin{aligned} &|+875 + 960 - 2.36X|| \lesseqgtr |+500| \\ &\text{i.e. } 1835 - 2.36X \lesseqgtr 500 \\ &or\ 2.36X \gtreqless 1335 \\ &\text{i.e.} X \gtreqless 566\ \text{psi} \end{aligned} \tag{AIB.52d}$$

The required range of prestress is therefore
For $\gamma = 0.5$, $244 \gtreqless X \gtreqless 632$ psi
For $\gamma = 0$, $304 \gtreqless X \gtreqless 566$ psi
This is clearly impossible with given values of internal pressure and temperature and cross fall.

For fixed internal pressure and vessel dimensions it is apparent that either the properties of the mix or the temperature cross fall must be altered. Therefore for further work the temperature cross fall was reduced to 20°C. The required range of prestress for this new cross fall is calculated below.

Since the stresses due to the temperature cross fall are proportional to the cross fall, the stresses due to the new cross fall are found by multiplying the temperature stresses Table (A) by $\frac{20}{30}$

For $\gamma = 0.15$

$$\sigma_{\text{ti}} = -1380 \times \tfrac{20}{30} = -920$$

$$\sigma_{\text{to}} = +1030 \times \tfrac{20}{30} = +687$$

For $\gamma = 0$,

$$\sigma_{\text{ti}} = -1180 \times \tfrac{20}{30} = -785$$

$$\sigma_{\text{to}} = +875 \times \tfrac{20}{30} = +583$$

Substituting in Eq. (AIB.52) we now get
For $\gamma = 0.15$

$$|-920 - 3.36| \lesseqgtr |-2200|$$

$$\text{i.e. } 3.30X \lesseqgtr 1280$$

$$\text{or } X \lesseqgtr 380 \text{ psi}$$

For $\gamma = 0$, the critical condition is

$$|-785 - 3.36| \lesseqgtr |-2200|$$

$$\text{i.e. } 3.36X \lesseqgtr 1415$$

$$\text{or } X \lesseqgtr 421 \text{ psi}$$

Equation (viii) now gives
For $\gamma = 0.15$

$$|+687 + 960 - 3.36X| \lesseqgtr |+5000|$$

$$\text{i.e. } 1647 - 2.36\,X \lesseqgtr 500$$

$$\text{or } 2.36X \gtreqless 1147$$

$$\text{i.e. } X \gtreqless 485 \text{ psi}$$

For $\gamma = 0$, the critical condition is

$$|+583+960-2.36X| \stackrel{-}{<} |+500|$$
$$\text{i.e. } |1543-2.36X| \stackrel{-}{<} 500$$
$$\text{or } 2.36X \stackrel{-}{>} 1043$$
$$X \stackrel{-}{>} 442 \text{ psi}$$

The required range of prestress is therefore
For $\gamma = 1.5$ $380 \stackrel{-}{>} X \stackrel{-}{>} 485$ psi
For $\gamma = 0$ $421 \stackrel{-}{>} X \stackrel{-}{>} 442$ psi
This case also gives an impossible value for the prestress. However, the difference is relatively small. As the calculation was based on a constant value of Young's Modulus and in practice Young's modulus fall of 20°C may be considered satisfactory. This is substantiated by that the formulae devised in Appendix A show that the effect of the varying E is to reduce the magnitudes of the stresses on the inner and outer faces of the wall. The computer analysis verifies that a cross fall of 20°C is acceptable.

Penetrations Liner Under Buckling

Again refer to Plates B.1 and B.2 finite-element technique to determine buckling and instability has been briefly formulated in the text. The European have given an improved design recommendations.

The European Prestandards ENV 1993-I-6 for the strength and stability of shell structures (eventually called 'shell eurocode') gives for the present buckling interaction case and the expansion is.

$$\left(\frac{\sigma_{xd}}{\sigma_{x\mathrm{Rd}}}\right)^{k_x} \left(\frac{\tau_d}{\tau_{\mathrm{Rd}}}\right)^{k_\tau} \leq 1 \qquad \text{(AIB.53)}$$

With the boxed (i.e. provisionally recommended) values

$$k_x = 1.25, \;\; k_\tau = 200 \qquad \text{(AIB.54)}$$

From the collected set of test and the own tests and from the numerical investigations carried out so far, the following tentative conclusions may be drawn:

(a) The concave interaction design Eqs. (AIB.53) and (AIB.54) seems to be somewhat unconservative in the range of very thin-walled cylindrical shells that buckle purely elastically.
(b) For medium-thick shells, the exponents according to Eq. (AIB.54) seem to be reasonable.
(c) For thick shells of which the buckling is dominated by plasticity influences, the exponents according to Eq. (AIB.54) tend to deliver overconservative design predictions.

Based on shell theory, the load deflection curves based on 'half wave', 'whole wave' and 'half penetration sectional model' are plotted viewing load factor v.s displacement. They are shown in Fig. (a) of Plate AIB.3. In all penetrations the lower end is assumed as fixed and the upper end is stiffened by a rigid diaphragm. The basic displacement δ_{mb} is taken to be tateral as

$$\delta_{mb} = F_b L^3 EI \tag{AIB.55}$$

Where $F_b = (\sigma_y - \frac{R}{A})\frac{I}{LR}$ (AIB.55a)

The final buckling behaviour is given in Plate AIB.3 while using three different models. The non-linear finite-element analysis results are plotted on the diagram (a) of Plate AIB.3 while keeping in mind the deformation of the welded stud, if any. Program ISOPAR is used to determine the buckling patterns indicated in Plates I, II and III. The maximum load factor indicates a 20% reduction in the buckling load due to an imperfection. The load factor over the non-linear bifraction load is 0.8 of the reduction is small, a large increase in the number of buckling waves. The buckling load in general is progressive but suddenly appears as computer outputs indicated 90° and 180°. Here multimodal initial imperfection is considered.

Boiler Penetration of Multi activity Vessels

The same analytical procedures are adopted as stated in the text and the non-linear finite-element analysis given in the text under various loading conditions the boiler cavities are examined. Plates AIB.4 and AIB.5 summarized the results in the forms of stress-trajectories. These boiler holes or cavities have been discussed already in the text.

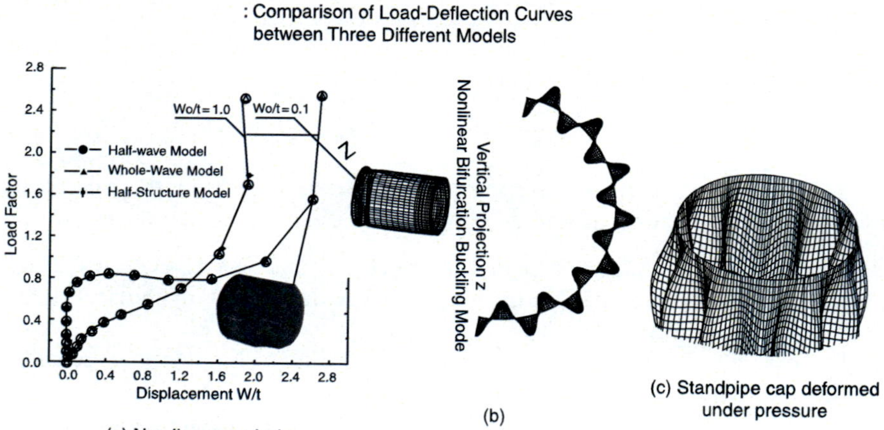

(a) Non linear analysis

(b)

(c) Standpipe cap deformed under pressure

Plate AIB.4

Stresses Across

Stresses at crossection under pressure and prestress + temperature
Pressure from presstress = 4.67 MN/m^2, Internmal pressure = 0

All values MN/m^2

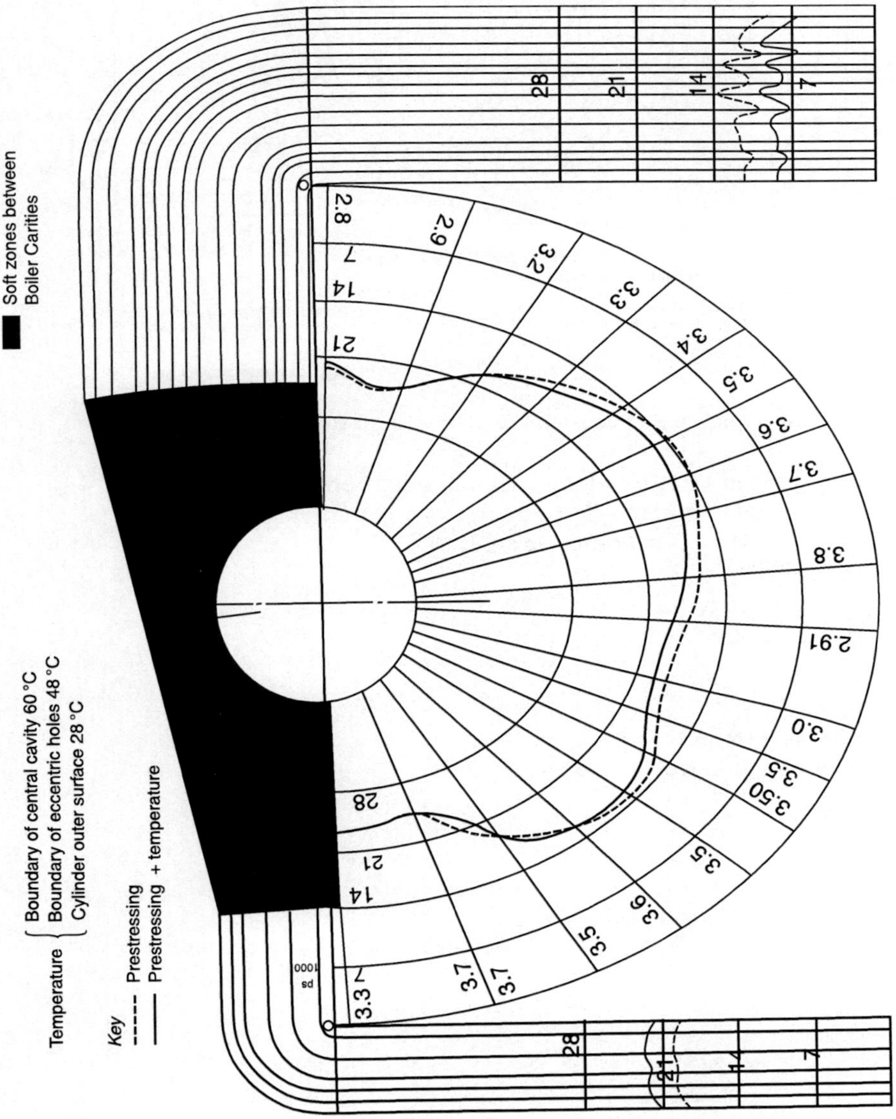

Plate AIB.5 Stresses across section

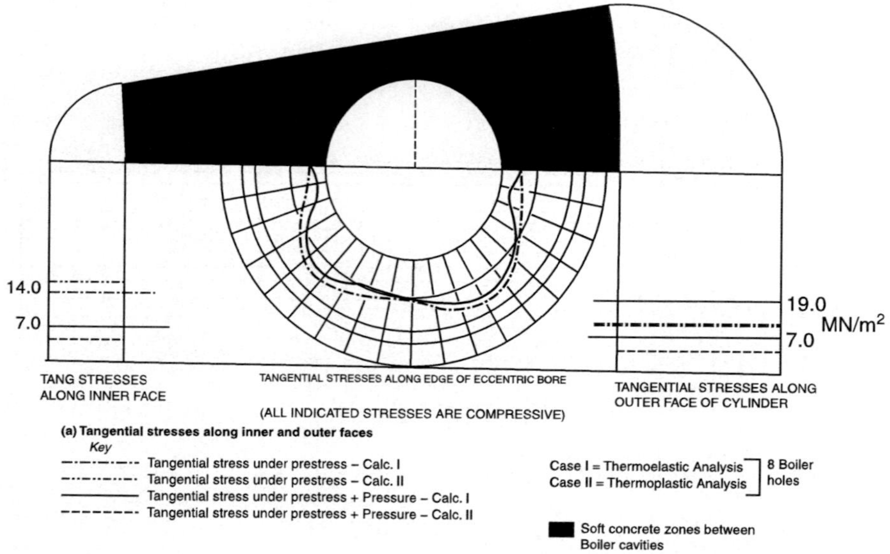

Plate AIB.6 Stresses around penetrations Thermoelastic stresses

(a) Tangential stress along liner and outer forces
(b) Case I = thermo elestic analysis
    Case II = thermo-plastic analysis

Concrete Part

$\nu_c = \gamma_c = 0.15$ to 0.2
$E_c = 20\ GN/m^2$

Steel Part

$V_s = 0.3 = \gamma_s$
$E_s = 200\ GN/m^2$
P = Internal pressure = $4.45\ MN/m^2$

Appendix C
Movements in Prestressed Concrete Reactor and Containment Vessels

General Note

A major part forming this appendix has been published by the author in 'Nuclear Engineering and Design' 50 (1978) pp. 463–473 by North Holland Publishing Company, under the same title. The same method can be applied to containment vessels for PWR where it can be proposed to place concrete using lift and bay arrangement for greater diameters and greater heights.

Reinforced and prestressed concrete reactor vessels are constructed by placing concrete using lift and bay arrangement. The vessels are built up in a number of individual bays. The size and shape of these bays and the order in which they are cast have a significant effect on the size and sense of construction movements. Lack of information about factors such as age of any bay before prestress, the time of year of casting and the time interval between adjacent vertical and horizontal bays would lead to costly design improvements and could lead to unforeseen technical problems both under operational and ultimate conditions. An attempt has been made to establish a philosophy and a rational method of assessing fairly accurately these movements. Various movements and their causes and effects are discussed in detail. A 3D finite-element, analysis discussed already in the text, is developed to predict these movements and to assess the behaviour of the vessel parameters under operational conditions. A computer programs F. BANG and ISOPAR, are developed and are tested against experimental and measured results.

Introduction

Prestressed concrete reactor vessels (PCRVs) have been recently adopted as primary containments for both the advanced and high temperature gas-cooled reactors. Among all the problems associated with this structure, the knowledge of predicted movements in certain parts of the vessel is still ambiguous. This chapter attempts to create a basis for the design of many plant components in relation to their built-in tolerances. Although symmetric and absolute relative movements are straightforward to calculate, the lack of knowledge concerning the complexity involved in evaluating the asymmetric movement can lead to costly design features which may be either cast into the vessel concrete or placed within the vessel itself. A 3D analysis using isoparametric elements is used to evaluate these movements under constructional and operational conditions. These movements, and the corresponding strains occurring in three dimensions, are built in to the ultimate load analysis. Under increasing gas pressure the vessel is analysed for pressure–displacement relationship with and without the initial built-in movements. In some cases the results have proved to be quite significant.

A computer program has been developed to determine these movements under constructional and operational conditions. A separate computer program is written to plot these movements using an plotter associated with IBM 7500. Finally, the theoretical model is tested against short-term partial experimental results obtained from Dungeness, Oldhury and Hartlepool vessels. The results for the Oldbury vessel are given in the text.

Philosophy of Movements

The object of this section is to discuss the philosophy of the causes, developments and the significance of movements and their effects on the safe operation of the vessel. In order to initiate such a discussion, it is logical to classify these movements on the basis of their presence at important stages during construction and operation.

Types of Movements

Constructional Movements

Constructional movements are defined as any movement of the pressure vessel concrete from the time of casting up to the application of prestress to the vessel. These movements are caused [1, 2, 3] by

(1) changes in ambient temperature
(2) shrinkage of the concrete
(3) deformation under load
(4) liberation and loss of heat of hydration

The constructional movements may he required in any one of the following two forms.

Absolute Construction Movements

The absolute construction movement of a point is defined as the movement of that point in the concrete relative to a fixed datum outside the vessel, from the moment of casting until prestress is applied to the vessel.

The absolute movements of any part of the vessel are given with respect to a datum on that vessel. The datum for vertical movements is taken as lie underside of the bottom cap. The vertical axis is taken as the datum for horizontal radial movements.

Relative Construction Movements

The relative construction movements of two points are the relative movements between the two points from the instant the second point is cast (the first point having already been cast) until prestress is applied to the vessel. The movements are given in the 3D coordinate systems which are discussed later in the development of the theoretical model. These movements can either be symmetrical or asymmetrical. When symmetry is considered, these movements occur within a vessel in which the concrete properties are entirely uniform and the loading of the vessel geometry are symmetrical about the vessel axis. The relative movement of two points on the vessel is the vertical movement of one point with respect to the other as datum. A similar definition holds for the relative radial movement of two points in the same vertical plane. For initial calculations it is assumed that the change in the distance between two points on the same horizontal plane is calculated from their absolute radial movements when the vessel is assumed symmetrical, it is safe to assume that no movement perpendicular to any radial line occurs.

The asymmetrical relative movements depend on variations which can be applied to some properties of concrete and to some loadings, and to the vessel geometry. The assessment of asymmetrical relative movements will be different to the one suggested for the symmetry. The vertical relative movements of two points will then be the vertical movement of one point with respect to the other as datum. The relative radial movement of two points on the same horizontal plane is the difference in the changes in their distances from the vessel axis. Similarly, the relative radial movement of two points in the same vertical plane is the difference in the changes in their distances from the vessel axis plus the difference in the changes in the distances of the vessel axis from its original position in the space at the levels of the two points under consideration. On the other hand the relative movements in the circumferential directions of the two points the algebraic sum of their movements are perpendicular to the radial lines through their original positions. In calculating these movement it is always advantageous to draw out the absolute movements of the two points.

Operational Movements

The operational movements are obtained by analysing the vessel under operational conditions. The reactor vessels [4, 13, 14–16] are analysed for several loading cases for these movements. Computer runs are also required in which the vessel properties, loading values, and summation factors are chosen so as to give the desired operation conditions. These independent computer runs are also necessary since they differ from the stress cases in that only the probable and maximum range of movements are calculated at the various stages in the vessel's life.

The probable movements are obtained by obtaining the expected elastic properties and the normal anticipated loads. It is suggested that the extreme movements should be the maximum inward and outward movements that are obtained by allowing extreme variation in elastic properties, shrinkage, cyclic expansion, prestress loss, pressure and coefficient of expansion, as well as for non-symmetrical conditions in the vessel. In addition to these maximum values, associated minimum values should also be calculated. They are obtained by considering the vessel subjected to the same loadings as with the maximum values but with the non-symmetry effects taken as being in the opposite direction to those used for the maximum values. These minimum values are used in the calculation of the relative movements.

Factors and Causes of Movements

Each of the individual movement factors may include general components as well as components special to the case being considered. Each of the many movement factors are components special to the case being considered. Each of the many movement factors is now discussed.

Effect of Creep

This is considered first since it is not only involved directly as a measure of the vessels stiffness but also indirectly because of its effect on prestress losses. Experimental work [2, 8, 10, 17] carried out to obtain specific strain curves indicates that a suitable relationship exists between the elastic modulus and creep rates and, of course, how much softening effect is there due to temperature.

The movements produced by externally applied loads are directly proportional to the specific strain values associated with the loading being considered. There is one difficulty, namely the complicated past history of loading and temperature. The calculation of specific strain values to be used are highly involved. Simplifying conservative assumptions have therefore been made in the theoretical model as regards the various ages of loading, time under loads and ruling temperatures. In the theoretical model the extreme value of the probable specific strain is considered as being that of a vessel loaded for 30 years at the normal working temperature, thus neglecting any occasions when the vessel is at a lower temperature, e.g. during shutdown periods. Will there be any significant effects, during shut down periods is extremely difficult to assess.

In certain cases it is also difficult to judge which combination of affects will produce the desired results. For the maximum inward movements it is obvious that the specific strain must be as high as possible so as to obtain the greatest prestress movement. However, when maximum on toward movements are considered, pressure is involved and it is possible that a larger outward movement will occur with a low-specific strain value. These values have been incorporated in the input. Perhaps they will compensate for the muse caused by the shutdown effects.

Prestress Loads

The effect of prestress on the movement ranges are the prestress losses, variation of prestress and asymmetric effects. The prestress losses are in turn affected by specific strain values, temperature, concrete stresses, tendon stresses, successive tensioning during the prestressing operation, shrinkage and relaxation of the steel. In addition, high-specific strain values cause high losses by virtue of the larger deflections incurred. Temperature, in addition to increasing specific strains, also produces high relaxation losses in the tendons. The level of concrete stresses determines the amount of deformation likely to occur which is directly relatable to the tendon stresses and strains to determine the loss after prestress encountered by vessel deflections. During the prestressing stage, the tendons lose some force due to the stressing of the successive cables, and allowance must be made as compensation. The relaxation of the steel itself must be taken into account since it is being affected by the temperature. Table AIC.1 is examined.

Shrinkage is another factor tending to increase the losses and changes after prestress must be used in calculating its effects. Some allowance is made for any asymmetric effects in the vessel, the prestress layout and during the jacking operation these are usually allowed for by some overall factor. This aspect is considered in the ultimate load analysis.

Pressure Loadings

The pressure is a more determinable and controllable loading than prcstress and the only special allowances are for the asymmetry of the vessel and a small factor to cover fluctuations, i.e. the gas incremental behaviour.

Temperature Loadings

The vessel temperatures are controlled fairly accurately so no great deviation from normal working temperatures should be considered in the theoretical model. For movements, either the *hot* or the *cold* vessel is used with sonic factor to cover variations in the expansion coefficient. The author emphasise this phenomenon to cover for the hot shutdown affects as stated earlier. By considering it, additional safety is maintained.

Case	Combined items
Prestress only	Elastic shortening Friction
Proof test	Steel relaxation (short term) Creep (short term) Elastic shortening Friction Internal proof pressure (1.15 X design pressure
Normal operation	Steel relaxation (long term) Creep (tong term) Elastic shortening Friction Thermal effects Internal design pressure (1.1 X working pressure)
Hot shut down	Steel relaxation (long term) Creep (long term) Elastic shortening Friction Thermal effects

Table AIC.1 Loading cases under operational conditions

Shrinkage

Upper and lower bands of shrinkage are used in the computer program input which allow for both age and temperature effects. Programs ISOPAR and F. BANG are equipped with such shrinkages.

Cyclic Expansion

Following a number of temperature cycles of typical Wylfa limestone concrete [2, 8, 10, 17], it is essential that a permanent set should occur, and allowance should be made for this. However, for some concrete the effects could be much less.

Combination and Presentation of Movements

In the form of an output the movements required in a vessel as described above fall into two groups, namely absolute and relative movements. To obtain the range of possible concrete movements for presentation, movements due to the individual effects already mentioned are summed algebraically taking due account of the construction programme: for instance, the maximum negative strain of one point relative to another point or datum, due to a change of ambient temperature, does not necessarily coincide in time with the maximum negative strain due to shrinkage. In this case the maximum combined strain due to these two defects may not be the sum of the two individual maxima, but less than this sum, and should be calculated. The maximum range of movements which could occur is normally presented in the vertical, radial and circumferential directions.

For absolute movements Table AIC.2 gives the maximum anticipated inward and outward movements at the various sections used in the computer analysis. In the majority of cases one definite set of loading factors suffices for obtaining a set of movements but locally, due to vessel shape, prestress arrangements or method of support, a slightly different loading combination may be required, but in these cases the movements involved usually tend to be small.

For maximum outward movement at the centre of the top pressure and cyclic expansion are required to be among the factors included, but along the same axis at the bottom pressure and shrinkage are required, this difference being due to the relative positions of the points and supports. Another important point to be considered is that certain factors may be interdependent and so cannot be combined arbitrarily with their extreme ranges. For example, the minimum prestress losses occur with lowest specific strain values, but for maximum inward movement a vessel with high-specific strain is usually used. The method of presentation of the results depends to what use they are to be put. To indicate the significance of the various factors at a few selected points on the vessel, a diagrammatic representation of the movements in the form of a bar chart could be used.

The relative movements are more difficult to present and various methods are available. One way is to present the tables giving the maximum movements and their associated minimum movements together with a series of general rules

DISPLACEMENT		CONCRETE DISPLACEMENTS FROM POSITIONS BEFORE PRESTRESS: INWARD MOVEMENT	OUTWARD MOVEMENT	P	P/S	T	∝	S	S/S	C	CO.
LOADING CONDITIONS CAUSING INWARD DISPLACEMENT	MIN.	P/S, C (net)		O	MIN.	AMB.	–	O	MIN.	MAX.	AVERAGE
	PROB.	S, P/S, C (net)		O	PROB.	AMB.	–	PROB.	PROB.	PROB.	MIN.
	MAX.	S, P/S (net)		O	MAX.	AMB.	–	MAX.	MAX.	MAX.	O
LOADING CONDITIONS CAUSING OUTWARD DISPLACEMENT	MIN.	P/S (net)	P, T, S	D.P.	MAX.	WORKING	MIN.	MAX.	CHECK MIN/MAX	MIN.	MAX.
	PROB.	(net)	P, T, C, P/S, S	W.P.	PROB.	WORKING	PROB.	PROB.	PROB.	PROB.	AVERAGE
	MAX.		P, C, T, (net) P/S	D.P.	MIN.	WORKING	MAX.	MIN.	CHECK MIN/MAX	MAX.	O

P = PRESSURE P.S.I.G
D.P. = DESIGN PRESSURE P.S.I.G
W.P. = WORKING PRESSURE
P/S = PRESTRESS.
T = TEMPERATURE.
∝ = COEFFICIENT OF EXPANSION.
S = SHRINKAGE.
S/S = SPECIFIC STRAIN.
C = CYCLIC SRAIN.
(hatched) = NET DISPLACEMENT.
PROB. = PROBABLE.
AMB. = AMBIENT TEMPERATURE.
MIN. = MINIMUM.
MAX. = MAXIMUM.
CO = CARBONATION.

Table AIC.2 Chart of inward and outward displacements

for working out the relative movements. This method has its difficulties. It may be more prudent to use the second method. i.e., the relative movements arc calculated as individual problems so that, although the general rules are used, special allowances or treatments can be incorporated.

Having assessed the variation of material, prestress and temperature properties for the vessel, it is now possible to compute the symmetrical absolute movements to the accuracy of the assumed concrete properties, values of prestress losses, etc. To obtain these movements, it is necessary first to decide which combination of the variations in vessel properties will yield the maximum and minimum vessel movements.

Two combination factors are possible for maximum and minimum outward movements. It will be necessary to calculate each to establish the maximum and minimum outward movements at any one point on the vessel. It is suggested that in the light of the above arguments the combination of movements should be as indicated in Table AIC.2, which will be discussed later, is assimilating the input data on the lines suggested in Table AIC.2.

Three-dimensional Analysis of Movements

Three-Dimensional Analysis Techniques

A 3D analysis is carried out using isoparametric 20-node solid elements for concrete (Table AIC.3) with cubic displacement expansion in platform and a

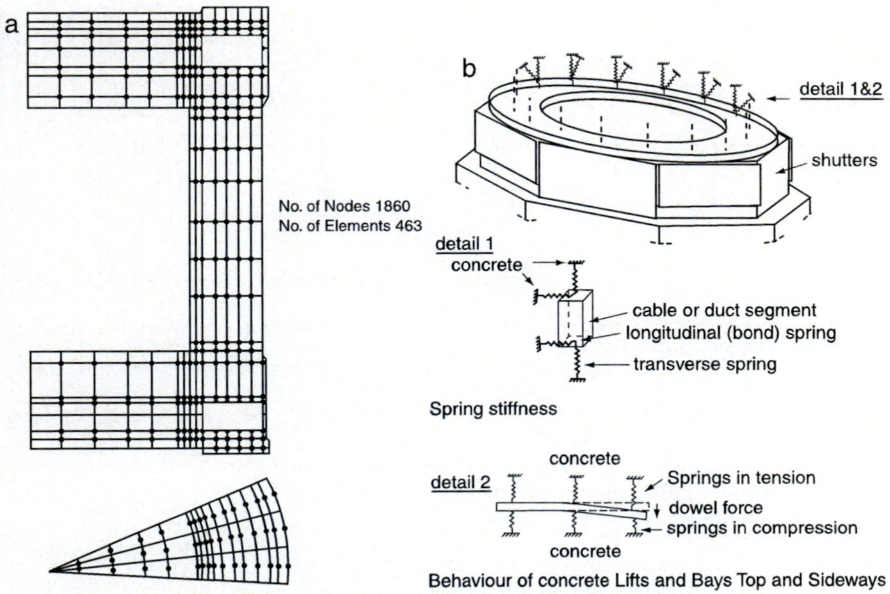

Fig. AIC.1 **(a)**,**(b)** Finite element idealisation of Oldbury vessel 20-node element, complete bay with spring locations

linear variation through the thickness of each lift and bay. The mutual response of concrete layers for lift and bays is represented by a linkage element (Fig. AIC.1b). The stiffness matrix for this element is given in table. It is assumed that while using these linkage elements the degrees of freedom of the isoparametric elements are retained at their boundaries so that the interconnection between concrete layers can be made at their interfaces. The stiffness assigned to the bond linkage is the slope of the bond-slip curve multiplied by the embedded item surface area A_s tributary to a "kink" to form in the embedded items such as ducts or reinforcement bars. The shear at the kink acts as a dowel force and is represented by vertical linkages.

When the vertical linkages are in tension, their stiffness is controlled for the most part by adhesion between the embedded items and the concrete and could

Table AIC.3 Matrix $[K_L]$ for linkage elements

$$\begin{bmatrix} k_H C^2 + k_V S^2 & k_H SC - k_v SC & -k_H C^2 - k_H S^2 & -k_H SC + k_V SC \\ h_H SC - k_V SC & k_H S^2 + k_V C^2 & -k_H SC + k_v SC & -k_H S^2 - k_v C^2 \\ -k_H C^2 - k_v S^2 & -k_H SC + k_v SC & k_H C^2 + k_v S^2 & k_H SC - k_v SC \\ -k_H SC + k_v SC & -k_H S^2 - k_v C^2 & k_H SC - k_v SC & k_H S^2 + k_v C^2 \end{bmatrix}$$

where $S = \sin\beta$, $C = \cos\beta$.
Sign convention + positive movement and tensile force between the connected.

Table AIC.4 Variable modulus and Poisson's ratio stiffness matrix [D]

$$\begin{bmatrix} {}^*K_{11} = \frac{(1-\nu_{23}\nu_{32})}{\bar{\nu}}E_1 & {}^*K_{12} = \frac{(\nu_{12}+\nu_{12}\nu_{32})}{\bar{\nu}}E_2 & {}^*K_{13} = \frac{(\nu_{13}+\nu_{12}\nu_{23})}{\bar{\nu}}E_3 & {}^*K_{14} = 0 & {}^*K_{15} = 0 & {}^*K_{16} = 0 \\ {}^*K_{21} = \frac{(\nu_{21}+\nu_{23}\nu_{31})}{\bar{\nu}}E_1 & {}^*K_{22} = \frac{(1-\nu_{13}\nu_{31})}{\bar{\nu}}E_2 & {}^*K_{23} = \frac{(\nu_{23}+\nu_{13}\nu_{21})}{\bar{\nu}}E_3 & {}^*K_{24} = 0 & {}^*K_{25} = 0 & {}^*K_{26} = 0 \\ {}^*K_{31} = \frac{(\nu_{31}+\nu_{21}\nu_{32})}{\bar{\nu}}E_1 & {}^*K_{32} = \frac{(\nu_{32}+\nu_{12}\nu_{31})}{\bar{\nu}}E_2 & {}^*K_{33} = \frac{(1-\nu_{12}\nu_{21})}{\bar{\nu}}E_3 & {}^*K_{34} = 0 & {}^*K_{35} = 0 & {}^*K_{36} = 0 \\ {}^*K_{41} = 0 & {}^*K_{42} = 0 & {}^*K_{43} = 0 & {}^*K_{44} = G_{12} & {}^*K_{45} = 0 & {}^*K_{46} = 0 \\ {}^*K_{51} = 0 & {}^*K_{52} = 0 & {}^*K_{53} = 0 & {}^*K_{54} = 0 & {}^*K_{55} = G_{23} & {}^*K_{56} = 0 \\ {}^*K_{61} = 0 & {}^*K_{62} = 0 & {}^*K_{63} = 0 & {}^*K_{64} = 0 & {}^*K_{65} = 0 & {}^*K_{66} = G_{31} \end{bmatrix}$$

where $\bar{\nu} = 1 - \nu_{12}\nu_{21} - \nu_{13}\nu_{31} - \nu_{23}\nu_{32} - \nu_{12}\nu_{23}\nu_{31} - \nu_{21}\nu_{13}\nu_{32}$

Table AIC.5 CABCO prestressing tendon main data

			Tendon Forces (kN)				
Anchorage	System	Tendon*	P_k	$0.8P_k$	$0.7P_k$	Anchorage size mm	Sheath internal dia. mm
		4/13 STD.	660.0	528.0	462.0	130	42
U1	Cabco	4/13 SUP.	736.0	588.8	515.2	"	"
		4/13 DYF.	836.0	668.8	585.2	"	"
		4/15 STD.	908.0	726.4	635.6	175	51
		4/15 SUP.	1000.0	800.0	700.0	"	"
		4/15 DYF.	1200.0	960.0	840.0	"	"
U2	Cabco	7/13 STD.	1155.0	924.0	808.5	"	"
		7/13 SUP.	1288.0	1030.4	901.6	"	"
		7/13 DYF.	1453.0	1170.4	1024.1	"	"
		4/18 STD.	1480.0	1184.0	1036.0	175	51
		4/18 DYF.	1520.0	1216.0	1064.0	"	"
		7/15 STD.	1589.0	1271.2	1112.3	215	75
		7/15 SUP.	1750.0	1400.0	1225.0	"	"
	Cabco	7/15 DYF.	2100.0	1680.0	1470.0	"	63
U3		12/13 STD.	1980.0	1584.0	1386.0	215	81
		12/13 SUP.	2208.0	1766.4	1545.6	215	"
	Multiforce	12/13 STD.	1980.0	1584.0	1386.0	215	"
		12/13 SUP.	2208.0	1766.4	1545.6	215	"
		7/18 STD.	2590.0	2072.0	1813.0	245	81
		7/18 DYF.	2666.0	2128.0	1862.0	"	75
U4	Cabco	12/13 DYF.	2508.0	2006.4	1755.6	"	75
		12/15 STD.	2724.0	2179.2	1906.8	"	81
		12/15 SUP.	3000.0	2400.0	2100.0	"	81
	Cabco	12/15 DYF.	3600.0	2880.0	2520.0	265	81
		19/13 STD.	3135.0	2508.0	2194.5	"	"
U5		19/13 SUP.	3496.0	2796.8	2447.2	"	"
	Multiforce	12/15 STD.	2724.0	2179.2	1906.8	245	"
		12/15 SUP.	3000.0	2400.0	2100.0	245	"
		12/15 DYF.	3600.0	2880.0	2520.0	265	"
		25/13 STD.	4125.0	3300.0	2887.5	300	90
U6	Multiforce	25/13 SUP.	4600.0	3680.0	3220.0	"	"
		13/15 DYF.	3900.0	3120.0	2730.0	"	84
		31/13 STD.	5115.0	4092.0	3580.0	335	100
		31/13 SUP.	5704.0	4563.2	3992.8	"	"
U7	Multiforce	19/15 STD.	4313.0	3450.0	3019.1	"	"
		19/15 SUP.	4750.0	3800.0	3325.0	"	"
		19/15 DYF.	5700.0	4560.0	3990.0	"	"

Table AIC.5 (continued)

Anchorage	System	Tendon*	Tendon Forces (kN) P_k	$0.8P_k$	$0.7P_k$	Anchorage size mm	Sheath internal dia. mm
U8	Multiforce	19/18 DYF.	7220.0	5776.0	5054.0	"	"
		5/18 STD.	1776.0	1480.0	1295.0	362 × 89	114 × 25
		5/18 DYF.	1824.0	1520.0	1330.0	"	"
	Strand-force	10/18 STD.	3530.0	2960.0	2590.0	362 × 171	" twin
		10/18 DYF.	3648.0	3040.0	2660.0	"	" twin

* STD. Standard
SUP. Super Strand
DYF. Dyform Strand

be set equal to zero. The stiffness of these vertical linkages that are in compression will depend on such factors as the geometry, the concrete cover, and the presence or absence of the embedded items in that zone of the vessel concrete. The stiffness matrix for the solid element is derived using the orthotropic material matrix given in Table AIC.4. The procedure for the stiffness matrix is summarized in Table AIC.5. The program ISOPAR is provided with two solution procedures, namely the B LOCKING TECHNIQUE and the FRONTAL SOLUTION.

Temperature, creep and shrinkage effects are included in the general formulation of the strain matrix. The details of this program are given in the text.

Concrete Stress–Strain Criteria

Concrete stress–strain criteria, plasticity and cracking [14, 15, 16, 17, 19, 20, 21, 22] have been assessed. In this work the hypoelastic approach is adopted for the multiaxial state of stress, the details of which are given in Ref. [23].

Summary of Results

Movement parameters for the Oldbury vessel, obtained from the theoretical model and from site measurements, are plotted in Fig. AIC.2 and AIC.3 for critical locations of this vessel.

They are in good agreement for many critical locations. However, at certain locations variations in results exist between the theoretical model and those obtained on site. Although net differences in value are insignificant in terms of the overall achievement of tolerances, a suitable explanation can be given.

Finite element data

$$\text{Strain} = \{\varepsilon\} = [B_i]\{\delta\} \qquad \text{(AIC.1)}$$

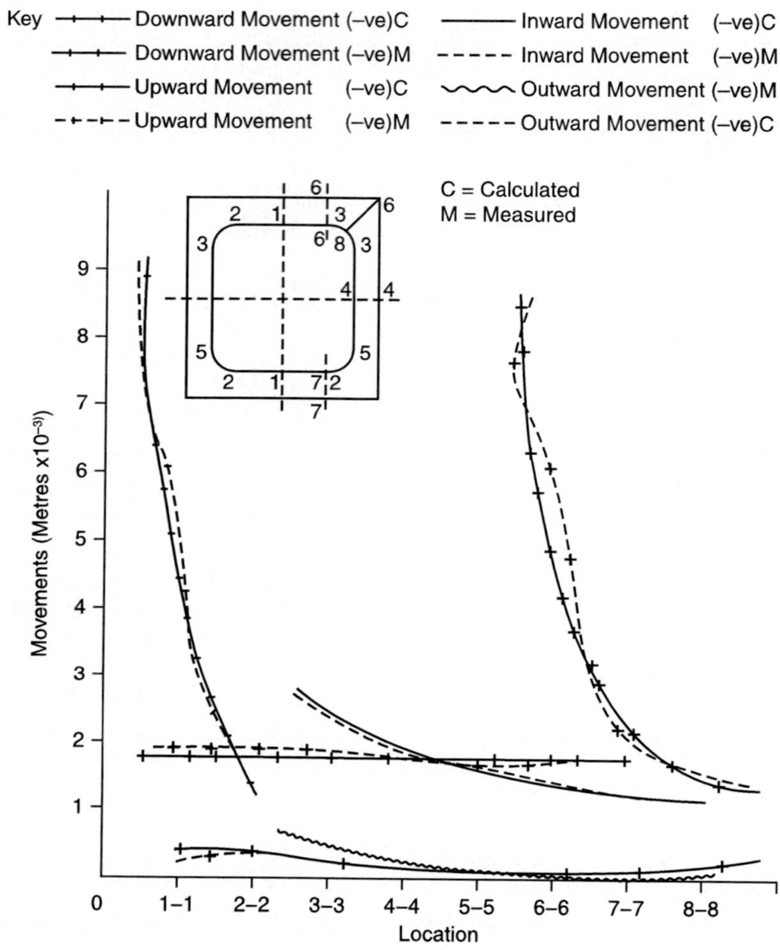

Fig. AIC.2 Absolute movements of Oldbury reactor vessel

where

$$[B_i] = \begin{bmatrix} \frac{\partial \rho i}{\partial x} & 0 & 0 \\ 0 & \frac{\partial \rho i}{\partial y} & 0 \\ 0 & 0 & \frac{\partial \rho i}{\partial z} \\ \frac{\partial \rho i}{\partial y} & \frac{\partial \rho i}{\partial x} & 0 \\ 0 & \frac{\partial \rho i}{\partial z} & \frac{\partial \rho i}{\partial y} \\ \frac{\partial \rho i}{\partial z} & 0 & \frac{\partial \rho i}{\partial x} \end{bmatrix} = \frac{1}{\det |J|} \begin{bmatrix} B_{1,} & \cdots\cdots & B_{1,n} \\ \vdots & & \vdots \\ & & \\ \vdots & & \vdots \\ B_{m,1} & \cdots\cdots & B_{m,n} \end{bmatrix}_{m \times n = 6 \times 6} \tag{AIC.2}$$

$$\text{Stress} = \{\sigma\} = [D](\{\varepsilon\} - \{\varepsilon_0\}) + \{\sigma_0\} \tag{AIC.3}$$

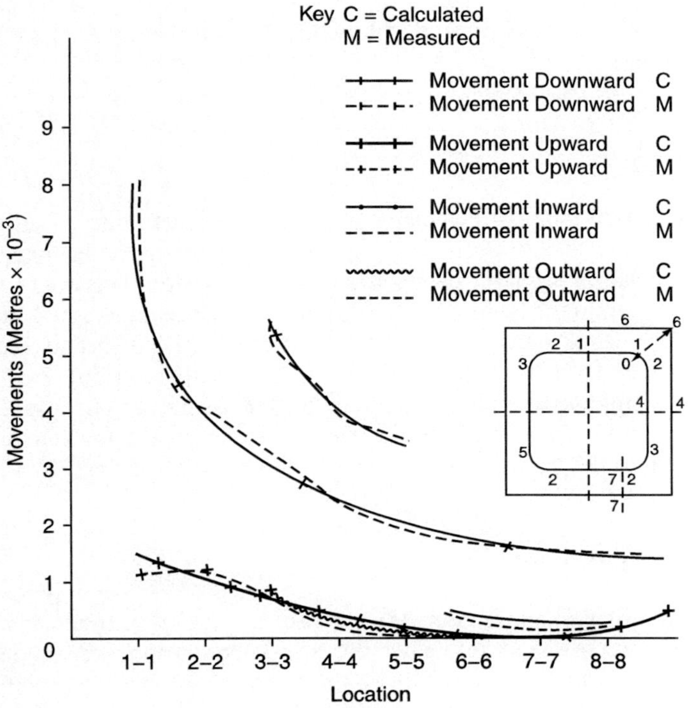

Fig. AIC.3 Relative movements of the Oldbury reactor vessel

where $\in_0$ and σ_0 = initial strain resp. stress and ρ = shape function.
$[K]_{6\times 6}$ = element stiffness matrix

$$= \int_{-1}^{+1}\int_{-1}^{+1}\int_{-1}^{+1} [B]^{T''}_{6\times n}[D]_{6\times 6}[B]_{6\times n} \det [J]\, d\xi\, d\eta\, d\xi. \qquad \text{(AIC.4)}$$

Forces:

$$\{F\} = \int_V [B]^{T''}[D]\{\varepsilon\} d\xi\, d\mu\, d\zeta \qquad \text{(AIC.5)}$$

T″ = Transpose

$$\{R_j\} = \underbrace{[K]^e\{\delta\}}_{\text{elastic}} + \underbrace{\{F_x\}^e_b}_{\text{bond}} + \underbrace{\{F_x\}^e_f}_{\text{body}} + \underbrace{\{F\}^e_{\sigma_\circ} + \{F\}^e_{\varepsilon_\circ}}_{\text{initial}} + \underbrace{\{F\}^e_t}_{\text{temperature}} \qquad \text{(AIC.6)}$$

Heat of Hydration

The effect on vessel movements of the heat of hydration of cement on settling is mainly felt during construction in the theoretical model; this was taken to have no further effect occur 100 days after any bay is cast. Although the figure of 100 days assumed is quite normal, this may not be so in the contractor's construction program.

Shrinkage

In order to establish the rate of increase or instantaneous values of shrinkage for any part of the vessel, it is necessary to know when a particular bay is poured and how long after the pour the temperature was applied. This might have even not been known to the contractors as different time interval is a time-dependent quantity and is extremely difficult to measure. In the theoretical model the shrinkage of concrete is related to time and temperature loading. For input the minimum shrinkage line starts from 0 to 100 microstrains. The initial gradient of the maximum shrinkage line is then continued until the temperature is applied and it rises linearly to a maximum of 400 microstrains.

Temperature Regions

This is a complex phenomenon. Movements are mainly caused by a uniform temperature across the section since the shrinkage rates are dependent on this uniform temperature. This may not be the case for temperature crossfall which causes large stresses and strains with negligible movements. The combined effect of these has been assumed in the analysis.

The age of the concrete at the time of different measurements may vary, which also has an effect on the value of the coefficient of thermal expansion.

Creep Rate

The creep rate is dependent on the type of aggregate, mixed proportion and temperature. The average values and rate increases are subject to a variation of $\pm 30\%$ due to the different quality of aggregate from any one pit and the daily variation of the mix. Only one specific rate is taken in the theoretical model on the basis of the available information. The vessel may have a different creep rate at different parts and they may be dependent also on the age of the concrete at loading. Although such variations have not been considered in the present input of the computer program, they can easily be handled by the theoretical model.

Boundary Restraints

The definition and provision of true partial restraints in the program at the level of skirt line is rather difficult. Owing to the limited computer storage, the restraint defined is the rollers in the horizontal direction and linkage effect in the vertical dimension. A rigid base or a base fully restrained was not considered since they would prevent movements and would unnecessarily cause stresses in the concrete. In fact, in practice the bottom of the cap is not rigidly jointed; instead it rests on the pads placed at a pitched circle diameter of the supporting ring column Structure.

Data for Reactor Vessels

In order to validate the theoretical model, data for the Oldburv reactor vessel are considered. This reactor vessel is a vertical concrete cylinder with top and bottom concrete flat caps. The prestressing tendons are arranged in the walls in a helical system at 45° with 11 layers running clockwise and 11 anticlockwise. Each layer contains 160 tendons. Some tendons are diverted and brought back to position since a systematic pattern of tendons cannot be followed at penetration through the walls. The tendons are anchored in the top and bottom stressing galleries creating even pitch around a circle for each layer. In both the top and bottom slabs the tendons are arranged in horizontal layers.

General Data

Tendons: PSC type 12 no/15 mm diameter having nominal high tensile steel strands with minimum g.u.t.s. 277 t per tendon.

Gas design pressure = 3.45 MN/m^2.

The main dimensions are

$$2R_{\text{I}} = 18.3\text{m} \qquad 2H_{\text{I}} = 23.5\,\text{m}$$
$$t = 4.6\,\text{m}; \qquad D_{\text{I}} = D_2 = 6.7\,\text{rn}$$

where R_1 = internal radius, H_1 = internal height, t = wall thickness and D_1, D_2 = cap thickness top and bottom. The minimum cube strength of concrete at 28 days is 41.37 KN/mm^2.

Conventional reinforcement

Primary 40 mm diameter
Secondary 12 mm diameter

Prestressing losses:

Vertical tendons: initial 3.5% commissioning 7.4%, 30 years 16.7%

Temperature: inside 59°C, outside 20°C
Elastic and creep ratio for concrete: 0.18.
Lift of concrete to be placed: 1.34 m.
E_s = Young modulus of steel = 200 KN/mm^2.
E_c = Young modulus of concrete = 41.4 KN/mm^2.
v = Poisson's ratio for steel = 0.3

Coefficient of linear thermal expansion:

Concrete 8.0 μM/m°C,
Steel 10.0 μM/m°C.

Thermal conductivity:

Steel = 41.6 W/m °C,
Concrete = 1.75 W/m °C.

Liner thickness = 12 mm.
Studs = 12 × 16.3 mm pitch not exceeding 305 mm, 324 fuelling pipes and 81 control rods with outer diameter 190 mm.

Provision for holes and equipment for man access penetration, BCD penetrations, corrosion and monitoring sample tubes, CO_2 inlet penetration thermo couples and gas or circulator ducts. Total weight of appliances, 36 66 750 kg, suitably distributed at an assumed 35 number of lifts based on their locations.

Conclusions

At the beginning of this appendix various questions were raised regarding the accuracy of determining movements under constructional and operational conditions. As is evident from the above discussions, results from the proposed theoretical model compare quite well with those measured in the Old bury vessel. Unfortunately, at the time of completing this appendix no results were available from the Hartlepool and Dungeness vessels, which are under construction, and therefore, theoretical results could not be compared.

In the Oldbury vessel with a complicated tendon geometry, the analysis for constructional movements shows that the fuelling line in the top cap is not straight and fuelling pipes in the stand pipe zone have moved individually from 3.3 to 12.5 mm from their original marked positions measured from the datum of the vessel. These figures have been improved when the vessel is prestressed longitudinally and circumferentially and correspondingly they range from 1.7 to 6.13 mm. Obviously these figures need consideration since the fuelling machine will be relatively out of alignment with the stand pipes. The comparative results show that these have been considered.

It is suggested that this method will give a useful guide to the other important areas of equipment where movements may take up excessive tolerances and thus introduce costly design improvements. As and when more data are released, a great deal of sophistication can be achieved in these results by using the suggested theoretical model. The reader may attempt to work on the results of Hartlepool and Dungeness B measured results.

Acknowledgements The author is grateful to the Central Electricity Generating Board for providing data.

References

1. Bangash, Y. Proceeding of the Institution Civil of Engineering, Paper 7710, Part 2, 57(1974)437;59(1975) 195.
2. Bangash, Y. Proceeding of Institution of Civil Engineering, Suppl. Paper 7478S (1972) 157; Suppl. xviii, 301.
3. Bangash, Y. Concrete 8(6) (1974) 46.
4. Bangash, Y.Journal of Institute of Nucleas Engineering. 13 (4) (1972) 108.
5. BNES, Symposium on Model Technique for Prestressed Concrete Pressure Vessels (Institute of Civil Engineer, London, 1969).
6. Burrow, R. D. et al., Nucleas Engineering Institute. 14 (162) (Nov.1969).
7. CECil, Private communication (1974).
8. Clough, R.W. Proceedings on Symposion on application of finite element methods in civil engineering, American Society of Engineering, 1969 p. 1.
9. Corum, J.E. et al. Proceeding on Symposion on application of finite element methods in civil enginceriniz, American Society for Civil Engineering 1969, 63.
10. Fjeld, S.A. Finite Element Method in Thrcc-Dimcnsional Theory of Elasticity, Technical University of Norway, Trondhcim, 1968.
11. Gardener, N.J. Proceeding of American Concrete Institute 66 (2) (1969) 136.
12. annant, D.J. Proceeding of American Concrete Institute. 66 (5) (1969) 391.
13. Hill, R. The Mathematical Theory of Plasticity Oxford University Press. 1956 p. 34.
14. ICE, Proceeding of Conference on Prestressed Concrete Pressure Vessels. Institute of Civil Engineer, London, 1968, papers 1, 2 and 3.
15. Langham, D. Proceedings 1st SM1RT Conference vol. 4, Part II, 1971 Paper 113/4.
16. Lannav, P. et al. Proceedings 1st SMiRT Conference, vol. 4, Part 11 1971, Paper 1-11/3.
17. Marcal, P.V. and King, l.P. Institute of Mechinical Science. 9, 1967, 143.
18. Newman, J.B. Ph.D. thesis, University of London, 1973.
19. Vos, B.I. Proceedings 1st SMiRT Conference, vol. 4, Part H 1971, Paper 112/S.
20. York, C.P. et al., American Concrete Institute. In Seminar on Concrete Reactors, Berlin, 1970.
21. Zienkicwicz, O.C. The Finite Element Method in Engineering Science, McGraw-Hill, London, 1971.
22. Iron, B.M. In: Finite element Techniques in Structural Mechanics (H. Tottcnhani and C. Brebbia eds.), Southampton University Press. 1970, p. 328.
23. Bangash, Y. Proceeding of Institute of Civil Engineering, 61 (Mar.) 1976, 59.

AIC.1 Appendix: ISOPAR And F. BANG

The complete program for the computing movements and Elasto—plastic analysis is given in a separate Appendix and the flowchart is given here in.

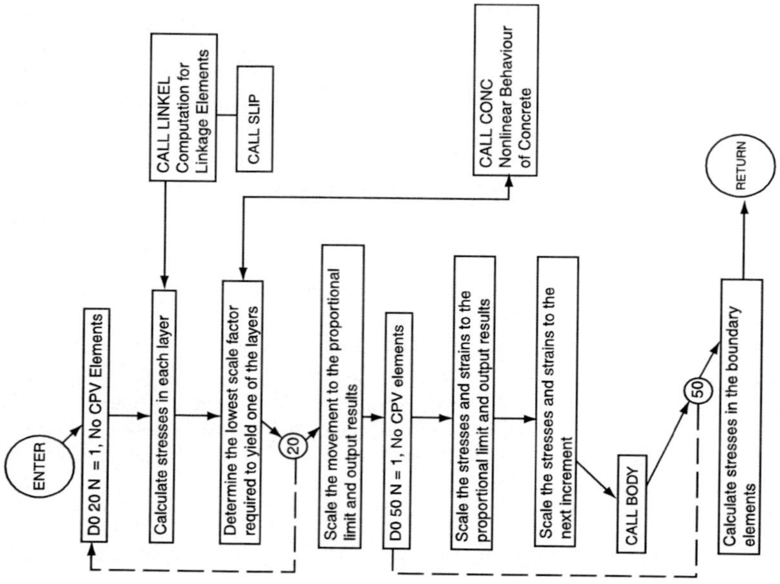

Flow chart for Sub Program LAY

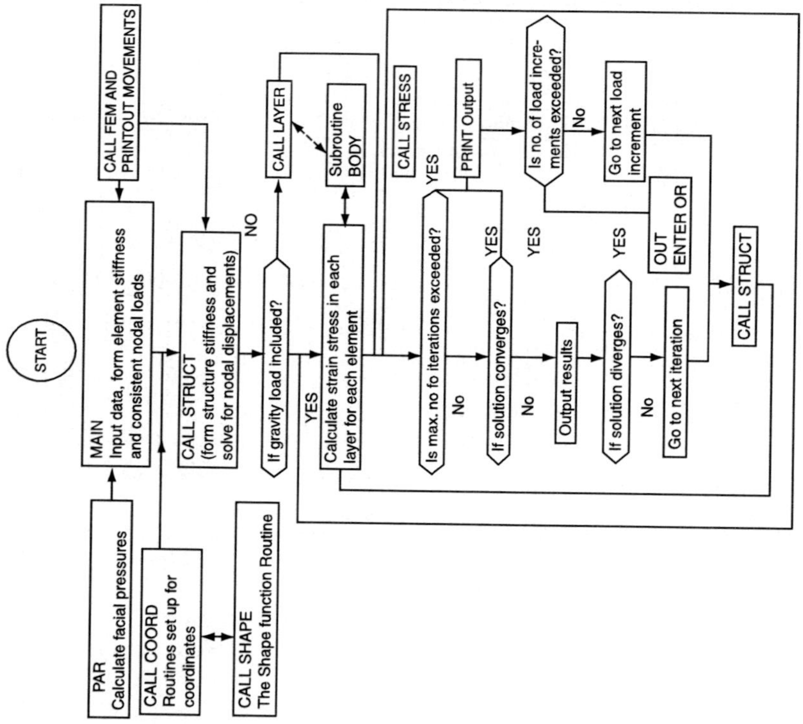

Flow chart for Program MAIN

Appendix D
Safe Analysis for Cooling Pipes for Reactor And Containment Vessels

General Note

Major part forming this appendix has been published by the author in the *Nuclear Engineering and Design,* 55(1979), pp. 305–313 by North Holland Publishing Company under the same title. The same can be applied to containments for PWR where cooling pipes are needed. The parameters will be different and care is taken to simulate data for Program ISOPAR or similar computer packages. The major analysis will still be applied to any kind of vessels.

A step-by-step analysis is given for the direct computation of 2D heat flow and safe pitching of the cooling pipes. Two models of the cooling system have been selected and calculations have been carried out for an existing vessel. On the one model this analysis is compared with the 3D finite-element analysis for obtaining insulation conductance for various cooling pipes. A safety factor is established.

Introduction

The cooling system is located behind the steel liner which is keyed back to the concrete by an extensive array of lugs. The design features of a cooling system depend on the reactor vessel lay out and the method of anchoring of the liner. In this chapter a conventional type of reactor vessel, namely Dungeness B, is chosen which has common design features of other vessels [1–3] and in which the heat exchangers are contained within the main pressure envelop. The heat flow through the liner and the concrete is therefore divided into two parts:

(a) Flow in the liner
(b) Flow in the concrete

General equations are required for the heat flow along the liner to the cooling pipe and in the main concrete. These equations derived in this Chapter which gives the values of maximum temperature insulation thicknesses and

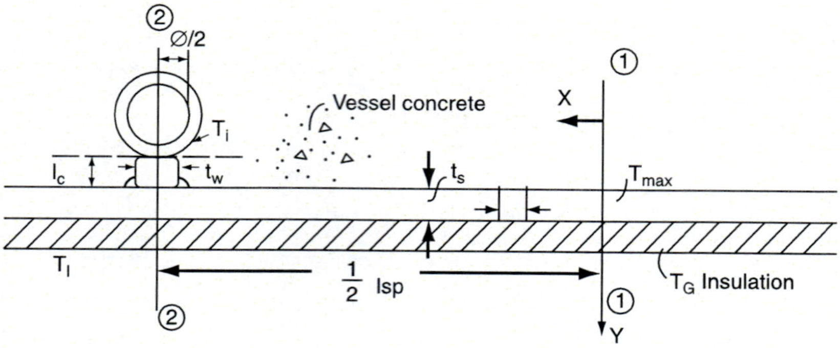

Fig. AID.1 Reactor liner insulation and concrete

conductances. Using these equations and knowing values of certain coefficients, the safe pitch value between any two cooling pipes is evaluated.

To produce a fully optimised design, further work is carried out on the arrangements of cooling pipes in regions where nuclear heating varies and which forms a large part of the heat loading of the system. All the temperature distributions derived have been calculated on the basis of a 3D isoperimetric finite-element analysis using frontal solutions [4]. For the vessel liner a basic mesh of 12 mm is used with an adiabatic boundary at a depth of 37 mm.

The overall width of the mesh is taken to be half the pipe pitch, extending from the centre liner of the cooling water pipe to a point midway between adjacent pipes. The geometry of the cooling system adopted in the calculations consists of two 19 mm I.D and 25 mm O.D. mild steel pipes welded to the liner as shown in Fig. AID.1. A computer program ISOPAR [7] is written to carry out various solutions to the finite-element analysis.

Analysis for Heat Flow in the Liner

Heat flow equations are derived using a model insulation system given in Fig. AID.1.

Section (I)–(I). Heat flow equation

$$-Kt_s(d^2T/dx^2)dx = [T_G - \theta]C_i dx + q dx.I.t_s \qquad \text{(AID.1)}$$

or

$$T = Ae^{\lambda x} + Be^{-\lambda x} + T_G + qt_s/C_i = Ae^{\lambda x} + Be^{-\lambda x} + T_G + q_o/\mu C_i \qquad \text{(AID.2)}$$

Where A, B are constants and $\lambda = \sqrt{C_i/Kt_s}$.

When the following boundary conditions are applied, constants A and B are computed:
$x = 0$; dT/dx = 0, at x = 0, A = B, at $x = \frac{1}{2} l_{sp}$; $T = T_1$

$$A = [T_l - T_G - q_o/\mu C_i] / \left[\exp\left(\frac{1}{2}\lambda l_{sp}\right) + \exp\left(-\frac{1}{2}\lambda l_{sp}\right)\right]$$

Substituting in Eq. (AID.1), T_{max} at $x = 0$ is given by

$$T_{max} = \frac{-T_G + T_l - q_o/\mu C_i}{\cos h \frac{1}{2}\lambda l_{sp}} + T_G + q_o/\mu C_i \tag{AID.3}$$

Net heat flow along (AID.2)–(AID.3) gives

$$-K\frac{1}{2}\phi\, \mathrm{dT/dy} = -K\frac{1}{2}\phi[(T_\imath - T_l)/l_c] = -Kt_s\, \mathrm{dT/dx} \tag{AID.4}$$

But

$$\mathrm{dT/dx} = (T_l - T_G q_o/\mu C_i)\lambda \tanh \lambda \frac{1}{2} l_{sp}$$

Therefore

$$T_\imath = \frac{\frac{1}{2}K\phi T_\imath/l_c + Kt_s\lambda \tan h(\frac{1}{2}\lambda l_{sp})(T_G + q_o/\mu C_i)}{\frac{1}{2}K\phi/l_c + Kt_s \tan h(\frac{1}{2}\lambda l_{sp})} \tag{AID.5}$$

Knowing values of T_{max}, T_i and T_G using Eqs. (AID.3) and (AID.4), the values of l_{sp} is computed.

Analysis for Heat Flow in Concrete

Heat flow equations are derived in this section and reference is made to an idealised situation given in Fig. AID.2.

For 2D gas it is assumed that the heat flow is between two parallel boundaries. The nuclear heating is represented by

$$q = q_o e^{-\mu x} \tag{AID.6}$$

Considering 3 strip of unit depth and thickness

$$\text{at } x = 0, T = T_1; \text{at } x = t : T = T_\ell \tag{AID.7}$$

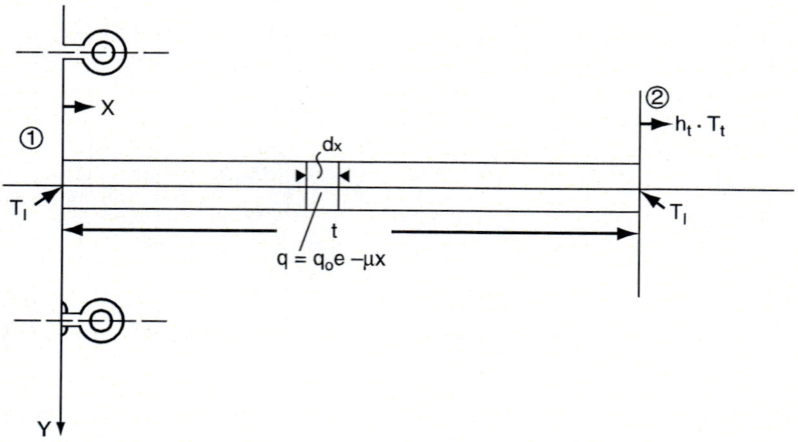

Fig. AID.2 Temperature on concrete

Both heat flow across and that generated in the element are summed up and the total heat flow equation is represented by

$$-K(\mathrm{d}^2T/\mathrm{d}x^2)\mathrm{d}x = q_o e^{-ux}\mathrm{d}x$$

Upon integration, the value of T is computed as

$$T = -(q_o/K\mu^2)e^{-\mu x} + Cx + D \tag{AID.8}$$

Applying boundary conditions given in Eq. (AID.7) values of C and D are computed and the following equation is derived:

$$\begin{aligned} D &= T_1 + q_o/K\mu^2 \\ T_l &= -(q_o/K\mu^2)e^{-\mu t} + Ct + T_1 + q_o/K\mu^2 \end{aligned} \tag{AID.9}$$

At $x = t$, $T = T_l$ heat flow

$$K\mathrm{d}T/\mathrm{d}x = (q_o/K\mu)e^{-\mu t} + C \tag{AID.10}$$

where

$$\begin{aligned} C &= 1/t[(T_l - T_1) + (q_o/K\mu^2)(e^{-\mu t} - 1)] \\ D &= T_1 + q_o/K\mu^2 \end{aligned} \tag{AID.11}$$

Since μt is large at $x = t$ and $e^{-\mu t} \to 0$.
$\therefore \mathrm{dT/dx} = \mathrm{C}$
Heat flow to ambient temperature $= h_t(T_t - T_l) = K.C.$

Hence

$$T_l \frac{h_t T_t + K_t.T_1 + q_o/t\mu^2}{h_t + K/t} \tag{AID.12}$$

For $T_{\max}, \mathrm{d}T/\mathrm{d}x = 0 = (q_o/K\mu)e^{-\mu x} + C$, hence C is determined, i.e. $\mu x = t_n - q_o/CK\mu$,

$$\eta = -q_o/CK\mu \tag{AID.13}$$

$$T_{\max} = \frac{C}{\mu}\left(1 + \frac{q_o}{C\mu}\right) + D = \frac{C}{\mu}(1 + t_n.\eta) + D$$

Analysis of Conductance and Insulation Thickness

Independent analysis of conductance: Mean temperature of the liner, using mean value theorem give

$$\theta_m = \frac{2}{l_{\rm sp}} \int_0^{1/2 l_{\rm sp}} (T_G + A\cos h\lambda x + B\sin h\lambda x)\mathrm{d}x \tag{AID.14}$$

$$\begin{aligned} &= \left({}^2/_{l_{\rm sp}}\right)\left[\frac{1}{2} T_G l_{\rm sp} + (A/\lambda)\sin h\frac{1}{2}\lambda l_{\rm sp}\right] \\ &= \left[T_{\rm G} + \frac{2}{\lambda^2 l_{\rm sp}} \frac{\frac{1}{2}t_\omega(T_w - T_1')}{t_{\rm s}.l_{\rm c}}\right] \end{aligned} \tag{AID.15}$$

Where T_1' the average temperature between the cooling pipe web and the liner.

$$T_1' + \left\{\frac{l_c t_s \lambda T_{\rm G} + \frac{1}{2}t_\omega.T_W \cot h(\frac{1}{2}\lambda l_{\rm sp})}{l_c t_s \lambda + \frac{1}{2}t_\omega \cot h(\frac{1}{2}\lambda l_{\rm sp})}\right\} \tag{AID.16}$$

Eliminating T_1' between Eqs. (AID.15) and (AID.16) and rearranging the terms, the value of C_i is given by

$$C_i = \frac{12Kt_s(\theta_m - T_G)}{l_{sp}\left(l_{sp} + 6t_s.l_c/\frac{1}{2}t_\omega\right)(T_G - \theta_m)} \tag{AID.17}$$

Insulation thickness

$$t_i = (K/C_i'')[(T - \theta_m)/(T_G - \theta_m)] \tag{AID.18}$$

If k varies with mean temperature, the approximate relationship is given by

$$K = 0.002670\theta'_m + 1.533 \quad \text{(AID.19)}$$

Where θ'_m *is* $100 < T < 700°C$ and

$$\theta'_m = \frac{1}{2}(T + \theta_m) \quad \text{(AID.20)}$$

C''_i = insulation conductance giving a mean temperature 60°C

Three-Dimensional Finite Element Analysis

A more refined analysis is required for hear conduction, conductance and insulation thickness in 3D space. Assuming θ represent θ_m and T'_1, then at any point the temperature θ satisfies an equation of heat conduction, provided initial and boundary conditions are given

$$\rho C_p.\partial\theta/\partial T' = \text{div}\,(\bar{K}\,\text{grade}\,\theta) + q, \quad \text{(AID.21)}$$

where $\rho, C_P, K, q, \partial T'$ denote density, specific heat, conductivity, heat generation rate and incremental time, respectively.

For a convective heat transfer from the surface, the heat flux, f, is related as

$$f = \bar{K}.\frac{\partial\theta}{\partial\hat{n}} = h_i\ (T - \theta) \quad \text{(AID.22)}$$

where $\hat{n}$ is the direction of the outward normal from the surface. For an insulated surface $f = 0$ and for steady-state condition $\partial\theta/\partial T' = 0$ but in other cases the temperature is allowed unboundedly though at a constant rate. An approximate solution is obtained by dividing the whole vessel and particularly the cooling system area into a series of isoparametric elements. In this case a 20-node isoparametric (Fig. AID.3) is chosen. One assumption is made and that is if the finite set of unknown temperatures consists of the temperature θ at the centroid of the element, then the following representation of the above equations for any element is true:

$$C'_i\frac{d\theta}{dT'} = \sum_{j=1}^{n} b_{ik}(\bar{\theta}_k - \theta_i) = q_i \quad \text{(AID.23)}$$

where θ_j and $\bar{\theta}_k$ the finite element and boundary temperatures, θ_j is the temperature of the ith element, C'_i is the heat capacity of an element equal to ρC_p, and a's, b's are constant coefficients known as 'admittances'. The a's measure

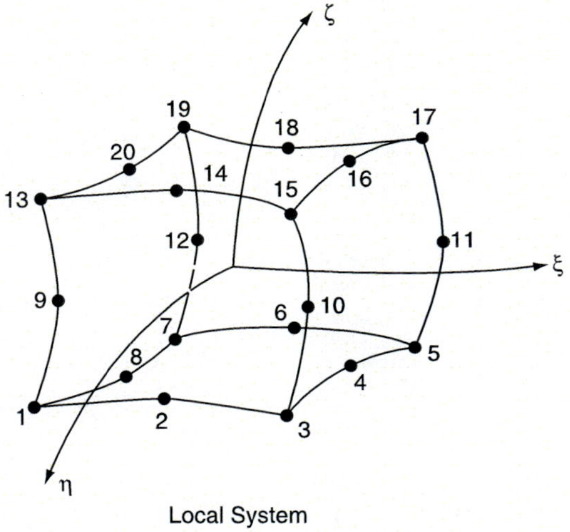

Fig. AID.3 Three-dimensional isoparametric 20 node hexahedron

the ease with which heat can pass between the one isoparametric element and another and the b s measure the ease with which heat can pass between any isoparametric element an a boundary. The exact form of admittances depends on the particular 3D geometrical configuration.

In the finite-element process, the left-hand side of Eq. (AID.23) is approximated by Crank–Nicholson method and is given by

$$\{\theta_m(T' = \Delta T') - \theta_m(T')\}/\Delta T' \tag{AID.24}$$

and the right-hand side by means of its value at time T' and $T' + \Delta T'$. A set of equations is obtained for the centroidal temperatures $\theta_m(T' + \Delta T')$ in terms known values $\theta_m(T')$. These equations are solved by iteration. Choosing a small enough time step, the iteration converges rapidly. For a steady-state condition, the right-hand side of Eq. (AID.21) must be equal to zero. A more detailed finite element version of Eqs. (AID.23) and (AID.24) is given in the text.

During the solution of Eq. (AID.23), Eqs. (AID.17), (AID.18), (AID.19) and (AID.20) are examined for conductance and insulation thickness. Appendix E gives a general layout of the subrouti terma of Program ISOPAR written for CDC 6600/7630. A provision is made in this program for determining conductance with and without nuclear heating.

Application to Dungeness B Vessel

In order to test the above analysis, data given below for the Dungeness B vessel is used.

Data

Cooling water (T_w)	
Maximum temperature	35°C
Velocity	4ft/s (1.22m/s)
K values (thermal conductivity):	
Steel	26 CHU ft/ft^2 °C (45 W/m°C)
Concrete	0.84 CHU ft/ft^{2}h °C (1.445 W/m °C)
Insulation	2.00 CHU ft/ft^{2}r °C (0.29 W/m°C)
Concrete heat transfer coefficient (h_i) At outer face 1.5 CHU/ft^{2}hr °C	(4.733 Wm^{2}h °C)

Gas temperature in vessel local to wall (TG)	450°C to 750°C
Ambient temperature at outside wall of vessel	25°C
Liner thickness (t_s) or t_w	½ in to $^3/_4$ in (12–19 mm)
Volumetric heating in vessel liner	152.0 CHU/ft^3 h (5663.4 CHU/m^3 h)
Volumetric heating in vessel concrete	34.36 e$^{-1.953}$ CHU/ft^3 h (1280 e$^{-1.953}$kJ/m^3)
Average temperature (θ_m)	60°C
Cooling pipe pitch $(I)_{sp}$	6 in. (150 mm). × 10 in. (250 mm) 14 in. (350 mm) and 18 in. (450 mm) CHU/ft^2 h °C.
Insulation conductance (C_i)	(W/m °C)
Thickness of concrete (t)	12 ft 6 in. (3.81 m)

Initial Specimen Calculations

For the Dungeness vessel parameters, the specimen calculations are carried out for maximum temperatures developed in the liner and in the concrete in the presence of cooling pipes at suitable pitches. Figures AID.1 and AID.2 are used. For the liner, using t_s= 19 mm. T_1 = 29.4°C, C_1 = 0.4, T$_G$ = 750°C the following parameters are obtained:

$$\lambda = 0.496, q_o/\mu = 59.6, l_c = 0.875, \tag{AID.25}$$

$$T_{\max} = 46°C, l_{sp} = 277\,mm\,(10.9''). \tag{AID.26}$$

For concrete, the following parameters are obtained:

$$T_t = 26.67°C, T_1 = 82.23°C, h_t = 1.5, \tag{AID.27}$$

$$q_o = 71.6,\ K = 0.84,\ T_2 = 31.5°C, T_{\max} = 93.2°C. \tag{AID.28}$$

Table AID.1 Vessel liner results

Pitch					
	Insulation conductance C_1	0.4	0.6	0.8	1.0
6 in.	Mean membrane temp. θ_m	54.7	63.2	71.3	79.1
(150 mm)	Max. membrane temp. T_{max}	55.9	64.9	73.5	81.8
	Insulation conductance. C_i	0.2	0.4	0.6	0.8
10 in.	Mean membrance temp. θ_m	52.0	65.5	78.1	89.7
(250 mm)	Max. membrane temp. T_{max}	53.6	68.2	82.0	94.7
	Insulation conductance C_i	0.3	0.4	0.5	0.6
14 in.	Mean membrane temp. θ_m	68.1	76.8	85.1	93.1
(350 mm)	Max. membrane temp. T_{max}	72.1	81.9	91.2	100.1
	Insulation conductance C_i	0.2	0.3	0.4	
18 in.	Mean membrane temp. θ_m	66.4	77.9	89.3	
(450 mm)	Max membrane temp. T_{max}	71.2	84.5	97.5	

Values of the mean and maximum liner temperature for each combination of pipe pitch and insulation conductance are given in Table AID.1. The values of conductance are based on Eqs. (16) and (17). In the Dungeness **B** vessel pipe pitches greater than 150 mm are not practicable, the reason being that the space

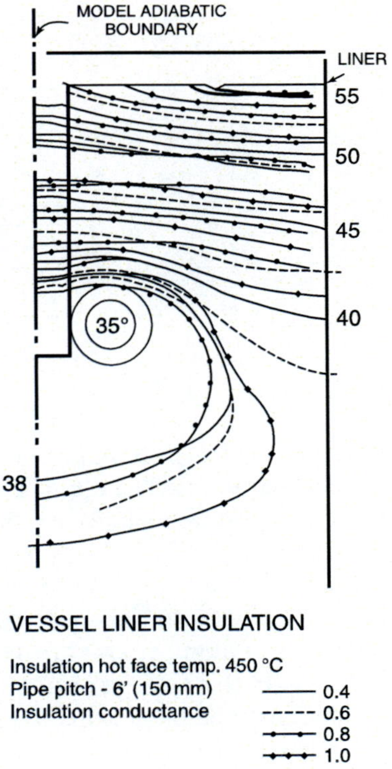

Fig. AID.4 Isotherm pattern

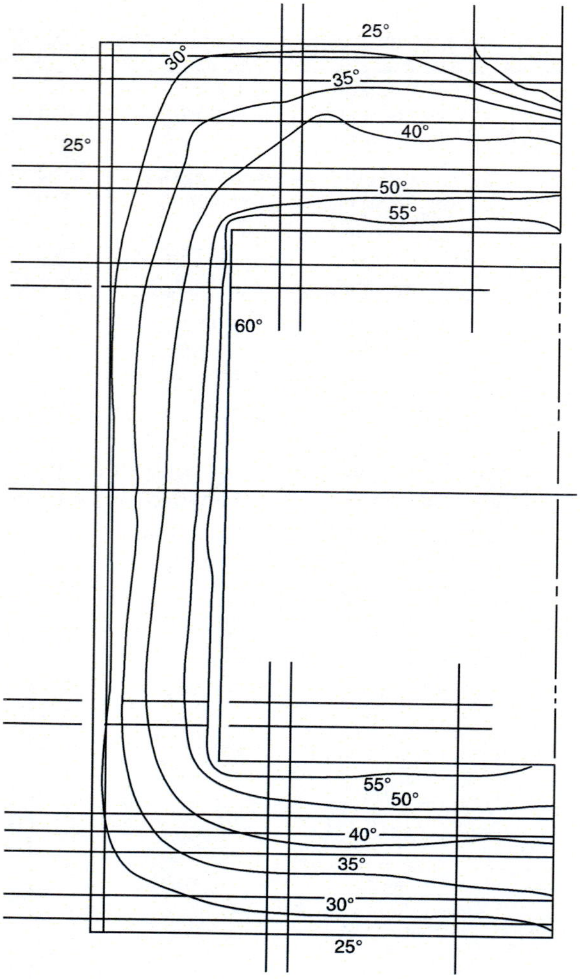

Fig. AID.5 Dungeness B type vessel temperature distribution

available for the insulation is limited to about 75 mm rather than only on economic grounds. The estimate of the thickness of the insulation for a given mean temperature at a given cooling pipe pitch is given by Eqs. (AID.18) and (AID.20).

Using three-dimensional finite element analysis and considering the cooling pipe zone in the vessel where cooling tubes lie on each side of the lug attached to the liner, isotherm patterns are drawn for insulation conductance varying from 0.2 to 1.0. In this case the insulation hot face temperature is taken to be 450°C. The cooling pipe pitch is maintained as 150 mm. Nuclear heating in the liner and concrete is included in the analysis. Figures AID.4 and AID.5 show such isotherm pattern using program ISOPAR with subroutine TERMA.

Comparison of Two Methods of Calculation

To make a comparison on the basis of neglecting nuclear heating in both methods of calculation, the insulation thicknesses at 450°C hot face temperature appropriate to liner/concrete wall are converted back to conductance using the assumed conductivity of 2 CHU in./ft^2 h °C (0.29 W/m °C). The comparative values of C_i, the insulation conductance, are given in Table AID.2 for different cooling pipe pitches and with and without nuclear heating. The results are in good agreement among themselves and with those obtained from prototype data (1, 6).

Table AID.2 Comparison of insulation conductances

	Insulation conductance C_i (CHU/ft^2 h °C) for $T_m = 60$°C and $T_G = 450$°C		
	Direat approach $C_f = 0.832/p(p + 4)$	Finite element	
Cooling pipe pitch	Nuclear heating neglected	Nuclear heating included	Without nuclear heating
6 in. (150 mm)	0.370	0.528	0.356
10 in. (250 mm)	0.207	0.310	0.215
14 in. (350 mm)	0.138	0.203	0.159
18 in. (450 mm)	0.101	0.146	0.115

Conclusions

A simplified approach for computing temperature distribution in the steel liner and concrete of prestressed concrete reactor vessels has been presented. A provision is made in this analysis for estimating the cooling pipe pitches, insulation thicknesses and conductances. A 3D finite-element analysis has been carried out for comparison and validation of the above analysis.

Notation

C_i = conductance of insulation
h_t = heat transfer coefficient from outer concrete surface to ambient temperature
k = thermal conductivity
l_e = length of cooling pipe from the liner
I_{sp} = pitch between cooling pipes
T' = time $\Delta T'$ = time increment
T = temperature; T'' transpose
t = thickness of concrete
t_i = thickness of insulation
t_s = liner thickness
θ_m = mean temperature; θ_n = temperature at nodes
φ = equivalent diameter of cooling pipe
λ, μ = = coefficients defined in the text
x = distance measured from the liner concrete interface

P = shape function
ρ = density

Appendix D.1
3-D Finite Element Formulation
Basic Thermal Diffusion Equation
Equations (D.21), D.22 and (D.23) are referred basic thermal diffusion equation

$$[\bar{C}_p]\left[\dot{\theta}\right] + [K]\{\theta\} = \{\bar{q}\} = \begin{Bmatrix} q_x \\ q_y \\ q_z \end{Bmatrix} \tag{D.1}$$

Heat flow temperature virtual work

$$\{\bar{q}\} = [D][\dot{\theta}] \tag{D.2}$$

where

$$[D] = \begin{bmatrix} \bar{K}_{xx} & 0 & 0 \\ 0 & \bar{K}_{yy} & 0 \\ 0 & 0 & \bar{K}_{zz} \end{bmatrix}$$

$$\dot{\theta} = \text{temperature vector} = \begin{bmatrix} \frac{\partial\theta}{\partial x}(x,y,z) \\ \frac{\partial\theta}{\partial y}(x,y,z) \\ \frac{\partial\theta}{\partial z}(x,y,z) \end{bmatrix}$$

$$\delta w = \text{virtual work} = \int_v \{\delta\dot{\theta}\}^{T''}\{\bar{q}\}\,\text{dvol} = \{\delta\theta_n\}^{T''} \tag{D.3}$$

$$\int_v \{B\}^{T''}[D][B]\text{dvol}\,\{\theta_n\} \tag{D.4}$$

where

$$[B] = \text{strain displacement matrix} = \begin{bmatrix} \left[\frac{\partial p}{\partial x}\right]^{T''} \\ \left[\frac{\partial p}{\partial y}\right]^{T''} \\ \left[\frac{\partial p}{\partial z}\right]^{T''} \end{bmatrix}$$

$\{\theta_n\}$ = nodal or centroidal temperature; $\{p\}$*shape function*

$$\theta(x,y,z) = [P]^T\{\theta_n\}$$

The virtual work associated with convective surfaces
Equation (D.22) is referred to internal

$$\begin{aligned} \delta w_1 &= \int_A \delta \Delta q \theta_n d \text{ (area)} \\ \Delta\theta &= \theta(x,y,z) - T \end{aligned} \tag{D.5}$$

$$q_{\hat{n}} = \text{vector direction normal to the surface } = h_t \Delta\theta$$

Hence

$$\delta w = \{\delta\theta_n\}^{T'''} \dddot{q} \int_v [P] \,\text{dvol} \tag{D.6}$$

$\dddot{q}$ = rate of heat generation/unit volume. The virtual work associated with a change stored energy in the light of Eq. (D.2) is written as

$$\begin{aligned} \delta w &= \int_v \delta\theta_{(x,y,z)} \rho C_p \frac{\delta\theta_{(x,y,z)}}{\delta T'} \text{dvol} \qquad (D.7a) \\ &= \{\delta\theta_n\}^{T'''} \rho C_p \int_v \{P\}\{P\}^{T'''} \text{dvol} \,[\dot{\theta}_n] \end{aligned}$$

where

$$[\dot{\theta}_n] = \delta\{\theta_n\}/\delta T'$$

Total element conductivity and specific heat material conductivity matrix

$$[K_c]_c = h_t \int_A \{Pl_\theta\}\{Pl_\theta\}^{T'''} dA \tag{D.8}$$

Specific heat matrix

$$[K]_{cp} = \rho C_p \int_A \{P\}\{P\}^{T'''} dvol$$

The element heat flow matrix

$$\begin{aligned} \{q\}_c &= h_t T \int_A \{Pl_\theta\} \,\text{dA} \\ \{q\}_g &= \dddot{q} \int_v \{P\} \text{dvol} \\ \text{dvol} &= \det\{J\} \text{d}\xi \,\text{d}\eta \,\text{d}\zeta \end{aligned}$$

Time integration scheme quadratic

$$\begin{aligned} &\left[\frac{2\Delta T'_o + \Delta T'_1}{\Delta T'_d} \frac{1}{\Delta T'_o} \{\bar{C}_p\} + [\bar{K}]\right]\{\theta\} \\ &= [\bar{q}](T') + \{\bar{C}_p\}\left(\frac{\Delta T'_d}{\Delta T'_o \Delta T'_1}\{\theta_{T-1}\} - \frac{\Delta T'_o}{\Delta T'_d - \Delta T'_1}\{\theta_{T-2}\}\right) \end{aligned}$$

where T'_o = initial time, T'_1 = previous considered time, T'_2 = previous time considered, $T'_d = T'_o - T'_2$, $\{\theta_{T-1}\}$, etc. = temperature at previous time steps.

Appendix D.2
Subroutine terma

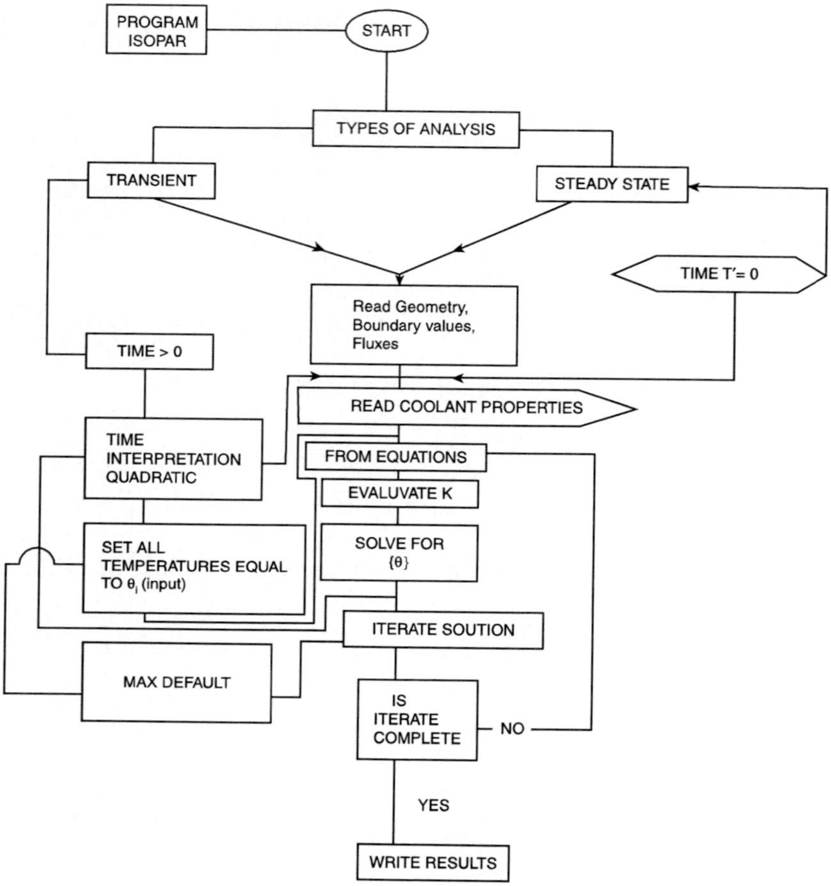

References

1. Pretressed concrete pressure vessel, Int. Conf. Inst. Civil Engrs., London, 1967, Papers Al to A3.
2. Bangash, Y. Journal of Institute of Nuclean Engineering. 13 1972 108–116.
3. Bangash, Y. Proceeings of Institute of Civil Engineering, 1972 Suppl. 155–173 and 301–308.
4. Irons, B.M. International Journal of Numerical Methods in Engineering, 2, 1970 5–32.
5. Hughes, J.W. et al. Prestressed concrete pressure vessels. International Conference of Institute of Civil Engineers, London, 1967, Paper 60, 703–713.
6. Turton, P.J. et al. Prestressed concrete pressure vessels. International Conference on Institute of Civil Engineers, London, 1967, paper 61, 715–723.
7. Bangash, Y. Nuclear Engineering and Design 50 (1978) 463–473.
8. Bangash, Y. Manual of Numerical Methods in Concrete, Thomas Telford, London 2001.

Appendix E
Programs Subroutines (For ISOPAR and F. BANG)

E.1 Introduction

This appendix gives short repetitive notes first associated with programs ISOPAR and F. BANG. These are then followed by computer listings of these programs.

E.2 Mass Matrices and Stiffness Matrices and Others

Table AIE.1 Stiffness and mass matrices (courtesy STRUCOM, London)
Orders of degree of freedom

The stiffness matrix in element co-ordination

$$[K_1] = \frac{AE}{L}\begin{bmatrix} C_1 & 0 & 0 & -C_1 & 0 & 0 \\ 0 & 0 & 0 & 0 & 0 & 0 \\ 0 & 0 & 0 & 0 & 0 & 0 \\ -C_1 & 0 & 0 & C_1 & 0 & 0 \\ 0 & 0 & 0 & 0 & 0 & 0 \\ 0 & 0 & 0 & 0 & 0 & 0 \end{bmatrix}$$

Where

A = element cross-sectional area
E = Youngs's modulus
L = element Length
C_1 = value given in the below

$$[S_1] = \frac{F}{L}\begin{bmatrix} 0 & 0 & 0 & 0 & 0 & 0 \\ 0 & C_2 & 0 & 0 & -C_2 & 0 \\ 0 & 0 & C_2 & 0 & 0 & -C_2 \\ 0 & 0 & 0 & 0 & 0 & 0 \\ 0 & -C_2 & 0 & 0 & C_2 & 0 \\ 0 & 0 & -C_2 & 0 & 0 & C_2 \end{bmatrix}$$

Table AIE.1

Value of stiffness coefficient (C_i)	
Previous iteration resulted in a tensile strain	Previous iteration resulted in a compressive stress
1.0	0.0
1.0	1.0×10^{-6}
0.0	1.0
1.0×10^{-6}	1.0

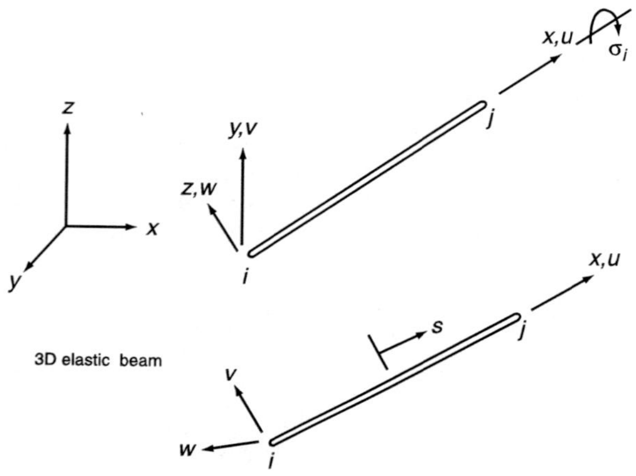

Fig. AIE.1 Three-dimensional elastic beam

Stiffness and mass matrices (courtesy STRUCOM, London)
Orders of degrees of freedom
The stiffness matrix in element co-ordinates is:

$$
[K_1] = \begin{bmatrix}
AE/L & & & & & & & & & & & \\
0 & a_z & & & & & & & & & & \\
0 & 0 & a_y & & & & & & & & & \\
0 & 0 & 0 & GJ/L & & & & & & \text{Symmetric} & & \\
0 & 0 & d_y & 0 & e_y & & & & & & & \\
0 & c_z & 0 & 0 & 0 & e_z & & & & & & \\
-AE/L & 0 & 0 & 0 & 0 & 0 & AE/L & & & & & \\
0 & b_z & 0 & 0 & 0 & d_z & 0 & a_z & & & & \\
0 & 0 & b_y & 0 & c_y & 0 & 0 & 0 & a_y & & & \\
0 & 0 & 0 & -GJ/L & 0 & 0 & 0 & 0 & 0 & GJ/L & & \\
0 & 0 & d_y & 0 & f_y & 0 & 0 & 0 & c_z & 0 & e_y & \\
0 & c_z & 0 & 0 & 0 & f_z & 0 & d_y & 0 & 0 & 0 & e_z
\end{bmatrix}
$$

Where

$$\begin{bmatrix} A = \text{cross}-\text{sectional area} & M_t = (\rho A+m)L(1-\varepsilon^{in}) \\ E = \text{Young's modulus} & \rho = \text{density} \\ L = \text{element length} & m = \text{added mass} \\ G = \text{shear modulus} & \varepsilon^{\text{in}} = \text{prestrain} \\ r_y = \sqrt{\frac{I_{yy}}{A}} = \text{radius of gyration} & \begin{array}{c} A_z = A\left(r_z,\phi'_y\right) \\ A_y = A\left(r_y,\phi'_z\right) \end{array} \\ r_y = \sqrt{\frac{I_{yy}}{A}} = \text{radius of gyration} & B_z = B\left(r_z,\phi'_y\right) \\ [\mathrm{M}_1] = M_t & F_y = F\left(r_z,\phi'_y\right) \end{bmatrix}$$

$$[K_1] = \begin{bmatrix} 1/3 \\ 0 & A_z & & & & & & F_y = F\left(r,\phi'_z\right) \\ 0 & 0 & A_y \\ 0 & 0 & 0 & J_x/3A & & & & & & & \text{Symmetric} \\ 0 & 0 & -C_y & 0 & E_y \\ 0 & C_z & 0 & 0 & 0 & E_z \\ 1/6 & 0 & 0 & 0 & 0 & 0 & 1/3 \\ 0 & B_z & 0 & 0 & 0 & D_z & 0 & A_z \\ 0 & 0 & B_y & 0 & -D_y & 0 & 0 & 0 & A_y \\ 0 & 0 & 0 & -J_x/6A & 0 & 0 & 0 & 0 & 0 & J_x/3A \\ 0 & 0 & D_y & 0 & F_y & 0 & 0 & 0 & C_y & 0 & E_y \\ 0 & D_z & 0 & 0 & 0 & F_z & 0 & -C_z & 0 & 0 & 0 & E_z \end{bmatrix}$$

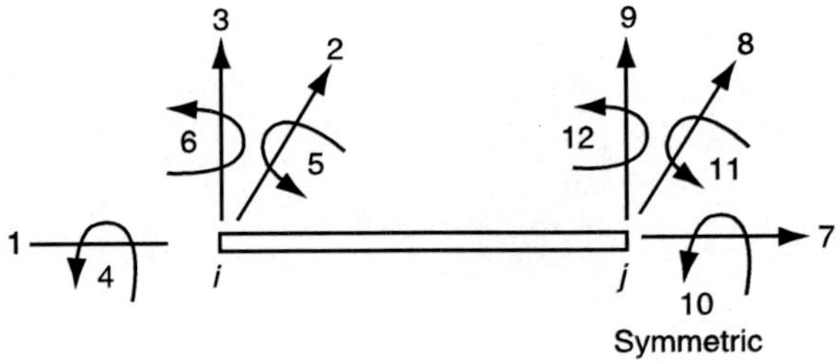

Fig. AIE.2 Order of degree of freedom

and where

$$A(r,\phi') = \frac{13/35 + 7/10\phi' + 1/3\phi'^2 + 6/5(r/L)^2}{(1+\phi')^2}$$

$$B(r,\phi') = \frac{9/70 + 3/10\phi' + 1/6\phi'^2 - 6/5(r/L)^2}{(1+\phi')^2}$$

$$C(r,\phi') = \frac{(11/210 + 11/120\phi'^2 + 1/24\phi'^2 + (1/10 - 1/2\phi')(r/L)^2)L}{(1+\phi')^2}$$

$$D(r,\phi') = \frac{(13/420 + 3/40\phi' + 1/24\phi'^2 - (1/10 - 1/2\phi')(r/L)^2)L}{(1+\phi')^2}$$

$$E(r,\phi') = \frac{(1/105 + 1/60\phi' + 1/120\phi'^2 + \left(2/15 + 1/6\phi' + 1/3\phi'^2\right)(r/L)^2)L^2}{(1+\phi')^2}$$

$$F(r,\phi') = \frac{(1/140 + 1/60\phi' + 1/120\phi'^2 + (1/30 + 1/6\phi' - 1/6\phi'^2)(r/L)^2)L^2}{(1+\phi')^2}$$

$$J = \text{torsional moment of inertia} = \left\{\begin{matrix} J_x & \text{If } I_x = 0 \\ I_x & \text{if } I_x \neq 0 \end{matrix}\right\}$$

I_x = input as I_{XX}
I_x = Polar moment of inertia = $I_y + I_z$

AIE.2.2 The element mass matrix in element co-ordinates

$$[M_1] = (\rho A + m)L\left(1 - \varepsilon^{\text{in}}\right)$$
$$\times \begin{bmatrix} 1/3 & 0 & 0 & 1/6 & 0 & 0 \\ 0 & A(r,\phi') & C(r,\phi') & 0 & B(r,\phi') & -D(r,\phi') \\ 0 & C(r,\phi') & E(r,\phi') & 0 & D(r,\phi') & -F(r,\phi') \\ 1/6 & 0 & 0 & 1/3 & 0 & 0 \\ 0 & B(r,\phi') & D(r,\phi') & 0 & A(r,\phi') & -C(r,\phi') \\ 0 & -D(r,\phi') & -F(r,\phi') & 0 & -C(r,\phi') & E(r,\phi') \end{bmatrix}$$

where
ρ = density
m = added m
ε^{in} = Prestrain

$$A(r,\phi') = \frac{13/35 + 7/10\phi' + 1/3\phi'^2 + 6/5(r/L)^2}{(1+\phi')^2}$$

$$B(r,\phi') = \frac{9/70 + 3/10\phi' + 1/6\phi'^2 - 6/5(r/L)^2}{(1+\phi')^2}$$

$$C(r,\phi') = \frac{(11/210 + 11/120\phi'^2 + 1/24\phi'^2 + (1/10 - 1/2\phi')(r/L)^2)L}{(1+\phi')^2}$$

$$D(r,\phi') = \frac{(13/420 + 3/40\phi' + 1/24\phi'^2 - (1/10 - 1/2\phi')(r/L)^2)L}{(1+\phi')^2}$$

$$E(r,\phi') = \frac{(1/105 + 1/60\phi' + 1/120\phi'^2 + \left(2/15 + 1/6\phi' + 1/3\phi'^2\right)(r/L)^2)L^2}{(1+\phi')^2}$$

$$F(r,\phi') = \frac{(1/140 + 1/60\phi' + 1/120\phi'^2 + (1/30 + 1/6\phi' - 1/6\phi'^2)(r/L)^2)L^2}{(1+\phi')^2}$$

$$r = \sqrt{\frac{1}{A}} = \text{radius of gyration}$$

AIE.2.3 Element types, shape functions, derivatives, stiffness matrices

$$\begin{bmatrix}
\frac{EA}{L} \\
0 & \frac{12EI\xi}{L^3(1+\bar\tau_\eta)} \\
0 & 0 & \frac{12EI\xi}{L^3(1+\bar\tau_\eta)} \\
0 & 0 & 0 & \frac{GJ}{L} \\
0 & 0 & \frac{-6EI_\eta}{L^2(1+\tau_\xi)} & 0 & \frac{(4+\bar\tau)EI_\eta}{L(1+\bar\tau_\eta)} \\
0 & \frac{6EI\xi}{L^2(1+\bar\tau_\eta)} & 0 & 0 & 0 & \frac{(4+\bar\tau)EI_\xi}{L(1+\bar\tau_\eta)} \\
\frac{-EA}{L} & 0 & 0 & 0 & 0 & 0 & \frac{AE}{L} \\
0 & \frac{12EI\xi}{L^3(1+\bar\tau_\eta)} & 0 & 0 & 0 & \frac{-6EI_\xi}{L^2(1+\bar\tau_\eta)} & 0 & \frac{12EI_\xi}{L^3(1+\bar\tau_\eta)} \\
0 & 0 & \frac{-12EI_\eta}{L^3(1+\bar\tau_\xi)} & 0 & \frac{6EI_\eta}{L^2(1+\bar\tau_\xi)} & 0 & 0 & 0 & \frac{12EI_\eta}{L^3(1+\bar\tau_\eta)} \\
0 & 0 & 0 & \frac{-GJ}{L} & 0 & 0 & 0 & 0 & 0 & \frac{GJ}{L} \\
0 & 0 & \frac{-6EI_\eta}{L^2(1+\bar\tau_\eta)} & 0 & \frac{(2-\bar\tau_\eta)EI_\eta}{L(1+\bar\tau_\xi)} & 0 & 0 & 0 & \frac{6EI_\eta}{L^2(1+\bar\tau_\xi)} & 0 & \frac{(4+\bar\tau_\xi)EI_\eta}{L(1+\bar\tau_\xi)} \\
0 & \frac{6EI\xi}{L^2(1+\bar\tau_\eta)} & 0 & 0 & 0 & \frac{(2-\bar\tau_\eta)EI_\xi}{L(1+\bar\tau_\eta)} & 0 & \frac{-6EI_\xi}{L^2(1+\bar\tau_\eta)} & 0 & 0 & 0 & \frac{(4+\bar\tau_\eta)EI_\xi}{L(1+\bar\tau_\eta)}
\end{bmatrix}$$

$$\bar{\tau}_\eta = \frac{12EI_\xi}{GA_{s\eta}L^2} = 24(1+v)\frac{A}{A_{sn}}\left(\frac{T_\xi}{L}\right)^2 \quad \bar{\tau}_\xi = \frac{12EI_\eta}{GA_{s\xi}L^2} = 24(1+v)\frac{A}{A_{s\xi}}\left(\frac{T_\eta}{L}\right)^2$$

A_s = shear area T = torsional moment of inertia P = density ξ, η local axes are parallel to Z and Y axes

The element pressure load vector in element coordinates IS

$$\{F_1^{pe}\} = \lfloor P_1\, P_2\, P_3 P_4\, P_5\, P_6 \rfloor^{\mathrm{T}}$$

Table AIE.2

Value of stress stiffness coefficient (C_2) Previous iteration resulted in a tensile stress	Previous iteration resulted in a compressive stress
1.0	0.0
1.0	$\frac{AE}{F \times 10^6}$
0.0	1.0
$\frac{AE}{F \times 10^6}$	1.0

$$F = \begin{cases} \text{for the first iteration : } AE\varepsilon^{\text{in}} \\ \text{for all subsequent iterations : the axial force} \\ \text{in the element as computed in the previous} \\ \text{stress pass of the element (output quantity)} \\ \text{FORC} \end{cases}$$

C_2 = Value given in the table above

AIE.2.4 The matrix for the tension-only or compression-only spar

$$[M_l] = \frac{M_t}{2}\begin{bmatrix} 1 & & & & & & & & & & & \\ 0 & 1 & & & & & & & & & & \\ 0 & 0 & 1 & & & & & & & & & \\ 0 & 0 & 0 & 0 & & & & & & & & \\ 0 & 0 & 0 & 0 & 0 & & \text{Symmetric} & & & & & \\ 0 & 0 & 0 & 0 & 0 & 0 & & & & & & \\ 0 & 0 & 0 & 0 & 0 & 0 & 1 & & & & & \\ 0 & 0 & 0 & 0 & 0 & 0 & 0 & 1 & & & & \\ 0 & 0 & 0 & 0 & 0 & 0 & 0 & 0 & 1 & & & \\ 0 & 0 & 0 & 0 & 0 & 0 & 0 & 0 & 0 & 0 & & \\ 0 & 0 & 0 & 0 & 0 & 0 & 0 & 0 & 0 & 0 & 0 & \\ 0 & 0 & 0 & 0 & 0 & 0 & 0 & 0 & 0 & 0 & 0 & 0 \end{bmatrix}$$

Table AIE.3 Chain rule

$$
\begin{bmatrix} \frac{\partial u}{\partial X} \\ \frac{\partial u}{\partial Y} \\ \frac{\partial u}{\partial Z} \\ \frac{\partial v}{\partial X} \\ \frac{\partial v}{\partial Y} \\ \frac{\partial v}{\partial Z} \\ \frac{\partial W}{\partial X} \\ \frac{\partial u}{\partial Y} \\ \frac{\partial u}{\partial Z} \end{bmatrix} = \frac{1}{\det J} \begin{bmatrix} C_{11} & C_{12} & C_{13} & 0 & 0 & 0 & 0 & 0 & 0 \\ C_{21} & C_{22} & C_{23} & 0 & 0 & 0 & 0 & 0 & 0 \\ C_{31} & C_{32} & C_{33} & 0 & 0 & 0 & 0 & 0 & 0 \\ 0 & 0 & 0 & C_{11} & C_{12} & C_{13} & 0 & 0 & 0 \\ 0 & 0 & 0 & C_{21} & C_{22} & C_{23} & 0 & 0 & 0 \\ 0 & 0 & 0 & C_{31} & C_{32} & C_{33} & 0 & 0 & 0 \\ 0 & 0 & 0 & 0 & 0 & 0 & C_{11} & C_{12} & C_{13} \\ 0 & 0 & 0 & 0 & 0 & 0 & C_{21} & C_{22} & C_{23} \\ 0 & 0 & 0 & 0 & 0 & 0 & C_{31} & C_{32} & C_{33} \end{bmatrix} \begin{bmatrix} \frac{\partial u}{\partial \xi} \\ \frac{\partial u}{\partial \eta} \\ \frac{\partial u}{\partial \varsigma} \\ \frac{\partial v}{\partial \xi} \\ \frac{\partial v}{\partial \eta} \\ \frac{\partial u}{\partial \varsigma} \\ \frac{\partial w}{\partial \xi} \\ \frac{\partial w}{\partial \eta} \\ \frac{\partial w}{\partial \varsigma} \end{bmatrix}
$$

where

$$
C_{11} = \frac{\partial Y}{\partial \xi}\frac{\partial Y}{\partial \xi} - \frac{\partial Z}{\partial \eta}\frac{\partial Y}{\partial \xi} \quad C_{12} = \frac{\partial Z}{\partial \xi}\frac{\partial Y}{\partial \varsigma} - \frac{\partial Z}{\partial \xi}\frac{\partial Y}{\partial \varsigma}
$$

$$
C_{13} = \frac{\partial Y}{\partial \xi}\frac{\partial Z}{\partial \varsigma} - \frac{\partial Z}{\partial \xi}\frac{\partial Y}{\partial \varsigma} \quad C_{21} = \frac{\partial Z}{\partial \xi}\frac{\partial X}{\partial \varsigma} - \frac{\partial X}{\partial \xi}\frac{\partial Z}{\partial \varsigma}
$$

$$
C_{22} = \frac{\partial X}{\partial \xi}\frac{\partial Z}{\partial \varsigma} - \frac{\partial Z}{\partial \xi}\frac{\partial X}{\partial \varsigma} \quad C_{23} = \frac{\partial Z}{\partial \xi}\frac{\partial X}{\partial \eta} - \frac{\partial X}{\partial \xi}\frac{\partial Z}{\partial \eta}
$$

$$
C_{31} = \frac{\partial X}{\partial \eta}\frac{\partial Y}{\partial \varsigma} - \frac{\partial Y}{\partial \eta}\frac{\partial X}{\partial \varsigma} \quad C_{32} = \frac{\partial Y}{\partial \xi}\frac{\partial X}{\partial \varsigma} - \frac{\partial X}{\partial \xi}\frac{\partial Y}{\partial \varsigma}
$$

$$
C_{33} = \frac{\partial X}{\partial \xi}\frac{\partial Y}{\partial \eta} - \frac{\partial Y}{\partial \xi}\frac{\partial X}{\partial \eta}
$$

det $[J]$ = the determinant of the Jacobian matrix

Table AIE.4 Stress and strain transformation matrices

$\{T''_{\sigma}\}\{T''_{\varepsilon}\}$

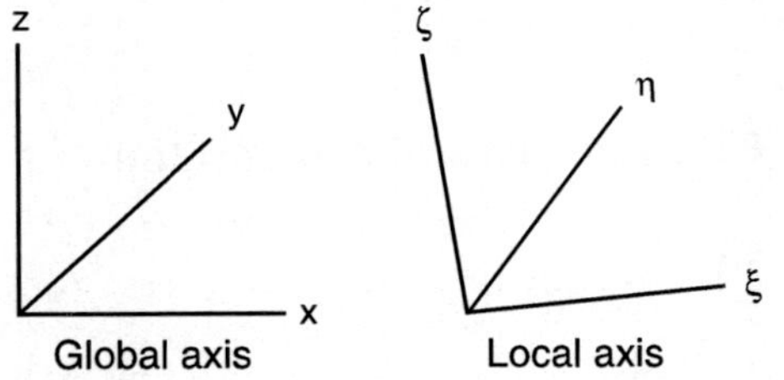

	x	y	z
ξ	l_1	m_1	n_1
η	l_2	m_2	n_2
ζ	l_3	m_3	n_3

Direction cosincs of the two axes are given by

$$
l_1 = \cos(\xi, x) \quad m_1 = \cos(\xi, y) \quad n_1 = \cos(\xi, z)
$$

$$
l_2 = \cos(\eta, x) \quad m_2 = \cos(\xi, y) \quad n_2 = \cos(\xi, z)
$$

$$
l_3 = \cos(\varsigma, x) \quad m_3 = \cos(\varsigma, y) \quad n_3 = \cos(\varsigma, z)
$$

The following relationships can be written for local and global strain and sterss vectors:

$$\{\varepsilon_x'\} = [T_\varepsilon]\{\varepsilon_x\} \ \{\sigma_x\} = [T_\varepsilon]^{T''}\{\sigma_x'\}$$

and also

$$\{\sigma_x'\} = [T_\sigma]\{\sigma_x\} \ \{\varepsilon_x\} = [T_\sigma]^{T''}\{\varepsilon_x'\}$$

$$[T_\varepsilon''] = \begin{bmatrix} l_1^2 & m_1^2 & n_1^2 & l_1m_1 & m_1n_1 & l_1n_1 \\ l_2^2 & m_2^2 & n_2^2 & l_2m_2 & m_2n_2 & l_2n_2 \\ l_3^2 & m_3^2 & n_3^2 & l_3m_3 & m_3n_3 & l_3m_3 \\ 2l_1l_2 & 2m_1m_2 & 2n_1n_2 & l_1m_2 + l_2m_1 & m_1n_2 + m_2n_1 & l_1n_2 + l_2n_1 \\ 2l_2l_3 & 2m_2m_3 & 2n_2n_3 & l_2m_3 + l_3m_2 & m_2n_3 + n_2m_3 & l_2n_3 + l_3n_2 \\ 2l_1l_3 & 2m_3m_3 & 2m_1n_3 & l_1m_3 + m_1l_3 & m_1n_3 + m_3n_1 & l_1n_3 + n_1l_3 \end{bmatrix}$$

$$[T_\varepsilon''] = \begin{bmatrix} l_1^2 & m_1^2 & n_1^2 & 2l_1m_1 & 2m_1n_1 & 2l_1n_1 \\ l_2^2 & m_2^2 & n_2^2 & 2l_2m_2 & 2m_2n_2 & 2l_2n_2 \\ l_3^2 & m_3^2 & n_3^2 & 2l_3m_3 & 2m_3n_3 & 2l_3m_3 \\ l_1l_2 & m_1m_2 & n_1n_2 & l_1m_2 + l_2m_1 & m_1n_2 + n_1m_2 & l_1n_2 + l_2n_1 \\ l_2l_3 & m_2m_3 & n_2n_3 & l_2m_3 + l_3m_2 & m_2n_3 + n_2m_3 & l_2n_3 + l_3n_2 \\ l_1l_3 & m_1m_3 & n_1n_3 & l_1m_3 + l_3m_1 & m_1n_3 + m_3n_1 & l_1n_3 + n_1l_3 \end{bmatrix}$$

E.3 Criteria for Convergence and Acceleration

Convergence Criteria

To ensure convergence to the correct solution by finer sub-division of the mesh, *the* assumed displacement function must satisfy the convergence criteria given below:

(a) Displacements must be continuous over element boundaries.
(b) Rigid body movements should be possible without straining.
(c) A state of constant strain should be reproducible.

Euclidean norm $\psi_1/R_1 \leq C$. The term ψ_1 represents the unbalanced forces and the norm of the residuals. With the aid of the iterative scheme described above, the unbalanced forces due to the initial stresses θ_1 become negligibly small. As a measure of their magnitude, the norm of the vector $\|\psi_1\|$ is used. The Euclidean norm and the absolute value of the largest component of the vector are written as

$$\|\psi\| = \left(|\psi_1|^2 + \cdots + |\psi_n|^2\right)^{1/2}$$
$$\|R_i\| = \left(\left|\{R_i\}^{T''}\{R_i\}\right|\right)^{1/2} \quad \text{(AIE.1)}$$

where T″ = Transpose
The convergence criterion adopted is

$$\|\psi\| = \max_i |\psi_i| < C = 0.001 \quad \text{(AIE.2)}$$

E.4 Uniform *Acceleration*

Various procedures are available for accelerating the convergence of the modified Newton–Raphson iterations. Figure AIE.1 shows the technique of computing individual acceleration factors, δ_1 and δ_2 are known. Then, assuming a constant slope of the response curve, and from similar triangles, the value of δ_3 is computed:

$$\frac{\delta_1}{\delta_2} = \frac{\delta_2}{\delta_3}\ \delta_3 = \delta_2 \frac{\delta_2}{\delta_1} \quad \text{(AIE.3)}$$

When δ_3 is added to δ_2, then the accelerated displacement δ_2' is expressed as

$$\delta_2' = \delta_2 + \delta_3 = \delta_2\left(1 + \frac{\delta_2}{\delta_1}\right) = \alpha\delta_2 \quad \text{(AIE.4)}$$

where the acceleration factor α is

$$\alpha = 1 + \frac{\delta_2}{\delta_1} \quad \text{(AIE.5)}$$

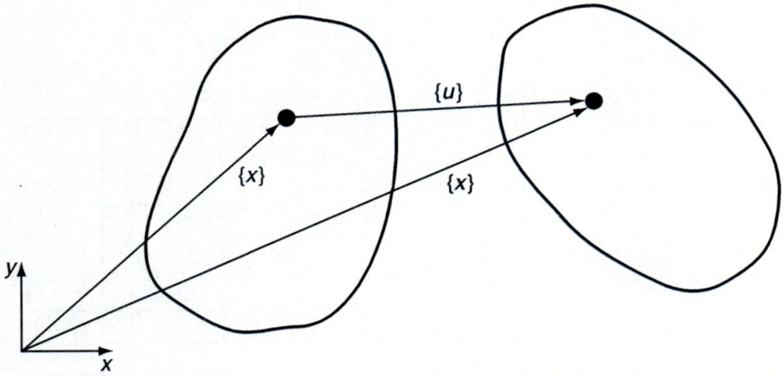

Fig. AIE.3 Position vectors and motion of a deforming body

Fig. AIE.4(a) Newton-Raphson method

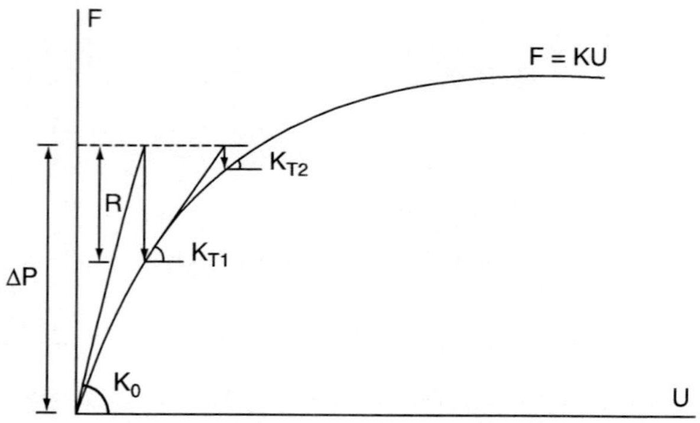

Fig. AIE.4(b) Initital stress method. Note: ΔP is a specific value of F

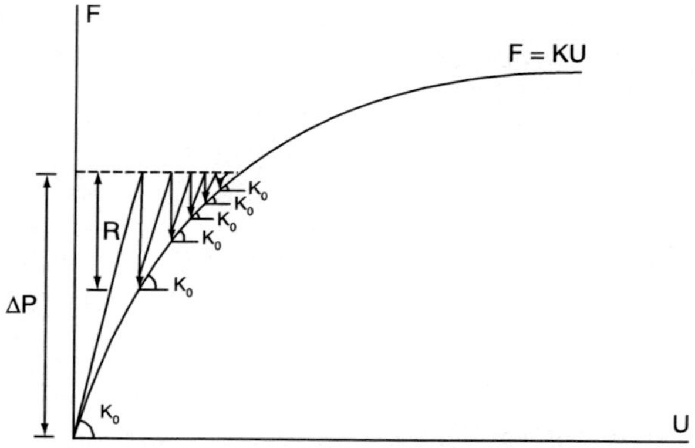

Fig. AIE.4(c) Technique of computing acceleration factors

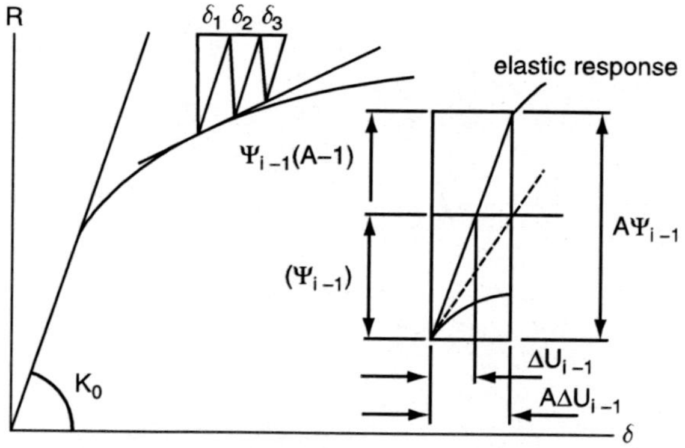

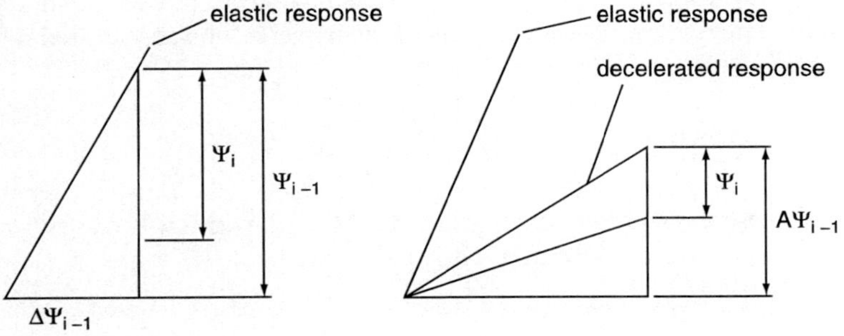

Fig. AIE.4(d) Graphical representation

Fig. AIE.4(e) Linear acceleration and load assumptions of the Wilson θ method

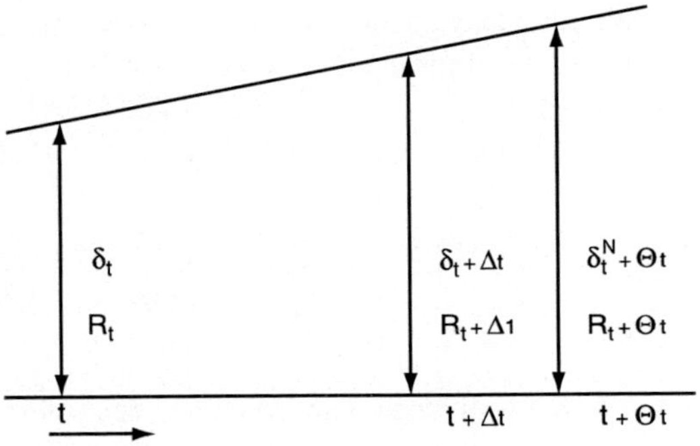

Fig. AIE.4(f) Quadralic and cubic variation of velocity and displacement assumptions of the Wilson θ method

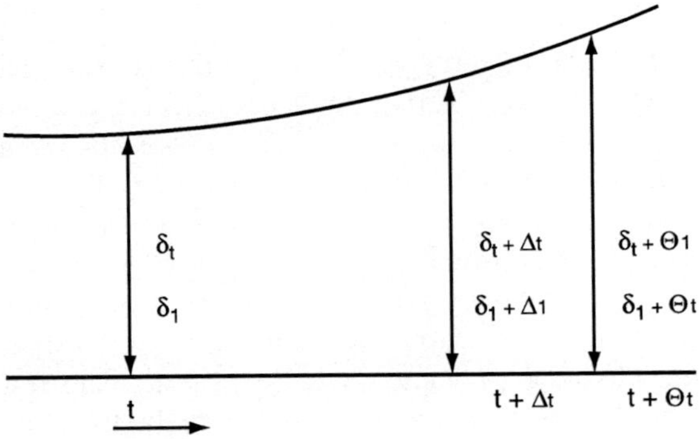

Generally the range of α is between 1 and 2. The value of $\alpha = 1$ for zero acceleration, and the value of a reaches the maximum value of 2 when the slope of the δ —R curve approaches zero.

The acceleration factor α is computed individually for every degree of freedom of the system. The displacement vector obtained from the linear stiffness matrix $[k_0]$ is then multiplied by the $[\alpha]$ matrix having the above constants on its diagonals. The remaining components of $[\alpha]$ are zero. The accelerated displacement vector is then expressed as follows:

$$\{\Delta u_i'\} = [a_{i-1}]\{+\Delta u_j\} \tag{AIE.6}$$

From these accelerated displacements $\{\Delta u_i'\}$the initial stresses $\{\sigma_0\}$ are found and they are equilibrated with the forces $\{\psi_1\}$ They are then used for the next solution

$$\{\Delta \bar{u}_j\} = [K_0]^{-1}\{\psi_i\} \tag{AIE.7}$$

which results in a new set of acceleration factors. Now an estimate for the displacement increment is made in order to find the incremental stresses and total stresses.

The residual forces needed to re-establish equilibrium can now easily be evaluated

$$\{\hat{\psi}_i\} = \int [B]^T\{\sigma_{0_\tau}\}dV - \{R_i\} \tag{AIE.8}$$

where $\{R_i\}$ represents the total external load; dV is the volume.

A new displacement now results from

$$\{\Delta u_{i+1}\} = -[K_0]^{-1}\{\hat{\psi}_i\} \tag{AIE.9}$$

In order to carry out these iterative steps, numerical integration is required. First of all the evalanation of $\{\hat{\psi}_1\}$ from the initial stresses is required, and this requires integration over the elastic—plastic region only. The value of $\{\hat{\psi}_1\}$ is computed by carrying out the integration over the entire domain of the analysis. Since these kinds of accelerated steps unbalance the equilibrium, therefore it has to be re-established by finding the residual forces $\{\hat{\psi}_1\}$. Since the state of stress produced by the accelerated displacements is not in balance with the residual forces of the previous iteration, the new residual forces $\{\hat{\psi}_1\}$ of Eq. (AIE.9) must balance $\{\sigma_T\}$ and $\{R_1\}$. Here the acceleration scheme is needed to preserve equilibrium, which will eventually make the equivalent forces over the whole region unnecessary. This is achieved by applying a uniform acceleration, i.e. the same acceleration factor $\bar{A}$ to all displacements, found by averaging the individual factors α_i

$$\bar{A} = \frac{1}{n}\sum_{i=1}^{n} \alpha_i \tag{AIE.10}$$

The force–displacement equation is then written by multiplying both sides with the scalar quantity $\bar{A}$ without disturbing the equilibrium.

$$\bar{A}\{\Delta u_i\} = [K_0]^{-1}\bar{A}\{\psi_i\} \qquad \text{(AIE.11)}$$

Now to evaluate $\{\psi_{i+1}\}$}, the previous value of $\{\psi_1\}$ must be multiplied by $\bar{A}$, and the previously accelerated forces from the initial stresses $\{\sigma_o\}$ must be included such that

$$\{\psi_{i+1}\} = \int_V [B]^{T''}\{\sigma_0\}dV - (A-1)\{\psi_{\mathrm{i}-1}\} \qquad \text{(AIE.12)}$$

Superelement and Substructuring

In general terms, such formulations are described by the following:

$$\left[\begin{array}{c|c} K & K_R \\ \hline K_R^{T''} & K_{RR} \end{array}\right]\left\{\begin{array}{c} U \\ \hline U_R \end{array}\right\} = \left\{\begin{array}{c} F \\ \hline F_R \end{array}\right\} \qquad \text{(AIE.13)}$$

The subscript R represents reaction forces. The top half of Eq. (AIE.13) is used to solve for $\{U\}$:

$$\{U\} = -[K]^{-1}[K_R]\{U_R\} + [K]^{-1}\{F\} \qquad \text{(AIE.14)}$$

The reaction forces $\{F_R\}$ are computed from the bottom half of the equation as

$$\{F_R\} = [K_R]^{\mathrm{T}''}\{U\} + [K_{RR}]\{U_R\} \qquad \text{(AIE.15)}$$

Equation (AIE.14) must be in equilibrium with Eq. (AIE.15).

For large structures with complicated features, a substructure (super element) may be adopted on the lines suggested in Eq. (AIE.13). This super element may then be used as a reduced element from the collection of elements. If subscripts r and r' represent the retained and removed degrees of freedom of the equations partitioned into two groups, then the expressions in Eq. (AIE.13) can be written as

$$\left[\begin{array}{c|c} K_{rr} & K_{rr'} \\ \hline K_{r'r} & K_{r'r'} \end{array}\right]\left\{\begin{array}{c} U_r \\ \hline U_{r'} \end{array}\right\} = \left\{\begin{array}{c} F_r \\ \hline F_{r'} \end{array}\right\} \qquad \text{(AIE.16)}$$

Equation (AIE.16) when expanded assumes the following form:

$$\{F_r\} = [K_{rr}]\{U_r\} + [K_{rr'}]\{U_{r'}\} \qquad \text{(AIE.17)}$$

$$\{F_{r'}\} = [K_{r'r}]\{U_r\} + [K_{r'r'}]\{U_{r'}\} \qquad \text{(AIE.18)}$$

When a dynamic analysis is carried out, the subscript r (retained) represents the dynamic degrees of freedom.

When Eq. (AIE.18) is solved, the value of $U_{r'}$ is then written, similarly to Eq. (AIE.14).

$$\{U_{r'}\} = [K_{r'r'}]^{-1}\{F_{r'}\} - [K_{r'r'}]^{-1}[K_{r'r}]\{U_r\} \tag{AIE.19}$$

Substituting $\{U_{r'}\}$ into Eq. (AIE.17) gives

$$\Big[\big[[K_{rr}] - [K_{r'r}][K_{rr'}]^{-1}[K_{r'r}]\big]\Big]\{U_r\} = \Big[\{F_r\} - [K_{rr'}][K_{rr'}]^{-1}\{F_{r'}\}\Big] \tag{AIE.20}$$

or

$$[\bar{K}]\{\bar{U}\} = \{\bar{F}\} \tag{AIE.21}$$

where

$$[\bar{K}] = [K_{rr}] - [K_{rr'}][K_{r'r'}]^{-1}[K_{r'r}] \tag{AIE.22}$$

$$\{\bar{F}\} = \{F_r\} - [K_{rr'}][K_{r'r'}]^{-1}\{F_{r'}\} \tag{AIE.23}$$

$$\{\bar{U}\} = \{U_r\} \tag{AIE.24}$$

and $[\bar{K}]$ and $\{\bar{F}\}$ are generally known as the substructure stiffness matrix and load vector, respectively.

In the above equations, the load vector for the substructure is taken as a total load vector. The same derivation may be applied to any number of independent load vectors. For example, one may wish to apply thermal, pressure, gravity and other loading conditions in varying proportions. Expanding the right-hand sides of Eqs. (AIE.15) and (AIE.16) gives:

$$\{F_r\} = \sum_{i=1}^{n}\{F_{ri}\} \tag{AIE.25}$$

$$\{F_{r'}\} = \sum_{i=1}^{n}\{F_{r'i}\} \tag{AIE.26}$$

where

$[G_i]$ = matrix of shape function derivatives
$[\tau_i]$ = matrix for the current Cauchy stresses i.e. true stresses σ_i

$$[F_i^{nr}] = \int [B_i]^{T''}\{\sigma_i\}d(Vol) \tag{AIE.27}$$

The increment of large strain $[\Delta\varepsilon_n]$ is then written as

$$[\Delta\varepsilon_n] = [R_{1/2}]^{1T''}[\Delta\bar{\varepsilon}_n][R_{1/2}] \tag{AIE.28}$$

where

$[R_{1/2}]$ = rotational matrix from the polar decomposition of the deformation gradient at the midpoint configuration

$$[F_{1/2}] = [R_{1/2}][U_{1/2}] \tag{AIE.29}$$

where

$$[F_{1/2}] = [I] + \frac{\partial\{U_{1/2}\}}{\partial\{X\}} \tag{AIE.30}$$

Large deformation and large strains are obtained where these formulations are linked to the main analyses, static dynamic and blast type of loading are encountered.

Introduction Interface or Gap Elements in Current Computer Programs

The potential of interface elements in elevators and escalators is vital. They play a very important role in bringing about cPose relations between various mechanical components in static and dynamic conditions. Examples are cited in lifts/elevators escalators and travelators where components rolling over each others or some static and others moving over the static ones. Various contact or gap elements have been explained in various finite-element packages : Some of them are cited below:

(a) Haliquist et al. Method
(b) DELFT Interface Traction Type Element
(c) ANSYS contact Elements
(d) ABACUS Gap/contact Elements
(e) LS-DYNA Gap/contact Elements

With program ISOPAR Haliquist et al. method has been adopted as this method clearly establishes the contact elements which suite the scope of the book. The method is described below:

where n = the number of independent load *vectors*.

Substituting into Eq. (AIE.12)

$$\{\bar{F}\} = \sum_{i=1}^{n}\{F_{rr'}\} - [K_{rr'}][K_{r'r'}]^{-1}\sum_{i=1}^{n}\{F_{r'i}\} \tag{AIE.31}$$

Large Strain

When the strains in a material exceed more than a few percent, the changing geometry due to this deformation can no longer be neglected. Analyses which include this effect are called large strain, or finite strain, analyses. A large strain analysis is performed in a static and dynamic transient analysis.

The theory of large strain computations can be addressed by defining a few basic physical quantities (motion and deformation) and the corresponding mathematical relationship. The applied loads acting on a body make it move from one position to another. This motion can be defined by studying a position vector in the 'deformed' and 'undeformed' configurations. Say the position vectors in the 'deformed' and 'undeformed' state are represented by $\{x\}$ and $\{X\}$, respectively, then the motion (displacement) vector $\{u\}$ is computed by

$$\{u\} = \{x\} - \{X\} \tag{AIE.32}$$

The deformation gradient $[F_d]$ is defined as

$$[F_d] = \frac{\partial\{x\}}{\partial\{X\}} + [\mathrm{I}] = [R][u] \tag{AIE.33}$$

where [I] = identity matrix.

The volume change at a point

$$\frac{dV}{dV_0} = \det[F] \tag{AIE.34}$$

Where

V_0 = Original volume
V = current volume
[u] = right stretch shape change matrix

Largestrain ε is written as

$$\{\varepsilon\} = \ln[\mathrm{u}] = \sum \ln(\lambda_{\mathrm{i}})\{e_i\}\{e_i\}^{\mathrm{T}''} \tag{AIE.35}$$

where

λ_i = eigenvalues of [u]
e_i= eigenvalues of [u]
R = rotation matrix ($[\mathrm{R}]^{\mathrm{T}''}[\mathrm{R}] = [\mathrm{I}]$)

Element formulation using Lagrangian formulation technique

$$[\,\mathrm{K}_j]\Delta u_j = [F^a] - [F_{\mathrm{i}}^{\mathrm{nr}}] \tag{AIE.36}$$

where

$$
\begin{aligned}
[\mathrm{K_i}] &= \text{tangent matrix} = [\mathrm{K_i}] + [S_i] \\
[\mathrm{K_i}] &= \text{usual stiffness matrix} \\
&= \int [B_i]^{\mathrm{T}''} [D_i][B_i]\mathrm{d(vol)}
\end{aligned}
\qquad \text{(AIE.37)}
$$

$$
\begin{aligned}
[\mathrm{S_i}] &= \text{stress stiffness contribution or geometric stiffness} \\
&= \int [\mathrm{G_i}]^{\mathrm{T}} [\tau_{\mathrm{i}}][\mathrm{G_i}]\, \mathrm{d(vol)}
\end{aligned}
\qquad \text{(AIE.38)}
$$

Haliquist et *al.* method

Haliquist *et al.* developed a useful concept of master and slave nodes sliding on each other. As shown in Fig. AIE.1 slave nodes are constrained to slide on master segments after impact occurs and must remain on a master segment until a tensile interface force develops. The zone in which a slave segment exists is called a slave zone. A separation between the slave and the master line is known. as void. The following basic principles apply at the interface:

(a) update the location of each slave node by finding its closest master node or the one on which it lies,
(b) for each master segment, find out the first slave zone that overlaps,
(c) show the existence of the tensile interface force.

Constraints are imposed on global equations by a transformation of the nodal displacement components of the slave nodes along the contact interface. Such a transformation of the displacement components of the slave nodes will eliminate their normal degrees of freedom and distribute their normal force components to the nearby master nodes. This is done using explicit time integration, as described in the finite-element solution procedures. Thereafter impact and release conditions are imposed. The slave and master nodes are shown in Fig. AIE.2. Hallquist et al. gave a useful demonstration of the identification of the contact point which is the point on the master segment to the slave node n_s and which finally becomes non-trivial during the execution of the analyses. When the master segment I is given the parametric representation and t is the position vector drawn to the slave node n_s, the contact point co-ordinate must satisfy the following equations:

$$
\begin{aligned}
\frac{\partial \hat{r}}{\partial \xi}(\xi_c, \eta_c) \times [\hat{i} - \hat{r}(\xi_c, \eta_c)] &= 0 \\
\frac{\partial \hat{r}}{\partial \eta}(\xi_c, \eta_c) \times [\hat{t} - \hat{r}(\xi_c, \eta_c)] &= 0
\end{aligned}
\qquad \text{(AIE.39)}
$$

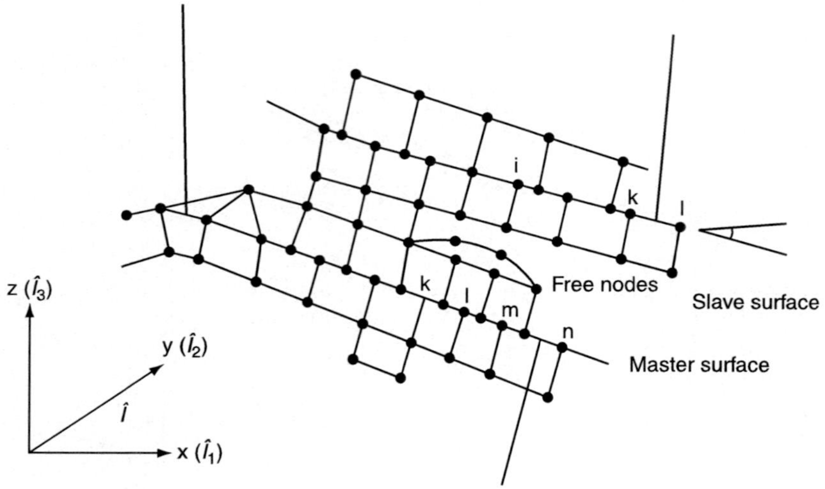

Fig. AIE.5 Hallquist contact method (modified by Bangash)

where (ξ_c, η_c) are the co-ordinates on the master surface segment S_i. Where penetration through the master segment S_i occurs, the slave node n_s (containing its contact point) can be identified using the interface vector f_s

$$f_s = -lk_i n_i \qquad \text{if } l<0 \tag{AIE.40}$$

to the degree of freedom corresponding to n_s, and

$$f_m^i = N_i(\xi_c \eta_c) f_s \qquad \text{if } l<0 \tag{AIE.41}$$

where

$$l = \hat{n}_i[\hat{t} - \hat{r}(\xi_c, \eta_c)] < 0 \tag{AIE.42}$$

A unit normal

$$\hat{n}_i = \hat{n}_i(\xi_c, \eta_c); \quad \hat{t}_i = \hat{n}_i \sum_{j=1}^{n} N_j (F_1)^j(t) \tag{AIE.43}$$

$$k_i = f_{si} K_i A_i^2 / V_i \tag{AIE.44}$$

where

$(F_I)^j(t)$ = impact at the jth node
K = stiffness factor

K_i, V_i, A_i = bulk modulus, volume and face area, respectively
f_{si} = scale factor normally defaulted to 0.10
$N_i = \frac{1}{4}(1 + \xi\xi_i)(1 + \eta\eta_i)$ for a 4-node linear surface

Bangash extended this useful analysis for other shape functions, such as N_i for 8-noded and 12-noded elements. On the basis of this theory and owing to the non-availability of the original computer source, a new sub-program CONTACT was writ-ten in association with the program ISOPAR. The subprogram CONTACT is in three dimensions

Gap element

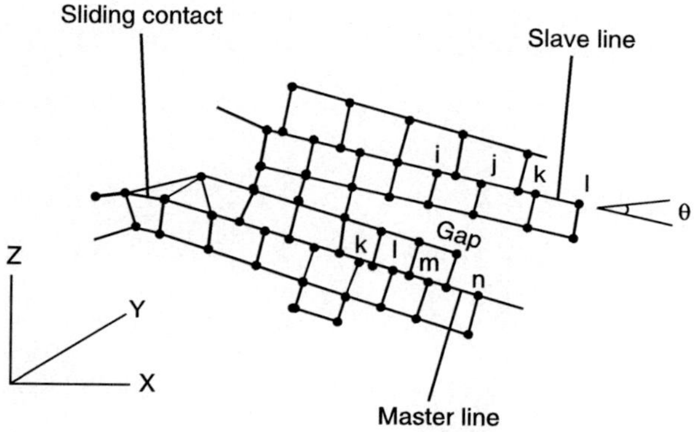

$$\left\{\hat{\vec{F}}_N\right\} = [K_{\gamma\gamma}]_i\{U_\gamma\}_i = [\Sigma K]\{\Sigma\Delta_i, \Delta_j \ldots\} = \left\{\hat{\vec{F}}_{i,j}, \cdots\right\} + \left\{\pm\mu\hat{\vec{F}}_n \ldots \pm k\Delta_i\right\}$$

$$\Delta_{sl} = \text{distance of sliding} \qquad \text{(AIE.45)}$$

$$= (\Delta_j - \Delta_i) - \frac{\mu\left|\hat{\vec{F}}_N\right|}{[K_{\gamma\gamma}]}$$

$$\mu = \text{friction} \qquad \text{(AIE.46)}$$

$$\{\hat{\vec{F}}_{SN}\} \le \mu\{\hat{\vec{F}}_N\} \text{ no sliding}$$

$$\ge \mu\{\hat{\vec{F}}_N\} \text{ sliding}$$

$$= 0 \text{ contact broken} \qquad \text{(AIE.47)}$$

$$\theta = \cos^{-1}\frac{X}{r} \text{ or } \sin^{-1}\frac{Y}{r}$$

Appendix F
Simplified Dynamic Analysis of Towers, Chimneys And Frames As Auxiliary Structures

AIF.1 The Determination of the Fundamental Frequency of a Tower Structure

A schematic diagram of the tower is shown in the Figure AIF.1. It consists of a circular tapering, hollow concrete tower L_1 in height, surmounted by a steel mast L_2 high. 'Tower' will be used to refer to the concrete structure and the mass of the steel must lumped at its end.

The tower tapers according to a linear relationship between diameter D, and x, measured from the ground:

$$D = D_o - D_s x \tag{AIF.1}$$

And the moment of inertia, I, varies according to a third-order polynomial in x:

$$I(x) = I_o - I_1 x + I_2 x^2 - I_3 x^3 \tag{AIF.2}$$

The mass per unit length, $W(x)$, over the tower height varies according to a linear relationship:

$$W(x) = Wf - Zx \tag{AIF.3}$$

Fig. AIF.1 The tower structure

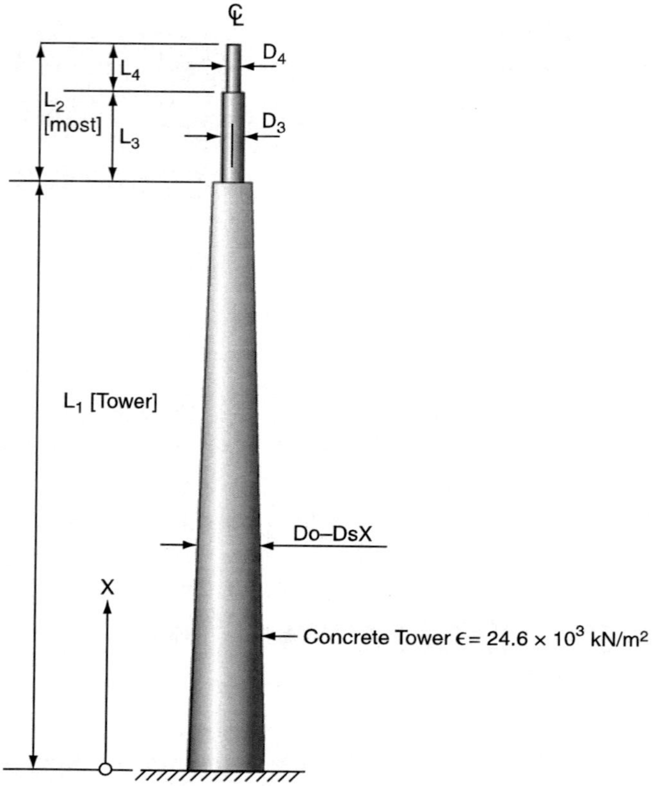

Table AIF.1 Characteristics of the tower

Length (m)	Mass (weight) and mass (wt) coeffs kg (KN)
$L_1 = 275.0$	$W_2 = 71121.0\ (697.7)$
$L_2 = 60.7$	$W_3 = 46206.0\ (453.3)$
$L_3 = 35.1$	$W_4 = 24915.0\ (244.4)$
$L_4 = 25.6$	$W_f = 97986.9\ \mathrm{m}^{-1}\ (961.3)\ \mathrm{m}^{-1}$
	$Z = 299.237\mathrm{m}^{-2}\ (2.9)\ \mathrm{m}^{-2}$
Diameters and coeffin (m)	Second moments of area and coeffs (m^4)
$D_3 = 3.60$	$I_{00} = 3044.86$
$D_4 = 2.27$	$I_1 = 38.7142\ \mathrm{m}^{-1}$
$D_{00} = 24.40$	$I_2 = 0.172539\ \mathrm{m}^{-2}$
$D_s = 0.067\ \mathrm{m}^{-1}$	$I_s = 0.00026\ \mathrm{m}^{-3}$

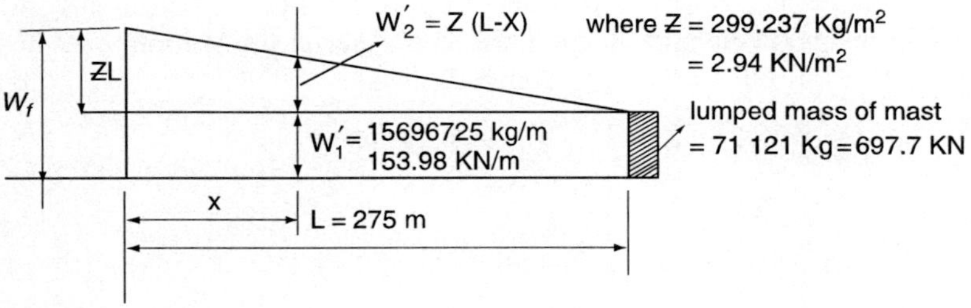

Fig. AIF.1(a) Base Tower parameters-I

The mass per unit length along the tower can be divided into a uniformly distributed load W_1' and a triangular distribution of W_2' as shown above.

Lengths of the tower are lumped into masses

$$MA = \int_{x_1}^{x_2} (WF - Zx)dx$$
$$= (x_2 - x_1)\left[WF - \frac{Z}{2}(x_2 + x_1)\right] \quad \text{(AIF.4)}$$

Where x_1 and x_2 are the cords of the start and end of the length under consideration.

The bending moments due to self-weight can be shown to be

$$M_0(x) = \frac{W_1}{2}(L-x)^2 + \frac{Z}{6}(L-x)^3 + MM(L-x) \quad \text{(AIF.5)}$$

The second moment of area of the tower varies according to

$$I_0(x) = I_{00} - I_1x + I_2x^2 - I_3x^3 \quad \text{(AIF.6)}$$

The bending moment due to a unit load at the lumped mass distance Y from the free end is

$$M_1(x) = (L - x - Y) \quad \text{(AIF.7)}$$

In all cases x is measured from the support.

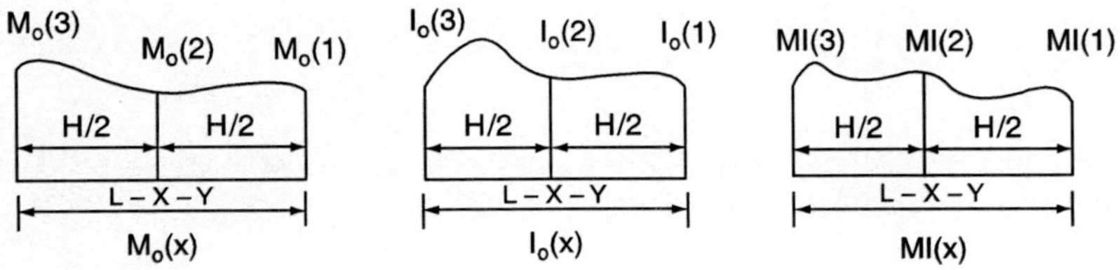

The static deflection of the contilever (tower) are computed by a combination of Simpson's Rule and the area–moment–inertia procedure.

The deflection DEL under any lumped mass

$$\begin{aligned}\text{DEL} &= \frac{1}{E}\int \frac{M_0(x).M_1(x)}{I_0(x)}dx \\ &= \frac{H}{6E}[M_0(1)M_1(1) \div I_0(1) + 4M_0(2)M_1(2) \div I_0(2) + M_0(3)M_1(3) \div I_0(3)]\end{aligned} \quad \text{(AIF.8)}$$

E being the modulus of elasticity of concrete.

The fundamental frequency of the tower is computed by the Rayleigh method. Hence the dynamic deflection is assumed, not to differ appreciably from the static deflection 'DEL'

$$\text{Max. strain energy of Oscillating masses} = \text{Potential energy}$$

$$= \sum \frac{\text{MA} \times \text{DEL}}{2} \quad \text{(AIF.9)}$$

$$\text{max. KE } \mathit{of} \text{ masses} = \frac{\omega^2}{2g}\sum MA \times \text{DEL}^2$$

Equating SE and KE one gets

$$\begin{aligned}\sum \text{MA} \times DEL &= \tfrac{\omega^2}{2g}\sum \text{MA} \times \text{DEL}^2 \\ \omega^2 &= g\frac{MA \times \text{DEL}}{MA \times \text{DEL}^2} = g \times \frac{\text{NUM}}{\text{DEN}}\end{aligned} \quad \text{(AIF.10)}$$

$$\therefore \omega = \left[\text{g} \times \frac{\text{NUM}}{\text{DEN}}\right]^2 \quad \text{(AIF.11)}$$

$$\text{Frequency } f = \frac{\omega}{2\pi} \quad \text{(AIF.12)}$$

g = acceleration due to gravity.

A Typical Case Study

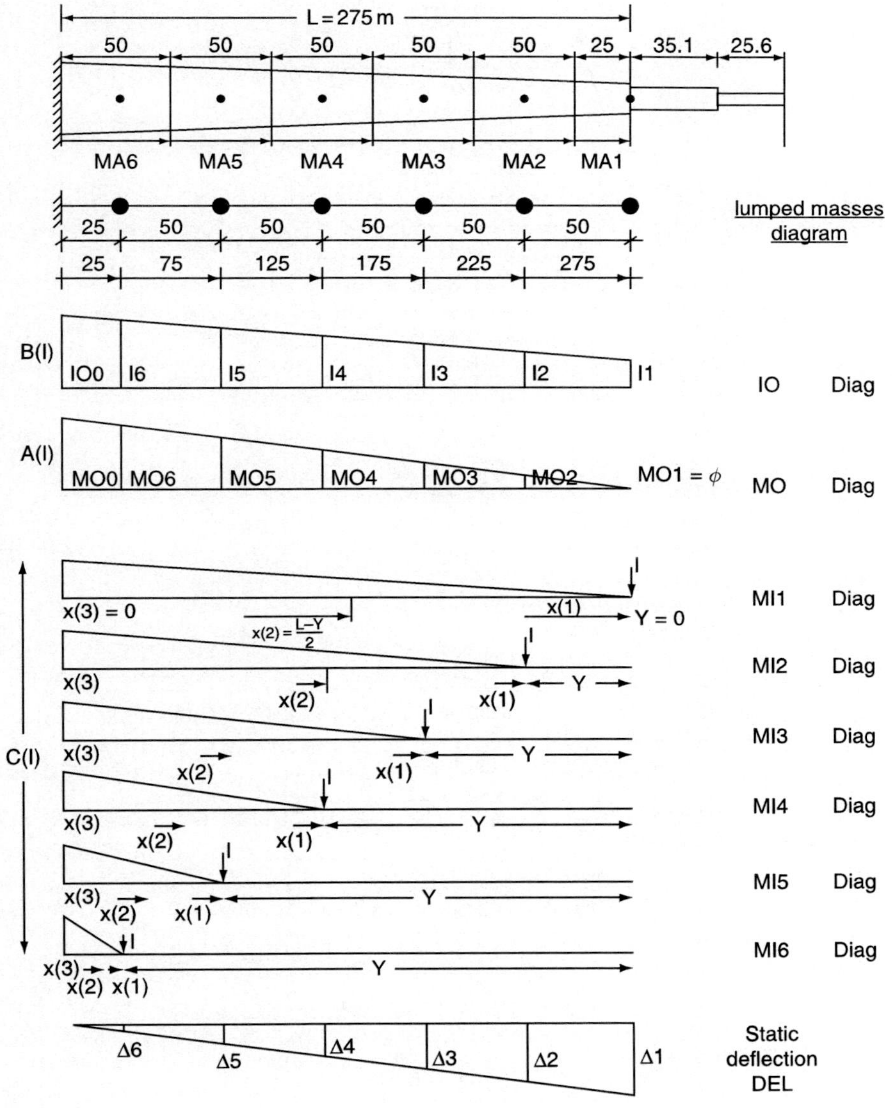

$$\Delta = \frac{1}{E}\int^{\text{Tower}} \frac{\text{MO MI } dx}{\text{IO}}$$

The Flow Chart

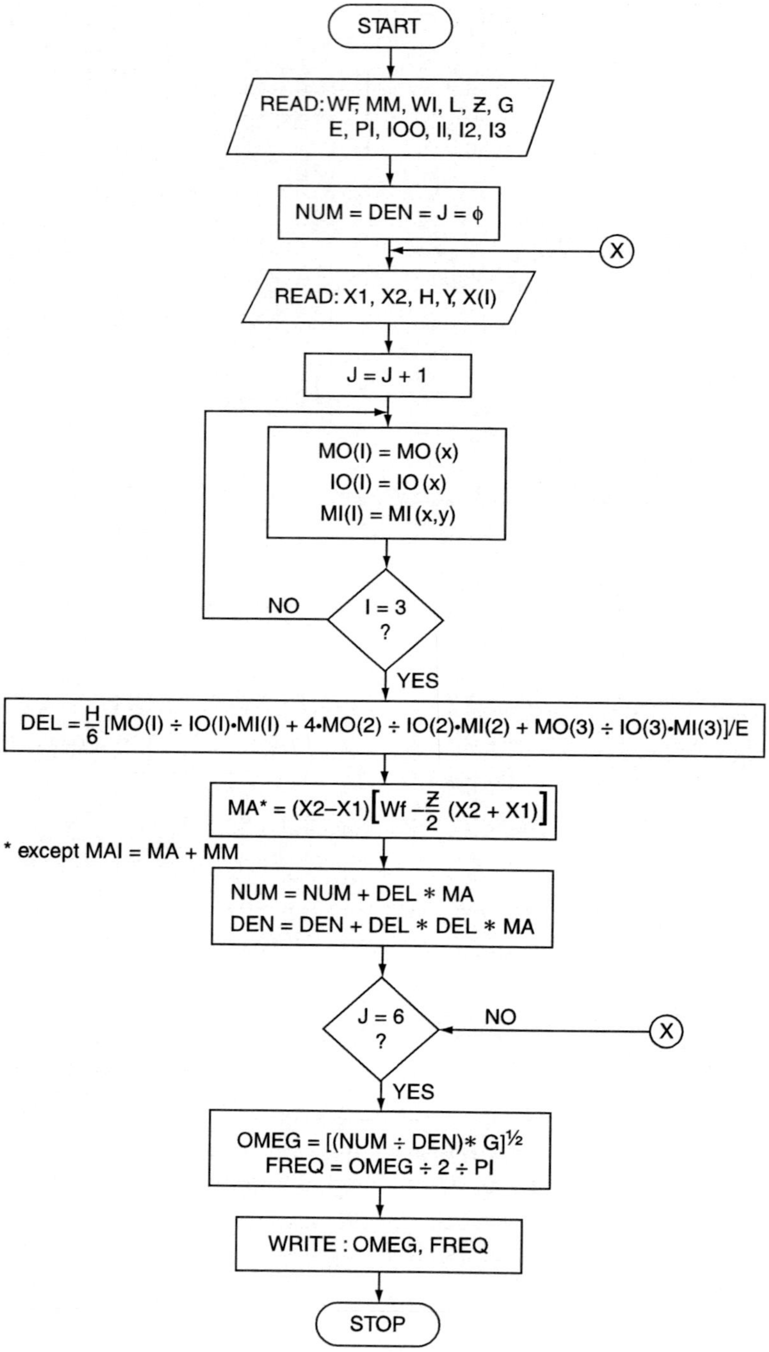

Output

The fundamental circular frequency of the structure (Tower 1) is $1.26828 \simeq 1.3$ rad per sec.

Comments and Conclusion

The Rayleigh method for natural frequencies was chosen for the computation of the fundamental frequency of a tower structure because of its considerable accuracy and ease. However, this method can be modified for the computation of a few of the lower modes frequencies.

Other methods have the advantage of computing the characteristic shapes of the various modes of vibration accurately, but are exceedingly cumbersome and require high computational times.

Since, one is only interested in the fundamental frequency, there is no justification for the adaptation of other methods.

The fundamental mode of vibration is that in which the tower and mast are in phase. In this exercise the mass of the mast, being very small compared with the mass of the tower, was lumped at the free end of the tower.

The value at $\omega = 1.3$ rad per sec is the same as that obtained by the use of Lagrange Equation (Ref: Vibration reduction in Tower Structures, by K.C. Johns & C.L. Kink).

Therefore, it can be concluded that the Rayleigh method is a very accurate method for the determination of the fundamental frequency. Other methods stated in the text can also be applied. They can be expensive and yet very elaborate result wise.

AIF.2 Cooling Tower and Cable net Tower shell Shapes for Nuclear Power Stations Hyperboloid of Revolution

When the rotating curve is a *hyperbola,* the generated surface is known as a *hyperboloid of revolution.* A single-sheet *hyperboloid of revolution* is so termed as it is a single-piece surface.

Two types of hyperboloid of revolution – i.e. a natural, draught cooling tower cable net tower shells – will be discussed below.

Natural draught cooling tower

For large power station blocks that are completely dependent on cooling tower natural draught coolers are the only types being built at present. The capital exjditure involved is relatively low and they require neither energy nor maintenance mechanical equipment. They have a correspondingly high level of reliability.

If the ground conditions are adequate, an annular foundation is usually laid individual foundations, e.g. pile foundations, are required, the number of pain columns should be kept to a minimum. In this case, it is also advisable to prest the lower shell periphery. The column framework provides the free cross-sec necessary for air intake in the exterior wall and transmits the reaction forces of the cooler shell into the foundation. The construction and installation of the fraction work require great care in order to avoid unnecessary bending stresses in the section. The column cross-section is rectangular or circular: the shape may be influence aerodynamic requirements.

For the large cooling towers built in the last few decades, the thin reinforced crete shell, reinforced in two axes on both surfaces, has proved eminently suitatable. For structural, thermodynamic and flow optimization reasons rotationally hyperbolic shapes are desirable, and occasionally also cylinders. These shapes are at same time aesthetically pleasing because they make the gigantic dimensions of the structures seem tolerable.

The cooler shell can be very thin. The minimum wall thickness in the cer and upper portions in almost all towers is 14 cm; in other words, in relation to diameter it is thinner than an eggshell. In the lower portion, the shell becomes thickness until at the point of transition to the bottom edge element of the shell it is at 60–80 cm thick.

At the top of the shell, there is a study ring (top edge element) which counter the tendency of the shell to ovalize under high wind loads, and thus also yhe risk distortions without strain in the shell. This ring can also serve as a surrounding gas if aviation obstruction lighting is required on the cooling tower for safety reason this case, ladders or staircases must also be provided.

Cable Net Cooling Towers

They have considerable advantages with respect to safety during construction in service as compared to a concrete tower. There is almost no limit to its size is most favourable in the case of bad soil conditions and earthquakes. Shapes be obtained with different materials, such as fabrics – which have been discussed. 3′, of the inclined straight line 3–3′, along two circular paths with uniform angular speed (Fig. AIF 2.1(a)), such that the generating line 3–3′, when produced, never cuts the line (O_2 O_1) passing through the centres of the circular paths.

Consider Figure AIF 2.1(b) in which the planes P_1 and P_2 may be parallel, but have smooth closed boundary curves.

In Figure AIF 2.1(b) a_1a_2, b_1b_2, etc., are straight lines, and the envelope due movement of these straight lines can be seen. Now,

1. If P_1 and P_2 are circular discs and parallel, and if $O_1a_1a_2O_2OO_1$ is a twisted (Hypar, Type II), i.e. the points a' on a_1a_2 and O on O_1O_2, never coincide, the

generated surface as a *hyperboloid of revolution* due to the revolution line a_1a_2 about O_1O_2.

2. If P_1 and P_2 are parallel circular discs, and if $O_1A_1A_2O_2O_1$ is a flat plan A_1A_2 forming a hyperbolic arc, we get a *hyperboloid of revolution*.
3. If the angles $A_1O_1a_1$ and $A_2O_2A_2$ lie in the parallel planes P_1 and P_2, arc parallel and are equal in magnitude, i.e. $\theta_1 = \theta_1$, then $O_1a_1a_2O_2O_1$ become plane consisting of two triangles O_1a_1O and O_2a_2O meeting at O on the O_1O_2. We now have a pair of conical surfaces meeting at a common verte: boundary curve A_1A_2 is transferred into a broken line A_1O–OA_2.

Membrane or cable net skin replaces the reinforced concrete shell of conven towers. A detailed analysis can be studied under fabric or net structures in thickness.

AIF.3 Hyperboloid of Revolution of One Sheet

The hyperboloid of revolution of one sheet has a very graceful appearance and profitably exploited for structural purposes such as a structure for the cooling towards great advantage emerges from the facts that the surface is generated by two factors of intersecting lines and the formwork can be achieved by straight boards with only slightly over their lengths. It is also interesting to note that the interset of grid straight lines form rhombuses of intersection. The shell surface can be built of precast rhombic elements, which can be repeated along the con circumference at a fixed height. Considerable utilization of this surface is being in the construction of cooling towers.

The surface is the locus of the equation

$$\frac{x^2}{C^2} + \frac{y^2}{A^2} - \frac{z^2}{B^2} = 1 \qquad \text{(AIF.13)}$$

in three-dimensional Cartesian coordinates. It is quadratic because it is the of a second-degree equation and is doubly ruled; each point lies on more that straight line of the surface.

The particular case of hyperboloid of revolution with $C = A$ in Eq. (AIF.13) discussed here for the surface and is defined by:

$$\frac{(x^2 + y^2)}{A^2} - \frac{z^2}{B^2} = 1 \qquad \text{(AIF.14)}$$

In Figure AIF.2.1(a) all sections (in planes) perpendicular to the axis of revolution of the z axis) are circles. The surface is also the locus described by a hyperbola revolation around its conjugate axis. Furthermore, the surface is also the locus described straight generatrix revolving around a non-parallel axis (Fig. AIF.2.1(b)). The project of this line on a parallel coaxial plane is one of the asmyptotes of the hyperbolic sections contained on the plane.

Consider in Figure AIF.2 an intersection by vertical plane $y = A$ which being introduced in Eq. (AIF.14) gives:

$$\frac{x^2}{A^2} = \frac{z^2}{B^2} \tag{AIF.15}$$

Therefore, the curve of intersection consists of a pair of straight lines have equations:

$$x = \pm \frac{A}{B} z \tag{AIF.16}$$

and slope, tan $D = B/A$.

Since the hyperboloid is a surface of revolution, the tangent plane to any point of the waist circle will give a similar pair of straight lines on the hyperboloid so that there exist two families of straight lines each of which covers the completely.

Referring to Figures AIF.3 and the following equations may be establis

$$x = A(\sec\ T)\ \cos\theta \tag{AIF.17}$$

$$y = A(\sec\ T)\ \sin\theta \tag{AIF.18}$$

$$z = B(\tan\ T) \tag{AIF.19}$$

and

$$R_1 = -\frac{D^2 3}{AB} \tag{AIF.20}$$

$$R_2 = \frac{AD}{B} \tag{AIF.21}$$

$$\text{Sin}\phi = \frac{B(\sec T)}{D} \tag{AIF.22}$$

$$\frac{d\phi}{dT} = \frac{AB(\sec T)}{D^2 2} \tag{AIF.23}$$

where

$$D = (A^2(\tan T)^2 + B^2(\sec T))^2 0.5 \tag{AIF.24}$$

Considering the unit weight of the dead load

$$x = 0,\ y = \sin\phi \text{ and } z = -\cos\phi$$

Substituting the auxiliary variables as:

$$G = \frac{A(\sec T)^2}{D} N_\phi \tag{AIF.25}$$

and

$$H = S(\sec T)^2 \tag{AIF.26}$$

(where $S = N_{\theta\phi} = N_{\phi\theta}$).

The self-weight stresses are given by:

$$\frac{\partial G}{\partial T} = -\frac{A}{B}\frac{\sqrt{B^2 + A^2 \sin^2 T}}{\cos^3 T} \tag{AIF.27}$$

$$H = 0 \tag{AIF.28}$$

$$N_\theta = \frac{A^2}{D^2} N_\phi - \frac{A^2}{B}\tan(T) \tag{AIF.29}$$

Boundary conditions

The planes of the top and bottom in Figure 6.48 are to be denoted by $T = T_1$ and $T = T_2$. Since the top is a free edge the stresses vanish there and therefore:

$$G = 0, \text{ i.e. } N_\phi = 0 \text{ and } H = 0, \text{ i.e. } N_{\theta\phi} = 0$$

for all values of 0 on $T = T_1$. (These conditions are sufficient for integration.)

By numerical integration with the boundary conditions the meridian force N_ϕ and hoop force N_θ can be obtained at all sections of the shell, as illustrated in the results of the computer program for NIZAM–4 cooling tower. Figure AIF.2 and its calculations indicate briefly the generalized stresses due to wind pressure at different heights when the cooling tower shapes are asymmetrical. Table AIF.1 gives loading systems and stresses per unit circumference of the tower shell surface.

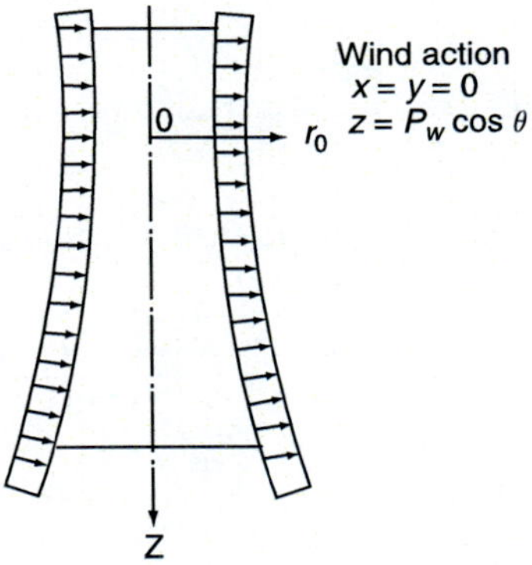

Fig. AIF.2 Asymmetrical membrane state of stress

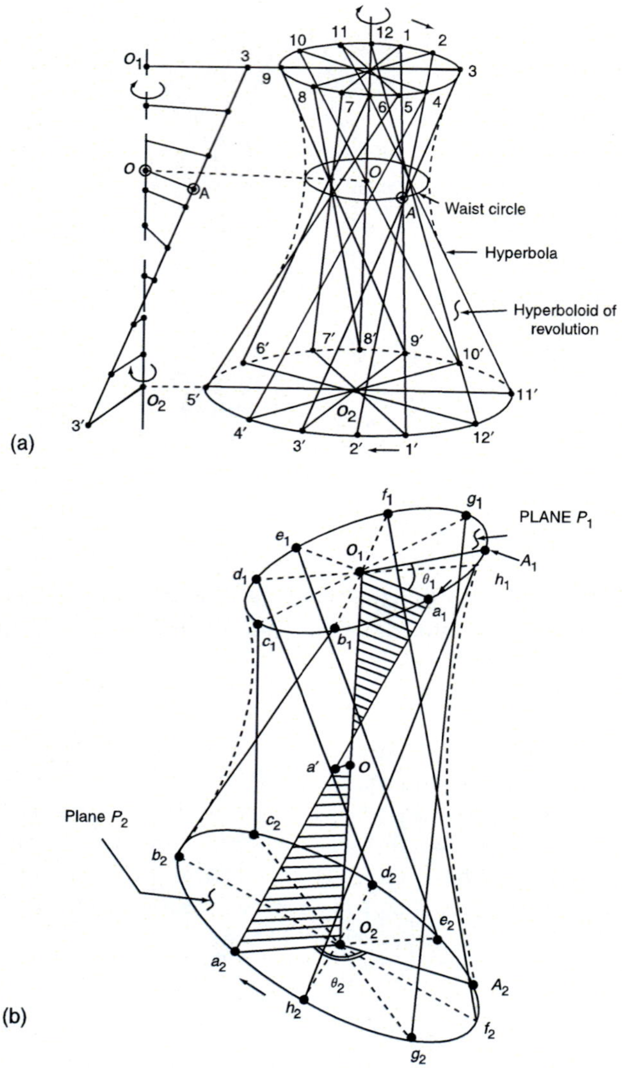

Fig. AIF.2b Generalized forms of rotational surface

Shell of revolution (refering to Fig. AIF.2)

$$N_\phi = \frac{P_W a \tan A}{6\rho^2 \sin\phi}\left[\frac{\tan^2 A - 2}{3\tan^2 A}(\rho^3 - \rho_0^3) + \rho_0 - \rho - \eta_0\rho_0(\eta - \eta_0) + \eta \ln\frac{\eta + \rho}{\eta_0 + \rho_0}\right]\cos\theta \qquad \text{(AIF.30)}$$

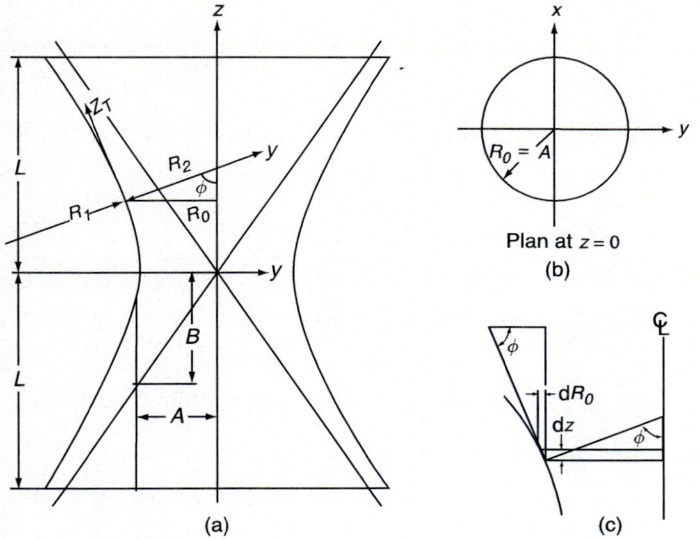

Fig. AIF 2.3 Cooling towers of equal radii

$$N_{\phi\theta} = N_{\theta\phi} = \frac{P_W a}{2} \quad \frac{\tan A}{\rho^3}\Big[\eta_0\rho_0 - \eta\rho - \eta\rho^2(\rho - \rho_0)$$

$$-\frac{\tan^2 A - 2}{3\tan^2 A}\eta(\rho^3 - \rho_0^3) - \ln\frac{\eta + \rho}{\eta_0 + \rho_0}\Big] \sin\theta \qquad \text{(AIF.31)}$$

$$N_\theta = -N_\phi \frac{r_2}{r_1} - P_w r_2 \cos\theta \qquad \text{(AIF.32)}$$

where $\rho = \frac{r_0}{a}, \eta = \frac{z}{b}$ ρ_0, η_0 value at upper edge ($\phi = \phi_0$)

Steel Frame Supporting a Rotating Machinery

Problem Example

A frame supporting a rotating machinery in a nuclear laboratory exerts a typical horizontal force $F(t)$ at the girder level as shown in Fig. AIF.3. In a normal practice the value of $F(t)$ is $F(t) = 200 \sin (5.3)t$ (*KN*) evaluation dynamic shear and bending moment an essentially demanded.

Data

$L = 15\,m$, $E = 200$ KN/m^2, I = 0.02m^4
ξ = damping = 0.05
The girder is rigid

Table AIF.1 Various loading systems on cooling towers and shell forces

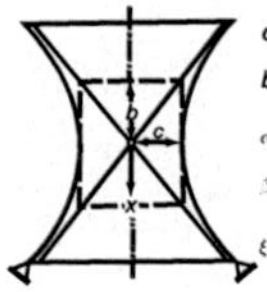

c throat radious

b imaginary hyperbolic distance

$\left.\begin{array}{l}\alpha = \frac{c}{b} \\ \beta = \sqrt{\xi^2+\xi}\end{array}\right\}$ dimensionless parameters

$\xi = \frac{x}{c}$ dimensionless coordinate

	Shell dimensions	
	$a = c\sqrt{1+\alpha^2\xi^2}$	$\bar{\xi} = \frac{\xi - \xi_0}{\sqrt{1-\beta^2\xi_0^2}}$
	$r_\phi = c\sqrt{1+\beta^2\xi^2}$	$r_{\phi 0} = c\sqrt{1+\beta^2\xi_0^2}$
	$r_\theta = \frac{b^2}{c}\sqrt{(1-\beta^2\xi^2)^3}$	$\chi_0 = \sqrt[4]{3(1-\mu^2)\frac{r_{\phi 0}^2}{t^2}}$

System loading	Imposed load	Shell forces	Deformation
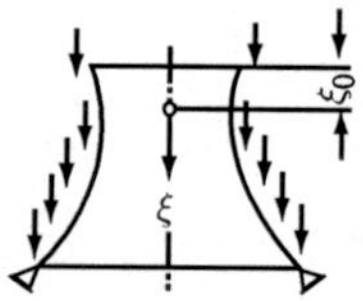	$p_x = p_E\sqrt{\frac{1+\alpha^2\xi^2}{1+\beta^2\xi^2}}$ $p_z = p_E\frac{\alpha^2\xi}{\sqrt{1+\beta^2\xi^2}}$	$N_\theta = \frac{p_E c\sqrt{1+\beta^2\xi^2}}{2\;\;1+\alpha^2\xi^2}\Big[\xi\sqrt{1+\beta^2\xi^2} - \xi_0\sqrt{1+\beta^2\xi_0^2} + \frac{1}{\beta}(\text{arsh}\,\beta\xi - \text{arsh}\,\beta\xi_0)\Big]$ $N_\phi = N_\theta\frac{\alpha^2}{1+\beta^2\xi^2} - p_E c\gamma^2\xi,\ T = 0$	$u' = \frac{1}{Et}(N_\theta - \mu N_\phi)$ $\bar{\omega} = -\frac{c}{Et}\sqrt{1+\alpha^2\xi^2}(N_\phi - \mu N_\theta)$ $\chi = \frac{\alpha^2\xi}{Et}\Big[\frac{1+\mu}{\sqrt{1+\alpha^2\xi^2}}(N_\theta - N_\phi) - \frac{d}{d\xi}(N_\phi - \mu N_\theta)\Big]$
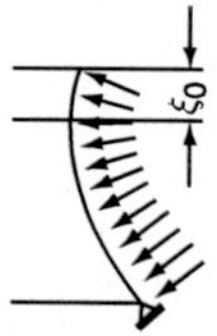	$p_z = p$	$N_\theta = -\frac{pc\alpha^2\sqrt{1+\beta^2\xi^2}}{2\;\;1+\alpha^2\xi^2}[\xi^2 - \xi_0^2]\ T = 0$ $N_\phi = N_\theta\frac{\alpha^2}{1+\beta^2\xi^2} - pc\sqrt{1+\beta^2\xi^2}$	
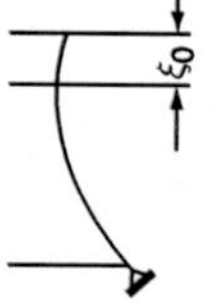	$p_z = \gamma c(\xi - \xi_0)$	$N_\theta = \gamma c^2 - \frac{\alpha^2\sqrt{1+\beta^2\xi^2}}{1+\alpha^2\xi^2}\Big[\frac{1}{3}\xi^3 - \frac{1}{2}\xi_0\xi^2 + \frac{1}{6}\xi_0^3\Big]$ $N_\phi = N_\theta\frac{\alpha^2}{1+\beta^2\xi^2} - \gamma c^2\sqrt{1+\beta^2\xi^2}(\xi - \xi_0)\ T = 0$	

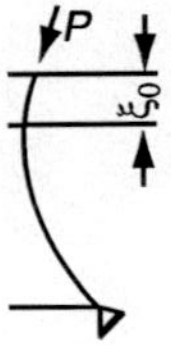	Concentrated load P inclined	$N_\theta = -P\dfrac{1+\alpha^2\xi_0^2}{1+\alpha^2\xi^2}\sqrt{\dfrac{1+\beta^2\xi^2}{1+\beta^2\xi_0^2}}$ $N_\phi = N_\theta\dfrac{\alpha^2}{1+\beta^2\xi^2},\ T=0$	$u' = \dfrac{1}{Et}(N_\theta - \mu N_\phi)$ $\bar{\omega} = -\dfrac{c}{Et}\sqrt{1+\alpha^2\xi^2}\times(N_\phi - \mu N_\theta)$ $\chi = \frac{\alpha^2\xi}{Et}\left[\frac{1+\mu}{\sqrt{1+\alpha^2\xi^2}}(N_\theta - N_\phi) - \frac{\mathrm{d}}{\mathrm{d}\xi}(N_{\phi-\mu N_\theta})\right]$
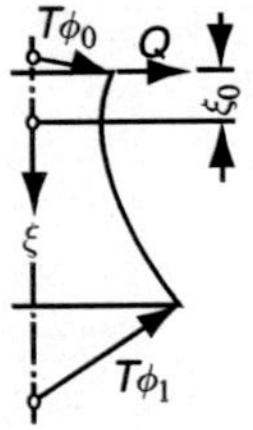	Horizontal shear load Q at top	$M_0 = Q\sqrt{\dfrac{1+\alpha^2\xi^2}{1+\beta^2\xi^2}\dfrac{r_{\phi_0}}{x_0}}e^{-x_0\bar{\xi}}\sin x_0\bar{\xi}$ $M_\phi = Q\dfrac{\alpha^2\xi}{\sqrt{1+\beta^2\xi^2}}\dfrac{r_{\phi_0}^2}{x_0^2}e^{-x_0\xi}\times[\cos\gamma 0\bar{\xi} + \sin\gamma 0\bar{\xi}] - \mu M_\theta$ $Q_\theta = Q\sqrt{\dfrac{1+\alpha^2\xi^2}{1+\beta^2\xi^2}}e^{-x\bar{\xi}}\times[\cos x_0\bar{\xi} - \sin x_0\bar{\xi}]$ $N_\theta = -Q_\theta\dfrac{\alpha^2\xi}{\sqrt{1+\alpha^2\xi^2}}\quad T=0$ $N_\phi = 2Q\sqrt{1+\alpha^2\xi^2}\dfrac{x_0}{r_{\phi_0}}e^{-x_0\bar{\xi}}\cos x_0\bar{\xi}$	$u' = -\dfrac{\mu Q}{2Bc}\dfrac{(1+\alpha^2\xi^2)^{\frac{3}{2}}}{(1+\beta^2\xi^2)^2}\dfrac{r_{\phi_0}^3}{x_0^3}$ $\times e^{-x_0\bar{\xi}}\cos x_0\bar{\xi}$ $\bar{\omega} = -\dfrac{Q}{2B}\dfrac{1+\alpha^2\xi^2}{1+\beta^2\xi^2}\dfrac{r_{\phi_0}^3}{x_0^2}$ $\times e^{-x_0\bar{\xi}}\cos x_0\bar{\xi}$ $\chi = \dfrac{Q}{2B}\sqrt{\dfrac{1+\alpha^2\xi^2}{1+\beta^2\xi^2}\dfrac{r_{\phi_0}^2}{x_0}}$ $\times e^{-x_0\bar{\xi}}[\cos x_0\bar{\xi} + \sin x_0\bar{\xi}]$
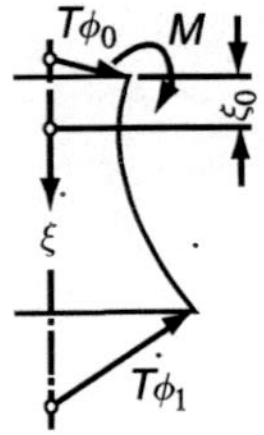	Moment M at top	$M_\theta = Me^{-x_0\bar{\xi}}[\cos x_0\bar{\xi} + \sin x_0\bar{\xi}]$ $M_\phi = \dfrac{M}{c}\frac{\alpha^2\xi}{\sqrt{(1+\alpha^2\xi^2)(1+\beta^2\xi^2)}}e^{-x_0\bar{\xi}}\times\cos x_0\bar{\xi} - \mu M_\theta$ $Q_\theta = -2M\dfrac{x_0}{r_{\phi_0}}e^{-x_0\bar{\xi}}\sin x_0\bar{\xi}$ $N_\theta = -Q_\theta\frac{\alpha^2\xi}{\sqrt{1+\alpha^2\xi^2}}\quad T=0$ $N_\theta = 2M\frac{x_0^2}{r_{\phi_0}^2}e^{-x_0\bar{\xi}}[\cos x_0\bar{\xi} - \sin x_0\bar{\xi}]$	$u' = -\frac{\mu M}{2Bc}\frac{1+\alpha^2\xi^2}{(1+\beta^2\xi^2)^{3/2}}\frac{r_{\phi_0}^2}{x_0^2}$ $\times e^{-x_0\bar{\xi}}[\cos x_0\bar{\xi} - \sin x_0\bar{\xi}]$ $\bar{\omega} = -\frac{M}{2B}\sqrt{\dfrac{1+\alpha^2\xi^2}{1+\beta^2\xi^2}\dfrac{r_{\phi_0}^2}{x_0^2}}$ $\times e^{-x_0\bar{\xi}}[\cos x_0\bar{\xi} - \sin x_0\bar{\xi}]$ ϕ $\chi = \frac{M}{B}\frac{r_{\phi_0}}{x_0}e^{-x_0\bar{\xi}}\cos x_0\bar{\xi}$

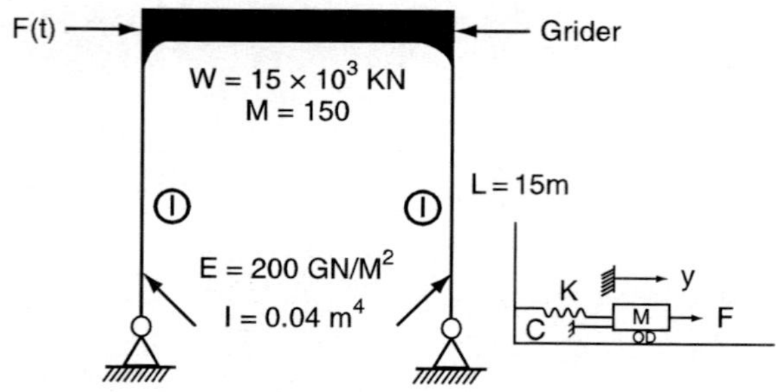

Fig. AIF.3.1 A frame supporting a rotating machinery

Solution

$$K = 3E^{(2I)}/_{L^3}$$
$$= 948.15KN/m$$

$$Y_{st} = \frac{F_0}{K} = \frac{200}{948.15}0.211\,\text{m}$$

$$w = \text{circular frequency}$$
$$= \sqrt{k/m}$$
$$= \sqrt{\frac{948(9.81)}{150}} = 7.874 \text{ rad/ sec}$$

Steady-state amplitude

$$r = \frac{\bar{w}}{w} = \frac{5.3}{7.874} = 0.673$$

$$Y = \frac{y_{st}}{(1 - r^2)^2 + (2r\xi)^2} = 0.38236\,\text{m}$$

The max. bending moment in the column $= V_{\max} L$

$$lV_{\max(\text{dyn})} = \frac{3EF}{L^3}.Y$$
$$= \frac{3 \times 200 \times 0.04}{(15)^3}(0.38263) = 2719\,\text{KN}$$

$$M_{\text{dyn}} = 2719 \times 15 = 40785\,\text{KNm}$$

AIF.4. Cooling Towers of Equal Radii

Stress resultants when cooling towers have top and bottom radii that are equally bounded by a smooth hyperboloid. The hyerbola is rotated by the z axis

$$\frac{B}{A} = \text{the slope of the asymptotes to the hyperbola}$$
$$A = R_0 \text{ at } Z = 0 \tag{AIF.33}$$
$$R_0 = R_2 \sin\phi$$

With the surface intersection to a vertical plane parallel to the x–z plane are

$$\frac{x^2 + A^2}{A^2} - \frac{Z^2}{B^2} = 1 \text{ or } \frac{x^2}{A^2} = \frac{Z^2}{B^2} \tag{AIF.34}$$

which is a straight line (A being the throat radius)

$$Z = \frac{B}{Z}x \tag{AIF.35}$$

A vertical plane tangent to the waist circle ($R_0 = A$) will intersect the surf straight lines of the same slope.

Equations are now rewritten in the following form:

$$\begin{aligned}
&\frac{\partial N_\theta}{\partial\theta} + N_{\theta\phi}\frac{\partial R_0}{R_1\partial\phi} + \frac{\partial(N_{\phi\theta}R_0)}{R_1\partial\phi} + \omega_\theta R_0 = 0\\
&\frac{\partial(N_\theta R_0)}{R_1\partial\phi} - N_\theta\frac{\partial R_0}{R_1\partial\phi} + \frac{\partial(N_{\theta\phi})}{\partial\theta} + \omega_\phi R_0 = 0\\
&\frac{N_\phi}{R_1} + \frac{N_\theta}{R_2} + \omega_y = 0
\end{aligned} \tag{AIF.36}$$

When z is the vertical coordinate, the radii will then become:

$$\begin{aligned}
&\frac{1}{R_1} = \frac{-\partial^2 R_0/\partial^2\partial_1^2}{[1 + (\partial R_0/\partial x)^2]^{(3/2)}}\\
&R_2 = \frac{R_0}{\sin\phi} \text{ and } R_0 = \frac{A}{B}(B^2 + Z^2)^{(1/2)}\\
&\frac{\partial R_0}{\partial Z} = \frac{A^2}{B^2}\frac{Z}{R_0}; \quad \frac{\partial^2 R_0}{\partial Z^2} = \frac{AB}{(B^2 + Z^2)^{\frac{3}{2}}}
\end{aligned} \tag{AIF.37}$$

Substituting into Eq. (AIF.37) the value of R_1 will become

$$R_1 = -A^2B^2\left(\frac{R_0^2}{A^4} + \frac{Z^2}{B^4}\right)^{(3/4)} \tag{AIF.38}$$

therefore

$$R_2 = A\left\{1 + \left(\frac{1}{B^2} + \frac{A^2}{B^4}\right)z^2\right\}^{(1/2)} \tag{AIF.39}$$

and hence

$$R_1 = -\frac{B^2}{A^2} R_2^3 \tag{AIF.40}$$

Gravity loads

$$\omega_2 = -\omega \sin \phi;\ \omega_y = -\omega \cos \phi \tag{AIF.41}$$

(Note the tangential axix Z_T along which ω_z is measured)

$$\begin{aligned} N_\phi &= \frac{\omega}{2\pi R_2 \sin^2 \phi} = -\frac{1}{2\pi R_2 \sin^2 \phi} \int_0^\phi \omega 2\pi R_0 R_1 \ \mathrm{d}\phi \\ &= -\frac{1}{R_2 \sin^2 \phi} \int_0^\phi \omega R_1 R_2 \ \sin\ \phi \ \mathrm{d}\phi \\ R_1 \ \mathrm{d}\phi &= \mathrm{d}s = \frac{\mathrm{d}z}{\sin \phi} \\ \mathrm{d}z &= R_1 \sin \phi \, \mathrm{d}\phi \end{aligned} \tag{AIF.42}$$

Eliminating all polar terms and saving extensive analytical work, the value of N_ϕ and N_θ will be evaluated as

$$N_\theta = -\omega_y R_2 + \frac{A^4}{B^2 R_2^2} N_\phi \tag{AIF.43}$$

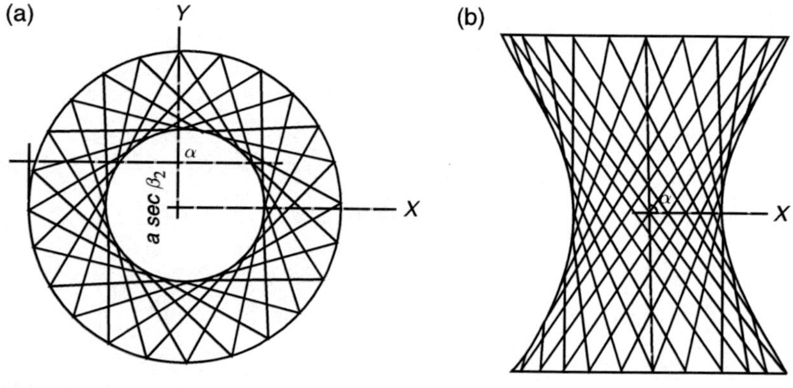

Fig. AIF.4 Cooling tower

A solution for the cooling tower existing in the Trojan Nuclear Power Plant of the Portland General Electric Company (GEC) built in 1974.

Data: $A = 9.1$ m, $B = 21.3$ m; $z_{\text{top}} = 9.1$ m; $z_{\text{bottom}} = 30.5$ m Intensity: Dead load 23.6kN/m^2. The stresses due to dead loads are desired strsses.

Notation for Figure AIF.5

$$\frac{R^2}{a^2} - \frac{z^2}{b^2} = 1 \tag{AIF.44}$$

In which R is the horizontal radius of the parallel circle; z is the vertical coordination along the axis of revolution; and a, b are characteristic curvature dimensions.

The principal radii of curvature are:

$$R_1 = -\frac{a \sin^2 \alpha \cos \alpha}{(\cos^2 \alpha - \cos^2 \theta)^{(3/2)}} \tag{AIF.45}$$

$$R_2 = \frac{R}{\sin \theta} \tag{AIF.46}$$

In which $\cot \alpha = a/b$ and $\tan \theta = (R/Z) \tan^2 \alpha$; R_1 and R_2 are the meridiona circumferential radii of curvature, respectively, and θ = the angle between the node to the surface and the axis of revolution.

Geometry of Hyperboloid of Revolution

The dependent variable can be separated using Fourier expansion. For example, displacements may be written as

$$u = \sum_{n=0}^{a} hu \cos n\phi. \tag{AIF.47}$$

$$v = \sum_{n=0}^{a} hv \sin n\phi \tag{AIF.48}$$

$$w = \sum^{a} hw \cos n\phi \tag{AIF.49}$$

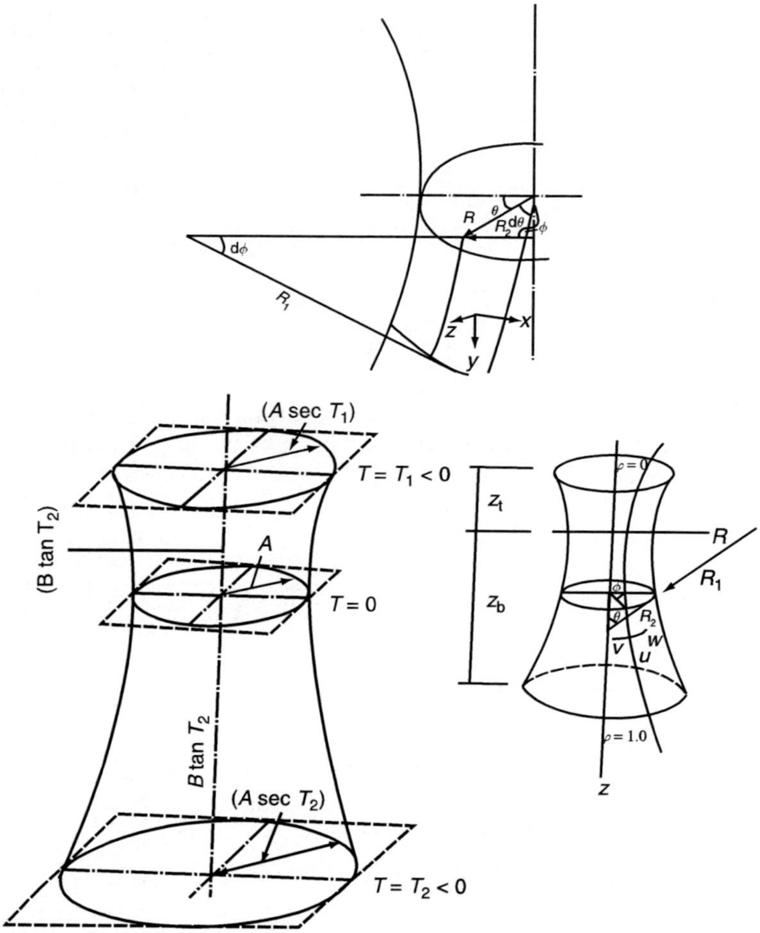

Fig. AIF.25 Geometry and stresses of cooling towers

Solution

Design factors

$$R_0(\text{at base}) = A\left\{1+\frac{z^2}{B^2}\right\}^{(1/2)} = 15.894\ \text{m} \qquad \text{(AIF.50)}$$

$$\tan\phi = \frac{B}{A}\left\{\frac{R_0}{R_0^2 - A^2}\right\}^{(1/2)} = 2.855;\ \ \cos\phi = \frac{A}{(A^2+B^2)^{(1/2)}}\phi^1 \qquad \text{(AIF.51)}$$

where $\phi^1 = 0.8415, f(\phi') = 8.277, R_0 = 9.95\ \text{m}; \tan\phi = 2.6245$ (AIF.52)

$$\cos\phi = \left\{\frac{4}{\sqrt{A^2+B^2}}\right\} \times 0.8415 = 0.33041$$

$$N\phi = -\tfrac{W}{4}B^2\{A^2+B^2\}^{(1/2)}\frac{\{1-\phi'^2\}^{(1/2)}}{(A^2+B^2-A^2\phi'^2)}[f(\phi')-f(\phi'_0)]$$

At base $= 13.58\times10^4$ kN/m

$$N\phi = \frac{WA^2}{\{A^2+B^2\}^{(1/2)}}\frac{\phi'}{\{1-\phi'^2\}^{(1/2)}} + N\phi\frac{A^2}{B^2}(1-\phi'^2) \tag{AIF.53}$$

At base $= 7344$ kN/m^2

$$N_{\phi(\mathrm{Top})} = 0 \text{ since } f(\phi'_0) - f'(\phi'_0) = 0$$

$$N_{\theta(\mathrm{Top})} = \frac{-WA^2}{\{A^2+B^2\}^{(1/2)}}\frac{\phi'_0}{\{1-\phi_0'^2\}^{(1/2)}} = 12.3\,\mathrm{kN/m/m} \tag{AIF.54}$$

$$\frac{N_\phi}{R_1} + \frac{N_\phi}{R_2} = -z = w\cos\phi \tag{AIF.55}$$

$$R_2\,(\text{at base}) = \frac{R_0}{\sin\phi} = \frac{R_0}{\tan\phi\cos\phi} = \frac{15.894}{2.855\times0.3306} = 16.84$$

$$R_1 = \frac{R_2^3}{(A^2/B^2)\times A^2} = \frac{(16.84)^3}{(9.1)^4}\times(21.3)^2 = 316 \tag{AIF.56}$$

Other values for R can be computed where variations exist. The same applied ZZ values, i.e. $w\cos\phi$ where w can be from wind, live loads and others. For de### $w\cos\phi = -23.6\times0.33041 = -7.8\,\mathrm{kN/m^2}$. Similarly, stresses for any load corrleation can be determined and the results are tabulated for the longitudinal and tragetic stresses along the entire height of the tower.

AIF.5 Damage Scenario of the Cooling Tower

A crack pattern and a map of meridional stresses are shown in Figures AIF.6 and AIF.7 using ISOPAR and DIANA software with a fracture simulation technics.

The entire shell was discredited with degenerated finite shell elements. Eight noded isoperimetric layer shell elements were used. With a special preprocessor, geometric imperfection and large deformations were taken into account. Smeared cracking option is used. Two grids of embedded steel bars reinforce the tower shell. Plastic flow of the reinforcing bars is considered. Bilinear softening is also employed. The analysis is considered using the shell stiffened and unstiffened. Wind directions with different intensities and distributions due to interference were considered. The maximum wind assumed is 150 miles/h, The results presented are based on this wind integrity based on BS 6399. A reference is made to the author's text: *Prototype Building Structures – Analysis and Dedign*. Thomas Telford, 1999.

Fig. AIF.6 Finite element analysis

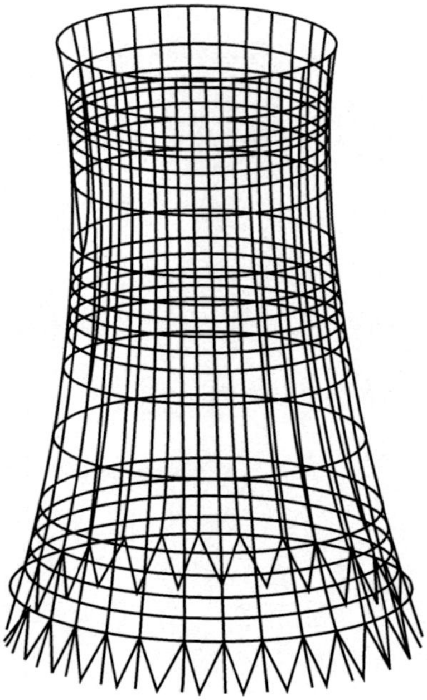

$$u = \frac{T_{1c} l_c}{2E b_c d_c \sin^2 \alpha_0}$$

$$v = \frac{S_c l_c}{2E b_c d_c \cos^2 \alpha_0}$$

$$w = \frac{\bar{Q}_{1c} l_c^3}{6EI} + \frac{M_{1c} l_c^2}{2EI \sin^2 \alpha_0}$$

$$\psi = \frac{\bar{Q}_{1c} l_c^2}{4EI \sin \alpha_0} + \frac{M_{1c} l_c}{2EI \sin^2 \alpha_0}$$

in which $I = b_c d_c^3/12$ and the subscript c applied to forces and moments indica summation over the frame width ($2I_c \cos \alpha_0$).

Notation

a, b	= characteristic curvature dimensions
b_c	= breadth of column
C	= elastic constant
C_0	= extensional stiffnesses of the shell
C_1, C_{12}	= extensional meridional and shear stiffnesses respectively
D_0	= flexural stiffnesses of shell

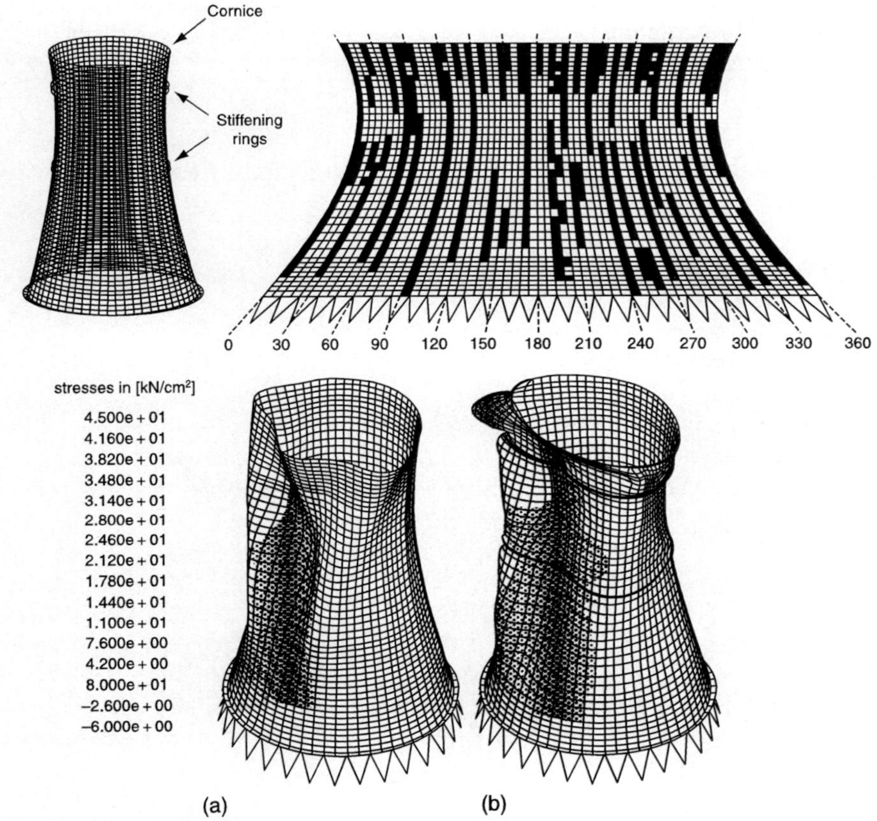

Fig. AIF.7 Damage scenario of a cooling tower shell. Finite element mesh schemes. (**a**) Unstiffened shell. (**b**) Stiffened Shell. (DYNA-3D and ISOPAR-5 programs.)

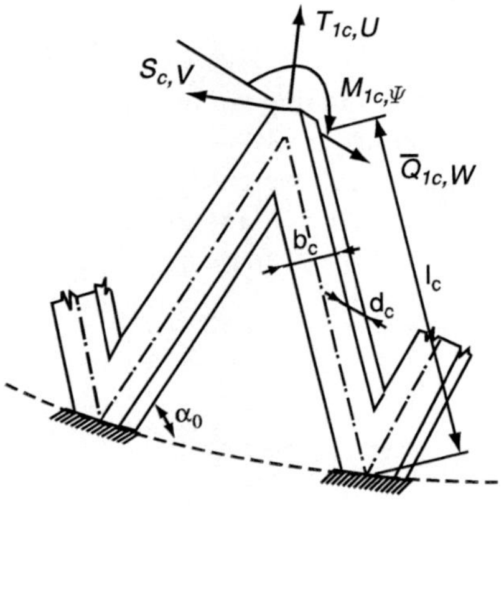

Fig. AIF.8 V Supports for cooling towers

D_l, D_{12} = flexural meridional and twisting stiffnesses respectively;
d_c = depth of column
E = Young's modulus

AIF.6 Quick Manual Solution Based on American Practices from the Equation of the Surface Defined

a = 70′; b = 181.6′
x = 74.5; z = −65.5 when $2x$ = 149′
h_{top} = 200′ dia = 149′; β = 20°
h = 214.3′ dia = 140′; β = 0
h = 20.6′ dia = 161.6′; β = 30°
h = 0′ dia = 251.8′; β = 50°
t = thickness = 6″
w = self weight/unit area of membrane surface = 71.9 lbs/ft^2
At base of the tower = $\tan\beta = \frac{z}{b} = \frac{214.3}{181.6}$
β = 50°

$$\alpha = \sqrt{a^2\tan^2\beta + b^2\sec^2\beta} = 293.58$$

Several values of	
β	$\bar{X}$
−15°	−78
−5°	−71
15°	−78
25°	−89
35°	−131
45°	−209
Summation	−722

The value of $\bar{X}$ = −722×72×0.1745 rad = −9080
Stresses $N_\phi = \frac{\bar{X}\alpha}{a\sec^2\beta} = \frac{9080\times 293.58}{70\times 2.4} = \frac{15900}{72} = 15900$ lbs/ft
corresponding comp stress $N_\theta = \frac{a^2}{\alpha^2}N_\phi - \frac{a^2}{b^2}\tan\beta = 939$ lbs/ft
σ (comp. stress) = 13 psi
Due to self weight Nominal steel will be required.

When other loads are applied the results from them will be algebraically added. Those loads may be wind and seismic loads.

Computer program: Dynamic wind pressure on cooling tower (NAZAM-4)

```
C *********************************************
C * Prepared by Liu, checked by Y. Bangash
C * THE SUBPROGRAM FINDS THE DYNAMIC WIND PRESSURE AT *
C * DIFFERENT HEIGHTS OF THE COOLING TOWER UNDER THE *
C * BASIC WIND SPEED 100 MPH (45 M/S).               *
C * THE BASIC WIND SPEED (V) IS THE 3-SECOND GUEST   *
C * SPEED ESTIMATED TO BE EXCEEDED ON THE AVERAGE    *
C * ONCE IN 100 YEARS.                               *
C * TOPOGRAPHICAL FACTOR = S1                        *
C * SHAPE FACTOR = S                                 *
C * STATISTICAL FACTOR = S2                          *
C * FORCE COEFFICIENT = CF                           *
C * DESIGN WIND SPEED (VS) = V*S1*S*S2               *
C * DYNAMIC WIND PRESSURE (PW) = CF*0.613*(VS**2)    *
C *********************************************
C
C   FUNCTION SUBPROGRAM
    SUBROUTINE WIND (PW,A,B,T,ZE)
    V=45.0
    S1=1.0
    S3=1.0
    Z=B*(TAN (T*0.01745))
    C=(ABS(Z/B))**2.0
    XX=(A*((1.0+C)**0.5))
    H=ZE-Z
C   HEIGHT/BREADTH==RATIO
    CALL SHAPE (S,H)
    SP=S1*S*S3
    VS = (V*SP)
    Q = 0.613 * ((VS)**2.0)
    HB=H/(XX+XX)
    CALL COEF (CF,HB)
    PW=(CF*Q)/1000
    RETURN
    END'
    WRITE (6,4)
4   FORMAT (///, 2X, 'THETA', 8X, 'BETA', 8X,
     'MERIDI NAL', 8X, *'HOOP', 13X, 'WIND')
    WRITE (6,5)
5   FORMAT(27X,'FORCES', 11X,'FORCES', 11X,'PRESSURE')
    WRITE(6,6)
```

```
6     FORMAT (/, 2X,' (Deg.)' , 7X,' (Deg.)' , 7X,' (KN/m)' ,
      11X,'*' (KN/m)' , 10X, (KN/m**2)' /2X, 6('*' ), 7X,
      6('*' ) , 6X, 9('*' ), * 8X, 10('*' ), 7X, 9('*' ))
      DO 100 T=-15,45,10
      CALL WIND (PW,A,B,T,ZE)
      AA=(X/Z)*(y**2)
      Q=ATAN (AA)
      H=SIN (Q)
      K=COS (Q)
      R1=(Z**3)/((B**2)*(y**2)*(K**3))
      R2=X/H
      P=X/A
      PO=RU/A
      ZO=-(HE-ZE)
      N=Z/B
      NO=ZO/B
      BB=(Y**2)/((P**2)*H)
      YP= ((Y**2.0)-2.0)*(P**3.0-PO**3.0)/(3.0*(Y**2.0))
      PD=PO-P
      OPN=(NO*PO)*(N-NO)
      IL=N*(ALOG((N+P)/(NO+PO)))
C     MERIDINAL FORCES DUE TO WIND LOADING = ANPP
      ANPP = ((PW*A)/6.0)*BB*(YP+PD-OPN+IL)*(COS(AN))
C     HOOP FORCES DUE TO WIND LOADING = ANXX
      ANXX = (-1)*(ANPP)*(R2/R1)-(PW*R2*(COS(AN)))
C     TOTAL MERIDINAL FORCES = MSUM
      MSUM=ANPP+ANP
C     TOTAL HOOP FORCES = HSUM
      HSUM=ANXX+ANX
      WRITE (6,111) T,F,MSUM,HSUM,PW
111   FORMAT (1X, F6.2, 6X, F7.2, 6X, E10.3, 8X, E10.3,
      8X, F6.3)
101   CONTINUE
100   CONTINUE
      STOP
      END

C   ***********************************************
C   * SUBROUTINE FOR FORCE COEFFICIENT
C   ***********************************************
      SUBROUTINE COEF (CF, HB)
      IF ((HB.GT.0.0) .AND.(HB.LE.0.5))   CF=0.70
      IF ((HB.GT.0.5) .AND.(HB.LE.1.0))   CF=0.70
      IF ((HB.GT.1.0) .AND.(HB.LE.2.0))   CF=0.80
      IF ((HB.GT.2.0).AND.(HB.LE.4.0))    CF=0.80
```

```
      IF ((HB.GT.4.0) .AND.(HB.LE.6.0))   CF=0.80
      IF ((HB.GT.6.0) .AND.(HB.LE.10.))   CF=0.90
      IF ((HB.GT.10.0) .AND.(HB.LE.20.)) CF=0.90
      IF (HB.GT.20.0)                     CF=1.00
      RETURN
      END

*********************************************************

      SUBROUTINE SELF(AND ANY ZE, A, B, DENS ANG)
      H=1/(COS(TT))**2
C     LET D BE ALPHA
      D=((A**2)*((ZE/B)**2) + (B**2)*(H))**0.5
      SUM=0.0
      DEF=-(A*(B**2+A**2*(SIN(T*0.01745))**2)**0.5) /
      *(B*(COS(T*0.01745)**3))
      SUM=SUM+DEF
      SG=SUM*ANG*DENS
      ANP= (SG*D) / (A*H)
      ANX= (((A**2)/(D**2))*ANP-(A**2/B)*(ZE/B))
      RETURN
      END

C
C     ***********************************************
C     * SUBROUTINE FOR SHAPE FACTOR                  *
C     ***********************************************
      SUBROUTINE SHAPE (S,H)
      IF ((H.EQ.0.0) .AND. (H.LE.3.0)) S=0.73
      IF ((H.GT.3.0) .AND. (H.LE.5.0)) S=0.78
      IF ((H.GT.5.0) .AND. (H.LE.10.0)) S=0.90
      IF ((H.GT.10.0) .AND. (H.LE.15.0)) S=0.94
      IF ((H.GT.15.0) .AND. (H.LE.20.0)) S=0.96
      IF ((H.GT.20.0) .AND. (H.LE.25.0)) S=0.98
      IF ((H.GT.25.0) .AND. (H.LE.30.0)) S=1.00
      IF ((H.GT.30.0) .AND. (H.LE.35.0)) 5=1.02
      IF ((H.GT.35.0) .AND. (H.LE.40.0)) S=1.03
      IF ((H.GT.40.0) .AND. (H.LE.45.0)) S=1.05
      IF ((H.GT.45.0) .AND. (H.LE.50.0)) S=1.06
      IF ((H.GT.50.0) .AND. (H.LE.60.0)) S=1.08
      IF ((H.GT.60.0) .AND. (H.LE.70.0)) S=1.10
      IF ((H.GT.70.0) .AND. (H.LE.80.0)) S=1.11
      IF ((H.GT.80.0) .AND. (H.LE.100.0)) S=1.13
      IF ((H.GT.100.0) .AND. (H.LE.120.0)) S=1.15
      IF ((H.GT.120.0) .AND. (H.LE.140.0)) S=1.17
```

```
IF ((H.GT.140.0) .AND. (H.LE.160.0)) S=1.19
IF ((H.GT.160.0) .AND. (H.LE.180.0)) S=1.20
IF ((H.GT.180.0) .AND. (H.LE.500.0)) S=1.21
RETURN
END
```

l = non-dimensional length of the meridian
l_c = length of column
M_1, M_2 = meridional and circumferential bending moment
n = harmonic number in circumferential direction
$\bar{Q}_1$ = effective meridional transverse shear force
R = radius of parallel circle
R_1, R_2 = meridional and circumferential radii, respectively
S = membrane shear force
T_1, T_2 = meridional and circumferential membrane forces, respectively
U, V, W = nondimensional displacements, u/h, v/h, w/h, respectively
u, v, w = meridional circumferential and normal displacements, respectively
Z = vertical coordinate along axis of revolution
$\alpha = \cot^{-1}(a/b)$
α_0 = inclination of columns of horizontal in their own plane
β = nondimensional factor of order of 10
θ = angle between normal to surface and vertical axis
μ = Poisson's ratio
ϕ = circumferential angle
ψ = meridional angle of rotation

Subscript c refers to summation over circumferential length ($2l_c \cos\alpha_0$).

Equivalence method

Considering the supporting system of columns, it can be shown that its equivalent uniform stiffnesses are

$$\text{Membrane force, } T_1;\ C_1 = E\frac{b_c d_c \sin^3\alpha_0(1-\mu^2)}{l_c\cos\alpha_0} \tag{AIF.57}$$

$$\text{Membrane shear, } S;\ C_{12} = E\frac{b_c d_c \sin\alpha_0\cos\alpha_0(1-\mu^2)}{l_c} \tag{AIF.58}$$

$$\text{Bending moment, } M_1;\ D_1 = E\left[\frac{I\sin^3\alpha_0}{l_c\cos\alpha_0} + 4\frac{i}{l_c}\frac{\sin\alpha_0\cos\alpha_0}{(1+\mu)}\right] \tag{AIF.59}$$

$$\text{Torsion, } H;\ D_{12} = E\left[\frac{4I}{l_c}\sin\alpha_0\cos\alpha_0 + \frac{8}{(1+\mu)}\frac{i}{l_c}\right] \tag{AIF.60}$$

$$C_1/C_0 = 1.16; \quad D_1/D_0 = 43.7$$
$$C_{12}/C_D = 0.154; \quad D_{12}/D_0 = 175$$

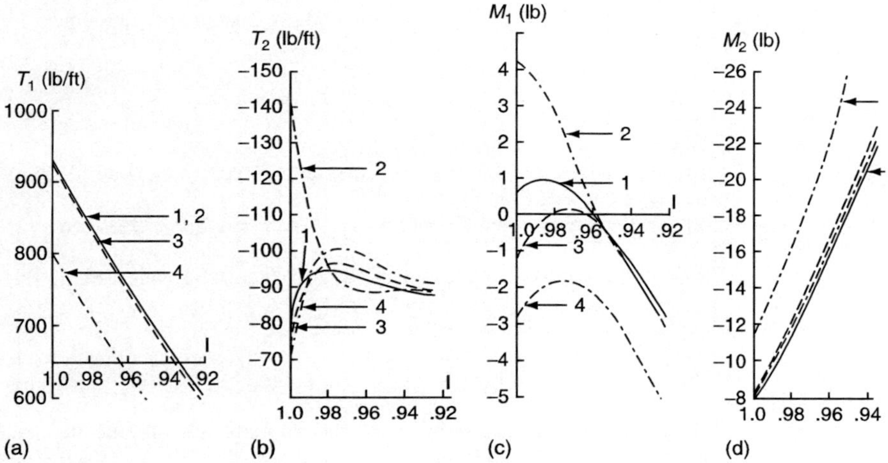

Table AIF.2

Relative stiffness (1)	Case number 1 (Equivalence) (2)	2 (3)	3 (4)
(C_1/C_0)	1.16	1.19	0.12
(C_{12}/C_0)	0.154	0.159	0.159
(D_1/D_0)	43.7	1.60	16.0
(D_{12}/D_0)	175.0	0.0	0.0

References

1. Zielinskl, J. and Rowe, R.E. An investigation of the stress distribution in the anchorage zones of post-tensioned concrete members. Res. Rep. No. 9. Cement and Concrete Association, September 1960.
2. Zielinski, J. and Rowe, R.E. The stress distribution associated with groups of anchorages in post-tensioned concrete members. Research Report No. 13. Cement and Concrete Association, October 1962.
3. Taylor, S.J. A failure criterion for concrete at cable anchorages. Ph.D. thesis, University of London, 1967.
4. Bortsch, R. Die Spannungen in Wälzgelenkquardern. Beton und Eisen, 1935, 35, No. 4 (Feb.) 61–66.
5. Bortsch, R. Walzgelenke und Steizenlager aus Eisenbeton. Beton and Eisen, 1938, 37, No. 19 (Oct.) 315–318 and 1938, No. 20 (Oct.) 328–332.
6. Lenschow, R.J. and Sozen, M.A. Practical analysis of the anchorage zone problem in prestressed beams. Journal of American Concrete Institute, 1965, 62 (Nov.) 1421–1439.

7. Newman, K. The structure and engineering properties of concrete. International Symposion on Theory Arch Dams, Southamptni, 1965.
8. Hsu, T. et al. Microcracking of plain concrete and the shape of the stress–strain curve. Journal of American Concrete Institute, 1963, 60, No. 2 (Feb.) 209–224.
9. Robinson, G.S. Methods of detecting the formation and propagation of microcracks in concrete. Int. Conf. Struct. Concr., London, Sep. 1965.
10. Jones, R. A method of studying the formation of cracks in a material subjected to stress. British Journal Applied Physics, 1952, 3, No. 7 (July), pp. 229–232.
11. Alexander, K.M. et al. Paste strength, aggregate bond strength, and the strength of concrete. mt. Conf. Struct. Concr., London, Sept. 1965.
12. Vile, U.W.D. Strength of concrete under short-term static biaxial stress. International Conference Structural Concrete, London, Sept. 1965.
13. Morsch, E. Uber die Berechnung der Gelenkquader. Beron und Eisen, 1924, No. 12, 156–161.
14. Taylor, S.J. The failure of concrete at cable anchorages. Structural Engineering (in the press).
15. Vile, G.W.D. A combined stress testing machine for concrete. The Engineer, 1965, (Jul.).
16. Newman, K. and Lachance, L. The testing of brittle materials under uniform uniaxial compressive stress. Proceedints of American Society for Testing and Material, 1965, 65.
17. Mchenry, D. The effect of uplift pressure on the shearing strength of concrete. 3rd Congress on Large Dams, Stockholm, 1948.
18. Akroyd, T.N.W. Failure mechanism of saturated concrete. Engineering, 1961, 191 (May 12) 568–659.
19. Magnel, G. Prestressed concrete, 2nd edn,. Concrete Publications Ltd, London, 1950.
20. Vile, U.W.D. Behaviour of concrete under simple and combined stresses. Ph.D. thesis, University of London, 1965, pp. 356–365.
21. Bleich, F. Der gerade Stab mit Rechteckquerschnitt als ebenes Problem. *Der Bauingenieur,* 1923, No. 9, *255–259* and 1923, No. 10, 304–307.
22. Guyon, Y. Prestressed Concrete, 1st edn,. London Contractors Record & Municipal Engineering, 1953.
23. Guyon, Y. Contraintes dans les pièces prismatiques soumises a des forces appliquées sur leurs bases, au voisinage de ces bases. Int. Ass. Bridge Struct. Engng Publications, 1951, 11, 195–226.
24. Sievers, H. Die Berechnung von Auflagerbanken und Auflagerquarden von Brückenpfeilern. Der Baningenleur, 1952, 27, No. 6 (Jun.) 209–213.
25. Sievers, H. Uber den Spannungszustand im Bereich der Ankerplatten von Spanngliedern vorgespannter Stahl-betonkonstruktionen. Der Bauingenieur, 1 956, 31, No. 4 (Apr.) 134–135.
26. Douglas, D. J. and Trahir, N.S. An examination of the stresses in the anchorage zone of a post tensioned prestressed concrete beam. Magazine on Concrete Research, 1960, 12, No. 34 (Mar.) 9–18.
27. Rydzewski, J.R. and Whitbread, F.J. Short end blocks for prestressed beams. Structural Engineering 1963, 41, No. 2 (Feb.) 41–53.
28. Christodoulides, S.P. A two-dimensional investigation of the end anchorages of post-tensioned concrete beams. Structural Engineering, 1955, 33, No. 4 (Apr.) 120–133.
29. Christodoulides, S.P. Three-dimensional investigation of the stresses in the end anchorage blocks of a prestressed concrete gantry beam. Structural Engineering, 1957, 35, No. 9 (Sep.), 349–356.
30. Ban S. et al. Anchorage zone stress distributions in post-tensioned concrete members. Proceeding of the World Conference on Prestress Concrete, San Francisco, July 1957, 16.1–16.14.

31. Ra5heeduzzafar. An investigation of the stress distribution in the anchorage zones of post-tensioned concrete beams. Ph.D. thesis, University of London, 1964.
32. Taylor, M.A. and Broms, B.B. Shear bond strength between coarse aggregate and cement paste or mortar. Journal American Concrete Institute, 1964, 61, No. 8 (Aug) pp. 939–956.
33. Sargius, M. Beitrag zur Ermittlung der Hauptzugspannungen am Endauflager vorgespannter Betonbalten. Dr Eng. thesis, University of Stuttgart, 1960.

Index